WITHDRAWN
UTSA LIBRARIES

Biogeochemical Investigations of Terrestrial, Freshwater, and Wetland Ecosystems across the Globe

Biogeochemical Investigations of Terrestrial, Freshwater, and Wetland Ecosystems across the Globe

Guest Editors:

R. KELMAN WIEDER
Department of Biology, Villanova University, Villanova, Pennsylvania, USA

MARTIN NOVÁK
Czech Geological Survey, Prague, Czech Republic

MELANIE A. VILE
*Patrick Center for Environmental Research, The Academy of Natural Sciences,
Benjamin Franklin Parkway, Philadelphia, Pennsylvania, USA*

Reprinted from *Water, Air, & Soil Pollution: Focus*
Vol. 4, Nos 2–3 (2004)

A C.I.P. Catalogue record for this book is available from the Library of Congress.

ISBN 1-4020-1814-2

Published by Kluwer Academic Publishers
P.O. Box 17, 3300 AA Dordrecht, The Netherlands

Sold and distributed in North, Central and Latin America
by Kluwer Academic Publishers
101 Philip Drive, Norwell, MA 02061, U.S.A.

In all other countries, sold and distributed
by Kluwer Academic Publishers
P.O. Box 322, 3300 AA Dordrecht, The Netherlands

Cover image:
Josef Šíma: *Landscape*, 1930
Reproduced with kind permission of the National Gallery in Prague

Printed on acid-free paper

All rights reserved
©2004 Kluwer Academic Publishers
and copyrightholders as specified on appropriate pages within.
No part of this work may be reproduced, stored in a retrieval system, or transmitted in any form or by any means, electronic, mechanical, photocopying, microfilming, recording or otherwise, without written permission from the Publishers, with the exception of any material supplied specifically for the purpose of being entered and executed on a computer system, for exclusive use by the purchser of the work.

Printed in the Netherlands.

TABLE OF CONTENTS

Preface   1
Acknowledgments   3

I: KEYNOTE ADDRESS

G.E. LIKENS / Biogeochemistry: Some Opportunities and Challenges for the Future   5–24

II: CATCHMENT AND REGIONAL-SCALE MODELLING

J. AHERNE, M. POSCH, P.J. DILLON and A. HENRIKSEN / Critical Loads of Acidity for Surface Waters in South-Central Ontario: Canada: Regional Application of the First-Order Acidity Balance (FAB) Model   25–36

J. AHERNE, T. LARSSEN, P.J. DILLON and B.J. COSBY / Effects of Climate Events on Elemental Fluxes from Forested Catchments in Ontario, Canada: Modelling Drought-Induced Redox Processes   37-48

M. ALVETEG / Projecting Regional Patterns of Future Soil Chemistry Status in Swedish Forests using SAFE   49–59

A.L. HORN, R.-A. DÜRING and S. GÄTH / Sorption of Cd in Soils: Pedotransfer Functions for the Parameters of the Freundlich Sorption Isotherm   61–71

M. HUGHES, D.D. HORNBY, H. BENNION, M. KERNAN, J. HILTON, G. PHILLIPS and R. THOMAS / The Development of a GIS-Based Inventory of Standing Waters in Great Britain together with a Risk-Based Prioritisation Protocol   73–84

Ø. KASTE / Simulation of Nitrogen Dynamics and Fluxes in Contrasting Catchments in Norway by Applying the Integrated Nitrogen Model for Catchments (INCA)   85–96

M. KERNAN, M. HUGHES, D. HORNBY, H. BENNION, J. HILTON, G. PHILLIPS and R. THOMAS / The Use of a GIS-Based Inventory to Provide a Regional Risk Assessment of Standing Waters in Great Britain Sensitive to Acidification from Atmospheric Deposition   97–112

A. LEPISTÖ, K. GRANLUND and K. RANKINEN / Integrated Nitrogen Modeling in a Boreal Forestry Dominated River Basin: N Fluxes and Retention in Lakes and Peatlands   113–123

T. LARSSEN, B.J. COSBY and T. HØGÅSEN / Uncertainties in Predictions of Surface Water Acidity using the MAGIC Model   125–137

F. MOLDAN, V. KRONNÄS, A. WILANDER, E. KARLTUN and B.J. COSBY / Modelling Acidification and Recovery of Swedish Lakes   139–160

K. RANKINEN, A. LEPISTÖ and K. GRANLUND / Integrated Nitrogen and Flow Modelling (INCA) in a Boreal River Basin Dominated by Forestry: Scenarios of Environmental Change   161–174

C. VAN DER SALM, G.J. REINDS and W. DE VRIES / Assessment of the Water Balance of European Forests: A Model Study   175–190

III: MONITORING AND MANIPULATIONS

C. BEIER, I.K. SCHMIDT and H.L. KRISTENSEN / Effects of Climate and Ecosystem Disturbances on Biogeochemical Cycling in a Semi-Natural Terrestrial Ecosystem   191–206

N.B. DISE and P. GUNDERSEN / Forest Ecosystem Responses to Atmospheric Pollution: Linking Comparative with Experimental Studies   207–220

X. FENG, J.W. KIRCHNER and C. NEAL / Spectral Analysis of Chemical Time Series from Long-Term Catchment Monitoring Studies: Hydrochemical Insights and Data Requirements   221–235

M. FERM and H. HULTBERG / Neutralisation of Sulphur Dioxide Deposition in a Coniferous Canopy   237–245

D. HOULE, C. GAGNON, S. COUTURE and A. KEMP / Recent Recovery of Lake Water Quality in Southern Québec Following Reductions in Sulfur Emissions   247–261

D.W. JOHNSON, R.B. SUSFALK, T.G. CALDWELL, J.D. MURPHY, W.W. MILLER and R.F. WALKER / Fire Effects on Carbon and Nitrogen Budgets in Forests   263–275

T. NAKAJI, T. KOBAYASHI, M. KUROHA, K. OMORI, Y. MATSUMOTO, T. YONEKURA, K. WATANABE, J. UTRIAINEN and T. IZUTA / Growth and Nitrogen Availability of Red Pine Seedlings under High Nitrogen Load and Elevated Ozone   277–287

S.A. NORTON, I.J. FERNANDEZ, J.S. KAHL and R.L. REINHARDT / Acidification Trends and the Evolution of Neutralization Mechanisms Through Time at the Bear Brook Watershed in Maine (BBWM), U.S.A.   289–310

R.L. REINHARDT, S.A. NORTON, M. HANDLEY and A. AMIRBAHMAN / Dynamics of P, Al, and Fe during High Discharge Episodic Acidification at the Bear Brook Watershed in Maine, U.S.A.   311–323

J.B. SHANLEY, P. KRÁM, J. HRUŠKA and T.D. BULLEN / A Biogeochemical Comparison of Two Well-Buffered Catchments with Contrasting Histories of Acid Deposition   325–342

IV: NITROGEN TRANSFORMATIONS AND PROCESSES

B. BERG and N. DISE / Validating a New Model for N Sequestration in Forest Soil Organic Matter   343–358

C.J. CURTIS, B.A. EMMETT, B. REYNOLDS and J. SHILLAND / Nitrate Leaching from Moorland Soils: Can Soil C:N Ratios Indicate N Saturation?   359–369

D. HOPE, M.W. NAEGELI, A.H. CHAN and N.B. GRIMM / Nutrients on Asphalt Parking Surfaces in an Urban Environment   371–390

C. HUBER, M. BAUMGARTEN, A. GÖTTLEIN and V. ROTTER / Nitrogen Turnover and Nitrate Leaching after Bark Beetle Attack in Mountainous Spruce Stands of the Bavarian Forest National Park   391–414

H. MEESENBURG, A. MERINO, K.J. MEIWES and F.O. BEESE / Effects of Long-term Application of Ammonium Sulphate on Nitrogen Fluxes in a Beech Ecosystem at Solling, Germany   415–426

S.E. MACHEFERT, N.B. DISE, K.W.T. GOULDING and P.G. WHITEBREAD / Nitrous Oxide Emissions from Two Riparian Ecosystems: Key Controlling Variables   427–436

D.S. REAY, K.A. SMITH and A.C. EDWARDS / Nitrous Oxide in Agricultural Drainage Waters Following Field Fertilisation   437–451

P. SCHLEPPI, F. HAGEDORN and I. PROVIDOLI / Nitrate Leaching from a Mountain Forest Ecosystem with Gleysols Subjected to Experimentally Increased N Deposition   453–467

V: STABLE AND RADIOGENIC ISOTOPES IN THE ENVIRONMENT

S.N. CHILLRUD, R.F. BOPP, J.M. ROSS, D.A. CHAKY, S. HEMMING, E.L. SHUSTER, H.J. SIMPSON and F. ESTABROOKS / Radiogenic Lead Isotopes and Time Stratigraphy in the Hudson River, New York   469–482

D. FOWLER, U. SKIBA, E. NEMITZ, F. CHOUBEDAR, D. BRANFORD, R. DONOVAN and P. ROWLAND / Measuring Aerosol and Heavy Metal Deposition on Urban Woodland and Grass Using Inventories of ^{210}Pb and Metal Concentrations in Soil   483–499

D. HOULE, R. CARIGNAN and J. ROBERGE / The Transit of $^{35}SO_4^{2-}$ and ^3H_2O Added *In Situ* to Soil in a Boreal Coniferous Forest   501–516

M. NOVÁK, R.L. MICHEL, E. PŘECHOVÁ and M. ŠTĚPÁNOVÁ / The Missing Flux in a ^{35}S Budget for the Soils of a Small Polluted Catchment   517–529

I. ROBERTSON, N.J. LOADER, D. McCARROLL, A.H.C. CARTER, L. CHENG and S.W. LEAVITT / δ^{13}C of Tree-Ring Lignin as an Indirect Measure of Climate Change   531–544

L. ROCK and B. MAYER / Isotopic Assessment of Sources of Surface Water Nitrate within the Oldman River Basin, Southern Alberta, Canada   545–562

VI: MERCURY AND METALS

G. BEASLY and P.E. KNEALE / Assessment of Heavy Metal and PAH Contamination of Urban Streambed Sediments on Macroinvertebrates   563–578

C.E. JOHNSON, R.J. PETRAS, R.H. APRIL and T.G. SICCAMA / Post-Glacial Lead Dynamics in a Forest Soil   579–590

Ü. MANDER, A. KULL and J. FREY / Residual Cadmium and Lead Pollution at a Former Soviet Military Airfield in Tartu, Estonia   591–606

J. MUNTHE and H. HULTBERG / Mercury and Methylmercury in Runoff from a Forested Catchment – Concentrations, Fluxes, and their Response to Manipulations   607–618

T. NAVRÁTIL, M. VACH, P. SKŘIVAN, M. MIHALJEVIČ and I. DOBEŠOVÁ / Deposition and Fate of Lead in a Forested Catchment, Lesni Potok, Central Czech Republic   619–630

E. VAVOULIDOU, E.J. AVRAMIDES, P. PAPADOPOULOS and A. DIMIRKOU / Trace Metals in Different Crop/Cultivation Systems in Greece   631–640

VII: PHOSPHORUS

H.P. JARVIE, C. NEAL and R.J. WILLIAMS / Assessing Changes in Phosphorus Concentrations in Relation to In-Stream Plant Ecology in Lowland Permeable Catchments: Bringing Ecosystem, Functioning into Water Quality Monitoring   641–655

J. VYMAZAL / Removal of Phosphorus in Constructed Wetlands with Horizontal Sub-Surface Flow in the Czech Republic   657–670

VIII: SCALING OF BIOGEOCHEMICAL PROCESSES

C. AKELSSON, J. HOLMQVIST, M. ALVETEG, D. KURZ and H. SVERDRUP / Scaling and Mapping Regional Calculations of Soil Chemical Weathering Rates in Sweden   671–681

M. RANTAKARI, P. KORTELAINEN, J. VUORENMAA, J. MANNIO and M. FORSIUS / Finnish Lake Survey: The Role of Catchment Attributes in Determining Nitrogen, Phosphorus, and Organic Carbon Concentrations   683–699

IX: SOIL ORGANIC MATTER AND DOC

I. CAÇADOR, A.L. COSTA and C. VALE / Carbon Storage in Tagus Salt Marsh Sediments   701–714

M. STARR and L. UKONMAANAHO / Levels and Characteristics of TOC in Throughfall, Forest Floor Leachate and Soil Solution in Undisturbed Boreal Forest Ecosystems   715–729

X: BIOGEOCHEMISTRY OF RESTORED ECOSYSTEMS

A.E. LUGO, W.L. SILVER and S.M. COLÓN / Biomass and Nutrient Dynamics of Restored Neotropical Forests   731–746

List of Reviewers   747–748

PREFACE

Fifteen years have passed since a small group of researchers at the Czech Geological Survey boldly convened a conference called *GEOMON*, *Geochemical Monitoring in Representative Basins*, held in Prague in 1987. The focus of the original *GEOMON* conference was rather narrow – monitoring of element pools and fluxes on a small catchment scale. Signaling a desire to broaden the focus to a more biogeochemical orientation, the 1993 meeting, also in Prague, was renamed *BIOGEOMON*. To foster wider international participation and cooperation, in 1997 *BIOGEOMON* was held at Villanova University in Pennsylvania. The most recent iteration of *BIOGEOMON* was held at the University of Reading in the United Kingdom and was the largest of the series of *BIOGEOMON* meetings to date. At Reading, *BIOGEOMON* hosted 43 invited speakers, 96 contributed talks and over 150 poster presentations. Over 260 delegates came to Reading in August 2002 from 25 countries around the world.

At Reading, themes that always have been strong at *BIOGEOMON* were continued: catchment monitoring and manipulations, catchment and regional-scale modeling, nitrogen transformations and processes, and stable and radiogenic isotopes in the environment. Beyond these traditionally emphasized themes, other sessions focused on mercury and metal dynamics, phosphorus, scaling of biogeochemical processes, terrestrial DOC and soil organic matter, rhizosphere biogeochemistry, biogeochemistry of restored ecosystems, and archives of global change on the continents. Most of these themes are represented in this Special Issue, a collection of peer-reviewed articles.

As human population growth continues into the 21st century, stresses on ecosystem function and compromising of life-essential ecosystem services will increase, locally, regionally, and globally. From monitoring through process-based investigation through modeling – a spectrum of research effort that is well represented in this Special Issue – biogeochemistry and ecosystem science will play an increasingly critical role in providing the scientific basis for informing public policy and decision making. In a world where local and regional political stability seems uncertain, it is our hope that international cooperation and collaboration in biogeochemistry and ecosystem science can contribute to peace and stability. We are committed to organizing future *BIOGEOMON* meetings at regular intervals.

MARTIN NOVÁK
MELANIE A. VILE
R. KELMAN WIEDER

July 2003

ACKNOWLEDGEMENTS

This Special Issue would not have come into being without the dedication and enthusiasm of many individuals. Hosting a conference such as *BIOGEOMON*, without the support of a particular professional or scientific society, is a challenge. Many people from the University of Reading worked tirelessly to make *BIOGEOMON*, The Fourth International Symposium on Ecosystem Behaviour, a smoothly run and highly successful conference. Special kudos to Hannah Prior as the *BIOGEOMON* Organising Committee Chair, to Paul Whitehead, Chair of the Local Arrangements Committee, and to Heather Browning, who handled the administration and financial aspects of the meeting. Others at the University of Reading deserving special thanks include Nicola Flynn, Rew Islam, J. Steve Robinson, Gemma Turner, and Patricia Whittaker. Special thanks go to the abstract volume committee (Merritt Turetsky, University of Alberta, Chair; Susan Crow, Oregon State University; Hannah Prior; Gemma Turner), who spent hours prior to the conference editing and formatting abstracts, and putting together the conference *Book of Abstracts*. The academic programme of *BIOGEOMON* was organized by Martin Novák and Tomáš Pačes (Czech Geological Survey), Melanie Vile (Princeton University), R. Kelman Wieder (Villanova University), and Hannah Prior.

All individuals who presented research at *BIOGEOMON* had the option of submitting a manuscript for inclusion in this Special Issue. We had 80 manuscripts submitted, a record for *BIOGEOMON*, of which 50 appear here. We are especially grateful to the 154 individuals around the world who helped us by reviewing manuscripts; we wish that we could do more to thank them than to include a list of reviewers as a part of this Special Issue. Megan McGroddy (Princeton University) and Kimberli Scott (Villanova University) contributed in an outstanding way through assisting in reviewer selection and through careful and thorough final editorial checking of manuscripts. We also are very grateful for the assistance, support, and patience of Betty van Herk of Kluwer Academic Publishers in working with us to make this Special Issue happen.

Finally, our respective institutions, The University of Reading, the Czech Geological Survey, and Villanova University provided generous support in ways unseen to make *BIOGEOMON* happen once again. Additional support was provided by the Aquatic Environments Research Centre, Lloyds TSB Bank Plc., and The Simon Barratt Fund.

MARTIN NOVÁK, Guest Editor
HANNAH PRIOR, *BIOGEOMON* Organising Committee Chair
MELANIE A. VILE, Guest Editor
PAUL WHITEHEAD, *BIOGEOMON* Local Arrangements Chair
R. KELMAN WIEDER, Guest Editor

BIOGEOCHEMISTRY: SOME OPPORTUNITIES AND CHALLENGES FOR THE FUTURE

GENE E. LIKENS
Institute of Ecosystem Studies, Millbrook, New York 12545, U.S.A.
(* author for correspondence, e-mail: LikensG@ecostudies.org)

(Received 20 August 2002; accepted 6 April 2003)

Abstract. There are major opportunities for big, important questions to drive biogeochemical research in the future. Some suggestions are presented, such as: what are the controls on N loss and retention in watershed-ecosystems; what are the rates and controls on biological N fixation and denitrification in diverse ecosystems; how does scale (temporal and spatial) control biogeochemical flux and cycling; what controls the apparent and actual weathering rates in terrestrial ecosystems and what is the fate of the weathered products; how can biogeochemical function best be integrated on regional to global scales; and what are the quantitative interrelationships between hydrologic cycles and biogeochemical cycles? Some brief examples and approaches to address such questions, for example, the value of multidisciplinary teams for addressing complicated questions, and the use of sophisticated tools (e.g., stable isotopes, spatial statistics, remote sensing), are presented.

Keywords: acid rain, antibiotics, biogeochemistry, eutrophication, hydrology, legacies, N and Ca cycling, scale, weathering

1. Introduction

Upon learning that the Desdemona he had just murdered in a jealous rage was indeed without guilt, Othello calls down extreme torments upon himself – devils, winds, hot sulphur, liquid fire (Othello, the Moor of Venice, Act V – Scene II, Shakespeare, 1564–1616). While the windborne sulfur deposited on land and water of eastern North America and western and central Europe over the past century or so may be less dramatic than is called to mind by this Shakespearean prose, its continuing addition to these ecosystems is neither trivial nor without ecological 'torment'.

Indeed, the atmospheric deposition of 'hot sulphur' in acid deposition has never killed anyone directly, and thus its guilt is subtle (difficult to detect), and its serious, long-term effects on terrestrial and aquatic ecosystems have been difficult to monitor and ultimately control. Despite governmental attempts to reduce acid deposition by reducing the precursor emissions of NO_x and particularly SO_2, acid deposition continues and its impacts persist at the ecological and biogeochemical level. Now, however, data are accumulating that show major direct human health effects of the particulate matter emitted as part of the overall air pollution problem associated with acid deposition (e.g., Pyn, 2002; Kaiser, 2000). Ironically, some

TABLE I

Some major questions and challenges for biogeochemistry

1.	What are the specific effects and relationships of the increasing size of the human population on flux and cycling of elements, and what are the biogeochemical effects of forcing functions often incongruent in space and time associated with these changes?
2.	What controls fluxes of N and P to and from natural and human-dominated (cities, agricultural) ecosystems? – effect of disturbance/nutrient saturation/instream-watershed retention/denitrification/ interactions with other element cycles; – effects of atmospheric deposition, and flows of fertilizer, food, waste.
3.	What is and what controls C sequestration in diverse ecosystems (e.g., forest, ocean, lakes, wetlands) on variable temporal and spatial scales?
4.	What controls weathering rates, and what are the fates of the weathered products, including nutrient loss in terrestrial ecosystems?
5.	What is the qualitative and quantitative role of non-human animals in the flux and cycling of nutrients, and what are the long-term effects of these fluxes (e.g., guano and other waste products)?
6.	How do the flux and cycling of antibiotics, steroids, hormones and pharmaceuticals affect element flux and cycling?
7.	What is the quantitative linkage between biogeochemistry and species richness, species extinction and invasion of alien species?
8.	What are the effects of lags and legacies on current and future biogeochemical fluxes and cycles?
9.	What are the quantitative interrelationships between hydrology and biogeochemistry?
10.	How can a better synoptic understanding of the biogeochemical flux, cycling and interaction of elements among air, land and water (including ocean) systems be achieved?
11.	What are the critical linkages and feedbacks among major nutrient and toxic element fluxes and cycles?
12.	What are the potential impacts of bioterrorism on biogeochemical fluxes and cycles, and human welfare that depends on these cycles?

estimates of the death toll related to such air pollution particles are controversial because of a glitch in the statistical software; nevertheless, morbidity estimates in the U.S. appear to be large (Kaiser, 2002).

My objective here is to identify some of the current major challenges for biogeochemistry where intensive research and creative thinking could be especially rewarding (Table I). I expand briefly on four of these: two (flux and cycling of N, and weathering release of Ca) that have been studied extensively for long periods and two (impacts of legacies, and the flux, cycling and effects of antibiotics, steroids, hormones, etc.) that have been studied less well from a biogeochemical point of view. My focus will be on biogeochemical flux and cycling. Flux is defined as a flow or movement across a boundary, real or specified, and cycling is movement and interchange within a boundary.

2. Challenges and Opportunities

Biogeochemistry is a vibrant, robust and growing field. From the early writings of Vernadsky (1944, 1945) and Hutchinson (1944, 1950) to in-depth analyses of lake (Schindler, 1980) and forest ecosystems (Likens et al., 1977) the field now has matured and become more focused on problems at regional to global scales and those directly relevant to humans (e.g., Schlesinger, 1997; Burke et al., 1998).

What are the 'big', important questions driving inquiry in biogeochemistry? I have no special crystal ball in this regard, but will offer some observations. I won't describe here current progress and discoveries. That has been done well in *Biogeomon 2002* and in this special issue.

Clearly our field, and others, is being pushed toward questions and programs requiring large, multidisciplinary teams (Likens, 1998, 2001a, b). This trend is fostered by the scope and complexity of the questions posed and addressed, as well as by the requirements of various funding agencies. This approach to big and relevant questions provides an important opportunity to make quantum leaps in our understanding as questions are tackled at large scales, by combining the talents of diverse disciplines, with powerful, new tools of analysis, statistics and communication, and by addressing directly the complexities involved.

Nevertheless, there often is a mismatch between the need for teamwork and the difficulties and insufficient training to organize a multidisciplinary research team and to make it work. We have opportunities for major advances in this area. Based on my experience as part of a team effort over several decades, I have emphasized the importance of trust in structuring efficient and productive teams (Likens, 1998, 2001a, b). Edmondson, in analyzing the effectiveness of hospital, medical teams and manufacturing work teams has found that psychological safety (PS), '... a shared belief held by members of a team that the team is safe for interpersonal risk taking...' strongly influences learning and performance of teams (Edmondson, 1999). Because most of us care about what other people think of us, Edmondson finds that PS is fundamental to leadership and thus characterizes the effectiveness and performance of group or team effort (e.g., Edmondson, 1999; Edmondson et al., 2001).

At a somewhat less lofty level, but possibly more important, is the value of individual intelligent inquiry, combined with serendipity, and often referred to as investigator-initiated research. Prior giants in our field, C. Darwin and G. E. Hutchinson, combined an extremely inquisitive approach with brilliant observational skills, characterized by keeping their eyes, ears and mind open to new thoughts and ideas (Likens, 2002). As a result they made seminal contributions to the field of biogeochemistry. Darwin isn't usually thought of as a biogeochemist, but many of his contributions were biogeochemical in nature (e.g., Darwin, 1881). A recent account of his valuing what probably was the first ecological experiment, including the chemical analysis of plants and soil in the development of his ideas, is especially noteworthy (Hector and Hooper, 2002).

I would argue that the truly major challenges for biogeochemistry in the future relate to aspects of human-accelerated environmental change, particularly their linkages and feedbacks (Likens, 1991, 1998), namely, global climate change, stratospheric ozone depletion, toxification of the biosphere (pollution of air, land and water), loss of biodiversity, and the all-pervasive effects of land-use changes. But, what are the specific effects and relationships of the increasing size of the human population on the flux and cycling of elements?

The effect of climate change on biogeochemical flux and cycling is poorly studied and in many cases is largely characterized by speculation, at least at the global scale. Some major questions related to climate change include: What controls carbon sequestration in different systems and on variable spatial and time scales, from stomates to ocean sediments?; What controls the fluxes of N to and from natural and human-dominated ecosystems? The effects of climate change (particularly from changes in atmospheric CO_2 concentration, regional changes in temperature, moisture and quality of precipitation) on biogeochemical flux and cycling, provide particularly important challenges and opportunities for research. Large-scale (plot/macrocosms) studies with elevated CO_2 (e.g., the Free-Air CO_2 Enrichment [FACE] network in the U.S.; Hendrey *et al.*, 1999; DeLucia *et al.*, 1999), and combining soil warming with experimental manipulations of moisture and nitrogen availability in subhectare- to hectare-sized plots (e.g., Rustad *et al.*, 2001) have great potential, but are expensive, difficult to maintain for long periods and require team efforts.

How can a better synoptic understanding of the biogeochemical flux, cycling and interaction of elements among air, land and water (including ocean) systems be obtained? Satellite-borne sensors and other types of remote sensing have been instrumental in the exciting development of synoptic approaches over a broad range of spatial and temporal scales (e.g., Matson and Ustin, 1991; Aber *et al.*, 1993; Martin and Aber, 1997; Burke *et al.*, 1998; Martin *et al.*, 1998; see Wessman and Asner, 1998). For example, the Airborne Visible InfraRed Imaging Spectrometer (AVIRIS) is making significant progress in assessing the N content of foliage on a regional scale (Ollinger *et al.*, 2002). Chlorophyll concentration maps are now available for entire areas of the ocean leading to better estimates of phytoplankton biomass and calculations of NPP for the entire ocean (Morel and Antoine, 2002). We are becoming much more skilled at addressing large-scale, biogeochemical questions with these new and developing technologies.

Nevertheless, much more attention needs to be given to integrating spatial and temporal patterns of global change to predict better the anthropogenic influence on ecosystem and biogeochemical processes. Generally, patterns of disturbance are not uniform (spatially or temporally) as area (scale) increases and processes become more incongruent in space and time. Recently, however, biogeochemistry has evolved from examining the flux and cycling of single elements to a greater focus on element interactions (e.g., Likens, 1981; Sterner *et al.*, 1992; Caraco *et al.*, 1993; Burke *et al.*, 1998). But, what are the critical linkages and feedbacks among major nutrient and toxic element cycles?

What is the linkage between biogeochemistry and species richness, species extinction and invasion of alien species? The effects of species richness, loss of species and invasion of alien species on biogeochemical flux and cycling are poorly known. An interesting example is the invasion of zebra mussels (*Dreissena polymorpha*) into the Hudson River (Strayer *et al.*, 1996, 1999; Caraco *et al.*, 1997). These alien species invaded the River in 1991 and by 1993 were the dominant filter feeder in the system, having reduced the native mussel population by 60%. There have been many ecological effects of this invasion, but the biogeochemical responses have included a decrease in dissolved oxygen by about 15% and increases in dissolved inorganic N and soluble reactive P in the River (Caraco *et al.*, 1997, 2000; Strayer *et al.*, 1999).

Most studies of community and population dynamics in ecosystems have not been integrated with biogeochemical analyses, but some recent studies have demonstrated the importance and value of such integration (e.g., Wedin and Tilman, 1990; Cronan and Grigal, 1995; Schimel *et al.*, 1996; Finzi *et al.*, 1998; Hooper and Vitousek, 1998; Tilman, 1998; Lovett *et al.*, 2002).

Better integration of chemistry, geology, biology and hydrology is needed in biogeochemical studies of watersheds and regions. For example, what are the quantitative interrelationships between hydrology and biogeochemistry; and what have been and will be the biogeochemical effects of extreme human impacts on hydrology, for example dams (less flashy runoff and sedimentation/sink for chemicals), paving (more flashy runoff), wetland drainage (more flashy runoff), reducing groundwater levels?

What will be the quantitative effect of El Niño – Southern Oscillations (ENSO) on biogeochemical flux and cycling, as influenced by global warming, on regional to global scales via the impact on flood-drought cycles and the resulting effects on biological and chemical cycles? There are strong spatial and temporal characteristics and responses to these climate anomalies (e.g., Molles and Dahm, 1990; www.cdc.noaa.gov/ENSO).

Moreover, understanding hydrological pathways can be very important for evaluating the biogeochemical response of soil, and groundwater, and surface waters to various environmental changes, including climate change, land-use disturbances, etc. (e.g., Hooper and Shoemaker, 1986; Church, 1997; Cirmo and McDonnell, 1997; Creed and Band, 1998; Kendall and McDonnell, 1998; Hill *et al.*, 2000; Mitchell, 2001).

Now, for a brief look at four major problems:

3. The Flux and Cycling of Nitrogen

Humans have disrupted the flux and cycling of N for a very long time (e.g., Vitousek et al., 1997), by the transfer of N from long-term storage pools: for example, by the release of N in fossil fuels to the atmosphere by combustion; and the transfer of N in guano deposits to agricultural fields. Through the Haber-Bosch process, huge amounts of N_2 in the atmosphere have been converted to biologically-reactive N (fertilizer) and applied to soils and waters worldwide (e.g., Vitousek et al., 1997; Galloway and Cowling, 2002b). Vitousek et al. (1997) estimated that human activity has approximately doubled the rate of nitrogen input to the global, terrestrial nitrogen cycle. Howarth et al. (2002) estimated that anthropogenic N inputs (e.g., inorganic fertilizer, atmospheric emissions of NO_x) to the U.S. doubled between 1961 and 1997.

Notwithstanding the importance of N to human societies, e.g., for food production, and the vast amount of research on the biogeochemistry of N, there still remain major uncertainties about the magnitude and controls on fluxes of N to and from natural and human-dominated ecosystems, especially at large scales (Table I). For example, ecological controls on N-fixation are poorly understood yet this is a critical biogeochemical flux, affecting primary production in both aquatic and terrestrial ecosystems (e.g., Howarth et al., 1999; Vitousek and Field, 1999; Marino et al., 2002; Vitousek et al., 2002). Even the potential role of N limitation for primary production, which has been studied extensively, is poorly understood in temperate forest ecosystems. The factors affecting the onset, magnitude and continuation of N-saturation in forests are variable spatially and temporally (e.g., Stoddard, 1994; Aber et al., 1998, 2002). Yet these factors may control retention and output of N from watersheds. A current 'hot topic' in N biogeochemistry is watershed retention and specifically the role of stream ecosystems and riparian areas in this retention (e.g., Fisher et al., 1998; Alexander et al., 2000; Peterson et al., 2001; Bernhardt et al., 2002). Likewise, our limited understanding of denitrification – where it occurs, how much occurs and what controls rates – is critical to understanding N retention in watersheds.

Caraco et al. (2003) found that relatively simple models, based on human-population density only, predicted variations of NO_3^- export for large rivers (watersheds >10 000 km^2), but these models lost predictive power at smaller scales and explained only 8% of the 1000-fold variation in NO_3^- export for watersheds <100 km^2. A somewhat more complex model using various loading factors, explained about 60 to 80% of the variance of this watershed export. Consideration of N retention and gaseous loss would increase the predictive power of these models, especially for small watersheds.

Based on volume-weighted concentrations and fluxes of dissolved inorganic N (DIN: $NH_4^+ + NO_3^-$) in precipitation at the Hubbard Brook Experimental Forest (HBEF) in New Hampshire, atmospheric inputs increased from about 200 mol ha^{-1} yr^{-1} in 1964–1965 to ∼700 mol ha^{-1} yr^{-1} in 1973–1974, and then fluc-

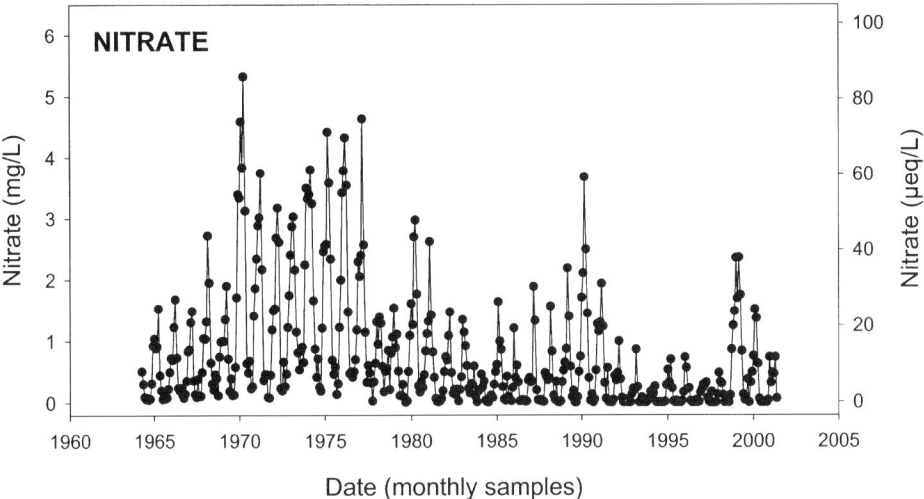

Figure 1. Monthly, volume-weighted concentrations of NO_3^- in stream water for Watershed 6 of the HBEF.

tuated around 500–600 mol N ha^{-1} yr^{-1} to the present. Streamwater fluxes of DIN, mostly NO_3^-, were relatively low (100–200 mol ha^{-1} yr^{-1}) during 1964–1965 to 1968–1969, increased to highest values (300–500 mol ha^{-1} yr^{-1}) during 1969–1976, and then have declined to the lowest on record (<100 mol ha^{-1} yr^{-1}) in 1993–1994 (Likens, 2001a, b). Increased streamwater fluxes during 1969–1970 to 1976–1977, 1979–1989 to 1980–1981 and 1989–1990 may have been related to soil freezing (in 1970, 1974, 1980, 1991–1993; Likens and Bormann, 1995; Mitchell *et al.*, 1996; Fitzhugh *et al.*, 2003), insect defoliation (in 1969–1971; Bormann and Likens, 1979); prior drought conditions (Aber *et al.*, 2002), and ice-storms (in 1998; Houlton *et al.*, 2003). Since forest biomass accumulation has been negligible since 1982 (Likens *et al.*, 1994; Likens, 2001a) it is surprising that streamwater fluxes of DIN have been so low since 1982.

A highly regular seasonal pattern is apparent in concentrations of streamwater NO_3^- at HBEF (Figure 1), with highest values during the winter and lowest during summer. The magnitude of the seasonal values has varied for different periods throughout the past 37 yr in response to the annual pattern (Figure 2). Only about 5–15% of annual N mineralization by soil heterotrophs occurs during the period of snow cover at the HBEF, typically from late November to mid April (Groffman *et al.*, 2001). Thus, the highest monthly streamwater values for NO_3^- during the dormant period for this largely deciduous forest are more likely due to reduced vegetation uptake than to microbial activity, but why have the maximum values occurred earlier with time (Figure 3)? The explanation for this change in timing is not clear, but may be related to anecdotal evidence that snowmelt began earlier and was more likely to be accompanied by rain after about 1983. April air tem-

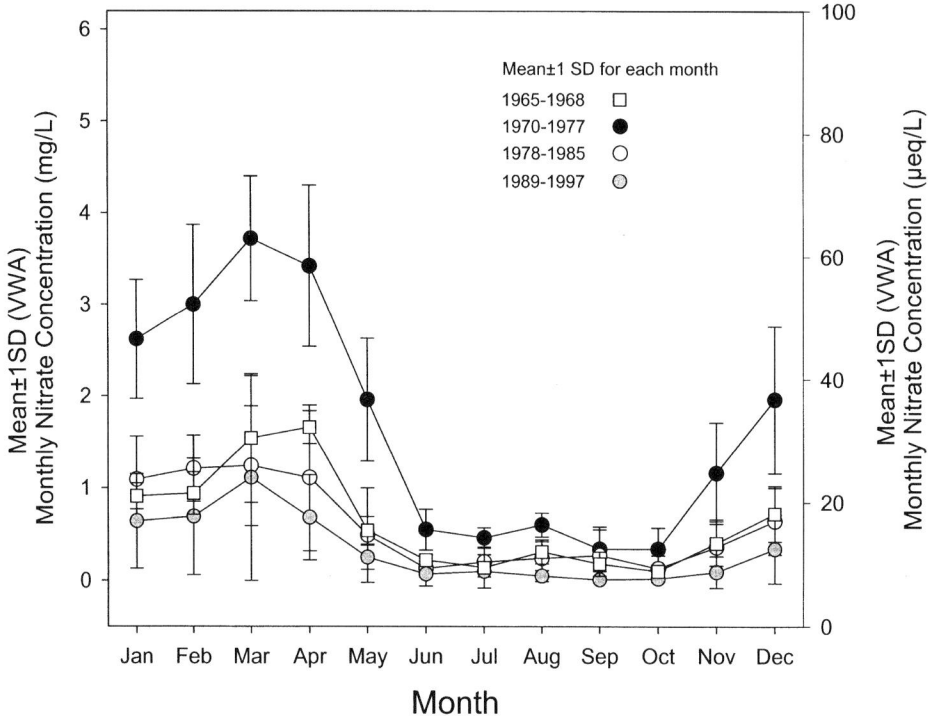

Figure 2. Mean (± SD) monthly streamwater concentrations for Watershed 6 of the HBEF during four different periods of the long-term record.

peratures also have increased significantly with time during this period (Likens, 2000). Because of the thaw concentration effect when the snowpack begins to melt (Hornbeck *et al.*, 1977), NO_3^- concentrations would be highest in stream water at that time. Thus, even after long periods of intensive study, interesting and important questions remain about the flux and cycling of N in natural systems.

Recently, an entire issue of *Ambio* (2002, Vol. 31, No. 2) was devoted to the biogeochemistry of reactive N, and various synthesis volumes are emerging (e.g., Boyer and Howarth, 2002; Neal, 2002). Complicated and exciting large-scale questions are posed by these efforts, such as, what are the effects of projected future changes in amounts of atmospheric deposition of reactive N throughout the globe and especially in developing countries and, what are the effects of reactive nitrogen flows as a result of production, transport and consumption of food containing N (Galloway and Cowling, 2002b). The location, extent (urban sprawl) and magnitude of future urban agglomerations (e.g., Likens 2001b) are particularly relevant to these flows and sinks (e.g., Baker *et al.*, 2001), and as such, represent major biogeochemical questions of great importance to human societies.

Galloway and Cowling (2002a) suggest two critical topics for research in N biogeochemistry: What is the fate of reactive N released to the environment by

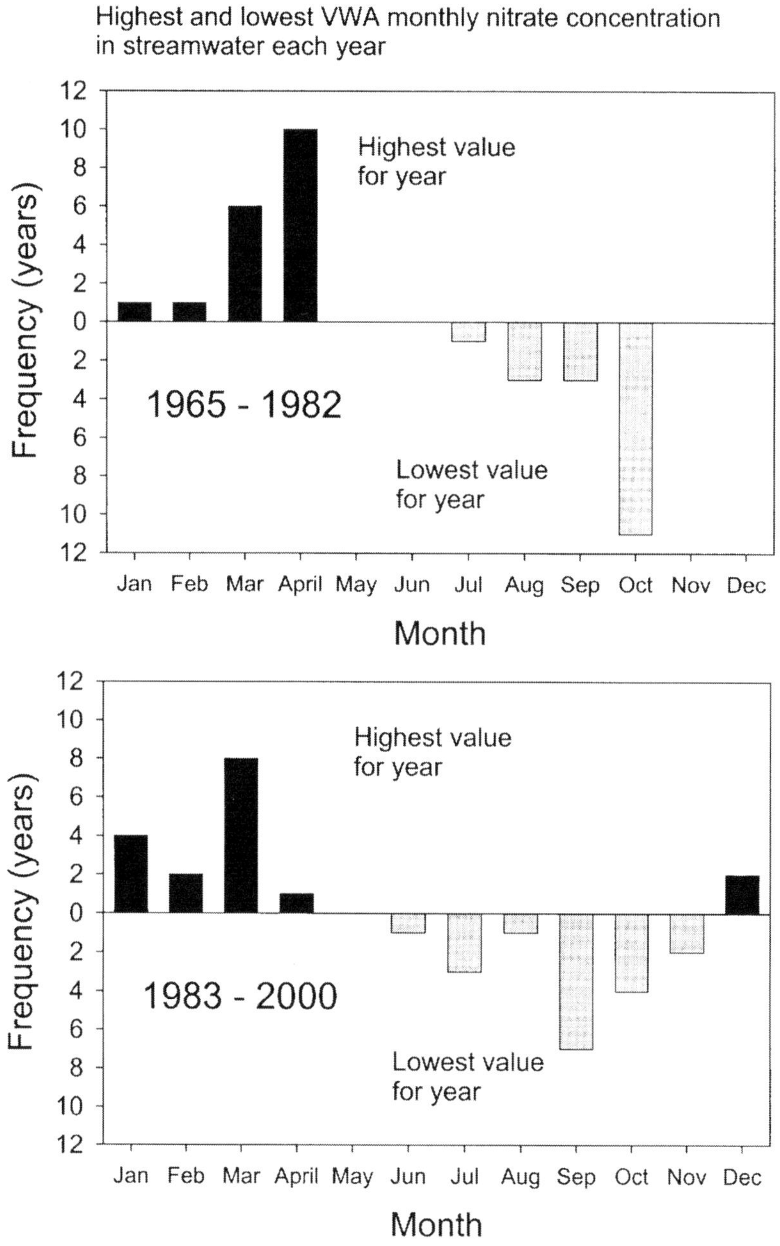

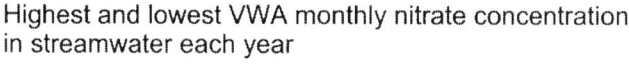

Figure 3. Highest and lowest, volume-weighted average streamwater NO_3^- concentrations for Watershed 6 of the HBEF during 1964–2000.

human action? And, how can this increasing amount of reactive N produced by energy production and food production be decreased? Given this call 'to optimize nitrogen management in food, fiber and energy production, and to minimize detrimental human and environmental impacts', they propose four opportunities: '– Increase nitrogen-use efficiency in food production; – Recycle reactive nitrogen within agroecosystems and managed forests; – Decrease NO_x emissions from fossil-fuel combustion by capturing NO_x for other uses or by eliminating its formation; – To the extent that the above are not possible, convert reactive nitrogen back to nonreactive N_2 before it is lost to the environment'. Clearly, these are critical and 'big' questions for understanding and managing the flux and cycling of reactive N relative to human societies.

4. Weathering Release

What is the difference between and what controls apparent and actual weathering rates in terrestrial ecosystems, and what is the fate of the weathered products? Very sophisticated tools, e.g., stable isotopes, are now available for studies of weathering release of nutrients, but the challenges are to ask the penetrating question, aided by the use of these tools, and to interpret the results obtained.

Using a mass-balance approach, Likens *et al.* (1998) estimated that about 50 mol Ca ha^{-1} a^{-1} was released by weathering (based on plagioclase weathering), compared to $\sim$30 mol Ca ha^{-1} a^{-1} input in atmospheric deposition. The weathering release supplied $\sim$25% of annual streamwater export and $\sim$3% of the annual gross plant uptake of Ca^{2+} during 1987–1992 at the HBEF. Recently, Blum *et al.* (2002) used Sr isotopes and Sr Ca^{-1} ratios to estimate that weathering of apatite by mycorrhizal 'mining' could account for significant Ca^{2+} in stream water and could provide an additional 'direct' source of Ca^{2+} for some trees at the HBEF. (Internal rock 'mining' by ectomycorrhizal mycelia is a developing 'hot topic' in weathering and plant nutrition – e.g., van Breemen *et al.*, 2000). Blum *et al.* (2002) estimated that the current sources of Ca^{2+} in stream water were $\sim$30% from the atmosphere, $\sim$35% from silicate minerals and $\sim$35% from apatite. Also using Sr isotopes, Bailey *et al.* (1996) suggested that vegetation obtained about 4% of its annual Ca^{2+} uptake from atmospheric inputs and weathering release represented $\sim$30% of streamwater output in a nearby watershed. Kennedy *et al.* (2002), however, used Sr isotopes to conclude that atmospheric input provided >90% of the Ca^{2+} used by vegetation in a remote Chilean forest and that weathering inputs of Ca^{2+} to plant nutrition were minimal.

Even though weathering has been studied for as long as any biogeochemical flux, uncertainties still remain in spite of the use of powerful tools. There are major biotic, geologic and atmospheric differences between these two systems (HBEF; remote Chilean forest) in opposite hemispheres. For example, the Chilean system is an old-growth forest, dominated by 'clean', atmospheric inputs, whereas the New Hampshire system is a relatively young forest, disturbed during the last century by logging, hurricanes and acid rain. There also are major differences in depth and type (till vs. bedrock) of weatherable materials (e.g., Likens *et al.*, 1977; Kennedy *et al.*, 1998) and cloudwater inputs (e.g., Weathers *et al.*, 2000). Do these factors explain the differences in results/conclusions relative to Ca^{2+} biogeochemistry, or are there even more fundamental differences? How do these differences affect the interpretation of effects due to air pollution and sustainable management of forests in both regions, and particularly as atmospheric inputs may change with time?

5. Legacies and Lags

So-called 'chemical time bombs' or legacies of past human actions or material displacements can profoundly impact current biogeochemical flux and cycling. Moreover, how have biogeochemical fluxes and cycles changed over geologic and historical time? What clues and tracers indicate their rates and magnitudes over time? Lags and legacies affecting questions of biogeochemical flux and cycling may be among the most challenging and interesting, to understand quantitatively. There are many examples known, and many more to be explored. Of the more obvious are avian guano deposits mined in Peru and Chile and deposited in the northern hemisphere as fertilizer; evapotranspiration to produce crops, with N transferred via the harvest to intensive utilization and waste generation areas, such as urban agglomerations and animal feedlots (such transfers often result in the transfer of toxic elements, e.g., pesticides); groundwater contaminants, such as chromium, can be used to track water and pollution movement and remediation of this resource over long times (Blowes, 2002); transfer of N, S and C condensed in fossil resources (coal and oil) to a dispersed form in the atmosphere with combustion and then returned at some distance from the source, via wet and dry deposition. Many biogeochemical legacies, e.g., effects on soil fertility, may be very old and widespread, yet are important to modern ecosystems as shown recently by anthropogical data from Brazilian rainforests (Moffat, 2002).

6. Antibiotics, Hormones, Steroids and other Biologically-Active Compounds

The occurrence, persistence, flux and cycling, and thus the biogeochemical importance of substances such as antibiotics, steroids, hormones and pharmaceuticals, are very poorly known in natural ecosystems. These substances are increasingly used by human societies, and potentially active forms are released widely to soils, to surface and ground waters, and to the atmosphere (e.g., Mallin, 2000). Unfortunately, there are few reliable data on the amounts of antimicrobials used in animal production in the U.S. One estimate suggests that about 11 200 metric tons per year of antimicrobials are used for nontherapeutic purposes on livestock, some 8 times more than used in human medicines (Mellon *et al.*, 2001). Worse, there are no quantitative data on the release of antimicrobials to the environment. Data

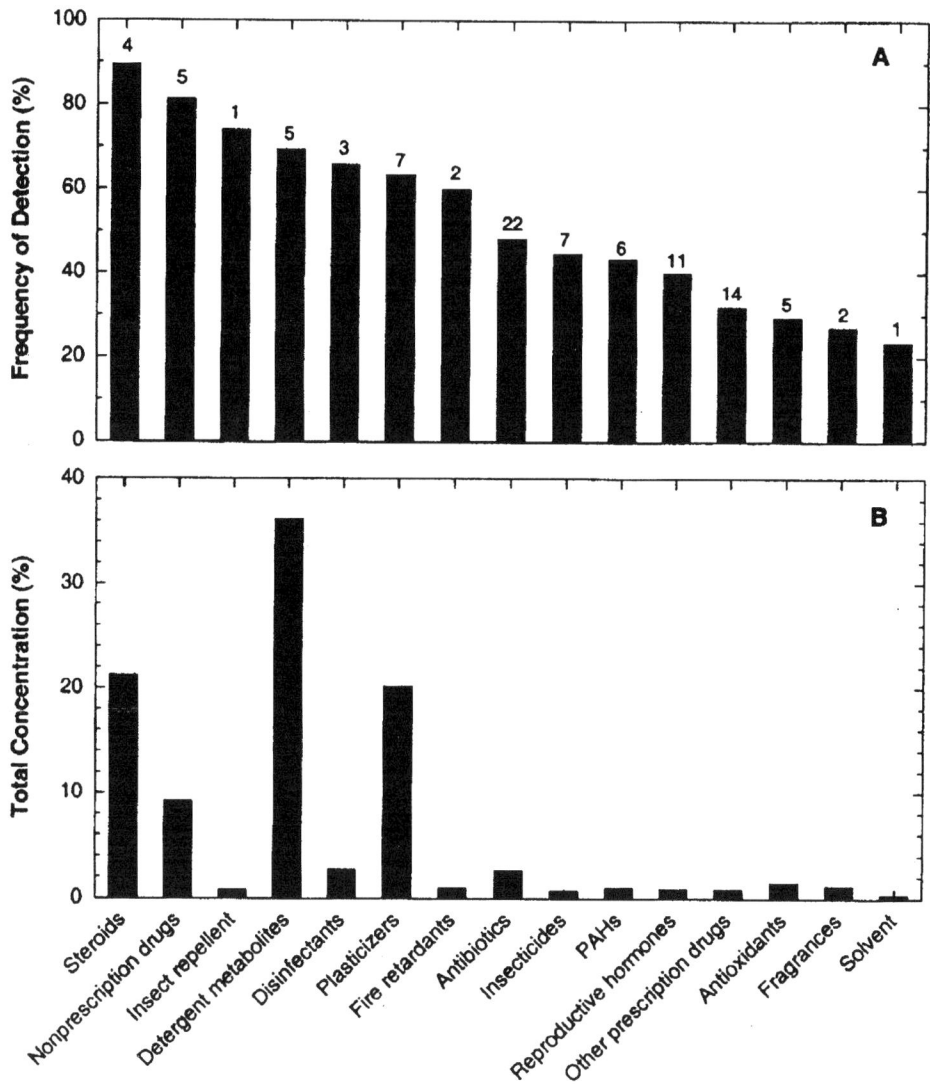

Figure 4. Frequency of detection and percent of total measured organic wastewater contaminants in 139 streams in 30 U.S. states during 1999 and 2000 (from Kolpin *et al.*, 2002).

on the release of hormonally-active compounds are even more scarce, or at least not available publicly (J. P. Meyers, personal communication). Apparently, many of these contaminants can pass through wastewater treatment plants and domestic septic systems into the environment relatively intact (Halling-Sorensen *et al.*, 1998; Kolpin *et al.*, 2002). A recent study (Kolpin *et al.*, 2002) found such organic contaminants in 80% of 139 streams sampled across the United States. 'The most frequently detected compounds were coprostanol (fecal steroid), cholesterol (plant and animal steroid), N, N-diethyltoluamide (insect repellant), caffeine (stimulant), triclosan (antimicrobial disinfectant), tri(2-chloroethyl)phosphate (fire retardant) and 4-nonylphenol (nonionic detergent metabolite)' (Kolpin et al., 2002). A median of 7 and as many as 38 of these contaminants were found in single water samples (Figure 4).

A significant impediment to studies of such components is the lack of appropriate or widely available methodology for their detection. Such contaminants often exist at very low concentrations. Also currently, the cost of such analyses is prohibitive to routine monitoring. Kolpin and colleagues (2002) needed to develop or modify 5 methods for their study. The development of specific molecular and microbial tools may be appropriate and particularly helpful for such studies in the future.

These biologically active compounds probably have major direct and indirect impacts on biogeochemical cycles. I believe that this subject deserves much attention in the future, particularly given the rapid development of high-density animal agriculture (e.g., Batie, 1993) around the world.

7. Final Thoughts

Biogeochemical information is critical for identifying the control points for management of environmental problems. 'Acid rain' (Odén, 1968; Likens et al., 1972) is a good example, as is the cultural eutrophication of freshwater lakes or coastal marine areas where control is managed largely by limiting the inputs of S, N or P (e.g., Hasler, 1947; Vollenweider, 1968; Schindler, 1977; Howarth, 1993; Vitousek and Howarth, 1991; Smith, 1998; Boyer and Howarth, 2002). Based on an understanding of these relationships, a need and an opportunity develop for direct human intervention in managing biogeochemical cycles to alleviate environmental problems, as may be required by managers and decision makers. Such widespread environmental problems persist and are likely to continue for the foreseeable future. Two media headlines during July 2002 proclaimed that acid rain damage is worse than 'previously believed/thought' based on studies of Ca^{2+} availability and plant uptake (Schaberg et al., 2001; Kennedy et al., 2002). It is always difficult to know what was previously 'thought', but clearly the ecological and biogeochemical impact of acid rain has not gone away and this environmental problem provides many challenges and research opportunities for the future. Using computer simulation approaches, Driscoll et al. (2001) and Evans et al. (2001) have suggested that surface waters in areas sensitive to acid rain in the northeastern U.S. and in the United Kingdom, respectively, would not recover to positive acid-neutralizing capacity for many decades under current emission reduction protocols. Resolving such complicated issues necessitates, as is always the case, dealing with the details, and 'the devil is in the details'. But, scientists thrive on studying, understanding, and integrating details to resolve the biogeochemical 'torments' of importance to human societies.

8. Conclusions

The dominant and enlarging role of humans in altering biogeochemical cycles, particularly fluxes – >100% of natural mobilization of N, S and probably P (Schlesinger, 1997); somewhat less to intermediate mobilization of others, e.g., B (Park and Schlesinger, 2002) and a smaller proportional impact on for example, C (but of great import relative to climate), throughout the planet is now obvious.

Controversial environmental issues and problems associated with large-scale disruption of natural biogeochemical cycles are continuing or increasing and are coupled with a need for increased understanding of the effects of direct human intervention (e.g., fertilizing ocean with Fe to combat global increases in atmospheric CO_2 (e.g., Chisholm et al., 2001; Johnson et al., 2002); hypoxia in coastal waters from agricultural N runoff (e.g., Rabalais and Turner, 2001); regional degradation of surface waters, forests and soils from acid rain (e.g., Likens et al., 1979; Driscoll et al., 2001; Evans et al., 2001; Stoddard et al., 1999), but areal 'hot spots' remain, such as intensive animal agriculture (e.g., Mallin, 2000).

Long-term data and sustained study are critical for understanding patterns and trends in biogeochemical problems.

There is a critical role for multidisciplinary teams in tackling complicated biogeochemical problems, but training of team leaders, team members and the teams themselves is largely lacking or grossly deficient. There appear to be major opportunities for improvement in team efficiency and productivity (e.g., Likens, 2001b; Edmondson, 1999). There is a need to sustain a strong and healthy mix of investigator-initiated research and large, team projects. There is a need for more large-scale, biogeochemical experiments to examine multiple stressors simultaneously. Such experiments provide powerful and dynamic opportunities for unraveling biogeochemical problems.

Several 'old', but still important biogeochemical problems persist in terms of scientific understanding and utilization and integration of knowledge in the management of these problems, e.g., eutrophication of aquatic ecosystems and acid rain.

There is a need for more and better synoptic study, and integration and interpretation of results over large scales (regional to global).

There is a need for better use of new and powerful tools (e.g., stable isotopes, spatial statistics, fractal analysis (e.g., Kirchner et al., 2000), genomics and other molecular techniques) in biogeochemical studies, but development of clear questions and interpretation of results remain a challenge. A 'tool' that potentially has great value for future biogeochemical studies is the archiving of samples for future analysis. Museum collections have been used in a variety of ways for biogeochemical studies and should be in the future. At the HBEF we established a formal archive of soil, water, wood, etc. samples in the early 1990s. Samples are bar coded and procedures have been established for protection and use of these samples. Alewell *et al.* (1999 and 2000) successfully measured ^{34}S in stored water samples, collected over several decades, to elucidate critical aspects of S biogeochemistry at the HBEF.

There is a need for greater understanding of the quantitative role of microorganisms in biogeochemical cycles.

Currently we live in a world characterized by significant bioterrorism (e.g., Likens, 2002). What are the potential impacts of bioterrorism on biogeochemical flux and cycling and on human welfare that depends on these cycles?

Acknowledgements

I thank J. Aber, E. S. Bernhardt, N. Caraco, J. J. Cole, R. Goldberg, J. P. Meyers, M. J. Mitchell, W. H. Schlesinger and J. S. Warner, for stimulating discussions and suggestions. D. Buso assisted with illustrations and P. Likens and H. Dahl with manuscript preparation. Two anonymous reviewers provided helpful comments. Financial support was provided by The Andrew W. Mellon Foundation, National Science Foundation and the Institute of Ecosystem Studies. This is a contribution to the program of the Institute of Ecosystem Studies.

References

Aber, J. D., Driscoll, C. T., Federer, C. A., Lathrop, R., Lovett, G. M., Melillo, J. M., Steudler, P. and Vogelmann, J.: 1993, 'A strategy for the regional analysis of the effects of physical and chemical climate change on biogeochemical cycles in northeastern (U.S.) forests', *Ecol. Model.* **67**, 37–47.

Aber, J. D., McDowell, W. H., Nadelhoffer, K. J., Magill, A., Berntson, G., Kamakea, M., McNulty, S. G., Currie, W., Rustad, L. and Fernandez, I.: 1998, 'Nitrogen saturation in temperate forest ecosystems: Hypotheses revisited', *BioScience* **48**, 921–934.

Aber, J. D., Ollinger, S. V., Driscoll, C. T., Likens, G. E., Holmes, R. T., Freuder, R. J. and Goodale, C. L.: 2002, 'Inorganic N losses from a forested ecosystem in response to physical, chemical, biotic and climatic perturbations', *Ecosystems* **5**, 648–658.

Alewell, C., Mitchell, M. J, Likens, G. E. and Krouse, R. H.: 1999, 'Sources of stream sulfate at the Hubbard Brook Experimental Forest: Long-term analyses using stable isotopes', *Biogeochemistry* **44**, 281–299.

Alewell, C., Mitchell, M. J., Likens, G. E. and Krouse, R.: 2000, 'Assessing the origin of sulfate deposition at the Hubbard Brook Experimental Forest', *J. Environ. Qual.* **29**, 759–767.

Alexander, R. B., Smith, R. A. and Schwarz, G. E.: 2000, 'Effect of stream channel size on the delivery of nitrogen to the Gulf of Mexico', *Nature* **403**, 758–761.

Bailey, S. W., Hornbeck, J. W., Driscoll, C. T. and Gaudette, H. E.: 1996, 'Calcium inputs and transport in a base-poor forest ecosystem as interpreted by Sr isotopes', *Water Resour. Res.* **32**, 707–719.

Baker, L. A., Hope, D., Ying Xu, Edmonds, J. and Lauver, L.: 2001, 'Nitrogen balance for the central Arizona-Phoenix ecosystem', *Ecosystems* **4**, 582–602.

Batie, S. S.: 1993, *Soil and Water Quality, An Agenda for Agriculture*, National Academy Press, Washington, D.C..

Bernhardt, E. S., Hall Jr., R. O. and Likens, G. E.: 2002, 'Whole-system estimates of nitrification and nitrate uptake in streams of the Hubbard Brook Experimental Forest', *Ecosystems* **5**, 419–430.

Blowes, D.: 2002, 'Tracking hexavalent Cr in groundwater', *Science* **295**, 2024–2025.

Blum, J. D., Klaue, A., Nezat, C. A., Driscoll, C. T., Johnson, C. E., Siccama, T. G., Eagar, C., Fahey, T. J. and Likens, G. E.: 2002, 'Mycorrhizal weathering of apatite as an important calcium source in base-poor forest ecosystems', *Nature* **417**, 729–731.

Bormann, F. H. and Likens, G. E.: 1979, *Pattern and Process in a Forested Ecosystem*, Springer-Verlag, Inc., New York, pp. 253.

Boyer, E. and Howarth, R. W. (eds): 2002, *The Nitrogen Cycle at Regional to Global Scales. Report of the International SCOPE Nitrogen Project*, Kluwer Academic Publishers, Dordrecht, Boston, London, pp. 519.

Burke, I. C., Lauenroth, W. K. and Wessman, C. A.: 1998, 'Progress in understanding biogeochemical cycles at regional to global scales', in M. L. Pace and P. M. Groffman (eds), *Successes, Limitations, and Frontiers in Ecosystem Science*, Springer-Verlag, New York, pp. 165–194.

Caraco, N. F., Cole, J. J. and Likens, G. E.: 1993, 'Sulfate control of phosphorus availability in lakes: A test and re-evaluation of Hasler and Einsele's model', *Hydrobiologia* **253**, 275–280.

Caraco, N., Cole, J. J., Raymond, P. A., Strayer, D. L., Pace, M. L., Findlay, S. E. G. and Fischer, D. T.: 1997, 'Zebra mussel invasion in a large, turbid river: Phytoplankton response to increased grazing', *Ecology* **78**, 588–602.

Caraco, N. F., Cole, J. J., Findlay, S. E. G., Fischer, D. T., Lampman, G. G., Pace, M. L. and Strayer, D. L.: 2000, 'Dissolved oxygen declines in the Hudson River associated with the invasion of the zebra mussel (*Dresissena polymorpha*)', *Environ. Sci. Technol.* **34**, 1204–1210.

Caraco, N. F., Cole, J. J., Likens, G. E., Lovett, G. M. and Weathers, K. C.: 2003, 'Variation in NO_3^- export from flowing waters of vastly different sizes: Does one model fit all?' *Ecosystems* **6** 344–352.

Chisholm, S. W., Falkowski, P. G. and Cullen, J. J.: 2001, 'Dis-crediting ocean fertilization', *Science* **294**, 309–310.

Church, M. R.: 1997, 'Hydrochemistry of forested catchments', *Annu. Rev. Earth Planet Sci.* **25**, 23–59.

Cirmo, C. P. and McDonnell, J. J.: 1997, 'Linking the hydrologic and biogeochemical controls of nitrogen transport in near-stream zones of temperate-forested catchments: A review', *J. Hydrol.* **199**, 88–120.

Creed, I. F. and Band, L. E.: 1998, 'Exploring functional similarity in the export of nitrate-N from forested catchments: A mechanistic modeling approach', *Water Resour. Res.* **34**, 3079–3093.

Cronan, C. S. and Grigal, D. F.: 1995, 'Use of calcium/aluminum ratios as indicators of stress in forest ecosystems', *J. Environ. Qual.* **24**, 209–226.

Darwin, C.: 1859, *On the Origin of Species by Means of Natural Selection, or the Preservation of Favored Races in the Struggle for Life*, John Murray, London.

Darwin, C.: 1881, *The Formation of Vegetable Mould, through the Actions of Worms, with Observations on Their Habits*, John Murray, London.

DeLucia, E. H., Hamilton, J. G., Naidu, S. L., Thomas, R. B., Andrews, J. A., Finzi, A., Lavine, M., Matamala, R., Mohan, J. E., Hendrey, G. R. and Schelsinger, W. H.: 1999, 'Net primary production of a forest ecosystem with experimental CO_2 enrichment', *Science* **284**, 1174–1179.

Driscoll, C. T., Lawrence, G. B., Bulger, A. J., Butler, T. J., Cronan, C. S., Eagar, C., Lambert, K. F., Likens, G. E., Stoddard, J. L. and Weathers, K. C.: 2001, 'Acidic deposition in the northeastern United States: Sources and inputs, ecosystem effect, and management strategies', *BioScience* **51**, 180–198.

Edmondson, A. C.: 1999, 'Psychological safety and learning behavior in work teams', *Admin. Sci. Quart.* **44**, 350–383.

Edmondson, A. C., Bohmer, R. M. and Pisano, G. P.: 2001, 'Speeding up team learning', *Harvard Bus. Rev.* **79**, 125–132.

Evans, C., Jenkins, A., Helliwell, R., Ferrier, R. and Collins, R.: 2001, *Freshwater Acidification and Recovery in the United Kingdom*, Centre for Ecology and Hydrology, The Macaulay Institute, Natural Environment Research Council, Wallingford, U.K., pp. 80.

Finzi, A. C., van Breemen, N. and Canham, C. D.: 1998, 'Canopy tree-soil interactions within temperate forests: Species effects on soil carbon and nitrogen', *Ecol. Appl.* **8**, 440–446.

Fisher, S. G., Grimm, N. B., Marti, E., Homes, R. M. and Jones Jr., J. B.: 1998, 'Material spiraling in stream corridors: A telescoping ecosystem model', *Ecosystems* **1**, 19–34.

Fitzhugh, R. D., Likens, G. E., Driscoll, C. T., Mitchell, M. J., Groffman, P. M., Fahey, T. J. and Hardy, J. P.: 2003, 'Role of soil freezing events in interannual patterns of stream chemistry at the Hubbard Brook Experimental Forest', *Environ. Sci. Tech.* **37**, 1575–1580.

Galloway, J. N. and Cowling, E. B.: 2002a, 'Reactive nitrogen', *Ambio* **31**, 59.

Galloway, J. N. and Cowling, E. B.: 2002b, 'Reactive nitrogen and the world: 200 years of change', *Ambio* **31**, 64–71.

Groffman, P. M., Driscoll, C. T., Fahey, T. J., Hardy, J. P., Fitzhugh, R. D. and Tierney, G. L.: 2001, 'Effects of mild winter freezing on soil nitrogen and carbon dynamics in a northern hardwood forest', *Biogeochemistry* **56**, 191–213.

Halling-Sorensen, B., Nielson, S. N., Lanzky, P. F., Ingerslev, F., Holten Lutzhoft, J. and Jorgenson, S. E.: 1998, 'Occurrence, fate and effects of pharmaceutical substances in the environment – A review', *Chemosphere* **36**, 357–394.

Hasler, A. D.: 1947, 'Eutrophication of lakes by domestic drainage', *Ecology* **28**, 383–395.

Hector, A. and Hooper, R.: 2002, 'Darwin and the first ecological experiment', *Science* **295**, 639–640.

Hendrey, G. R., Ellsworth, D. S., Lewin, K. F. and Nagy, J.: 1999, 'A free-air enrichment system for exposing tall forest vegetation to elevated atmospheric CO_2', *Global Change Biol.* **5**, 209–293.

Hill, A. R., Devito, K. J., Campagnolo, S. and Sanmugadas, K.: 2000, 'Subsurface denitrification in a forest riparian zone: Interactions between hydrology and supplies of nitrate and organic carbon', *Biogeochemistry* **51**, 193–223.

Hooper, D. U. and Vitousek, P. M.: 1998, 'Effects of plant composition and diversity on nutrient cycling', *Ecol. Monogr.* **68**, 121–149.

Hooper, R. P. and Shoemaker, C. A.: 1986, 'A comparison of chemical and isotopic hydrograph separation', *Water Resour. Res.* **22**, 1444–1454.

Hornbeck, J. W., Likens, G. E. and Eaton, J. S.: 1977, 'Seasonal patterns in acidity of precipitation and the implications for forest-stream ecosystems', *Water, Air, Soil Pollut.* **7**, 355–365.

Houlton, B. Z., Driscoll, C. T., Fahey, T. J., Likens, G. E., Groffman, P. M., Bernhardt, E. S. and Buso, D. C.: 2003, 'Nitrogen dynamics in ice storm-damaged forest ecosystems: Implications for nitrogen limitation theory', *Ecosystems* **6**, 431–443.

Howarth, R. W.: 1993, 'The role of nutrients in coastal waters', in Report from the National Research Council Committee on Wastewater Management for Coastal Urban Areas, *Managing Wastewater in Coastal Urban Areas*, National Academy Press, Washington, D.C., pp. 177–202.

Howarth, R. W., Chan, F. and Marino, R.: 1999, 'Do top-down and bottom-up controls interact to exclude nitrogen-fixing cyanobacteria from the plankton of estuaries? An exploration with a simulation model', *Biogeochemistry* **46**, 203–231.

Howarth, R. W., Boyer, E. W., Pabich, W. J. and Galloway, J. N.: 2002, 'Nitrogen use in the United States from 1961–2000 and potential future trends', *Ambio* **31**, 88–96.

Hutchinson, G. E.: 1944, 'Nitrogen in the biogeochemistry of the atmosphere', *Am. Sci.* **32**, 178–195.

Hutchinson, G. E.: 1950, *Survey of Contemporary Knowledge of Biogeochemistry. III. The Biogeochemistry of Vertebrate Excretion.* Bull. Amer. Mus. Nat. History, **96**, pp. 554.

Johnson, K. S. and Karl, D. M.: 2002, 'Is ocean fertilization credible and creditable'? (Response to Chisholm *et al*). *Science* **296**, 468.

Kaiser, J.: 2000, 'Evidence mounts that tiny particles can kill', *Science* **289**, 22–23.

Kaiser, J.: 2002, 'Software glitch threw off mortality estimates', *Science* **296**, 1945–1947.

Kendall, C. and McDonnell, J. (eds): 1998, *Isotope Tracers in Catchment Hydrology*, Elsevier, The Netherlands.

Kennedy, M. J., Chadwick, O. A., Vitousek, P. M., Derry, L. A. and Hendricks, D. M.: 1998, 'Changing sources of base cations during ecosystem development, Hawaiian Islands', *Geology* **26**, 1015–1018.

Kennedy, M. J., Hedin, L. O. and Derry, L. A.: 2002, 'Decoupling of unpolluted temperate forests from rock nutrient sources revealed by natural $^{87}Sr/^{86}Sr$ and ^{84}Sr tracer addition', *Proc. Natl. Acad. Sci. U.S.A.* **99**, 9639–9644.

Kirchner, J. W., Feng, X. H. and Neal, C.: 2000, 'Fractal stream chemistry and its implications for contaminant transport in catchments', *Nature* **403**, 524–527.

Kolpin, D. W., Furlong, E. T., Meyer, M. T., Thurman, E. M., Zaugg, S. D., Barber, L. B. and Buxton, H. T.: 2002, 'Pharmaceuticals, hormones, and other organic wastewater contaminants in U.S. streams, 1999–2000: A national reconnaissance', *Environ. Sci. Technol.* **36**, 1202–1211.

Likens, G. E. (ed.): 1981, *Some Perspectives of the Major Biogeochemical Cycles*, Vol. 17, SCOPE IVth General Assembly, Stockholm, Sweden. John Wiley & Sons, Ltd., Chichester, pp. 170.

Likens, G. E.: 1991, 'Human-accelerated environmental change', *BioScience* **41**, 130.

Likens, G. E.: 1998, 'Limitations to intellectual progress in ecosystem science', in M. L. Pace and P. M. Groffman (eds.), *Successes, Limitations and Frontiers in Ecosystem Science, 7th Cary Conference, Institute of Ecosystem Studies*, Millbrook, New York, Springer-Verlag, New York, Inc., pp. 247–271.

Likens, G. E.: 2000, 'A long-term record of ice-cover for Mirror Lake, NH: Effects of global warming'? *Verh. Int. Ver. Limnol.* **27**, 2765–2769.

Likens, G. E.: 2001a, 'Biogeochemistry, the watershed approach: Some uses and limitations', *Mar. Freshwat. Res.* **52**, 5–12.

Likens, G. E.: 2001b, 'Ecosystems: Energetics and biogeochemistry'. in W. J. Kress and G. W. Barrett (eds), *A New Century of Biology*, Smithsonian Institution Press, Washington and London, pp. 53–88.

Likens, G. E.: 2002, 'A thimbleful of powder', *BioScience* **52**, 547.

Likens, G. E. and Bormann, F. H.: 1995, *Biogeochemistry of a Forested Ecosystem*, 2nd ed., Springer-Verlag, New York, pp. 159.

Likens, G. E., Bormann, F. H. and Johnson, N. M.: 1972, 'Acid rain', *Environment* **14**, 33–40.

Likens, G. E., Bormann, F. H., Pierce, R. S., Eaton, J. S. and Johnson, N. M.: 1977, *Biogeochemistry of a Forested Ecosystem*, Springer-Verlag, New York, pp. 146.

Likens, G. E., Wright, R. F., Galloway, J. N. and Butler, T. J.: 1979, 'Acid rain', *Sci. Am.* **241**, 43–51.

Likens, G. E., Driscoll, C. T., Buso, D. C., Siccama, T. G., Johnson, C. E., Ryan, D. F., Lovett, G. M., Fahey, T. and Reiners, W. A.: 1994, 'The biogeochemistry of potassium at Hubbard Brook', *Biogeochemistry* **25**, 61–125.

Likens, G. E., Driscoll, C. T., Buso, D. C., Siccama, T. G., Johnson, C. E., Lovett, G. M., Fahey, T. J., Reiners, W. A., Ryan, D. F., Martin, C. W. and Bailey, S. W.: 1998, 'The biogeochemistry of calcium at Hubbard Brook', *Biogeochemistry* **41**, 89–173.

Lovett, G. M., Weathers, K. C. and Arthur, M. A.: 2002, 'Control of N loss from forested watersheds by soil C:N ratio and tree species composition', *Ecosystems* **5**, 712–718.

Mallin, M. A.: 2000, 'Impacts of industrial animal production on rivers and estuaries', *Am. Sci.* **88**, 26–37.

Marino, R., Chan, F., Howarth, R. W., Pace, M. and Likens, G. E.: 2002, 'Ecological and biogeochemical interactions constrain planktonic nitrogen fixation in estuaries', *Ecosystems* **5**, 719–724.

Martin, M. E. and Aber, J. D.: 1997, 'Estimation of forest canopy lignin and nitrogen concentration and ecosystem processes by high spectral resolution remote sensing', *Ecol. Appl.* **7**, 431–443.

Martin, M. E., Newman, S. D., Aber, J. D. and Congalton, R. G.: 1998, 'Determining forest species composition using high spectral resolution remote sensing data', *Remote Sens. Environ.* **65**, 249–254.

Matson, P. A. and Ustin, S. L.: 1991, 'Special feature: The future of remote sensing in ecological studies', *Ecology* **76**, 1917.

Mellon, M., Benbrook, C. and Benbrook, K. L.: 2001, *Hogging It! Estimates of Antimicrobial Abuse in Livestock*, Union of Concerned Scientists, Cambridge, Massachusetts, pp. 109.

Mitchell, M. J.: 2001, 'Linkages of nitrate losses in watersheds to hydrological processes', *Hydrol. Process.* **15**, 3305–3307.

Mitchell, M. J., Driscoll, C. T., Kahl, J. S., Likens, G. E., Murdoch, P. S. and Pardo, L. H.: 1996, 'Climatic control of nitrate loss from forested watersheds in the northeast United States', *Environ. Sci. Technol.* **30**, 2609–2612.

Moffat, A. S.: 2002, 'South American landscapes: Ancient and modern', *Science* **296**, 1959–1961.

Molles, M. C. and Dahm, C. N.: 1990, 'A perspective on El Niño and La Niña: Global implications for stream ecology', *J. N. Am. Benthol. Soc.* **9**, 68–76.

Morel, A. and Antoine, D.: 2002, 'Small critters – Big effects', *Science* **296**, 1980–1982.

Neal, C.: 2002, 'Assessing nitrogen dynamics in catchments across Europe within an INCA modelling framework', *Hydrol. Earth System Sci.* **6**, 1–615.

Odén, S.: 1968, 'Nederbördens Och Luftens Försurning - Dess Orsaker, Förlopp Och Verkan 1 Olka Miljöer', *Ekologikommittén Bull.* **1**, 1–86.

Ollinger, S. V., Smith, M. L., Martin, M. E., Hallett, R. A., Goodale, C. L. and Aber, J. D.: 2002, 'Regional variation in foliar chemistry and N cycling among forests of diverse history and composition', *Ecology* **83**, 339–355.

Park, H. and Schlesinger, W. H.: 2002, 'The global biogeochemical cycle of boron', Unpublished manuscript, 39 pp.

Peterson, B. J., Wollheim, W. M., Mulholland, P. J., Webster, J. R., Meyer, J. L., Tank, J. L., Marti, E., Bowden, W. B., Valett, H. M., Hershey, A. E., McDowell, W. H., Dodds, W. K., Hamilton, S. K., Gregory, S. and Morrall, D. D.: 2001, 'Control of nitrogen export from watersheds by headwater streams', *Science* **292**, 86–89.

Pyn, S.: 2002, 'Small particles add up to big disease risk', *Science* **295**, 1994.

Rabalais, N. N. and Turner, R. E. (eds): 2001, *Coastal Hypoxia: Consequences for Living Resources and Ecosystems*, Coastal and Estuarine Studies 58, American Geophysical Union, Washington, D.C., pp. 454.

Rustad, L. E., Campbell, J. L., Marion, G. M., Norby, R. J., Mitchell, M. J., Hartley, A. E., Cornelissin, J. H. C. and Gurevitch, J.: 2001, 'A meta-analysis of the responses of soil respiration, net nitrogen mineralization, and aboveground plant growth to experimental ecosystem warming', *Oecologia* **126**, 543–562.

Schaberg, P. G., DeHayes, D. H. and Hawley, G. J.: 2001, 'Anthropogenic calcium depletion: A unique threat to forest ecosystem health?' *Ecosystem Health* **7**, 214–228.

Schimel, J. P., Cleve, K. V., Cates, R. G., Clausen, T. P. and Reichert, P. B.: 1996, 'Effects of balsam poplar (*Populus balsamifera*) tannins and low molecular weight phenolics on microbial activity in taiga floodplain soil: Implications for changes in N cycling during succession', *Can. J. Bot.* **74**, 84–90.

Schindler, D. W.: 1977, 'Evolution of phosphorus limitations in lakes', *Science* **195**, 260–262.

Schindler, D. W.: 1980, 'Evolution of the experimental lakes project', *Can. J. Fish. Aquat. Sci.* **37**, 313–319.

Schlesinger, W. H.: 1997, *Biogeochemistry: An Analysis of Global Change*, 2nd ed., Academic Press, San Diego, pp. 588.
Smith, V. H.: 1998, 'Cultural eutrophication of inland, estuarine, and coastal waters', in M. L. Pace and P. M. Groffman (eds), *Successes, Limitations, and Frontiers in Ecosystem Science*, Springer-Verlag, New York, pp. 7–49.
Sterner, R. W., Elser, J. J. and Hessen, D. O.: 1992, 'Stoichiometric relationships among producers, consumers and nutrient cycling in pelagic ecosystems', *Biogeochemistry* **17**, 49–67.
Stoddard, J. L.: 1994, 'Long-term changes in watershed retention of nitrogen', in L. A. Baker (ed.), *Environmental Chemistry of Lakes and Reservoirs*, American Chemical Society Advances in Chemistry Series, No. 237, Americal Chemical Socierty, Washington, D.C., pp. 223–284.
Stoddard, J. L., Jeffries, D. S., Lükewille, A., Clair, T. A., Dillon, P. J., Driscoll, C. T., Forsius, M., Johannessen, M., Kahl, J. S., Kellogg, J. M., Kemp, A., Mannlo, J., Monteith, D., Murdoch, P. S., Patrick, S., Rebsdorf, A., Skjelväte, B. L., Stainton, M. P., Trasen, T., Van Dam, H., Webster, K. E., Wieting, J. and Wilander, A.: 1999, 'Regional trends in aquatic recovery from acidification in North America and Europe', *Nature* **401**, 575–578.
Strayer, D. L., Powell, J., Ambrose, P., Smith, L. C., Pace, M. L. and Fischer, D. T.: 1996, 'Arrival, spread and early dynamics of a zebra mussel (*Dreissena polymorpha*) population in the Hudson River estuary', *Can. J. Fish. Aquat. Sci.* **53**, 1143–1149.
Strayer, D. L., Caraco, N. F., Cole, J. J., Findlay, S. and Pace, M. L.: 1999, 'Transformation of freshwater ecosystems by bivalves. A case study of zebra mussels in the Hudson River', *BioScience* **49**, 19–27.
Tilman, D.: 1998, 'Species composition, species diversity, and ecosystem processes: Understanding the impacts of global change', in M. L. Pace and P. M. Groffman (eds), *Successes, Limitations, and Frontiers in Ecosystem Science*, Springer-Verlag, New York, pp. 452–472.
van Breemen, N., Finlay, R., Lundström, U., Jongman, A. G., Giesler, R. and Olsson, M.: 2000, 'Mycorrhizal weathering: A true case of mineral plant nutrition?', *Biogeochemistry* **49**, 53–67.
Vernadsky, W. I.: 1944, 'Problems in biogeochemistry. II', *Trans. Conn. Acad. Arts Sci.* **35**, 493–494.
Vernadsky, W. I.: 1945, 'The Biosphere and the noösphere', *Am. Sci.* **33**, 1–12.
Vitousek, P. M. and Field, C. B.: 1999, 'Ecosystem constraints to symbiotic nitrogen fixers: A simple model and its implications', *Biogeochemistry* **46**, 179–202.
Vitousek, P. M. and Howarth, R. W.: 1991, 'Nitrogen limitation on land and in the sea: How can it occur?', *Biogeochemistry* **13**, 87–115.
Vitousek, P. M., Aber, J. D., Howarth, R. W., Likens, G. E., Matson, P. A., Schindler, D. W., Schlesinger, W. H. and Tilman, D. G.: 1997, 'Human alteration of the global nitrogen cycle: Sources and consequences', *Ecol. Appl.* **7**, 737–750.
Vitousek, P. M., Cassman, K., Cleveland, C., Crews, T., Field, C. B., Grimm, N. B., Howarth, R. W., Marino, R., Tartinelli, L., Rastetter, E. B. and Sprent, J. I.: 2002, 'Towards an ecological understanding of biological nitrogen fixation', *Biogeochemistry* **57/58**, 1–45.
Vollenweider, R. A.: 1968, 'Scientific Fundamentals of Lake and Stream Eutrophication, with Particular Reference to Phosphorus and Nitrogen as Eutrophication Factors', *Tech. Report DAS/DSI/68.27*, OECD, Paris, France.
Weathers, K. C., Lovett, G. M., Likens, G. E. and Caraco, N.: 2000, 'Cloudwater inputs of nitrogen to forest ecosystems in southern Chile: Forms, fluxes and sources', *Ecosystems* **3**, 590–595.
Wedin, D. and Tilman, D.: 1990, 'Species effects on nitrogen cycling: A test with perennial grasses', *Oecologia* **84**, 433–441.
Wessman, C. A. and Asner, G. P.: 1998, 'Ecosystems and problems of measurement at large spatial scales', in M. L. Pace and P. M. Groffman (eds), *Successes, Limitations and Frontiers in Ecosystem Science*, Springer-Verlag, New York, pp. 346–371.

CRITICAL LOADS OF ACIDITY FOR SURFACE WATERS IN SOUTH-CENTRAL ONTARIO, CANADA: REGIONAL APPLICATION OF THE FIRST-ORDER ACIDITY BALANCE (FAB) MODEL

J. AHERNE[1]*, M. POSCH[2], P. J. DILLON[1] and A. HENRIKSEN[3]

[1] *Environmental and Resource Studies, Trent University, Peterborough, ON K9J 7B8, Canada;*
[2] *National Institute for Public Health and the Environment (RIVM), Bilthoven, The Netherlands;*
[3] *Norwegian Institute for Water Research (NIVA), Oslo, Norway*

(* *author for correspondence, e-mail: julian.aherne@ucd.ie, phone: +1 705 748 1011 ext. 5351, fax: +1 705 748 1569*)

(Received 20 August 2002; accepted 1 April 2003)

Abstract. Major sulphur emission control programs have been implemented in North America, resulting in current emissions being ~30% less than those in 1980. However, the level of acidic deposition remaining is still unlikely to promote widespread recovery of aquatic ecosystems. The First-order Acidity Balance (FAB) model has been applied to south-central Ontario (285 lakes in the Muskoka River Catchment) to evaluate the need for further reductions in emissions. As a result of the past decline in deposition, the proportion of lakes with critical loads exceedance has dropped substantially; however, further reductions in sulphur and nitrogen emissions are required to eliminate critical loads exceedance. Based on bulk deposition of sulphate and nitrogen (41.1 mmol$_c$ m^{-2} yr^{-1} and 62.5 mmol$_c$ m^{-2} yr^{-1}, respectively) for the period 1995–1999, 166 lakes (58.3%) exceed critical loads. Even with full implementation of SO$_2$ abatement programs in Canada (achieved in 1994) and the United States (legislated for 2010), critical loads will be exceeded in a large proportion (46.6%) of the study lakes.

Keywords: acid deposition, acidification, critical load, exceedance, FAB model, lakes, Muskoka, nitrogen, sulphate

1. Introduction

During the 1970s and 1980s, the acidification of surface waters by atmospherically deposited sulphur (S) became a major international concern. Major S emission control programs were implemented in North America, resulting in current emissions being ~30% less than in 1980 (Jeffries *et al.*, 2000). Consequently, S deposition in south-central Ontario has decreased by ~40% in the past two decades, while nitrogen (N) deposition has remained unchanged. In response, S concentrations in many lakes have declined (Jeffries *et al.*, 1995; McNicol *et al.*, 1998). However, the level of acidic deposition remaining is still unlikely to promote widespread recovery of aquatic ecosystems (Jeffries *et al.*, 2000; Henriksen *et al.*, 2002).

Critical loads have been widely accepted in Europe as a basis for the development of air pollution control strategies, as evidenced by the Second Sulphur Protocol and the Gothenburg Protocol (see Gregor *et al.*, 2001). Similarly, the critical load concept has been used in Canada to design an emission reduction programme (RMCC, 1990; Jeffries *et al.*, 1993). At present two steady-state models are widely used for calculating critical loads for surface waters (Henriksen and Posch, 2001): the Steady-State Water Chemistry (SSWC) model and the First-order Acidity Balance (FAB) model. The SSWC model requires weighted annual mean water chemistry, or an estimate thereof, and annual runoff to calculate critical loads of acidity, CL(A). The FAB model allows the simultaneous calculation of critical loads of acidifying S and N deposition similar to the Simple Mass Balance (SMB) model widely used for computing forest soil critical loads (Sverdrup and De Vries, 1994). In addition to processes in the terrestrial catchment soils, such as uptake, immobilisation, and denitrification, the FAB model also takes into account the in-lake retention of S and N.

Due to its modest data requirements, the SSWC model has been used in regional critical load assessments across Europe (see Posch *et al.*, 2001), North America (Snucins *et al.*, 2001; Henriksen *et al.*, 2002) and Asia (Duan *et al.*, 2000). A recent application to ~1500 lakes in south-central Ontario indicated that the reductions in S deposition have resulted in reduced critical loads exceedance (Henriksen *et al.*, 2002). However, to achieve a more complete chemical recovery even larger reductions are necessary, and the deposition of N has to be taken into account. Critical loads of acidifying S and N deposition for 285 lakes in the Muskoka River Catchment, south-central Ontario are presented in this paper. The advantage of the FAB model is that it can provide estimates of the reductions needed for S and N deposition, and more importantly, the ability to play one off against the other. The objectives were to evaluate the exceedance of critical loads using the FAB model and to better quantify the need for further reductions in S and N deposition.

2. Materials and Methods

2.1. STUDY AREA

The Muskoka River Catchment extends into four regions (Muskoka, Haliburton, Nipissing, and Parry Sound) in south-central Ontario (Figure 1). All regions are underlain by impermeable Precambrian granitic bedrock covered by thin glacial till. The land cover is largely mixed forest, with deciduous forests dominating where the soil is thicker and coniferous forests dominating where the soils are thin. Cottage developments (seasonal and permanent) are common in all regions. The climate of the study area is north temperate, with long-term average precipitation ranging between about 800–1100 mm, about one quarter to one third of which falls as snow. The mean monthly temperatures for January and July are −10 and 19 °C, respectively, and the long-term annual average temperature is approximately 5.0 °C.

Bulk deposition has been measured in Muskoka and Haliburton during the past two decades (Figure 2). Either four (1980–1989) or three (1990-present) stations

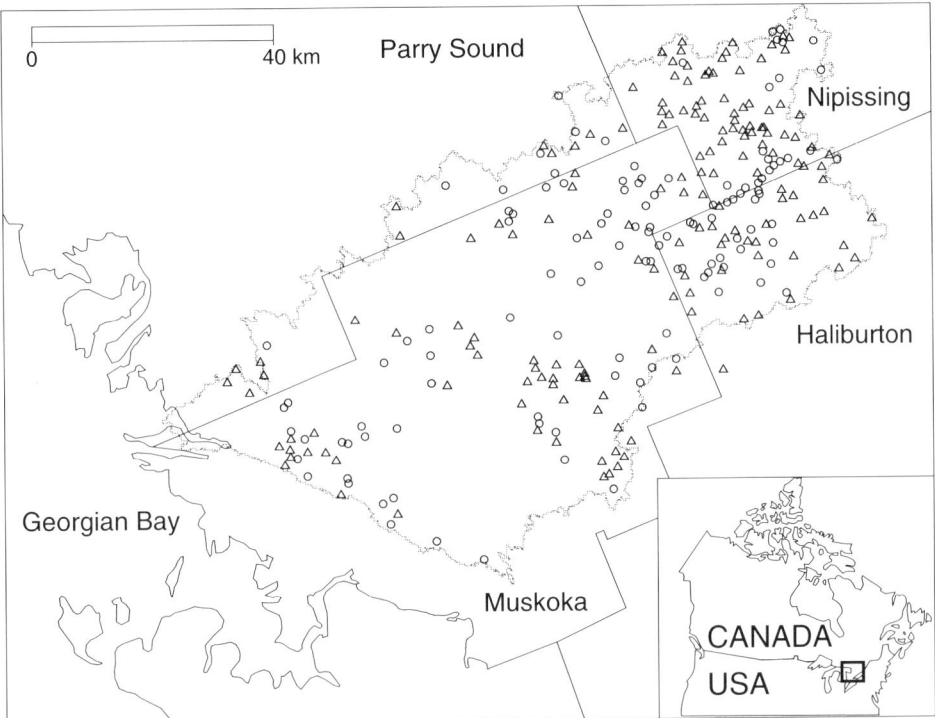

Figure 1. Location of the study area and the 285 study lakes (circles and triangles) in the Muskoka River Catchment (dotted line), south-central Ontario, Canada.

located in this region were used to generate annual deposition for the Muskoka-Haliburton region, which is assumed to be representative of deposition to the study area. The annual average bulk depositions (1995–1999) for S and N (NO_3^- and NH_4^+) are 41.1 mmol$_c$ m^{-2} yr^{-1} and 62.5 mmol$_c$ m^{-2} yr^{-1}, respectively, hereafter referred to as 'current' deposition. Independent measurements of wet and dry S deposition in the region suggest that bulk deposition corresponds reasonably well to total deposition (Dillon *et al.*, 1988). Although this may be true for lake surfaces, throughfall measurements in the region (Neary and Gizyn, 1994) indicate that total deposition to forests is significantly greater than bulk deposition. As such, exceedances may be underestimated in the current study. Estimated S bulk deposition for the year 2010 is 36.4 mmol$_c$ m^{-2} yr^{-1}, which corresponds to the year when all currently legislated pollution controls in both Canada and the United States will be fully implemented. The forecasted deposition is based on modelled annual S wet deposition to Muskoka-Haliburton from the Acid Deposition and Oxidant Model (ADOM; Michael D. Moran, Meteorological Service of Canada, personal communication). Nitrogen deposition is assumed to remain at current levels (62.5 mmol$_c$ m^{-2} yr^{-1}; 58.5% as NO_3^- and 41.5% as NH_4^+) as there has been no significant trend or change in N deposition during the past two decades (Figure 2). Furthermore,

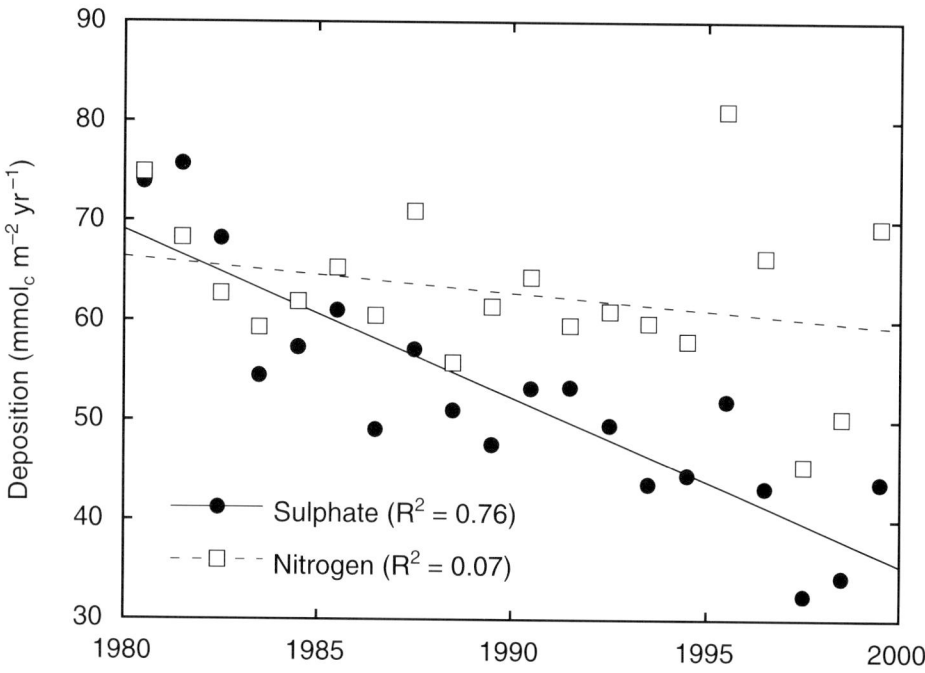

Figure 2. Sulphate and nitrogen (NO_3^- and NH_4^+) bulk deposition for Muskoka-Haliburton, Ontario, during the period 1980–1999. Lines represent least squares linear regression. Note: data are presented as hydrological years, i.e., 1980 represents the average deposition from 1 June 1980–31 May 1981.

any mandated large source emission reductions of N pollutants are predicted to be offset by increases from other sectors (Michael D. Moran, Meteorological Service of Canada, personal communication).

Chemical data for lakes in south-central Ontario have been collected during the course of a number of different research projects and lake surveys during the period 1981–1995, with most of the data collected between 1983 and 1988. All chemical analyses were carried out at the Ontario Ministry of the Environment laboratories (for methods see: OME, 1983). Mean annual runoff coefficients were taken from a provincial runoff map generated from measured 30 yr average runoff at all long-term hydrologic gauging stations throughout Ontario. Catchment and lake areas, as well as the fraction of forest and wetland, were derived from digitised maps (Ontario Base Maps 1:50 000 scale). Area data were generated for 859 catchments (lakes >5 ha) from the 1000+ located within the Muskoka River Catchment. From the combined water chemistry, runoff and catchment area databases, 285 lake catchments have data suitable for the FAB model (Figure 1). These lakes represent a subset of the regional data set (n $\sim$ 1370 for Muskoka, Haliburton, Nipissing, and Parry Sound) used for a previous SSWC CL(A) assessment (Henriksen *et al.*, 2002). In general, they are slightly smaller and more acid sensitive, as evidenced by the median statistics (Muskoka River Catchment: lake area = 25.5 ha, runoff =

510 mm, Ca^{2+} = 132.1; regional data set: lake area = 28.1 ha, runoff = 479 mm, Ca^{2+} = 144.7). The total combined area of these catchments is largely forested (86.1%) with some wetlands (4.5%) and the remaining attributed to grasslands (9.4%).

2.2. THE FAB MODEL

The current application follows the most recent published model description (Henriksen and Posch, 2001) and, as such, will not be repeated here. However, in brief, the simultaneous treatment of S and N does not allow the calculation of a single critical load value, but rather, the concept of a Critical Load Function (CLF) must be introduced. Every pair of N and S deposition, (N_{dep}, S_{dep}), satisfying Equation (1) defines a CLF. The area below the CLF defines deposition pairs for which there is no exceedance. The 'maximum' critical load of S and N ($CL_{max}(S)$ and $CL_{max}(N)$, respectively), i.e., the critical load for either S_{dep} or N_{dep} alone (i.e., setting the other to zero), define the extent of the CLF:

$$(1 - \rho_S) \times S_{dep} + (1 - \rho_N) \times b_N \times N_{dep} = (1 - \rho_N) \times M_N + CL(A), \quad (1)$$

where the dimensionless coefficient b_N and M_N depend on the denitrification fraction in soils, the net uptake of N, and the net immobilisation of N and are a function of the magnitude of N_{dep}. The in-lake retention coefficients for S and N (ρ_S and ρ_N, respectively) are modelled by a kinetic equation (Kelly et al., 1987) making them a function of runoff, the lake:catchment ratio, and the net mass transfer coefficients for S and N (s_S and s_N, respectively). Finally, CL(A) is the acidity critical load estimated by the SSWC model. It is assumed that the lakes and their catchments are small enough to be properly characterised by average soil and lake-water properties; furthermore, all lakes are treated as headwater lakes.

The denitrification fraction is modelled as a linear function of the fraction of wetlands in a catchment (see Posch et al., 1997). The net uptake of N refers to the net removal through harvesting of forests, which has been set to zero in the current study. Tree harvesting is a common occurrence in south-central Ontario (Watmough and Dillon, 2002); however, more research is needed to quantify adequately the removal of biomass through harvesting. The assumption of zero net uptake will result in higher critical loads. Immobilisation refers to the long-term storage of N in soil organic matter. A default value of 14.3 $mmol_c$ m^{-2} yr^{-1} has been used, which is in agreement with recent studies on critical loads for forest soils in the region (Watmough and Dillon, 2002). This value falls within the range for coniferous and deciduous forests (7.1–21.4 $mmol_c$ m^{-2} yr^{-1} depending on warm-cold climate) suggested by Hornung et al. (1995). However, under current conditions with increased N cycling due to elevated N deposition, this range may be higher. Net N mass transfer coefficients (s_N) for total inorganic nitrogen (TIN) have been estimated recently for eight long-term study lakes in Muskoka-Haliburton (Kaste and Dillon, 2003); the median value (6.3 m yr^{-1}) has been selected for the

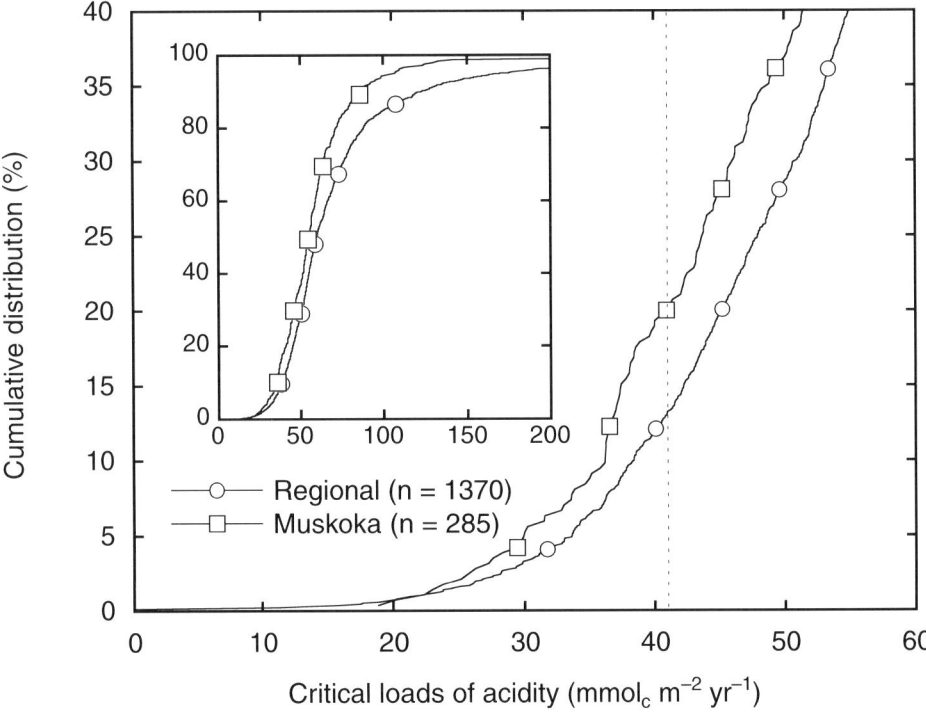

Figure 3. Cumulative distribution function for critical loads of acidity <60 mmol$_c$ m^{-2} yr^{-1} for the study lakes in the Muskoka River Catchment and lakes in the regional data set. The inset depicts the cumulative distribution function for critical loads of acidity <200 mmol$_c$ m^{-2} yr^{-1}. The dotted line represents sulphate bulk deposition for the period 1995–1999 (41.1 mmol$_c$ m^{-2} yr^{-1}).

study lakes. A default value of 0.5 m yr^{-1} is typically used for s_S (*cf.* Posch *et al.*, 1997; Hindar *et al.*, 2001). However, estimates for the eight long-term study lakes suggest values that are not significantly different from zero (P. J. Dillon, unpublished data). As such, it was assumed that $s_S = 0$ for the study lakes. The current net mass transfer coefficients are based on long-term averages and are the best available estimates of long-term steady-state retention for the region (Kaste and Dillon, 2003).

3. Results and Discussion

The SSWC CL(A) for south-central Ontario previously have been estimated in a companion paper (Henriksen *et al.*, 2002). The 285 study lakes have lower critical loads than the regional data set; current S deposition suggests that 20% of the study lakes exceed SSWC CL(A) compared to ∼13% for the regional data set (Figure 3). The 5th percentile CL(A) of 30.0 mmol$_c$ m^{-2} yr^{-1} (Table I) for the Muskoka River Catchment is similar to the regional data set estimate (33.7 mmol$_c$

TABLE I

Summary catchment and critical load data (lake:catchment ratio, SSWC critical loads of acidity, maximum critical loads for S and N, and the fraction of N retained in the terrestrial catchment and the lake[a]) for the study lakes (n = 285) in the Muskoka River Catchment

	r	CL(A)	$CL_{max}(S)$	$CL_{max}(N)$	N_{terr}	N_{lake}
		($mmol_c\ m^{-2}\ yr^{-1}$)			(%)	
5 percentile	0.03	30.0	30.0	89.5	20.9	19.2
25 percentile	0.08	43.9	43.9	131.4	24.4	35.5
50 percentile	0.15	54.8	54.8	175.2	26.3	47.2
75 percentile	0.21	67.9	67.9	242.1	28.8	55.0
95 percentile	0.32	106.2	106.2	424.0	30.9	63.5
Mean	0.15	60.5	60.5	203.4	26.3	44.4

[a] Presented as a percentage of annual average nitrogen bulk deposition, 62.5 $mmol_c\ m^{-2}\ yr^{-1}$ (1995–1999).

$m^{-2}\ yr^{-1}$: Henriksen et al., 2002) and to previous critical load studies in central Ontario (33.3 $mmol_c\ m^{-2}\ yr^{-1}$: Jeffries and Lam, 1993; Jeffries et al., 1999). For further details on the regional data set, the SSWC model and its application to south-central Ontario (see Henriksen et al., 2002).

The CLFs for each of the 285 study lakes have been aggregated by computing percentile functions (Table I and Figure 4). All combinations of N_{dep} and S_{dep} lying below the p-th percentile function protect 100-p percent of the lakes. Approximately 60% of the lakes exceed critical load (40% are protected) based on current S and N deposition (see cross in Figure 4). Unlike the SSWC method, there is no unique amount of S and N to be reduced to reach non-exceedance; for example, from current deposition, 95% of the lakes could be protected by reducing S deposition alone, or alternatively by reducing a combination of S and N deposition (path between cross and 5% line in Figure 4). For the Muskoka River Catchment, S reductions are more effective than N reductions, as the 5th percentile CLF cannot be reached by N reductions alone.

$CL_{max}(S)$ is equivalent to CL(A) as in-lake S retention has been set to zero ($s_S = 0$). $CL_{max}(N)$ varies substantially, reflecting differences in land cover and lake:catchment ratio. Much of the deposited N is retained in the lake (N_{lake}: median = 47% of the current N deposition) compared to the terrestrial catchment (N_{terr}: median = 26%). The variation in N_{lake} is almost entirely explained by the lake:catchment ratio (Hindar et al., 2001). Retention of N in the terrestrial catchment is considerably more static as it depends on a constant N immobilisation plus a denitrification fraction based on the proportion of wetland area in the catchment. Lakes with a high in-lake N retention have a lower sensitivity to acidity (Hindar et al., 2001). In total, 166 lakes (58.3%, shown as triangles in Figure 1) exceed critical

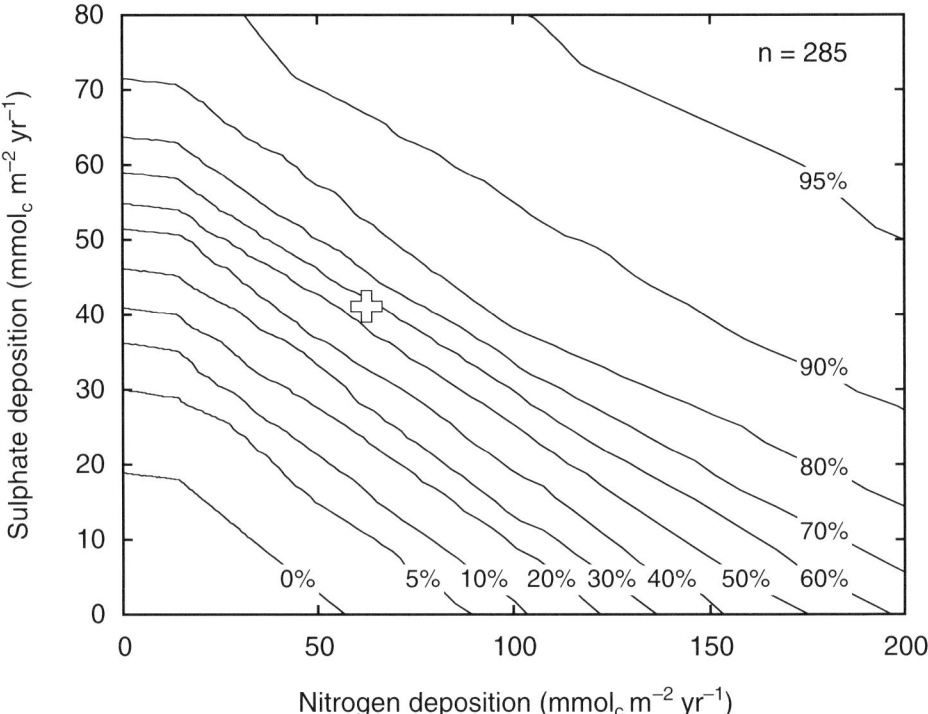

Figure 4. Percentile distributions of the Critical Loads Functions for the study lakes in the Muskoka River Catchment. Also shown is the annual average bulk deposition of sulphate and nitrogen (41.1 mmol$_c$ m^{-2} yr^{-1} and 62.5 mmol$_c$ m^{-2} yr^{-1}, respectively) for the period 1995–1999 (cross).

loads under current deposition. The majority of lakes are located at the intersection of the four regions (Muskoka, Haliburton, Nipissing, and Parry Sound), with the greatest proportion of exceeded lakes in Nipissing. Similar regional patterns were observed in the SSWC assessment (Henriksen *et al.*, 2002).

Sulphate deposition in Muskoka shows a clear downward trend during the study period ($R^2 = 0.76$; Figure 2), with the majority of the decrease occurring pre-1985. The decrease in deposition is consistent with decreases in S emissions in eastern Canada of 46% over the past two decades (McNicol *et al.*, 1998). As a result of the declining deposition, the proportion of lakes with critical loads exceedance has dropped substantially, from ∼90% in the late 1970s to 70% in the late 1990s (Figure 5). However, due to the considerable year-to-year variation in deposition (especially between 1995–1999, Figure 5), it is more reasonable to use the four-year moving average as an indicator of current exceedance; as such, ∼50% of the lakes still have exceedances.

In the year 2010, we estimate that 133 lakes (46.6%) will exceed critical loads. Clearly, even with full implementation of SO_2 abatement programs in Canada (achieved in 1994) and the United States (legislated for 2010), critical loads are exceeded in a large proportion of the study lakes. The relative effectiveness of

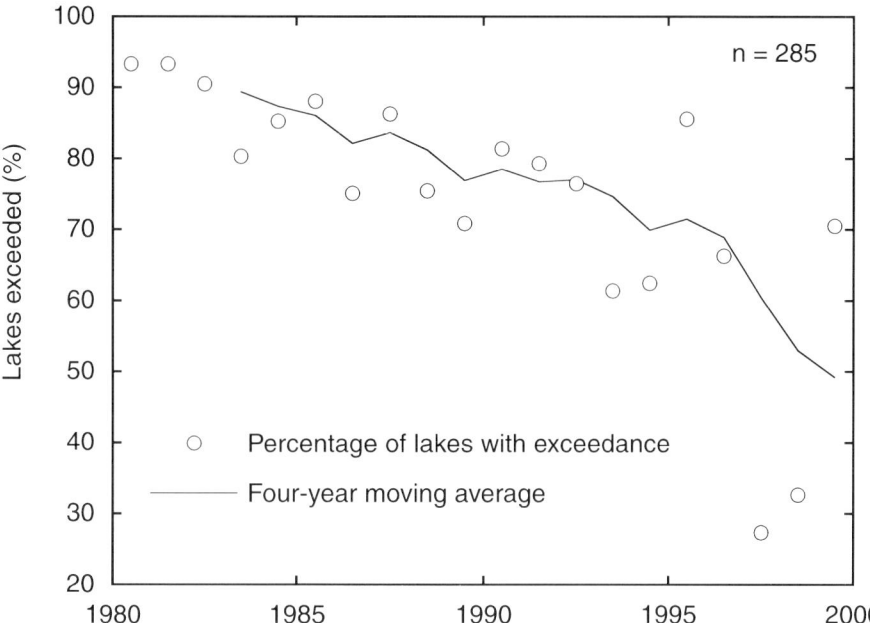

Figure 5. Proportion (%) of study lakes in the Muskoka River Catchment with exceedance of critical load during the period 1980–1999 (circles). Also shown is the four-year moving average for the percentage of study lakes with exceedance.

future S and N reductions may be investigated by re-plotting the percentile CLFs (or protection isolines, Figure 4) with 2010 deposition as the origin and percentage N and S reduction on the axes (Figure 6). In relative terms, S reductions are more effective than N reductions. For example, approximately 85% of the lakes can be protected by a 42% reduction in S deposition compared to 75% of the lakes for a 42% reduction in N deposition. However, to achieve non-exceedance for all lakes, reductions in both quantities are necessary.

The current model application treats all study lakes as headwater lakes, but in reality only 146 (51%) are headwater lakes. To investigate the effects of this simplification, we applied an alternative method to the non-headwater lakes. Hindar *et al.* (2001) presented an extension of the FAB model ('big-lake' approach) to account for lake systems (lakes linked by streams), which treats all lakes in the system as a single lake, situated in the combined catchment. Unlike Hindar *et al.* (2001), the two approaches produce very similar critical loads for the 139 non-headwater lakes (Table II). The 'big-lake' approach estimates a marginally lower $CL_{max}(N)$ for the lower percentiles of the distribution, due to the slightly increased in-lake N retention. This equates to a slightly lower exceedance: 79 lakes (56.8%) versus 81 lakes (58.3%). Hindar *et al.* (2001) attributed increased in-lake N retention to increases in the lake:catchment ratio and changes in land cover, the latter affecting N input to the lake. However, land cover was reasonably static between the two

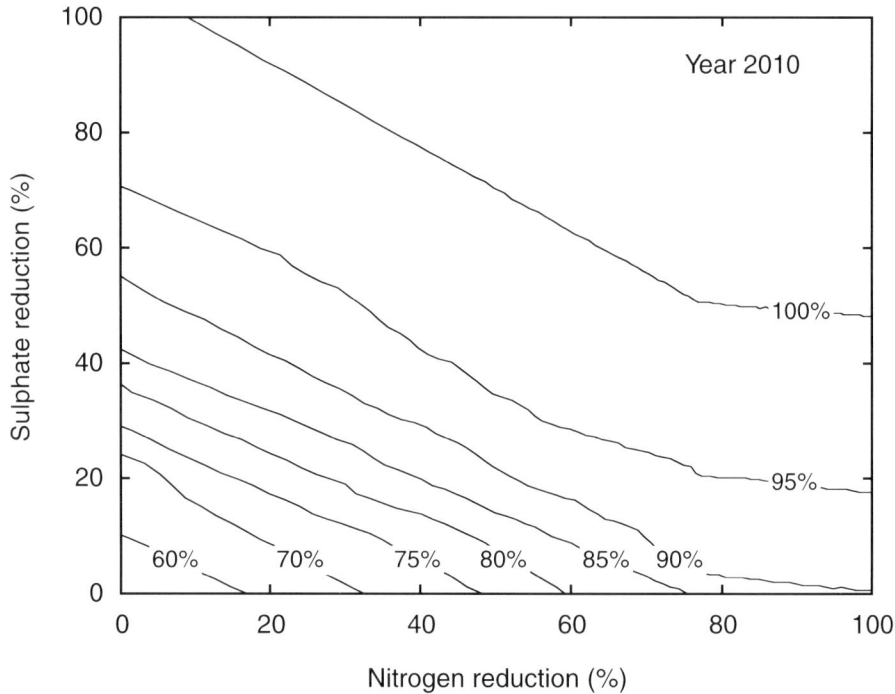

Figure 6. Isolines of protection percentages for study lakes in the Muskoka River Catchment for all possible uniform percentage reductions starting from 2010 bulk deposition for sulphate and nitrogen (36.4 mmol$_c$ m^{-2} yr^{-1} and 62.5 mmol$_c$ m^{-2} yr^{-1}, respectively).

TABLE II

Summary critical load and catchment data (maximum critical loads for S and N, and the fraction of N retained in the terrestrial catchment and the lake[a] and the lake:catchment ratio) for all non-headwater lakes (n = 139) estimated using the standard method compared with the 'Big-lake' method

	Standard method					Big-lake method			
	$CL_{max}(S)$	$CL_{max}(N)$	N_{terr}	N_{lake}	r	$CL_{max}(N)$	N_{terr}	N_{lake}	r
	(mmol$_c$ m^{-2} yr^{-1})			(%)		(mmol$_c$ m^{-2} yr^{-1})		(%)	
5 percentile	35.0	80.0	21.6	14.3	0.02	91.2	23.9	26.5	0.05
25 percentile	45.8	126.4	25.2	31.2	0.06	136.8	26.1	36.2	0.08
50 percentile	55.9	166.9	27.2	44.5	0.13	169.7	27.5	42.9	0.12
75 percentile	70.0	231.7	29.1	52.9	0.19	209.5	28.7	48.9	0.16
95 percentile	99.9	351.9	31.1	60.8	0.29	329.9	30.4	56.3	0.23
Mean	59.8	189.0	27.0	41.6	0.13	181.7	27.3	42.3	0.13

[a] Presented as a percentage of annual average nitrogen bulk deposition, 62.5 mmol$_c$ m^{-2} yr^{-1} (1995–1999).

approaches, as is evident in the similarity of the N_{terr} distributions (Table II). While the effects of increased in-lake N retention are important for certain individual lake systems, the effects are negligible when studying the Muskoka lakes in a regional context.

The greatest potential error in the current study is whether the specified long-term terrestrial N sinks are underestimated or whether the current high levels of N retention are sustainable. Further, Kaste and Dillon (2003) suggest that if terrestrial catchments move toward N saturation, in-lake retention rates will increase owing to the higher input. The current N retention rates are based on the best available data, and until further information becomes available it seems reasonable to base the FAB model on the precautionary principle: current terrestrial N sinks are unsustainable; and all N inputs not retained by assumed long-term sustainable N sinks will be leached as NO_3^- sometime in the future.

4. Conclusions

In this study, critical loads of acidifying S and N deposition using the FAB model have been computed for 285 lakes in the Muskoka River Catchment. Although the study focused on a sub-set of acid-sensitive lakes from the regional data set, the results present a conservative estimate of critical load exceedance in south-central Ontario as calculations are based on bulk deposition and ignore the effect of forest harvesting. The study shows that even with full implementation of SO_2 abatement programs in Canada and the United States, further reductions in S and N emissions are required to protect the majority of lakes in south-central Ontario.

Acknowledgements

This work was supported by grants from Ontario Power Generation Inc. and the Natural Sciences and Engineering Research Council of Canada to P. J. Dillon, and by the Ontario Ministry of the Environment. The authors would like to thank Caryn Perry (Ministry of Natural Resources, Peterborough, Canada) for generating the area-specific catchment data from the Ontario Base Maps and Dr. Michael D. Moran (Meteorological Service of Canada, Toronto, Canada) for providing 2010 deposition estimates.

References

Dillon, P. J., Lusis, M., Reid, R. A. and Yap, D.: 1988, 'Ten-year trends in sulphate, nitrate and hydrogen deposition in central Ontario', *Atmos. Environ.* **22**, 901–905.

Duan, L., Hao, J., Xie, S. and Du, K.: 2000, 'Critical loads of acidity for surface waters in China', *Sci. Tot. Environ.* **246**, 1–10.

Gregor, H.-D., Nagel, H.-D. and Posch, M.: 2001, 'The UN ECE international programme on mapping critical loads and levels', *Water, Air, Soil Pollut.: Focus* **1**, 5–19.

Henriksen, A. and Posch, M.: 2001, 'Steady-state models for calculating critical loads of acidity for surface waters', *Water, Air, Soil Pollut.: Focus* **1**, 375–398.

Henriksen, A., Dillon, P. J. and Aherne, J.: 2002, 'Critical loads of acidity to surface waters in south-central Ontario, Canada: Regional application of the steady-state water chemistry model', *Can. J. Fish. Aquat. Sci.* **59**, 1287–1295.

Hindar, A., Posch, M. and Henriksen, A.: 2001, 'Effects of in-lake retention of nitrogen on critical load calculations', *Water, Air, Soil Pollut.* **130**, 1403–1408.

Hornung, M., Sutton, M. A. and Wilson, R. B. (eds): 1995, *Mapping and Modelling of Critical Loads for Nitrogen: A Workshop Report*, Institute of Terrestrial Ecology, United Kingdom, pp. 207.

Jeffries, D. S. and Lam, D. C. L.: 1993, 'Assessment of the effect of acidic deposition on Canadian lakes: Determination of critical loads for sulphate deposition', *Water. Sci. Technol.* **28**, 183–187.

Jeffries, D. S., Clair, T. A., Dillon, P. J., Papineau, M. and Stainton, M. P.: 1995, 'Trends in surface water acidification at ecological monitoring sites in southeastern Canada (1981–1991)', *Water, Air, Soil Pollut.* **85**, 577–582.

Jeffries, D. S., Lam, D. C. L., Moran, M. D. and Wong, I.: 1999, 'The effect of SO_2 emission controls on critical load exceedance for lakes in southeastern Canada', *Water. Sci. Technol.* **39**, 165–171.

Jeffries, D. S., Lam, D. C. L., Wong, I. and Moran, M. D.: 2000, 'Assessment of changes in lake pH in southeastern Canada arising from present levels and expected reductions in acidic deposition', *Can. J. Fish. Aquat. Sci.* **57**, 40–49.

Kaste, Ø. and Dillon, P. J.: 2003, 'Inorganic nitrogen retention in acid-sensitive lakes in southern Norway and southern Ontario, Canada – A comparison of mass balance data with an empirical N retention model', *Hydrol. Process.* (in press).

Kelly, C. A., Rudd, J. W. M., Hesslein, R. H., Schindler, D. W., Dillon, P. J., Driscoll, C. T., Gherini, S. A. and Hecky, R. E.: 1987, 'Prediction of biological acid neutralization in acid-sensitive lakes', *Biogeochemistry* **3**, 129–140.

McNicol, D. K., Mallory, M. L., Laberge, C. and Cluis, D. A.: 1998, 'Recent temporal patterns in the chemistry of small acid sensitive lakes in central Ontario, Canada', *Water, Air, Soil Pollut.* **105**, 343–351.

Neary, A. J. and Gizyn, W. I.: 1994, 'Throughfall and stemflow under deciduous and coniferous forest canopies in south-central Ontario', *Can. J. For. Res.* **24**, 1089–1110.

OME: 1983, 'Handbook of Analytical Methods for Environmental Samples', *Technical Report*, Ontario Ministry of the Environment, Rexdale, Ontario, Canada, pp. 246.

Posch, M., Kämäri, J., Forsius, M., Henriksen, A. and Wilander, A.: 1997, 'Exceedance of critical loads for lakes in Finland, Norway and Sweden: Reduction requirements for acidifying nitrogen and sulfur deposition', *Environ. Manage.* **21**, 291–304.

Posch, M., De Smet, P. A. M., Hettelingh, J.-P. and Downing, R. J. (eds): 2001, 'Modelling and Mapping of Critical Thresholds in Europe', *Status Report 2001*, RIVM, Bilthoven, The Netherlands, pp. 188.

RMCC: 1990, *The 1990 Canadian Long-range Transport of Air Pollutants and Acid Deposition Assessment Report, Part 4: Aquatic Effects*, Federal/Provincial Research and Monitoring Coordinating Committee, Ottawa, Ontario, pp. 151.

Snucins, E., Gunn, J., Keller, B., Dixit, S., Hindar, A. and Henriksen, A.: 2001, 'Effects of regional reductions in sulphur deposition on the chemical and biological recovery of lakes within Killarney Park, Ontario, Canada', *Environ. Monit. Assess.* **67**, 179–194.

Sverdrup, H. and De Vries, W.: 1994, 'Calculating critical loads for acidity with the simple mass balance method', *Water, Air, Soil Pollut.* **72**, 143–162.

Watmough, S. A. and Dillon, P. J.: 2002, 'Do critical load models adequately protect forests in south-central Ontario?', *Can. J. For. Res.* **33**, 1544–1556.

EFFECTS OF CLIMATE EVENTS ON ELEMENTAL FLUXES FROM FORESTED CATCHMENTS IN ONTARIO, CANADA: MODELLING DROUGHT-INDUCED REDOX PROCESSES

J. AHERNE[1]*, T. LARSSEN[2], P. J. DILLON[1] and B. J. COSBY[3]
[1] *Environmental and Resource Studies, Trent University, Peterborough, ON K9J 7B8, Canada;*
[2] *Norwegian Institute for Water Research (NIVA), Oslo, Norway;* [3] *Department of Environmental Science, University of Virginia, Charlottesville, Virginia, U.S.A.*
(* author for correspondence, e-mail: julian.aherne@ucd.ie, phone: +1 705 748 1011 ext. 5351, fax: +1 705 748 1569)

(Received 20 August 2002; accepted 9 May 2003)

Abstract. One of the principal influences on elemental fluxes from forested catchments in south-central Ontario is the atmospheric deposition rate of strong acids. While sulphate deposition has decreased by ~40% in the past two decades, nitrate deposition has remained unchanged and is now equivalent to sulphate deposition. Sulphate concentrations in headwater lakes and their inflows have decreased, but much less than expected based on the anticipated direct response of the catchments. Reduction-oxidation (redox) processes occurring in wetlands have been identified as the reason for delayed recovery, and climate events as controlling these redox processes. A new version of the biogeochemical model MAGIC (model of acidification of groundwater in catchments) with a wetland compartment that incorporates redox processes driven by climate events has been generated. The application of MAGIC to a subcatchment of Plastic Lake in south-central Ontario indicates that the basic structure of the model appears to be consistent with the observed data. Moreover, the wetland component was essential in reproducing the observed trends, which include sulphate retention in non-drought years and re-oxidation of previously stored (reduced) sulphur in drought years.

Keywords: acid deposition, acidification, MAGIC model, Plastic Lake, reduction-oxidation processes, sulphate, wetlands

1. Introduction

One of the principal influences on elemental fluxes from forested catchments in south-central Ontario is the atmospheric deposition rate of strong acids. While sulphate (SO_4^{2-}) deposition has decreased by ~40% in the past two decades, nitrate (NO_3^-) deposition has remained unchanged and now equals that of SO_4^{2-}. Consequently, SO_4^{2-} concentrations in headwater lakes and their inflows have decreased, but much less than expected based on the anticipated direct response of the catchments (Jeffries *et al.*, 1995; McNicol *et al.*, 1998; Stoddard *et al.*, 1999). As a further consequence, recovery of alkalinity and pH has been slow. Reduction-oxidation (redox) processes occurring in wetlands have been identified as the reason for delayed recovery, and climate events as controlling these redox processes (Dillon *et al.*, 1997; Dillon and Evans, 2000; Eimers and Dillon, 2002).

A number of studies on Plastic Lake catchment in south-central Ontario have demonstrated that SO_4^{2-} in a stream draining a wetland portion of the catchment increased substantially after periods of drought (Dillion and LaZerte, 1992; LaZerte, 1993; Devito and Hill, 1999). These drought periods resulted in lower water tables in the wetland, oxidation of previously stored (reduced) sulphur (S) compounds, and subsequent efflux of S in acid form.

A number of dynamic process-oriented models have been developed to investigate the response of surface waters to acid deposition (see Tiktak and Van Grinsven, 1995). An important characteristic of such models is that they include descriptions of all chemical and physical processes that are important in controlling the response of a catchment. A new version of the biogeochemical model MAGIC (model of acidification of groundwater in catchments: Cosby *et al.*, 1985, 2001) with a wetland compartment that incorporates redox processes driven by climate events has been generated.

This paper presents the application of MAGIC to a subcatchment of Plastic Lake. The objective of the study was to test if the process description included in the wetland component is capable of explaining the observed chemistry at Plastic Lake catchment.

2. Materials and Methods

2.1. STUDY AREA

Plastic Lake (32.1 ha headwater lake) is located on the Precambrian Shield in Haliburton County, south-central Ontario (45°11′N, 78°50′W), in an area that has received substantial levels of acid deposition for decades (Dillon *et al.*, 1987). Plastic Lake's catchment (95.5 ha) is drained by seven small streams, six of which are ephemeral. The largest of the seven subcatchments (PC1, 23.3 ha, see Figure 1) flows year-round except in very dry years. Five of the seven subcatchments, including PC1, drain wetlands. The upland soils of PC1 (and it's gauged subcatchment PC1-08, 3.5 ha, see Figure 1) are sandy, shallow, humo-ferric or ferro-humic podzols (Lozano, 1987). Quartz, plagioclase, potassium feldspar, and amphibole dominate the soil mineralogy (Kirkwood and Nesbitt, 1991). The PC1 upland forest cover is primarily coniferous, dominated (80% of basal area in 1999: Watmough and Dillon, 2001) by white pine (*Pinus strobus* L.), red maple (*Acer rubrum* L.), eastern hemlock (*Tsuga canadensis* (L.) Carrière), and red oak (*Quercus rubra* L.). The average age of the PC1-08 stand is ~90 yr with a maximum of 200 yr. The wetland (*Sphagnum*-conifer swamp, 2.2 ha, see Figure 1) soils are humic mesisols dominated by white cedar (*Thuja occidentalis* L.). The catchment altitude ranges from 380–420 m a.s.l., annual rainfall averages ~1000 mm and the mean January and July air temperatures are –9.4 and 18.6, respectively.

The chemistry and volume of bulk precipitation have been measured since 1976 (Dillon *et al.*, 1988; Eimers and Dillon, 2002). In addition, the chemistry and

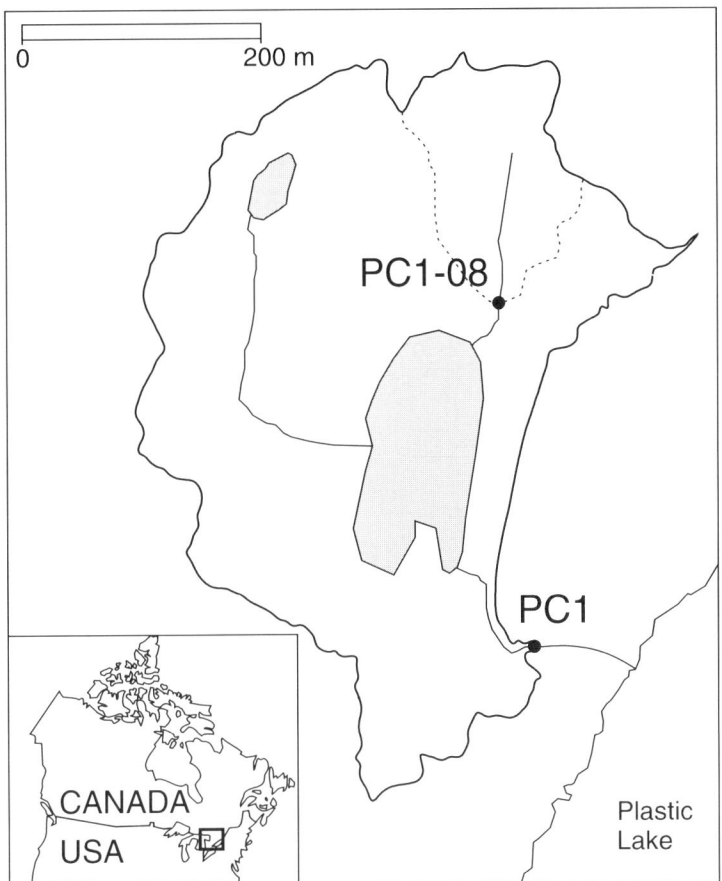

Figure 1. Map of Plastic Lake gauged catchment No. 1 (PC1) showing the gauged subcatchment (PC1-08; broken line) and wetland areas (grey). Note: gauged catchment outlets (sampling stations) are indicated by filled circles.

volume of throughfall was measured between 1983 and 1986 (Neary and Gizyn, 1994). Throughfall data can, if interpreted with care, be used to approximate total deposition of some ions (Hultberg and Grennfelt, 1992; Ferm and Hultberg, 1995). In combination with historic emission rates (Husar, 1994; Galloway, 1995; Lefohn et al., 1999; EPA, 2000), annual average total deposition sequences from ~1850-present have been generated (see Figure 2 for SO_4^{2-} sequence). Stream discharge and chemistry has been measured for the PC1 subcatchment (see Figure 1) since 1979. Between 1986 and 1995 stream discharge and chemistry also were monitored at PC1-08 above the wetland (see Figure 1). Water was collected from the B-horizon (average depth 38 cm) using zero tension lysimeters within (or just outside) PC1-08 between 1986 and 1995 (LaZerte and Scott, 1996). The field, hydrologic, and analytical methods are described in detail elsewhere (OME, 1983; Scheider et al., 1983; Locke and Scott, 1986; methods have remained unchanged). During

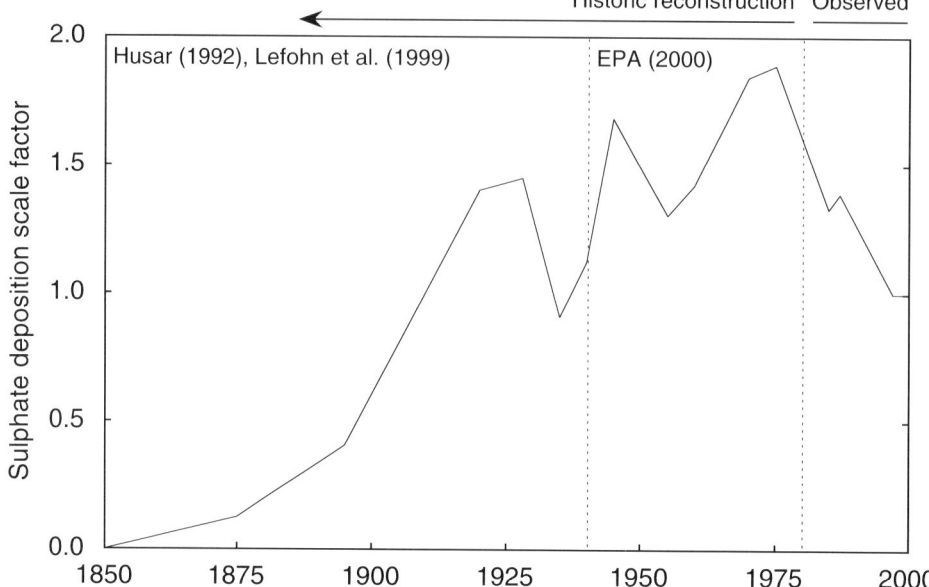

Figure 2. Estimated historic sulphate deposition (scale factor; 1997 deposition (41 μmol_c L^{-1}) is equivalent to a factor of 1.0) for the period 1850–1997. The broken lines separate the different data sources used to construct the deposition sequence.

1983, soil samples were collected from 15 pits located in the upland part of the catchment (Lozano, 1987); sampling was repeated in 1999 using adjacent soil pits (Watmough and Dillon, 2003). Bulk density and exchangeable cations were determined for both surveys using standard methods (OME, 1983). Similarly, forest biomass and cation pools at PC1 were surveyed in 1983 (Lozano and Parton, 1987) and 1999 (Watmough and Dillon, 2003). Forest biomass is currently in steady state (Watmough and Dillon, 2001); forest re-growth since ∼1900 was described using a Michaelis-Menten equation.

2.2. THE MAGIC MODEL

MAGIC is a lumped-parameter model of intermediate complexity, developed to predict the long-term effects of acidic deposition on soils and surface water chemistry. The model was first described by Cosby *et al.* (1985) and developments are reviewed by Cosby *et al.* (2001) and, as such, will not be repeated here. In brief, the model predicts the monthly and annual average concentrations of the major ions for soil solution and surface water chemistry. MAGIC represents the watershed with aggregated, uniform soil compartments (one or two), and a surface water compartment that can be either a lake or a stream. Time series inputs to the model include annual or monthly estimates of: deposition of ions from the atmosphere (wet plus dry deposition); discharge volumes and flow routing within the catchment; biological production, removal and transformation of ions; internal sources and sinks of ions from weathering or precipitation reactions; and climate data. Constant parameters in the model include physical and chemical characteristics of the soils and surface waters, and thermodynamic constants. The model is calibrated using observed values of surface water and soil chemistry for a specified period.

MAGIC has been in use for more than 15 years and has been applied extensively in North America and Europe (see Appendix I in Cosby *et al.*, 2001). Several refinements or additions to MAGIC have been proposed or implemented through the years as a result of the many applications of the model. These changes address inadequacies in the original structure and incorporate new processes that subsequent research has indicated are of increasing importance in natural systems. The latest version of MAGIC (7.77) includes a wetland compartment that incorporates redox processes driven by climate events. The wetland compartment is similar to the existing soil compartment(s) except that SO_4^{2-} adsorption characteristics are replaced by S reduction and oxidation characteristics. Consequently, four new input parameters are required: monthly S reduction (mmol m^{-2}); monthly S oxidation (mmol m^{-2}); a monthly discharge (Q) threshold (cm); and the initial reduced S pool (mol m^{-2}). The MAGIC 'wetland' module must be run on a monthly time step as redox dynamics are driven by the monthly Q threshold, i.e., annual resolution data are not sensitive enough to define periods of drought. When wetland monthly discharge is less than the specified Q threshold only, S oxidation occurs; S reduction can be specified to occur at all times or only when discharge is greater than the Q threshold.

2.3. CALIBRATION OF MAGIC

The PC1 catchment was represented in MAGIC using three compartments (soil, wetland and stream). The soil compartment represents the lumped horizons of the upland soils with 95% of its discharge routed through the wetland before entering the stream. The model was manually calibrated against stream-water observations for the period 1980–1997, rather than a single reference year as is typically carried out. The calibration was carried out in two stages. The first calibration focused only on the PC1-08 subcatchment (excluding the wetland component) using observed inputs and outputs for the period 1986–1995. This enabled the soil compartment to be fully parameterised before introducing the wetland. The second calibration focused on the PC1 catchment (Table I).

Both calibrations consisted of a number of sequential stages. Firstly, soil, wetland and stream physical and chemical characteristics for which data were available were specified (fixed parameters, see Table I). Secondly, uptake rates of nitrogen in the catchment were set based on the input-output budgets, i.e., an empirical rather than process-oriented description was selected in MAGIC. Thirdly, the SO_4^{2-} concentrations were calibrated against stream-water observations by adjusting the

TABLE I

Input parameters (fixed and calibrated) for the three-compartment (soil, wetland and stream) MAGIC model application to Plastic Lake catchment No. 1 (PC1)

Parameter	Units	Soil	Wetland	Stream
Relative area	%	–	10.0	0.01
Depth	m	0.4	1.0	–
Porosity	%	45	45	–
Bulk density	kg m^{-3}	1000	800	–
Cation exchange capacity (CEC)	mmol$_c$ kg^{-1}	80[1]	150	–
SO_4^{2-} adsorption half saturation	mmol$_c$ m^{-3}	3500[2]	–	–
SO_4^{2-} adsorption maximum capacity	mmol$_c$ kg^{-1}	7.5[2]	–	–
Aluminium solubility constant[a]	log	9.2	7.5	7.5
Temperature[b]	°C	6.7	6.7	6.7
Partial pressure of CO_2[a, b]	atm	0.005	0.005	0.0005
Dissolved organic carbon (DOC)[a, b]	mmol m^{-3}	55	100	70
Ca^{2+} uptake[b]	mmol$_c$ m^{-2} yr^{-1}	13.6	–	–
Mg^{2+} uptake[b]	mmol$_c$ m^{-2} yr^{-1}	1.6	–	–
K^+ uptake[b]	mmol$_c$ m^{-2} yr^{-1}	2.7	–	–
NH_4^+ uptake[b]	% of input	100.0	100.0	–
NO_3^- uptake[b]	% of input	99.7	85.0	–
SO_4^{2-} undefined soil source[a, b]	mmol$_c$ m^{-2} yr^{-1}	13.0	–	–
Ca^{2+} weathering[a]	mmol$_c$ m^{-2} yr^{-1}	19.5	–	–
Mg^{2+} weathering[a]	mmol$_c$ m^{-2} yr^{-1}	13.5	–	–
K^+ weathering[a]	mmol$_c$ m^{-2} yr^{-1}	1.0	–	–
Na^+ weathering[a]	mmol$_c$ m^{-2} yr^{-1}	10.5	–	–
Exchangeable Ca^{2+} in 1847[a] (1983 target)	% of CEC	17.6 (7.5)	4.5	–
Exchangeable Mg^{2+} in 1847[a] (1983 target)	4.7 (1.5)	2.0	–	
Exchangeable K^+ in 1847[a] (1983 target)	% of CEC	2.3 (1.6)	0.5	–
Exchangeable Na^+ in 1847[a] (1983 target)	% of CEC	2.0 (1.2)	0.0	–
Wetland module: sulphur redox parameters				
S reduction[a]	mmol m^{-2} mo^{-1}	–	12	–
S oxidation[a]	mmol m^{-2} mo^{-1}	–	85	–
Q threshold[a]	cm mo^{-1}	–	1.5	–
S initial pool	mol m^{-2}	–	10	–

Data source: [1] LaZerte and Findeis (1994); [2] Catherine Eimers, Waterloo University (unpublished data).
[a] Parameters obtained through calibration or slight adjustment of initial input data.
[b] Parameters that also require inputs at monthly resolution. Note: base cation uptake by trees is based on a specified hindcast uptake sequence.

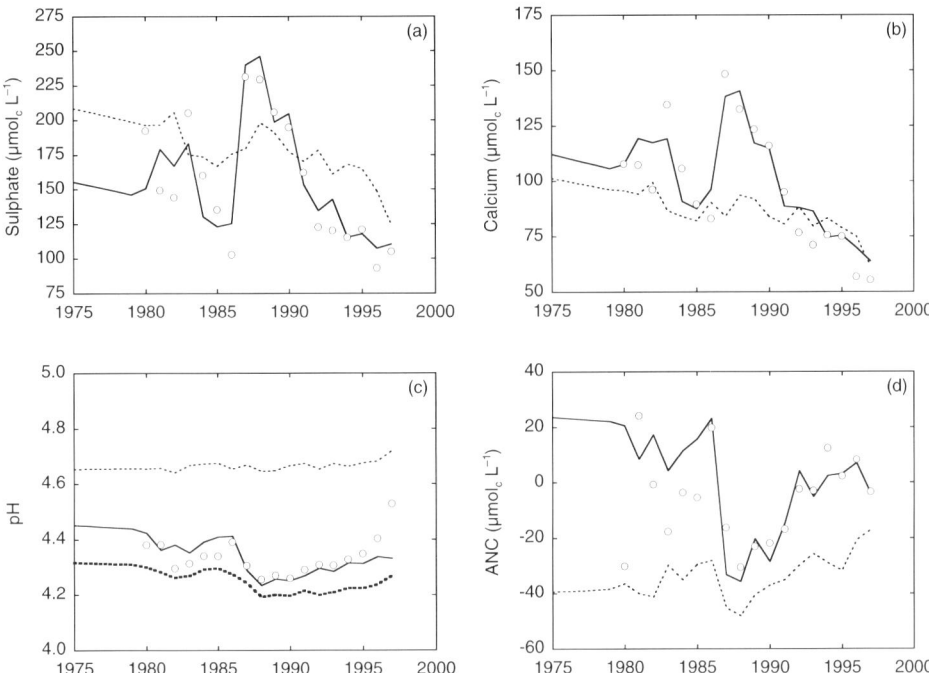

Figure 3. Simulated (line) and observed (circle) volume-weighted annual average stream-water concentrations of: (a) sulphate; (b) calcium; (c) pH; and (d) ANC (acid neutralising capacity; sum of base cations minus the sum of acid anions) at PC1. Note: simulated volume-weighted annual average soil solution (broken line) and wetland (heavy broken line) concentrations are also shown. In all graphs, except pH, soil solution and wetland concentrations are approximately equivalent and, as such, only soil solution concentrations are shown.

wetland parameters (or addition of an undefined source to the soil compartment, see Table I). Fourthly, weathering rates and initial base saturation levels were calibrated using observed stream-water and soil base saturation data. Finally, having calibrated the model to simulate the correct concentrations of strong acid anions and base cations in surface waters, the hydrogen ion, total aluminium and organic anion concentrations in the surface water were calibrated by adjusting parameters controlling weak inorganic and organic acid speciation and aluminium solubility.

Observations and model output were compared using standard goodness-of-fit indices (Janssen and Heuberger, 1995; Alewell and Manderscheid, 1998); linear regression of observed against predicted (R^2), and Normalised Mean Absolute Error (NMAE). An R^2 of 1.0, or an NMAE of zero, indicate absolute agreement of predicted with observed data.

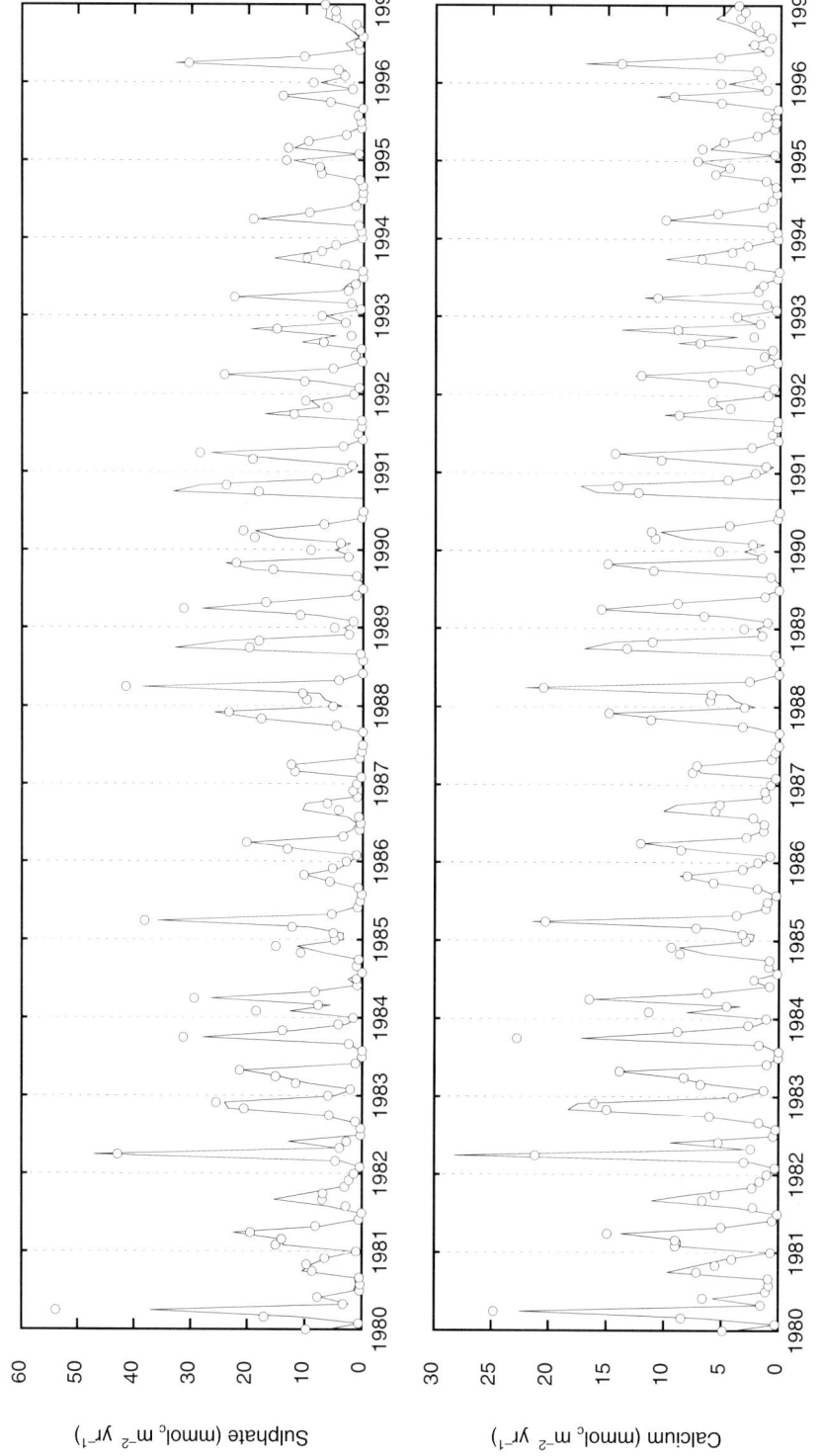

Figure 4. Simulated (line) and observed (circle) monthly average stream-water fluxes of sulphate and calcium at PC1, January 1980–December 1997.

3. Results and Discussion

The output of the calibrated model compare well with observed annual and monthly data (see Figures 3 and 4). The observed trends in annual average SO_4^{2-} and calcium (Ca^{2+}) concentrations are successfully reproduced by the model (Figure 3: $R^2 = 0.81$, NMAE = 0.11 and $R^2 = 0.84$, NMAE = 0.09, respectively). Similarly, the model successfully reproduces the monthly fluxes of SO_4^{2-} and Ca^{2+} (Figure 4: $R^2 = 0.91$, NMAE = 0.21 and $R^2 = 0.93$, NMAE = 0.18, respectively).

The calibrated parameters (Table I) are in general agreement with previous studies, i.e., Watmough and Dillon (2003) estimated weathering to range between 32–42 $mmol_c$ m^{-2} (calibrated value: 44.5 $mmol_c$ m^{-2}); LaZerte and Scott (1996) estimated the log of the aluminium solubility constant in the soil B-horizon to be 9.2 ± 0.5 (calibrated value: 9.2). An unknown source of SO_4^{2-} was required to balance SO_4^{2-} outputs from PC1-08 in the initial calibration of the upland soils. This discrepancy in SO_4^{2-} input-output budgets has been observed elsewhere in North America (Driscoll et al., 1995). Driscoll et al. (1998) proposed two explanations for the unmeasured S sources: a significant underestimate of dry deposition; or internal catchment S sources, such as weathering or mineralisation of soil organic S. Based on current understanding of total deposition at PC1, the unmeasured source was assumed to be an internal catchment source (Table I).

In contrast, the wetland retains SO_4^{2-} during non-drought years and then exports SO_4^{2-} during drought years (1983, 1987, 1988, 1989 and 1990; see Figure 3a). LaZerte (1993) estimated the annual average net SO_4^{2-} retention during 1984–1986 (non-drought years) to be 22 $mmol_c$ SO_4^{2-} m^{-2}, which agrees with the simulated value of 20 $mmol_c$ SO_4^{2-} m^{-2}. The pulse of SO_4^{2-} during drought years is balanced mostly by an export of cations (see Figure 3b). The retention of SO_4^{2-} in non-drought years, however, is not balanced by a retention of cations (compare Figures 3a, b) but by their reduced export (compared to drought years) and by a decrease in proton export (see Figure 3c). This is also clearly seen in ANC values (Figure 3d); the large difference between soil solution and stream-water ANC during non-drought years is explained by the net retention of SO_4^{2-} and net export of base cations from the wetland. In general, there is a net export of base cations from the wetland during the entire simulation. The model simulations agree with previous studies (Dillion and LaZerte, 1992; LaZerte, 1993) that demonstrate the importance of the PC1 wetland in controlling the runoff chemistry.

The results of the simulation clearly demonstrate that drought years have a substantial effect on wetland dynamics. The apparent discrepancy between simulated and observed stream-water chemistry at the beginning of the calibration period (1980–1997) can be explained by a drought that occurred in 1977. The absence of this drought from the calibration data-set produces a mis-match between observed and simulated stream-water concentrations prior to the calibration period. This highlights the importance of high-resolution (monthly) long-term stream-water discharge, which is an important driver of the wetland compartment.

4. Conclusions

Catchment solute models are used for two purposes, to predict and to explain. As such, it is essential that they include descriptions of all process that are important in controlling the chemical response of a catchment. The MAGIC model has been tested extensively during the past 15 years using data from a variety of sites. This is the first test of a new version of MAGIC that includes a wetland module. The application to PC1 indicates that the basic structure of the model appears to be consistent with the observed data. Moreover, the model proved to be very successful in reproducing the observed trends under a highly constrained calibration, i.e., in general, all necessary input parameters are site measured and calibrated parameters are in good agreement with previous research findings. The model results support the hypothesis that wetland redox processes explain much of the inter-annual variation in SO_4^{2-} and base cation concentrations.

Acknowledgements

This work was supported by grants from Ontario Power Generation Inc. and the Natural Sciences and Engineering Research Council of Canada to P. J. Dillon, and by the Ontario Ministry of the Environment.

References

Alewell, C. and Manderscheid, B.: 1998, 'Use of objective criteria for the assessment of biogeochemical ecosystem models', *Ecol. Model.* **107**, 213–224.

Cosby, B. J., Hornberger, G. M., Galloway, J. N. and Wright, R. F.: 1985, 'Modeling the effects of acid deposition: Assessment of a lumped parameter model of soil water and streamwater chemistry', *Water Resour. Res.* **21**, 51–63.

Cosby, B. J., Ferrier, R. C., Jenkins, A. and Wright, R. F.: 2001, 'Modelling the effects of acid deposition: refinements, adjustments and inclusion of nitrogen dynamics in the MAGIC model', *Hydrol. Earth Syst. Sci.* **5**, 499–517.

Devito, K. J. and Hill, A. R.: 1999, 'Sulphate mobilization and pore water chemistry in relation to groundwater hydrology and summer drought in two conifer swamps on the Canadian shield', *Water, Air, Soil Pollut.* **113**, 97–114.

Dillon, P. J. and LaZerte, B. D.: 1992, 'Response of the Plastic Lake catchment, Ontario, to reduced sulphur deposition', *Environ. Pollut.* **77**, 211–217.

Dillon, P. J. and Evans, H. E.: 2000, 'Long-term changes in the chemistry of a soft-water lake under changing acid deposition rates and climate fluctuations', *Verh. Int. Verein. Limnol.* **27**, 2615–2619.

Dillon, P. J., Reid, R. A. and De Grosbois, E.: 1987, 'The rate of acidification of aquatic ecosystems in Ontario, Canada', *Nature* **329**, 45–48.

Dillon, P. J., Lusis, M., Reid, R. A. and Yap, D.: 1988, 'Ten-year trends in sulphate, nitrate and hydrogen deposition in central Ontario', *Atmos. Environ.* **22**, 901–905.

Dillon, P. J., Molot, L. A. and Futter, M.: 1997, 'The effect of El Niño-related drought on the recovery of acidified lakes', *Environ. Monit. Assess.* **46**, 105–111.

Driscoll, C. T., Postek, K. M., Kretser, W. and Raynal, D. J.: 1995, 'Long-term trends in the chemistry of precipitation and lake water in the Adirondack region of New York, U.S.A.', *Water, Air, Soil Pollut.* **85**, 583–588.

Driscoll, C. T., Likens, G. E. and Church, M. R.: 1998, 'Recovery of surface waters in the northeastern U.S. from decreases in atmospheric deposition', *Water, Air, Soil Pollut.* **105**, 319–329.

Eimers, M. C. and Dillon, P. J.: 2002, 'Climate effects on sulphate flux from forested catchments in south-central Ontario', *Biogeochemistry* **61**, 337–355.

EPA: 2000, *National Air Pollutant Emission Trends: 1900–1998*, EPA-454/R-00-002, United States Environmental Protection Agency, North Carolina (online: www.epa.gov/ttn/chief/trends/trends98).

Ferm, M. and Hultberg, H.: 1995, 'Method to estimate atmospheric deposition of base cations in coniferous throughfall', *Water, Air, Soil Pollut.* **85**, 2229–2234.

Galloway, J. N.: 1995, 'Acid deposition: Perspectives in time and space', *Water, Air, Soil Pollut.* **85**, 15–24.

Hultberg, H. and Grennfelt, P.: 1992, 'Sulphur and seasalt deposition as reflected by throughfall and runoff chemistry in forested catchments', *Environ. Pollut.* **75**, 215–222.

Husar, R.: 1994, 'Sulphur and nitrogen emission trends for the United States: An application of the materials flow approach', in R. U. Ayres and U. E. Simonis (eds), *Industrial Metabolism: Restructuring for Sustainable Development*, United Nations University Press, pp. 390 (online: www.unu.edu/unupress/unupbooks/80841e/80841e00.htm).

Janssen, P. H. M. and Heuberger, P. S. C.: 1995, 'Calibration of process-oriented models', *Ecol. Model.* **83**, 55–66.

Jeffries, D. S., Clair, T. A., Dillon, P. J., Papineau, M. and Stainton, M. P.: 1995, 'Trends in surface water acidification at ecological monitoring sites in southeastern Canada (1981–1991)', *Water, Air, Soil Pollut.* **85**, 577–582.

Kirkwood, D. E. and Nesbitt, H. W.: 1991, 'Formation and evolution of soils from an acidified watershed: Plastic Lake, Ontario, Canada', *Geochim. Cosmochim. Acta* **55**, 1295–1308.

LaZerte, B. D.: 1993, 'The impact of drought and acidification on the chemical exports from a minerotrophic conifer swamp', *Biogeochemistry* **18**, 153–175.

LaZerte, B. D. and Findeis, J.: 1994. 'Acidic leaching of a podzol Bf horizon from the Precambrian Shield, Ontario, Canada', *Can. J. Soil Sci.* **75**, 43–54.

LaZerte, B. D. and Scott, L.: 1996, 'Soil water leachate from two forested catchments on the Precambrian Shield, Ontario', *Can. J. For. Res.* **26**, 1353–1365.

Lefohn, A. S., Husar, J. D. and Husar, R. B.: 1999, 'Estimating historical anthropogenic global sulfur emission patterns for the period 1850–1990', *Atmos. Environ.* **33**, 3435–3444.

Locke, B. A. and Scott, L. D.: 1986, 'Studies of lakes and watersheds in Muskoka-Haliburton, Ontario: Methodology (1976–1985)', *Data Report DR 86/4*, Dorset Research Centre, Ontario Ministry of the Environment, Dorset, Ontario, Canada.

Lozano, F.: 1987, 'Physical and chemical properties of the soils at the southern biogeochemistry study site', *Report BGC-018*, Dorset Research Centre, Ontario Ministry of the Environment, Dorset, Ontario, Canada.

Lozano, F. and Parton, W. J.: 1987, 'Forest cover characteristics of the Harp #4 and Plastic #1 subcatchments at the southern biogeochemical study site', *Report BGC-006*, Dorset Research Centre, Ontario Ministry of the Environment, Dorset, Ontario, Canada.

McNicol, D. K., Mallory, M. L., Laberge, C. and Cluis, D. A.: 1998, 'Recent temporal patterns in the chemistry of small acid sensitive lakes in central Ontario, Canada', *Water, Air, Soil Pollut.* **105**, 343–351.

Neary, A. J. and Gizyn, W. I.: 1994, 'Throughfall and stemflow under deciduous and coniferous forest canopies in south-central Ontario', *Can. J. For. Res.* **24**, 1089–1110.

OME: 1983, 'Handbook of analytical methods for environmental samples', *Technical Report*, Ontario Ministry of the Environment, Rexdale, Ontario, Canada, pp. 246.

Scheider, W. A., Reid, R. A., Locke, B. A. and Scott, L. D.: 1983, 'Studies of lakes and watersheds in Muskoka-Haliburton, Ontario: Methodology (1976–1982)', *Data Report DR 83/1*, Dorset Research Centre, Ontario Ministry of the Environment, Dorset, Ontario, Canada.

Stoddard, J. L., Jeffries, D. S., Lükewille, A., Clair, T. A., Dillon, P. J., Driscoll, C. T., Forsius, M., Johannesen, M., Kahl, J. S., Kellogg, J. H., Kemp, A., Mannio, J., Monteith, D. T., Murdoch, P. S., Patrick, S., Rebsdorf, A., Skjelkvåle, B. L., Stainton, M. P., Traaen, T., Van Dam, H., Webster, K. E., Wieting, J. and Wilander, A.: 1999, 'Regional trends in aquatic recovery from acidification in North America and Europe', *Nature* **401**, 575–578.

Tiktak, A. and Van Grinsven, H. J. M.: 1995, 'Review of sixteen forest-soil-atmosphere models', *Ecol. Model.* **83**, 35–53.

Watmough, S. A. and Dillon, P. J.: 2001, 'Base cation losses from a coniferous catchment in central Ontario, Canada', *Water, Air, Soil Pollut.: Focus* **1**, 507–524.

Watmough, S. A. and Dillon, P. J.: 2003, 'Major fluxes from a coniferous catchment in central Ontario, 1983–1999', *Biogeochemistry* (in press).

PROJECTING REGIONAL PATTERNS OF FUTURE SOIL CHEMISTRY STATUS IN SWEDISH FORESTS USING SAFE

MATTIAS ALVETEG

Department of Chemical Engineering, P.O. Box 124, Lund University, SE 22100 Lund, Sweden
(e-mail: Mattias.Alveteg@chemeng.lth.se, phone: +46 46 222 3627, fax: +46 46 149156)

(Received 20 August 2002; accepted 23 May 2003)

Abstract. As part of the Abatement Strategies for Transboundary Air Pollution (ASTA) research program, the dynamic soil chemistry model SAFE was used to make hindcasts and future projections of soil solution chemistry for 645 Swedish forest soils between 1800 and 2100. The data needed were derived from different databases of different spatial resolution ranging from site-specific measurements of soil and stand characteristics from the Swedish Forest Inventory to species-specific nutrient content ranges based on literature surveys. The time-series of nutrient uptake and atmospheric deposition needed were created using the MAKEDEP model and the future scenarios were based on the 1999 Gothenburg protocol. The version of MAKEDEP used included nutrient content elasticity, and the modelled biomass nutrient content thus varies between regions as well as over time. The results were analysed by dividing the sites into three different regions (southwest, central and north) as well as nationally. It was shown that acidification remains a severe environmental problem in the southwest region even after implementation of the 1999 Gothenburg protocol, whereas in the north the problem is far less pronounced.

Keywords: acidification, critical loads, MAKEDEP, modelling, recovery

1. Introduction

The emission of sulphur dioxide peaked around the 1980's in Europe (Mylona, 1993) and has declined considerably since. The decline is to large degree a result of international negotiations within the UN-ECE Convention on Long-Range Transboundary Air Pollution (LRTAP). The latest protocol to the LRTAP convention was signed in Gothenburg, 1999 and was based on estimates of cost-effective reductions of emissions with the aim of mitigating acidification, eutrophication and ground level ozone. The environmental targets were set using critical loads of acidity, critical loads of nutrient nitrogen, and critical levels of ozone.

A problem with the critical loads concept is that it does not give any information about the time delays of the system. Studying only the critical loads of acidity for forest soils may lead to slightly misleading conclusions, because there might be a considerable time delay in the recovery from acidification. Several researchers in Europe are therefore working with the dynamic aspects of acidification, with the aim of including these aspects in future work within the LRTAP convention and the Clean Air For Europe (CAFE) program (Munthe *et al.*, 2002).

The aims of this study are to (1) show how dynamic aspects of acidification and recovery from acidification can be studied on the regional level using the Swedish national forest inventory database, national deposition modeling, and the SAFE and MAKEDEP models, and (2) to demonstrate the difficulties in interpreting such results when using discrete forest management events.

2. Model Setup

The model setup was adapted from earlier Swiss studies (Alveteg *et al.*, 1998a; Kurz *et al.*, 1998) using updated versions of the SAFE and MAKEDEP models. The SAFE model (Warfvinge and Sverdrup, 1992; Alveteg, 1998) is a dynamic, multi-layer, soil chemistry model, developed with the objective to study the effects of acid deposition on soils and groundwater. The largest difference between the SAFE model and other similar models (Posch *et al.*, 2003) such as MAGIC, SMART, and VSD is that SAFE uses soil mineralogy, simulated soil water chemistry, and transition state theory to calculate the mineral weathering rate, whereas the other models rely either on calibration against stream water chemistry or on the user specifying the weathering rate (optional in SMART and VSD). In this study, concentration-dependent nitrogen immobilisation and denitrification were implemented using the same model formulation as in the regionalised version of PROFILE (Bertills and Lövblad, 2002). The initial conditions used to start the SAFE model are calculated by a steady-state version of the model, the initSAFE model.

The MAKEDEP model (Alveteg *et al.*, 1998c) uses general deposition trend curves and site specific data to create site-specific time-series of deposition needed by dynamic soil chemistry models. The MAKEDEP model used in this study (Alveteg *et al.*, 2002) includes nutrient content elasticity.

3. Data Acquisition

Much of the input data needed in this study were derived using the same methods as in earlier assessment of Swedish critical loads (Warfvinge and Sverdrup, 1995; Bertills and Lövblad, 2002), which in turn rely heavily on the National Forest Inventory (NFI). In a dynamic assessment, however, additional input is needed: exchangeable cations (i.e., cation exchange capacity and base saturation) as well as time series of deposition, nutrient cycling, and nutrient uptake.

Information on exchangeable cations was taken from the publicly available NFI (http://www.slu.se), where samples have been taken from up to 5 different soil horizons. Sites with measurements of exchangeable cations in less than three horizons were excluded from this study. Consequently, the soil was divided into three to five layers.

TABLE I

Deposition reduction factors (2010 compared to 1997) as estimated by Gun Lövblad, IVL, for six different regions in Sweden and their approximate midpoint. For each site, the deposition scenario for the closest region midpoint was used

Longitude	Latitude	SO_4^{2-} deposition	NO_x deposition	NH_4^+ deposition
15	56	0.573	0.660	0.892
15	57	0.571	0.548	0.913
12.5	57	0.577	0.623	0.971
18	59	0.752	0.731	1.006
15	63	0.770	0.632	1.015
18	66	0.876	0.717	1.044

The deposition values (1997) used in this study are identical to earlier studies on critical loads of acidity and nitrogen (Bertills and Lövblad, 2002). The deposition standard curves needed by MAKEDEP were the same as in an earlier study (Alveteg et al., 1995) up until 1997. For 1997–2010 deposition standard curves were based on estimates made by Gun Lövblad, IVL (Table I) of the effect of the Gothenburg protocol, and after 2010 they were held constant.

For standing biomass data, the NFI data used were tree species composition, mean tree age, and biomass (based on Hugin model calculations) for stem, branch, and canopy. Data were selected from both the first and the second NFI measurement campaigns (during the 1980's and 1990's, respectively), favouring data with mean tree age older than 30 yr to avoid unnecessary uncertainty in the MAKEDEP calculations. Information on tree species composition was used together with information on nutrient content ranges for different tree species (De Vries and Kros, 1991; Heintzenberg et al., 2002) to deduce a reasonable nutrient content range at each site.

3.1. FOREST MANAGEMENT

The importance of forest management on predicted recovery has been demonstrated earlier (Heliwell et al., 1998b). For base cations, the importance is obvious, especially since the current removal rate through harvesting might be larger than the sustainable removal rate (Holmqvist et al., 2002).

In Sweden, clear-cutting is the dominant harvesting practice. There are two fundamentally different ways to incorporate clear-cut management in a regional assessment. The focus may be either on the individual stands or on larger forests. If the focus is on individual forest stands, clear-cut harvesting is a discrete event in time where the entire stand is removed and replanted. On the other hand, if we focus on larger forest areas ignoring some of the heterogeneity in time and space, clear-cut harvesting of individual stands could be seen as continuous harvesting, i.e. harvest always occurs somewhere in the large forest. In this study, however, the chosen focus is on simulating the dynamics of individual stands. Harvesting is thus treated as a series of discrete events, thinning 10% when the forest is 10, 20 and 30 yr old and cutting down the entire forest after one rotation period.

On the local level, information on rotation periods, timing and extent of forest thinning etc. may be available from the forest owner. On the regional level in Sweden, the only readily available information is the mean age of the trees and typical rotation periods at different latitudes. In this study, the length of the rotation period was assumed to be 80 yr in southern Sweden and 200 yr in the very northern part of the country, increasing linearly with latitude. It is likely that the actual rotation periods are somewhat shorter, but the available information on average tree age of the individual forest stands precluded the use of shorter rotation periods.

4. Results

The necessary input data to run MAKEDEP and SAFE could be derived for a total of 710 sites out of the 1883 sites in the critical loads database. The sites without all necessary input usually lacked information on exchangeable cations and/or standing biomass. MAKEDEP failed to reconstruct nutrient uptake and atmospheric deposition for 65 sites, i.e. the atmospheric deposition of nitrogen as reconstructed by MAKEDEP was not enough to build up the specified nitrogen storage in biomass in these 65 sites. SAFE was thus applied to 645 sites.

The results clearly indicate that acidification remains a problem in all but the northern part of Sweden even if the Gothenburg protocol will be fulfilled (Figure 1). Further reductions are thus needed if the goal of protecting 95 of all ecosystems should be fulfilled. The results also indicate that recovery from soil acidification is a slow process and even if deposition is reduced below the critical load at all sites, the problem with acidified soils will prevail for several decades.

On site level, changes in land use and discrete events such as harvesting seem to be of tremendous importance in the short term (Figure 2). The variation in soil solution chemistry for individual sites during the rotation period raises the question of how the chemical threshold should be defined. Setting the minimum Bc: Al ratio to 1, for example, does not uniquely define a deposition target as the minimum Bc: Al ratio will vary within the rotation period. Is the proper threshold the average ratio throughout a rotation period or the minimum value of the rotation period?

5. Discussion

The south-western part of Sweden clearly stands out in this study as a region where it is likely that we will see negative effects of acidification into the foreseeable

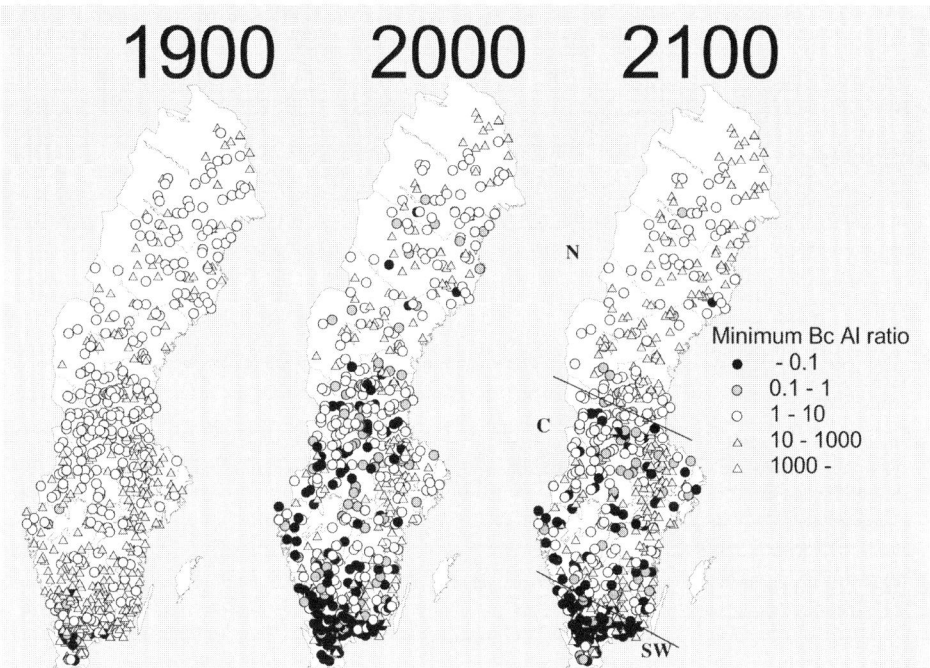

Figure 1. Simulated minimum Bc:Al ratio for three different years and the division of Sweden into three regions, south-west, central and north.

future (Figure 1). The assessed chemical recovery from acidification is likely to be a prerequisite for biological recovery. The timing of the biological response is however difficult to estimate, due to competition between different species, for example.

Comparing the result of this study with earlier studies is difficult since most other studies emphasize stream water chemistry rather than soil chemistry (Helliwell *et al.*, 1998a; Sefton and Jenkins, 1998; Collins and Jenkins, 1998; Evans *et al.*, 1998). Soil solution chemistry has slower dynamics than surface water chemistry, which is why the delay times in this study are longer than those simulated for Swedish lakes (Moldan *et al.*, 2004). Another problem in comparing results with surface water studies is the fact that the soil in those studies often is treated as one compartment only, something which leads to different dynamics and excludes the effect of changes in nutrient cycling (Alveteg *et al.*, 1998b) on soil solution chemistry. In multi-layer assessments, the recovery from acidification is typically faster in the upper horizons and slower in the lower horizons (De Vries *et al.*, 1995; Alveteg *et al.*, 1998a).

One of the underlying principles within LRTAP and CAFE is the link between emissions and deposition: deposition is calculated with the EMEP and RAINS models using meteorological conditions such as prevailing wind directions and

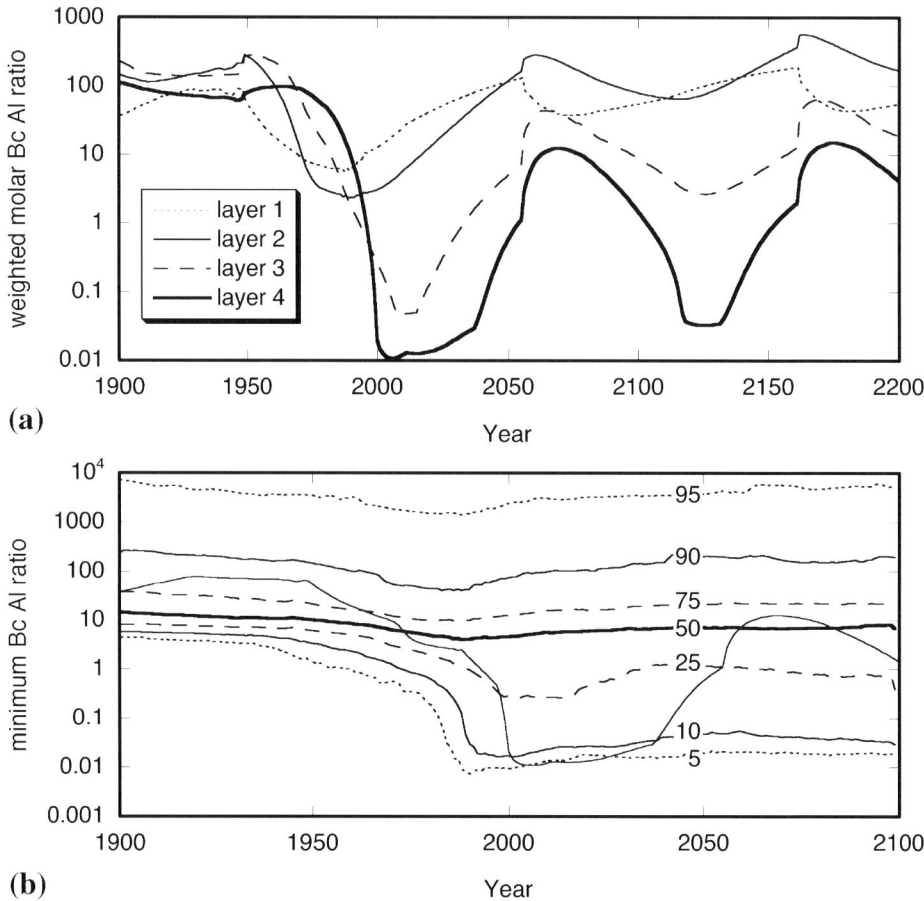

Figure 2. (a) Weighted molar Bc:Al ratios for site NFI 83/5773/2031 at longitude 12, latitude 58 which is assumed to be harvested in, 1947 and to have a rotation period of 107 yr. (b) Percentile traces of simulated minimum Bc:Al ratio. As an illustration, the result for the site in subfigure A is drawn in grey.

regional land use maps. Changes in land use therefore change the deposition trends and patterns. In this study, the MAKEDEP model modified the future deposition trends based on the evolution of the canopy biomass. For an individual forest plot, scaling deposition with canopy biomass might be worthwhile because that will indicate how the acidification and recovery processes are affected by changes in land use and harvesting practices. At the European level, scaling deposition trends with site-specific evolution of canopy biomass might introduce inconsistencies in the assessment. If one forest is cut down, deposition to that area decreases, but because the overall emission-deposition mass balance still must be satisfied, deposition must increase somewhere else. Large-scale land use change scenarios must therefore be treated already at the EMEP modelling level and great care is needed to make consistent assessments of the effects.

Furthermore, for dynamic simulations to be useful within the LRTAP and CAFE framework, the future projections should preferably be monotonous. If not, the choice of mitigation strategy might be sensitive to the target year at which the results are evaluated. Due to discrete harvesting events used in this study the future projections of soil chemistry are, however, not monotonous (Figure 2). I would therefore suggest to use discrete events only in assessments directed towards model evaluation and small scale assessments and averaged descriptions of land use when the assessment is to be used as a basis for negotiations on emission reductions.

Theoretically, it makes no difference for the long-term goal if acidification and recovery from acidification are evaluated with steady-state models or dynamic models. In real assessments, however, discrepancies may arise due to differences in (1) model formulation, (2) selection of site, and (3) data acquisition. Differences in model formulation may arise from slightly different modelling objectives or from implementation problems. Differences in the selection of sites are often unavoidable due to the more input needed by dynamic models. Finally, as dynamic models require time-series of input and steady-state models only long-term averages, the input data to the two kinds of models might produce different results. After a regional assessment of the dynamic aspects of acidification there is thus a need to harmonise the dynamic and steady-state assessments, for example by recalculating the critical loads.

The differences between the long-term results in this study and steady-state calculations were assessed by comparing the minimum Bc:Al ratio in 2100 and calculations with the regionalised version of PROFILE both for the sites at which dynamic simulations were made and for the entire critical loads database of 1883 sites (Figure 3). Due to the discrete harvesting events, the dynamic simulations may be far from steady-state in 2100 (Figure 2). The comparison nevertheless suggests that the selected sites well represent the full Swedish critical loads data set but that there are discrepancies between the steady-state and dynamic results, especially in the south-western part of Sweden.

6. Conclusions

The recovery from soil acidification is a slow process and may take several decades. Further reductions of acidifying pollutants, sulphur dioxide, nitrogen oxides and ammonia, are needed if 95% of the ecosystems should be protected.

The objectives of a regional assessment of dynamic aspects of acidification should always be clearly stated, as one intended use may preclude another. If the effects of local scale changes in harvesting practices are studied, the assessment will not be suitable for use at an international level due to possible violations of the large scale emission-deposition mass balance. Furthermore, the use of discrete

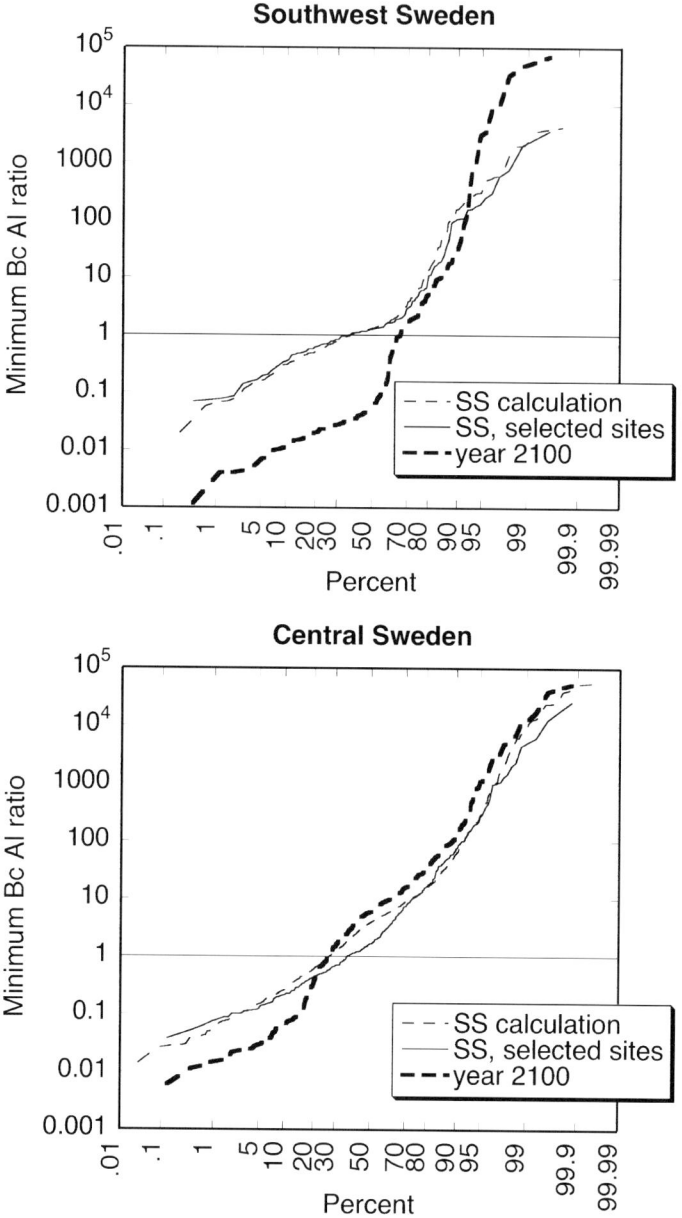

Figure 3. Comparison between minimum Bc: Al ratio as (1) calculated using steady-state model PROFILE for all sites, (2) calculated using steady-state model for same sites as in the dynamic assessment and (3) minimum Bc:Al ratio as simulated in year 2100 using SAFE. The three regions are defined in Figure 1. In south-western, central and northern Sweden 65, 26 and 2% of the modelled sites have a minimum Bc:Al ratio below 1 in the year 2100.

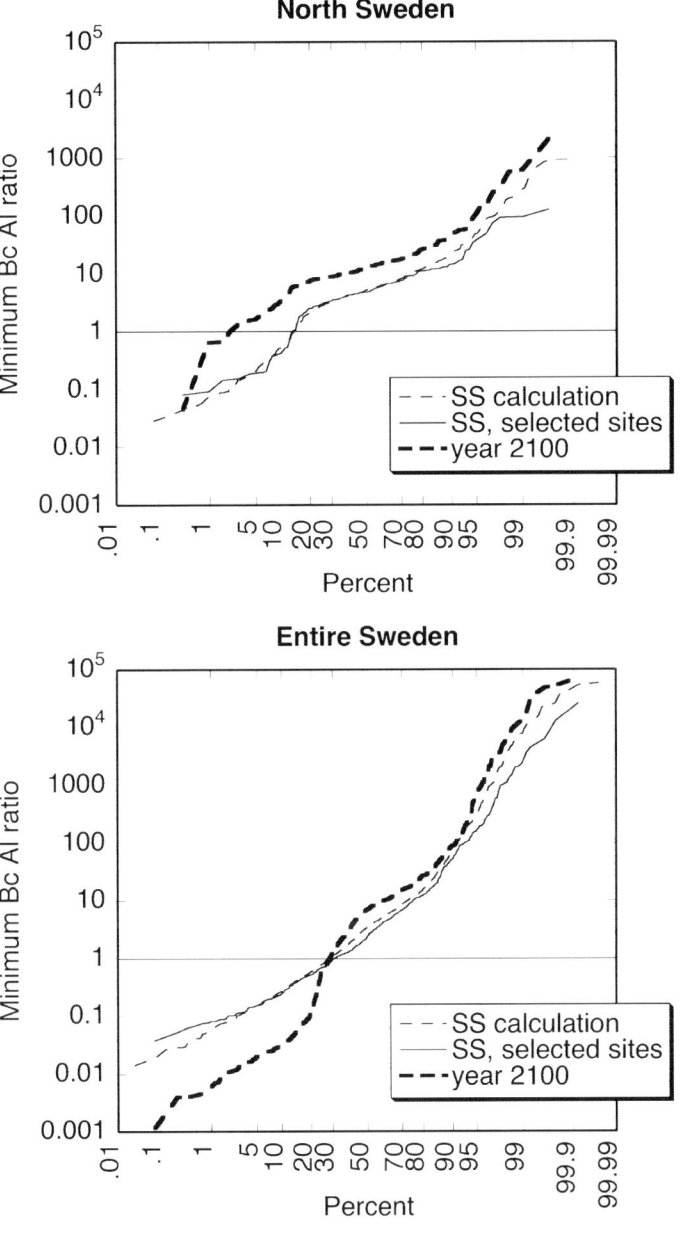

Figure 3. (continued)

events in the forest management scenarios, creates problems in assessing acceptable target loads since some sites might fluctuate between exceeded/not exceeded over the rotation period.

Acknowledgements

This work was funded by the MISTRA foundation through the ASTA and SUFOR programs and uses databases provided by the Swedish Agricultural University (SLU) in Uppsala, IVL and the Swedish Meterological Institute (SMHI).

References

Alveteg, M.: 1998, 'Dynamics of forest soil chemistry', *Doctoral Thesis*, Lund University, Sweden.
Alveteg, M., Sverdrup, H. and Warfvinge, P.: 1995, 'Regional assessment of the temporal trends in soil acidification in southern Sweden, using the SAFE model', *Water, Air, Soil Pollut.* **85**, 2509–2514.
Alveteg, M., Kurz, D. and Sverdrup, H.: 1998a, 'Integrated assessment of soil chemical status. 1. Integration of existing models and derivation of a regional database for Switzerland', *Water, Air, Soil Pollut.* **105**, 1–9.
Alveteg, M., Walse, C. and Sverdrup, H.: 1998b, 'Evaluating simplifications used in regional applications of the SAFE and MAKEDEP models', *Ecol. Model.* **107**, 265–277.
Alveteg, M., Walse, C. and Warfvinge, P.: 1998c, 'Reconstructing historic atmospheric deposition and nutrient uptake from present day values using MAKEDEP', *Water, Air, Soil Pollut.* **104**, 269–283.
Alveteg, M., Kurz, D. and Becker, R.: 2002, 'Incorporating nutrient content elasticity in the MAKEDEP model', in *Sustainable Forestry in Temperate Regions – Proceedings from the SUFOR International Workshop, 7–9 April 2002 in Lund, Sweden*, Reports in Ecology and Environmental Engineering 1, Department of Chemical Engineering II, Lund University, 2002, Lund, Sweden, pp. 52–67.
Bertills, U. and Lövblad, G. (eds): 2002, *Kritisk belastning för svavel och kväve, Rapport 5174*, Stockholm: Naturvårdsverket.
Collins, R. and Jenkins, A.: 1998, 'Regional modelling of acidification in Wales; calibration of a spatially distributed model incorporating land use change', *Hydrol. Earth Syst. Sci.* **2**, 533–541.
De Vries, W. and Kros, J.: 1991, 'Assesment of critical loads and the impact of deposition scenarios by steady state and dynamic soil acidification models', *Technical Report 46*, Winand Staring Centre.
De Vries, W., Kros, H. and Van der Salm, C.: 1995, 'Modelling the impact of acid deposition and nutrient cycling on forest soils', *Ecol. Model.* **79**, 231–254.
Evans, C. D., Jenkins, A., Helliwell, R. C. and Ferrier, R. C.: 1998, 'Predicting regional recovery from acidification; the MAGIC model applied to Scotland, England and Wales', *Hydrol. Earth Syst. Sci.* **2**, 543–554.
Heintzenberg, F., Jönsson, A. M., Karlsson, P. E., Nihlgård, B., Pleijel, H., Rosengren, U., Schlyter, P., Selldén, G., Skärby, L., Sonesson, K., Stjernquist, I., Sverdrup, H., Thelin, G., Uddling, J. and Welander, N. T.: 2002, 'Forest vitality and stress implications', in H. Sverdrup and I. Stjernquist (eds), *Developing Principles and Models for Sustainable Forestry in Sweden*, Kluwer Academic Publishers, Dordrecht, The Netherlands, pp. 197–272.
Helliwell, R. C., Ferrier, R. C., Evans, C. D. and Jenkins, A.: 1998a, 'A comparison of methods for estimating soil characteristics in regional acidification models; an application of the MAGIC model to Scotland', *Hydrol. Earth Syst. Sci.* **2**, 509–520.
Helliwell, R. C., Ferrier, R. C. and Jenkins, A.: 1998b, 'A two-layer application of the MAGIC model to predict the effects of land use scenarios and reductions in deposition on acid sensitive soils in the U.K.', *Hydrol. Earth Syst. Sci.* **2**, 497–507.

Holmqvist, J., Thelin, G., Rosengren, U. and Sverdrup, H.: 2002, 'Assessment of nutrient sustainability Asa Forest Research Park', in H. Sverdrup and I. Stjernquist (eds), *Developing Principles and Models for Sustainable Forestry in Sweden*, Kluwer Academic Publishers, Dordrecht, pp. 381–426.

Kurz, D., Alveteg, M. and Sverdrup, H.: 1998, 'Integrated assessment of soil chemical status. 2. Application of a regionalised model to 622 forested sites in Switzerland', *Water, Air, Soil Pollut.* **105**, 11–20.

Moldan, F., Kronnäs, V., Wilander, A., Karltun, E. and Cosby, B. J.: 2004, 'Modelling acidification and recovery of Swedish lakes', *Water, Air, Soil Pollut.: Focus* **4**(2–3), 139–160.

Munthe, J., Grennfelt, P., Sverdrup, H., Sundqvist, G., Alveteg, M., Bishop, K., Bergkvist, Helliwell, R. C., Ferrier, R. C., Evans, C. D. and Jenkins, A., Falkengren-Grerup, U., Hanssonc, H. C., Moldan, F., Karlsson, P. E., Näsholm, T., Pleijel, H. and Westling, O.: 2002, 'ASTA new concepts and methods for effect-based strategies on transboundary air pollution, Synthesis report, April 2002', *Technical Report*, IVL Swedish Environmental Research Institute.

Mylona, S.: 1993, 'Trends of sulphur dioxide emissions, air concentrations and depositions of sulphur in Europe since 1880', *Technical Report EMEP/MSC-W 2/93*, The Norwegian Meteorological Institute.

Posch, M., Hettelingh, J.-P. and Slootweg, J.: 2003, 'Manual for dynamic modelling of soil response to atmospheric deposition', *Technical Report 259101012*, RIVM.

Sefton, C. E. M. and Jenkins, A.: 1998, 'A regional application of the MAGIC model in Wales: Calibration and assessment of future recovery using a Monte-Carlo approach', *Hydrol. Earth Syst. Sci.* **2**, 521–531.

Warfvinge, P. and Sverdrup, H.: 1992, 'Calculating critical loads of acid deposition with PROFILE – A steady-state soil chemistry model', *Water, Air, Soil Pollut.* **63**, 119–143.

Warfvinge, P. and Sverdrup, H.: 1995, 'Critical loads of acidity to Swedish forest soils', *Reports in Ecology and Environmental Engineering 5, Chemical Engineering II*, Lund University.

SORPTION OF Cd IN SOILS: PEDOTRANSFER FUNCTIONS FOR THE PARAMETERS OF THE FREUNDLICH SORPTION ISOTHERM

A. L. HORN[1]*, R.-A. DÜRING[2] and S. GÄTH[2]

[1] *Ecology Centre, Department of Hydrology and Water Resources Management, CAU Kiel, Olshausenstr. 40, D-24098 Kiel, Germany;* [2] *Institute of Landscape Ecology and Resources Management, Division of Waste Management and Environmental Research, Justus-Liebig-University, Heinrich-Buff-Ring 26C, D-35392 Giessen, Germany*
(* author for correspondence, e-mail: ahorn@hydrology.uni-kiel.de; phone: +49-(0)431-8801268; fax: +49-(0)431-8804607)

(Received 20 August 2002; accepted 18 April 2003)

Abstract. Using a data set derived from 480 soil samples, pedotransfer functions for the coefficient (K) and the exponent (m) of the Freundlich sorption isotherm were developed. The functions were validated with data from 135 independent soil samples. We found high goodness-of-fit for the approaches of the Freundlich coefficient, whereas the efficiency of the models for predicting the independent samples should be improved. For the Freundlich exponent, the mean value of the parameterisation data set ($m = 0.88$) served as a better predictor than any of the developed functions. These findings contribute to the interpretation stated by previous authors that variation of the Freundlich exponent is of minor importance and therefore negligible in the course of sorption modelling.

Keywords: cadmium, Freundlich coefficient, Freundlich exponent, parameterisation, validation

1. Introduction

Risk assessment of large-scale environmental pollution by heavy metals has become a major task of environmental sciences in recent years (e.g., Ingwersen *et al.*, 2000; Keller *et al.*, 2001). Because many assessment tools are based on deterministic models, they require information on the sorption behaviour of heavy metals in soil, which mainly determines the environmental fate of the metals. Laboratory studies do not adequately satisfy the demand for sorption data at the scale of interest because the respective analyses would be intolerably expensive and time-consuming. Alternatively, pedotransfer functions can provide an adequate means to derive the required soil information (McBratney *et al.*, 2002).

An early pedotransfer function approach in the context of sorption analysis was presented by Chardon (1984) and van der Zee and van Riemsdijk (1987). The authors replaced the coefficient of the Freundlich sorption isotherm by a power function and included soil organic carbon and pH as independent variables. Other authors (Anderson and Christensen, 1988; Christensen, 1989; Reinds *et al.*, 1995; Streck and Richter, 1997; Kookana and Naidu, 1998; Römkens and Salomons,

1998; Springob and Böttcher, 1998; Tiktak *et al.*, 1998, 1999; Wilkens *et al.*, 1998; Elzinga *et al.*, 1999; Ingwersen, 2001) followed this approach using the same type of mathematical extension, but with individual combinations of independent variables.

In the course of parameter estimation according to the concept of Chardon (1984) and van der Zee and van Riemsdijk (1987), the exponent of the Freundlich equation receives a single constant value. Following the interpretation of van der Zee and van Riemsdijk (1987), based on a literature review, this result is conclusive as the Freundlich exponent 'is relatively constant over a wide range of experimental conditions'. However, other experimental studies found considerable variation of the Freundlich exponent between different soils (Buchter *et al.*, 1989; Wilkens, 1995; Springob and Böttcher, 1998a). Regarding this effect, Buchter *et al.* (1989), Schug *et al.* (1999) and Thiele and Leinweber (2001) investigated pedotransfer functions for the Freundlich exponent and found medium to low goodness-of-fit in cases when more than 10 samples were included in the analysis.

Considering the controversial evaluation of the importance of Freundlich exponent variation, as well as the unsatisfactory goodness-of-fit during its prediction in the preceding works, the aim of this study was to reappraise the significance of the relationship between the Freundlich exponent and other soil properties. For this purpose, we recalculated the pedotransfer function approaches using a larger set of sorption data than the previous authors. Further, we derived pedotransfer functions for the Freundlich coefficient in order to obtain a complete transfer equation of the Freundlich sorption isotherm. The approaches were validated using independent data sets derived by identical analytical methods.

2. Materials and Methods

The study areas, soil sampling and soil analysis are described in detail by Schug (2000), Szibalski (2001), and Sauer (2002). In the following sections, only a brief summary of the materials and methods is given.

2.1. SOIL SAMPLES

Nine study areas in different federal states of Germany (Hesse, Saxony, Thuringia, Rhineland-Palatinate) were selected for soil sampling. The sampling sites, where 615 soil samples were taken, represent a broad variety of soil properties (Table I) and landscape characteristics (land use, geology, geomorphology).

2.2. SORPTION EXPERIMENTS

The sorption characteristics of the soils were determined by batch experiments as described by Schug *et al.* (2000). A suspension of soil and a 0.01 M $Ca(NO_3)_2$ solution (soil/solution ratio 1:2.5 dry wt/vol), which contained 4 kBq ^{109}Cd and

TABLE I

Descriptive statistics of soil properties (SOC = soil organic carbon; OXIDE = sum of dithionite-extractable Fe and Mn) and parameters of the Freundlich sorption isotherm

	CLAY [g $(100g)^{-1}$]	SOC [g $(100g)^{-1}$]	pH [-]	OXIDE [g kg^{-1}]	K [L^m mg^{1-m} kg^{-1}]	m [-]
Minimum	0.2	0.1	2.9	0.06	0.23	0.51
Maximum	70.8	14.1	7.8	54.56	1294.44	1.01
Mean	18.6	1.7	4.2	12.99	101.95	0.88
Median	17.9	1.1	5.2	10.98	24.37	0.89

variable amounts of stable Cd (6.0 ng – 2.2 mg Cd as CdCl$_2$), was shaken horizontally for 16 h at 175 rpm, centrifuged (15 min at 8360 × g) and filtered (0.45 μm pore size). The supernatant was analysed for radioactivity by a NaI scintillation counter. The distribution of radioactive Cd between the solid phase and the soil solution was assumed to represent the sorption behaviour of total mobile non-radioactive Cd in soil. Total mobile Cd was defined as the sum of added Cd and Cd extractable from soil with EDTA in an independent experiment (Schug, 2000). The Cd concentration in the soil solution (C) was calculated as the product of total mobile Cd and the share of radioactivity in the supernatant divided by the volume of the added solution. The Cd concentration in the solid phase (S) was derived in a similar way as the product of total mobile Cd and the share of radioactivity in the solid phase divided by the amount of soil used in the experiment.

For each sample, the model of the Freundlich sorption isotherm was fitted to the experimental data:

$$S = KC^m \tag{1}$$

where S (mg kg^{-1}) is the Cd concentration in the solid phase, C (mg L^{-1}) is the Cd concentration in soil solution, K (L^m mg^{1-m} kg^{-1}) is Freundlich coefficient, and m (dimensionless) is the Freundlich exponent. Table I shows descriptive statistics of the estimated Freundlich parameters.

2.3. DERIVATION OF PEDOTRANSFER FUNCTIONS

According to Chardon (1984), the following pedotransfer function approach was used for the Freundlich coefficient (model signature: Cd-K):

$$K = \beta_0 \cdot \prod_{i=1}^{k} (P_i)^{\beta_i} \tag{2}$$

where β_0 and β_i are parameters, P_i indicates a soil property relevant for Cd sorption, and k is the number of relevant soil properties.

For the Freundlich exponent, we considered two pedotransfer functions. First, a power function was defined (model signature Cd-POW(m)):

$$m = \beta_0 \cdot \prod_{i=1}^{k} (P_i)^{\beta_i} \tag{3}$$

with β_0 and β_i indicating the parameters.

Second, a linear relationship between the exponent and the independent variables was assumed (model signature Cd-LIN(m)):

$$m = \beta_0 \cdot \sum_{i=1}^{k} \beta_i P_i \tag{4}$$

In all approaches, the P_i were defined by the variables H^+ (or pH in Equation (4)), SOC, CLAY and OXIDE. Parameter estimation was performed by multiple linear regression (SPSS 10, SPSS Inc.), which required logarithmic transformation of the power functions. Data from 480 soil samples (soil properties, Freundlich coefficients and Freundlich exponents) were used for the regression analysis.

Goodness-of-fit was evaluated by the adjusted coefficient of determination (R^2_{adj}). In the results section, two model types are presented for each approach. First, the model with highest R^2_{adj} including all significant variables is shown. In order to indicate this model type in the results section, an M (= maximum number of variables) is appended to the model signature, e.g. Cd-K-M. Alternatively, a model with a reduced number of variables is presented. This model type receives an R as appendix of model signature and is considered advantageous in case of limited data availability during regionalisation. The type R model was derived from the respective type M model by exclusion of the least significant variables. The criterion for termination of variable exclusion was the R^2_{adj} of the type R model, which should experience a maximum relative decrease of 5% in comparison to the R^2_{adj} of the type M model. This loss in goodness-of-fit was considered to be acceptable for practical purposes.

The data of the 135 independent soil samples not included in parameterisation were used for validation of the pedotransfer functions. The efficiency of the predictions was evaluated by the geometric mean error ratio (GMER) and the geometric standard deviation of the error ratio (GSDER) according to Tietje and Hennings (1996):

$$GMER = \exp\left[\frac{1}{n}\sum_{i=1}^{n} \ln\left(\frac{M_i}{O_i}\right)\right] \tag{5}$$

$$GSDER = \exp\left[\left(\frac{1}{n-1}\sum_{i=1}^{n}\left[\ln\left(\frac{M_i}{O_i}\right) - \ln(GMER)\right]^2\right)^{0.5}\right] \tag{6}$$

where n (dimensionless) is number of data sets, M_i is the ith predicted value (*e.g.*, concentration of sorbed Cd, Freundlich coefficient or Freundlich exponent), and O_i is the ith observed value.

If predicted values equal observed values, GMER and GSDER reach unity. A GMER > 1.0 indicates overestimation and GMER < 1.0 shows underestimation of the mentioned values. The GSDER is a measure of the scatter of the error ratio around the mean.

Further, the coefficient for efficiency (E) was applied for evaluation of the predictions (Nash and Sutcliffe, 1970):

$$E = 1 - \frac{\sum_{i=1}^{n}(O_i - M_i)^2}{\sum_{i=1}^{n}(O_i - \bar{O})^2} \tag{7}$$

where $\bar{O}$ is the mean of the observed values. E ranges from minus infinity to 1.0 with higher values indicating better agreement.

3. Results and Discussion

3.1. PARAMETERISATION

The parameters and the goodness-of-fit measures of the pedotransfer functions are presented in Table II. The relationships for the Freundlich coefficient achieved a high statistical significance and the equations for the Freundlich exponent obtained medium goodness-of-fit measures.

The coefficient of determination for the transfer functions of the Freundlich exponent is in the range of the studies of Buchter *et al.* (1989), Schug *et al.* (1999) and Thiele and Leinweber (2001), who achieved an R^2 (not adjusted) of 0.44 to 0.69 in cases where more than 10 samples were considered in the regression analysis. The lack in goodness-of-fit can not be attributed to the choice of the mathematical relationship, *i.e.* to the assumptions on the underlying processes controlling the sorption parameter. The differences between the goodness-of-fit measures of the linear equation and the power function are small and the statistical consistency of both approaches (normal distribution of the residuals, homoscedasticity) was not violated. The values of R^2 rather can be interpreted by the low sensitivity of the Freundlich exponent with respect to soil properties. For example, the range of pH and SOC values in the samples used for parameterisation is 2.9 to 7.8 and 0.1 to 14.1, respectively, whereas the range of the Freundlich exponent is only 0.51 to 1.01. Under these conditions, small analytical errors in the input data of the Freundlich exponent, which can not be excluded completely, have a stronger

TABLE II

Parameters, goodness-of-fit measures and validation results of the pedotransfer functions for the Freundlich coefficient and exponent

Model	Parameters (dimensionless)					Statistics			
	β_0	β_i				R^2_{adj}	Validation		
		H^+	SOC	OXIDE	CLAY		GMER	GSDER	E
Cd-K-M	2.432E-3	−0.625	0.574	0.175	0.276	0.928	0.54	2.10	0.46
Cd-K-R	5.943E-3	−0.645	0.627			0.916	0.47	2.05	0.44
Cd-POW(m)-M	1.107E-0	1.7E–2	n.s.[a]	−6.7E–3	n.s.[a]	0.640	0.97	1.03	−0.46
Cd-POW(m)-R	1.088E-0	1.7E–2				0.635	0.97	1.03	−0.37
Cd-LIN(m)-M	1.081E-0	−3.5E–2[b]	n.s.[a]	−6.4E–4	n.s.[a]	0.653	0.97	1.03	−0.43
Cd-LIN(m)-R	1.070E-0	−3.5E–2[b]				0.641	0.97	1.03	−0.40

[a] Not significant ($p > 0.05$)
[b] The negative logarithm of the proton activity (pH) was implemented in the regression equation.

negative effect on the goodness-of-fit during regression analysis in comparison to regression with variables of high sensitivity.

Comparing the type M models and the type R models, the derivation of functions with a reduced number of variables proves to be an efficient concept. Despite of the definition of a critical value regarding the loss in goodness-of-fit, a reduced model could be found for each approach. Therefore, data demand is restricted to a maximum of two variables, which can be derived from laboratory analysis with minor effort.

The reduced model for the Freundlich coefficient contains proton activity and SOC. This combination complies with earlier definitions of the pedotransfer function (van der Zee and van Riemsdijk, 1987; Streck and Richter, 1997; Ingwersen, 2001).

The reduced pedotransfer functions for the Freundlich exponent include only the proton activity (or pH value) as the independent variable. This result is in accordance with the definition of the pedotransfer function by Buchter et al. (1989). The complete models for the Freundlich exponent also consider the variable OXIDE in addition to proton activity, whereas SOC and CLAY are not significant. The model of Schug et al. (1999) is similar to this variant. However, the authors found pH and SOC as significant variables, whereas dithionite-extractable Fe and silt content showed no significant influence. The results of Thiele and Leinweber (2001) do not correspond with any of the above-mentioned models. The authors found no significant correlation between the Freundlich coefficient and other soil properties for samples from topsoil. They established a functional relationship between the Freundlich coefficient and the Freundlich exponent for this reason. In the subsoil, they found that the total Cd content and the pH were significant, but the number of samples was rather low ($n = 9$).

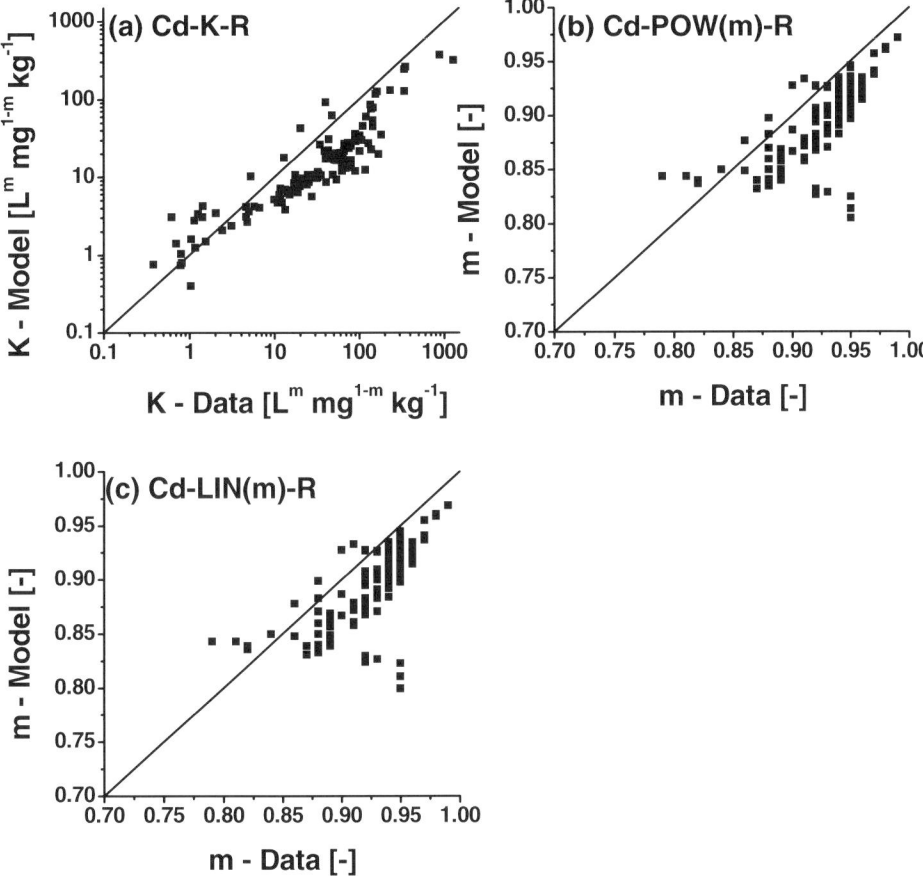

Figure 1. Comparison of measured and predicted data for (a) the reduced model type of the Freundlich coefficient pedotransfer functions, (b) the reduced model type of the power function approach for the Freundlich exponent, and (c) the reduced model type of the linear function for the Freundlich exponent.

3.2. VALIDATION

The validation of the pedotransfer functions for the Freundlich coefficient (Table II) shows that the models underestimate the independent data (GMER < 1.0) and that the standard deviation of the predictions is high (GSDER > 2.0). Figure 1a illustrates the relationships between measured and predicted values.

For the models of the Freundlich exponent, the GMER and the GSDER show a good approximation to the optimum value (GMER = GSDER = 1.0). However, the illustrations of the reduced model types (Figure 1, b and c) indicate still considerable differences between the observed and the predicted values. The negative value of the coefficient of efficiency (E) supports this interpretation as it points out that the mean of the observed Freundlich exponents, would be a better predictor than the

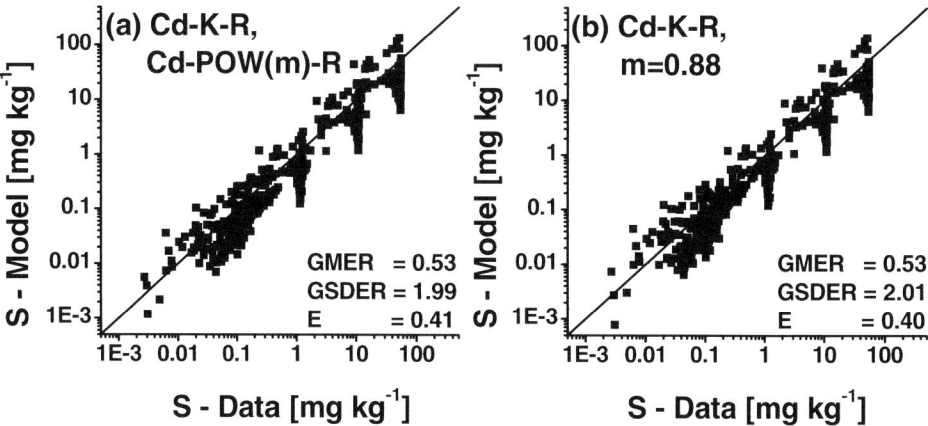

Figure 2. Comparison of measured and predicted concentrations of sorbed Cd in soil (S); predictions by (a) a pedotransfer function of the Freundlich sorption isotherm combined of Cd-K-R and Cd-POW(m)-R, and (b) a Freundlich isotherm approach implementing Cd-K-R and a constant value of the Freundlich exponent (mean of data used for parameterisation, $m = 0.88$).

model output (Legates and McCabe, 1999). These findings confirm the argument of van der Zee and van Riemsdijk (1987), who concluded from a literature review that a constant Freundlich exponent can be used in the regression analysis. Further, they are in accordance with the results of Springob and Böttcher (1998b) who reported that fixation of the Freundlich exponent did not essentially change the quality of their pedotransfer functions for Cd sorption in soil.

The use of complete pedotransfer functions for the Freundlich sorption isotherm, *i.e.* combinations of functions for the Freundlich coefficient with transfer equations for the Freundlich exponent, led to similar findings. We tested the validity of the approaches using the independent measurements of the concentration of sorbed Cd in the 135 validation samples. Figure 2a shows the results for the best model, which was constructed from Cd-K-R and Cd-POW(m)-R. The values of the validation criteria were almost identical to the results of the separate Cd-K-R model, thus indicating that a strong influence on prediction by the pedotransfer function for the Freundlich coefficient exists and that the definition of a functional relationship for the Freundlich exponent is of minor importance. These findings were confirmed if a constant Freundlich exponent, i.e. the mean value of the data set for parameterisation, was implemented in the model (Figure 2b). The changes of the validation criteria were small in this case.

The mean value for the Freundlich exponent in this study supports the range of $m = 0.82 - 0.88$ concluded from previous studies (van der Zee and van Riemsdijk, 1987; Reinds *et al.*, 1995; Temminghoff *et al.*, 1995; Streck and Richter, 1997; Springob and Böttcher, 1998b; Tiktak, 1999; Ingwersen, 2001). However, this range can only serve as a rough estimate in cases where no data on the Freundlich exponent are available. Some authors, however, report of distinctly higher or lower values for the parameter (Streck and Richter, 1997; Filius *et al.*, 1998; Tiktak *et al.*, 1998; Wilkens *et al.*, 1998; Tiktak *et al.*, 1999; Ingwersen *et al.*, 2000; Ingwersen, 2001).

4. Conclusions

The results of this study give additional support that functional relationships for the Freundlich exponent are of minor importance for the derivation of conclusive pedotransfer functions describing the sorption of Cd in soil. A mean value for the Freundlich exponent ($m = 0.88$ according to the data of this study) offers a good approximation for practical purposes.

The accuracy of the pedotransfer functions is determined mainly by the transfer equation of the Freundlich coefficient. The derivation of respective functional relationships showed that a high goodness-of-fit during parameterisation must not result in a successful validation of the model. This observation indicates that validation contributes valuable information on the performance of the models, and for this reason, it should become an essential step in the course of model development.

Acknowledgements

This project is part of the Collaborative Research Programme 299 'Land Use Options for Peripheral Regions' granted by the German Research Foundation. We thank Bettina Schug, Christian Friedrich, Ute Rost and Daniela Sauer for support of the study.

References

Anderson, P. R. and Christensen, T. H.: 1988, 'Distribution coefficients of Cd, Co, Ni, and Zn in soils', *J. Soil Sci.* **39**, 15–22.

Buchter, B., Davidoff, B., Amacher, M. C., Hinz, C., Iskandar, I. K. and Selim, H. M.: 1989, 'Correlation of Freundlich Kd and n retention parameters with soils and elements', *Soil Sci.* **148**, 370–379.

Chardon, W. J.: 1984, 'Mobiliteit van cadmium in de bodem', *Ph.D. Thesis*, Wageningen Universiteit (The Netherlands).

Christensen T. H.: 1989, 'Cadmium soil sorption at low concentrations: VIII, Correlation with soil parameters', *Water, Air, Soil Pollut.* **44**, 71–82.

Elzinga, E. J., van Grinsven, J. J. M. and Swartjes, F. A.: 1999, 'General purpose Freundlich isotherms for cadmium, copper and zinc in soils', *Europ. J. Soil Sci.* **50**, 139–149.

Filius, A., Streck, T. and Richter, J.: 1998, 'Cadmium sorption and desorption in limed topsoils as influenced by pH: isotherms and simulated leaching', *J. Environ. Qual.* **27**, 12–18.

Ingwersen, J., Streck, T., Utermann, J. and Richter, J.: 2000, 'Ground water preservation by soil protection: Determination of tolerable total Cd contents and Cd breakthrough times', *J. Plant Nutr. Soil Sci.* **163**, 31–40.

Ingwersen, J.: 2001, 'The environmental fate of cadmium in the soils of the waste water irrigation area in Braunschweig – measurement, modelling and assessment', *Ph.D. Thesis*, Technische Universität Braunschweig (Germany).

Keller, A., von Steiger, B., van der Zee, S. E. A. T. M. and Schulin, R.: 2001, 'A stochastic empirical model for regional heavy-metal balances in agroecosystems', *J. Environ. Qual.* **30**, 1976–1989.

Kookana, R. S. and Naidu, R.: 1998, 'Effect of soil solution composition on cadmium transport through variable charge soil', *Geoderma* **84**, 235–248.

Legates, D. R. and McCabe Jr., G. J.: 1999, 'Evaluating the use of "goodness-of-fit" measures in hydrologic and hydroclimatic model validation', *Water Resour. Res.* **35**, 233–241.

McBratney, A. B., Minasny, B., Cattle, S. R. and Vervoort, R. W.: 2002, 'From pedotransfer functions to soil inference systems', *Geoderma* **109**, 41–73.

Nash, J. E. and Sutcliffe, J. V.: 1970, 'River flow forecasting through conceptual models. I. A discussion of principles', *J. Hydrol.* **10**, 282–290.

Reinds, G. J., Bril, J., de Vries, W., Groenenberg, J. E. and Breeuwsma, A.: 1995, 'Critical loads and excess loads of cadmium, copper and lead for European forest soils', *DLO Winand Staring Centre Report 96*, Wageningen (The Netherlands).

Römkens, P. F. A. M. and Salomons, W.: 1998, 'Cd, Cu and Zn solubility in arable and forest soils: consequences of land use changes for metal mobility and risk assessment', *Soil Sci.* **163**, 859–871.

Sauer, D.: 2002, 'Genese, Verbreitung und Eigenschaften periglaziärer Lagen im Rheinischen Schiefergebirge – anhand von Beispielen aus Westerwald, Hunsrück und Eifel', *Ph.D. Thesis*, Justus-Liebig-Universität Giessen (Germany).

Schlichting, E., Blume, H.-P. and Stahr, K.: 1995, *Bodenkundliches Praktikum: eine Einführung in pedologisches Arbeiten für Ökologen, insbesondere Land- und Forstwirte und für Geowissenschaftler*, Blackwell Wissenschafts-Verlag, Berlin.

Schug, B., Hoß, T., Düring, R. and Gäth, S.: 1999, 'Regionalization of sorption capacities for arsenic and cadmium', *Plant Soil* **213**, 181–187.

Schug, B.: 2000, 'Entwicklung von Pedotransferfunktionen zur Regionalisierung des Retentionspotenzials von Böden für Cadmium, Blei und Zink', *Ph.D. Thesis*, Justus-Liebig-Universität Giessen (Germany).

Schug, B., Düring, R. and Gäth, S.: 2000, 'Improved cadmium sorption isotherms by the determination of initial contents using the radioisotope ^{109}Cd', *J. Plant Nutr. Soil Sci.* **163**, 197–202.

Springob, G. and Böttcher, J.: 1998a, 'Parameterization and regionalization of Cd sorption characteristics of sandy soils, I, Freundlich type parameters', *J. Plant Nutr. Soil Sci.* **161**, 681–687.

Springob, G. and Böttcher, J.: 1998b, 'Parameterization and regionalization of Cd sorption characteristics of sandy soils, II, Regionalization: Freundlich k estimates by pedotransfer functions', *J. Plant Nutr. Soil Sci.* **161**, 689–696.

Streck, T. and Richter, J.: 1997, 'Heavy metal displacement in a sandy soil at the field scale: I, Measurements and parameterization of sorption', *J. Environ. Qual.* **26**, 49–56.

Szibalski, M.: 2001, 'Großmaßstäbige Regionalisierung labiler Bodenkennwerte in standörtlich hochdiversen Kulturlandschaften', *Ph.D. Thesis*, Justus-Liebig-Universität Giessen (Germany).

Temminghoff, E. J. M., van der Zee, S. E. A. T. M. and de Haan, F. A. M.: 1995, 'Speciation and calcium competition effects on cadmium sorption by sandy soil at various pHs', *Europ. J. Soil Sci.* **46**, 649–655.

Thiele, S. and Leinweber, P.: 2001, 'Parameterization of Freundlich adsorption isotherms for heavy metals in soils from an area with intensive livestock production', *J. Plant Nutr. Soil Sci.* **164**, 623–629.

Tietje, O. and Hennings, V.: 1996, 'Accuracy of the saturated hydraulic conductivity prediction by pedo-transfer functions compared to the variability within FAO textural classes', *Geoderma* **69**, 71–84.

Tiktak, A., Alkemade, J. R. M., van Grinsven, J. J. M. and Makaske, G. B.: 1998, 'Modelling cadmium accumulation at a regional scale in the Netherlands', *Nutr. Cycling. Agroecosyst.* **50**, 209–222.

Tiktak, A.: 1999, 'Modeling non-point source pollutants in soils; Applications to the leaching and accumulation of pesticides and cadmium', *Ph.D. Thesis*, Universiteit van Amsterdam (The Netherlands).

Tiktak, A., Leijnse, A. and Vissenberg, H.: 1999, 'Uncertainty in a regional-scale assessment of cadmium accumulation in the Netherlands', *J. Environ. Qual.* **28**, 461–470.

van der Zee, S. E. A. T. M. and van Riemsdijk, W. H.: 1987, 'Transport of reactive solute in spatially variable soil systems', *Water Resour. Res.* **23**, 2059–2069.

Wilkens, B. J.: 1995, 'Evidence for groundwater contamination by heavy metals through soil passage under acidifying conditions', *Ph.D. Thesis*, Universiteit Utrecht (The Netherlands).

Wilkens, B. J., Brummel, N. and Loch, J. P. G.: 1998, 'Influence of pH and zinc concentration on cadmium sorption in acid, sandy soils', *Water, Air, Soil Pollut.* **101**, 349–362.

THE DEVELOPMENT OF A GIS-BASED INVENTORY OF STANDING WATERS IN GREAT BRITAIN TOGETHER WITH A RISK-BASED PRIORITISATION PROTOCOL

M. HUGHES[1]*, D. D. HORNBY[2], H. BENNION[1], M. KERNAN[1], J. HILTON[2], G. PHILLIPS[3] and R. THOMAS[4]

[1] *Environmental Change Research Centre, University College London, London, WC1H 0AP, U.K.;*
[2] *Centre for Ecology and Hydrology, Winfrith Technology Centre, Dorchester, Dorset, DT2 8ZD, U.K.;* [3] *Environment Agency, National Centre for Risk Analysis Options Appraisal, Reading, RG1 8DQ, U.K.;* [4] *Environment Agency, National Centre for Environmental Data, Bath, Avon, BA2 9ES, U.K.*
(* author for correspondence, e-mail: m.hughes@ucl.ac.uk, phone: +44 020 7679 5522, fax: +44 020 7679 7565)

(Received 20 August 2002; accepted 2 April 2003)

Abstract. An inventory of standing waters (freshwater lakes and lochs) was derived from Ordnance Survey digital map data at a scale of 1:50 000 and represents the most comprehensive survey of its kind for Great Britain. The inventory includes 43 738 water bodies in England, Scotland, Wales and the Isle of Man and contains basic physical data such as location, surface area, perimeter and altitude. Catchment areas were computed for water bodies with a surface area larger than 1 ha from a digital terrain model (DTM) using customised routines in a geographical information system (GIS). The resulting polygons were then used to derive catchment-related information from a variety of national datasets including population density, livestock density, land cover, solid and drift geology, meteorological data, freshwater sensitivity status, acid deposition and conservation status. Using data derived from the inventory a risk-based prioritisation protocol was developed to identify standing waters at risk of harm from acidification and eutrophication. This information is required by the Environment Agency, Scottish Environmental Protection Agency and the U.K. statutory conservation bodies to co-ordinate actions and monitor change under international, European and national legislation.

Keywords: acidification, catchment delineation, eutrophication, GIS, Great Britain, risk-assessment, standing waters

1. Introduction

The water bodies of Great Britain are an important resource with local, national and international significance. They support a wide range of activities including water supply, nature conservation, fisheries, tourism, leisure and scientific research and are important habitats for many plant and animal species. In 1979, Smith and Lyle published a paper describing the distribution of Great Britain's water resource based on 1:250 000 scale maps. Since then there has been no comparable survey published for Great Britain based on more detailed map data that has become available.

Recent EU legislation together with the growing demands of Great Britain's statutory environmental protection and conservation agencies has led to the need for a comprehensive computer-based inventory of standing waters. The Water Framework Directive (EU, 2000), which came into force in December 2000, requires the setting of water quality objectives for all water bodies (not just those designated by the member state) and provides for a classification of water bodies according to ecological quality status (DEFRA, 2001). The implementation of U.K. Biodiversity Action Plans for both habitats (specifically eutrophic and mesotrophic standing waters) and species also require data on the extent, distribution and physical, chemical and biological status of standing waters across the U.K. (U.K. Biodiversity Group, 1998).

A consortium of stakeholders was brought together to build and populate the inventory and to develop a risk-based prioritisation protocol to help assess environmental harm from nutrients and acid deposition. The protocol requires catchment-based data for several environmental parameters and so it was necessary to compute catchment polygons. These were used in a GIS with appropriate national datasets to derive the environmental data required to apply the protocol.

This paper describes how the inventory was built, gives an account of the distribution of standing waters in Great Britain and outlines the development of the prioritisation protocol.

2. Methods

2.1. Assembling the inventory

During the last 100 years there have been various attempts to describe the extent and distribution of standing waters in Great Britain (e.g. Smith and Lyle, 1979; Barr *et al.*, 1994; Fuller *et al.*, 1994; Haines-Young *et al.*, 2000), Scotland (e.g., Murray and Pullar, 1910; Lyle and Smith, 1994) and Northern Europe (e.g. Henriksen *et al.*, 1998). For Great Britain as a whole, the most comprehensive of these was the Smith and Lyle survey, which was based on a visual inspection of 1:250 000 paper maps from Ordnance Survey (OS). At this scale the lower level for inclusion is about 4 ha. The survey also undertook sample counts from 1:63 360 OS maps and it was estimated that the number of lochs was more than seven times the number found in the original survey, the additional lochs being small water bodies not shown at the 1:250 000 scale. Clearly, at larger scales there will be a greater number of water bodies defined and the lower level for inclusion will get smaller.

In choosing the data source for the inventory the main consideration was the degree of detail needed. The basic requirements were for an outline of the water body from which the co-ordinates of its centroid, surface area and perimeter could be derived. It was decided at an early stage to concentrate on water bodies with a surface area of at least 1 ha. OS Land-Form PANORAMA™ contour data at

1:50 000 contain features representing contours, spot heights, ridge lines, coastline and lake outlines as seen on the 1:50 000 Landranger paper map series. Lakes in PANORAMA are defined as 'bodies of inland water, including ponds, lakes, lochs and the lower parts of some rivers' (Ordnance Survey, 2001). Each lake outline has an associated elevation attribute (to the nearest meter) and all line objects have a quoted spatial accuracy of 3 m root mean square error (Ordnance Survey, 2001). A visual inspection of the data indicates that water bodies with surface areas as small as 0.5 ha are accurately represented and although smaller water bodies do exist in the dataset (the smallest being 0.02 ha), their representation is somewhat generalised. The dataset remains relatively manageable in terms of computer processing whereas data at a larger scale (such as OS Land-Line™) would have introduced problems of data manageability and unnecessary complexity (see Hughes and Fisher, 2000).

Each of the 812 400 km^2 PANORAMA tiles was processed individually to extract the lake features (OS feature code 0202) using 3 SPARC Ultra-5 workstations and ESRI software. Tiles were converted to ArcInfo coverages using ESRI's MapManager software and subsequently lake features were extracted automatically, converted to polygons using custom scripts in ArcInfo and assembled as a single coverage. Each lake polygon was then assigned a pair of geometric centroid co-ordinates (to the nearest metre) and basic physical parameters (surface area in hectares, altitude and perimeter both to nearest metre, number of islands).

The data were error-checked using a variety of methods, both automatic and manual. Lake features that did not form whole polygons were closed automatically using an iterative process where snapping distance was gradually increased from 2 to 10 m. Open polygons with gaps larger than 10 m were edited manually with reference to the paper map, resulting in more than 700 additional water bodies. The conversion to polygon process occasionally produced slivers (very small polygons adjacent to 'real' polygons) and these were searched for automatically (by size) but checked visually against the map. Some 300 polygons were rejected as being slivers. The original dataset contained large rivers coded as lakes and split into sections appearing as long chains of adjacent rectangular polygons. These were removed manually, resulting in the loss of a further 2200 polygons. The final number of polygons in the database was 46 570; of which 43 738 were lakes and the remaining 2832 were islands.

Additional parameters were derived for each lake polygon. OS-style grid-references, geographic coordinates, distance to sea, shoreline development index (a measure of shoreline complexity), and length and bearing of line of maximum fetch were all computed using Microsoft Excel, ArcView and ArcInfo. Attribute data were managed in a Microsoft Access database and linked to the GIS using unique identification codes. The lake centroids were used in a series of overlays to identify co-occurrence with a range of national datasets (Table I). Measured depth data for approximately 5% of the water bodies >1 ha were collected from a wide range of sources (including Murray and Pullar, 1910) and used in a simple multiple

TABLE I

Data available for lake centroids	Data available for catchments
Agency (EA/SEPA) Region co-occurrence	LCM90 Landcover class (Fuller et al., 1994)
1995–1997 Acid Deposition (CEH)	Drift Geology (1:625 000)
Freshwater Sensitivity (Hornung et al., 1995)	Solid Geology (1:625 000)
Protected areas co-occurrence (includes National Park, Forest Park, National Nature Reserve (NNR), RAMSAR, Special Area of Conservation (SAC), Special Protection Area (SPA), Site of Special Scientific Interest (SSSI))	1995–1997 Acid Deposition (CEH)
	Freshwater Sensitivity (Hornung et al., 1995)
	Animal stocking density (MAFF)
	Modelled hindcast P load (Johnes et al., 2000)
	Modelled current P load (Hilton et al., 1999)
	1991 Population (Bracken and Martin, 1995)
EN Character/Natural Area co-occurrence	Mean annual runoff (CEH)
OS Landranger Map sheet	

regression model to predict mean and maximum lake depths (for the calculation of volume and retention time) based on surface area, altitude and perimeter. Separate models were developed for England, Wales and Scotland – giving better results than a global model. The model used was a least squares regression which, despite large residuals, did at least give a good prediction of depth if broad depth classes (such as those used in the Water Framework Directive) were used.

2.2. CATCHMENT DELINEATION AND OVERLAY

Catchment areas were derived for all water bodies with a surface area >1 ha. The lake polygons ($n = 14\,353$) were extracted and processed with a flow grid derived from the Institute of Hydrology digital terrain model (DTM) (Morris and Flavin, 1990) to generate catchment polygons. This 50 m resolution DTM is based on digitised contours from the Ordnance Survey 1:50 000 map (the same source as the lake polygons used in the inventory), but has been adjusted to conform to a digitised stream network for greater accuracy. An assessment of DTM quality for hydrological applications (Wise, 2000) found that this particular model gave the lowest RMSE error statistics and was well suited for hydrological applications due to its low number of sinks (groups of cells that do not flow into any surrounding cells).

Each lake polygon was used to select grid cells from the flow grid and the cell with the greatest value (i.e. maximum flow) was selected as the pour point (i.e. the outflow cell for the watershed). ArcView's Spatial Analyst Watershed function was used to generate a catchment outline from the pour point, which was saved as a polygon. The catchment polygons were subsequently processed to calculate their area, perimeter and lake to catchment ratio. Catchment polygons were then used in the GIS to extract data from national datasets (Table I).

The catchment overlay process for each dataset took one of two forms depending on the data type. A catchment-weighted procedure was used for overlay with gridded maps of distributed data at varying resolutions (such as acid deposition (1 km) and P load (5 km)) whereby a mean value was found by calculating the proportion of each gridded data cell overlaid by the catchment polygon. For datasets containing categorical data in discrete units (such as geology and land cover), the proportion of each category was calculated as an actual area (in ha) and percentage of catchment. (Note: many of the national datasets did not cover the Isle of Man therefore these water bodies were excluded from the risk prioritisation exercise.)

2.3. RISK PRIORITISATION

The purpose of the risk prioritisation exercise was to assess the risk of harm from nutrients (eutrophication) and acid deposition (acidification) to ecological condition, which in turn would allow a prioritisation of water bodies for action. The approach follows DETR guidelines for environmental risk assessment (DETR, 2000) whereby risk is placed in an objective framework with multiple tiers, ensuring that actions are focused where they are most beneficial to society (Pollard et al., 2000). A detailed account of the protocol can be found in Bennion et al. (2002) but in short, the prioritisation system for lakes is based on three essentially independent properties: importance – or value to society; hazard – posed to a lake from sources of nutrients and acidity; and sensitivity – of a lake to change in water quality.

The first tier of the protocol takes as its input the entire population of lakes larger than 1 ha in surface area and applies a simple set of risk screening criteria under the three headings: importance, hazard and sensitivity. Lakes that meet these criteria then fall through to tier 2 where a more detailed risk assessment is performed. Lakes that are output from this tier are then carried forward into the third and final tier where a detailed quantitative risk assessment can be carried out on a site-specific level. The protocol results in a list of water bodies ranked by importance, exposure to hazard and sensitivity, which can be used by managers to coordinate and prioritise actions. It is also hoped that this process may help identify previously overlooked water bodies with conservation value.

Importance criteria were selected in consultation with stakeholders to satisfy the key requirements of EU legislation (Water Framework Directive), U.K. Biodiversity Action Plans (specifically the Lakes Habitat Action Plans) and the agency's (EA/SEPA) eutrophication strategies. The criteria were size (>50 ha), conservation status (RAMSAR, SAC or SPA) and designation under EU Bathing Waters Directive.

At tier 1, an assessment of the risk of harm from eutrophication is made from estimated nutrient loads for water bodies using the GIS-derived catchment polygons

TABLE II
Numbers of water bodies by country and surface area

	<1 ha[a]	1–5 ha	5–10 ha	10–50 ha	50–100 ha	>100 ha	Total
England	10738	4260	710	625	64	51	16448
Scotland	17727	5294	1195	1205	168	171	25760
Wales	894	394	88	90	10	17	1493
Isle of Man	26	9	0	2	0	0	37

[a] The dataset contains no water bodies <0.02 ha and the number between 0.02 and 0.2 are almost certainly under-represented.

and national datasets of land cover (Fuller et al., 1994) and population (Bracken and Martin, 1995) and published phosphorus export coefficients (Hilton et al., 1999). At tier 2, the eutrophication risk assessment is improved by the addition of measured chemical and biological data on current trophic status and an estimate of enrichment based on hindcast models (e.g. Bennion et al., 1996; Johnes et al., 1996; Ferrier et al., 1997). The sensitivity of a lake to eutrophication was assessed using an estimate of retention time (or flushing rate) derived from measured or modelled lake depths.

For acidification, the main hazard is regarded as atmospheric deposition and this was estimated at tier 1 using the GIS-derived catchment polygons and national maps of acid deposition (provided by the Centre for Ecology and Hydrology). At tier 2 the acidification scheme is enhanced with calculations of critical loads and exceedances (e.g. Curtis et al., 2000) and pH hindcast models (e.g. Jenkins et al., 1990; Jones et al., 1993). Sensitivity was assessed using a national map of sensitivity of freshwaters to acidification (Hornung et al., 1995).

3. Results and Discussion

3.1. Lakes inventory – Summary data

The inventory contains 43 738 water bodies in England, Scotland, Wales and Isle of Man. A breakdown of distribution by surface area and country is given in Table II. The majority of water bodies in each country have a surface area smaller than 1 ha with less than 10% having a surface area larger than 10 ha. The total surface area of standing waters in the inventory is 213 911 ha – covering approximately 1% of the land surface of Great Britain.

In their survey of Scottish lochs, Lyle and Smith (1994) grouped water bodies larger than 25 ha into logarithmic area classes to investigate the relationships between numbers of lochs, accumulated area and volume. They found that there was a natural order in the frequency of occurrence based on surface area, which is confirmed by the present study for Great Britain as a whole. The relationship

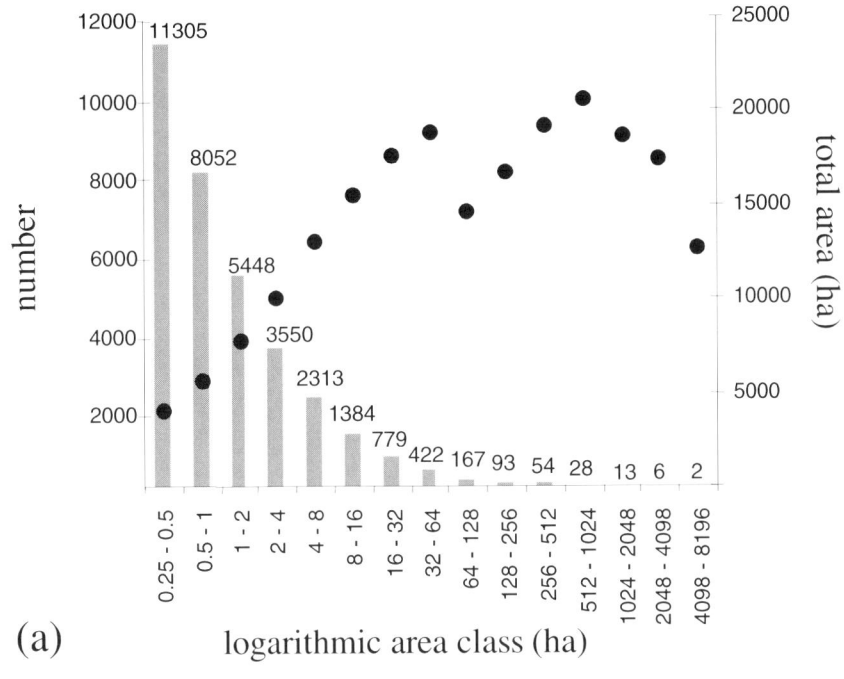

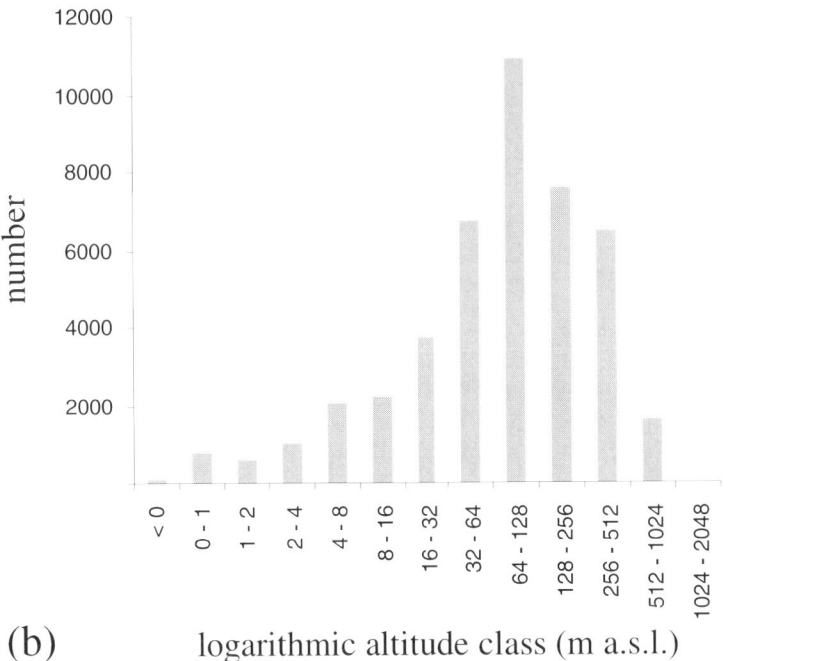

Figure 1. (a) Numbers of water bodies (grey bars) and total areas (black dots) for logarithmic area classes; (b) numbers of water bodies for logarithmic altitude classes.

TABLE III

Numbers of water bodies by country and altitude (metres above sea level)

	<10	10–50	50–100	100–300	300–500	500–750	>750
England	2686	4826	4757	3866	245	66	2
Scotland	1828	4651	4481	9001	4205	1327	267
Wales	183	180	213	468	345	102	2
Isle of Man	4	21	2	10	0	0	0

between numbers of water bodies and total area for each logarithmic area classes is shown in Figure 1 and shows that although the number of water bodies increases as the logarithmic area class decreases, the total area starts to decrease – something which Lyle and Smith predicted but were unable to demonstrate with their dataset. It is useful to look at this relationship because it gives an idea of the level at which water bodies, in a strategic sense at least, become unimportant in terms of surface area and volume. By extrapolating the relationship between logarithmic area class and total surface area of water bodies (e.g. polynomial curve-fitting) it can be estimated that although the number of very small (<0.25 ha) water bodies in Great Britain almost certainly exceeds one hundred thousand their accumulated area is probably no more than 5% of the total for GB and their accumulated volume less than 1%. The estimated total surface area of standing water for Great Britain based on these results is ~220 000 ha. This includes the water bodies in the inventory and estimates for the smaller water bodies not shown at 1:50 000 and is somewhat larger than the recent Countryside 2000 survey estimate of 190 000 ha, but smaller than Smith and Lyles 1979 estimate of 240 400 ha. The Countryside 2000 survey (Haines-Young et al., 2000) was based on a stratified random sampling of 1 km^2 squares and despite the lower total area claims to include lowland ponds, giving a total number of standing water bodies at 400 000 (although they do not give a lower limit of inclusion).

Table III and Figure 1 show the distribution of water bodies by altitude and country and by logarithmic altitude class, respectively. There are two different distributions influenced by topography – England with the majority of its water bodies at lower altitudes and Scotland and Wales with the majority at slightly higher altitudes. In terms of the Water Framework Directive 'System A' ecoregions, 75.8% of water bodies (by number) occur in the lowland ecoregion (<200 m), 23.8% in the mid-altitude ecoregion (200–800 m) and the remaining 0.4% in high ecoregion (>800 m). The figures for distribution by surface area are similar except for the high ecoregion, which only has 0.06% of the total surface area for Great Britain.

3.2. Risk prioritisation protocol

Two schemes were developed for assessing the risk to standing waters from eutrophication and acidification. Both relied on data generated by overlay of catchment polygons and national datasets. The protocol was tested on a subset of 30 lakes for which measured data were available with the primary objective of assessing the reliability of the importance, hazard and sensitivity measures as indicators of risk. Detailed results can be found in Bennion *et al.* (2002).

For eutrophication, the protocol reliably models the risk of enrichment and the likelihood of response to restoration in most cases. Modelled current nutrient loads are well estimated when inputs from humans, land cover and livestock are considered and there are no point sources of nutrient input, which the inventory does not include. Results suggest that the lake depth model is not performing well and consistently overestimates depth. For shallow lowland lakes this is significant because depth is used to predict stratification class suggesting a better response to restoration than would be expected in reality. A more rigorous model of lake depth is planned for the future based on morphometric analysis using surrounding slope angles derived from contours.

The acidification scheme was tested using palaeolimnological data (diatom assemblages) to confirm high risk of acidification and lakes predicted as being at low risk due to agricultural improvements in the catchment are confirmed. The use of critical load and exceedance data to improve risk assessments was found to be beneficial, especially the diatom critical load since the diatom community is regarded as the most sensitive aspect of aquatic biota.

This work considers just two possible risk-assessments based on eutrophication and acidification, but the protocol could easily be adapted to account for any other type of environmental pressure provided that sufficient datasets were available.

3.3. Errors and uncertainty

The main drawback of using PANORAMA data as the basis for the inventory is that it is a static dataset. OS has not updated it since data capture took place in 1983. This means that water bodies created or modified since 1983 will not appear in the dataset in their present form. On the whole, this is not a problem since water bodies do not change much over time. A visual comparison of all lakes >1 ha and current 1:50 000 maps indicates that around 4% of water bodies in the inventory have been significantly modified or no longer exist, however the majority of these occurrences appear to be related to cartographic errors in both the original data (e.g. features incorrectly coded as water features) and the current map. Real cases of lake loss are concentrated in areas around extractive industries, quarries, urban areas, docks and dunes. Conversely, it can be expected that many new lakes have been created since the PANORAMA data capture in 1983 and this is confirmed by other surveys, especially in lowland regions (e.g. Haines-Young *et al.*, 2000).

In several cases, catchment-derived data for certain lakes were suspect, leading to an incorrect risk assessment. Closer inspection revealed that the catchment delineation process had failed to produce an accurate watershed for the lake concerned. The majority of these 'errors' occur when a small water body is close to a large river. The cell that gets selected as the pour point for the catchment is the one with the highest accumulated flow and if a lake cell coincides with a river cell then the river catchment will actually be computed. In the risk assessment protocol this leads to a small water body being erroneously attributed large nutrient and acid deposition loads. Another type of error that may occur is poor catchment delineation in areas of low relief – such as in the Fens or the Norfolk Broads. The vertical resolution of the DTM is insufficient in these cases to allow the accurate mapping of a lake's catchment. A confounding factor in these areas is that lake catchments often are influenced by man-made drainage features, which are not represented in the DTM. Clearly, these lowland water bodies represent a challenge. Higher resolution LiDAR data could be used as a means to more accurately delineate their catchments (Lloyd and Atkinson, 2002).

4. Conclusions

Despite being based on 25 year old data, the inventory presented in this paper provides the most comprehensive survey of Great Britain's standing water resource published to date. Together with the catchment-derived data and risk-prioritisation protocol, it represents a valuable resource for GB's statutory environmental protection and conservation agencies. Potential applications include use in stock-at-risk assessments (e.g. Kernan *et al.*, 2004), the development of a lake classification scheme or 'typology' as required by the Water Framework Directive, and investigations into the relationships between catchment-related parameters such as rainfall, erosion, sediment loads, ecological conditions and catchment morphology. Already the inventory (together with additional datasets) is being used as part of a project to predict in-lake phosphorus from diffuse and point sources. The catchment delineation routine needs refining for water bodies in areas of low relief and to take account of man-made drainage before catchment-based data for these lakes can be used with confidence in the risk-assessment protocol.

Acknowledgements

The authors would like to thank all those who provided data for the inventory, especially; Ian Fozzard and Duncan Taylor (Scottish Environmental Protection Agency), Willie Duncan and Mary Hennessy (Scottish Natural Heritage), Catherine Duigan (Countryside Council for Wales), Allan Stewart (English Nature) and Oliver Jarratt (Environment Agency National Centre for Environmental Data).

Also, thanks to those who contributed to the development of the risk prioritisation protocol, in particular, Colin Reynolds (CEH Windermere), Laurence Carvalho (CEH Edinburgh), Judy Clark and Chris Curtis (University College London). This project was jointly funded by the Environment Agency (EA), English Nature, Countryside Council for Wales and the Scotland and Northern Ireland Forum for Environmental Research (SNIFFER) under EA contract number P2-239.

References

Barr, C. J., Howard, D. C. and Benefield, C. B.: 1994, *Countryside Survey 1990. Inland Water Bodies*, Department of the Environment, London (Countryside 1990 Series).

Bennion, H., Hilton, J., Hughes, M., Clark, J., Hornby, D., Kernan, M. and Simpson, G.: 2002, 'Development of a risk based prioritisation protocol for standing waters in Great Britain based on a georeferenced inventory – Phase 2', Environment Agency R&D, *Technical Report P2-260/2/TR1*.

Bennion, H., Juggins, S. and Anderson, N. J.: 1996, 'Predicting epilimnetic phosphorous concentrations using an improved diatom-based transfer function and its application to lake eutrophication management', *Environ. Sci. Technol.* **30**, 2004–2007.

Bracken, I. and Martin, D.: 1995, 'Linkage of the 1981 and 1991 U.K. Censuses using surface modelling concepts', *Environ. Plann. A* **27**, 379–390.

Curtis, C. J., Allott, T. E. H., Hughes, M., Hall, J., Harriman, R., Helliwell, R., Kernan, M., Reynolds, B. and Ullyett, J.: 2000, 'Critical loads of sulphur and nitrogen for fresh waters in Great Britain and assessment of deposition reduction requirements with the First-order Acidity Balance (FAB) model', *Hydrol. Earth Syst. Sci.* **4**, 125–140.

DEFRA (Department for Environment, Food and Rural Affairs): 2001, *First Consultation Paper on the Implementation of the EC Water Framework Directive (2000/60/EC)*, DEFRA.

DETR (Department of the Environment, Transport and the Regions): 2000, *Guidelines for Environmental Risk Assessment and Management*, Revised Departmental Guidance.

European Union: 2000, *Directive of the European Parliament and the Council 2000/60/EC Establishing a Framework for Community Action in the Field of Water Policy*, Luxembourg.

Ferrier, R. C., Owen, R., Edwards, A. C., Malcolm, A. and Morrice, J. G.: 1997, 'Hindcasting of phosphorus concentrations in Scottish Standing Waters', in P. J. Boon and D. L. Howell (eds), *Freshwater Quality: Defining the Indefinable?*, HMSO, Edinburgh.

Fuller, R. M., Groom, G. B. and Jones, A. R.: 1994, 'The land cover map of Great Britain: An automated classification of Landsat Thematic Mapper data', *Photogramm. Eng. Remote Sensing* **60**, 553–562.

Haines-Young, R. H., Barr, C. J., Black, H. I. J., Briggs, D. J., Bunce, R. G. H., Clarke, R. T., Cooper, A., Dawson, F. H., Firbank, L. G., Fuller, R. M., Furse, M. T., Gillespie, M. K., Hill, R., Hornung, M., Howard, D. C., McCann, T., Morecroft, M. D., Petit, S., Sier, A. R. J., Smart, S. M., Smith, G. M., Stott, A. P., Stuart, R. C. and Watkins, J. W.: 2000, *Accounting for Nature: Assessing Habitats in the U.K. Countryside*, DETR, London.

Henriksen, A., Skjelvale, B. L., Mannio, J., Wilander, A., Harriman, R., Curtis, C., Jensen, J. P., Fjeld, E. and Moiseenko, T.: 1998, 'Northern European Lake Survey, 1995', *Ambio* **27**, 80–91.

Hilton, J., Irons, G. P. and McNally, S.: 1999, 'Pilot Catchment Study of Nutrient Sources – Control Options and Costs', Environment Agency, *Technical Report*, pp. 345.

Hornung, M., Bull, K. R., Cresser, M., Ullyett, J., Hall, J. R., Langan, S. and Loveland, P. J.: 1995, 'The sensitivity of surface waters of Great Britain to acidification predicted from catchment characteristics', *Environ. Pollut.* **87**, 207–214.

Hughes, M. and Fisher, P.: 2000, 'Evaluating the derivation of sub-urban land-cover data from ordnance survey land-line', in P. Atkinson and D. Martin D. (eds), *GIS and Geocomputation: Innovations in GIS 7*, Taylor and Francis, London.

Jenkins, A., Whitehead, P. G., Cosby, B. J. and Birks, H. J. B.: 1990, 'Modelling long-term acidification: A comparison with diatom reconstructions and the implications for reversibility', *Philos. Trans. R. Soc. Lond., Series B* **327**, 209–214.

Johnes, P. J., Curtis, C., Moss, B. Whitehead, P., Bennion, H. and Patrick, S.: 2000, 'Trial Classification of Lake Water Quality in England and Wales: A proposed approach', Research and Development *Technical Report E53*, Environment Agency, Bristol.

Johnes, P. J., Moss, B. and Phillips, G. L.: 1996, 'The Determination of water quality by land use, livestock numbers and population data – Testing of a model for use in conservation and water quality management', *Freshwat. Biol.* **36**, 451–473.

Jones, V. J., Flower, R. J., Appleby, P. G., Natkanski, J., Richardson, N., Rippey, B., Stevenson, A. C. and Battarbee, R. W.: 1993, 'Palaeolimnological evidence for the acidification and atmospheric contamination of lochs in the Cairngorm and Lochnagar areas of Scotland', *J. Ecol.* **81**, 3–24.

Kernan, M., Hughes, M. Hornby, D., Bennion, H., Hilton, J., Phillips, G. and Thomas, R.: 2004, 'The use of a GIS-based inventory to provide a regional assessment of standing waters in Great Britain sensitive to acidification from atmospheric deposition', *Water, Air, Soil Pollut.: Focus* **4**, 97–112.

Lloyd, C. D. and Atkinson, P. M.: 2002, 'Deriving DSM's from LiDAR data with kriging', *Int. J. Remote Sens.* **23**, 2519–2524.

Lyle, A. A. and Smith, I. R.: 1994, 'Standing waters' in P. S. Maitland, P. J. Boon and D. S. McLusky (eds), *The Fresh Waters of Scotland*, John Wiley & Sons.

Morris, D. G. and Flavin, R. W.: 1990, 'A digital terrain model for hydrology', *Proc 4th International Symposium on Spatial Data Handling*, Vol. 1, 23–27 July, Zurich, pp. 250–262.

Murray, J. and Pullar, L.: 1910, *Bathymetrical Survey of the Fresh Water Lochs of Scotland*, Challenger Office, Edinburgh.

Ordnance Survey: 2001, 'Land-Form PANORAMA User Guide v3.0', Ordnance Survey, Southampton.

Pollard, S., Duarte-Davidson, R., Yearsley, R., Twigger-Ross, C., Fisher, J., Willows, R. and Irwin, J.: 2000, 'A Strategic Approach to the Consideration of Environmental Harm', NCRAOA, *Report Number 36*.

Smith, I. R. and Lyle, A. A.: 1979, *Distribution of Freshwaters in Great Britain*, Institute of Terrestrial Ecology, Edinburgh.

U.K. Biodiversity Group: 1998, *Tranche 2 Action Plans. Vol. 2: Terrestrial and Freshwater Habitats*, English Nature.

Wise, S.: 2000, 'Assessing the quality for hydrological applications of digital elevation models derived from contours', *Hydrol. Process* **14**, 1909–1929.

SIMULATION OF NITROGEN DYNAMICS AND FLUXES IN CONTRASTING CATCHMENTS IN NORWAY BY APPLYING THE INTEGRATED NITROGEN MODEL FOR CATCHMENTS (INCA)

ØYVIND KASTE

Norwegian Institute for Water Research, Southern Branch, Televeien 3, N-4879 Grimstad, Norway
(e-mail: oeyvind.kaste@niva.no, phone: +47 37295055, fax: +47 37044513)

(Received 20 August 2002; accepted 18 April 2003)

Abstract. The process-based INCA model was applied to Dalelva Brook (3.2 km^2) and the Bjerkreim River (685 km^2) including several subcatchments, in order to test the model's ability to simulate streamwater nitrate (NO_3^-) dynamics and output fluxes under highly contrasting climatic conditions and nitrogen (N) loading. The simulated runoff volumes and mean NO_3^- concentrations at Dalelva and Bjerkreim were within +2 to +10% of the measured average during 1993–1995 (–19 to +31% within individual years). INCA to a great extent also reproduced the observed streamwater flow dynamics at both study sites (coefficient of determination, $r^2 > 0.70$). Temporal variation of streamwater NO_3^- during 1993–1995 was captured quite well by the model, especially at small catchments with a distinct seasonal NO_3^- pattern ($r^2 = 0.46$–0.68). At the Bjerkreim River outlet, the relationship were somewhat weaker ($r^2 = 0.26$, $p < 0.01$). Despite a few situations where the model failed to capture the streamwater NO_3^- dynamics, INCA proved to be a quite robust tool for simulating NO_3^- dynamics and output fluxes in the two study catchments.

Keywords: catchment, integrated modelling, nitrate leaching, nitrogen deposition, nitrogen dynamics, nitrogen transformation

1. Introduction

Human activities such as fossil fuel combustion, fertiliser application, cultivation of N fixing crops, and discharge of domestic and industrial effluents have caused N enrichment of many terrestrial and aquatic ecosystems (Galloway *et al.*, 1995). This has increased the importance of N in acidification of upland lakes and rivers (Henriksen and Brakke, 1988) and has resulted in eutrophication of many sensitive estuaries and coastal areas (Howarth *et al.*, 1996). N concentrations and output fluxes in river systems reflect the integration of several diffuse or point sources within the catchment and various terrestrial and aquatic N retention processes. To deal with this large complexity in river basin management, integrated and spatially distributed catchment models might be useful tools.

The process-oriented INCA model attempts to integrate these factors by linking hydrology and N inputs from atmospheric deposition, agriculture and populated areas with the microbial processes controlling N behaviour in soils and river reaches (Whitehead *et al.*, 1998a, b; Wade *et al.*, 2002a). By computing a mass

TABLE I

Catchment characteristics at Dalelva and Bjerkreim with sub-catchments, showing the percentages of the six land cover classes modelled within this INCA application

	Size km^2	Land cover class (%)					
		Forest	Heath	Peat	Pasture	Arable	Lakes
Dalelva catchment	3.2	20	61	4	–	–	15
Bjerkreim catchment	685.2	18	64	1	4	2	11
Svela sub-catchment	0.6	61	39	–	–	–	–
Øygard sub-catchment	2.5	4	83	6	–	–	7
Apeland sub-catchment	1.7	60	9	–	23	10	–

balance for all sources and sinks of N in up to six land cover types within a catchment, the model assesses the contribution of multiple sources to catchment N pools and river NO_3^- and ammonium (NH_4^+) concentrations. To provide a generic tool for management of water quality across Europe, the model presently is under validation and modification through investigations in a wide range of aquatic ecosystem types across eight European countries (Wade *et al.*, 2002b). Two Norwegian catchments with extensive data are included in the joint project, the arctic Dalelva Brook and the southern boreal Bjerkreim River.

The main objectives of this paper are to: (1) test the ability of the INCA model to simulate streamwater NO_3^- dynamics and export in two catchments with highly contrasting climatic conditions and N loading, and (2) simulate realistic mass balances for N sources and sinks in the study catchments.

2. Materials and Methods

The study catchments, Dalelva Brook (69°41′N, 30°23′E) and the Bjerkreim River (58°28′N, 5°59′E) are located near the northeastern and southwestern borders of Norway, respectively (Figure 1). Dalelva has a small catchment (3.2 km^2), whereas Bjerkreim is a larger (685 km^2) and more complex river system with four main branches and several relatively large lakes. Dalelva is characterised by a cold and relatively dry arctic climate with low N deposition, 2–3 kg N ha^{-1} yr^{-1}, whereas the Bjerkreim area has a milder, humid climate with much higher N deposition, 14–22 kg N ha^{-1} yr^{-1}. Heathlands and mountains dominate both catchments, but in Bjerkreim there are also some farmed areas in the river valleys (Table I) and about 2300 inhabitants. For a more detailed description of the study sites, see Kaste *et al.* (1997) and Kaste and Skjelkvåle (2002).

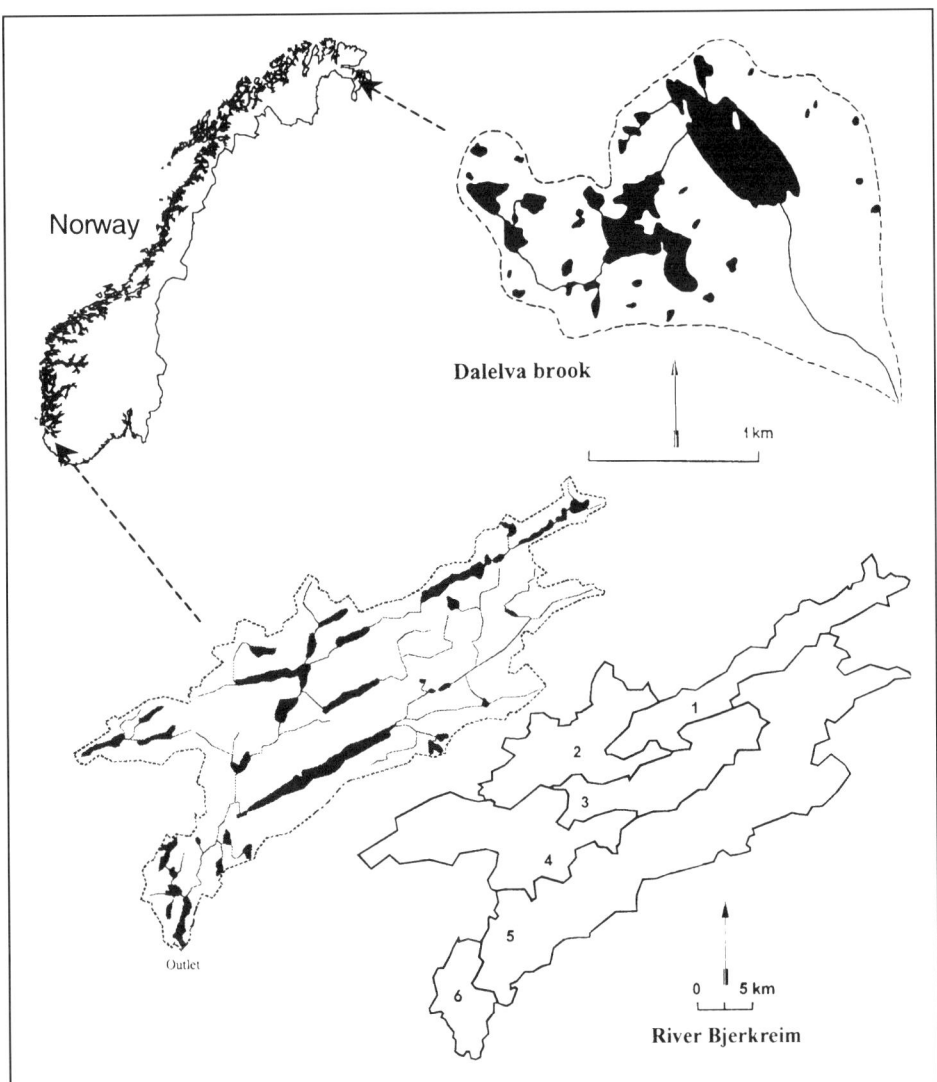

Figure 1. Map of Norway showing locations of Dalelva Brook and the Bjerkreim River, including the six river reaches defined within the model application. The Øygard and Svela subcatchments are located within reach 4, Apeland within reach 5.

The INCA model requires input of daily time series of air temperature (AT), actual precipitation (P), soil moisture deficit (SMD), and hydrologically effective rainfall (HER; the fraction of P that contributes directly to runoff). The data can be integrated for the whole catchment (lumped) or ascribed to individual subcatchments (semi-distributed). In addition, the model requires information about subcatchment structure (number, size, reach length), physical properties of the selected subcatchments, and inputs of N from atmospheric deposition, fertiliser application, and effluent discharges. A full description of the model, which includes the data requirements and the equations used to simulate the N dynamics in the plant/soil system and in the stream, is given by Wade *et al.* (2002a).

Water chemistry data from the modelling period 1993–1995 comprise weekly samples at Dalelva and fortnightly samples from three small subcatchments (Svela, Øygard, Apeland) and six river reaches (including the outlet) in Bjerkreim (Figure 1). Water flow was recorded hourly at the outlet of Dalelva, Øygard and Svela, and on a daily basis near the Bjerkreim River outlet. Daily data on AT, P and snow depth were obtained from the closest monitoring stations operated by the Norwegian Meteorological Institute. As a rough approximation, the AT values at each subcatchment were corrected by $-0.75\ °C$ for every 100 m elevation above the closest monitoring station (Gottschalk and Killingtveit, 1997). N deposition data were obtained from the closest monitoring stations operated by the Norwegian Institute for Air Research. During 1993–1995, precipitation amounts and atmospheric N inputs were estimated for the entire Bjerkreim River system, including several sub-catchments (Tørseth and Semb, 1997). For a more detailed description of input data and methodology, see Kaste *et al.* (1997) and Kaste and Skjelkvåle (2002).

The daily change in SMD was calculated from an estimated evapotranspiration rate (ET) minus P. Evapotranspiration was expressed as a function of AT, and according to long-term water balances for Norwegian catchments (Otnes and Ræstad, 1978) annual ET amounts to roughly 0.15 mm $°C^{-1}\ d^{-1}$ in the two regions. The volumes and dynamics of simulated vs. observed flow indicate that this ET factor was appropriate for both catchments during 1993–1995. The maximum SMD estimated was 68 mm at Øygard during the summer of 1995. The time series of HER was calculated as:

$$HER = (P + M) - ET - S, \qquad (1)$$

where P is liquid precipitation, M is snowmelt water, ET is evapotranspiration and S is soil water storage. For periods with saturated soils (SMD = 0), S will be zero. Water inputs from melting snow were estimated by a separate snow accumulation and snow melt model (Vehviläinen, 1992; Rekolainen and Posch, 1993).

When calibrating the model, procedures recommended by Wade *et al.* (2002a) were applied. After including the appropriate initial values, INCA was set up to simulate the actual hydrology, in terms of both dynamics and absolute flow, before any parameters controlling N storage, transformations or transport were adjusted. Secondly, the parameters controlling land phase and in-stream N transformation rates were adjusted such that annual process loads were within the ranges reported in the literature and a reasonable match between simulated and observed streamwater NO_3^- concentrations was obtained.

In the relatively large and complex Bjerkreim watershed, the model was first calibrated for small, homogenous subcatchments (0.5–2.5 km^2). These were Svela (61% forest), Øygard (83% heathlands and mountains), and Apeland (33% pasture

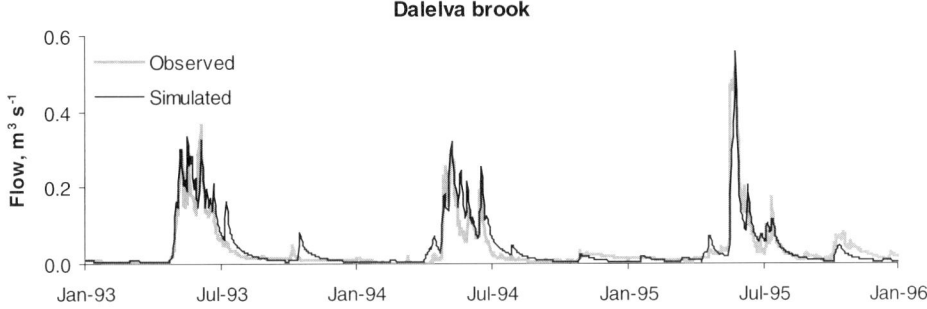

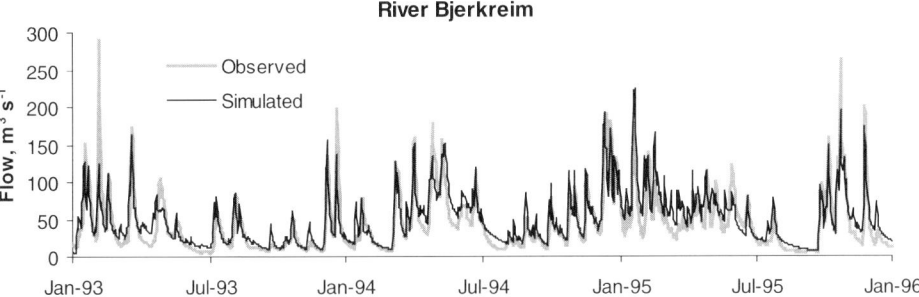

Figure 2. Simulation of streamwater flow at Dalelva Brook and the Bjerkreim River. Note the different scales.

and arable land). When scaling up to the entire Bjerkreim catchment (685 km^2), the main river was first divided into six reaches (Figure 1). Hydrological time series (SMD, HER, AT, P) were then assigned to each of the individual reaches, and hydrological parameters, such as storage volumes and velocity/flow relationships, were calibrated. Further, N process parameters from the small catchments were applied to the corresponding land cover classes in the main catchment. At Dalelva the whole catchment was treated as one single reach.

3. Results

The simulated 1993–1995 average runoff at Dalelva Brook and Bjerkreim River showed relatively small deviations (+10 and +2%, respectively) from the observed values (Table II). Within single years, however, the variation between simulated and observed runoff volumes was larger. This was especially evident at the Dalelva catchment, where the annual simulated runoff deviated between −16 and +31% from the observed values (Table II). In this catchment, where the snowpack depth during winter often exceed 1 m, even small gradients in AT relative to the adjacent monitoring station can cause erroneous estimates of snow accumulation and melting within specific calendar years. INCA to a great extent reproduced the observed

TABLE II

Modelling of streamwater discharge and NO_3^- concentrations at Dalelva Brook and the Bjerkreim River (incl. three sub-catchments) 1993–1995. Ratio of simulated to observed mean values and the relationship between simulated and observed temporal patterns; individual years (ranges) and whole period (parentheses)

1993–1995	Streamwater discharge		NO_3^- concentration	
	Mean (sim./obs. ratio)	Temporal pattern (r^2)	Mean (sim./obs. ratio)	Temporal pattern (r^2)
Dalelva Brook	0.84–1.31 (1.10)	0.80–0.88 (0.83)	1.01–1.18 (1.10)	0.54–0.74 (0.64)
Bjerkreim River	0.93–1.07 (1.02)	0.68–0.82 (0.73)	0.99–1.07 (1.02)	0.20–0.35 (0.26)
Svela sub-catchment	0.97–1.29 (1.11)	0.38–0.52 (0.47)	0.90–1.23 (1.00)	0.71–0.80 (0.68)
Øygard sub-catchment	0.86–1.09 (1.02)	0.55–0.79 (0.66)	0.90–1.12 (1.03)	0.43–0.71 (0.46)
Apeland sub-catchment	1.03–1.27 (1.10)	0.33–0.56 (0.49)	0.81–1.30 (1.02)	0.18–0.26 (0.18)

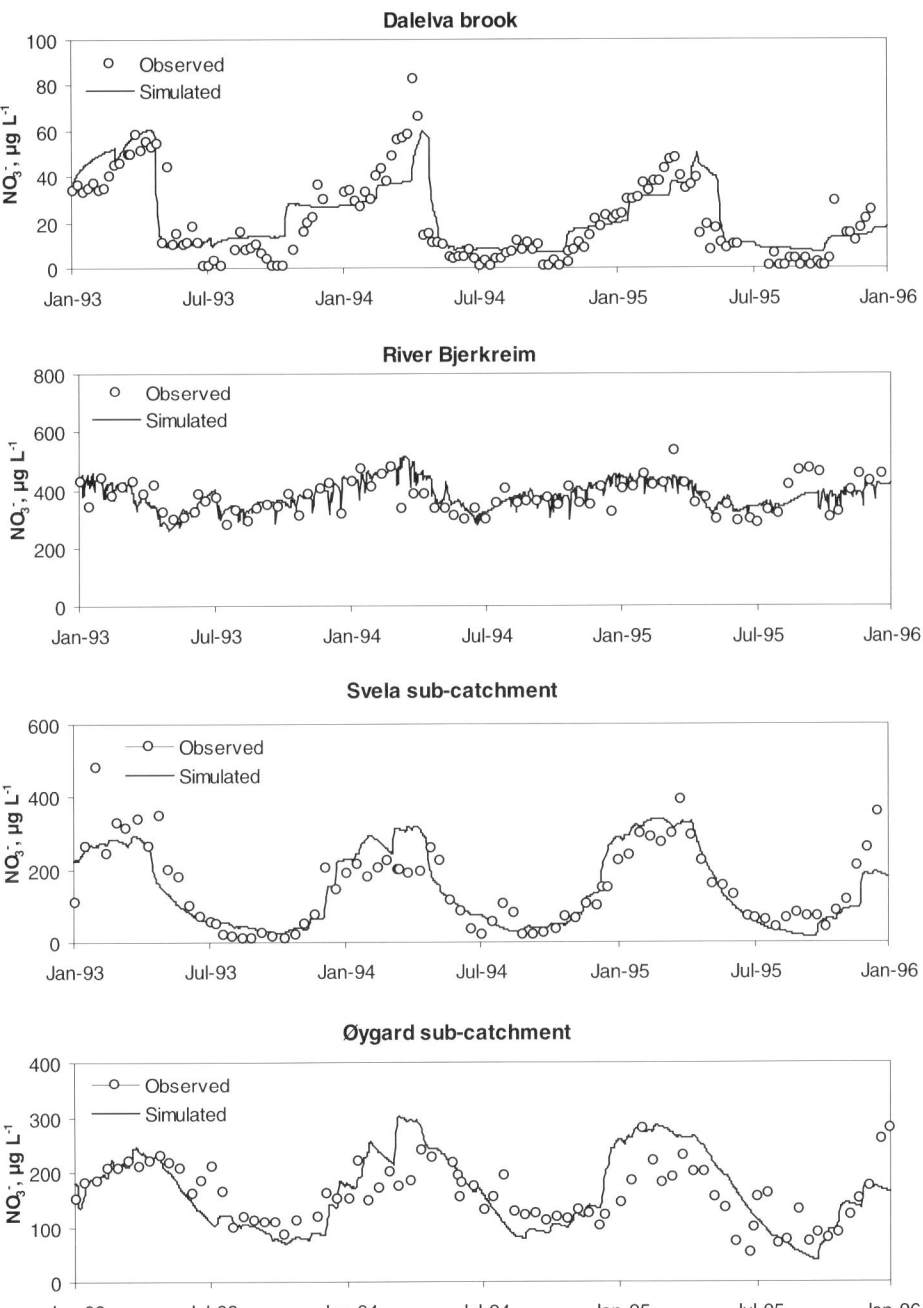

Figure 3. Simulation of streamwater NO_3^- concentrations at Dalelva Brook, Bjerkreim River and two small subcatchments within the Bjerkreim River basin. Note the different scales.

streamwater flow dynamics at the study sites (Figure 2). Based on the coefficient of determination (r^2) from daily simulations vs. observations, the model captured over 70% of the variance in streamwater flow during 1993–1995 at both Dalelva Brook and the Bjerkreim River (individual years: 68–88%; Table II). The correlation was somewhat weaker at the small catchments in Bjerkreim where streamwater flow is highly variable.

The simulated 1993–1995 arithmetic mean NO_3^- concentrations, based on daily-simulated values, were in good agreement with those based on observed data (Table II). Within single years, the predicted mean NO_3^- concentrations deviated between −19 and +30% from the observation-based averages. The greatest between-year variations were found at the Apeland catchment, which is heavily influenced by agricultural activities. INCA captured the temporal variation of streamwater NO_3^- quite well during 1993–1995, especially at Dalelva, Svela, and Øygard, where streamwater NO_3^- concentrations follow a distinct seasonal pattern. The coefficients of determination (r^2) for simulated vs. observed NO_3^- concentrations in these catchments were 0.64, 0.68 and 0.46, respectively (Table II). In the larger, lake-influenced Bjerkreim River basin and at Apeland, where streamwater NO_3^- concentrations are more variable due to complex interactions between hydrology and multiple N sources, the relationships between simulated and observed NO_3^- values were weaker ($r^2 = 0.26$ and 0.18, respectively; $p < 0.01$).

The model simulations are based on annual N mass balances for the land phase and in the stream. Table III summarises the calibrated process loads on a catchment scale, which integrates the contribution from all the actual land cover types. Whereas both N inputs and outputs are closely connected to field observations, the internal N transformation rates are calibrated values held within the ranges measured at the actual sites or reported from comparable ecosystems. The catchment N budgets illustrate the large difference in N inputs and outputs at the two study catchments. A dominant fraction of the total simulated NO_3^- retention takes place within the terrestrial catchment and in lakes, whereas only a minor fraction is ascribed to in-stream processes. The denitrification rates applied were considerably higher in lakes than in the terrestrial part of the catchment. As an example, the denitrification rates applied to lakes within the Bjerkreim catchment were around 30 kg N ha^{-1} (lake surface) yr^{-1}, whereas values of 1–1.5 kg ha^{-1} yr^{-1} were applied for areas covered with forest or heathland/mountains. Due to fertiliser inputs of N, the estimated NO_3^- leaching losses from farmed areas within Bjerkreim were much higher (15–18 kg N ha^{-1} yr^{-1}) than from forested and heathland/mountainous areas that leached 3 and 8 kg N ha^{-1} yr^{-1}, respectively.

4. Discussion

In larger catchments, such as the Bjerkreim River basin, there are often significant climatic gradients and a complex composition of various land cover types, lakes,

TABLE III

Simulated land phase and in-stream N balances for Dalelva Brook and the Bjerkreim River during 1993–1995. Sources (+) and sinks (−) are given in kg N ha^{-1} yr^{-1}. The internal N transformation rates are calibrated values

	Dalelva Brook		Bjerkreim River	
	NO_3^-	NH_4^+	NO_3^-	NH_4^+
Land phase[a]:				
Deposition	+1.43	+1.30	+9.83	+8.79
Fertilisation			+3.55	+3.54
Mineralisation		+3.13		+26.01
Nitrification	+0.50	−0.50	+5.58	−5.58
Denitrification	−0.75		−5.84	
Plant uptake	−1.06	−1.29	−7.35	−1.78
Immobilisation		−2.64		−29.47
δ Storage (+/−)	−0.05	+0.02	+1.38	−0.03
Leaching	−0.05	−0.02	−7.15	−1.48
In-stream:				
Effluents			+0.21	+0.21
Nitrification	+0.02	−0.02	+1.53	−1.53
Denitrification	−0.01		−0.09	
δ Storage (+/−)			−0.04	+0.07
Output	−0.06	−0.00	−8.76	−0.23

[a] Including lakes.

and river reaches with variable hydrological properties and N retention abilities. Hence, the small catchments Svela, Øygard and Apeland, with distinct land cover characteristics and a relatively simple hydrological functioning were used as a basis for model calibration to the Bjerkreim River catchment. Such an aggregation procedure undoubtedly includes uncertainties regarding spatial variability in factors such as soil physical properties, nutrient status, vegetation types, forest age, fertilisation practises, and in-stream/in-lake processes. However, supported by high-quality N input/output data and also some within-site N process measurements, this approach may produce a more realistic simulation of the catchment N cycle than by treating the whole Bjerkreim catchment in a single calibration.

Heathland/mountains and forests, as represented by the Øygard and Svela catchments, respectively, cover over 80% of the Bjerkreim catchment. The calibrated N mineralisation and N uptake rates within the Svela forest (42 and 8 kg N ha^{-1} yr^{-1}, respectively) were somewhat lower than values previously measured within the same catchment (50–98 and 17 N ha^{-1} yr^{-1}, Mulder *et al.*, 1997). At Øygard, measurements of net N mineralisation, net nitrification, and denitrification conducted in heathland and wetland plots suggest ranges of 1–12, 0–3, and 0.3–0.4 kg N ha^{-1} yr^{-1}, respectively (O. J. Kjønaas, unpublished data). The net mineralisation and net nitrification rates applied for the heathlands/mountains and wetlands at Øygard were relatively close to these values (10–15 and 0.3–0.7 kg N ha^{-1} yr^{-1}, respectively), whereas the denitrification rates used for wetlands were higher (up to 8 kg N ha^{-1} yr^{-1}). In the absence of N process data from the Apeland catchment, the simulated N retention and losses from pasture and arable land are based solely on mass balance approaches and literature data (cf. Whitehead *et al.*, 1998b).

The calibrated in-lake denitrification rates used at Dalelva and Bjerkreim (7 and 32 kg N ha^{-1} (lake surface) yr^{-1}, respectively) are difficult to evaluate due to large spatial variation in lake physical properties, hydrologic conditions, and nutrient loading. In comparison, in-lake N retention rates estimated from input/output budgets were in the range of 1–25 kg N ha^{-1} yr^{-1} in five southern Norwegian acid-sensitive lakes with highly contrasting N inputs (Kaste and Dillon, 2003). The calibrated in-stream denitrification rates at Dalelva and Bjerkreim accounted for 8 and <1% of the annual NO$_3^-$ output, respectively. Although in-stream retention rates are highly variable between sites, values in the range 3–20% of the annual total N export commonly are reported (Hill, 1979; Jansson *et al.*, 1994; Howarth *et al.*, 1996).

In the arctic Dalelva catchment, low external N inputs, low soil temperatures, and short growing season lead to much lower N transformation rates than in Bjerkreim (Table III). Field measurements indicate that N mineralisation rates in arctic ecosystems usually are low (1–6 kg N ha^{-1} yr^{-1}) compared to soils of lower latitudes (Nadelhoffer *et al.*, 1992). The relative importance of N fixation vs. mineralisation, however, tends to increase toward colder regions, where rates of both are commonly within the range of 0.2–2.5 kg N ha^{-1} yr^{-1} (Chapin and Bledsoe, 1992). Nitrification has been reported in arctic soils, even though cold and wet conditions are not considered favourable for nitrifiers (Nadelhoffer *et al.*, 1992). Denitrification also occurs, especially during soil freezing and thawing cycles (Goodroad and Keeney, 1984).

As a generic model developed for a wide range of ecosystem types and climatic regions, INCA inevitably has its limitations when it comes to special cases or events. At the Norwegian study sites, there were a few situations where the model failed to capture the streamwater NO$_3^-$ dynamics. Among these were the simulation of: (1) the dampened seasonal NO$_3^-$ signal, often occurring downstream of large lakes (Kaste *et al.*, 2003), (2) elevated NO$_3^-$ concentrations during baseflow conditions in summer, and (3) rapid changes in streamwater NO$_3^-$ during events. Despite these few shortcomings, INCA proved to be a quite robust model for simulating the retention and transport of N under the contrasting climatic conditions and N loading prevailing at the Norwegian sites. From this, it appears that the model may provide a useful management tool that can be used to assess the impacts of possible changes in land management, N deposition, and climate on future streamwater N concentrations and loads.

Acknowledgements

This work has been supported by the European Commission (INCA Project EVK1-1999-00011), the Research Council of Norway (NFR) through the research project *N retention and acidification in mountains and heathlands*, and the Norwegian Institute for Water Research (NIVA). Chemical data have been collected as part of the *Nitrogen from Mountains to Fjord* programme (funded by NFR and NIVA) and the *Norwegian monitoring programme for long-range transported air pollutants* (funded by the Norwegian Pollution Control Authority [SFT]). I thank Richard F. Wright, NIVA, for valuable comments on the manuscript.

References

Chapin, D. M. and Bledsoe, C. S.: 1992, 'Nitrogen fixation in arctic plant communities', in F. S. Chapin III, R. L. Jeffries, J. F. Reynolds, G. R. Shaver and J. Svoboda (eds), *Arctic Ecosystems in a Changing Climate – An Ecophysiological Perspective*, San Diego, Academic Press, pp. 301–319.

Galloway, J. N., Schlesinger, W. H., Levy II, H., Michaels, A. and Schnoor, J. L.: 1995, 'Nitrogen fixation: Anthropogenic enhancement – Environmental response', *Global Biochem. Cycles* **9**, 235–252.

Goodroad, L. L. and Keeney, D. R.: 1984, 'Nitrous oxide emissions from soils during thawing', *Can. J. Soil Sci.* **64**, 187–194.

Gottschalk, L. and Killingtveit, Å.: 1997, *Textbook in Hydrology*, Institute for Geophysics, University of Oslo, Norway (in Norwegian).

Henriksen, A. and Brakke, D. F.: 1988, 'Increasing contributions of nitrogen to the acidity of surface waters in Norway', *Water, Air, Soil Pollut.* **42**, 183–201.

Hill, A. R.: 1979, 'Denitrification in the nitrogen budget of a river ecosystem', *Nature* **281**, 291–292.

Howarth, R. W., Billen, G., Swaney, D., Townsend, A., Jaworski, N., Lajtha, K., Downing, J. A., Elmgren, R., Caraco, N., Jordan, T., Berendse, F., Freney, J., Kudeyarov, V., Murdoch, P. and Zhao-Liang, Z.: 1996, 'Regional nitrogen budgets and riverine N and P fluxes for the drainages to the North Atlantic Ocean: Natural and human influences', *Biogeochemistry* **35**, 75–139.

Jansson, M., Leonardson, L. and Fejes, J.: 1994, 'Denitrification and nitrogen retention in a farmland stream in Southern Sweden', *Ambio* **23**, 326–331.

Kaste, Ø. and Dillon, P. J.: 2003, 'Inorganic nitrogen retention in acid-sensitive lakes in southern Norway and southern Ontario, Canada – A comparison of mass balance data with an empirical N retention model', *Hydrol. Process.* **17**, 2393–2407.

Kaste, Ø. and Skjelkvåle, B. L.: 2002, 'Nitrogen dynamics in runoff from two small heathland catchments representing opposite extremes with respect to climate and N deposition in Norway', *Hydrol. Earth Syst. Sci.* **6**, 351–362.

Kaste, Ø., Henriksen, A. and Hindar, A.: 1997, 'Retention of atmospherically-derived nitrogen in subcatchments of the Bjerkreim River in Southwestern Norway', *Ambio* **26**, 296–303.

Kaste, Ø., Stoddard, J. L. and Henriksen, A.: 2003, 'Implication of lake water residence time on the classification of Norwegian surface water sites into progressive stages of nitrogen saturation', *Water, Air, Soil Pollut.* **142**, 409–424.

Mulder, J., Nilsen, P., Stuanes, A. O. and Huse, M.: 1997, 'Nitrogen pools and transformation in forest ecosystems with different atmospheric inputs', *Ambio* **26**, 273–281.

Nadelhoffer, K. J., Giblin, A. E., Shaver, G. R. and Linkins, A. E.: 1992, 'Microbial processes and plant nutrient availability in arctic soils', in F. S. Chapin III, R. L. Jeffries, J. F. Reynolds, G. R. Shaver and J. Svoboda (eds), *Arctic Ecosystems in a Changing Climate – An Ecophysiological Perspective*, Academic Press, San Diego, pp. 281–300.

Otnes, J. and Ræstad, E.: 1978, *Hydrology in Practice*, Engineering Publishers (Ingeniørforlaget), Oslo, Norway, pp. 314 (in Norwegian).

Rekolainen, S. and Posch, M.: 1993, 'Adapting the CREAMS model for Finnish conditions', *Nordic Hydrol.* **24**, 309–322.

Tørseth, K. and Semb, A.: 1997, 'Atmospheric deposition of nitrogen, sulphur and chloride in two watersheds located in Southern Norway', *Ambio* **26**, 258–265.

Vehviläinen, B.: 1992, 'Snow Cover Models in Operational Watershed Forecasting', *Ph.D. Thesis*, National Board of Waters and the Environment, Helsinki, Finland, pp. 112.

Wade, A. J., Durand, P., Beaujouan, V., Wessel, W. W., Raat, K. J., Whitehead, P. G., Butterfield, D., Rankinen, K. and Lepistö, A.: 2002b, 'A nitrogen model for European catchments: INCA, new model structure and equations', *Hydrol. Earth Syst. Sci.* **6**, 559–582.

Wade, A. J., Whitehead, P. G. and O'Shea, L. C. M.: 2002a, 'The prediction and management of aquatic nitrogen pollution across Europe: An introduction to the Integrated Nitrogen in European Catchments project (INCA)', *Hydrol. Earth Syst. Sci.* **6**, 299–314.

Whitehead, P. G., Wilson, E. J. and Butterfield, D.: 1998a, 'A semi-distributed integrated nitrogen model for multiple source assessment in catchments (INCA): Part I – Model structure and process equations', *Sci. Total Environ.* **210/211**, 547–558.

Whitehead, P. G., Wilson, E. J., Butterfield, D. and Seed, K.: 1998b, 'A semi-distributed integrated nitrogen model for multiple source assessment in Catchments (INCA): Part II – Application to large river basins in South Wales and eastern England', *Sci. Total Environ.* **210/211**, 559–583.

THE USE OF A GIS-BASED INVENTORY TO PROVIDE A REGIONAL RISK ASSESSMENT OF STANDING WATERS IN GREAT BRITAIN SENSITIVE TO ACIDIFICATION FROM ATMOSPHERIC DEPOSITION

M. KERNAN[1]*, M. HUGHES[1], D. HORNBY[2], H. BENNION[1], J. HILTON[2], G. PHILLIPS[3] and R. THOMAS[4]

[1] *Environmental Change Research Centre, University College London, London, WC1H 0AP, U.K.;*
[2] *Centre for Ecology and Hydrology, Winfrith Technology Centre, Dorchester, Dorset, DT2 8ZD, U.K.;* [3] *Environment Agency, National Centre for Risk Analysis Options Appraisal, Reading, RG1 8DQ, U.K.;* [4] *Environment Agency, National Centre for Environmental Data, Bath, Avon, BA2 9ES, U.K.*
(* *author for correspondence, e-mail: mkernan@geog.ucl.ac.uk, phone: +44 20 7679 5523, fax: +44 20 7679 7565)*

(Received 20 August 2002; accepted 26 April 2003)

Abstract. Previous attempts to identify regions of Britain vulnerable to acidification have used sensitivity maps based on the distribution of soils, geology, and land cover across Great Britain. Additionally, a systematic survey of freshwaters undertaken as part of the U.K. critical loads mapping programme provides a regional assessment of both sensitivity (critical loads) and, in tandem with deposition data, potential impact (critical load exceedance). Both approaches, while useful for identifying regional patterns, do not enable estimates of the number of affected water bodies to be made. Recent EU legislation (e.g., The Water Framework Directive) requires member states to set water quality objectives for all water bodies. We developed a GIS-based inventory of standing water bodies in response to the need for legislation-driven assessments of the status of the U.K. lake population. This paper describes how the inventory can be used to assess the number of standing water bodies in Britain that are vulnerable to acid deposition (at current levels), building on the sensitivity mapping undertaken previously. Using this approach, approximately 31% of all standing waters in Great Britain (excluding the Shetlands and Orkney) larger than 0.02 ha are identified as 'at risk' from acidification. Higher proportions are vulnerable in Scotland and Wales. Additionally, large numbers of standing waters in areas designated for environmental protection purposes are also vulnerable.

Keywords: acidification, freshwater sensitivity, GIS, Great Britain Standing Waters Inventory, lakes, risk assessment

1. Introduction

Since the 1850's, there has been widespread acidification of freshwaters across upland areas of Britain, caused by the emission and subsequent deposition of elevated levels of anthropogenically derived compounds of sulphur (S) and nitrogen (N) (Battarbee, 1990). More recently there have been substantial declines in S and, to a lesser extent, oxidised N (NO_x) deposition (NEGTAP, 2001). Further, there is evidence (although this is not consistent) that there has been some chemical recovery of acidified systems; biological recovery also has been detected (Harriman et al., 2001; NEGTAP, 2001). However the spatial patterns of deposited S are such that the greatest reductions have been observed close to emission sources, such as the East Midlands of England. Much smaller reductions have been observed in the west coast uplands of England, Wales and Scotland. The soils and geology characterising these upland areas generally are base-poor and the surface waters are poorly buffered against acidified precipitation. Given that deposition levels in these areas have not been significantly reduced (and in combination with other confounding factors such as climate change, N breakthrough and forestry), it is evident that freshwater acidification remains a significant threat to sensitive areas of Britain (Monteith and Evans, 2000).

Previous attempts to identify regions of Britain vulnerable to surface water acidification have been based on sensitivity mapping, using the distribution of soils, geology and land cover types across Britain (Edmunds and Kinniburgh, 1986; Hornung et al., 1995; Langan and Wilson, 1992) to differentiate between areas of low, medium and high sensitivity to acid deposition. A systematic survey of freshwaters (ca. 1500 lakes and streams) undertaken as part of the U.K. critical loads mapping programme (CLAG Freshwaters, 1995) also provides a regional assessment of sensitivity through the use of maps of freshwater critical loads. These can be use in tandem with maps of acid deposition to identify areas where sensitive freshwaters may have been impacted. Both approaches, while useful for identifying regional patterns, do not enable estimates of the number of affected water bodies to be made. The former does not consider water bodies explicitly and the latter uses a grid-based approach whereby a single water body in the most sensitive part of each 10×10 km grid square was sampled.

The critical loads approach was developed in response to the United Nations Economic Commission for Europe (UNECE) requirement for protocols aimed at reducing S and N emissions. The grid-based mapping approach enables the spatial distribution of sensitive freshwaters to be highlighted and the areas where deposition levels are likely to lead to exceedance of critical loads to be identified. However, more recent EU legislation, including the Water Framework Directive (EU, 2000), requires water quality objectives to be set for all water bodies necessitating a classification according to ecological quality status (DEFRA, 2001). Additionally data on the extent, distribution and physical, chemical, and biological status of standing waters across the U.K. are required for implementation of U.K. Biodiversity Action Plans (U.K. Biodiversity Group, 1998). Consequently, there is a need for a comprehensive inventory of U.K. freshwaters to enable assessments to be made encompassing the population of U.K. standing water bodies and, more generally, to provide environment protection agencies and conservation bodies with a means of quantifying 'stock at risk' from environmental damage.

Within this context, an inventory of standing water bodies has been developed (Hughes et al., 2004) to facilitate a risk-based prioritisation protocol so that impacts from elevated nutrient levels and acid deposition can be assessed. This paper describes how this inventory can be used to assess the number of standing water bodies in Britain that are vulnerable ('at risk') to acid deposition (at current levels), building on the sensitivity mapping exercises undertaken previously.

2. Datasets Used

2.1. GREAT BRITAIN STANDING WATERS INVENTORY

The Great Britain Standing Waters Inventory ('The Inventory') has been developed, using customised GIS routines, as part of an ongoing assessment of the status of U.K. freshwaters (Hughes *et al.*, 2004). All standing water bodies larger than 0.02 ha are included in the Inventory and are represented both by a digital polygon of the lake boundary and as a geometric centre point derived from U.K. Ordnance Survey PANORAMA™ digital data (Ordnance Survey, 2001). Each water body is coded according to a unique numerical identifier that enables data from a wide variety of sources to be linked once these have been harmonised with the Inventory. Additionally, all standing waters with a surface area greater than 1 ha have digital catchment boundaries which were derived from a 50 m resolution Digital Elevation Model (DEM) (Hughes *et al.*, 2004). Smaller water bodies tend to have more poorly defined catchments, increasing the potential errors in the catchment delineation process. A total of 43 738 standing waters are contained within the Inventory (Figure 1). Of these, 14 353 are >1 ha and have digital catchment boundaries. The remaining 29 385 are represented by the centroid of the water body. Due to scaling differences, it is likely that the number of water bodies between 0.02 and 0.2 ha is almost certainly under-represented (Hughes *et al.*, 2004).

2.2. SECONDARY DATA SOURCES

The Inventory has been linked to a number of national datasets enabling each standing water body to be characterised according to a range of catchment attributes (described below). Two methods are used to establish the links. For standing waters with digital catchment boundaries, the bounding polygons are overlaid onto national digital maps allowing the proportion of each 'class' in the national dataset present in the catchment to be determined. For standing waters <1 ha the centroid of the water body is overlaid onto the national dataset and the 'class' at that point is used to describe the standing water in terms of that particular coverage. The linked datasets have been combined in an MS ACCESS database. Table I lists the secondary datasets used in this assessment, their source and how they are represented in the Inventory database.

The Freshwater Sensitivity (FWS) map (Hornung *et al.*, 1995) was derived using national digital maps of soils and geology to indicate areas susceptible to acidification under different flow regimes. Where possible, the FWS map was used to ascribe a sensitivity class to each standing water body in the Inventory. The map

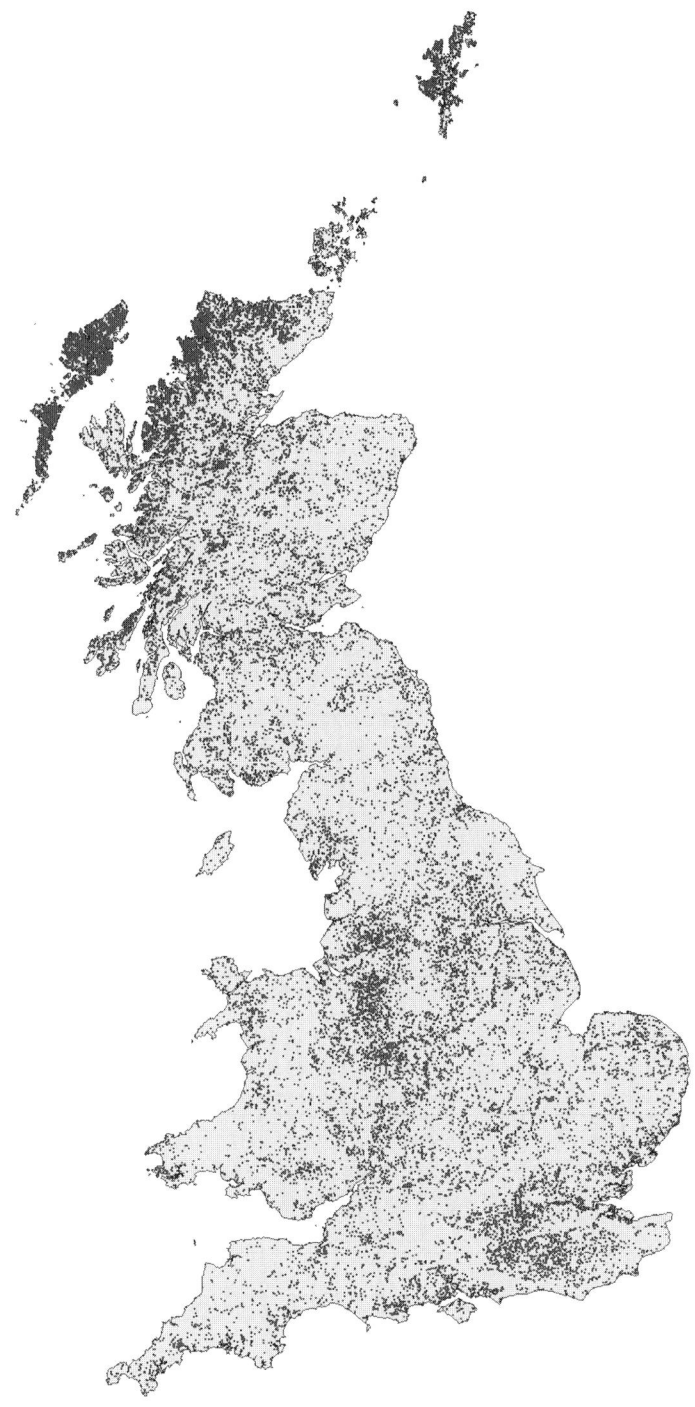

Figure 1. Distribution of standing water bodies across Great Britain.

TABLE I

Datasets used in the GB Lakes Inventory

Dataset	Source	Representation in database
Freshwater Sensitivity (FWS)	Hornung et al. (1995)	>1 ha – % of each FWS Class in catchment <1 ha – FWS at lake centroid
Protected areas co-occurrence – National Park (NP), National Nature Reserve (NNR), RAMSAR, Special Area of Conservation (SAC), Special Protection Area (SPA), Site of Special Scientific Interest (SSSI)	Centre for Ecology and Hydrology (CEH)	Standing Water centroid overlaid onto digital maps of protected area boundaries –
National Deposition Data	CEH	>1 ha – 5 km deposition grid overlaid onto catchment boundary to produce weighted value <1 ha – 5 km deposition grid value at centroid
LCM90 Land Cover Classification	Fuller et al. (1994)	>1 ha – % of each land cover class in catchments <1 ha – Dominant land cover class in 1 km grid square in which water body is located

does not cover The Shetlands and Orkney and these have been omitted from this exercise. For standing waters >1 ha the dominant FWS class was used. For smaller sites the FWS class at the centroid of the water body was used. The Inventory was also matched with the national deposition map, which comprises deposition values mapped onto a 5×5 km grid. A net acidifying deposition (NO_x + NH_x + non-marine sulphate + non-marine chloride – non-marine calcium + non-marine magnesium for 1995–1997 period in keq ha^{-1} yr^{-1}) value for each site was derived either using an area-weighted calculation (for sites >1 ha) or a point based approach using the value of the grid square where the standing water centroid is located. For land cover, the Inventory was laid onto the digital LCM90 Land Cover map of the U.K. (Fuller et al., 1994). The percentage of each land cover class was determined for those sites with catchment polygons and for sites <1 ha, the dominant land cover class in the 1×1 km square in which the site is located was used. It was also

TABLE II

Number and percentage of standing water bodies in each freshwater sensitivity class (Hornung *et al.*, 1995), broken down by country

Freshwater Sensitivity Class		England[a]		Wales		Scotland		Total[b]	
		No.	%	No.	%	No.	%	No.	%
1	High (acid waters will occur all flows)	335	2	363	26	15898	69	16596	41
2	Medium-high (acid waters at very high flows)	594	4	151	11	2830	12	3575	9
3	Medium-low (acid waters are unlikely)	967	6	163	12	1636	7	2766	7
4	Low (acid waters are very unlikely)	3196	20	147	10	1232	5	4575	11
5	Non-sensitive (acidic waters will not occur)	10609	68	591	42	1395	6	12590	31
Total		15701		1415		22991		40102	

[a] Includes Isle of Man.
[b] No freshwater sensitivity data for Shetland and Orkney.

possible to identify those standing waters occurring within the boundaries of protected areas by overlaying the Inventory onto digital maps of National Parks (NP), National Nature Reserves (NNR), RAMSAR sites, Special Areas of Conservation (SAC), Special Protection Areas (SPA), and Sites of Special Scientific Importance (SSSI). Thus, with the exception of the Shetlands and Orkney, it is possible to characterise each standing water body in England, Wales and Scotland according to freshwater sensitivity, deposition, land cover, and conservation designation.

3. Methods and Results

To assess the numbers of standing water bodies vulnerable to acidification, a multi-layer screening exercise was undertaken using the Inventory in combination with the linked national datasets. The screening exercise is illustrated in Figure 2, a flow diagram describing the iterative procedure used to determine sites at risk from acidification.

3.1. Freshwater sensitivity

To identify sites sensitive to acidification, the Freshwaters Sensitivity Map was used. Table II shows the number and percentage of standing waters in each of the FWS classes, broken down by country.

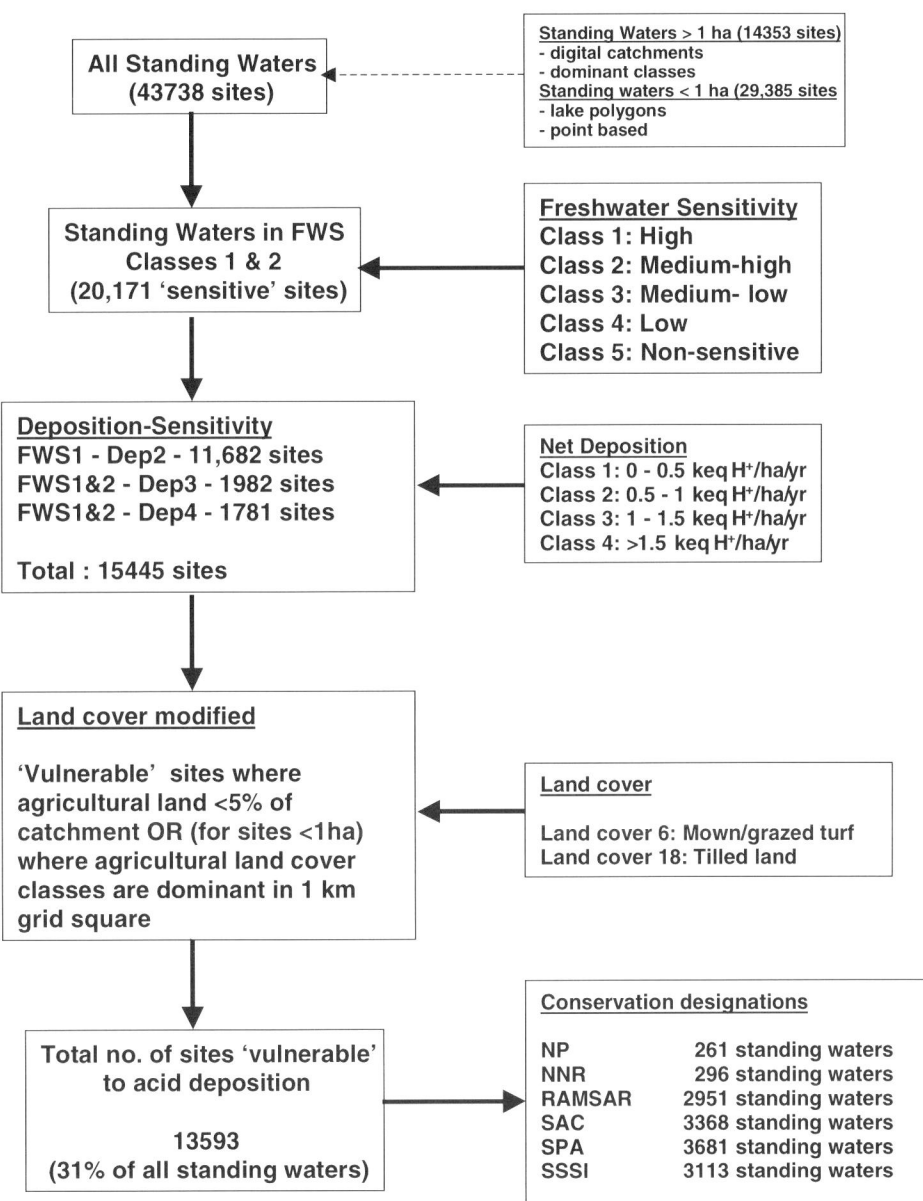

Figure 2. Iterative procedure used to determine the number of standing waters 'at risk' from acidification.

TABLE III

Scheme combining freshwater sensitivity class with deposition class to identify sites 'at risk' from acidification

Net Deposition class	FWS class 'at risk'	No. of sites (in FWS 1–2)
1 (0–0.5 keq H^+ ha^{-1} yr^{-1})	None	3375
2 (0.5–1 keq H^+ ha^{-1} yr^{-1})	1	11682
3 (1–1.5 keq H^+ ha^{-1} yr^{-1})	1 and 2	1982
4 (>1.5 keq H^+ ha^{-1} yr^{-1})	1 and 2	1782
	Total 'at risk'	15446

Across Great Britain, approximately 40% of standing waters fall into the most sensitive sensitivity class (Class 1), while 31% are classed as non-sensitive (Class 5). The majority of the most sensitive sites are in Scotland, where 69% of standing waters fall into this category. Conversely, in England, most standing waters (68%) are in non-sensitive areas. In Wales, standing waters are more evenly distributed across the sensitivity classes although the highest numbers are classed as non-sensitive. This breakdown by sensitivity class is to be expected given the distribution of sensitive soils and geology across Britain. These are dominant in the upland areas of northern and western Britain. The converse is true in England, where the moderate to well-buffered geological strata dominate in southern and eastern England giving rise to soils with high base saturation and pH levels (Hornung et al., 1995). Additionally, in Scotland, the highest concentrations of standing waters bodies are to be found in the highly sensitive north-west region (see Figure 1).

The first level of screening differentiates between 'sensitive' standing waters and those located in less sensitive areas. Sites ascribed FWS classes 1 (acid waters will occur at all flows) and 2 (acid waters will occur at very high flows) were considered as sensitive, a threshold also used previously by Reynolds et al. (2001). Sites in FWS classes 3–5 (acid waters unlikely, very unlikely, and will not occur, respectively) were omitted from further consideration. Table II shows that 20 171 sites are deemed 'sensitive' according to this criterion.

3.2. NET DEPOSITION

The freshwater sensitivity map identifies areas that are likely to be sensitive to acidification as a result of the limited buffering capacity of underlying soils and geology. To assess whether these sites are 'at risk' from acidification, it is necessary to consider the levels of acid deposition falling at each site. While there are no data available at this resolution, the national deposition map provides a guide to the spatial patterns of acidifying deposition across Britain (NEGTAP, 2001). This paper uses a scheme combining FWS class and deposition to highlight areas where the combination of these might lead to surface water acidification. A net deposition classification based on the critical loads exceedance classes used in U.K. critical loads mapping (CLAG Freshwaters, 1995) was produced so that areas of high sensitivity were deemed at risk under relatively low deposition levels. With decreasing sensitivity, the deposition level required for a site to be considered at risk increases. This scheme is shown in Table III and results in 15 445 standing waters being classed as at risk of acidification on the basis of sensitivity and mapped deposition value.

3.3. LAND COVER MODIFICATION

There are two digital versions of the FWS Map, the one used in this exercise and a second map that incorporates the effects of land use on freshwater sensitivity. The ITE Land Classification Database (Bunce and Heal, 1984) was used to identify those areas where arable or intensive grassland are dominant, the assumption being that these areas are subject to liming, which raises the pH and base saturation of the soil, increasing the buffering capacity. Prior to amalgamation with the geology map, the soil map was modified to reflect the increased buffering capacity of soils receiving regular additions of lime (Hornung *et al.*, 1995). For this study, land use modification was undertaken using the LCM90 Land Cover Classification Dataset (Fuller *et al.*, 1994). Rather than omit all sites with agricultural land in the catchment, standing waters with more than 5% of tilled land or mown/grazed turf were deemed not at risk (Allott *et al.*, 1995a). This cut-off was employed to allow for misclassifications within the land cover data (Fuller *et al.*, 1994) and for uncertainties associated with the resolution of the data. For sites where no digital catchment outlines were available, the dominant land cover in the 1 × 1 km grid square within which the site is located was used to determine whether the site should be excluded from further consideration on the basis of land cover. Following the application of the land cover criteria, the total number of sites at risk of acidification is 13 593 (31% of the population). Of these, 12 774 are in Scotland (49% of all Scottish standing waters), 507 in England (3%), and 312 in Wales (20%) (Figure 3). Large numbers of standing waters in England and Wales in non-sensitive parts of the country are not vulnerable (*cf.* Figure 1) and, reflecting the patterns identified in sensitivity mapping exercises, the areas most at risk are in northern and south-west Scotland, the Lake District, upland Wales and the Pennines.

3.4. CONSERVATION DESIGNATION

By overlying the standing waters at risk of acidification onto the digital boundaries for all the protected areas listed in Section 2.2 it is possible to estimate the number of standing waters vulnerable to acidification in designated protection areas. This is summarised by country in Table IV. There are large numbers of standing waters within the boundaries of protected areas that may be at risk from acid deposition. In Scotland and Wales, the proportion at risk in protected areas is very high. While not

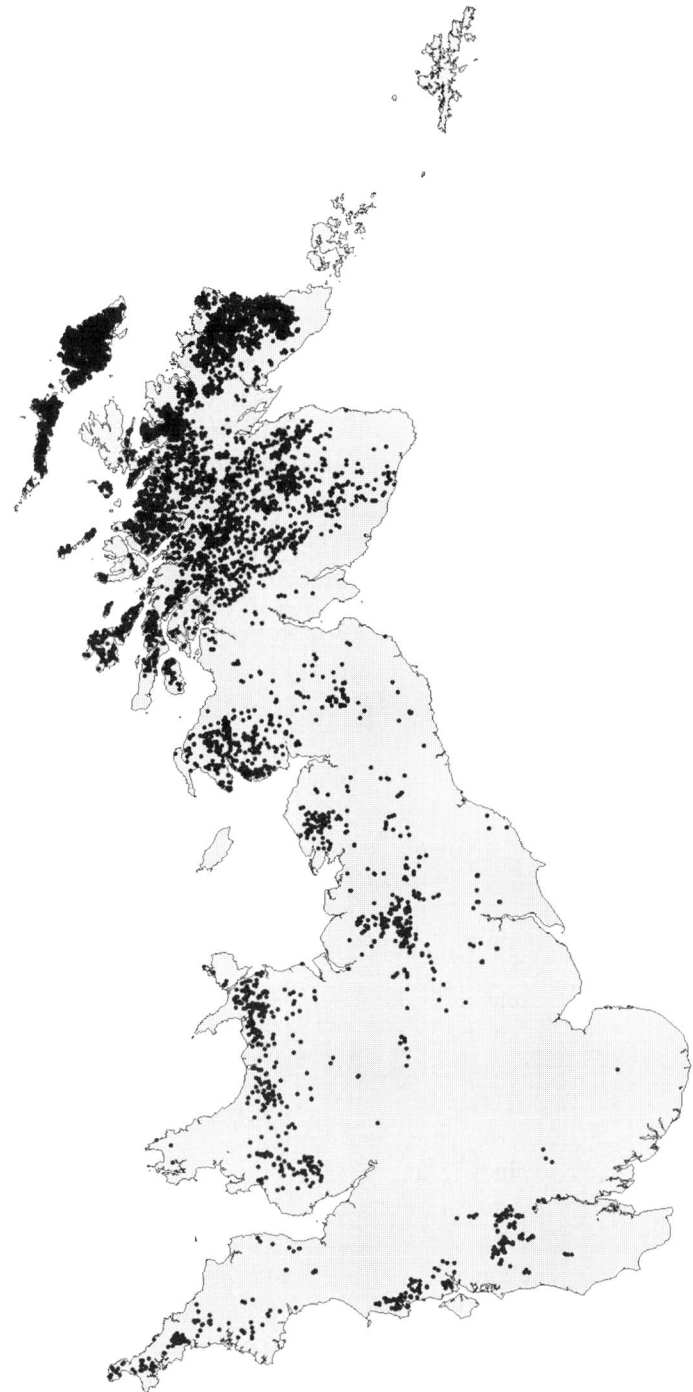

Figure 3. Distribution of standing waters 'at risk' from acidification.

TABLE IV

Total number of standing waters within the boundaries of protected areas and the number and % 'at risk' (AR) from acidification (NB Does not include recently designated Loch Lomond and Trossachs National Park)

Protection designation	England			Wales			Scotland			GB		
	No.	AR	%	No.	AR	%	No.	AR	%	No.	AR	%
SSSI	1592	137	9	265	99	37	5074	2877	57	6931	3113	45
SAC	543	114	21	133	69	52	4706	3185	68	5382	3368	63
SPA	645	93	14	39	23	59	5091	3565	70	5775	3681	64
RAMSAR	545	51	9	18	2	11	4033	2898	72	4596	2951	64
NNR	160	19	12	32	5	16	545	275	50	737	299	41
NP	637	98	15	377	163	43	0	0		1014	261	26

all protected areas have been designated because of their freshwater features, numerous SSSI's and SACs, particularly in Scotland, include standing water habitats as primary reasons for their designations.

4. Discussion

4.1. 'Stock-at-risk' from acidification

The regional assessment presented here suggests, using the criteria described, that there are large numbers of standing waters 'at risk' of acidification. Approximately 31% of the total population of standing waters >0.02 ha are vulnerable (33% of those with sites freshwater sensitivity data). As Figure 3 shows, these are primarily in the sensitive upland areas of Britain and reflect both the spatial patterns of the Freshwater Sensitivity Map (Hornung et al., 1995) and the distribution of lakes in Britain. Thus, the highest concentrations of standing waters in Britain (in the Outer Hebrides and north west Scotland) coincide with areas of high sensitivity. Although deposition levels are relatively low, the soils and geology in these areas are highly sensitive. Critical load values in the area are low (CLAG Freshwaters, 1995) and are exceeded at numerous sites (Allott et al., 1995b). However, it has been argued that these exceedances are an artefact of the critical loads methodology, which may be inappropriate for catchments with historically high inputs of neutral sea-salt sulphate (Harriman et al., 1995).

Further south, in the central Scottish uplands, the density of standing waters is reduced, but soils and geology remain highly sensitive and deposition levels are higher. Other regions at risk include south-west Scotland, where poorly weathered granite formations along with high deposition levels mean that standing waters in this area are potentially vulnerable to the acidifying effects of acid deposition. Upland areas of England and Wales are also at risk due to the combination of sensitive soils and geology and a closer proximity to the major deposition sources. Large numbers of standing waters in the Lake District, the Pennines, and throughout upland Wales are vulnerable. In southern England, standing waters on acid heathlands (e.g. in Surrey and Dorset, in the New Forest) and on the granite outcrops of the southwest are also vulnerable. Many of the water bodies vulnerable in these regions are in areas protected under a variety of designations.

These 'vulnerable' areas have been highlighted before both by sensitivity mapping exercises (Langan and Wilson, 1992; Hornung *et al.*, 1995) and by maps of critical loads for freshwater acidity (CLAG Freshwaters, 1995). However, the development of the first comprehensive digital inventory of British standing waters (Hughes *et al.*, 2004) enables the numbers of standing waters 'at risk' to be quantified. This is important within the context of recent European legislation that requires member states to consider the status of all standing waters. However, the designation of a standing water as 'at risk' from acidification is subject to a number of caveats based on both the resolution of the data employed and the assumptions made regarding vulnerability.

4.2. DATA RESOLUTION ISSUES

The Freshwater Sensitivity Map is used to represent the soil and geology present in the catchments of each lake in the Inventory. These are recognised as the main controls of freshwater chemistry at the regional scale (Hornung *et al.*, 1990). However, at the catchment scale, small areas of soil or geology that might be observed at higher resolution are not mapped at national or regional scales. This is particularly important where, for example, rocks with calcite-bearing veins might be present within an otherwise acidic catchment, yet are not included in national scale mapping. The deposition data employed in this exercise are modelled values, interpolated across a 5×5 km grid. Given that these are averaged for altitude, it is probable that in upland and forested areas, grid values underestimate the amount received at a given site. The LCM90 land cover data are based on satellite imagery and are available at 25 m and 1 km resolution. For this exercise only those areas where intensive management practices do not occur (e.g. tilled land) are considered. However, there are classification errors with these data and it is likely that the screening exercise has resulted in numbers of sites being either omitted or included erroneously in the 'at risk' category.

4.3. ASSUMPTIONS UNDERPINNING 'SENSITIVITY' AND 'STOCK-AT-RISK' STATUS

In addition to the problems posed by mapping resolution of the national datasets, a number of assumptions have been made with regard to the combinations of deposition and freshwater sensitivity that are likely to lead to acidification problems. The lowest deposition levels used are 0.5 keq H^+ ha^{-1} yr^{-1}. Although relatively low, these levels are found in areas with very little buffering capacity present in soils and geology and it may be that even these low deposition levels are sufficient to cause acidification of surface waters. The approach used here can be manipulated to produce estimates of standing waters 'at risk' under a variety of future deposition scenarios and, by selecting different combinations of freshwater sensitivity class and deposition thresholds, the exercise can be as precautionary as required.

Underpinning the approach is the assumption that the national FWS and deposition datasets can be used to gauge sensitivity and impact at each site. Previous authors have used existing chemistry data to validate the use of the FWS map to predict freshwater sensitivity. A regional survey of water quality in Wales showed that almost all sites defined as highly sensitive were located in areas predicted to exhibit acid waters from a sensitivity map using agricultural land use, soils, and geology data (Hornung *et al.*, 1990). A similar exercise using a modified version of the FWS Map indicated that predictions of acid sensitivity holds for larger catchments, (>20 ha), but that the large ranges in water quality in smaller catchments are not predicted (Reynolds *et al.*, 2001). Other researchers have examined the relationships between sensitivity as defined by the FWS Map and freshwater critical loads values for individual sites (Hall *et al.*, 1995; Kernan, 1995) and have highlighted the difficulties inherent in using national datasets to predict conditions at individuals sites. Additionally, examination of national critical loads data shows that critical load values and exceedances vary considerably in areas classed as highly sensitive according to the FWS Map (CLAG Freshwaters, 1995, Curtis *et al.*, 1995). Regional studies confirm that, even within regions that are highly sensitive such as the Cairngorms in north-east Scotland (Helliwell *et al.*, 2002), Snowdonia in north Wales (Kernan *et al.*, 1999), Galloway in south-east Scotland (Ferrier *et al.*, 2001), and the Pennines in northern England (Evans *et al.*, 2000), the sensitivity of sampled sites exhibits considerable variation. These studies illustrate the problems of scale encountered when applying the National Freshwaters Sensitivity Map to individual sites.

The mechanisms and attributes that determine whether a standing water body is sensitive to acidification are highly variable both spatially and temporally. The surface and subsurface areas of catchments are generally heterogeneous in nature and fine-scale variations in soil and rock geochemistry across catchments are not captured on national scale maps (Reynolds *et al.*, 2001). Attempts at using higher resolution mapped data have proved inconclusive (Kernan *et al.*, 2001; Ullyett *et al.*, 2001). It is likely that the number of 'sensitive' standing waters are overestimated using the Freshwater Sensitivity Map. The presence of small areas of base rich geology within a catchment dominated by sensitive formations will have a much greater impact on the surface waters than the presence of small areas of acid soils and geology in a well buffered catchments (Hornung *et al.*, 1995). Conversely, it is probable that the gridded deposition map underestimates the levels of acid input in the most sensitive upland areas. Further work will seek to examine the relationships between water chemistry data, higher resolution deposition data and the FWS Map to quantify the confidence with which the map can be used in asessments of this kind.

5. Conclusions

Given the problems of scale, it is clear that for catchment scale management, the use of the Freshwater Sensitivity Map, in tandem with other national datasets, provides a basic guide as to whether a particular site may be sensitive to acidification or impacted. With the uncertainties inherent in using national datasets to identify standing waters 'at risk' from acidification it is not recommended that this approach be adopted to assess the status of individual standing water bodies. At a site-specific level, the most appropriate approach remains field sampling campaigns. However, it is not possible to do this for all standing waters. Nevertheless, U.K. Environmental Protection Agencies are required to respond to recent legislation at the European and National level (e.g., the Water Framework Directive, Lake Habitat Action Plan) to provide an assessment of the status of the population of standing water bodies. Additionally, U.K. conservation bodies are seeking to evaluate the condition of standing waters to prioritise activities and highlight problems in protected areas. This approach allows the numbers of water bodies 'at risk' to be estimated under a range of future deposition scenarios providing a regional assessment of the number of standing water bodies vulnerable to acidification. As such it offers a valuable screening tool to prioritise those areas most at risk. Using the classification parameters selected here, it is clear that these coincide with those areas where acid emission reductions are having the least effect on acid deposition (NEGTAP, 2001). The approach can be modified by changing the sensitivity criteria applied to reflect regional variations in sensitivity (e.g., the effects of neutral sea-salts in the north-west of Scotland) or by employing more up-to-date national datasets for deposition and land cover and as they become available. Additonally, with the development of the Standing Waters Inventory offering, for the first time, a U.K. database in electronic form, similar GIS-based approaches using different national datasets can be undertaken for other potential threats such as eutrophication (see Hughes *et al.*, 2004) and climate change.

Acknowledgements

This work has been funded by the Environment Agency under R&D project P2-239. Additional support provided by DEFRA (Contract EPG 1/3/83) with guidance from the Freshwaters Umbrella group. The authors would also like to thank Jackie Ullyet and Jane Hall at CEH Monks Wood for provision of data.

References

Allott, T. E. H., Curtis, C. J., Hall, J., Harriman, R. and Battarbee, R. W.: 1995a, 'The impact of nitrogen deposition on upland surface waters in Great Britain: A regional assessment of nitrogen leaching', *Water, Air, Soil Pollut.* **85**, 279–302.

Allott, T. E. H., Golding, P. N. E. and Harriman, R.: 1995b, 'A palaeolimnological assessment of the impact of acid deposition on surface waters in north-west Scotland, a region of high sea-salt inputs', *Water, Air, Soil Pollut.* **85**, 2425–2430.

Battarbee, R. W.: 1990, 'The causes of lake acidification, with special reference to the role of acid deposition', *Philos. Trans. R. Soc. Lond.* **B 327**, 33–347.

Bunce, R. G. H. and Heal, O. W.: 1984, 'Landscape evaluation and the impact of changing land-use on the rural environment: the problem and an approach', in R. D. Roberts and T. M. Roberts (eds), *Planning and Ecology*, Chapman and Hall, London, pp. 16–188.

CLAG Freshwaters: 1995, 'Critical loads and acid deposition for U.K. freshwaters', *A Report to the Department of the Environment from the Critical Loads Advisory Group, Freshwaters sub-group*, Environmental Change Research Centre, London, pp. 80.

Curtis, C., Allott, T. E. H., Battarbee, R. W. and Harriman, R.: 1995, 'Validation of the U.K. critical loads for freshwaters: Site selection and sensitivity', *Water, Air, Soil Pollut.* **85**, 2467–2472.

DEFRA (Department for environment, Food and Rural Affairs): 2001, *First Consultation Paper on the Implementation of the EC Water Framework Directive (2000/60/EC)*, DEFRA.

Edmunds, W. M. and Kinniburgh, D. G.: 1986, 'The susceptibility of U.K. groundwaters to acidic deposition', *J. Geol. Soc. Lond.* **143**, 707–720.

European Union: 2000, Directive of the European Parliament and the Council 2000/60/EC, *Establishing a Framework for Community Action in the Field of Water Quality*, Luxembourg.

Evans, C. D., Jenkins, A. and Wright, R. F.: 2000, 'Surface water acidification in the south Pennines I. Current status and spatial variability', *Environ. Pollut.* **109**, 11–20.

Ferrier, R. C., Helliwell, R. C., Cosby, B. J., Jenkins, A. and Wright, R. F.: 2001, 'Recovery from acidification of lochs in Galloway, south-west Scotland, U.K., 1979–1998', *Hydrol. Earth Syst. Sci.* **5**, 421–432.

Fuller, R. M., Groom, G. B. and Jones, A. R.: 1994, 'The land cover map of Great Britain: An automated classification of Landsat Thematic Mapper data', *Photogramm. Eng. Remote Sens.* **60**, 553–562.

Hall, J. R., Wright, S. M., Sparks, T. H., Ullyet, J., Allott, T. E. H. and Hornung, M.: 1995, 'Predicting freshwater critical loads from national data on geology, soils and land use', *Water, Air, Soil Pollut.* **85**, 2243–2448.

Harriman, R., Christies, A. E. G. and Watt, A. W.: 1995, 'Chemical and biological exceedances of critical loads: Are they compatible?', in R. W. Battarbee (ed.), *Acid Rain and its Impact: The Critical Loads Debate*, Proceedings of a Conference held at the Environmental Change Research Centre, University College London, 1993, ENSIS Publishing, London, pp. 108–114.

Harriman, R., Watt, A. W., Christies, A. E. G., Collen, P., Moore, D. W., McCartney, A. G., Taylor, E. M. and Watson, J.: 2001, 'Interpretation of trends in acid deposition and surface water chemistry in Scotland during the past three decades', *Hydrol. Earth Syst. Sci.* **5**, 407–420.

Helliwell, R. C., Wright, R. F., Ferrier, R. C., Jenkins, A. and Evans, C.: 2002, 'Acidification of lochs in the Cairngorm Mountains, NE Scotland. *Water, Air, Soil Pollut.: Focus* **2**, 4–59.

Hornung, M., Le-Grice, S. and Norris, D.: 1990, 'The role of geology and soils in controlling surface water acidity in Wales', in R. W. Edmunds, A. S. Gee and J. H. Stoner (eds), *Acid Waters in Wales*, Kluwer Academic Publishers, Dordrecht, pp. 55–66.

Hornung, M., Bull, K., Cresser, M., Ullyett, J., Hall, J. R., Langan, S., Loveland, P. J. and Wilson, M. J.: 1995, 'The sensitivity of surface waters of Great Britain to acidification predicted from catchment characteristics', *Environ. Pollut.* **87**, 204–214.

Hughes, M., Horney, D. D., Bennion, H., Kernan, M., Hilton, J., Phillips, G. and Thomas, R.: 2004, 'The development of a GIS-based inventory of standing waters in Great Britain together with a risk-based prioritisation protocol', *Water, Air, Soil Pollut.: Focus* (this volume).

Kernan, M.: 1995, 'The use of catchment attributes to predict surface water critical loads: a preliminary analysis', *Water, Air, Soil Pollut.* **85**, 2479–2484.

Kernan, M. and Allott, T. E. H.: 1999, 'Spatial variability of nitrate concentration in lakes in Snowdonia, North Wales, U.K.', *Hydrol. Earth Syst. Sci.* **3**, 395–408.

Kernan, M., Hall, J., Ullyet, J. M. and Allott, T. E. H.: 2001, 'Variation in freshwater critical loads across two upland catchments in the U.K.: Implications for catchment scale management', *Water, Air, Soil Pollut.* **130**, 116–1174.

Langan, S. J. and Wilson, M. J.: 1992, 'Predicting the regional occurrence of acid surface waters in Scotland using an approach based on geology, soils and land use', *J. Hydrol.* **138**, 515–528.

Monteith, D. T., and Evans, C. (eds): 2000, *U.K. Acid Waters Monitoring Network: 10 Year Report*, ENSIS Publishing, London, pp. 364.

NEGTAP: 2001, 'Transboundary air pollution: Acidification, eutrophication and ground level ozone in the U.K.', *Report prepared by the National Expert Group on Transboundary Air Pollution at CEH Ediburgh*, DEFRA Contract EPG 1/3/153, pp. 314.

Ordnance Survey: 2001, 'Land-form PANORAMA User Guide v3.0', Ordnance Survey, Southampton.

Reynolds, B., Neal, C. and Norris, D. A.: 2001, 'Evaluation of regional acid sensitivity predictions using field data: Issues of scale and heterogeneity', *Hydrol. Earth Syst. Sci.* **5**, 75–81.

U.K. Biodiversity Group: 1998, *Tranche 2 Action Plans. Vol. 2: Terrestrial and Freshwater Habitats*, English Nature.

Ullyet, J., Hall, J. R., Hornung, M. and Kernan, M.: 2001, 'Mapping the potential sensitivity of surface waters to acidification using measured freshwater critical loads as an indicator of acid sensitive areas', *Water, Air, Soil Pollut.* **130**, 1235–1240.

INTEGRATED NITROGEN MODELING IN A BOREAL FORESTRY DOMINATED RIVER BASIN: N FLUXES AND RETENTION IN LAKES AND PEATLANDS

AHTI LEPISTÖ*, KIRSTI GRANLUND and KATRI RANKINEN

Finnish Environment Institute, P.O. Box 140, FIN-00251 Helsinki, Finland
(* author for correspondence, e-mail: Ahti.Lepisto@ymparisto.fi, phone: +358 9 40300238, fax: +358 9 40300291)

(Received 20 August 2002; accepted 27 April 2003)

Abstract. Two models, N_EXRET and INCA, were applied to the Simojoki river basin (3160 km^2) in northern Finland in order to assess nitrogen retention in wetlands and lakes. N_EXRET is a spatial, export coefficient-based N export and retention model developed for large river basins. It utilizes remote sensing-based land use and forest classification, evaluated export coefficients, and data on areal N deposition and point sources of N. A new version (v1.7) of the Integrated Nitrogen in CAtchments model (INCA) is a semi-distributed, dynamic nitrogen process model, which simulates and predicts nitrogen transport and processes within catchments. Average retention of the gross total N load of 700 t a^{-1} to the river system was estimated using N_EXRET model as 17 t N a^{-1} to the wetlands and 77 t N a^{-1} to the lakes. A good fit was found between modeled and measured values along the river. Inorganic N fluxes simulated by the INCA model were compared with measured fluxes along the river Simojoki, with a good fit between modeled and measured NH_4^+-N fluxes, and an adequate fit for NO_3^--N fluxes. Both fluxes were overestimated at the first reach, below Lake Simojärvi. High percentage of peatlands led to high NH_4^+-N/NO_3^--N ratios derived from data, indicating negligible nitrification in large river subbasins and particularly in small research catchments.

Keywords: boreal river basin, dynamic modeling, Finland, INCA, N_EXRET, N modeling, N processes, N retention

1. Introduction

Nitrogen (N) concentrations and fluxes in rivers reflect the integration of catchment N sources (non-point sources, atmospheric N deposition, direct effluent discharges, mineralisation) and sinks (plant uptake, immobilisation, denitrification). Integrated nitrogen models provide valuable tools to integrate the effects of varying N sources at different spatial scales – river basins and small catchments – in order to test different scenarios of interest for environmental policy makers.

Typical features in the boreal zone in northern Finland are long winters (5–7 months) with continuous snow cover, deep soil frost, and ice cover over rivers and lakes. A snowmelt-induced spring flood in late April–May dominates the annual hydrological pattern. Smaller flow peaks occur in autumn due to rainfall. Most nutrient leaching occurs during these high flow periods. N uptake by vegetation, immobilisation and mineralisation in catchment soils, accumulation in the seasonal snowpack, and flow dynamics are the most important factors affecting seasonal variation of inorganic N in surface waters.

The sum of catchment N sources normally exceeds the total output fluxes from a catchment/river basin on a long-term basis. This discrepancy is often referred to as retention (e.g., Dillon *et al.*, 1990; Arheimer, 1998). Retention is a lumped expression for the net effect of various biogeochemical processes responsible for temporary or permanent N removal from the water phase, such as biological uptake (assimilation), sedimentation, and denitrification. Possibilities for retention are favoured where water is stored in the landscape; the key environments are riparian areas, wetlands, streams, and lakes (Arheimer, 1998). A number of interacting factors that correlate with a system's capacity to retain N have been identified, including C:N ratio of soil organic matter, soil texture, rate of biomass accumulation and past human land use (Vitousek *et al.*, 1997). Retention of N is usually high in northern N-limited forest ecosystems and is even higher in peatlands. In peatlands, NO_3^--N is used by plants during growth, in NO_3^- reduction, or removed through microbial reduction or denitrification. Because of this biogeochemical reactivity, NO_3^--N inputs are efficiently retained, and the concentrations in peatland waters are low. The concentrations of NH_4^+-N may be of the same order or higher than those of NO_3^--N (e.g., Saukkonen and Kortelainen, 1995; Joensuu *et al.*, 2002), due to inhibition of nitrification.

A considerable portion of N flux from boreal forest and peatland-dominated river basins may reach the sea in the form of organic N, e.g. 75–80% in the northern Bothnian Bay of the Baltic Sea (Pitkänen, 1994). Most of organic N occurs as dissolved organic N, DON. In ongoing studies in northern rivers in which both total and filtered samples have been analyzed, particulate fractions (PON) have been very minor, ~5% (T. Mattsson, unpublished). Dominance of dissolved N suggests low retention by sedimentation in fast-flowing northern rivers. However, a large part of DON may be bioavailable in the estuaries. For example in the northeastern United States, between 40 and 72% of the DON in two large rivers was utilized by estuarine bacteria (Seitzinger and Sanders, 1997).

To address these issues and to compare results from two models, N fluxes and retention in a boreal river basin were studied using an export/retention coefficient-based N_EXRET model (Lepistö *et al.*, 2001) and an N process-based INCA model (Wade *et al.*, 2002). First, modeled total N fluxes (N_EXRET model) and inorganic N fluxes (INCA model) were compared with available empirical data. Second, N retention estimates for wetlands were compared using both models, and N retention in lakes was estimated (N_EXRET model). Third, the impact of peatlands on NH_4^+-N/NO_3^--N ratios and nitrification in larger river subbasins and in small catchments was assessed.

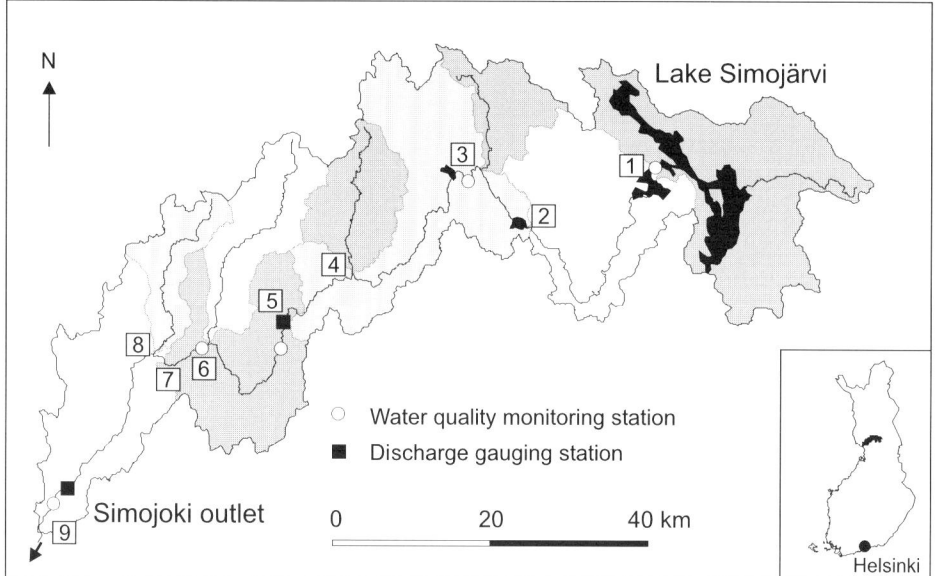

Figure 1. Map of the Simojoki river basin showing reach (1–9) boundaries, the subcatchment areas and the locations of monitoring sites.

2. Materials and Methods

2.1. Site description, hydrological monitoring and water sampling

The river Simojoki (3160 km^2) (Figure 1) discharges to the Gulf of Bothnia in the Baltic Sea. The river Simojoki is a salmon river in near-natural state, the dominant human impact being forestry, mainly forest drainage and cutting. The total length of the river between the outlet of Lake Simojärvi and the sea is 193 km. Peatlands and peatland forests are common in the region and an average of 0.5% of the total catchment area is felled annually. Agricultural fields cover only 2.7% of the catchment area (Perkkiö *et al.*, 1995).

The river Simojoki has two discharge gauging stations; one is located at the river outlet, the other at the Hosionkoski rapids. The N concentration data were obtained from the water quality database operated by the Finnish Environment Institute (SYKE) and regional environment centres. Water samples were taken at the outlet and at different sites of the Simojoki river (Figure 1) during the period of 1994–1996. Total N was analysed as NO_3^--N after oxidation with $K_2S_2O_8$. NH_4^+-N was analysed by a spectrophotometric method with hypochlorite and phenol, and NO_3^--N by the cadmium amalgam method.

2.2. N_EXRET MODEL

N_EXRET is a spatial, export coefficient-based N export and retention model developed for large river basins and described in detail by Lepistö *et al.* (2001). It utilizes remote sensing-based land use and forest classification (Vuorela, 1997), evaluated export coefficients, areal N deposition, and point sources of N. It simulates N fluxes by all the major land use types, and retention of once-leached N in the peatlands and lakes of the region concerned. N_EXRET is a simple conceptual model for simulating total N export and retention in large scales. It does not include N processes or allow simulation of inorganic N fractions as INCA does. On the other hand, simulation of long watercourses, including large lakes, is possible using N_EXRET, whereas INCA is a river basin model concentrating on the land-water interphase, and is not meant to be applied to basins with large lakes (simulation of in-lake processes not included).

2.3. INCA MODEL

On the basis of earlier work by Whitehead *et al.* (1998), a new version of the dynamic, process-based and semi-distributed INCA model was developed and is described in detail by Wade *et al.* (2002). N process equations are written in terms of loads rather than concentrations, thus allowing a robust tracking of mass conservation when using numerical integration. N process rate-coefficients (mineralization, nitrification, denitrification, plant uptake) depend on soil moisture and temperature (Wade *et al.*, 2002). The model integrates hydrology, catchment and river N processes, and simulates daily NO_3^--N and NH_4^+-N concentrations as time series at key sites, as profiles down the river system or as statistical distributions.

2.4. MODEL INPUT DATA

2.4.1. *N_EXRET Model and Sensitivity Analysis*

Proportions of 12 land-use classes within a 1×1 km grid were obtained from the above-mentioned satellite-based database. The export coefficients used were the so-called 'best available' values taken from the empirical studies assumed to be representative for the area, and were the same as those reported by Lepistö *et al.* (2001). The effects of retention coefficients *a* (peatlands) and *b* (lakes) were studied. The criterion was the difference between modeled and measured N fluxes in the outflow of the basin. The studied range of retention coefficients was 0.05–0.40 kg ha^{-1} a^{-1} for peatlands and 2.0–6.0 kg ha^{-1} a^{-1} for lakes.

2.4.2. *INCA Model*

The INCA model application utilises the most recent satellite image-based land use and forest classification data for Finland (Vuorela, 1997), supplemented with satellite image-based maps of final cuttings in mineral and organic soils, provided by the Finnish Forest Research Institute (FFRI). Inorganic N deposition for both

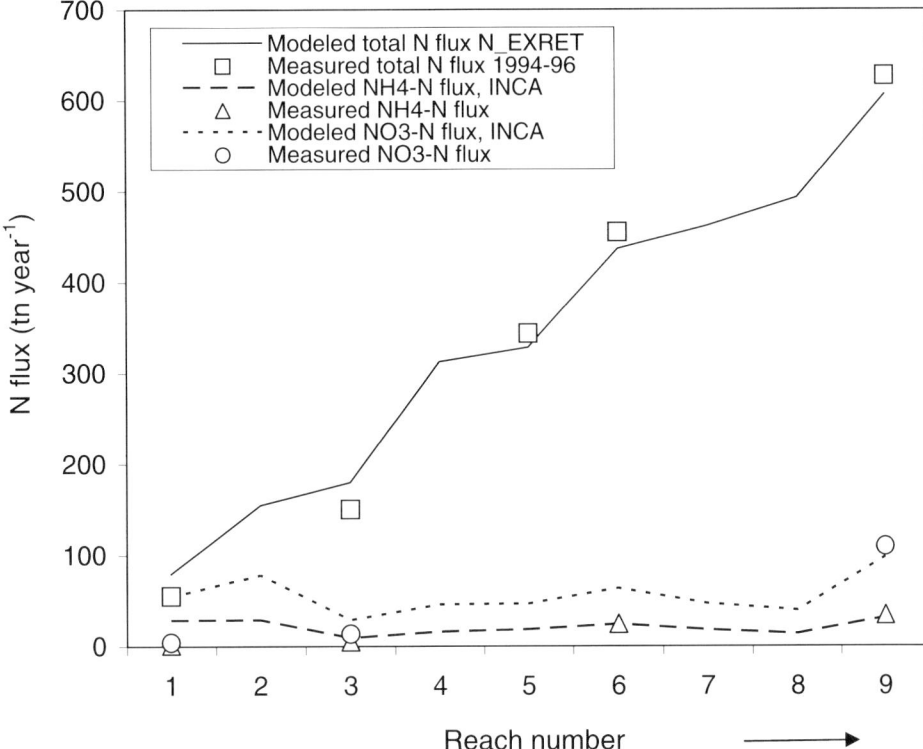

Figure 2. N fluxes calculated using the N_EXRET model in nine reaches (Figure 1) along the river Simojoki, compared with measured (1994–1996) flux in five sites. Inorganic N fluxes calculated using the INCA model compared with measured flux (reach 1 represents to Lake Simojärvi outlet and reach 9 the sea outlet).

INCA and N_EXRET models was estimated using the areal deposition model DAIQUIRI (Kangas and Syri, 2002). Calibration of the INCA hydrological sub-model to Simojoki river basin has been described earlier by Rankinen *et al.* (2002) and N process model calibration and N process fluxes for different land-use types elsewhere in this volume (Rankinen *et al.*, 2004).

3. Modeled and Measured N Fluxes in Different Sites along the River

3.1. N_EXRET MODELED TOTAL N FLUXES

The studied river basin has a closed N cycle with low input and output fluxes of inorganic N. During the period of 1982–2000, up to 80% of the total gross N load was organic N, 6% was NH_4^+-N, and 14% was NO_3^--N (Rankinen *et al.*, 2002).

Total N fluxes – including organic N – were simulated with the N_EXRET model. Figure 2 shows N fluxes calculated using the N_EXRET model in nine

TABLE I

Model sensitivity to the variation of retention parameters a and b. The criterion was the modeled N flux compared (%) with measured N flux at the outlet of the Simojoki river basin

Retention		b, lakes (kg ha^{-1} a^{-1})				
		2	3	4	5	6
a, peatlands (kg ha^{-1} a^{-1})	0.05	104	101	98	95	92
	0.10	103	100	97	94	91
	0.20	100	97	94	91	88
	0.30	98	94	91	88	85
	0.40	95	92	89	86	83

reaches along the river Simojoki. Modeled, accumulated total N flux was compared with measured (1994–1996) flux in those five sites along the river for which data were available. A good fit was found between modeled and measured values along the river, which means that estimated retention in wetlands and lakes of the basin is acceptable. Average retention to the wetlands was estimated as 17 t N a^{-1}, and to the lakes as 77 t N a^{-1}, representing 2.4% (wetlands) and 11% (lakes) of the total, gross N load of 700 t a^{-1} (non-point source loading, N deposition and point sources) to the river system. The 'best' fit was obtained with coefficients of 0.1 (wetlands) and 4.0 (lakes), when taking into account model performance both in the upper (Lake Simojärvi outlet) and lower parts (outlet to the sea) of the river basin. Modeled N flux to the sea was 606 t a^{-1} (97%), which compared well with the measured value (1994–1996) of 627 t a^{-1}.

The studied range of retention coefficients was 0.05–0.40 for wetlands and 2.0–6.0 for lakes, i.e., somewhat smaller but within the range of values given in Lepistö et al. (2001). Table I shows the sensitivity analysis by 25 model runs, with covariation of the coefficients and their effects on modeled N flux compared (%) with measured N flux at the river outlet. Retention coefficients of ≥ 0.40 for peatlands, and ≥ 5.0 for lakes did not give adequate results, whereas different combinations gave equally good results. Retention by the lakes was estimated to be over one order of magnitude higher than by the peatlands.

3.2. INCA MODELED INORGANIC N FLUXES

Correspondingly, inorganic N fluxes simulated by the INCA model were compared with measured fluxes along the river Simojoki (Figure 2). A good fit was found between modeled and measured NH_4^+-N fluxes, and an adequate fit for NO_3^--N fluxes. Both fluxes were overestimated at the first reach, below Lake Simojärvi. The INCA model does not include lake processes, although N retention in the lake has probably had an important role. There is a slight increase in inorganic N fluxes along the river, but not so pronounced as in the case of organic N. Particularly concerning NH_4^+-N, inputs to the river and losses as nitrification of NH_4^+ to NO_3^- in the river itself are balanced.

4. N Retention in Catchments and Implications for N Processes

4.1. RETENTION IN PEATLANDS

Retention was modeled by N_EXRET to vary in the range of 0.05–0.40 kg ha^{-1} a^{-1} in peatlands, with a best estimate of 0.10 kg ha^{-1} a^{-1}. These rates are very low, as is retention in the whole river basin. Considerable retention may already have occurred along the flow paths in mineral soils before water enters the peat areas in lowlands. However, these rates are comparable with estimated denitrification in organic soils by the INCA model, 0.24–0.36 kg ha^{-1} a^{-1} (Rankinen et al., 2004).

There exists huge variability between upstream waterlogged patches, different ombrotrophic and minerotrophic peatland types, and riparian peaty areas in the lower reaches of a river. This means that the average regional retention rates estimated here are extremely difficult to compare with measurements in certain, individual peatland areas. In peatlands, NO_3^- is utilized by plants during growth, in NO_3^- reduction, and in denitrification. Because of this biogeochemical reactivity, NO_3^--N inputs are efficiently retained and the concentrations of NO_3^- in peatland waters are normally low (Gorham et al., 1984).

4.2. RETENTION IN LAKES

Most N, once leached to a stream, is probably not retained before reaching the first lakes along the watercourse. Most of the retention in the study area occurs in the large Lake Simojärvi: 60% of the total lake retention (77 t N a^{-1}) within the river basin is estimated to occur in this lake. Retention by the lakes was estimated to vary in the range of 2.0–4.0 kg ha^{-1} a^{-1}, i.e. over one order of magnitude higher than by the peatlands.

Nitrogen in the Simojoki river basin is largely bound in humic material and the inorganic N concentrations are low. On average, 80% of the measured N flux to the sea was organic dissolved and particulate N. Lakes are important for long-term retention of N in these northern regions, with sedimentation being probably more important than denitrification. Ahlgren et al. (1994) studied Lake Norrviken, central Sweden, and found that sediment retention was 14% of external N loading, compared with 5–11% retention by denitrification. All of the sedimentation may not be permanent; for example, it has been reported that 54% of sedimented organic N was mobilised and released from the sediments of a humic lake (Jonsson and Jansson, 1997).

Retention of N was estimated to be negligible, even 'negative' in the river Simojoki basin based on an earlier study (Lepistö et al., 2001). In that study, the Simojoki basin was used for model testing with the same retention coefficients as in the more southern Oulujoki river basin. It seems probable that retention decreases towards the north, with colder climate, less denitrification, and slower N cycling. In this study, we attempted to estimate more accurate lake retention values (4 kg ha^{-1} a^{-1} compared with the earlier estimate of 8 kg ha^{-1} a^{-1}), specifically for the Simojoki basin.

4.3. IMPACT OF PEATLANDS ON NH_4^+-N AND N TRANSFORMATION PROCESSES

Peatland effects on regional inorganic N fluxes and N transformation were studied by comparing NH_4^+-N/NO_3^--N ratios and NH_4^+-N concentration levels with the percentage of peatlands in the catchment. Using the presented data by Saukkonen and Kortelainen (1995) from 29 small (0.7–56 km^2) catchments, it was determined that the ratio increased clearly with increasing peatland percentage (Figure 3a), with highest values between 5 and 6 in catchments strongly dominated by peatlands (>60%). Variability is high with an R^2 of only 0.36. In the large subbasins of the rivers Simojoki and Oulujoki (Lepistö et al., 2001), the same phenomenon was found to be smoothed, the ratio varying between 0.14 and 1.10, and with a considerably higher R^2 of 0.81 (Figure 3a). Further, NH_4^+-N concentrations increase as a function of peatland percentage (Figure 3b).

The high concentrations of NH_4^+-N and the high ratios may reflect anoxic acid conditions in peatlands, with an inhibitory effect on nitrification (Sikora and Keeney, 1983), and/or non-effective retention of NH_4^+ by the soil. The former interpretation is more probable in the ecosystems studied here because, in the boreal zone, nitrification even in mineral forest soils in natural state is typically very low (e.g. Priha and Smolander, 1995), and even lower in organic soils. Nitrification also may be severely inhibited by certain dissolved organic compounds typical in soil systems with high humic levels. Further evidence for the importance of NH_4^+-N is provided by the fact that comparable amounts of NH_4^+-N and NO_3^--N losses are estimated from Finnish forest land with high percentage of peatland forests (Kortelainen and Saukkonen, 1998). The same phenomenon can be detected for groundwater N, which is typically in the form of NH_4^+ close to the stream in boreal forest catchments studied in Canada, indicating reduced, anoxic conditions (Ford and Naiman, 1989).

5. Concluding Remarks

The river Simojoki basin has a relatively closed N cycle with minor inorganic N input and output fluxes. Nitrogen in the Simojoki river basin is largely bound in

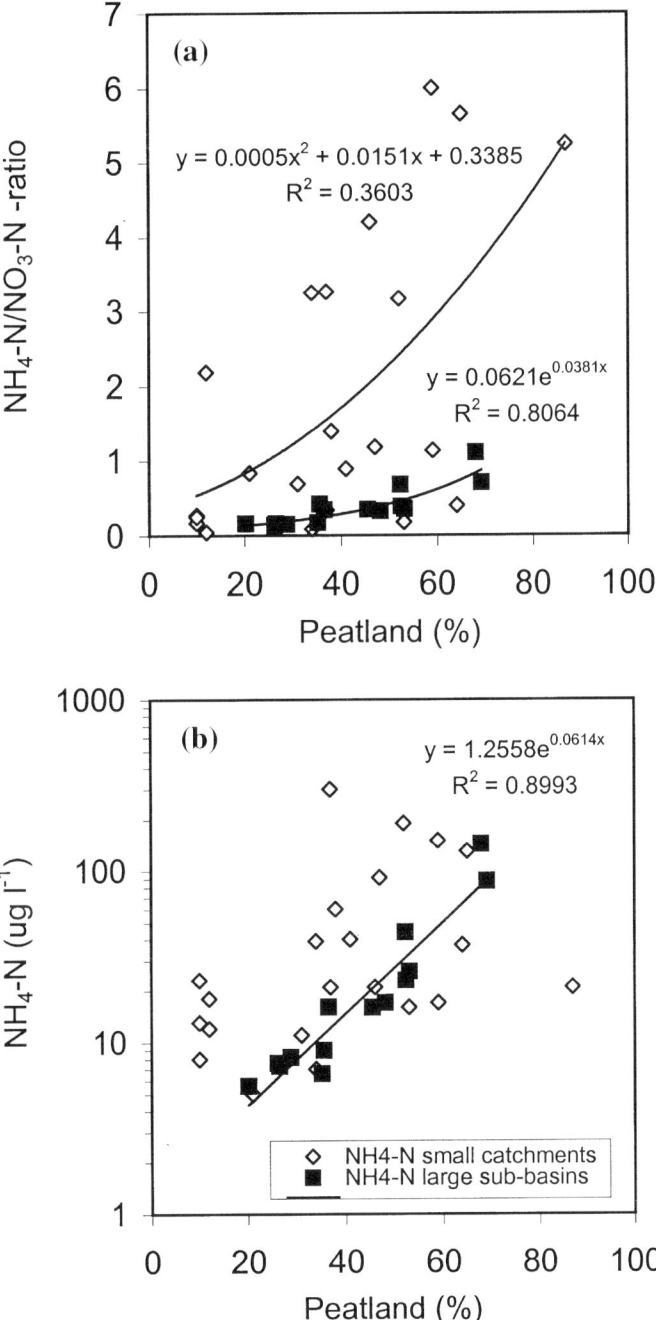

Figure 3. (a) NH_4^+-N/NO_3^--N ratios and (b) NH_4^+-N concentrations as a function of peatland percentages in small catchments (data from Saukkonen and Kortelainen, 1995) and in large subbasins of the rivers Simojoki and Oulujoki (data from Lepistö *et al.*, 2001).

humic material: on average, 80% of the measured N flux to the sea is organic dissolved and particulate N.

A good fit was observed between modeled (N_EXRET) and measured total N fluxes along the river, which means that the retention estimate for wetlands and lakes of the basin is acceptable. Retention by the lakes was estimated to be over one order of magnitude higher than by the peatlands. Major retention in the basin occurs in the large Lake Simojärvi. Below this lake, retention in Simojoki river itself is minor. Inorganic N fluxes simulated by the INCA model were compared with measured fluxes along the river Simojoki. A good fit was obtained between modeled and measured NH_4^+-N fluxes, and an adequate fit for NO_3^--N fluxes.

Peatlands have a strong effect on N transformation processes, detected as the highest NH_4^+-N/NO_3^--N ratios derived from the data of catchments with the highest percentages of peatland, indicating negligible nitrification particularly in small headwater catchments but also in large river sub-basins in the area.

Acknowledgements

This study was supported by the Commission of the European Union, the INCA project (EVK1-CT-1999-00011).

References

Ahlgren, I., Sörensson, F., Waara, T. and Vrede, K.: 1994, 'Nitrogen budgets in relation to microbial transformations in lakes', *Ambio* **23**, 367–377.
Arheimer, B.: 1998, 'Riverine nitrogen – Analysis and modelling under nordic conditions', *Ph.D. Thesis*, Linköping University, Sweden, Linköping Studies in Arts and Science, No. 185, pp. 64.
Dillon, P. J., Evans, R. D. and Molot, L. A.: 1990, 'Retention and resuspension of phosphorus, nitrogen, and iron in a central Ontario lake', *Can. J. Fish. Aquat. Sci.* **47**, 1269–1274.
Ford, T. E. and Naiman, R. J.: 1989, 'Groundwater-surface water relationships in boreal forest watersheds: Dissolved organic carbon and inorganic nutrient dynamics', *Can. J. Fish. Aquat. Sci.* **46**, 41–49.
Gorham, E., Bayley, S. E. and Schindler, D. W.: 1984, 'Ecological effects of acid deposition upon peatlands: A neglected field in 'acid-rain' research', *Can. J. Fish. Aquat. Sci.* **41**, 1256–1268.
Joensuu, S., Ahti, E. and Vuollekoski, M.: 2002, 'Effects of ditch network maintenance on the chemistry of run-off water from peatland forests', *Scand. J. For. Res.* **17**, 238–247.
Jonsson, A. and Jansson, M.: 1997, 'Sedimentation and mineralisation of organic carbon, nitrogen and phosphorus in a large humic lake, northern Sweden', *Arch. Hydrobiol.* **141**, 45–65.
Kangas, L. and Syri, S.: 2002, 'Regional nitrogen deposition model for integrated assessment of acidification and eutrophication', *Atmos. Environ.* **36**, 1111–1122.
Kortelainen, P. and Saukkonen, S.: 1998, 'Leaching of nutrients, organic carbon and iron from Finnish forestry land', *Water, Air, Soil Pollut.* **105**, 239–250.
Lepistö, A., Kenttämies, K. and Rekolainen, S.: 2001, 'Modeling combined effects of forestry, agriculture and deposition on nitrogen export in a northern river basin in Finland', *Ambio* **30**, 338–348.

Perkkiö, S., Huttula E. and Nenonen, M.: 1995, 'Water protection plan for the Simojoki river basin. (Simojoen vesistön vesiensuojelusuunnitelma)', Publication of the Water and Environment Administration – Series A 200, pp. 102 (in Finnish).

Pitkänen, H.: 1994, 'Eutrophication of the Finnish coastal waters: Origin, fate and effects of riverine nutrient fluxes', Publication of the Water and Environment Research Institute 18, Helsinki, pp. 44.

Priha, O. and Smolander, A.: 1995, 'Nitrification, denitrification and microbial biomass in soil from two N-fertilized and limed Norway spruce forests', *Soil Biol. Biochem.* **27**, 305–310.

Rankinen, K., Lepistö, A. and Granlund, K.: 2002, 'Hydrological application of the INCA model with varying spatial resolution and nitrogen dynamics in a northern river basin', *Hydrol. Earth Syst. Sci.* **6**, 339–350.

Rankinen, K., Lepistö, A. and Granlund, K.: 2004, 'Integrated nitrogen and flow modelling (INCA) in a boreal river basin dominated by forestry: Scenarios of environmental change', *Water, Air, Soil Pollut.: Focus* (this volume).

Saukkonen, S. and Kortelainen, P.: 1995, 'Metsätaloustoimenpiteiden vaikutus ravinteiden ja orgaanisen aineen huuhtoutumiseen', Finnish Environment, No. 2, pp. 15–32 (in Finnish).

Seitzinger, S. P. and Sanders, R. W.: 1997, 'Contribution of dissolved organic nitrogen from rivers to estuarine eutrophication', *Mar. Ecol. Prog. Ser.* **159**, 1–12.

Sikora, L. J. and Keeney, D. R.: 1983, 'Further aspects of soil chemistry under anaerobic conditions', in A. J. P. Gore (ed.), *Ecosystems of the World 4A, Mires: Swamp, Bog, Fen and Moor: General Studies*, Elsevier, Amsterdam, pp. 247–256.

Vitousek, P., Aber, J. D., Howarth, R. W., Likens, G., Matson, P. A., Schindler, D. W., Schlesinger, W. H. and Tilman, D. G.: 1997, 'Human alteration of the global nitrogen cycle: Sources and consequences', *Ecol. Applic.* **7**, 737–750.

Vuorela, A.: 1997, 'Satellite image based land cover and forest classification of Finland', in R. Kuittinen (ed.), *Proc. Finnish-Russian Seminar on Remote Sensing*, Helsinki, 29 August–1 September, 1994, Reports of the Finnish Geodetic Institute 97 2, pp. 42–52.

Wade, A. J., Durand, P., Beaujouan, V., Wessels, W. W., Raat, K. J., Whitehead, P. G., Butterfield, D., Rankinen, K. and Lepistö, A.: 2002, 'A nitrogen model for European ecosystems: INCA, new model structure and equations', *Hydrol. Earth Syst. Sci.* **6**, 559–582.

Whitehead, P. G., Wilson, E. J. and Butterfield, D.: 1998, 'A semi-distributed Integrated Nitrogen model for multiple source assessment in Catchments (INCA): Part I – Model structure and process equations', *Sci. Total Environ.* **210/211**, 547–558.

UNCERTAINTIES IN PREDICTIONS OF SURFACE WATER ACIDITY USING THE MAGIC MODEL

THORJØRN LARSSEN[1]*, B. JACK COSBY[2] and TORE HØGÅSEN[1]

[1] *Norwegian Institute for Water Research, P.O. Box 173, Kjelsås, 0411 Oslo, Norway;*
[2] *Department of Environmental Sciences, University of Virginia, Charlottesville, VA 22903, U.S.A.*
(* *author for correspondence, e-mail: thorjorn.larssen@niva.no, phone: +47 2218 5194, fax: +47 2218 5200)*

(Received 20 August 2002; accepted 6 April 2003)

Abstract. In this study, we have used the MAGIC model together with data from the Birkenes catchment in Norway, at which 27 years of data (1974–2000) are available. We calibrated the MAGIC model to the five year observed average chemistry around 1990, and then used the data from the five year period around 1980 to refine the calibration. From 1990, forecasts were run for the different sets of inputs and parameters, and the sets of inputs and parameters were further refined using observations for the period 1996–2000. Through an automatic calibration routine, the model was calibrated a large number of times with different sets of input data to account for the uncertainties in the observed data using a Monte Carlo set-up. The results show that the uncertainty in the model predictions decreases as more observed data from different points in time are used in the model calibration. The results also show that when using the time series data in calibration, the distribution of the forecast changed. The distribution of the predicted Acid Neutralisation Capacity (ANC) in the future is lower for the more refined model calibration. The 10 and 90 percentiles of predicted ANC in 2010 are −3 to 21 μeq L^{-1} when only a five-year average is used for calibration, but are −7 to 9 μeq L^{-1} when data from the three different time periods are used.

Keywords: acidification, Birkenes, modelling, Monte Carlo, Norway, prediction, recovery

1. Introduction

As acid deposition in Europe decreases as a result of emission reductions, there is an increased need to know whether these reductions are sufficient, as well as when ecosystems recover from acidification. To address these issues, dynamic acidification models are required. The predictive power of dynamic acidification modelling is a crucial issue for determining the usefulness of such models to support policy-making. Dynamic models for surface water acidification have been developed the last two decades (Warfvinge *et al.*, 1992). Models have been refined and enlarged through inclusion of additional processes as new data and new knowledge has become available (e.g., Cosby *et al.*, 2001). As more data become available, uncertainties in model simulations can be better quantified and included as a part of the presentation of modelling results to policy makers. Quantification of uncertainties in predictions is of increasing importance, as acid deposition decreases to levels closer to the critical loads.

The policy goals of acidification modelling have changed. Earlier the objective was often to illustrate the effects of emission reductions. Recently, as substantial reductions in emissions have been implemented, the focus has increasingly changed to issues such as: (a) *when* will waters and soils recover from acidification, and (b) are the latest internationally-agreed measures sufficient (Bull *et al.*, 2001). To address these questions policymakers have taken increasing interest in dynamic models, especially as the time component is lacking in the steady-state models (UN-ECE, 2002).

Uncertainty in dynamic modelling has been addressed early in the history of development of the models (Hornberger *et al.*, 1986, 1989). Renewed interest in uncertainty of model predictions has arisen because of the policy relevance of the results. The availability of the data from long-term monitoring programs facilitates systematic analysis of uncertainty. The extensive data available from a few calibrated catchments are of great value in exploring the predictive power of dynamic models and to evaluate the uncertainties in their predictions.

In this paper, we use the long-term data set from the Birkenes catchment in Norway and the dynamic acidification model MAGIC (Cosby *et al.*, 1985, 2001) in a Monte Carlo setup to explore how the predictions and related uncertainty relate to the temporal quality of the available observational data.

2. Material and Methods

Data from the Birkenes catchment are among the longest time series data of precipitation and surface water chemistry available in the world. Weekly data with major anions and cations are available since 1974 for both bulk deposition and runoff chemistry (SFT, 2001; Aas *et al.*, 2001). The data span the years with maximum acid deposition (1970's) and the period with 50–60% decrease in S deposition (1985 to 2000) (Figure 1).

MAGIC is a lumped-parameter model of intermediate complexity, developed to predict the long-term effects of acidic deposition on soils and surface water chemistry (Cosby *et al.*, 1985, 2001). The model simulates soil solution chemistry and surface water chemistry to predict annual average concentrations of the major ions in lakes and streams. The model has been extensively used at a range of different sites and applications (Wright, 2001). Model inputs are deposition amount and composition and physical and chemical parameters for soil and surface water. Weathering rates and initial (pre-industrial) base saturation values are calibrated by comparing model outputs and observations for observed water chemistry and soil base saturation.

In this study, we take into account the uncertainties in all model inputs. The range of uncertainty, as well as our ability to quantify Distributions are calculated from the observations when sufficient data are available. When data are lacking, a rectangular distribution is applied. Values for the inputs are drawn from the

UNCERTAINTIES IN PREDICTIONS OF SURFACE WATER ACIDITY 127

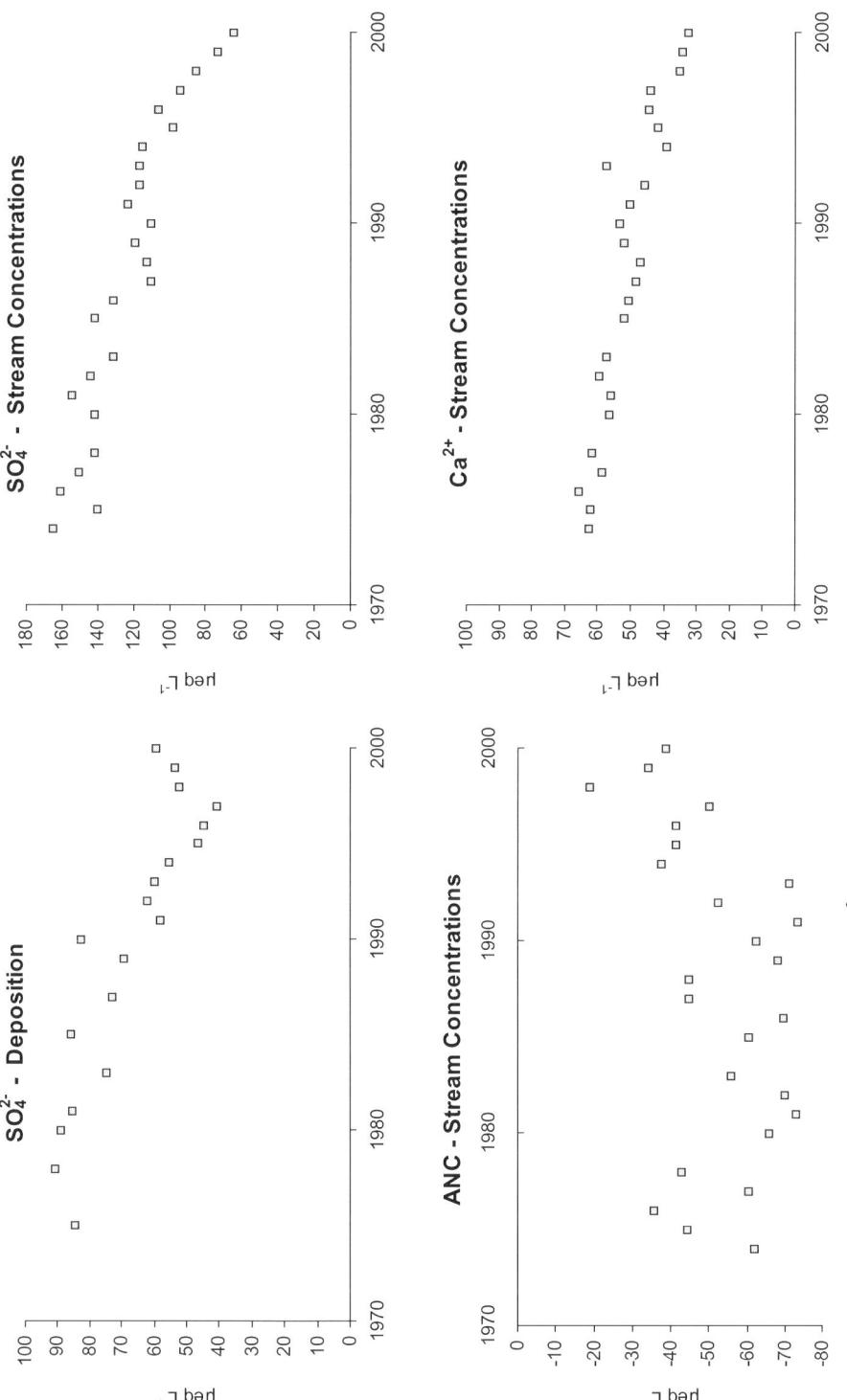

Figure 1. Observed annual average concentration of SO_4^{2-} in deposition and stream and ANC and Ca^{2+} concentration in the stream for the period 1973 to 2000.

distributions, and the model is calibrated to the given set of inputs. This procedure is repeated a large number of times, and hence a large number of calibrated sets of parameters are created.

2.1. ESTIMATE OF TOTAL DEPOSITION

The total deposition is split in the model inputs as wet and dry deposition. The wet deposition input is taken from the bulk chemistry measurements. The dry deposition fraction is calculated based on annual stream chemistry fluxes for chloride (Cl^-) and sulphate (SO_4^{2-}) and the average sea salt composition in the ocean. Cl^- is assumed to be conservative and hence having equal input and output fluxes each year. The factor between the output and input fluxes of Cl^- is used as dry deposition factor for the other sea salt ions. For SO_4^{2-} the dry deposition factor was increased in order to take into account the dry deposition of SO_2 so that the long-term input flux matches the long-term output flux.

Distributions for the input deposition data were calculated from the observations. To take the co-variation between the different inputs into account, we used linear regressions on Cl^- and SO_4^{2-} as a basis for calculating the distributions for the other ions. Regression on Cl^- was used for the mainly sea salt originating ions (sodium (Na^+), magnesium (Mg^{2+}) and potassium (K^+)), while calcium (Ca^{2+}) was regressed on SO_4^{2-}. This was done to avoid initialising the model with impossible deposition chemistry (e.g., low Cl^- and high Na^+ concentrations) and still keeping sufficient variation in the inputs.

For Cl^- the normal distribution of the mean of the volume weighted annual averages (1974–2000) was used as input. From the linear regressions the Na^+ and Mg^{2+} concentration distributions were drawn randomly from the 99.9% confidence intervals of the estimates.

The random variation from year to year in the SO_4^{2-} deposition was calculated using the variation around a linear regression line was used. The input distribution was drawn using Latin hypercube sampling from the from the 99.9% confidence interval around the 1990 estimate. The input distribution for Ca^{2+} was drawn randomly from the 99.9% confidence intervals of the estimates from the regression with SO_4^{2-}. The distributions used are shown in Table I.

2.2. SOIL PARAMETERS

Cation exchange capacity and bulk density were measured at 4 plots in the catchment at 4 depths at each plot in 1984 and 1992 (SFT, 1993). Data from both years (there is no significant difference between the years) and all plots and horizons were used to calculate the depth-weighted average bulk density and the mass-weighted average CEC. The shapes of their distributions cannot be derived from the data, as such, rectangular distributions were used with the upper and lower input limits set to ±10% of the average values. For the other soil parameters, fewer measurements are available and subjective judgements were used to set the limits

TABLE I

Average and standard deviations for the distributions for wet deposition data used as inputs in the model calibration (1990). See text for description of estimation of the distributions

Parameter	Average	Standard deviation
Ca^{2+}	10.9	0.5
Mg^{2+}	19	1.1
Na^+	82	4.9
K^+	4.7	0.3
SO_4^{2-}	70	2.7
NO_3^-	56	1.8
Cl^-	96	6.0

TABLE II

Ranges for soil and stream parameters used as model inputs. For all parameters in the table rectangular shape was used for the distributions

	Unit	Min	Max
Soil parameters			
Depth	m	0.30	0.50
Porosity	%	40	60
Bulk density	$kg\ m^{-3}$	695	850
Cation exchange capacity	$meq\ kg^{-1}$	95	117
CO_2-pressure	% of atm.	0.50	2.00
Solution organic charge	$\mu mol\ L^{-1}$	0	250
Stream characteristics			
CO_2-pressure	% of atm.	0.05	0.20
Solution organic charge	$\mu mol\ L^{-1}$	0	25

of the inputs. Rectangular distributions were used, as there are no data suggesting other distributions. The input data ranges are given in Table II.

2.3. MODEL CALIBRATION AND TARGETS

Observed surface water chemistry and the relative amount of each of the exchangeable base cations were used as targets in the model calibration (Table III), meaning

TABLE III

Observed stream chemistry and exchangeable base cations (expressed as % of CEC) and soil pH used as model calibration targets for 1990. The minimum and maximum values given are the limits used in model calibration

Parameter	Unit	Minimum	Maximum
Stream chemistry			
Ca^{2+}	(μeq L^{-1})	43.6	51.3
Mg^{2+}	(μeq L^{-1})	24.8	35.5
Na^+	(μeq L^{-1})	108.6	139.7
K^+	(μeq L^{-1})	2.4	6.3
SO_4^{2-}	(μeq L^{-1})	105.7	121.5
NO_3^-	(μeq L^{-1})	5.8	13.5
Cl^-	(μeq L^{-1})	104.4	166.1
Soil chemistry			
Ca^{2+}	(%)	2.9	4.9
Mg^{2+}	(%)	1.1	3.1
Na^+	(%)	0.1	2.1
K^+	(%)	1.8	3.8
pH		3.9	4.5

that model outputs were compared to the measured values and accepted as a successful run and stored if within the target window. Distributions for the target water chemistry for the different ions were estimated in the same way as described for the deposition input distributions. The principal calibration year was 1990. The model was calibrated to fit the average for the observed data 1988–1992 using an automatic optimising procedure. Input parameters were drawn from distributions by Latin hypercube sampling. Out of 10 000 sets of inputs, 8891 were successfully calibrated.

Sulfate adsorption was ignored in this application, as SO_4^{2-} input and output have the same slopes over time. The nitrogen (N) parameters, i.e. nitrification and uptake rates, were calibrated such that the observed total N input matched nitrate (NO_3^-) output. Base cation concentrations in the stream and on the soil ion exchanger were calibrated by automatically adjusting the weathering rates and the initial (i.e., pre-industrial) base saturation (Table IV). The pH and the aluminium concentration were calibrated using the charge of organic acidity and the dissolution constant for aluminium hydroxide in the soil and in stream water.

TABLE IV

Initial ranges for the calibrated parameters

Parameter	Unit	Min	Max
Soil weathering rates			
Ca^{2+}	meq m^{-2} a^{-1}	0	100
Mg^{2+}	meq m^{-2} a^{-1}	0	100
Na^+	meq m^{-2} a^{-1}	0	100
K^+	meq m^{-2} a^{-1}	0	100
Aluminum dissolution constants			
Soil	Log10	6	11
Stream	Log10	6	11
Initial soil base cation saturation			
Ca^{2+}	%	0.1	50
Mg^{2+}	%	0.1	50
Na^+	%	0.1	50
K^+	%	0.1	50
NO_3^- uptake	%	0	100

TABLE V

Limits used for surface water chemistry in the calibration refinement procedure. Units: μeq L^{-1}

		Ca^{2+}	SO_4^{2-}	ANC
1980	Minimum	54	135	−86
	Maximum	63	157	−39
1990	Minimum	a	a	−83
	Maximum	a	a	−40
2000	Minimum	26	52	−59
	Maximum	49	117	−14

[a] See Table III.

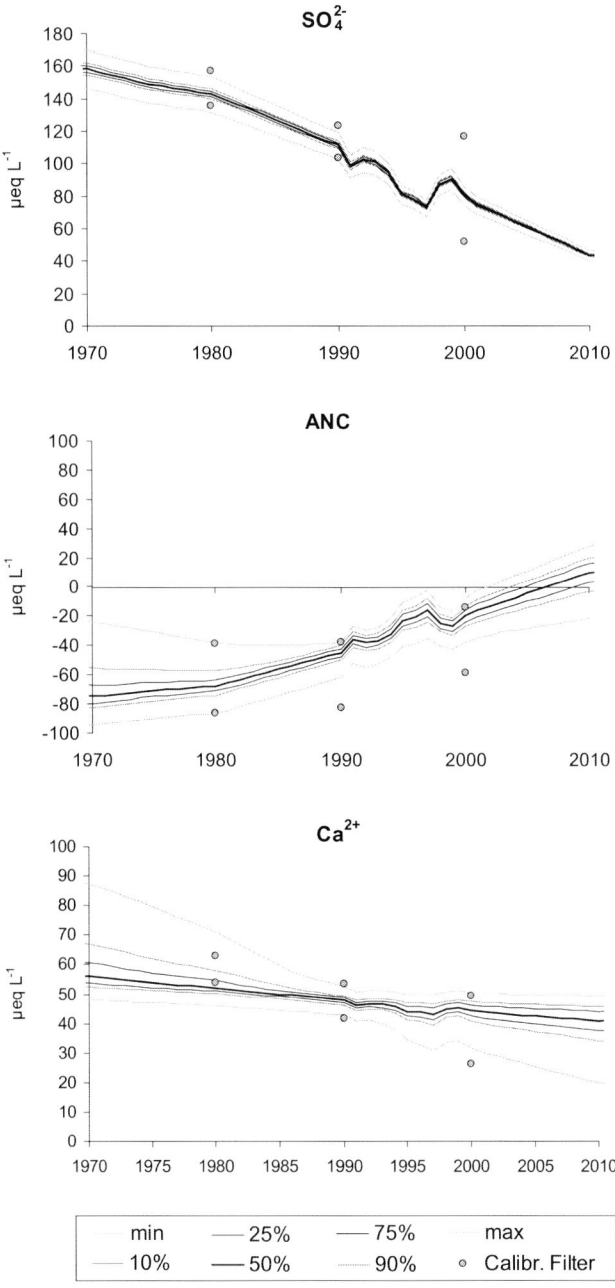

Figure 2. Comparison of limits used for calibration, post calibration filtering and modelled distributions of SO_4^{2-} concentrations (upper panels), ANC (middle panels) and Ca^{2+} concentration (lower panels). The limits used in calibration are circles and the modelled values are shown as lines for different percentiles of the modelled distribution. The left panels show results when using only the time period around 1990 for model calibration, while the right panels show the results after refinements using data for three discrete points in time.

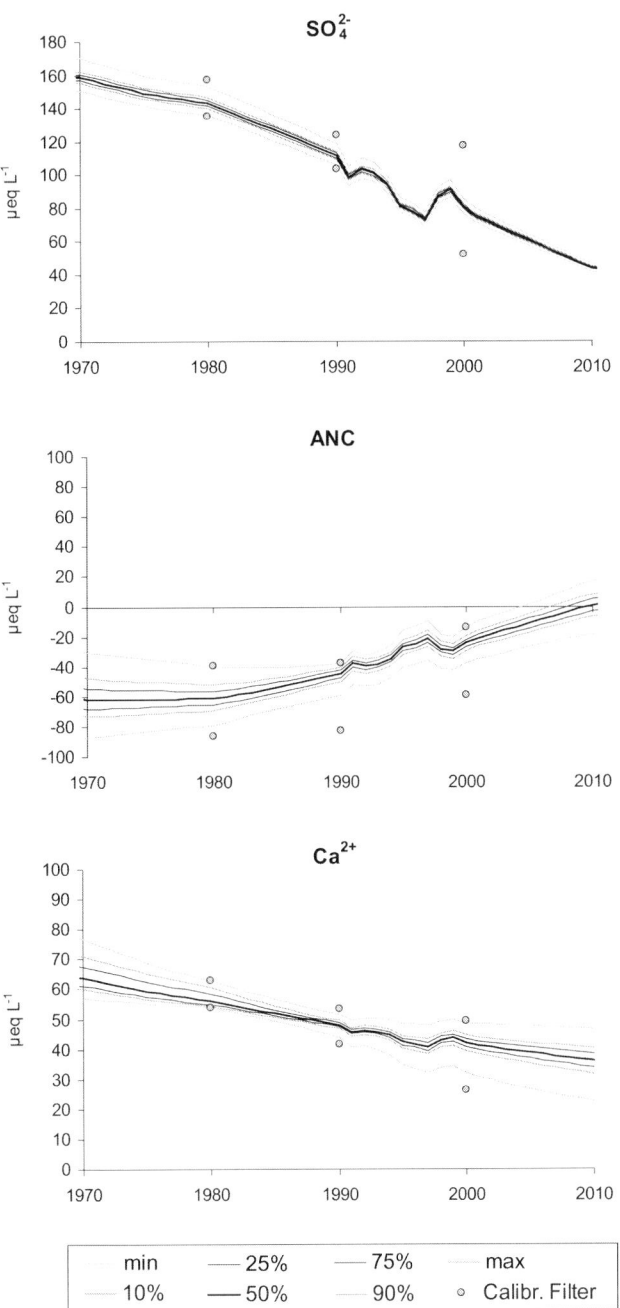

Figure 2. (Continued).

2.4. POST CALIBRATION REFINEMENT

For the period from 1991 to 2000, a model forecast was run driven by the observed annual deposition. The observed stream water chemistry average for 1996–2000 (denoted as 2000 in the following) was used to set a third window for refinement of the input distributions. After calibration to 1990 values, the accepted inputs were further refined based on observations around 1980 (1978–1982) and 2000 (1996–2000). This was done in a stepwise manner to understand how the prediction distributions changed as additional criteria were applied. ANC (acid neutralising capacity; the sum of base cations minus the sum of acid anions) for 1990 was used as an additional criterion to the calibration to the separate ions. SO_4^{2-}, Ca^{2+} and ANC were used as criteria for 1980 and 2000. The average for the relevant 5 yr periods, plus and minus 1.96 times the standard deviation, were used as upper and lower limits (Table V).

2.5. FORECAST DEPOSITION SCENARIO

As a forecast scenario, a reduction in S deposition in 2010 at 30% of the 2000 level was assumed. This is approximately the reduction in anthropogenic S deposition expected after implementation of the Gothenburg protocol. For N and the base cations the deposition was kept constant at their 1990 levels.

3. Results and Discussion

Of the original 10 000 sets of inputs, 8891 sets calibrated to match the 5 yr average around 1990. When adding the criteria to fit within the window for ANC, the number of sets was reduced to 8147. All of the sets that were rejected when applying this filter had ANC higher than the window (Table IV). This means that although all separate ions fit within their respective windows, their combination tended to give too high ANC, i.e. the base cations were skewed toward being higher than the observed mean or the acid anions skewed toward lower values than the observed mean.

When applying additional filters for ANC, SO_4^{2-} and Ca^{2+} in 1980, underestimation of Ca^{2+} concentration occurred for more than half of the parameter sets. This means that when only the one point in time was used for calibration, the slope of the decreasing Ca^{2+} concentration was commonly not met. When the filters for 2000 were applied the number of successful calibrations was reduced slightly further, but most of the refinement was done through the 1980 filter (Table VI).

The modelled development of ANC, SO_4^{2-} and Ca^{2+} concentrations are shown for the case where only the observations around 1990 were used as window, and the case where all three windows were used (Figure 2). The observed trends show substantial changes in the surface water chemistry with decreased concentrations of SO_4^{2-} and Ca^{2+} and a corresponding increase in the ANC. The response of the

TABLE VI

Number of sets of inputs falling outside the limits given in Table V and hence successively rejected in the post calibration refinement procedure. A total of 10 000 sets were run; of these 8891 calibrated and 2234 passed all three filters

	1990	Adding 1980 filters			Adding 2000 filters		
	ANC	SO_4^{2-}	ANC	Ca^{2+}	SO_4^{2-}	ANC	Ca^{2+}
Too high	744	0	2	190	0	14	2
Too low	0	7	1	5697	0	0	0

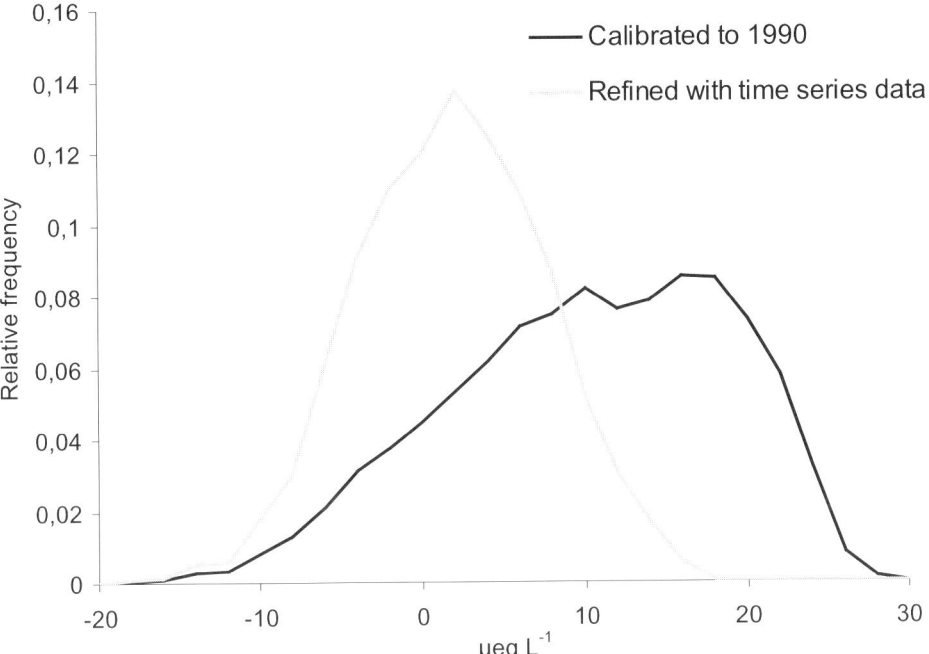

Figure 3. Distributions of predicted ANC in 2010 following the Gothenburg Protocol scenario. The solid black line shows the distribution when calibrating to only one point in time (5 yr average around 1990); the grey line shows the distribution using also the post calibration filters.

Ca^{2+} concentration to the change in the SO_4^{2-} concentration is dependent on the calibrated ion exchange constants. The post-calibration filters led to a narrower distribution in the modelled concentrations (Figure 2). Similarly, the distribution of the predicted ANC in 2010 is narrower when the 1980 and 2000 criteria are applied (Figure 3).

In addition, another interesting and less obvious feature is the shift in the distribution toward lower ANC values when the post-calibration filters were applied

(Figure 3). The median of the predicted ANC for 2010 is 10.5 μeq L^{-1} for the calibrated data set and 1.0 μeq L^{-1} when the post-calibration filters were applied. Hence, one would possibly make a too positive prediction if only using the one point in time for calibration. Again, the difference relates to the fact that many of the predicted Ca^{2+} concentrations for 1980 were lower than the observed value, and only the input parameter sets giving a steeper slope for the Ca^{2+} concentration (and hence lower slope for ANC) were accepted by all the filters.

The shift in the predicted distributions for 2010 illustrates the importance of time series data. For this particular catchment the predicted ANC is close to zero, and the differences between the two predictions for 2010 are important.

The results address the uncertainties in dynamic acidification modelling in a probabilistic manner. The outputs can be addressed in terms of probability of reaching a given effect-based target. The implication of the results for policymaking using this methodology could be to present the results as a probability of reaching, for example, an ANC of 10 μeq L^{-1} by 2010. The answer to this question would be about 50% with the distribution based only on the 1990 data, but only about 5% when the time-series-constrained data are used. The probability of reaching the goal would obviously increase with increased deposition reduction. Different emission reduction scenarios may in this way be analysed in a probabilistic way in the context of reaching a water quality target.

Acknowledgements

Julian Aherne is acknowledged for constructive comments on the manuscript. The work was carried out with financial contributions from the Norwegian Research Council Grant 140401/720, and the Norwegian Directorate for Nature Management Grant 01040648.

References

Aas, W., Tørseth, K., Solberg, S., Berg, T., Manø, S. and Yttri, K. E.: 2001, 'Overvåking av langtransportert forurenset luft og nedbør. Atmosfærisk tilførsel 2000', *Rapport 828/2001*, Statens forurensningstilsyn, Oslo, Norge.

Bull, K. R., Achermann, B., Bashkin, V., Chrast, R., Fenech, G., Forsius, M., Gregor, H.-D., Guardans, R., Haussmann, T., Hayes, F., Hettelingh, J. P., Johannessen, T., Krzyzanowski, M., Kucera, V., Kvæven, B., Lorenz, M., Lundin, L., Mills, G., Posch, M., Skjelkvåle, B. L. and Ulstein, M. J.: 2001, 'Coordinated effects monitoring and modelling for developing and supporting international air pollution control agreements', *Water, Air, Soil Pollut.* **130**, 119–130.

Cosby, B. J., Ferrier, R. C., Jenkins, A. and Wright, R. F.: 2001, 'Modelling the effects of acid deposition: Refinements, adjustments and inclusion of nitrogen dynamics in the MAGIC model', *Hydrol. Earth System Sci.* **5**, 499–518.

Cosby, B. J., Hornberger, G. M., Galloway, J. N. and Wright, R. F.: 1985, 'Modelling the effects of acid deposition: Assessment of a lumped parameter model of soil water and streamwater chemistry', *Water Resour. Res.* **21**, 51–63.

Hornberger, G. M., Cosby, B. J. and Galloway, J. N.: 1986, 'Modeling the effects of acid deposition: Uncertainty and spatial variability in estimation of long-term sulfate dynamics in a region', *Water Resour. Res.* **22**, 1293–1302.

Hornberger, G. M., Cosby, B. J. and Wright, R. F.: 1989, 'Estimating uncertainty in long-term reconstructions', in J. Kämäri, D. F. Brakke, A. Jenkins, S. A. Norton and R. F. Wright (eds.), *Regional Acidification Models*, Springer-Verlag, Berlin, pp. 279–290.

SFT: 1993, 'Overvåking av langtransportert forurenset luft og nedbør, Årsrapport 1992', *Statlig program for forurensningsovervåking Rapport 533/93*, Statens forurensningstilsyn, Oslo, Norway, pp. 296.

SFT: 2001, 'Overvåking av langtransportert forurenset luft og nedbør. Årsrapport – Effekter 2000', *Statlig program for forurensningsovervåking Rapport 834/01*, Statens forurensningstilsyn, Oslo, Norway, pp. 197.

UN-ECE: 2002, 'Convention on long-range transboundary air pollution', Executive Body for the convention. Informal document No. 1 for the Eighteenth Session, 28 November–1 December 2000, http://www.unece.org/env/eb/.

Warfvinge, P., Holmberg, M., Posch, M. and Wright, R. F.: 1992, 'The use of dynamic models to set target loads', *Ambio* **21**, 369–376.

Wright, R. F.: 2001, 'Use of the dynamic model 'MAGIC' to predict recovery following implementation of the Gothenburg protocol', *Water, Air, Soil Pollut.: Focus* **1**, 455–482.

MODELLING ACIDIFICATION AND RECOVERY OF SWEDISH LAKES

FILIP MOLDAN[1]*, VERONIKA KRONNÄS[1], ANDERS WILANDER[2],
ERIK KARLTUN[3] and BERNARD J. COSBY[4]

[1] Swedish Environmental Research Institute (IVL), Box 47 086, SE-402 58 Göteborg, Sweden;
[2] Department of Environmental Assessment, Swedish University of Agricultural Sciences, Box 7050, SE-750 07 Uppsala, Sweden; [3] Department of Forest Soils, Swedish University of Agricultural Sciences, Box 7001, SE-750 07 Uppsala, Sweden; [4] Department of Environmental Sciences, University of Virginia, Charlottesville, VA 22901 U.S.A.
(* author for correspondence, e-mail: filip.moldan@ivl.se; phone: +46 - 31-725 62 31; fax: +46-31-725 62 90)

(Received 20 August 2002; accepted 21 April 2003)

Abstract. The MAGIC model was calibrated to 143 lakes in Sweden, all of which are monitored in Swedish national monitoring programmes conducted by the University of Agricultural Sciences (SLU). Soil characteristics of the lake catchments were obtained from the National Survey of Forest Soils and Vegetation also carried out by SLU. Deposition data were provided by the Swedish Meteorological and Hydrological Institute (SMHI). The model successfully simulated the observed lake and soil chemistry at 133 lakes and their catchments. The fact that 85% of the lakes calibrated successfully without being treated in an individual way suggests that data gathered by the national monitoring programmes are suitable for modelling of soil and surface water recovery from acidification. The lake and soil chemistry data were then projected into the future under the deposition scenario based on emission reductions agreed in the Gothenburg protocol. Deposition of sulphur (sea salt corrected) was estimated to decrease from 1990 to 2010 by 65–73%; deposition of nitrogen was estimated to decrease by 53%. The model simulated relatively rapid improvements in lake water chemistry in response to the decline in deposition from 1990 to 2010, but the improvements levelled off once deposition stabilised at the lower value. There was a major improvement of simulated lake water charge balance acid neutralising capacity (ANC) from 1990 to 2010 in all lakes. The modelled lakes were divided into acidification sensitive and non-sensitive. The modelled sensitive lakes are representative of 20% of the most sensitive lakes in Sweden. By 2010, the ANC in the sensitive lakes was 10 to 50 μeq L^{-1} below estimated pre-industrial levels and did not increase much further from 2010 to 2040. Soils at the majority of the modelled catchments continued to lose base cations even after the simulated decline in acid deposition was complete, i.e. after the year 2010. Based on this model prediction, the acidification of the Swedish soils will in general not be reversed by the deposition reduction experienced over the last 10 years and expected to occur by the year 2010.

Keywords: acidification, dynamic modelling, lake, MAGIC, regional modelling, soil, Sweden

1. Introduction

There are many lakes in Sweden. About 9% of the country is covered by lakes. There are about 85,000 lakes with a surface area larger than 0.01 km^2. National surveys of water quality in the lakes started in 1972 (Bernes, 1986) and have been repeated five times since. In Sweden, deposition of sulphur (S) and nitrogen (N) peaked in the 1970s and remained high during the 1980s (Mylona, 1996). Between 1990 and 1998, deposition of S has decreased by about 50% and deposition of N by about 20%. Acidification of soils and surface waters was expected to slow down, stop or even reverse due to the much lower deposition. Since 1976, there has been a documented decline of sulphate (SO_4^{2-}) concentrations in lakes in southern Sweden (Forsberg et al., 1985). Bernes (1991) confirmed this result on a countrywide basis. Neither study found unequivocal evidence that declining SO_4^{2-} was accompanied by an increase in alkalinity or pH. Wilander (1997) assessed trends in lake water chemistry of 140 lakes between 1983 and 1994, showing increasing alkalinity for the first time. In 70% of the investigated lakes, the SO_4^{2-} decrease was accompanied by increasing alkalinity, although the increase in alkalinity was much smaller than the decrease in SO_4^{2-}. Wilander and Lundin (2000) reassessed the trends in lake chemistry of 140 lakes using data up to 1997. In 96% of the investigated lakes, SO_4^{2-} concentration declined over the period studied. The median change was 2 μeq L^{-1} a^{-1}. At the same time, the concentrations of base cations decreased in 87% of the lakes, with a median change that was approximately the same as that of SO_4^{2-}, i.e. 2 μeq L^{-1} a^{-1}. Therefore, the charge balance ANC has increased by much less and in fewer lakes (53%) (Wilander and Lundin, 2000). There was, however, an increase in pH in 85% of the lakes. The median change was 0.1 pH unit over the 10 years. These results are consistent with assessments of the lake chemistry trends in Finland, Norway and Sweden between 1990 and 1999 (Skjelkvåle et al., 2001). In the three Scandinavian countries, SO_4^{2-} concentration decreased in 69% of monitored lakes, non-marine base cations decreased in 26%, and ANC increased in 32%.

Monitoring of changes in soil properties is more difficult than monitoring changes in precipitation or lake water chemistries. This is because soils are more spatially heterogeneous and the expected rate of change in soil properties, such as soil base saturation, soil acidity, or the carbon/nitrogen (C/N) ratio, is relatively slow. However, soils in Sweden have acidified (Karltun, 1995) and acidification of surface waters and groundwater and effects on organisms are to a large extent consequences of the soil acidification in their catchments. Complete reversal of acidification in the surface waters is possible only if soil acidification is reversed. The National Survey of Forest Soils and Vegetation (Ståndortskarteringen, SK) carried out by the Swedish University of Agricultural Sciences (SLU) is a systematic forest soil inventory collecting soil samples for environmental monitoring purposes. Based on sampling of 23,500 soil plots since 1983, SK describes the current status of soils in Sweden.

The availability of these extensive studies permits an extrapolation of the observed situation into the future. Here we present the implementation of the MAGIC model and the simulation of lake water and soil chemistry up to the year 2040 under the currently agreed upon emission reductions within the Gothenburg protocol (UN/ECE, 1999). In total, 143 lakes included in a national monitoring programme carried out by SLU have been modelled (Figure 1).

Figure 1. Map of Sweden with soil sampling sites from SK, time-series lakes from SLU and three geographical regions.

2. Methods

2.1. MODEL DESCRIPTION

MAGIC is a lumped-parameter model of intermediate complexity, developed to predict the long-term effects of acidic deposition on surface water chemistry. The model simulates soil solution and surface water chemistry to predict the monthly and annual average concentrations of the major ions in these waters. MAGIC consists of: 1) a section in which the concentrations of major ions are assumed to be governed by simultaneous reactions involving SO_4^{2-} adsorption, cation exchange, dissolution-precipitation-speciation of aluminium (Al) and dissolution-speciation of inorganic carbon (C), and 2) a mass balance section in which the flux of major ions to and from the soil is assumed to be controlled by atmospheric inputs, chemical weathering, net uptake and loss in biomass and losses to runoff.

At the heart of MAGIC is the size of the pool of exchangeable base cations in the soil. As the fluxes to and from this pool change over time owing to changes in atmospheric deposition, the chemical equilibria between soil and soil solution shift to give changes in surface water chemistry. The degree and rate of change of surface water acidity thus depend both on flux factors and the inherent characteristics of the affected soils.

Cation exchange is modelled using equilibrium (Gaines-Thomas) equations with selectivity coefficients for each base cation and Al. Sulphate adsorption is represented by a Langmuir isotherm. Aluminium dissolution and precipitation is assumed to be controlled by equilibrium with a solid phase of Al trihydroxide. Aluminium speciation is calculated by considering hydrolysis reactions as well as complexation with SO_4^{2-} and fluoride. Effects of carbon dioxide on pH and on the speciation of inorganic C are computed from equilibrium equations. Organic acids are represented in the model as triprotic analogues. First-order rates are used for retention (uptake) of nitrate (NO_3^-) and ammonium (NH_4^+) in the catchment. Weathering rates are assumed to be constant (for details of the model, see Cosby *et al.*, 1985a–c, 2001).

2.2. MODEL IMPLEMENTATION

Atmospheric deposition and net uptake-release fluxes for the base cations and strong acid anions are required as inputs to the model. The volume lake discharge for the catchment, values for soil and lake temperature, partial pressure of carbon dioxide in the soil, and lake water and organic acid concentrations in soil water and lake water also must be provided to the model.

As implemented in this project, the model is a two-compartment representation of a lake catchment. Atmospheric deposition enters the soil compartment (and the lake compartment) and the equilibrium equations are used to calculate soil water chemistry. The water is then routed to the lake compartment, and the appropriate equilibrium equations are reapplied to calculate lake water chemistry. The model is implemented using average hydrologic conditions and meteorological conditions; median annual deposition, precipitation and lake discharge are used to drive the model.

The aggregated nature of the model requires it to be calibrated to observed data for each of the modelled lakes before it can be used to examine potential system response for that lake. Calibration is achieved by setting the values of certain parameters in the model (called 'fixed' parameters). The model is then run (using observed atmospheric and hydrologic inputs) and the outputs (lake water and soil chemical variables, called 'criterion' variables) are compared to observed values of these variables. If the observed and simulated values differ, the values of another set of parameters in the model (called 'optimised' parameters) are adjusted to improve the fit. After a number of iterations, the simulated-minus-observed values of the criterion variables usually converge to zero (within some specified tolerance). The model is then considered calibrated for that system. In this, study the model was calibrated to soil and lake water chemistry from 1997.

2.3. INPUT DATA

2.3.1. *Deposition*
The 1997 deposition of calcium (Ca^{2+}), magnesium (Mg^{2+}), sodium (Na^+), potassium (K^+), NO_3^-, NH_4^+, SO_4^{2-}, and chloride (Cl^-) have been estimated using the MATCH model (Robertson *et al.*, 1999, www.smhi.se) in a 20 x 20 km square grid across Sweden. Data on total deposition given as mass per area were divided by the precipitation volume to calculate rainwater concentrations. There were no data on dry deposition of Cl^- available, so data on dry and wet deposition of Na^+ and data on wet deposition of Cl^- were used to make a first approximation the dry deposition of Cl^-. The ratio between dry and wet deposition was assumed to be the same for Na^+ and Cl^-.

Modelled deposition provided by the SMHI was adjusted using the observed lake water chemistry to account for the local variation within the 20×20 km squares. The total deposition of Cl^- was adjusted at each site using lake water chemistry, so that the input flux of Cl^- was equal to the output flux at each lake (no sink or source of Cl^- in any of the catchments is assumed). Base cations, Na^+, K^+, Mg^{2+} and Ca^{2+}, were added or subtracted along with the Cl^- in sea salt proportions. The modelled SO_4^{2-} deposition was adjusted using the measured lake water chemistry to account for variations in dry deposition of SO_4^{2-}. It was assumed that, as a result of the declining SO_4^{2-} deposition in preceding years, part of the SO_4^{2-} in the output flux of the lakes in 1997 had been desorbed from catchment soils or, in lakes with large retention time, it originated from the lake water itself. Since the deposition of S was decreasing, this release from storage would make the output flux of S larger than the influx in 1997 by an estimated 35%. The estimated output flux excess of 35% was derived by requiring that the adjustments in deposition of SO_4^{2-} at each site that resulted from this correction procedure did not change the average SO_4^{2-} deposition over all the lakes or the trend in SO_4^{2-} deposition from the south to the north of Sweden. It also corresponds well with the SO_4^{2-} desorption of about 35%, which has been calculated at the monitored catchment at Gårdsjön, SW Sweden (Moldan and Ek, 2000). The modelled deposition of N species was adjusted to account for variations in dry deposition by assuming that the ratio between the adjusted deposition and the deposition given by SMHI was the same for the N species and SO_4^{2-} at each lake.

The adjusted 1997 deposition at each site was then scaled both to the past and to the future. Past deposition was calculated by scaling 1997 deposition to the history of emissions in Europe (Mylona 1996; Simpson *et al.*, 1997). The future deposition scenario was based on RAINS model calculations carried out by The

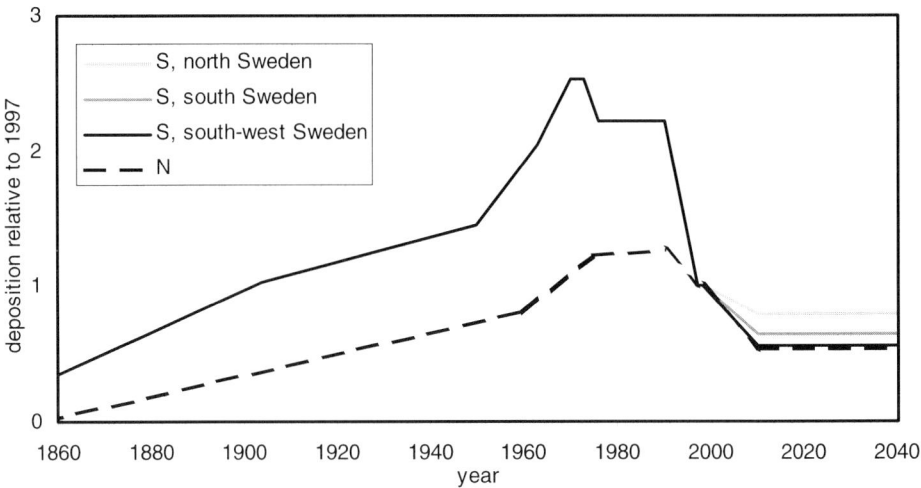

Figure 2. Deposition sequences of sulphur and nitrogen.

International Institute for Applied Systems Analysis (IIASA) using the predicted emissions according to agreements in the Gothenburg protocol (Gun Lövblad, IVL, pers. comm.). The country was divided into three regions, north, south, and south-west, adapted from Nordic Council of Ministers (1984) (Figure 1). The relative change in the future deposition of S was 56% of the 1997 level in the south-western region, 65% in the south, and 73% in the north (Figure 2).

2.3.2. *Soils*

The soil sampling sites from SK are distributed throughout Sweden. The variation in the soil parameters is greater in southern Sweden than in northern Sweden (Figure 1). This is reflected in the survey design by a larger number of sites per area in the south of Sweden than in the north. The soil sampling sites are generally not located within the catchments of the modelled lakes. The soil parameters used in the modelling are averages of parameters from soil sampling sites within a distance S to each lake, where S depends on the north coordinate X. Care must be taken that the distance S is neither too small (no soil pits included in the average for a lake) or too large (soil pits included in the average that are remote from the lake, and therefore not representative). An analysis of the SK data and their distribution suggested that S could be most reliably estimated as (Kronnäs *et al.*, 2002):

$$S(X) = \frac{S_0 \left.\frac{dN}{dX}\right|_{X=6.8\cdot 10^6}}{-6\cdot 10^{-13} X + 5\cdot 10^{-6}} = \frac{S_0 \cdot 9.2 \cdot 10^{-7}}{-6\cdot 10^{-13} X + 5\cdot 10^{-6}} \quad (1)$$

where S_0 is S for the middle of Sweden ($X = 6,800,000$ m). For the chosen S_0 of 25,000 m, the soil parameters are averages of approximately 15 soil sampling sites for each lake, regardless of where in Sweden it is.

Soil depth, amount of exchangeable Ca^{2+}, Mg^{2+}, Na^+ and K^+ per mass of soil, CEC, soil pH, and amount of C and N were vertically aggregated for the profiles of each soil sample included for a lake. Soil bulk densities were estimated by Karltun (1995) and averaged over the profiles. The values of these vertically aggregated soil parameters for each lake were then averaged across all soil profiles within distance S of the lake to provide inputs to MAGIC. Soil porosity was assumed to be 50% for all sites, soil temperature was assumed to be the same as lake water temperature, and soil water dissolved organic carbon (DOC) was assumed to be 8 mg L^{-1} for all catchments (based on data from monitoring in permanent forest plots in Sweden, ICP – level II).

Forest uptake of N was assumed to be directly proportional to N deposition, with the proportionality determined by observed input and output fluxes. For most lake catchments, the N uptake was complete, since there was no N leaching. For each catchment, the productivity of the forest and the fractions of spruce and pine were estimated from maps in National Atlas of Sweden (1996). Those productivity values were used to estimate the forest uptake of Ca^{2+}, Mg^{2+} and K^+ using cation concentrations in forest biomass (Jacobson and Mattsson, 1998). Some of the modelled lakes are situated in areas with very low productivity and for those forests the uptake had to be extrapolated. Forest uptake was assumed to be constant over time, i.e. the forests in the catchments were modelled as extensively managed so that the average age of the stand is approximately constant. The forest uptake of Na^+ and SO_4^{2-} was considered negligible compared to input.

2.3.3. *Lakes*

When the monitoring programme was designed, the lakes were selected to represent all types of lakes that commonly occur in Sweden, especially with respect to total organic carbon (TOC) concentrations, acidity, size and geographical position (SNV, 1993). Furthermore, the intensively monitored lakes are situated in areas with no significant agricultural activities and are generally not limed or polluted (e.g., by wastewater). The lakes have been sampled with regard to water chemistry (Ca^{2+}, Mg^{2+}, Na^+, K^+, SO_4^{2-}, Cl^-, ANC, NH_4^-, NO_3^-, TOC, and pH) several times per year since 1984. For modelling purposes, DOC was assumed equal to the measured TOC. Dissociation of the organic acids was calculated according to a triprotic dissociation model (Hruška *et al.*, 2001). Lake water temperature was estimated using both measurements from SLU and air temperatures from SMHI.

Catchment area, lake area, and specific runoff for 114 of the lakes were provided by SLU. For those 29 lakes lacking runoff data, averages of the runoff of geographically nearby lakes were used. Values of lake volume for 72 of the lakes were obtained from SMHI (www.smhi.se). SLU provided lake volumes for another 9 of the modelled lakes. Lake volumes were calculated from the average depth for another 28 lakes. From these known volumes, two linear regression equations against lake area were made, one for large lakes (area > 5 km^2) and one for smaller.

Lake volumes were approximated for the 34 lakes with unknown volumes. The regression equations were:

$$V = 6.75 \cdot A \qquad A \leq 5 \text{ km}^2 \ R^2 = 0.72 \qquad (2)$$

$$V = 9.74 \cdot A - 1.98, \qquad A > 5 \text{ km}^2 \ R^2 = 0.90 \qquad (3)$$

where V is the volume in 10^6 m^3. The average residence times of water in the lakes were calculated from catchment area, runoff and lake volume. Because high lake retention times might mean that in-lake processes (as opposed to the processes on the terrestrial parts of the catchments) would be more important in mediating the lake chemistry responses to changing deposition, only those lakes with retention times shorter than five years were modelled (Figure 1).

The model calibration and reference year was 1997. For lakes actually sampled in 1997, the median observed values of lakewater chemistry from 1997 were used for calibration. For 29 lakes, the measurements ended in 1995. For those lakes, linear trends were calculated for each water quality variable in each lake and estimated 1997 values were extrapolated from the trends. Similar extrapolations also were made for the lakes that had measurements in 1997, and the ratio was formed between the estimated 1997 value and the measured 1997 value for each variable and each lake. The average of these ratios for each variable over all lakes was then used to adjust the estimated values of the 29 lakes, to account for the yearly variation of 1997. The predicted value of a variable for which no observation was available in 1997 is thus:

$$C_{1pred}(T) = \tfrac{1}{\overline{k}} * C_{1trend}(T) = \overline{\tfrac{C_{2obs}(T=1997)}{C_{2trend}(T=1997)}} * \left(\left(\tfrac{\sum C_{1obs}}{n} - b \tfrac{\sum T}{n} \right) + bT \right)$$

$$\text{where} \quad b = \tfrac{n \sum T C_{1obs} - (\sum T)(\sum C_{1obs})}{n \sum T^2 - (\sum T)^2}$$

(4)

where T is the year, C_2 is a variable value from a lake that has measurements in 1997, C_1 is a variable value from one of the lakes that has no measurements in 1997, and n is the number of C_1-observations. The bar denotes averaging of k over all lakes with measurements in 1997.

The model was calibrated to the year 1997 (Table I). Lake and soil chemistry of each individual lake and its catchment was simulated from 1857 to 2040.

3. Results and Discussion

3.1. COMPARISON OF ADJUSTED MAGIC INPUT DATA AND ORIGINAL DATA

An analysis that requires the application of MAGIC (or any model) to a large number of sites will invariably encounter the problem of missing or inadequate data at some (or most) sites. As described above, procedures can be adopted to

TABLE I

MAGIC model parameters after optimisation

Parameter	Average value	Standard deviation	Units	Parameter	Average value	Standard deviation	Units
Fixed parameters				**Evaluation data, 1997**			
Soil depth	0.856	0.160	m	Exchangeable Ca^{2+} in soil	19.24	9.65	%
Porosity	50	0	%	Exchangeable Mg^{2+} in soil	6.18	2.78	%
Bulk density of soil	800	194	kg/m^3	Exchangeable Na^+ in soil	0.75	0.52	%
CEC in soil	69	50	meq kg^{-1}	Exchangeable K^+ in soil	2.77	0.98	%
DOC in soil	27	0	μeq L^{-1}	Base saturation in soil	30.29	11.57	%
Soil temperature	6.10	1.79	°C	Soil pH	5.40	0.42	–
Half saturation SO_4^{2-}	100	5.9	μeq L^{-1}	Lake concentration of Ca^{2+}	188	152	μeq L^{-1}
Max sulphate adsorption	0.79	1.05	meq kg^{-1}	Lake concentration of Mg^{2+}	86.5	48.3	μeq L^{-1}
Weathering rate of Ca^{2+}	61.6	53.3	meq m^{-2} a^{-1}	Lake concentration of Na^+	146	108	μeq L^{-1}
Weathering rate of Mg^{2+}	18.1	15.2	meq m^{-2} a^{-1}	Lake concentration of K^+	14.5	7.4	μeq L^{-1}
Weathering rate of Na^+	13.0	6.2	meq m^{-2} a^{-1}	Lake concentration of NH_4^+	1.32	1.37	μeq L^{-1}
Weathering rate of K^+	4.03	2.72	meq m^{-2} a^{-1}	Lake concentration of SO_4^{2-}	126	82	μeq L^{-1}
Selectivity coeff., Ca^{2+}	4.87	3.08	–	Lake concentration of Cl^-	133	120	μeq L^{-1}
Selectivity coeff., Mg^{2+}	5.34	3.01	–	Lake concentration of NO_3^-	3.64	4.68	μeq L^{-1}
Selectivity coeff., Na^+	2.29	1.86	–	Sum of base cations	435	259	μeq L^{-1}
Selectivity coeff., K^+	−2.41	1.71	–	Sum of acid anions	264	188	μeq L^{-1}
Runoff	0.384	0.162	m a^{-1}	ANC	184	184	μeq L^{-1}

TABLE I
continued

Parameter	Average value	Standard deviation	Units	Parameter	Average value	Standard deviation	Units
Relative area of lake	10.3	6.3	%	Lake pH	6.36	0.89	–
Retention time	1.56	1.23	Years				
Solubility of Al^{3+}	8.50	0.29	log10				
Lake temperature	6.1	1.8	°C				
CO_2 in lake	0.06	0	% atm				
DOC in lake	33.8	19.3	µeq L^{-1}				
Driving variables, 1997							
Precipitation	0.669	0.135	m a^{-1}	Forest uptake of Ca^{2+}	8.00	5.23	meq m^{-2} a^{-1}
Forest uptake of K^+	2.52	1.09	meq m^{-2} a^{-1}	Forest uptake of Mg^{2+}	1.16	0.74	meq m^{-2} a^{-1}
Concentrations in precipitation of:							
Ca^{2+}	10.34	2.84	µeq L^{-1}	NH_4^+	38.7	27.5	µeq L^{-1}
Mg^{2+}	15.77	13.1	µeq L^{-1}	SO_4^{2-}	47.2	26.5	µeq L^{-1}
Na^+	59.5	55.9	µeq L^{-1}	Cl^-	69.3	65.1	µeq L^{-1}
K^+	4.81	2.38	µeq L^{-1}	NO_3^-	39.3	25.1	µeq L^{-1}

estimate missing values or to adjust incomplete or inadequate data. Obviously, it is hoped that these procedures produce correct and unbiased estimates or corrections. To examine this question, the representativity for this application of MAGIC, the adjusted input data used for the modelling were compared to the original dataset.

3.1.1. *Deposition*

The built-in mass balance of all major solutes in the MAGIC model requires consistency of the catchment inputs (deposition and weathering) with the outputs (uptake to vegetation and runoff) and with the change in the size of pools of solutes in the soil, lake, and in vegetation. For that reason, the regionally estimated atmospheric deposition often had to be adjusted to the local conditions reflected in the measured lake water composition (see above). The adjusted site-specific deposition data have a larger variation than the averaged data from SMHI with a 20 km x 20 km resolution (Figure 3). Furthermore, the adjusted site-specific deposition of SO_4^{2-}, NO_3^-, NH_4^+, and base cations (BC) was on average higher than deposition in the corresponding 20 km × 20 km squares (estimated dry deposition of Cl^- was not available on 20 km × 20 km squares and therefore it could not be compared to the site-specific value calculated from the lake water Cl^- concentrations.) The higher deposition at the lake catchments could be a consequence of enhanced dry deposition because of relatively higher percentage of forest cover (65% forest cover at the lake catchments, compared to 54% for the whole country), because the lakes were selected such that they were preferably unaffected by agriculture. The largest differences were in the deposition of base cations (Figure 3). Base cations were added to or subtracted from the SMHI estimates at the sea salt ratios, when the deposition of Cl^- was calculated from the Cl^- concentrations in lakes. Lack of measured deposition at the sites, together with particular uncertainty involved in measuring or modelling the deposition of BC (Lövblad *et al.*, 2000), makes it difficult to compare the BC deposition used for the modelling with reality or to justify further adjustments. It should be noted that the deposition calculated by SMHI on the 20 km × 20 km square grids was on average slightly higher in the modelled areas than for the whole of Sweden. This is because the south of Sweden, where the deposition is highest, was over represented by selection of the modelled lakes.

3.1.2. *Soils*

The distributions of soil parameters are compared for: 1) the original 2900 SK sites with complete soil analysis, 2) the subset of selected 1600 SK sites used to calculate values for the lake catchments, and 3) the 143 calculated values used in the modelling. The spread in the distributions is approximately the same for the selected SK sites as for the whole of SK (Figure 4). One exception is soil pH, where the sites with the largest pH are not represented. The distributions of the soil parameters for the catchments of the 143 modelled lakes (calculated as averages of approximately 15 soil samples for each catchment) are smaller than

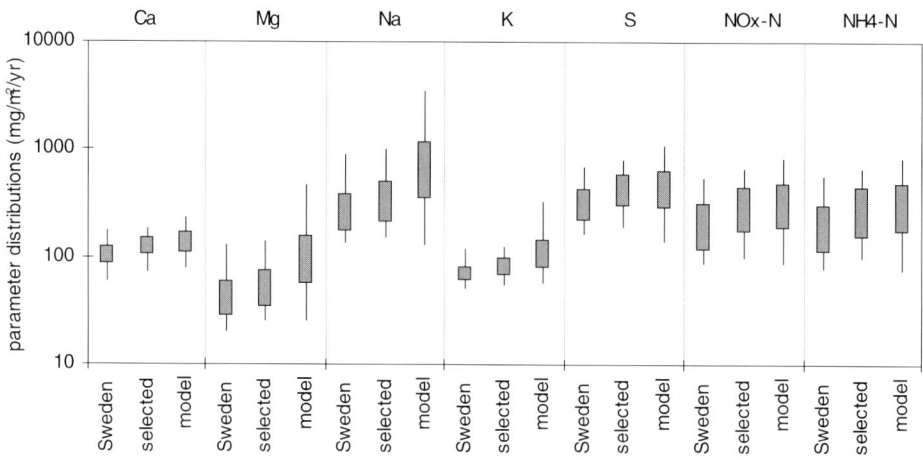

Figure 3. Distributions of deposition parameters, where *Sweden* denotes deposition values (SMHI) for the whole of Sweden, *Selected* denotes values of those squares (SMHI) where any of the modelled lakes is situated and *Model* denotes the values used in the modelling. The bars show the distribution between the 25- and 75-percentiles and the whiskers between the 5- and 95 percentiles.

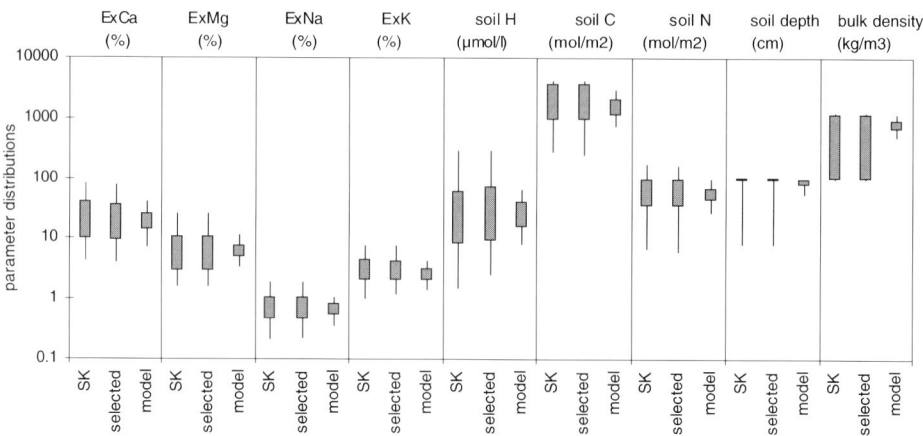

Figure 4. Distributions of soil parameter values at different sites, where *SK* denotes the whole SK, *Selected* denotes those sites of SK that were used to calculate values for lake catchments, and *Model* denotes the values calculated for the catchments. The bars show the distribution between the 25- and 75-percentiles and the whiskers between the 5- and 95 percentiles.

the distributions of 1600 selected SK sites. The fact that the distributions for the 2900 SK sites and the subset of 1600 SK sites are about equal, but the modelled catchment distributions are smaller, again shows that there is considerable variation of the soil parameters at a local scale.

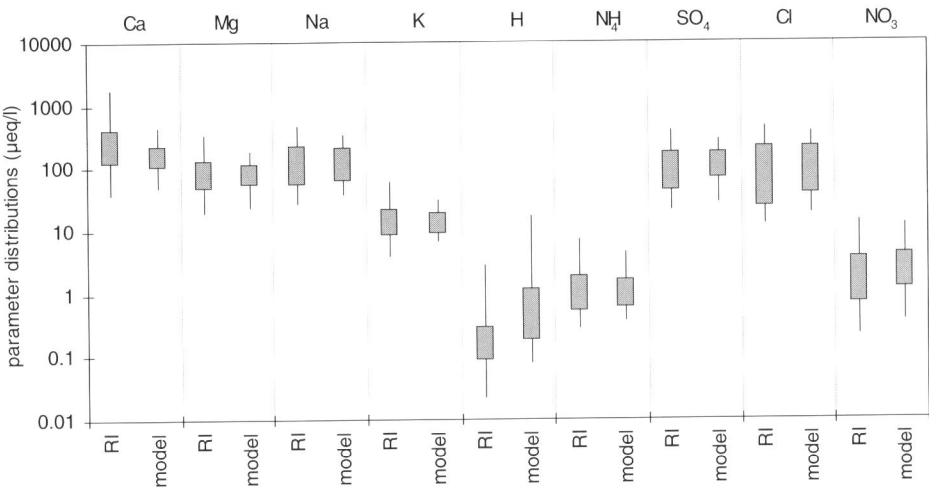

Figure 5. Distributions of lake parameter values for two sets of measured lakes, where *RI* denotes lakes in the National Surface Water Survey of 1995 (Riksinventeringen) and *Model* denotes the values for the modelled lakes. The bars show the distribution between the 25- and 75-percentiles and the whiskers between the 5- and 95 percentiles.

3.1.3. *Lakes*

Apart from the intensely monitored lakes modelled in this study, every fifth year several thousands of lakes are sampled by SLU in the National Surface Water Surveys, (Riksinventeringen, RI) (Wilander *et al.*, 1998). The lakes are randomly selected to represent all Swedish lakes. In 1995, 4,113 lakes were sampled. The distributions of selected water chemical parameters from RI and the distributions for the modelled lakes are reasonably comparable (Figure 5) except for pH, where the modelled lakes are more acid than the RI lakes. Such bias is to be expected since the former were selected to be lakes in forested areas and relatively susceptible to acidification.

On average, the 143 intensively monitored lakes had in 1995 a measured median pH of 6.3 and a median ANC of 128 μeq L^{-1}, while the 4113 lakes sampled the same year across the whole country had a median pH of 6.8 and a median ANC of 230 ueq L^{-1} (data from www.ma.slu.se). The median pH and ANC of the modelled lakes correspond to the 20th percentile values of the 4113 lakes, which were pH 6.4 and ANC 128 μeq L^{-1}. In other words, allowing for a certain degree of generalisation, 50% of the 143 modelled lakes (the low pH and low ANC lakes) are comparable to 20% of the most acidic lakes across the whole Sweden.

3.2. COMPARISON BETWEEN MODELLED AND MEASURED DATA

The results of the model calibration to the year 1997 data are shown in Figure 2. The tolerance for successful model calibrations was set for lake water concentrations as ± 10 μeq L^{-1} for SO$_4^{2-}$, ± 6 μeq L^{-1} for K$^+$ and NO$_3^-$, ± 4 μeq L^{-1}

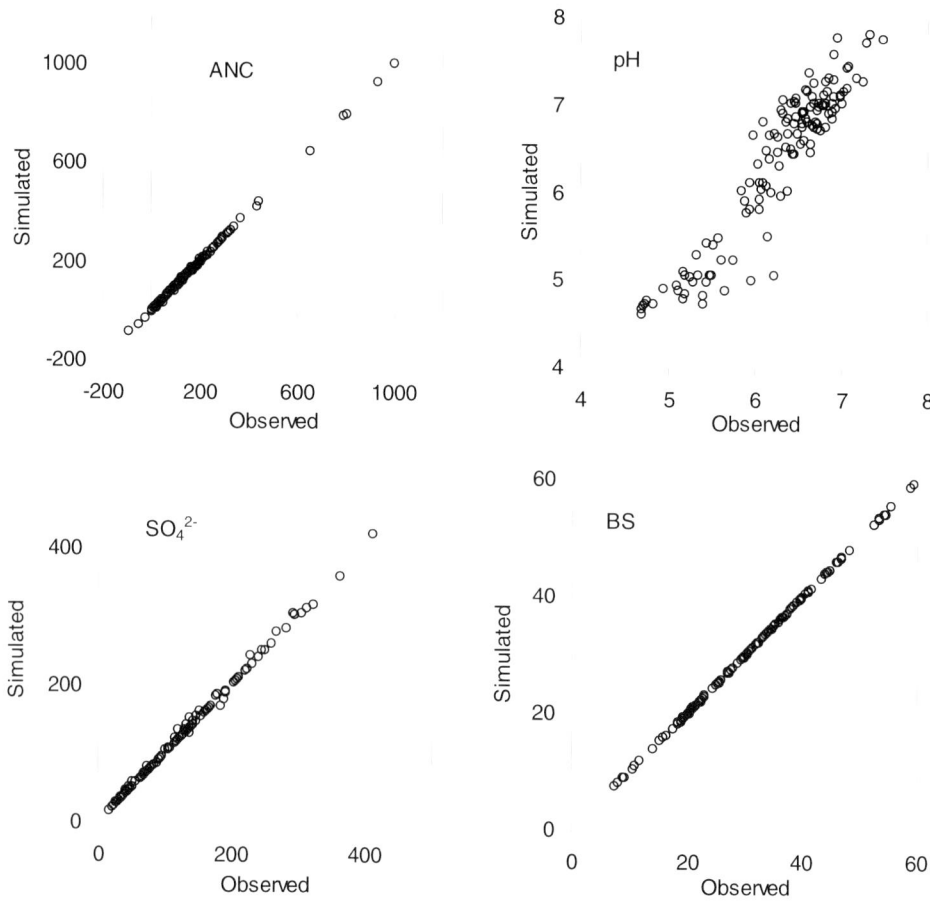

Figure 6. Simulated vs. observed values for 1997 of ANC, pH and SO_4^{2-} in the lakes and soil base saturation (BS) in the lake catchments.

for NH_4^+, ± 2 μeq L^{-1} for Ca^{2+}, Mg^{2+}, Na^+, and Cl^-, and ± 0.2% units for soil exchangeable stores of Ca^{2+}, Mg^{2+}, Na^+ and K^+. When provided with these criteria, the model predicted both lake water and soil chemistry very well for the calibration year 1997 in 133 cases (Figure 6). The remaining 10 lakes failed to calibrate without site specific adjustments different from the uniform procedures applied to the input data as described above.

Lake pH was not used for calibration. There was a generally good agreement between measured and modelled pH (Figure 6), although the model tended to underestimate pH at acid lakes and overestimate at more alkaline lakes. We decided not to attempt to improve the model performance with respect to pH because of the lack of data for measured Al concentrations in the lakes. The presence of Al interferes with the calculation of pH, especially in acid waters.

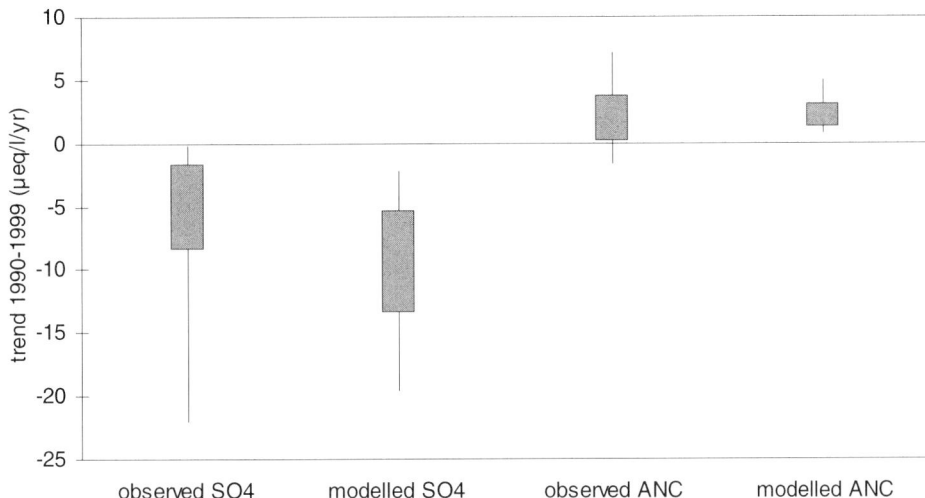

Figure 7. The 5, 25, 75 and 95 percentiles of trends (1990-1999) of ANC and SO_4^{2-} for the 63 lakes common to the modelling and the Skjelkvåle *et al.* (2001) investigation.

The Swedish lakes included in the recent Scandinavian trend analysis by Skjelkvåle *et al.* (2001) overlapped in 63 cases with lakes included in this MAGIC modelling exercise. Modelled and measured trends of lake water between 1990 and 1999 were compared for ANC and for SO_4^{2-} for these 63 lakes. There was both a modelled and a measured decrease in SO_4^{2-} concentrations and an increase of ANC in this group of lakes. The modelled decrease in lake water SO_4^{2-} and increase in ANC did not, however, fully cover the extremes of both minimum and maximum trends in the observed data. The modelled trends in SO_4^{2-} were on average slightly larger than observed. The trends in ANC were on average nearly the same (Figure 7).

3.3. GOTHENBURG PROTOCOL SCENARIO

The historical lake ANC (as simulated by MAGIC) has decreased (from 1860 to 1990) at practically all lakes (Figure 8). As a result of realised and planned efforts to control emissions, the model simulations included a major decrease in deposition of S between 1990 and 2010. As a result, there was an increase of ANC (Figure 8) in all modelled lakes over this period.

Soil base saturation has decreased historically from 1860 to 1990 at all modelled lake catchments (Figure 9) due to two causes: the leaching of base cations from the soils enhanced by the acidifying deposition and the historical patterns of land use, which was in most cases harvesting of spruce or pine plantations. However, in contrast to ANC, there was in general very little or no increase in soil BS between 1990 and 2010. Base saturation even continued to decrease at a number of modelled catchments.

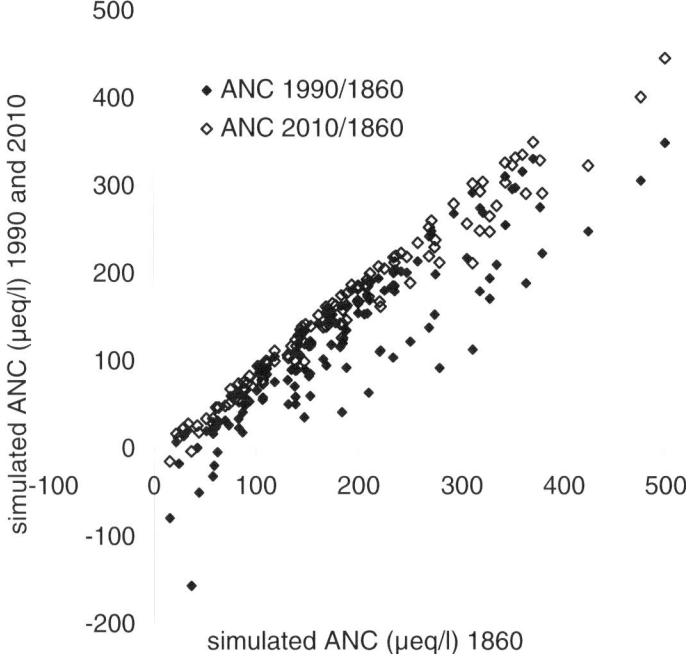

Figure 8. Simulated ANC in the lakes for 1997 and 2010 compared with simulated pre-industrial ANC (1860). Dots below the 1:1 line have acidified since 1860.

All of the 133 lakes have been affected by acid deposition, but many have never become heavily acidified (Figure 8). That is, however, not surprising considering the selection of the lakes modelled, which included all types of small lakes in Sweden. To investigate the outcome of the Gothenburg scenario specifically on sensitive lakes, the lakes were divided into two groups according to their measured ANC in 1995. The 'sensitive' lakes, with ANC <128 μeq L^{-1} (61 lakes), and 'non-sensitive' lakes with ANC >128 μeq L^{-1} (72 lakes). The 'sensitive' lakes should be approximately representative of 20% of most sensitive lakes in Sweden (see above); the 'non-sensitive' lakes are representative of the rest of the lake population with the exception of very large lakes.

The modelled 'non-sensitive' lakes have never experienced severe acidification and showed both a decrease in ANC during the historical acidification and an increase in ANC since the deposition peaked and started to decrease. This general pattern of change, however, was moderate, due either to the relatively low cumulative acid deposition or to the well-buffered surrounding soils (or to a combination of these two factors). The moderate acidification was at most of these lakes probably without any dramatic biological effects.

The changes in ANC at the 'sensitive' lakes were, however, in relative terms, much more pronounced both during the acidification and recovery phase (Figure 10). The median ANC decreased from 106 μeq L^{-1} in 1860 to 70 μeq L^{-1} in 1997,

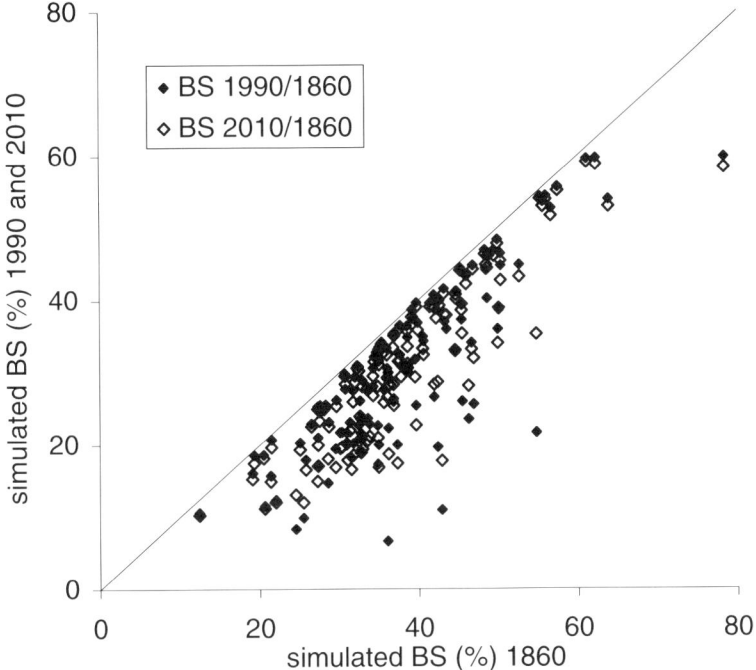

Figure 9. Simulated soil base saturation (BS) in the lake catchment soils for 1997 and 2010 compared with simulated pre-industrial BS (1860). Dots below the 1:1 line have acidified since 1860.

and then increased to 80 μeq L^{-1} by 2010. Further increases in ANC beyond 2010 were only small (2 μeq L^{-1} from 2010 to 2040). Acidity of the lakes in terms of pH (data not shown) developed in a similar way. Median pH decreased from pre-industrial (1860) value of 6.3 to 5.5 in 1997, then recovered to 6.0 and remained 6.0 with no further increase till the end of the modelled period in 2040.

The median soil base saturation of the sensitive lake catchments decreased from 34% in 1860 to 24% in 1997 (Figure 11). The decreasing deposition brought about very little or no increase in BS from 1997 to 2010 and to 2040 except for the catchments with lowest BS, where there was an increase of approximately 2% units.

3.4. COMPARISON TO OTHER STUDIES

MAGIC has been used to reconstruct the history of acidification and to simulate future recovery trends on a regional basis and in a large number of individual catchments in both North America and Europe (e.g., Beier *et al.*, 1995; Cosby *et al.*, 1990, 1995, 1996; Hornberger *et al.*, 1989; Jenkins *et al.*, 1990; Lepistö *et al.*, 1988; Whitehead *et al.*, 1988, 1997; Norton *et al.*, 1992). MAGIC has been used in regional modelling of lake acidification and recovery in Norway (Cosby and Wright, 1998) and in Great Britain (Evans *et al.*, 2001) and is currently being

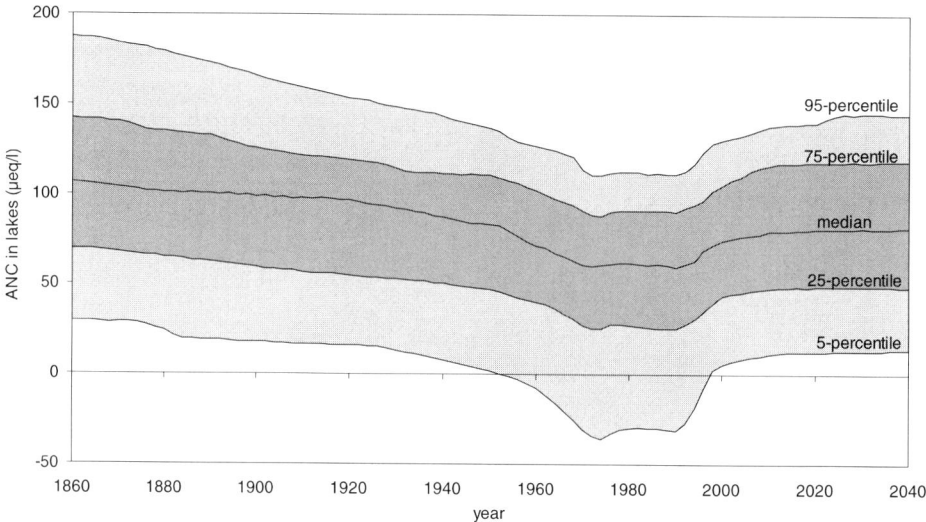

Figure 10. Modelled ANC at 63 sensitive lakes. Dark shaded band encompass 50 % of the lakes, the whole band 90% of the lakes.

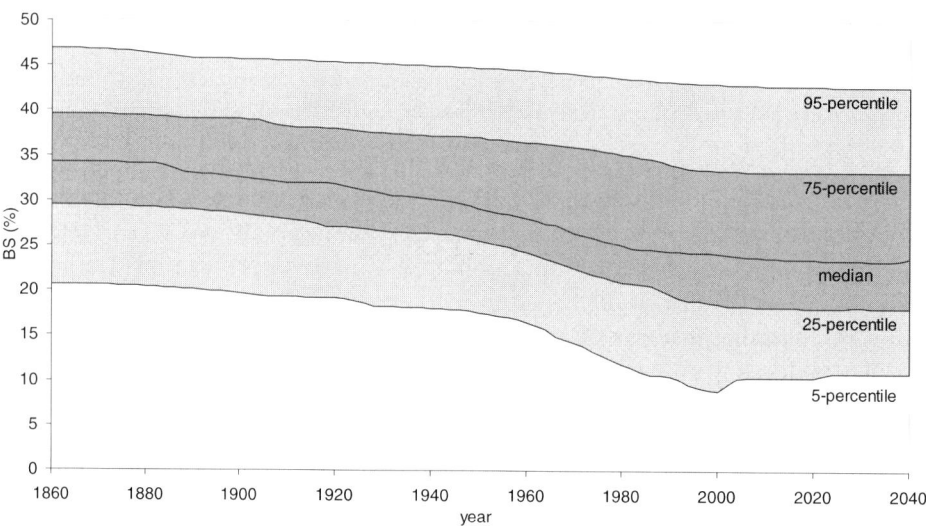

Figure 11. Modelled soil base saturation at 63 sensitive lakes. Dark shaded band encompass 50% of the lakes, the whole band 90% of the lakes.

applied to several other regions in Europe within the major EU project on recovery from acidification, RECOVER:2010 (Ferrier *et al.*, 2001). In Norway, data from three national lake surveys in the southern part of the country indicated that >65% of the lakes were acidic between 1974 and 1995, and that during the 1990's the average ANC of the lakes increased. After 10 years of sulphur deposition at 20% of the 1974 deposition, in 1995 >65 % of the lakes had positive ANC (Cosby and

Wright, 1998). These results appear both qualitatively and quantitatively similar to the results of the MAGIC model presented here.

The modelling of the Swedish lakes presented above demonstrated a relatively high degree of internal consistency of the monitoring data used for the modelling. The data for soils, deposition and lakes were gathered by the National Survey of Forest Soils and Vegetation and the lake monitoring programme with varying temporal and spatial resolution, with different aims and over different periods of time. However, these proved suitable for this kind of modelling. The model was successfully applied to a relatively large number of sites using the input data that were synthesised and aggregated in the same way for all sites. This opens possibilities of applying the model to even much larger numbers of lakes, refining the model calibration and testing multiple scenarios.

Although the average regional response of the lake chemistry was modelled well, the observed trends exhibited larger extremes than the modelled ones. The procedure in which the regional data are used for the modelling involves smoothing of extremes in several ways, a process that is related to the scale at which the data are available. Real lake systems are subject to a great deal of spatial heterogeneity. Small-scale differences in geology, land-use, deposition exposure, etc. can result in large effects on the chemistry of a particular lake. However, the spatial distribution of the data used for the modelling may not always capture that small-scale variability. The maps and surveys from which the modelling data are drawn capture the means and large-scale trends in variables, but small-scale differences are frequently not observable in the data sets used to drive the model simulations. Therefore, it is to be expected that the observed lake chemistry values and trends will exhibit larger extremes than model predictions, as is the case here.

The simulated increase of pH and ANC in the lakes demonstrates the benefits of the deposition reductions that have occurred from 1990 to 1997, the additional reductions since 1997, and reductions that are expected to occur by 2010 under the Gothenburg protocol. The simulations also showed that for the recovery of the catchment soils, there will not be much improvement by the year 2010. Recovery of the soils will require either much longer time or even greater reductions in deposition, and/or different land use or some combination thereof.

4. Conclusions

Both sensitive and non-sensitive lakes in Sweden have been acidified by acid deposition. At present they are recovering, and they will recover more as deposition continues to decrease. Model simulations suggest that, given the deposition reductions agreed under the Gothenburg protocol, lake ANC in the year 2010 will be 10 to 50 μeq L^{-1} below pre-acidification levels of ANC in 90% of the sensitive lakes, with little further recovery projected beyond 2010. Soils in Sweden also have been acidified, but so far they appear to have shown little recovery. Model simulations suggest that the deposition reductions expected under the Gothenburg protocol may not be sufficient to reverse the soil acidification on a regional basis. Continued loss of base cations, especially with forestry, puts the soils at high risk even with the large reductions in deposition achieved to date. The successful calibration of 133 out of 143 lakes without any of the lakes being treated in any individual way suggests that data gathered by the National Survey of Forest Soils and Vegetation and from the lake monitoring programme are indeed suitable for modelling of present and future soil and surface water recovery from acidification.

Acknowledgements

This work was funded by the Swedish ASTA programme (International and National Abatement Strategies for Transboundary Air Pollution), financed by MISTRA foundation with contribution from National board of forestry. The National Survey of Forest Soils and Vegetation and the monitoring of lakes are parts of the national environmental monitoring programme financed by the Swedish Environmental Protection Agency. We thank Brit Lisa Skjelkvåle, NIVA for Nordic lakes trend analysis data and Gun Lövblad, IVL, for estimates of future deposition.

References

Beier, C., Hultberg, H., Moldan, F. and Wright, R. F.: 1995, 'MAGIC applied to roof experiments (Risdalsheia, N; Gårdsjön, S; Klosterhede, DK) to evaluate the rate of reversibility of acidification following experimentally reduced acid deposition', *Water, Air, Soil Pollut.* **85**, 1745–1751.

Bernes, C.: 1986, *Sura och försurade vatten (Acid and Acidified Waters)*, Statens Naturvårdsverk, Solna, Monitor 1986 (in Swedish).

Bernes, C.: 1991, *Acidification and Liming of Swedish Freshwaters*, Swedish Environmental Protection Agency, Solna, Monitor 12.

Cosby, B. J., Wright, R. F., Hornberger, G. M. and Galloway, J. N.: 1985a, 'Modelling the effects of acid deposition: assessment of a lumped parameter model of soil water and streamwater chemistry', *Water Resourc. Res.* **21**, 51-63.

Cosby, B. J., Wright, R. F., Hornberger, G. M. and Galloway, J. N.: 1985b, 'Modelling the effects of acid deposition: estimation of long-term water quality responses in a small forested catchment', *Water Resourc. Res.* **21**, 1591–1601.

Cosby, B. J., Hornberger, G. M., Galloway, J. N. and Wright, R. F.: 1985c, 'Time scales of catchment acidification: A quantitative model for estimating freshwater acidification', *Environ. Sci. Technol.* **19**, 1144–1149.

Cosby, B. J., Jenkins, A., Miller, J. D., Ferrier, R. C. and Walker, T. A. B.: 1990, 'Modelling stream acidification in afforested catchments: Long-term reconstructions at two sites in central Scotland', *J. Hydrol.* **120**, 143–162.

Cosby, B. J., Wright, R. F. and Gjessing, E.: 1995, 'An acidification model (MAGIC) with organic acids evaluated using whole-catchment manipulations in Norway', *J. Hydrol.* **170**, 101–122.

Cosby, B. J., Norton, S. A. and Kahl, J. S.: 1996, 'Using a paired-catchment manipulation experiment to evaluate a catchment-scale biogeochemical model', *Sci. Total Environ.* **183**, 49–66.

Cosby, B. J. and Wright, R. F.: 1998, 'Modelling regional response of lake water chemistry to changes in acid deposition: the MAGIC model applied to lake surveys in southernmost Norway 1974–1986–1995', *Hydrol. Earth Syst. Sci.* **2**, 563–576.

Cosby, B. J., Ferrier, R. C., Jenkins A. and Wright, R. F.: 2001, 'Modelling the effects of acid deposition: refinements, adjustments and inclusion of nitrogen dynamics in the MAGIC model', *Hydrol. Earth Syst. Sci.* **5**,499–517.

Evans, C., Jenkins, A., Helliwell, R., Ferrier, R. and Collins, R.: 2001, *Freshwater Acidification and Recovery in the United Kingdom*, Centre for Ecology and Hydrology, Wallingford, United Kingdom.

Ferrier, R. C., Jenkins, A., Wright, R. F., Schöpp, W. and Barth, H.: 2001, 'Assessment of recovery of European surface waters from acidification 1970–2000: An introduction to the special issue', *Hydrol. Earth Syst. Sci.* **5**, 274–282.

Forsberg, C., Morling, G. and Wetzel, R. G.: 1985, 'Indications of the capacity for rapid reversibility of lake acidification', *Ambio* **14**, 164–166.

Hornberger, G. M., Cosby, B. J. and Wright, R. F.: 1989, 'Historical reconstructions and future forecasts of regional surface water acidification in southernmost Norway', *Water Resourc. Res.* **25**, 2009–2018.

Hruška, J., Laudon, H., Johnson, C.E., Köhler, S. and Bishop, K.: 2001, 'Acid/base character of organic acids in a boreal stream during snowmelt', *Water Resourc. Res.* **37**, 1043.

Jacobson, S. and Mattsson, S.: 1998, *'Snurran' – ett Excel-program som beräknar näringsuttag vid skörd av trädrester*, SkogForsk resultat No 1 1998, Uppsala, 4 pp.

Jenkins, A., Whitehead, P. G., Cosby, B. J. and Birks, H. J. B.: 1990, 'Modelling long-term acidification: a comparison with diatom reconstructions and the implications for reversibility', *Philos. Trans. R. Soc. Lond.* **B 327**, 435–440.

Karltun, E.; 1995, 'Acidification of forest soils on glacial till in Sweden', Report 4427, 1–76. Swedish Environmental Protection Agency, Solna, Sweden.

Kronnäs, V., Moldan, F., Wilander, A., Karltun, E and Cosby, B. J.: 2002, 'Swedish national monitoring data for regional modelling of acidification in 143 lakes', *Poster at Biogeomon 2002, 4th International Symposium on Ecosystem Behaviour*, The University of Reading, Reading, UK, August 17–21, 2002.

Lepistö, A., Whitehead, P. G., Neal, C. and Cosby, B. J.: 1988, 'Modelling the effects of acid deposition: estimation of longterm water quality responses in forested catchments in Finland', *Nordic Hydrol.* **19**, 99–120.

Lövblad, G., Persson, C. and Roos, E.: 2000, 'Deposition of base cations in Sweden', Report 5119, Swedish Protection Agency, Stockholm.

Moldan, F. and Ek, A.: 2000, 'Experimental studies in Sweden', in P. Warfvinge and U. Bertills (eds.), *Recovery from Acidification in the Natural Environment. Present Knowledge and Future Scenarios*, Report 5034, Swedish Protection Agency, Stockholm, pp. 37–52.

Mylona, S.: 1996, 'Sulphur dioxide emissions in Europe 1880–1991 and their effect on sulphur concentrations and depositions', *Tellus* **48B**, 662–689, and Corrigendum, *Tellus*, **49B**, 447–448.

National Atlas of Sweden: 1996, Skogen. Red. Nilsson, N.-E. (in Swedish).

Nordic Council of Ministers: 1984, Naturgeografisk regionindelning av Norden (in Swedish).

Norton, S. A., Wright, R. F., Kahl, J. S. and Scofield, J. P.: 1992, 'The MAGIC simulation of surface water acidification at, and 1st year results from, the Bear-Brook watershed manipulation, Maine, U.S.A.', *Environ. Pollut.* **77**, 2–3, 279–86.

Robertson, L., Langner, J. and Engardt, M.: 1999, 'An Eulerian limited-area atmospheric transport model', *J. Appl. Meteor.* **38**, 190–210.

Simpson, D., Olendrzynski, K., Semb, A., Støren, E. and Unger, S.: 1997, 'Photochemical oxidant modelling in Europe: Multi-annual modelling and source-receptor relationships', EMEP/MSC-W Report 3/97.

Skjelkvåle, B. L., Mannio, J., Wilander, A. and Andersen, T.: 2001, 'Recovery from acidification of lakes in Finland, Norway and Sweden 1990–1999', *Hydrol. Earth Syst. Sci.* **5**, 327–337.

SNV: 1993, 'Svensk nationell miljöövervakning', (Swedish national environmental monitoring programme) Program antaget av Naturvårdsverkets miljöövervakningsnämnd 7 juni 1993. SNV Rapport 4275 (in Swedish).

UN/ECE: 1999, 'The 1999 protocol to abate acidification, eutrophication and ground-level ozone', (2003, March 7, online) URL: http://www.unece.org/env/lrtap/multi_h1.htm.

Whitehead, P. G., Barlow, J., Haworth, E. Y. and Adamson, J. K.: 1997, 'Acidification in three Lake District tarns: Historical long term trends and modelled future behaviour under changing sulphate and nitrate deposition', *Hydrol. Earth Syst. Sci.* **1**, 197–204.

Whitehead, P. G., Reynolds, B., Hornberger, G. M., Neal, C., Cosby, B. J. and Paricos, P.: 1988, 'Modelling long term stream acidification trends in upland Wales at Plynlimon', *Hydrol. Process.* **2**, 357–368.

Wilander, A.: 1997, 'Referenssjöarnas vattenkemi under 12 år; tillstånd och trender. (Reference lakes water chemistry under 12 years; status and trends)', Naturvårdsverket Rapport 4652 (in Swedish).

Wilander, A. and Lundin, L.: 2000, 'Recovery of surface waters and forest soils in Sweden', in P. Warfvinge and U. Bertills (eds), *Recovery from Acidification in the Natural Environment: Present Knowledge and Future Scenarios*, Trelleborg Naturvårdsverket Report 5034, pp. 53–66.

Wilander, A., Johnson, R., Goedkoop, W. and Lundin, L.: 1998, 'National surface water survey: A synoptic study of water chemistry and benthic fauna in Swedish lakes and streams', (in Swedish with English summary), Swedish Environment Protection Agency Report 4813.

INTEGRATED NITROGEN AND FLOW MODELLING (INCA) IN A BOREAL RIVER BASIN DOMINATED BY FORESTRY: SCENARIOS OF ENVIRONMENTAL CHANGE

KATRI RANKINEN*, AHTI LEPISTÖ and KIRSTI GRANLUND

Finnish Environment Institute, P.O. Box 140, FIN-00251 Helsinki, Finland
(* author for correspondence, e-mail: Katri.Rankinen@ymparisto.fi, phone: +358 9 4030 0244, fax: +358 9 4030 0290)

(Received 20 August 2002; accepted 1 April 2003)

Abstract. A new version (v1.7) of the Integrated Nitrogen in CAtchments model (INCA) was applied to the northern boreal Simojoki river basin (3160 km^2) in Finland. The INCA model is a semi-distributed, dynamic nitrogen (N) process model which simulates N transport and processes in catchments. The INCA model was applied to model flow and seasonal inorganic N dynamics of the river Simojoki basin over the period 1994–1996, and validated for two more years. Both calibration and validation of the model were successful. The model was able to simulate annual dynamics of inorganic N concentrations in the river. The effects of forest management and atmospheric deposition on inorganic N fluxes to the sea in 2010 were studied. Three scenarios were applied for forestry practices and two for deposition. The effects of forest cutting scenarios and atmospheric deposition scenarios on inorganic N flux to the sea were small. The combination of the maximum technically possible reduction of N deposition and a decrease of 100% in forest cutting and peat mining areas decreased NO_3^--N flux by 6.0% and NH_4^+-N flux by 3.1%.

Keywords: dynamic modelling, forest management, N deposition, nitrogen, northern river basin, scenarios

1. Introduction

Nitrogen (N) plays an important role in controlling the trophic status of surface waters. Despite considerable effort, accurate modelling of N sources, transformations and sinks on the catchment scale still remains a challenge. Responses of catchment N storages and fluxes to changes in climate, atmospheric deposition and land use are of particular importance. Deterministic, dynamic models are suitable for predicting effects of environmental changes. In northern forest ecosystems these changes are typically due to changes in forest management and atmospheric deposition.

Forest management can affect many aspects of river basin hydrology and N dynamics. Total runoff generally increases with forest disturbance as a result of reduced interception and transpiration. Nitrogen is considered to be the growth-limiting factor in most boreal forest ecosystems and natural ecosystems are characterized by a tight internal cycling of N. Leaching losses and gaseous losses are generally less than a few kg N ha^{-1} yr^{-1} (Gundersen and Bashkin, 1994). High leaching losses may occur after clear-cutting due to enhanced mineralisation and nitrification, and reduced N uptake of vegetation. Lowering of the water table by forestry ditching also may increase gaseous losses from nutrient-rich peatlands, although it has only a minor effect in poor peatlands (Martikainen *et al.*, 1993).

Forestry is distributed evenly in Finland, but peat mining is concentrated in certain areas and has a long duration. Peat mining causes elevated nutrient and suspended solids leaching and concentrations in downstream waters (Heikkinen, 1990; Klöve, 2001).

Transboundary air pollution with its adverse environmental impacts has been a severe environmental problem in Europe and North America for decades. Excess N deposition causes eutrophication of soils and waters. In 1990, as much as 55% of the ecosystem areas in the present EU-15 countries were estimated to receive eutrophying N deposition in excess of their tolerance limits (Comission, 1999). In the northern boreal ecosystems the N deposition levels are low, but the receiving surface water ecosystems are typically acidic, oligotrophic and sensitive to any excess loading.

The emphasis of this study was on calibrating and validating the dynamic, semi-distributed INCA model to the river Simojoki basin in northern Finland. The long-term effects of forest management and atmospheric deposition on inorganic N fluxes to the sea were studied in order to evaluate the effects of present and future forestry practices (Maa- ja metsätalousministeriö, 1999) and the emission reduction obligations agreed within the UN/ECE (1999). Three forestry scenarios and two atmospheric N deposition scenarios were used to simulate changes in inorganic N fluxes in 2010. Combinations of these scenarios were tested in an attempt to fulfill the Water Protection Targets (Ministry of the Environment, 1998), aiming at 50% reduction in nutrient load to inland waters and to the Baltic Sea.

2. Materials and Methods

2.1. SIMOJOKI RIVER BASIN

The river Simojoki discharges into the Gulf of Bothnia in the Baltic Sea. The river basin (3160 km^2) is composed of nine sub-basins (Figure 1). Over the period 1961–1975, mean annual precipitation was 650–750 mm, mean annual evapotranspiration was about 330 mm and annual runoff 350–450 mm. There are about 170–180 winter days in year and the mean annual temperature is +0.5 – +1.5 °C. The duration of the snow cover is from the middle of November to early May (Perkkiö *et al.*, 1995). According to the Finnish Meteorological Institute the growing season started on average on 10 May in the years 1961–1990. The duration of the growing season was on average 140 days.

The river Simojoki is a salmon river in near-natural state, and the dominant human impacts are forestry and atmospheric N deposition (Perkkiö *et al.*, 1995).

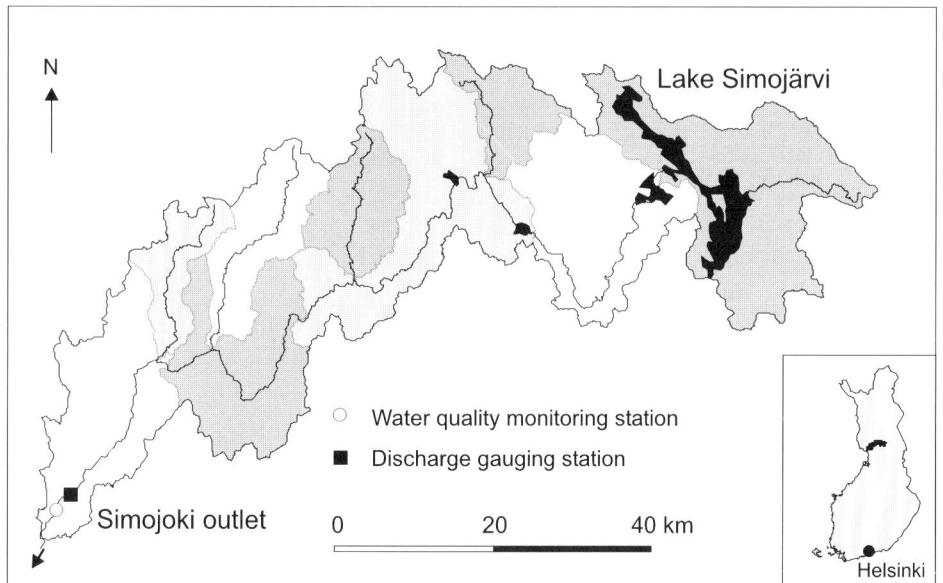

Figure 1. Location of the river Simojoki basin in northern Finland.

Peatlands and peatland forests are common in the region and an average of 0.5% of the total catchment area is felled annually. Forestry drainage was most intensive during the 1960s and 1970s, and by 1991 over 30% (i.e. 1000 km^2) of the total catchment area had been drained (Perkkiö *et al.*, 1995). Currently, supplementary drainage works are being conducted in these old drainage areas. Urban areas cover only 0.06% and agricultural fields cover 2.7% of the catchment area (Perkkiö *et al.*, 1995). Grass cultivation for animal husbandry is the most common form of agricultural production. In 1995, there were 1365 ha of peat mining area (0.43% of the catchment area), which by 1997 had increased to 1570 ha (0.5% of the catchment area).

The N concentration data for the river Simojoki were obtained from the water quality data base operated by the Finnish Environment Institute (SYKE) and regional environment centres. Sampling frequency at the outlet of the Simojoki river was 12–17 samples a^{-1} during the study period.

2.2. THE INCA MODEL

On the basis of an earlier work by Whitehead *et al.* (1998), a new version of the dynamic, process-based and semi-distributed INCA (Integrated Nitrogen in Catchments) model was developed and described in detail by Wade *et al.* (2002). This model integrates hydrology, catchment and river N processes, and simulates daily NO_3^--N and NH_4^+-N concentrations as time series at key sites, as profiles down the river system or as statistical distributions. Three components are included: the hydrological model, the catchment N process model and the river N process model.

Sources of N include atmospheric deposition, the terrestrial environment, urban areas and direct discharges. The model can simulate N processes in six land use classes.

INCA simulates plant uptake of mineral N, nitrification, denitrification, mineralization, and immobilisation within each sub-catchment and land use class. Rate coefficients of N processes are temperature- and moisture-dependent. A new version (v1.7) of the INCA model takes into account the isolating effect of the snow pack in calculating soil temperature and uses a temperature response function (Bunnel *et al.*, 1977) capable of describing temperature dependencies of N processes in northern areas.

2.3. CALIBRATION

The simulated N concentrations in both the terrestrial and in-stream components of the model depend on water volumes. Hence, the hydrological component of the INCA model was calibrated first before moving on to the catchment N process model work. Calibration of the hydrological sub-model and the effect of spatial resolution of the input were discussed earlier (Rankinen *et al.*, 2002a).

The N process sub-model was calibrated using input data for 1994–1996 because these years correspond to the GIS-based land use classification. Land use classes were derived from the satellite image-based land use and forest classification of Finland (Vuorela, 1997), and were supplemented with satellite image-based maps of final cuttings in mineral and organic soils, provided by the Finnish Forest Research Institute. In this application the six land-use classes are: forest on mineral soil (ForMin), cut forest on mineral soil (ForMinCu), forest on organic soil (ForOrg), cut forest on organic soil (ForOrgCu), agriculture (Arable) and open surface water, covering 35, 4, 52, 1, 2, and 6%, respectively, of the whole river basin area.

In parameterization, all the forest cutting areas on organic soil were assumed to be ditched. Ground vegetation was assumed to start to recover on forest cut areas. The N process parameters (Table I) were adjusted until the annual process loads were in the range reported in the literature (Table II). Simulated immobilisation was assumed to be about 62% of the gross mineralisation (Stottlemyer and Toczydlowski, 1999).

Initial NO_3^--N and NH_4^+-N concentrations in soil water and groundwater were adjusted according to values found in the literature and to observed river concentrations in the beginning of the simulation. Inorganic N deposition was calculated using the regional deposition model DAIQUIRI (Syri *et al.*, 1998). In agricultural areas, only conventional cultivation was assumed; no organic farming. Agricultural areas were assumed to be fertilised at the end of May (70 kg NH_4^+-N ha^{-1} and 30 kg NO_3^--N ha^{-1}) and at the end of June (30 kg NH_4^+-N ha^{-1} and 24 kg NO_3^--N ha^{-1}). Loads from peat-mining areas were added as point-sources, because nutrient con-

TABLE I

Parameter values of N processes used in calibration. ForMin stands for forest on mineral soil, ForMinCu for forest cut areas on mineral soil, ForOrg for forest on organic soil, ForOrgCu for forest cut areas on organic soil and Arable for grass cultivation

Process	Land use class	Value	Unit
Denitrification	ForMin	0.01	day^{-1}
	ForMinCu	0.03	day^{-1}
	ForOrg	0.01	day^{-1}
	ForOrgCu	0.05	day^{-1}
	Arable	0.015	day^{-1}
Nitrification	ForMin	0.001	day^{-1}
	ForMinCu	0.03	day^{-1}
	ForOrg	0.001	day^{-1}
	ForOrgCu	0.05	day^{-1}
	Arable	0.25	day^{-1}
Mineralisation	ForMin	0.5	kg N ha^{-1} day^{-1}
	ForMinCu	0.55	kg N ha^{-1} day^{-1}
	ForOrg	0.5	kg N ha^{-1} day^{-1}
	ForOrgCu	0.55	kg N ha^{-1} day^{-1}
	Arable	1.2	kg N ha^{-1} day^{-1}
N Fixation	ForMin	0.001	kg N ha^{-1} day^{-1}
	ForMinCu	0.001	kg N ha^{-1} day^{-1}
	ForOrg	0.0035	kg N ha^{-1} day^{-1}
	ForOrgCu	0.0035	kg N ha^{-1} day^{-1}
	Arable	0.25	kg N ha^{-1} day^{-1}
Immobilisation	ForMin	0.75	day^{-1}
	ForMinCu	0.6	day^{-1}
	ForOrg	0.45	day^{-1}
	ForOrgCu	0.5	day^{-1}
	Arable	0.065	day^{-1}

centrations in runoff waters were available (average inorganic N load in summer 1995 was 9.1 kg d^{-1}).

2.4. SCENARIOS OF ENVIRONMENTAL CHANGE IN 2010

The effects of forest management and atmospheric deposition on inorganic N fluxes to the sea in 2010 were studied. Three scenarios were applied for forestry practices

TABLE II

Catchment process fluxes: comparison of measured values within the published literature with simulated mean annual fluxes during the period 1994 to 1996 in the river Simojoki basin. Land use class abbreviations are described in Table I

Process/ land use	Simulated value (kg ha^{-1} a^{-1})	Measured value or range in values (kg ha^{-1} a^{-1})	Reference
N uptake			
ForMin	25–49	28–51	Mälkönen (1974)
ForMinCu	14–26	11–23	Mälkönen (1974)
ForOrg	24–44	26–42	Finér (1989)
ForOrgCu	24–37		
Arable	220–250	225[f]	
Mineralisation			
ForMin	77–120	15–120	Persson and Wirén (1995)[b]; Smolander et al. (1998a)
ForMinCu	86–132		
ForOrg	77–120		
ForOrgCu	100–132		
Arable	154–221	210	Sippola (2000); Lindén et al. (1992)
Nitrification			
ForMin	0.04–0.07	0–7	Martikainen (1984); Persson and Wirén (1995)[b]
ForMinCu	2.3–4.1	13	Persson and Wirén (1995)[b]
ForOrg	0.07–0.12		
ForOrgCu	5.1–6.9		
Arable	124–175		
Denitrification			
ForMin	0.25–0.4	<1	Gundersen and Bashkin (1994)
ForMinCu	1.2–1.6	<0.5	Smolander et al. (1998b)
ForOrg	0.24–0.36	0–4.7[a]	Regina et al. (1996)[a]; Martikainen et al. (1994)
ForOrgCu	2.4–3.3	<1	Nieminen (1998)
Arable	13–20	3–9	Gundersen and Bashkin (1994)

TABLE II
(continued)

Process/ land use	Simulated value (kg ha^{-1} a^{-1})	Measured value or range in values (kg ha^{-1} a^{-1})	Reference
N Fixation			
ForMin	0.32	0.31–3.8	Granhall and Lindberg (1978, 1980)
ForMinCu	0.32	0.27	Granhall and Lindberg (1980)
ForOrg	1.11–1.12	<1–2	Granhall (1981)
ForOrgCu	1.11–1.12	<1–2	Granhall (1981)
Arable	4	~4	Rekolainen *et al.* (1992)
Inorganic N leaching			
ForMin	0.45–0.65	0.38–0.59^c	Kortelainen *et al.* (1997); Lepistö (1996)
ForMinCu	0.88–1.2		
ForOrg	0.5–0.78	0.48^d	Kortelainen *et al.* (1997)
ForOrgCu	0.83–0.97	0.92	Lepistö (1996)
Arable	14.6–19.8	13.8–16.2^e	Vuorenmaa *et al.* (2002)
		1–38	Jaakkola (1984)

^{a} Including drained peatlands.
^{b} Low-leaching sites 6 (Farabol 1) and 8 (Skogaby) which could represent northern conditions.
^{c} Average of mineral soil catchments.
^{d} Average of peatland (>35%) catchments.
^{e} Specific loss from the arable land part of agricultural research catchments.
^{f} Based on statistical information of yields.

and two for deposition. Firstly, an increase of 20% in forest cut areas (cut +20%) is based on Finland's National Forest Programme (Maa- ja metsätalousministeriö, 1999). Secondly, a decrease of 100% in forest cut areas (cut-100%) was tested in an attempt to fulfill the Water Protection Targets (Ministry of the Environment, 1998), aiming at 50% reduction in nutrient load to inland waters and to the Baltic Sea. Thirdly, a decrease of 100% in both forest cutting and peat mining areas (cut and peat – 100%) was tested.

Expected future N deposition (UNE) in 2010 was calculated from the emission reduction obligations agreed within the UN/ECE (1999). The maximum technically possible reduction of N deposition was estimated using the Maximum Feasible Reductions (MFR) emission scenario compiled by IIASA (International Institute for Applied Systems Analysis). The average N deposition decreased from 2.3 kg ha^{-1} a^{-1} to 2 kg ha^{-1} a^{-1} (UNE scenario) and to 1.4 kg ha^{-1} a^{-1} (MFR scenario).

3. Results and Discussion

3.1. FLOW AND INORGANIC N DYNAMICS

Five years of monitoring data on flow and inorganic N concentrations enabled splitting of data for the calibration period 1994–1996 (Figure 2) and for the model validation years 1993 and 1997 (Figure 3). The model was able to simulate annual dynamics of inorganic N concentrations in the river during the calibration and validation periods. The observed concentration level in autumn and early spring was reached for NH_4^+-N and NO_3^--N, as well as the extremely low NH_4^+-N concentrations in summer. Simulated NO_3^--N concentrations remained somewhat higher than the observed concentrations in summer. Observed NO_3^--N and NH_4^+-N concentrations decreased rapidly in the beginning of May, but the decrease in the simulated concentrations was delayed. The difference was due to vegetation uptake of N starting in the beginning of May, when there is still a lot of water discharging through the soil. According to (Rankinen *et al.*, 2002b) the INCA model is more sensitive to hydrological input than to vegetation N uptake.

The seasonal variation of inorganic N in the river Simojoki followed the 'normal' seasonal pattern of a non-polluted and undisturbed northern river, with low or undetectable concentrations during summer and appreciably higher concentrations during the dormant season. In northern forests, inorganic N is usually retained effectively during the growing season, such that stream water concentrations are often negligible. The timing and magnitude of high flow peaks are very important when simulating N fluxes, because under the present climatic conditions most of the N export occurs during periods with high flow. Particularly, the early phase of the spring flood is a critical period due to flushing of inorganic N from catchment soils and from melting snow, followed by dilution processes (Arheimer *et al.*, 1996). Thus, a successful model calibration for these high flow periods is extremely important for estimating the total load of N to the rivers and coastal waters.

3.2. N PROCESS LOADS

Most of simulated annual N process loads were in the range reported in the literature (Table II). Special attention was paid to find literature studies carried out in boreal environments, but two sites from the study of Persson and Wirén (1995) were also included. They were assumed to be able to represent northern conditions, for example annual mineralisation was in the range reported by Smolander *et al.* (1998a). Annual nitrification at these two sites was up to 7 kg ha^{-1} a^{-1}, which is higher than usually found in boreal forests. Nitrification rates in boreal forests on

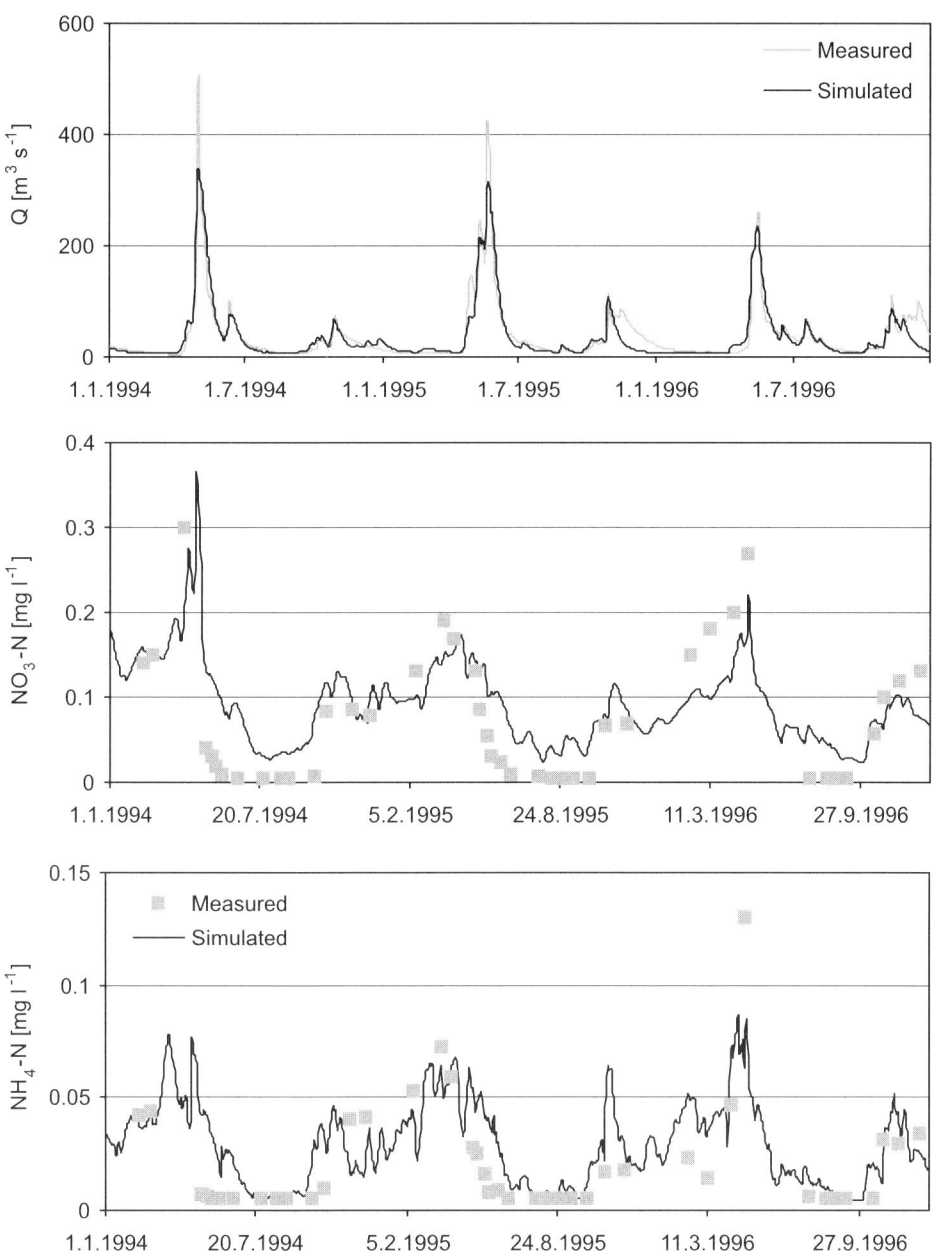

Figure 2. Simulated and observed discharge, NO_3^--N and NH_4^+-N concentrations at the outlet of the river Simojoki. Calibration years 1994–1996.

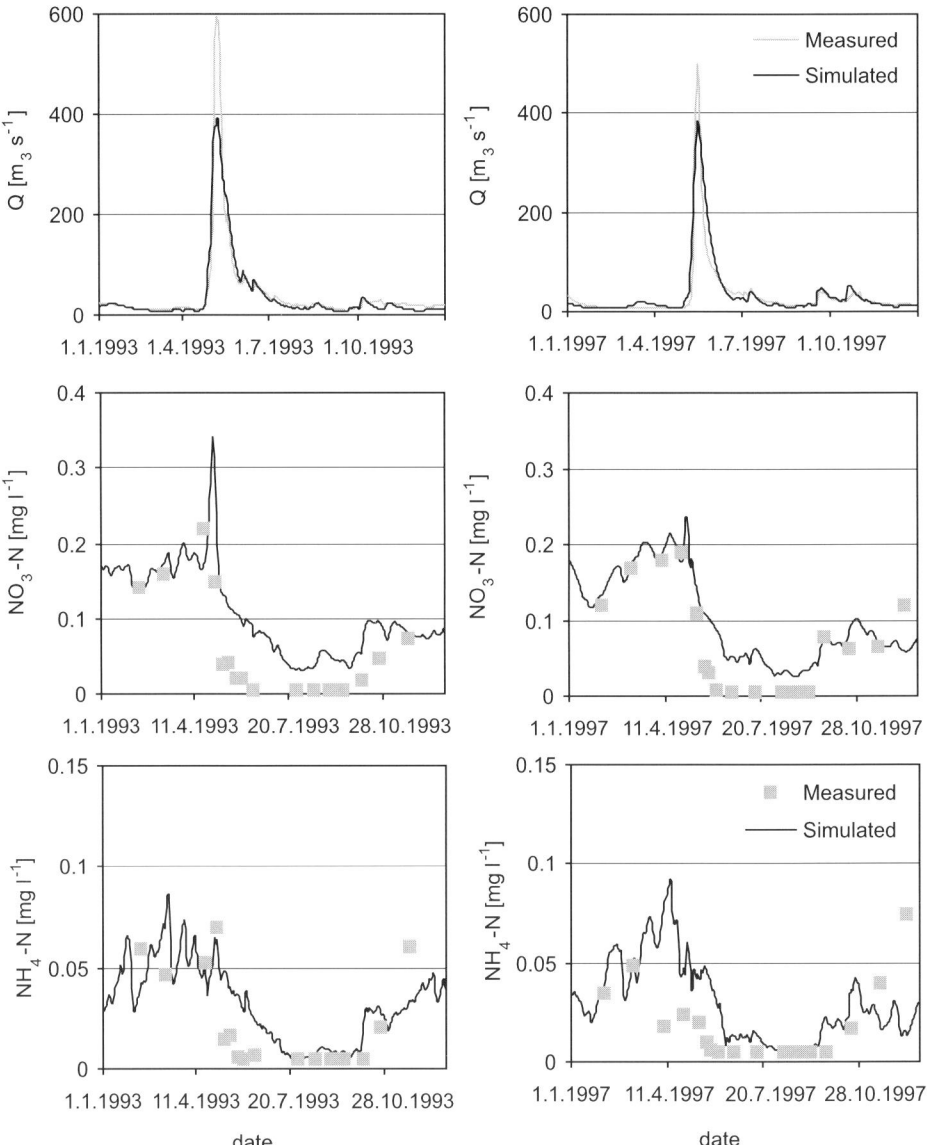

Figure 3. Simulated and observed discharge, NO_3^--N and NH_4^+-N concentrations at the outlet of the river Simojoki. Validation years 1993 and 1997.

mineral soil are reported to be negligible except in forest cut areas (Martikainen, 1994; Smolander *et al.*, 1998b).

The values presented by Granhall and Lindberg (1978, 1980) and Granhall (1981) were used to calibrate N fixation in forests and forest cut areas. Higher values were used for forest on organic soil than for forest on mineral soil, because Granhall and Lindberg (1978) found positive correlation between N fixation and organic matter content in soils. According to the latest studies of DeLuca *et al.* (2002), N fixation potential in forests of northern Scandinavia and Finland varied between 1.5–2.0 kg N ha^{-1} a^{-1}. This range is of the same order of magnitude as atmospheric N deposition in the study basin, which indicates that N-fixation has to be taken into account in N balances of these ecosystems.

The ground vegetation was assumed to start to recover on forest cut areas. Simulated vegetation uptake was compared to measurements of Mälkönen (1974), in the way that values for ground vegetation were used for forest cut areas, and values for total stands for forest on mineral soil.

Simulated inorganic N leaching from all land-use classes were of the same order of magnitude reported in the literature (Table II). Further, a relatively good fit was found when simulated and measured N fluxes were compared along the river (Lepistö *et al.*, 2004), suggesting that N leaching estimates were acceptable in the river basin scale.

3.3. SCENARIOS

The effects of forest cutting scenarios and atmospheric deposition scenarios on inorganic N flux to the sea were not high. Increasing forest cut areas by 20% (Maa-ja metsätalousministeriö, 1999) increased NO_3^--N flux to the sea by 0.7% and NH_4^+-N flux by 0.5%. Decreasing forest cut areas by 100% decreased NO_3^--N flux by 2.7% and NH_4^+-N flux by 1.7%. Decreasing both forest cut areas and peat mining areas by 100% decreased NO_3^--N flux to the sea by 3.3% and NH_4^+-N flux by 3.0%. Decreasing atmospheric deposition according to the UNE scenario decreased NO_3^--N flux by 0.8% but NH_4^+-N flux did not change. Decreasing atmospheric deposition according to the MFR scenario decreased NO_3-N flux by 2.6% and NH_4^+-N flux by 0.1%.

The combination of the maximum technically possible reduction of N deposition and a decrease of 100% in forest cutting and peat mining areas decreased NO_3^--N flux from 98.3 to 92.4 Mg yr^{-1} (6.0%) and NH_4^+-N flux from 31.6 to 30.6 Mg yr^{-1} (3.1%), i.e. the total inorganic N flux to the sea would decrease by about 5%. The organic N load from forestry could be assumed to decrease correspondingly, however. The mosaic of treatment areas in large river basins may result in a smoothed impact concerning both hydrology and nutrient leaching. The direct and local effect of forest treatments can be much higher (Ahtiainen and Huttunen, 1999).

Even though these changes in forestry and atmospheric deposition were drastic, they would not be sufficient to achieve the 50% load reduction target set by the Water Protection Targets (Ministry of the Environment, 1998). Climate change and agricultural scenarios which will be assessed later would probably have a more pronounced impact.

4. Concluding Remarks

The INCA model was applied to the river Simojoki basin to model flow and seasonal inorganic N dynamics over the period 1994–1996, and validated for two more years. Both calibration and validation of the model were successful. Most of simulated annual N process loads were in the range reported in the literature.

According to the model scenarios, a drastic future change in forest management practices (cut and peat–100%) would decrease inorganic N flux by 4.5%, while an increase of 20% in cut areas would increase the flux only by 1.2%. The maximum technically feasible reduction of atmospheric N deposition would decrease N flux by 2.8%. This decrease together with the highest decrease in cut areas would have a combined impact of ∼5% on inorganic N fluxes. There were no marked change in inorganic N leaching according to these scenarios. The area treated annually by forestry was small (∼0.5% of the catchment area), and the location of forest cut areas was scattered in the river basin. Atmospheric N deposition in the area is relatively low and forests are N limited, so a change in N deposition did not play a more important role in N leaching.

Acknowledgements

This study was supported by the Commission of the European Union, the INCA project (EVK1-CT-1999-00011).

References

Ahtiainen, M. and Huttunen, P.: 1999, 'Long-term effects of forestry managements on water quality and loading in brooks', *Boreal Environ. Res.* **4**, 101–114.

Arheimer, B., Andersson, L. and Lepistö, A.: 1996, 'Variation of nitrogen concentration in forest streams – influences of flow, seasonality and catchment characteristics', *J. Hydrol.* **179**, 281–304.

Bunnel, F. L., Tait, D. E. N., Flanagan, P. W. and Van Cleve, K.: 1977, 'Microbial respiration and substrate weight loss. I. A general model of the influences of abiotic variables', *Soil Biol. Biochem.* **9**, 33–40.

Comission, E.: 1999, 'Proposal for a Directive of the European Parliament and of the Council on National Emission Ceilings for Certain Atmospheric Pollutants', COM (1999) 125final, Brussels, Belgium.

DeLuca, T. H., Zackrisson, O., Nilsson, M.-C. and Sellstedt, A.: 2002, 'Quantifying nitrogen-fixation in feather moss carpets of boreal forests', *Nature* **419**, 917–920.

Finér, L.: 1989, 'Biomass and nutrient cycle in fertilized and unfertilized pine, mixed birch and spruce stands on a drained mire', *Acta For. Fenn.* **208**, 1–63.

Granhall, U.: 1981, 'Biological nitrogen fixation in relation to environmental factors and functioning of natural ecosystems', in P. Clark and T. Rosswall (eds), *Terrestial Nitrogen Cycles*, Vol. 33, Ecological Bulletin (Stockholm), pp. 131–144.

Granhall, U. and Lindberg, T.: 1978, 'Nitrogen fixation in some coniferous forest ecosystems', in U. Granhall (ed.), *Environmental Role of Nitrogen-fixing Blue-green Algae and Asymbiotic Bacteria*, Vol. 26, Ecological Bulletin (Stockholm), pp. 178–192.

Granhall, U. and Lindberg, T.: 1980, 'Nitrogen input through biological nitrogen fixation', in E. Persson (ed.), *Structure and Function of Northern Coniferous Forests – An Ecosystem Study*, Vol. 32, Ecological Bulletin (Stockholm), pp. 333–340.

Gundersen, P. and Bashkin, V. N. (eds): 1994. *Nitrogen Cycling*, John Wiley & Sons Ltd., Chichester, pp. 255–283.

Heikkinen, K.: 1990, 'Transport of organic and inorganic matter in river, brook and peat mining water in the drainage basins of the river Kiiminkijoki', *Aqua Fenn.* **20**, 143–155.

Jaakkola, A.: 1984, 'Leaching losses of nitrogen from a clay soil under grass and cereal crops in Finland', *Plant Soil* **76**, 59–66.

Klöve, B.: 2001, 'Characteristics of nitrogen and phosphorus loads in peat mining wastewaters', *Water Res.* **35**, 2353–2362.

Kortelainen, P., Saukkonen, S. and Mattson, T.: 1997, 'Leaching of nitrogen from forested catchments in Finland', *Global Biogeochem. Cycles* **11**, 627–638.

Lepistö, A.: 1996, 'Hydrological processes contributing to nitrogen leaching from forested catchments in Nordic conditions', *Monographs of the Boreal Environment Research*, 1–71.

Lepistö, A., Granlund, K. and Rankinen, K.: 2004, 'Integrated nitrogen modelling in a boreal forestry dominated river basin: N fluxes and retention in lakes and peatlands', *Water, Air, Soil Pollut.: Focus* **4**(2–3), 113–123.

Lindén, B., Lygstad, I., Sippola, J., Soegaard, K. and Kjellerup, V.: 1992, 'Nitrogen mineralisation during the growing season. 1. Contribution to the nitrogen supply of spring barley', *Swed. J. Agric. Res.* **22**, 3–12.

Maa- ja metsätalousministeriö: 1999, 'Kansallinen metsäohjelma 2010', Maa- ja metsätalousministeriö, MMM:n julkaisuja 2/1999, pp. 38.

Mälkönen, E.: 1974, *Annual Primary Production and Nutrient Cycle in some Scots Pine Stands*, Department of Soil Science, Finnish Forest Research Institute, Helsinki, pp. 87.

Martikainen, P. J.: 1984, 'Nitrification in two coniferous forest soils after different fertilization treatments', *Soil Biology Biochem.* **16**, 577–582.

Martikainen, P. J., Nykänen, H., Crill, P. and Silvola, J.: 1993, 'Effect of a lowered water table on nitrous oxide fluxes from northern peatlands', *Nature* **366**, 51–53.

Martikainen, P. J., Nykänen, H., Silvola, J., Alm, J., Lång, K., Smolander, A. and Ferm, A.: 1994, 'Nitrous oxide (N_2O) emissions from some natural environments in Finland', in *6th International Workshop on Nitrous Oxide Emissions*, Turku, Finland, 7–9 June, pp. 553–560.

Ministry of the Environment: 1998, *Water Protection Targets to 2005*, Ministry of the Environment. The Finnish Environment 226, Finnish Environment Institute, Helsinki, pp. 82.

Nieminen, M.: 1998, 'Changes in nitrogen cycling following the clearcutting of drained peatland forests in southern Finland', *Boreal Environ. Res.* **3**, 9–21.

Perkkiö, S., Huttula, E. and Nenonen, M.: 1995, 'Simojoen vesistön vesiensuojelusuunnitelma', *Vesi- ja ympäristöhallinnon julkaisuja-sarja A* **200**, 102.

Persson, T. and Wirén, A.: 1995, 'Nitrogen mineralization and potential nitrification at different depths in acid forest soils', *Plant Soil* **168–169**, 55–65.

Rankinen, K., Lepistö, A. and Granlund, K.: 2002a, 'Hydrological application of the INCA (Integrated Nitrogen in CAtchments) model with varying spatial resolution and nitrogen dynamics in a northern river basin', *Hydrol. Earth System Sci.* **6**, 339–350.

Rankinen, K., Lepistö, A. and Granlund, K.: 2002b, 'Sensitivity of the INCA model to N process parameters and hydrological input', *Integrated Assessment and Decision Support Proceedings of the 1st Biennial Meeting of the International Environmental Modelling and Software Society*, 24–27 June 2002, University of Lugano, Switzerland, Vol. 1, pp. 317–321.

Regina, K., Nykänen, H., Silvola, J. and Martikainen, P. J.: 1996, 'Fluxes of nitrous oxide from peatland as affected by peatland type, water table level and nitrification capacity', *Biogeochemistry* **35**, 401–418.

Rekolainen, S., Kauppi, L. and Turtola, E.: 1992, 'Maatalous ja vesien tila. MAVEROn loppuraportti [in Finnish]', Luonnonvarajulkaisuja 15. Luonnonvarainneuvosto, Maa- ja metsätalousministeriö, pp. 61.

Sippola, J.: 2000, 'Estimation of soil nitrate in the spring as a basis for adjustment of nitrogen fertiliser rates', *Agric. Food Sci. Finland* **9**, 71–77.

Smolander, A., Kukkola, M. and Mälkönen, E.: 1998a, 'Metsäekosysteemin toiminta typpikuormituksen alaisena', in E. Mälkönen (ed.), *Ympäristönmuutos ja metsien kunto. Metsien terveydentilan tutkimusohjelman loppuraportti*, Vol. 691, Metsäntutkimuslaitos, pp. 175–181.

Smolander, A., Priha, O., Paavolainen, L., Steer, J. and Mälkönen, E.: 1998b, 'Nitrogen and carbon transformations before and after clear-cutting in repeatedly N-fertilized and limed forest soil', *Soil Biology Biochem.* **30**, 477–490.

Stottlemyer, R. and Toczydlowski, D.: 1999, 'Nitrogen mineralization in a mature boreal forest, Isle Royale, Michigan', *J. Environ. Qual.* **28**, 709–720.

Syri, S., Johansson, M. and Kangas, L.: 1998, 'Application of nitrogen transfer matrices for integrated assessment', *Atmos. Environ.* **32**, 409–413.

UN/ECE: 1999, 'Protocol to the 1979 convention on long-range transboundary air pollution to abate acidification, eutrophication and ground-level ozone', *UN/ECE Document EB/AIR/1999/1*, United Nations, New York, Geneva.

Vuorela, A.: 1997, 'Satellite image based land cover and forest classification of Finland', *Reports of the Finnish Geodetic Institute* **97**, 42–52.

Vuorenmaa, J., Rekolainen, S., Lepistö, A., Kenttämies, K. and Kauppila, P.: 2002, 'Losses of nitrogen and phosphorus from agricultural and forest areas in Finland during the 1980s and 1990s', *Environ. Monit. Assess.* **76**, 213–248.

Wade, A., Durand, P., Beaujoan, V., Wessels, W., Raat, K., Whitehead, P. G., Butterfield, D., Rankinen, K. and Lepistö, A.: 2002, 'Towards a generic nitrogen model of European ecosystems: New model structure and equations', *Hydrol. Earth Syst. Sci.* **6**, 559–582.

Whitehead, P. G., Wilson, E. J. and Butterfield, D.: 1998, 'A semi-distributed integrated nitrogen model for multiple source assessment in catchments (INCA): Part I – Model structure and process equations', *Sci. Total Environ.* **210/211**, 547–558.

ASSESSMENT OF THE WATER BALANCE OF EUROPEAN FORESTS: A MODEL STUDY

CAROLINE VAN DER SALM*, GERT JAN REINDS and WIM DE VRIES

Alterra Green World Research, P.O. Box 47, 6700 AA Wageningen, The Netherlands
(* author for correspondence, e-mail: carolinevandersalm@wur.nl; phone: +31 317 474700; fax: +31 317 419000)

(Received 20 August 2002; accepted 9 May 2003)

Abstract. As part of the UN-ECE Intensive Monitoring Program, data on precipitation, throughfall and soil solution concentrations are measured on a regular basis in approximately 300 forest stands. These data were used to construct element budgets for European forests. To construct such budgets drainage fluxes have to be modeled. In this paper, the research chain from model selection to data derivation and application of the selected model to 245 of the 300 sites is described. To select a suitable hydrological model the Cl^- balance method, two capacity models (a multi and a single layer version) and a Darcy model have been applied to two forest sites. The results indicate that drainage fluxes calculated with the Darcy model are more accurate than fluxes derived with the capacity model, in particular in situations where water availability is limited. The Darcy model was applied to the sites using a mixture of generic data and site data. Despite the use of generic data, the calculated drainage fluxes appear feasible. Median transpiration fluxes were 350 mm and the lowest values are found in northern Europe and highest values are found in central Europe. Median drainage fluxes were 150 mm yr^{-1} with the highest values in areas with high rainfall. Uncertainty analyses indicate that the use of local instead of interpolated meteorological data leads to lower drainage fluxes at 70% of the sites. The median deviation in calculated drainage fluxes is 20 mm yr^{-1}. The use of local soil data had little impact on the calculated fluxes.

Keywords: capacity model, Cl^- budgets, Darcy model, evapotranspiration, forests, hydrology

1. Introduction

Knowledge on the fate of chemical substances often is derived from long-term mass balances. For example, important information on the fate of nitrogen and on relationships between nitrogen losses and soil chemical properties has been derived from mass balances of nitrogen for forests (Dise *et al.*, 1998; Gundersen *et al.*, 1998). As part of the UN-ECE Intensive Monitoring Program, data on precipitation, throughfall, and soil solution concentrations are measured regularly in approximately 300 forest stands (De Vries *et al.*, 2001). Mass balances of chemical substances in European forests may be constructed on basis of the available data. The obtained results provide a unique opportunity to increase the insight in the fate of the major chemical substances in European forests.

To construct mass balances of chemical substances in forests, both the input and output fluxes have to be obtained. Input fluxes can be derived from measured throughfall and precipitation fluxes. The assessment of output fluxes, however, is more difficult. A lot of the UN-ECE intensive monitoring sites are deeply drained. The measurement of soil water fluxes at such sites is extremely difficult and is therefore hardly ever carried out. The alternative is to apply a hydrological model.

In the current study, soil water fluxes have been derived for 245 forests in order to obtain element budgets. The forests occur at a broad range of soil types and climate conditions. Data availability for the sites is rather limited. To calculate water fluxes different types of models may be used, ranging from simple budget models to comprehensive mechanistic models (Bouten and Jansson, 1995; Arah and Hodnett, 1997; Granier et al., 1999; Boyle et al., 2000). Mechanistic models are generally applicable to a broader range of conditions compared to empirical models. However, when deriving fluxes for the 245 sites, the advantages of a mechanistic model may be overruled by uncertainties caused by the lack of input data. The aim of this paper is to show the procedure used to select the most suitable model and to derive water fluxes for the 245 sites. In this paper, a research chain consisting of the following steps has been used: first, a hydrological model has been selected that is most suitable to calculate fluxes for the 245 sites. This model should be able to calculate not only realistic long-term fluxes, but also fluxes for relatively short time periods (week/month) because soil solution concentrations are measured at these intervals. To select such a model, a number of models were compared on two intensive monitoring sites. After selection of the most suitable model, input data have been derived for the 245 sites. The data have been applied to all sites and the results have been evaluated. Finally the impact of uncertainties in a number of important input parameters has been evaluated.

2. Model Selection

2.1. Hydrological models considered

Hydrological fluxes generally are calculated using Darcy models, capacity models, or using the Cl balance method. In this study, four models are considered to calculate fluxes for the 245 sites: a Darcy model (SWATRE; Belmans et al., 1993; Groenenberg et al., 1995), a single-layer and a multi-layer capacity model (Hendriks et al., 1997), and the Cl balance method.

In Darcy models, water fluxes are calculated on basis of Darcy's law:

$$Q = -K(h)(\frac{dh}{dz} + 1)$$

where Q is the water flux (m^3 m^{-2} d^{-1}), $K(h)$ is the hydraulic conductivity, and dh/dz is the pressure head gradient in the vertical direction. This equation is combined with the continuity equation to yield the Richards equation to calculate changes in water content over time. In capacity models, it is assumed that outflow only occurs when the maximum storage capacity (field capacity) of a soil layer is exceeded. The output flux of a soil layer than equals the water excess during the time step:

$$q = F_{in} - F_{up} - \frac{dS}{dt}$$

where F_{in} is the input flux (m d^{-1}), F_{up} is the root uptake flux (m d^{-1}), and dS/dt is the change in water content (m d^{-1}). When the soil water content is less than the maximum soil water content, the water flux is zero and rainfall and evapotranspiration lead to changes in water content only. One of the main weaknesses of this model is that water fluxes only occur above field capacity. Moreover, this model ignores the capillary rise of (ground) water and is thus limited to well drained soils with a deep water table (Vanclooster and Boesten, 2000).

An alternative way to calculate water fluxes is the Cl$^-$ balance method. This method is based on the assumption that Cl$^-$ does not react with the soil. Water fluxes can than be calculated as:

$$q_o = \frac{q_i c_i}{c_o}$$

where c_i is the input concentration, c_o is the output concentration, q_i is the input flux, and q_o is the output flux. But this method can only be successfully applied to time steps at which there is no change in storage of water in the soil.

Differences between hydrological models also arise from differences in the way (evapo)transpiration and interception are dealt with. The capacity model and the Darcy model used in this study use the same (evapo)transpiration and interception sub-models. Interception evaporation was calculated using the Gash model (Gash, 1979; Gash et al., 1995). Potential evapotranspiration can be calculated in both models using either the Penman-Monteith (Monteith, 1981) or the Makkink equation (Makkink, 1957). Potential soil evaporation is calculated as a function of the potential evapotranspiration, the leaf area index, and the length of the dry period. Actual transpiration was calculated as a function of the water content (capacity model) or the pressure head (Darcy model).

The choice of model depends on the data availability, the time period considered and the desired accuracy of the calculated fluxes. An overview of the basic data needs of the models is given in Table I. The Cl$^-$ balance method only requires Cl$^-$ concentrations and throughfall fluxes. The model can be applied successfully when long-term average output fluxes (several years) have to be calculated. The capacity and Darcy models both require a substantial amount of data. The main difference is that the capacity model needs less information on soil physical characteristics, only the water content at field capacity has to be known.

TABLE I

Data requirements of the three types of models

Data	Model		
	Darcy'	Capacity	Chloride balance
Precipitation	X	X	
Throughfall			X
Parameters to calculate evapotranspiration	X	X	
Soil water retention curve	X	X^a	
Hydraulic conductivity	X		
Cl concentrations in throughfall and soil water			X

[a] Only the water content at field capacity is needed.

TABLE II

Forest-stand characteristics used in the model evaluation procedure

	Solling	Speuld
Country	Germany	The Netherlands
Location	51° 40' N, 9° 30' E	52° 13' N, 5° 39' E
Altitude	505 m	40 m
Soil type	Spodic Dystric Cambisol	Cambic Podzol
Texture	Loam to loamy silt	Loamy sand
Forest	Norway Spruce	Douglas Fir
Year of plantation	1888	1960
Water table	> 50 m	> 40 m
Average yearly precipitation (mm)	1037	804
Considered period	1976–1989	1985–1989

2.2. MODEL COMPARISON

2.2.1. Study sites

To evaluate the model performance, models were applied to an intensively monitored forest stand (Table II) in Germany (Solling) and in the Netherlands (Speuld). The Cl^- balance model has been applied only to the Solling site, as the monitoring period for the Speuld site was too short and because of the large spatial variability and the limited measurements of Cl^- concentrations during the summer month due to drought.

For both sites, comprehensive datasets (including meteorological, soil solution data, and soil physical data) are available (Tiktak *et al.*, 1995a; Tiktak *et al.*, 1995b). Input data for the models were derived from previous studies in which

TABLE III

Average yearly drainage fluxes[a] at 90 cm depth (mm a^{-1})

	Darcy model (SWATRE)	Capacity model (single layer)	Capacity model (multi layer)	Cl$^-$ balance method
Solling (12 yr)	409	409	418	395
Solling (4 yr)	438	411	417	440
Speuld (4 yr)	224	94	137	–

[a] Fluxes have been calculated for hydrological years (starting April 1st), to minimize year to year differences in fluxes due to storage of water.

the SWATRE model was applied to Solling (Groenenberg *et al.*, 1995) and Speuld (Tiktak *et al.*, 1995b; Kros, 2002). For the capacity models, parameters were directly translated from the data used for the Darcy model. All models were run on a daily time step basis. The thickness of the soil layers ranged from 2.5 cm in the topsoil to 10 cm in the subsoil, except for the single layer model in which a soil layer of 90 cm was considered.

3. Results

The average yearly drainage fluxes at 90 cm depth were quite comparable for the four models at the Solling site (Table III). The differences in average yearly fluxes were much more pronounced for Speuld. The drainage fluxes calculated by the Darcy model were more than two times those calculated by the single layer capacity model. The fluxes calculated by the multi-layer capacity models were also lower (39%) than fluxes calculated by the Darcy model.

The difference in fluxes between the Darcy model and the two capacity models were larger when relatively short time periods (four years or less) are considered (Table III). The difference between the models was relatively small (up to 10%) in years with an average rainfall (Figure 1). Large differences occurred in dry years such as 1976 and 1982. The capacity models calculated lower fluxes compared to the Darcy model in these dry years, due to the fact that drainage ceases completely as soon as the water content of the soil is lower than field capacity. The Cl$^-$ budget method strongly overestimated drainage fluxes in dry years, due to the fact that soil solution can not be sampled with lysimeter cups during the dry periods.

The seasonal distribution of the fluxes differed for the three models. This is clearly illustrated by the calculated monthly drainage fluxes in Speuld. In relatively dry years (e.g. 1985), the capacity models indicate that drainage fluxes were zero from February until October (Figure 2). In wet years (e.g. 1987), the capacity models calculated substantially higher fluxes during the summer, compared to the Darcy model due to the fact that in the capacity models water is lost by drainage as soon as the field capacity is exceeded.

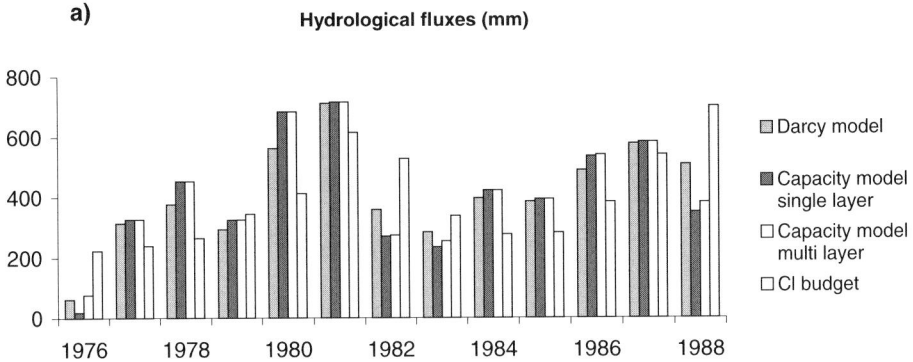

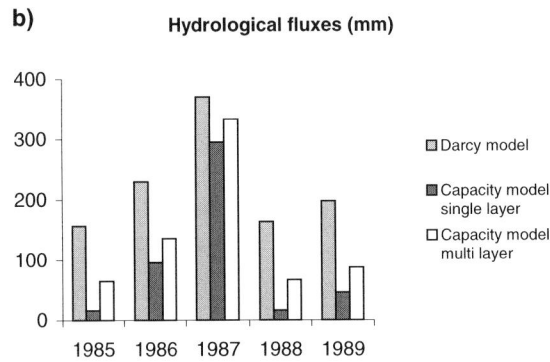

Figure 1. Yearly drainage fluxes (mm a^{-1}) at the Solling site (a) and the Speuld site (b).

The differences in calculated drainage fluxes are clearly reflected in the simulated water contents. In Speuld, water contents were measured continuously in 1989 using TDR (Tiktak and Bouten, 1992, 1994). The Darcy model simulated the measured water contents quite well (Figure 3). The capacity models failed to simulate the decline in water content during the dry summer month. These deviations are due to the fact that drainage ceases completely in the capacity models when the water content drops below field capacity.

3.1. SELECTION OF THE HYDROLOGICAL MODEL

The application of the three hydrological models to the above intensive monitoring sites makes clear that drainage fluxes calculated with the Darcy model are more reliable than fluxes derived with the capacity model. Differences between calculated drainage fluxes are most pronounced in situations where water availability is limited, whereas differences are minimal when precipitation is high. A disadvantage of the use of a Darcy model is the greater data need with respect to soil physical

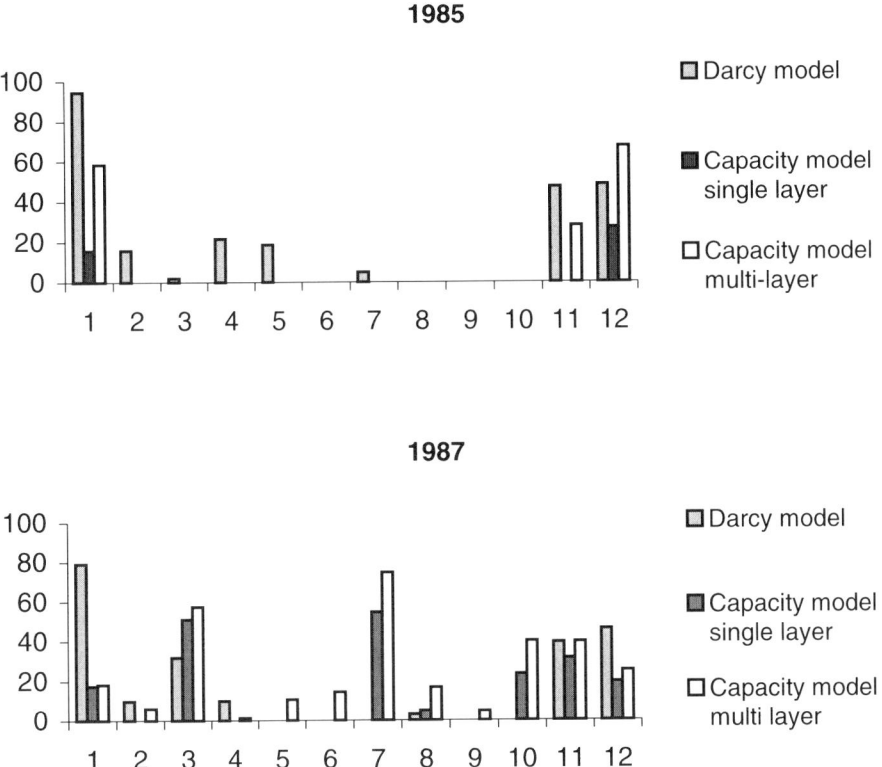

Figure 2. Calculated monthly fluxes (mm) at 90 cm depth at the Speuld site.

parameters. When measurements are lacking, these parameters may be obtained from available databases such as those presented by Wösten *et al.* (1999) or Leij *et al.* (1996). However, this may lead to less accurate results, thus overruling the advantages of a more complicated model.

The influence of the use of generic soil physical data (Wösten *et al.*, 1999) on the simulated output fluxes was evaluated for Speuld and Solling. At the Solling site, the use of generic soil physical data led to a slight overestimation of the drainage fluxes (Figure 4). The largest deviation was found for the Darcy model. With this model, the average drainage fluxes were 27 mm a^{-1} greater. At the Speuld site, the use of generic data for the Darcy model led to 26% lower average output fluxes compared to the use of measured data. The use of generic data had a smaller impact on the fluxes simulated with the capacity models but even with generic data the Darcy model gives a more realistic estimate of the fluxes than the capacity models.

On basis of the above results, the Darcy model was selected to calculate the fluxes for the 245 forest stands. For a limited number of sites, the fluxes derived on

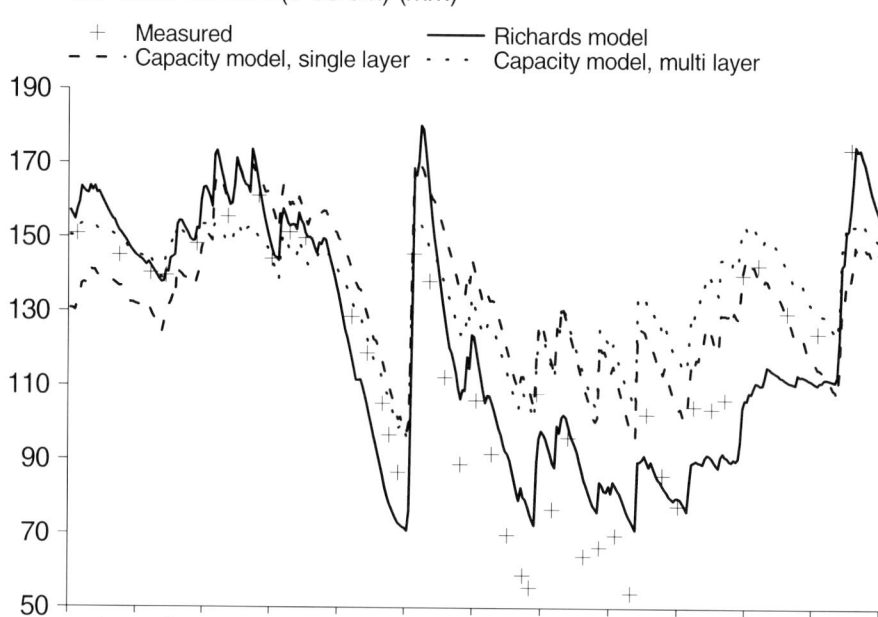

Figure 3. Measured and simulated soil water content (mm) for 1989 in the upper 90 cm of the soil in Speuld.

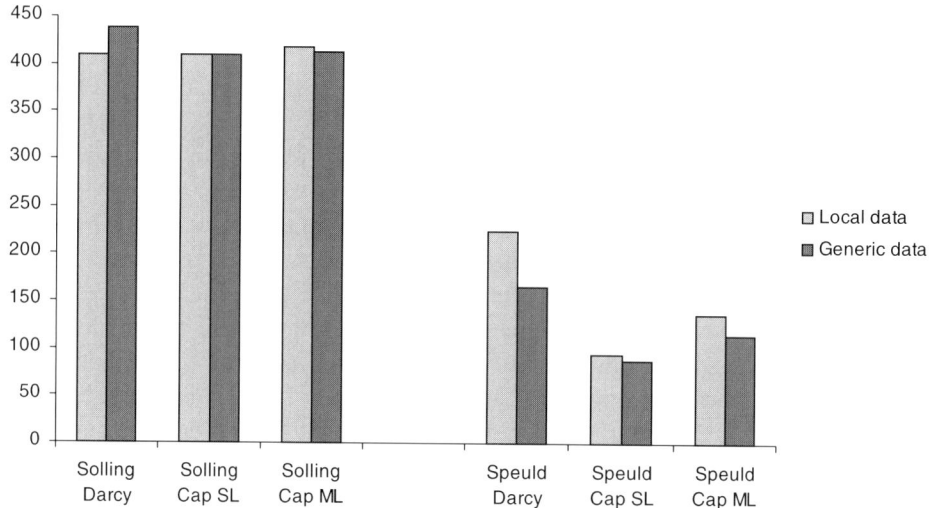

Figure 4. Average yearly drainage fluxes for the Solling and Speuld sites using local and generic soil physical data.

4. Model Application and Uncertainties

4.1. DATA DERIVATION

4.1.1. *Meteorological Data*
To calculate the water fluxes, daily data on precipitation, net radiation, temperature, wind speed, and relative humidity are required. Daily measurements were only available for a limited number (72) of plots. Therefore, to obtain consistent budgets, we did not use the local data, but used interpolated data for all plots.

Data on precipitation, net radiation, temperature, wind speed, and humidity were derived by interpolation between data from 1300 European meteorological stations. Interpolation to each site was performed using an inverse distance weighting procedure for four meteorological stations located in the surroundings of the plot (Klap *et al.*, 1997).

Precipitation may vary considerably over short distances, leading potentially to large differences in daily precipitation between interpolated site data and the actual measured precipitation at the site. At all sites, precipitation data are available on a weekly (47%), biweekly (27%), or monthly (26%) basis. To obtain a best estimate of the daily precipitation at the site, interpolated daily precipitation data were corrected using the measured biweekly or monthly values according to:

$$P_{i,site} = P_{i,int} \cdot \frac{P_{period,site}}{P_{period,int}}$$

where $P_{i,site}$ is the daily interpolated and corrected precipitation at the site, $P_{i,int}$ is the daily interpolated precipitation for the site, $P_{period,site}$ is the measured precipitation at the site during a given measurement period (two weekly or monthly), and $P_{period,int}$ is the interpolated precipitation during this period.

4.1.2. *Parameterization of the Interception Module*
Apart from meteorological data, the interception module (Gash) needs data on the storage and evaporation from the canopy. These data were not available and instead the parameters were calibrated on measured (weekly to monthly) throughfall data for the 245 plots using an automatic calibration procedure based on the Levenberg-Marquardt algorithm. A full description of the calibration procedure can be found in De Vries *et al.* (2001).

4.1.3. *Soil and vegetation parameters*
Soil-related parameters (such as number and thickness of the soil layers, soil physical characteristics) and vegetation-dependent parameters (e.g., crop factor,

TABLE IV
Derivation of the main input parameters for SWATRE

Parameter	Obtained from
Depth of the soil profile	Soil type and soil phase
Lower boundary conditions	Soil type
Texture of the layers	Directly from the Intensive Monitoring database, from the parent material or based on the FAO soil code
Soil physical characteristics	Based on texture data using transfer function (Wösten et al., 1999)
Root distribution	Depth of the soil profile and literature data (De Visser and de Vries, 1989)
Basic crop resistance	Literature values for different tree species (Hendriks et al., 1997)
Tree height	Intensive Monitoring database or estimated (Klap et al., 1997)
Interception parameters	Calibration on measured throughfall

evapotranspiration parameters) were not available in the Intensive Monitoring database and have to be derived indirectly using literature data and transfer functions (Table IV).

The Intensive Monitoring database does not (yet) contain profile descriptions of the various sites, and therefore the depth of the soil profile has been derived on basis of the soil type and soil phase. Information on the soil type was available for each plot in the database. The soil phase for each plot was derived by overlaying the map with plot locations with the soil map of the Soil Geographical Database of Europe at a 1:1 Million scale (Eurosoil, 1999). The depth of the profile was derived from a pedotransfer rule indicating the depth of soil profile (shallow (< 40 cm), moderate (40–80 cm), deep (80–120 cm) and very deep (> 120 cm)) as a function of soil code and phase (Pedotransfer rule 411, European Soil Bureau). A depth of 30, 60, 100 and 230 cm has been allocated to the respective depth classes. The soil profile has been divided into a maximum of 7 horizons, which were subdivided into layers ranging in thickness from 2.5 cm in the top to 25 cm at the bottom of the soil profile. The thickness of the organic layer was calculated from the measured weight of the organic layer (kg m^2) and the bulk density of that layer (kg m^3) calculated with a transfer function that relates bulk density to the measured C content of organic soils (van Wallenburg, 1988).

With respect to the lower boundary conditions, for all soil profiles free drainage of soil water at the bottom of the soil profile was assumed. Lithosols and Gleysols were not represented in the selection of 245 European sites.

TABLE V

Simulated median drainage fluxes (mm a^{-1}) and Cl$^-$ budgets (mol$_c$ ha^{-1} a^{-1}) for the 245 sites

Tree species	Number of sites	Number of years	Calculated fluxes[a] (mm a^{-1})					Cl$^-$ budget (mol$_c$ ha^{-1}a^{-1})
			P	E$_I$	E$_t$	E$_s$	D	
Pine	51	3	642	152	314	55	79	211
Spruce	98	3	963	290	385	32	205	167
Oak	24	3	725	177	338	105	123	− 3
Beech	45	3	891	241	356	94	138	113
Others	27	3	1122	300	389	81	278	102
All	245	3	860	240	356	64	152	135

[a] Average yearly fluxes over the monitoring period with P = precipitation, E$_i$ = interception evaporation, E$_t$ = transpiration, E$_s$ = soil evaporation and D= drainage flux.

Soil physical characteristics for each horizon were derived using transfer functions (Wösten et al., 1999), depending on the texture class and whether the layer is assumed to be a topsoil (A or E horizon) or a subsoil compartment (B and C horizons). Texture for the different horizons were either obtained directly from the Intensive Monitoring database (voluntary submissions) or derived from the FAO soil code (FAO Composition Rules, 1981). Soil physical characteristics for the organic layer were taken from data for a Douglas fir stand (Schaap et al., 1997), data for other tree species are not available in literature.

Root distribution was based on estimated soil depth and literature data for deciduous and coniferous forest (De Visser and de Vries, 1989). Basic crop resistance (r$_s$) values for the calculation of the potential evapotranspiration were derived from the literature (Hendriks et al., 1997). Values ranged from 50 s m^{-1} for oak species, 85 s m^{-1} for beech, 90 s m^{-1} for spruce and Douglas fir, to 100 s m^{-1} for pine. Tree height was based on the reported measurements in the intensive monitoring database. For plots where data for tree height was not reported (26 sites), it was estimated as a function of tree species, tree age, climatic zone and the C/N ratio (Klap et al., 1997).

4.2. RESULTS

The calculated transpiration and drainage fluxes showed plausible values and trends over Europe (Figure 5, Table V). Low transpiration fluxes were found in northern Scandinavia and locally in southern Europe. The highest fluxes were found in central and western Europe. In large parts of Europe drainage fluxes were between 150 and 250 mm yr^{-1}. Low values were found in areas with low precipitation in northeastern Germany, Sweden and Spain. The highest fluxes were found at sites with high precipitation such as mountainous areas and Western Europe.

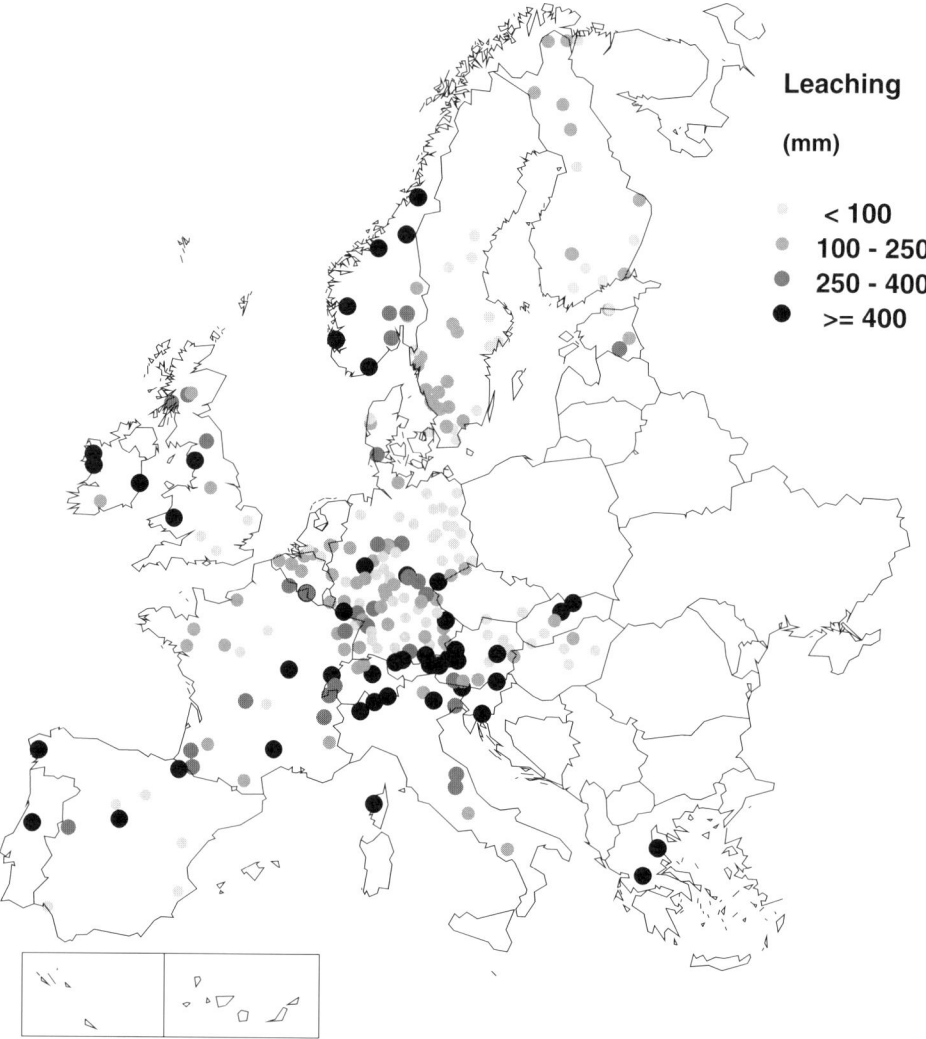

Figure 5. Average transpiration fluxes (a) and drainage fluxes (b) for the 245 monitoring plots.

Median transpiration fluxes ranged from 315 mm a^{-1} at the pine stands to 385 mm a^{-1} for spruce. Median transpiration fluxes were close to the mean value of 333 mm a^{-1} of Roberts (1983), who found that transpiration fluxes for European forest are in a very narrow range due to feedback mechanisms with soil and atmosphere. Soil evaporation fluxes were highest for the oak stands, which were generally less dense. The median drainage fluxes ranged from 79 mm on pine stands compared to 205 mm for Spruce. The drainage fluxes for pine are quite low because the median precipitation is lower on pine compared to spruce and beech.

Median values for the Cl^- budgets were generally positive, which might indicate that there is a slight tendency for an underestimation of the drainage fluxes. De-

TABLE VI

Simulated median drainage fluxes (mm a^{-1}) using local and generic data

Flux (mm a^{-1})	Impact of meteorological data (72 sites)		Impact of soil data (14 sites)		
	Local data	Interpolated data	Local texture and soil profile[a]	Local texture[b]	Generic data[c]
Precipitation	845	895	1251	1251	1251
Interception	228	244	432	432	432
Throughfall	609	618	819	819	819
Transpiration	340	384	445	450	448
Soil evaporation	68	74	31	31	31
Drainage flux	183	145	287	284	284

[a] Fluxes calculated using both the local soil profile descriptions and texture analysis.
[b] Fluxes calculated using local texture analysis.
[c] Fluxes calculated using data derived form the 1:1 million soil map.

viations of the Cl$^-$ budget from zero are quite common when only a limited number of years is considered. For example, at the Solling spruce stand, Cl$^-$ budgets for a four year period ranged from –227 to 130 mol$_c$ ha^{-1} a^{-1} with a median value of 57 mol$_c$ ha^{-1} a^{-1}.

4.3. UNCERTAINTIES IN CALCULATED FLUXES

Interpolated meteorological data are used instead of local data and soil data (including soil depth, texture, root depth, and soil physical characteristics) were derived indirectly from the soil type. To quantify the impact of uncertainty in the input data on the calculated fluxes the following two comparisons were carried out. First, the impact of meteorological data was evaluated for 72 sites for which both local and interpolated meteorological data were available. Second, for 14 sites in France, soil profile descriptions and texture analysis were available. These data were used instead of the generic data derived on basis of soil type. The soil profile data provided more accurate data on soil depth and root depth. The texture data were used to make better local estimates of the soil physical characteristics. The total impact of uncertainty in the soil physical characteristics could not be evaluated because local soil physical characteristics were not available for these sites.

The use of local meteorological data led to higher median drainage fluxes, caused by a lower median (evapo)transpiration (Table VI). This deviation is largely caused by higher wind speed in the interpolated dataset, which is probably due to the fact that meteorological stations are usually located in vast open areas. The cumulative frequency distribution shows that drainage fluxes were underestimated at 70% of the sites when generic data are used. The median deviation is approx-

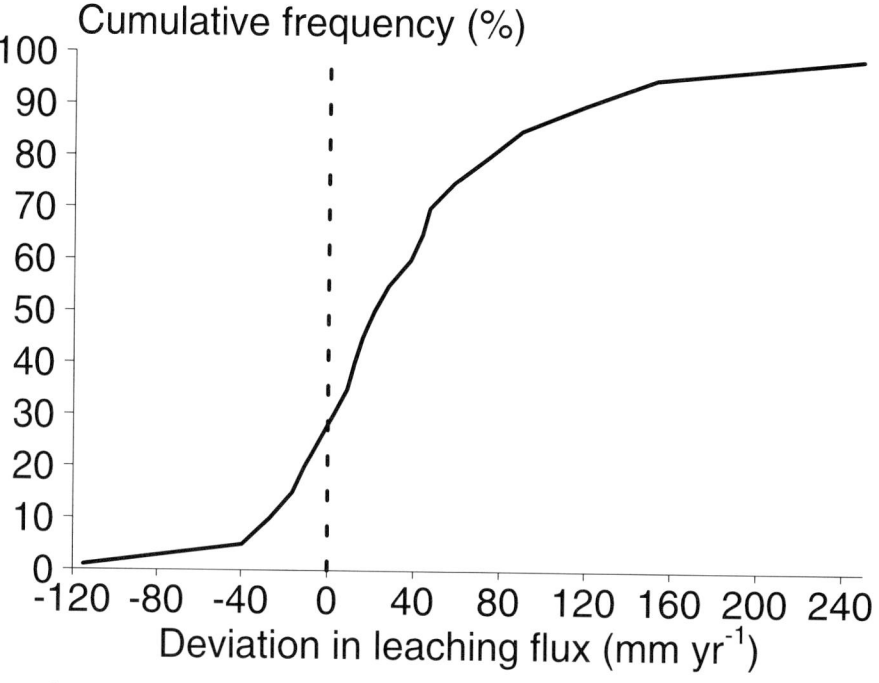

Figure 6. Cumulative frequency distribution of the deviation (local-interpolated) in simulated drainage fluxes when using local meteorological data instead of generic data.

imately 20 mm and 60% of the sites have a deviation of less than 40 mm (Figure 6).

The impact of the use of local soil data on the derived fluxes is very limited. This is partly due to the fact that the pedotransfer function used considers only a five-texture class. At almost 50% of French sites, the same texture class is assigned on basis of the parent material and the measured texture data. However, even when the texture data lead to another texture class the impact is generally small (median deviation in drainage flux of 12 mm).

5. Conclusions

To select a suitable model to calculate water fluxes for 245 forest stands four models were applied to two intensively monitored forest stands. This model comparison showed that the Darcy model was most suitable to calculate the drainage fluxes for a broad range of circumstances. The (multi-layer and single layer) capacity models performed quite well under wet conditions, but less under dry conditions. The Cl^- budget method performed quite well under wet conditions when an extensive period (> 4 yr) was considered.

The Darcy model was applied to all 245 sites and results showed that the magnitude and the geographic distribution of the drainage fluxes were plausible. However, median Cl⁻ budgets where slightly positive, indicating a slight underestimation of the drainage fluxes. The uncertainty analysis confirmed that the use of generic meteorological data leads to an underestimation of the drainage fluxes. This underestimation is mainly caused by too high wind speeds in the interpolated meteorological data set. The use of generic soil data had little influence of the calculated fluxes when pedotransfer functions are used to derive soil physical characteristics. However, the use of locally measured soil physical characteristics may have a larger impact on the simulated fluxes.

References

Arah, J. and Hodnett, M.: 1997, 'Approximating soil hydrology in agroforestry models', *Agroforestry Forum* **8**, 17–20

Belmans, C., Wesseling, J. G. and Feddes, R. A.: 1983, 'Simulation model of the water balance of a cropped soil providing different types of boundary conditions: SWATRE', *J. Hydrol.* **63**, 271–286.

Bouten, W. and Jansson, P. E.: 1995, 'Water balance of the Solling spruce stand as simulated with various forest-soil-atmosphere models', *Ecol. Model.* **83**, 245–253.

Boyle, G. M., Farrell, E. P., Cummins, T. and Nunan, N.: 2000, 'Monitoring of forest ecosystems in Ireland', Forest Ecosystem Research Group Report 48, University College Dublin, Ireland.

De Vissser, P. and de Vries, W.: 1989, 'De gemidelde jaarlijkse waterbalans van bos-, heide- en graslandvegetaties', *Rapport 2085*, Stichting voor Bodemkartering, Wageningen, The Netherlands.

Dise, N.B., Matzner, E. and Gundersen, P.: 1998, 'Synthesis of nitrogen pools and fluxes from European forest ecosystems', *Water, Air, Soil Pollut.* **105**, 143–154.

De Vries, W., Reinds, G. J., van der Salm, C., Draaijers, G. P. J., Bleeker, A., Erisman, J. W., Auee, J., Gundersen, P., van Dobben, H., de Zwart, D., Derome, J., Voogd, J. C. H. and Vel, E.: 2001, 'Intensive monitoring of forest ecosystems', *Technical Report 2001*, EC-UN/ECE 2001, Brussels, Geneva, 166 pp.

Eurosoil: 1999, 'Metadata: soil geographical data base of Europe v. 3.2.8.0', *Eurosoil*, Ispra, Italy.

FAO: 1981, 'FAO-Unesco soil map of the world 1:5.000.000', Volume V Europe, *UNESCO*-Paris, 199 pp.

Gash, J. H. C.: 1979, 'An analytical model of rainfall interception by forests', *Quart. J. R. Meteorol. Soc.* **105**, 43–55.

Gash, J. H. C., Lloyd, C. R. and Lachaud, G.: 1995, 'Estimating sparse forest rainfall interception with an analytical model', *J. Hydrol.* **170**, 79–86.

Granier, A., Bréda N., Biron, P. and Vilette, S.: 1999, 'A lumped water balance model to evaluate duration and intensity of drought constraints in forest stands', *Ecol. Model.* **116**, 269–283.

Groenenberg, J. E., Kros, van der Salm, J. C. and de Vries, W.: 1995, 'Application of the model NuCSAM to the Solling spruce site', *Ecol. Model.* **83**, 97–107.

Gundersen, P., Callesen, I. and de Vries, W.: 1998, 'Nitrate leaching in forest ecosystems is related to forest floor C/N ratios', *Environ. Pollut.* **102**, 404–407.

Hendriks, C. M. A., de Vries, W., van Leeuwen, E. P., Oude Voshaar, J. H. and Klap, J. M.: 1997, 'Assesment of the possibilities to derive relationships between stress factors and forest condition for the Netherlands', *Report 147*, DLO-Winand Staring Centre, Wageningen, The Netherlands.

Klap, J. M., de Vries, W., Erisman, J. W. and van Leeuwen, E. P.: 1997, 'Relationships between forest condition and natural and anthropogenic stress factors on the European scale; a pilot study', DLO-Winand Staring Centre, *Report 150*, Wageningen, The Netherlands.

Kros, J: 2002, 'Evaluation of biogeochemical models at local and regional scale', *Ph. D. Thesis*, Wageningen University, Wageningen, The Netherlands, 284 pp.

Leij, F. J., Alves, W. J., van Genuchten, M. Th. and Williams, J. R.: 1996, 'Unsaturated Soil Hydraulic Database, UNSODA 1.0 User's Manual', Report EPA/600/R96/095, U.S. Environmental Protection Agency, Ada, Oklahoma, 103 pp.

Makkink, G. F: 1957, 'Testing the Penman formula by means of lysimeters', *J. Inst. Water Eng.* **11**, 277–288.

Monteith, J. L.: 1981, 'Evaporation and surface temperature', *Quart. J. R. Met. Soc.* **107**, 1–27.

Roberts, J.: 1983, 'Forest transpiration: a conservative hydrological process?', *J. Hydrol.* **66**, 133–141.

Schaap, M. G., Bouten, W. and Verstraten, J. M.: 1997, 'Forest floor water content dynamics in a Douglas fir stand', *J. Hydrol.* **201**, 367–383.

Tiktak, A. and Bouten, W.: 1992, 'Modelling soil water dynamics in a forested ecosystem. III: Model description and evaluation of discretization', *Hydrol. Process.* **6**, 455–465.

Tiktak, A. and Bouten, W.: 1994, 'Soil water dynamics and long-term water balances of a Douglas fir stand in the Netherlands', *J. Hydrol.* **156**, 265–283.

Tiktak, A., Bredemeier, M. and van Heerden, C.: 1995a, 'The Solling dataset. Site characteristics, monitoring data and deposition scenarios', *Ecol. Model.* **83**, 17–34.

Tiktak, A., van Grinsven, J. J. M., Groenenberg, J. E., van Heerden, C, Janssen, P. H. M., Kros, J., Mohren, G. M. J., van der Salm, C., van de Veen, J. R. and de Vries. W.: 1995b, 'Application of three Forest-Soil-Atmosphere models to the Speuld experimental forest', National Institute of Public Health and Environmental Protection, Bilthoven, the Netherlands, Report 733001003.

Vanclooster, M. and Boesten, J. J. T. I.: 2000, 'Application of pesticide simulation models to the Vredepeel dataset I. Water, solute and heat transport', *Agric. Water Manage* **44**, 105–117.

van Wallenburg, C.: 1988, 'De bodemdichtheid van koopveen-, weideveen- en waardveengronden in relatie to bodemkenmerken', *Rapport 2040*, Stichting voor Bodemkartering, Wageningen, The Netherlands.

Wösten, J. H. M., Lilly, A., Nemes, A. and Le Bas, C.: 1999, 'Development and use of a database of hydraulic properties of European soils', *Geoderma* **90**, 169–185.

EFFECTS OF CLIMATE AND ECOSYSTEM DISTURBANCES ON BIOGEOCHEMICAL CYCLING IN A SEMI-NATURAL TERRESTRIAL ECOSYSTEM

CLAUS BEIER[1]*, INGER KAPPEL SCHMIDT[2] and
HANNE LAKKENBORG KRISTENSEN[3]

[1] *RISØ National Laboratory, P.O.B. 49, DK-4000 Roskilde, Denmark;* [2] *Danish Forest and Landscape Research Institute, Hørsholm Kongevej 11, DK-2970 Hørsholm, Denmark;* [3] *Danish Institute of Agricultural Sciences, Dept. of Horticulture, P.O.B. 102, DK-5792 Aarslev, Denmark*
(* *author for correspondence, e-mail: claus.beier@risoe.dk; phone: +45 4677 4161; fax: +45 4677 4160*)

(Received 8 May 2002; accepted 15 May 2003)

Abstract. The effects of increased temperature and potential ecosystem disturbances on biogeochemical cycling were investigated by manipulation of temperature in a mixed *Calluna*/grass heathland in Denmark. A reflective curtain covered the vegetation during the night to reduce the heat loss of IR radiation from the ecosystem to the atmosphere. This 'night time warming' was done for 3 years and warmed the air and soil by 1.1 °C. Warming was combined with ecosystem disturbances, including infestation by *Calluna* heather beetles (*Lochmaea suturalis* Thompson) causing complete defoliation of *Calluna* leaves during the summer 2000, and subsequent harvesting of all aboveground biomass during the autumn. Small increases in mineralisation rates were induced by warming and resulted in increased leaching of nitrogen from the organic soil layer. The increased nitrogen leaching from the organic soil layer was re-immobilised in the mineral soil layer as warming stimulated plant growth and thereby increased nitrogen immobilisation. Contradictory to the generally moderate effects of warming, the heather beetle infestation had very strong effects on mineralisation rates and the plant community. The grasses completely out-competed the *Calluna* plants which had not re-established two years after the infestation, probably due to combined effects of increased nutrient availability and the defoliation of *Calluna*. On the short term, ecosystem disturbances may have very strong effects on internal ecosystem processes and plant community structure compared to the more long-term effects of climate change.

Keywords: defoliation, ecosystem disturbance, experimental manipulation, ecosystem response, heathland, nitrogen cycling, temperature increase, warming

1. Introduction

Historical records show an increase in global mean temperatures of 0.6 °C over the last 100 years (Houghton *et al.*, 2001), which has co-occurred with elevated atmospheric CO_2 (Watson *et al.*, 1991; Luxmoore *et al.*, 1998). The increase over land has been due mainly to an increase in the diurnal minimum temperatures (T_{min}), which have increased twice as much as maximum temperatures (T_{max}), primarily because of increased cloudiness (Watson *et al.*, 1995). Scenarios for anthropogenic emissions of CO_2 and other greenhouse gases have been predicted to cause increased global temperatures of 1–3.5 °C and changes in precipitation patterns with more severe droughts and floods (Watson et al., 1995).

Temperature and water are the main drivers for many biological and chemical processes and climatic changes are therefore likely to have a large influence on the functioning of natural and semi-natural environments (Watson et al., 1995). Particularly, global warming is expected to stimulate mineralisation processes leading to increased loss of C and nutrients, but also stimulating plant growth leading to increased C fixation and nutrient immobilisation (Shaver et al., 2000). In a recent meta-analysis study synthesising results from a large number of warming experiments, a general warming-induced increase in soil respiration of 20% and in nitrogen mineralisation of 46% was found, while the effects on biomass accumulation were less clear (Rustad et al., 2001). The net result of these processes determines the overall feedback between carbon cycle and climate, and thereby the acceleration or deceleration of the global warming (Shaver et al., 2000). Increased mineralisation may lead to other effects, such as increased nutrient availability. If nutrient availability exceeds plant and microbial demand, nutritional constraints on the ecosystem will be altered and increase the potential, not only for plant growth, but also for changes in species composition (Heil and Bobbink, 1993) and nutrient losses by leaching to ground or surface waters (Lukewille and Wright, 1997). In this sense global warming may have cascading effects on biogeochemical processes silimar to those that have been described for increased atmospheric nitrogen input to terrestrial ecosystems (Galloway and Cowling, 2002).

Also, vegetation defoliation can cause disturbance to biogeochemical cycling in terrestrial ecosystems. In *Calluna* heathlands, infestation by heather beetles (*Lochmaea suturalis* Thompson) is a naturally occurring disturbance that can cause complete defoliation of the *Calluna* vegetation and have major effects on biogeochemical cycling in heathlands (Heil and Bobbink, 1993). Defoliation increases the transfer of litter and frass to the soil, increases light penetration, and reduces plant nutrient uptake leading to accelerated decomposition of the top organic soil layer, increased N mineralization, and leaching of soluble organic matter and nutrients to deeper soil layers (Kristensen and McCarty, 1999; Nielsen et al., 2000). Many of the biogeochemical processes affected by such ecosystem disturbances also are affected by climatic changes, and ecosystem disturbances therefore are likely to interact with climatic changes to amplify or reduce the effects. However, interactions between climate and ecosystem disturbances and the effects on biogeochemical cycling are little known.

The present paper describes a field scale manipulation experiment in a Danish heathland as part of the EU-projects CLIMOOR (Climate Driven Changes in the Functioning of Heath and Moorland Ecosystems) and VULCAN (Vulnerability Assessment of Shrubland Ecosystems in Europe under Climatic Changes). The treatments included simulation of global warming. After one year of experimental manipulation, the site was severely affected by infestation with heather beetles, followed by cutting of the vegetation in an attempt to restore the *Calluna* vegetation.

The aims of the present paper are to investigate the effects of warming on the main ecosystem processes driving changes in biogeochemical cycling and to contrast these climate driven changes with the effects caused the heather beetle infestation and cutting of vegetation.

2. Material and Methods

2.1. SITE DESCRIPTION

The study was conducted in a heathland area at Mols Bjerge (56°23 N; 10°57 W) 5 km from the east coast of Jutland in central Denmark. The site is 57 m above sea level in a hilly glacial sandy moraine from the late phase of the Würm glaciation. The climate is relatively dry with an average annual precipitation of 550 mm (1960–1990). During the three year study reported here, average annual air temperature and precipitation was 8.7 °C and 750 mm, respectively. The site is affected by the nearby sea and the surrounding agriculture with high deposition levels of sea salts and moderate levels of nitrogen (wet deposition 38.6 kg Cl^- ha^{-1} yr^{-1} and about 13 kg N ha^{-1} yr^{-1}, respectively). The soil type is a sandy podzol with a 1–4 cm organic mor layer on top. The soil contains 11 mg N g^{-1} and 0.34 mg P g^{-1} in the organic layer and 0.4 mg N g^{-1} and 0.2 mg P g^{-1} in the mineral layer.

The site was cultivated until the 1950's and was grazed by sheep and cattle until 1992. Since 1992, the area has been a nature reserve with no management actions except some selective removal of pine trees and bushes. The vegetation in the experimental area was generally dominated by the evergreen dwarf shrub *Calluna vulgaris (L.) Hull* until the 1960's, whereafter an increasing amount of grass, mainly *Deschampsia flexuosa (L.) Trin*, has been observed. When the warming experiment was initiated in 1999, the aboveground biomass was about 1050 g m^{-2}. The vegetation was dominated by a mixture of *Calluna* (about 45%) and grasses, mainly *D. flexuosa* (about 46%). A naturally occurring outbreak of heather beetles (*Lochmaea suturalis* Thompson) started in July/August 1999 and peaked during the summer of 2000, defoliating and killing the main part of the *Calluna* plants (>95%). To help restore the *Calluna* vegetation after the heather beetle infestation all aboveground vegetation was cut and removed from the experimental plots in September 2000. During the autumn of 2000, and especially the following year of 2001, the vegetation re-established with complete dominance of *D. flexuosa*.

2.2. TREATMENTS

A field warming experiment was conducted as 'night time warming' at three study plots of 5 m H 4 m each. The warming plots were covered by a light scaffolding of galvanised steel tubes carrying a reflective aluminium curtain (ILS ALU, AB Ludvig Svensson, Sweden). The curtains reflected 97% of the direct and 96% of the diffuse radiation and allowed transfer of water vapour. The study plots were open at all sides. The movement of the curtains was automatically controlled according to pre-set climatic conditions:

- Light intensity – at sunset, the curtains were drawn over the vegetation to reduce the IR reflection, thereby conserving energy and leading to an increased temperature in the heated plots. At sunrise, the curtains were automatically removed to keep the plots open during the day.
- Rain – during rain events at night the curtains were automatically removed.
- Wind – a high wind speeds (>10 m/s) during the night, the curtains were automatically retracted to avoid damage to the systems.

The warming treatment increased the monthly average midnight air temperatures by 0.6–1.9 °C (average yearly increase 1.1 °C) and midnight soil temperatures by 0.6–1.6 °C (average yearly increase 1.1 °C). Parallel to the warming treatment, three untreated control plots were operated for comparison. The control plots were covered by a similar light scaffolding as for the warming treatment, but without any curtain. The control and warming plots were placed in three blocks within an area of 25 m H 40 m in the heathland. Three additional experimental plots dominated by intact *Calluna* that had not been infested by heather beetles were included in the investigation of N mineralisation. These plots were situated within a distance of less than 50 m from the control and warming plots.

2.3. MEASUREMENTS

2.3.1. *Input and Climatic Measurements*

The input of water and nutrients were measured by two open air rain gauges sampled in monthly intervals and analysed for NH_4^+, NO_3^- and DON (Autoanalyzer 3, Bran+Luebbe, Norderstedt, Germany). The functioning of the curtains and the extent of the treatment were recorded and checked by bihourly measurements of air and soil temperature (110 Termocouple Reference Thermistor types – probe 107, Campbell scientific, Logan, Utah, USA), radiation balance (NR-lite, Kipp & Zonen, Delft, the Netherlands), air humidity (VAISALA humitter 50, Helsinki, Finland), wind speed (2 RISØ Cup anemometer P2546A, Roskilde, Denmark), and monthly measurements of water input to each plot (funnels). For more details on the experimental setup, see Beier *et al.* (in press).

2.3.2. *Litter Decomposition*

Litter decomposition was studied by the litterbag technique. Fresh litter was collected in September 1999, air-dried in the lab and put into polyethylene litterbags (10 g litter, mesh size 1 mm, three bags per plot and species and sampling time) and installed in the field in May 2000 at the starting date of the drought period. Litter from *Calluna* and *D. flexuosa* was incubated separately and incubations were terminated after 1, 6 and 18 months. After collection litterbags were air-dried and cleaned of soil and ingrown plants. The remaining litter was weighed for estimation of weight loss and analysed for total C, N (EA 1110 elemental analyser, CE instruments, Milan) and lignin, determined by ADF (Acid Detergent Fibre=cellulose+lignin) followed by acid digestion (van Soest, 1963).

2.3.3. *Soil Mineralisation*
Soil nitrogen mineralisation and nitrification were measured by the buried bag technique. Measurements were conducted seasonally during one year starting in winter 1999/2000. At each sampling date, two replicate paired intact soil cores, 5 cm in diameter were taken from the top soil horizon under *Calluna* in each study plot including plots with intact *Calluna* outside the treatment. One of each pair was analysed for initial NH_4^+ and NO_3^- content. The other core was placed in a polyethylene bag and replaced in the ground. After incubation in the field for 1–2 months, cores were removed and analysed similarly. All soil cores were analysed individually for soil moisture (105 °C in 48 hr), organic matter content by combustion, and NH_4^+ and NO_3^- by extraction in 1M KCl (1:10 soil to extract ratio) and the change in inorganic-N (net mineralisation) and NO_3^--N only (net nitrification) during the incubation period was calculated as the change in extractable nitrogen content per gram organic matter ($\mu g\ N\ g\ OM^{-1}$).

2.3.4. *Soil Respiration*
Soil respiration was measured biweekly or monthly during 2000 and 2001 at two permanent chamber-bases (100 mm diameter) installed on bare soil in each plot. Measurements were done by an infrared gas analyser (PP-systems – EGM-2, Hertfordshire, UK) with a standard chamber, which was fitted onto the permanent chamber bases in the plots during the measurements.

2.3.5. *Soil Solution*
Soil water was collected monthly at two soil depths. Water percolating through the organic layer was collected by a 10 cm H 20 cm PVC tray placed beneath the organic layer and connected to a sampling bottle placed in a 50 cm soil pit outside the plot. Three samples were taken from each plot and pooled before analyses. Seepage water was collected continuously and sampled monthly in each plot by three PTFE tension soil suction cups (PRENART super quarz, Copenhagen, Denmark) connected to the same sampling bottle placed in a cold and dark soil pit outside the plot. Soil water samples were analysed for NO_3^-, NH_4^+ and TON (as for precipitation) and TOC (Shimadzu TOC-5000 Analyzer, Duisburg, Germany). TON was only measured campaign-wise.

2.3.6. *Element Fluxes*
Water balances in the soil were estimated with the water model EVACROP (Olesen and Heidmann, 1990) using measured numbers of temperature, precipitation and soil moisture. The element fluxes were calculated by multiplying the average

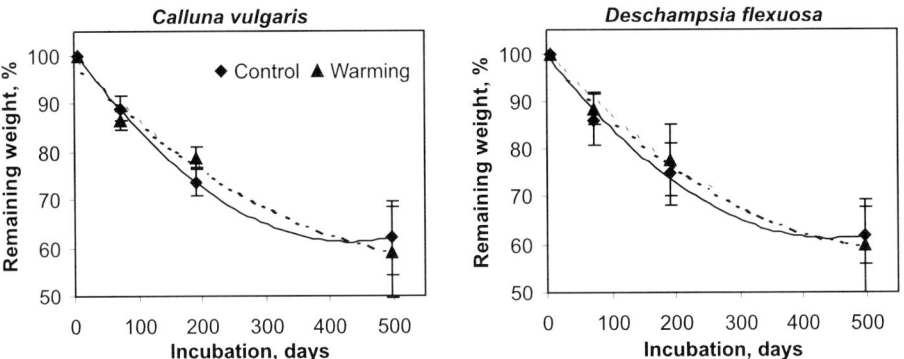

Figure 1. Decomposition of Deschampsia flexuosa and Calluna vulgaris litter in relation to treatment shown as remaining weight after incubation in litterbags for 30–500 days. Incubation started on 25th May 2000. Lines are best fitted 2nd order polynomials. Error bars indicate SE.

monthly concentrations in precipitation or soil solution with the average precipitation or modelled drainage in the soil.

The relative mobilization or retention in percentage was estimated as the difference between the estimated input and output fluxes (Equation 1).

$$\text{Retention} = (\text{Input}_{precipitation} - \text{Output}_{leachate} / \text{Input}_{precipitation}) \times 100 \quad (1)$$

2.4. STATISTICAL METHODS

A mean weight loss of litterbags was calculated for each plot, log transformed and rates of decomposition were analysed by a one-way ANOVA at each sampling period to determine the effect of warming. Repeated measures analyses of variance (rmANOVA) with temperature and block as main factors were applied to test the effects of the main factors on soil respiration and seasonal (e.g., winter, spring, early summer, late summer, autumn) mean concentrations of soil solution chemistry during the three years of measurement. Net N mineralisation and nitrification data were $\log(x+1)$ transformed before analysis in order to obtain homogeneous variances and analysed separately for each season (e.g., winter, spring, summer and autumn) by one-way ANOVA with temperature or vegetation type as main factor. All statistical analyses were performed with SAS using the GLM procedure and type II sum of squares. Differences were considered significant at $p < 0.05$ level.

3. Results

The decomposition of plant litter followed a general pattern with a fast initial phase followed by a slower phase (Figure 1). 40% of the litter material was decomposed

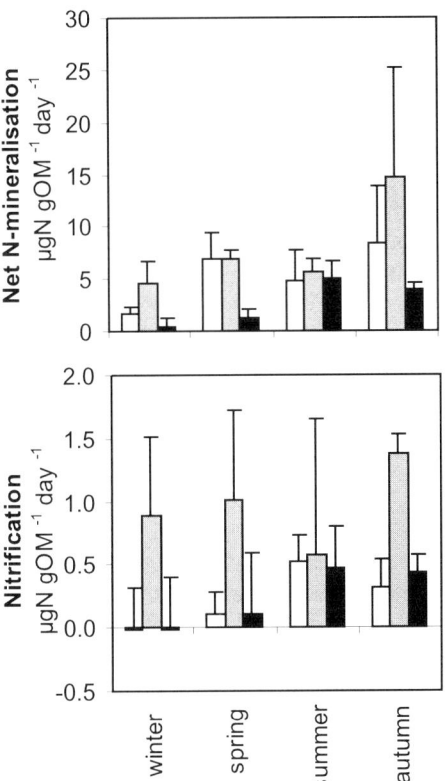

Figure 2. Net N-mineralisation and nitrification at Mols during winter 1999 to autumn 2000 (μg N g^{-1} OM^{-1} d^{-1} (mean + SE)). Control and warming plots were subject to heather beetle infestation throughout the mineralisation study. 'Intact' samples were taken in the same area but from plots not affected by heather beetles. The last sampling in autumn 2000 weas done before the vegetation was cut. Significance levels of * indicates $p < 0.05$.

within the 500 days of incubation independent of litter type and treatment. The decomposition rates were not significantly different among treatments and showed almost similar patterns for both treatments and plant species despite large differences in litter quality between *D. flexuosa* and *Calluna*, in particular lignin content being 7.8 % and 35.4% respectively (Figure 1).

The N-mineralisation was in the order of 0.5-15 μg N g^{-1} OM d^{-1} and was not significantly higher in the warming ($p < 0.0559$) and control ($p < 0.3082$) plots compared to the 'intact' *Calluna* plots outside the treatment plots (Figure 2). Mineralisation rates in the intact cores showed a seasonal pattern with low winter and spring rates and high summer rates. The control plots showed low mineralisation rates in the winter and more or less constant rates for the rest of the year with no distinct seasonal pattern (Figure 2). Warming tended to increase N-mineralisation ($p < 0.2342$) especially during the winter and autumn in the organic layer (Figure 2).

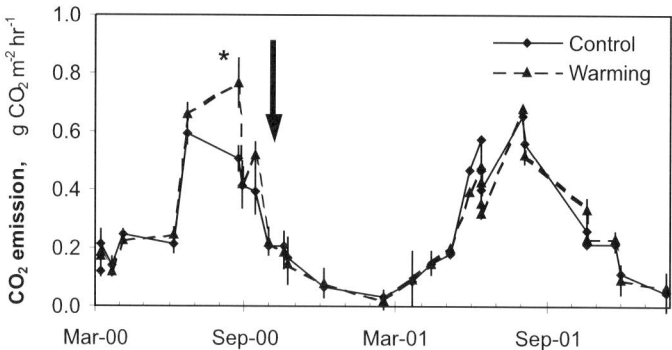

Figure 3. Soil respiration (g CO_2 m^{-2} hr^{-1}) during 2000 and 2001 in the warming and control plots at Mols. Error bars indicate SE. Arrow indicates time of vegetation removal from the plots. Significance levels of * indicates $p < 0.05$.

Nitrification rates were 10 times lower than the N-mineralisation rates with a seasonal pattern almost similar to net mineralisation although the control plots showed a seasonal pattern with highest rates in the summer (Figure 2). In contradiction to mineralization, the 'intact' cores and cores from the control plots generally showed similar rates. Nitrification in the warming treatment showed no seasonal trends but generally the rates tended to be higher relative to the control ($p < 0.1296$), significantly so in the autumn ($p < 0.0220$).

Soil respiration rates were 0.1-0.8 g CO_2 m^{-2} hr^{-1} with a clear seasonal pattern over the year with significantly higher rates during the summer compared to the winter (Figure 3). The soil respiration rates were almost identical for the control and warming plots for all sampling dates except from a significant effect of the warming treatment during the summer 2000 which did not appear again in 2001.

Soil water concentrations of dissolved organic matter in the organic layer were 20–30 mg L^{-1} at the start of the experiment increasing to 40–60 mg L^{-1} during the summer 1999 (Figure 4). DOC levels were almost identical and not significantly different in the warming treatment and the control and showed a similar pattern as for the N-mineralisation with a relatively constant level over year 2000 in the control and a slight increase in the summer 2000 in the warming (Figure 4). Beneath the root zone (90 cm depth) the DOC concentration was 10 times smaller than in the top soil, and showed no clear seasonal patterns (data not shown).

Soil water concentrations of nitrogen in both the organic layer and beneath the root zone were generally low and constant over the year although a slight seasonal pattern with lower N-concentrations in the peak of the growing seasons were seen below the root zone (Figure 5). The nitrogen retention in the soil was high (>80%) in the first treatment year and dropped significantly in the control plot to about 24% in 2000 when heather beetle infestation progressed (Table I). The increased N mineralisation caused by the warming resulted in increased leaching losses of

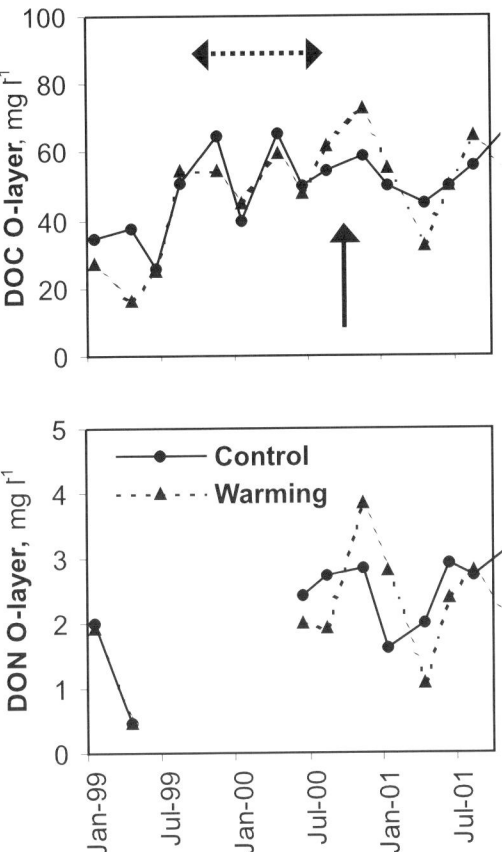

Figure 4. Soil water concentrations DOC and DON beneath the organic layer in the control (full line and bullets) and the warming treatment (dashed line and triangles). Times of infestation (dashed arrow) and cutting (solid arrow) are indicated.

NO_3^- and NH_4^+ beneath the organic layer, while in contrast, the nitrogen concentrations in the soil solution in the warming treatment decreased under the root zone compared to the control in the second year of treatment (Figure 5, Table I). The higher N retention after defoliation in the warming treatment indicates a much higher immobilization of N in soil and plants.

4. Discussion and Conclusion

At the beginning of the experiment in 1999, the heathland ecosystem at Mols was already subject to grass invasion. This is a result of natural heathland succession towards grass dominance on relatively nutrient rich soils (Holmsgaard, 1986; Gimingham, 1995). The heathland had received inputs of nitrogen from

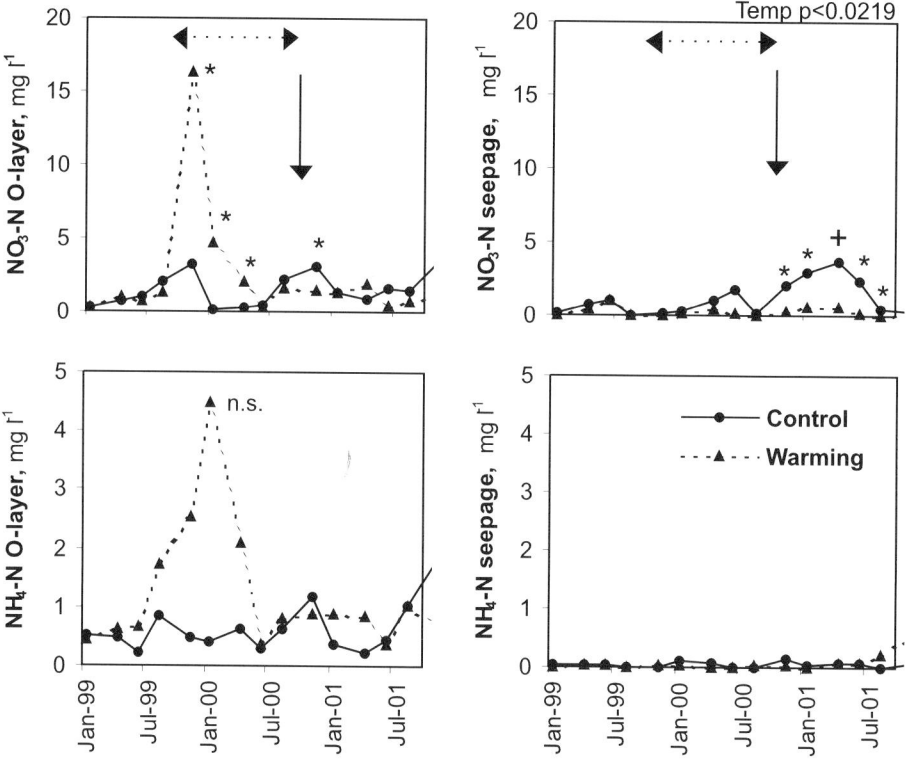

Figure 5. Soil solution concentrations of NO_3^- and NH_4^+ beneath the organic layer and in the seepage water in the control (full line and bullets) and the warming treatment (dashed line and triangles). Overall significant effects of the rmANOVA are given in top left corner. Seasonal effects are given for each time. Significance levels of + and * indicates $p \leq 0.1$ and $p \leq 0.05$, respectively. Times of infestation (dashed arrow) and cutting (solid arrow) are indicated.

atmospheric deposition, probably for decades, being close to the critical loads level defined for lowland dry heathlands of 15–20 kg N ha^{-1} yr^{-1} (EEA, 1999), and it is likely that the grass invasion has been stimulated by increased nitrogen availability. The mineralisation and nitrification rates found in the organic layer under intact *Calluna* are comparable to rates found in Dutch heathlands, which have been subjected to very high inputs of atmospheric N deposition (e.g., van Vuuren *et al.*, 1992), which further indicates a relatively high N availability compared to the very low or negative rates found in other northern *Calluna* heath (Rangeley and Knowles, 1988; Kristensen and Henriksen, 1998; Kristensen, 2001). On the other hand, the heathland at Mols is still capable of immobilising the incoming nitrogen by either microbial immobilisation as has been shown in another similar heathland (Kristensen and McCarty, 1999) or by increased plant uptake as the vegetation shifts to grasses that have a much lower C/N ratio.

TABLE I

Nitrogen fluxes (kg N ha^{-1} yr^{-1}) at the Danish heatland site at Mols following experimental temperature manipulation. In the second treatment year (2000), the *Calluna vulgaris* was completely defoliated by heather beetles followed by cutting to promote regrowth of *Calluna*. Input = wet deposition; OM-leaching = leaching from the organic horizon; Output = leaching under the root zone in 90 cm depth; Retention = Input-Output; % retention = % of input retained

Treatment	Control		Warming	
	NO_3^-	NH_4^+	NO_3^-	NH_4^+
1999				
Input	5.8	7.1	5.3	6.5
OM-leaching	2.9	1.8	11	5.3
Output	1.2	0.14	0.47	0.18
Retention	4.6	7.0	4.8	6.3
% Retention	80	98	91	97
2000				
Input	5.8	7.1	5.3	6.5
OM-leaching	3.4	2.2	10	9.8
Output	4.4	0.42	1.2	0.10
Retention	1.4	6.6	4.1	6.4
% Retention	24	94	79	99

The effects of the warming treatment on biogeochemical processes were generally moderate, which may be expected with the relatively small temperature increase of about 1 °C. Even though warming only led to relatively small increases in mineralisation rates and soil respiration with the largest effects in the summer, warming increased nitrogen leaching from the top soil layer 3–4 fold, mainly attributed to a strong increase in soil water N concentrations during the winter 1999/2000, one year after the start of the experiment (Figure 5). Correspondingly, increased N leaching in response to warming has also been reported in a forested catchment (Lukewille and Wright, 1997) and upland grassland (Ineson *et al.*, 1998). The increased mineralisation and leaching of N from the organic layer was not seen beneath the root zone as the mobilization of nitrogen was counteracted by a 30% increased plant production in the warmed plots (Torben Riis Nielsen, unpublished data) illustrating that interactions between the plants and the soil may complicate demonstration of climate induced effects on biogeochemistry, at least in the short term.

The infestation by the heather beetles in the late summer 1999 and subsequent defoliation and harvesting of all vegetation in all plots during the summer and autumn 2000 increased the mobilization of DOC and nitrogen from the organic soil layer but not below the root zone (Figures 4 and 5, Table I (year 2000)). This corresponds to findings in forest ecosystems, where insect infestations, defoliation, and clear-cutting have been shown to increase nitrogen fluxes in the soil water (Swank et al., 1981; Pedersen and Bille-Hansen, 1999) and in particular to findings from another Danish heathland (Nielsen et al., 2000) where leaching of DOC from the organic soil layer was doubled due to heather beetle infestation. but all of the released DOC was trapped again in the B-horizon. The mobilizing effect of the beetle infestation and vegetation removal on nitrogen leaching was much smaller in the warmed plots compared to the control, probably as a consequence of faster and stronger regrowth of the grass *Deschampsia flexuosa* in the warming treatment, rapidly re-established the plant sink for nitrogen supported by an observed 30 % increase in grass biomass in the summer 2000 in the heated plots (Torben Riis-Nielsen, unpublished data). So, warming seemed to reduce N losses by stimulating plant growth. However, the increase of about 1 °C in the warming treatment is on the low end of projected temperature rises (Houghton et al., 2001). A larger degree of warming might be expected to cause larger increases in N-mineralization, at least transient, and thereby potentially increase the risk of nitrogen leaching below the root zone in cases where the plants nitrogen demand is exceeded.

In general, the results indicate alterations in the soil processes, but the responses often were not significant. In contrast, plants integrate the small yearly mineralisation increases over several years. Similar patterns of low response in soil respiration, net mineralisation but larger increases in plant biomass have been shown in several warming experiments in the Arctic (Christensen et al., 1997; Jonasson et al., 1999; Schmidt et al., 2002).

The increase in soil solution nitrogen concentrations in the warmed plots during the first autumn and winter may be due alone to warming induced increased mineralisation of organic material which to some extent was supported by the mineralisation data. However, it may also be due to interaction between the heather beetle infestation and the warming. The seasonal measurements of nitrogen mineralisation and nitrification in the control plots infested by heather beetles indicate an increase in N mineralisation by more than 50% compared to the measurements from the intact areas outside the plots. This agrees well with previous findings showing that heather beetle infestations can alter nitrogen cycling in *Calluna* heathlands by switching soils from net immobilisation of NH_4^+ to net ammonification (Kristensen and McCarty, 1999). The increase of NH_4^+ availability caused by beetle infestation in Mols did, however, stimulate nitrification contrary to results from the previous study. The observed increase in soil solution N may therefore be due to a combination stimulation of mineralisation by both the heather beetle infestation and the warming treatment.

Subsequent to the heather beetle infestation and the removal of all vegetation, the study plots relatively quickly developed a thin grass cover over the autumn of 2000 and further a dense grass cover during the spring and summer of 2001. The *Calluna* plants did not re-establish within the first two growing seasons after the infestation and cutting. This crucial role of heather beetles in the conversion of heathlands to grasslands when the heathland is under pressure for grass invasion and increased nutrient availability has been described previously (Brunsting and Heil, 1985; Prins *et al.*, 1991). In the present study, the attempt to restore the *Calluna* vegetation by cutting and removing all aboveground plant biomass after the beetle infestation was not found to enable regrowth of *Calluna* within the first two years after cutting.

Based on the results of this study five important aspects have to be considered:

1) The area of *Calluna* heathlands in Europe is decreasing due to invasion by grasses and trees. This development is stimulated by increased nutrient availability due to increased atmospheric N deposition (Heil and Bobbink, 1993). The increased temperature was in the present study found to stimulate mineralisation and N availability in the heath ecosystem. This indicates that global warming may, at least during the time of transition add to the nutritional pressure on heathlands.

2) The occurrence of heather beetle infestation of *Calluna* is thought to be increased by increased N deposition and availability in heathlands (Heil and Bobbink, 1993). The problem may also increase the risk of herbivory leading to increased pressure or increased frequency of insect infestations (Penuelas *et al.*, in press). As seen in this study, the insect infestation in ecosystems already under pressure may have detrimental effects and change the ecosystem characteristics fundamentally.

3) Ecosystem disturbances will interact with increased temperatures to reduce or amplify the effects on biogeochemical cycling. The direction and extent of the interaction among disturbances and climatic changes depends on the relative extent of the stimulation of plant and soil processes. In this study, with moderate temperature increases the stimulation of the plant growth by warming caused rapid and efficient re-immobilization of nutrients immobilized by the ecosystem disturbance. At more extensive temperature increases the interactions may lead to loss of nutrients from the ecosystem.

4) The change in vegetation from *Calluna* bushes to grasslands may also affect carbon and nutrient budgets in the ecosystem. In the present study, a conversion from a pure *Calluna* cover to a pure grass cover reduces the aboveground carbon stock by about 40%, while aboveground nitrogen content will almost double in the grassland ecosystem as the C/N ratio is only half the *Calluna* plants. These aboveground effects have to be balanced with the adjacent belowground effects, which are uncertain at this level. One might expect the turnover rate of the organic matter in the soil to be faster in the grassland leading to reduced carbon storage, but the results from the warming study

are not conclusive yet. Recent findings on dry grasslands indicate an overall reduction in carbon storage in grasslands compared to shrublands under drier conditions (Jackson et al., 2002).

5) The measurements show that on the short term (year to year) ecosystem disturbances can have a much stronger impact on the biogeochemical cycling compared to moderate changes in climatic conditions. However, climatic conditions act on a much longer time scale, and the effects therefore have to be assessed on a long term. Furthermore, ecosystem disturbances may potentially act simultaneously with changes in climatic conditions and the frequency may even be increased leading to synergistic effects on ecosystem biogeochemistry.

Acknowledgments

The project was funded by EU under the projects CLIMOOR (Contract ENV4-CT97-0694) and VULCAN (Contract EVK2-CT-2000-00094) and the participating research institutes. We owe a lot of grateful thanks to all other institutes and researchers involved with the Climoor and Vulcan projects for inspiration and enthusiasm and in particular to the technical staff at our institutes for their skillful field and lab work. Further information about the project can be found on www.vulcanproject.com.

References

Brunsting, A. M. H. and Heil, G. W.: 1985, 'The role of nutrients in the interactions between a herbivorous beetle and some competing plant species in heathlands', *Oikos* **44**, 23–26.

Beier, C., Emmett, B., Gundersen, P., Tietema, A., Penuelas, J., Estiarte, M., Gordon, C., Gorissen, A., Llorens, L., Roda, F. and Williams, D.: 'Novel approaches to study climate change effects on terrestrial ecosystems in the field – drought and passive night time warming', *Ecosystems*, in press.

Christensen, T. R., Michelsen, A., Jonasson, S. and Schmidt, I. K.: 1997, 'Carbon dioxide and methane exchange of a subarctic heath in response to climate change related environmental perturbations', *Oikos* **79**, 34-44.

EEA: 1999, 'Nutrients in European ecosystems', *Environmental Assessment Report No. 4*, European Environmental Agency, Luxembourg, 156 pp.

Galloway, J. N. and Cowling, E. B.: 2002, 'Reactive nitrogen and the world: 2000 years of change', *Ambio* **31**, 64–71.

Gimingham, C. H.: 1995, 'Heaths and moorland: an overview of ecological change', in D. B. A. Thompson, A. J. Hester and M. B. Usher (eds), *Heaths and Moorland: Cultural Landscapes*, Scottish Natural Heritage, HMSO, Edinburgh, pp. 9–19.

Heil, G. W. and Bobbink, R.: 1993, 'Impact of atmospheric nitrogen deposition on dry heathlands. A stochastic model simulation competition between *Calluna* vulgaris and two grass species', in R. Aerts and G. W. Heil (eds), *Heathlands: Patterns and Processes in a Changing Environment*, Kluwer Academic Publishers, Dordrecht, The Netherlands, pp. 181–200.

Holmsgaard, J. E.: 1986, 'The heath of Nørholm. 5th report', *Det forstlige forsøgsvæsen* **40**, 273–329.

Houghton, J. T., Ding., Y., Griggs, D. J., Noguer, M. van der Lindern, P. J. and Xiaosu, D. (eds): 2001, *Climate Change 2001: The Scientific Basis*, Cambridge University Press, Cambridge.

Ineson, P., Benham, D. G., Poskitt, J., Harrison, A. F., Taylor, K. and Woods, C.: 1998, 'Effects of climate change on nitrogen dynamics in upland soils. 2. A soil warming study', *Global Change Biol.* **4**, 153–161.

Jackson, R. B., Banner, J. L., Jobbagy, E. G., Pockman, W. T. and Wall, D. H.: 2002, 'Ecosystem carbon loss with woody plant invasion of grasslands', *Nature* **418**, 623–626.

Jonasson, S., Michelsen, A., Schmidt, I. K. and Nielsen, E. V.: 1999, 'Responses in microbes and plants to changed temperature, nutrient, and light regimes in the arctic', *Ecology* **80**, 1828–1843.

Kristensen, H. L. and Henriksen, K.: 1998, 'Soil nitrogen transformations along a successional gradient from *Calluna* heathland to *Quercus* forest at intermediate atmospheric nitrogen deposition', *Appl. Soil Ecol.* **8**, 95–109.

Kristensen, H. L. and McCarty, G. W.: 1999, 'Mineralization and immobilization of nitrogen in heath soil under intact *Calluna*, after heather beetle infestation and nitrogen fertilization', *Appl. Soil Ecol.* **13**, 187–198.

Kristensen, H. L.: 2001, 'High immobilization of NH_4^+ in Danish heath soil related to succession, soil and nutrients: Implications for critical loads of N', *Water, Air, Soil Pollut.: Focus* **1**, 211–230.

Lukewille, A. and Wright, R. F.: 1997, 'Experimentally increased soil temperature causes release of nitrogen at a boreal forest catchment in southern Norway', *Global Change Biol.* **3**, 13–21.

Luxmoore, R. J., Hanson, P. J., Beauchamp, J. J. and Joslin, J. D.: 1998, 'Passive nighttime warming facility for forest ecosystem research', *Tree Physiol.* **18**, 615–623.

Nielsen, K. E., Hansen, B., Ladekarl, U. L. and Nørnberg, P.: 2000, 'Effects of N-deposition on ion trapping by B-horizons of Danish heathlands', *Plant Soil* **223**, 265–276.

Olesen, J. E. and Heidmann, T.: 1990, 'EVACROP – Et program til beregning af aktuel fordampning og afstrømning fra rodzonen, Version 1.00', (in Danish), Dept. of Agrometeorology, Research Centre Foulum, Tjele, Denmark.

Pedersen, L. B. and Bille Hansen, J.: 1999, 'Effects of nitrogen load to the forest floor in Sitka spruce stands (*Picea sitchensis*) as affected by difference in deposition and spruce aphid infestations', *Water, Air, Soil Pollut.* **85**, 1173–1178.

Penuelas, J., Gordon, C., Llorens, L., Nielsen, T., Tietema, A., Beier, C., Brunal, P., Emmett, B., Estiarte, M., Gorissen, T. and Williams, D.: 'Non-intrusive field experiments show different plant responses to warming and drought among sites, seasons and species in a North-South European gradient', *Ecosystems*, in press.

Prins, A. H., Berdowski, J. J. M. and Latuhihin, K. J.: 1991, 'Effect of NH_4^+ fertilization on the maintenance of a *Calluna vulgaris* vegetation', *Acta Bot. Neerland.* **40**, 269–279.

Rangeley, A. and Knowles, R.: 1988, 'Nitrogen transformations in a Scottish peat soil under laboratory conditions', *Soil Biol. Biochem.* **20**, 385–391.

Rustad, L. E., Campbell, J. L., Marion, G. M., Norby, R. J., Mitchell, M. J., Hartley, A. E., Cornelissen, J. H. C., Gurevitch, J. and GCTE-NEWS: 2001, 'A meta-analysis of the response of soil respiration, net nitrogen mineralization, and aboveground plant growth to experimental ecosystem warming', *Oecologia* **126**, 543–562.

Schmidt, I. K., Jonasson, S., Shaver, G. R., Michelsen, A. and Nordin, A.: 2002, 'Mineralization and distribution of nutrients in plants and microbes in four arctic ecosystems: responses to warming', *Plant Soil* **242**, 93–106.

Shaver, G. R., Canadell, J., Chapin, F. S., Gurevitch, J., Harte, J., Henry, G., Ineson, P., Jonasson, S., Melillo, J., Pitelka, L. and Rustad, L.: 2000, 'Global warming and terrestrial ecosystems: a conceptual framework for analysis', *BioScience* **50**, 871–882.

Swank, W. T., Waide, J. B., Crossley Jr., D. A. and Todd, R. L.: 1981, 'Insect defoliation enhances nitrate export from forest ecosystems', *Oecologia* **51**, 297–299.

van Soest, P. J.: 1963, 'Use of detergents in the analysis of fibrous feeds', *J. Assoc. Off. Anal. Chem.* **46**, 825–835.

van Vuuren, M. M. I., Aerts, R., Berendse, F. and de visser, W.: 1992, 'Nitrogen mineralization in heathland ecosystems dominated by different plant species', *Biogeochemistry* **16**, 151–166.

Watson, R. T., Rodhe, H., Oeschger, H. and Siegentaler, U.: 1991, 'Greenhouse gases and aerosols', in J. T. Houghton, G. J. Jenkins and J. J. Ephraums (eds), *Climate Change, The IPCC Scientific Assessment*, Cambridge University Press, Cambridge, pp. 1–40.

Watson, R. T., Zinyowera, M. C. and Moss, R. H. (eds): 1995, *Climate Change 1995: Impacts, Adaptations and Mitigation of Climate Change: Scientific-Technical Analyses*, Cambridge University Press, Cambridge, 878 pp.

FOREST ECOSYSTEM RESPONSES TO ATMOSPHERIC POLLUTION: LINKING COMPARATIVE WITH EXPERIMENTAL STUDIES

N. B. DISE[1,2]* and P. GUNDERSEN[3]

[1] *Department of Earth Science, The Open University, Milton Keynes, MK76AA, U.K.;*
[2] *Department of Biological Science, Villanova University, Villanova, Pennsylvania, 19085, U.S.A.;*
[3] *Danish Forest and Landscape Research Institute, Kongevej 11, DK-2970, Hoersholm, Denmark*
(* *author for correspondence, e-mail: N.B.Dise@open.ac.uk, phone: +44 1908 655075, fax: +44 1908 655151*)

(Received 20 August 2002; accepted 29 April 2003)

Abstract. The impact on an ecosystem of an environmental stress, such as climate change or air pollution, can be studied through experimentation, through comparisons of sites across a gradient of the stress, through long-term studies at a single site, or through theoretical or modelling approaches. Although the former three techniques often are used to develop and test models, it is much rarer to explicitly link experimental, comparative or long-term studies together. Here we present a concept for combining experimental and comparative research to assess the direction and rate of change, the expected long-term state, and the rate at which the long-term state is achieved after an ecosystem is exposed to an environmental stress. We do this by comparing the response of a forest in Denmark to experimentally increased N deposition with the expected long-term response based on a European database of forests exposed to different levels of N deposition over long time periods. The analysis suggests that if N deposition were to increase by 3-fold to about 50 kg N ha^{-1} a^{-1} at the Danish site, and remain at this level, the N concentration in needles would respond within 2–4 yr after the onset of the enhanced N deposition, and would rapidly plateau to an expected mean value of 18.0 mg N g^{-1} dry mass (95% confidence interval ± 2.5 mg g^{-1}). The N concentration of new litter also would respond rapidly (1–2 yr) to reach an expected value of 16.6 mg N kg^{-1} dry mass (± 3). The N concentration of the organic layer in the soil would increase much more slowly, but a significant increase would be expected within 5–10 yr. Mineral soil pH would take more than 7 yr to change. Finally, the flux of dissolved inorganic N in leachate would begin to increase immediately, but would take many years to reach the expected level of 22.4 kg N ha^{-1} a^{-1} (± 4).

Keywords: acidification, empirical studies, experimental, forest decline, forest ecosystems, gradient studies, nitrogen saturation

1. Introduction

How can we tell if an ecosystem has changed in response to a pollutant, how long it will continue to change, or how fast it will recover? These questions are particularly difficult in environmental science because in many cases the time for a full response to a disturbance may greatly exceed the duration of a research grant.

One approach is field or laboratory experimentation, in which a single factor is experimentally altered to the level of interest. If we observe a change in a response variable, we can infer with some confidence that the manipulated factor has caused the change, and we can say how long it takes for the system to respond. However, experimental studies are biased toward short-term responses, whereas important long-term changes may never occur within the usual funding lifetime of a research project. Furthermore, there is the risk that artefacts of the manipulation itself (Gundersen *et al.*, 1998) or site-specific factors might explain part of the response.

An alternate approach is through comparative studies. These often take the form of surveys, either through the literature or through field measurement, of sites along a gradient of the driving variable. Comparative studies can point out long-term responses, but they cannot tell us when those responses occurred after the onset of the stress. They also cannot prove causality. Long-term studies at a single site also can be used to detect environmental change, and all of these can be incorporated into ecosystem models (Carpenter *et al.*, 1995).

Many models, both conceptual and mathematical, incorporate more than one approach to building an understanding of how ecosystems respond to environmental perturbations (e.g., Aber *et al.*, 1998). Few if any attempts have been made, however, to explicitly link the conclusions of an experimental and a comparative study together to make specific predictions about the nature, timing, and direction of response of an ecosystem to an environmental stress. In this paper, we bring together a field manipulation study of N deposition on the Klosterhede forest in Denmark with a comparative database of N deposition effects on forests across Europe. Our aim is to illustrate that combining both approaches gives us insight into important questions that neither can answer alone: i.e., what is the long-term chemical change that an increase in N deposition will cause to the needles, litter, organic matter, soil pH and runoff water quality of a forest, and how long will it take each of the chemical changes to reach its long-term quasi-equilibrium state?

2. Materials and Methods

2.1. DATASETS

2.1.1. *Klosterhede Experimental Forest*

Klosterhede forest is located near Lemvig in Western Jutland (8°24'E, 56°29'N), Denmark. Deposition of inorganic N to Klosterhede is 18 kg N ha^{-1} a^{-1}. The best estimate is that N deposition increased from low levels (ca. 5 kg ha^{-1} a^{-1}) in the early 1950s to the current level by the early 1980s and has remained fairly constant since (Gundersen, 1989; Hovmand and Bille-Hansen, 1999). Experiments were carried out in a 75 yr old Norway spruce (*Picea abies* L. Karst) plantation, which is second generation after heathland. Mean tree height is approximately 20 m, basal area is 30 m^2 ha^{-1}, and stem density is 860 ha^{-1}. The soil is a coarse sandy nutrient-poor podzol (Typic Haplorthod) with low clay content (<4%). An organic layer of 7 cm (C/N ratio 32) has developed during the current rotation.

As part of the NITREX project (Wright and van Breemen, 1995), three control plots and one N addition plot were established in December 1990 (Gundersen and Rasmussen, 1995). N was experimentally added to precipitation to simulate a threefold increase in N deposition from the background levels of 18 kg N ha^{-1} a^{-1} to 53 kg N ha^{-1} a^{-1}. N was added by monthly hand spraying of dissolved NH$_4$NO$_3$, in which the total additional water was less than 0.3% of annual throughfall. The experiment began in 1992 and is ongoing, although there was no treatment between autumn 1997 and autumn 1999. Nitrogen concentrations were measured in major plant and soil pools each year from 1991 to 1995 (Gundersen, 1998) and again in 1999. Solute fluxes in leachate water were monitored monthly from 1991 through 1995. The results we describe are based on 4 years of data for seepage water, needles and litter, and 7 years for all other components.

Data were analysed by using repeated measurements analysis of variance, using multiple sampling sites within each plot as replicates and pre-treatment data as a covariate. Before analysis, data were tested for normality and log-transformed if necessary. Means comparisons were done by t-tests where appropriate. Differences were considered significant at the $p < 0.05$ level unless otherwise stated.

2.1.2. *IFEF*

IFEF (Indicators of Forest Ecosystem Functioning) is a compilation of data from published sources, data questionnaires, and existing databases, numbering about 200 forests across Europe. It has been used to analyse European-wide patterns in fluxes of dissolved N, Al, and Mg in runoff or leachate from forests, which can be related to meaningful environmental drivers (Dise *et al.*, 1998a, b, 2001; Armbruster *et al.*, 2002; MacDonald *et al.*, 2002).

For N, the database consists of input (as throughfall) and output (as runoff or seepage water) fluxes of dissolved inorganic N (NO$_3^-$ and NH$_4^+$, hereafter referred to as DIN) from 181 forests. For a smaller (and variable) number of these sites, data exist on fluxes of other major elements, soil and vegetation chemistry, and site characteristics such as tree age and species. A minimum of one year's data were used for element fluxes, and the period of data collection ranged between 1980 and 1998.

The data were passed through several filters designed to check for data quality. Site data were discarded if the output of Cl$^-$ was not within ± 50% of the Cl$^-$ input, if the output of NO$_3^-$ was larger than the input (indicating ecosystem disturbance), or if the B horizon pH of a site was greater than 6.5 (to restrict our analysis to acid-sensitive sites). For this study, tree age was restricted to stands >20 yr old, and only conifers were used, so as to have stands that were similar to Klosterhede.

2.2. DATA ANALYSIS

The analyses of data generated by the comparative approach and the experimental manipulation approach rest on major assumptions that are clarified in the following text and emphasised *in italics*.

The relationships between DIN fluxes in throughfall ('N_{in}', in kg N ha^{-1} a^{-1}) and hypothesised ecosystem response variables in IFEF were described using simple linear regressions. We used only response variables for which the relationship between DIN input and the variable was both highly significant ($p < 0.01$) and explained a reasonable proportion of the variability in the response variable data across the sites ($r^2 > 15\%$).

Five response variables fulfilled these criteria: (1) N concentration of current-year needles (mg g^{-1} dry mass), (2) N concentration of fresh litter (mg g^{-1} dry mass), (3) N concentration of the organic horizon ($O_f + O_h$) (mg g^{-1} dry mass), (4) soil pH in the B horizon, and (5) the flux of DIN in seepage water or runoff ('N_{out}', in kg N ha^{-1} a^{-1}). *The regression equation is considered a simple model of the underlying relation between the response variable and N deposition across Europe*. We call this the 'long-term response', in the sense of the time over which most of the forests have been exposed to elevated N deposition.

We recognise that for each forest, the level of the response variable will reflect not only N deposition, but also other factors like climate, management, age, tree species, etc. *We assume that the large number of sites spread out over Europe will reflect this variability in a random distribution of residuals around the regression line, and that this variability is reflected in the 95% confidence intervals (CIs) on predictions.*

To evaluate the changes in the experimental manipulation, we determined whether there was a significant change in each N status variable at Klosterhede over the period that N deposition was increased from 18 to 53 kg N ha^{-1} a^{-1}. The new value of each response variable was then compared to its predicted 'long-term' value from the regression. If the new value was within the 95% CI of the predicted long-term value, it was considered to be not significantly different from the long-term response. If the new value was outside the 95% CI, it was considered to have not reached the long-term state, and thus still be in transition. The 95% confidence intervals were weighted to reflect the fact that each value for Klosterhede is the mean of many observations rather than a single measurement. Thus we assumed that *the mean difference in a response variable between two groups of forests receiving 18 and 53 kg N ha^{-1} yr^{-1} DIN in throughfall in IFEF was the best estimate of the long-term change in that variable for Klosterhede should N pollution be increased to 53 kg N ha^{-1} yr^{-1} ('space for time' substitution).*

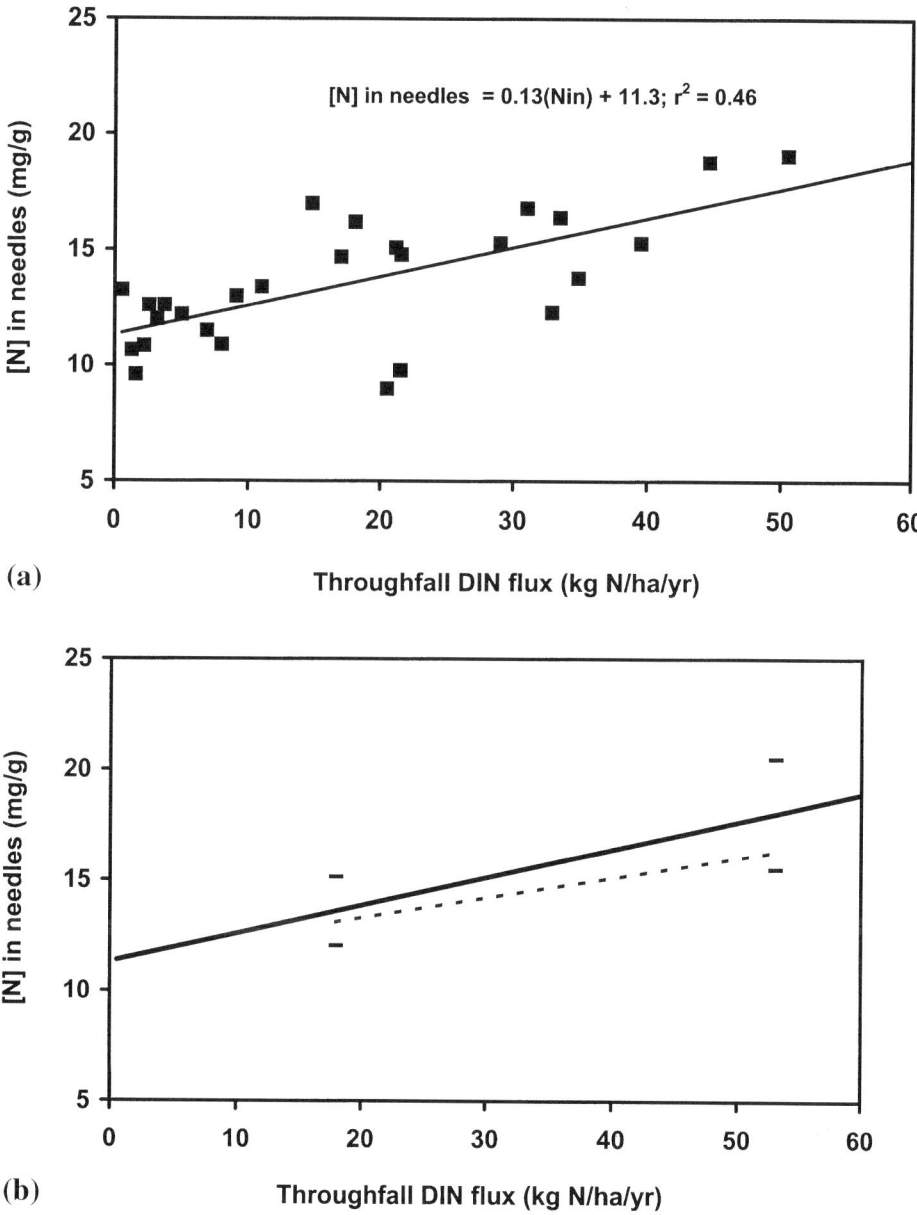

Figure 1. N concentration in current-year needles versus N input in throughfall: (a) Comparative relationship for coniferous forests in the IFEF database; (b) Response to experimental increased nitrogen input (18 to 53 kg N ha^{-1} a^{-1}) at Klosterhede (dotted line, data from Table I) compared to the predicted relationship from IFEF (solid black line) as well as the 95% confidence intervals on the predicted values at inputs of 18 and 53 kg N ha^{-1} a^{-1}. All relationships are significant except where noted (NS).

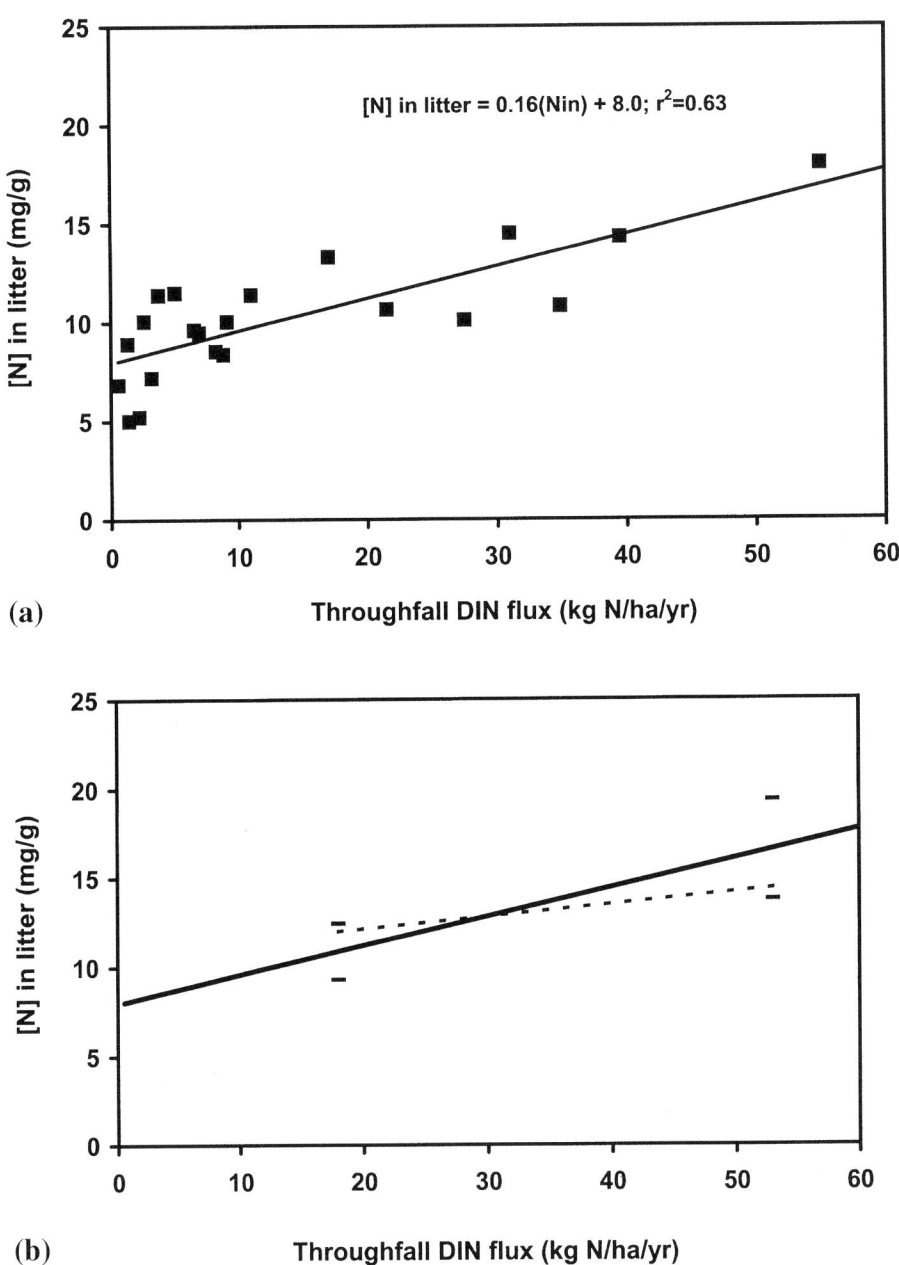

Figure 2. N concentration in needle litter versus N input in throughfall. For details see Figure 1 legend.

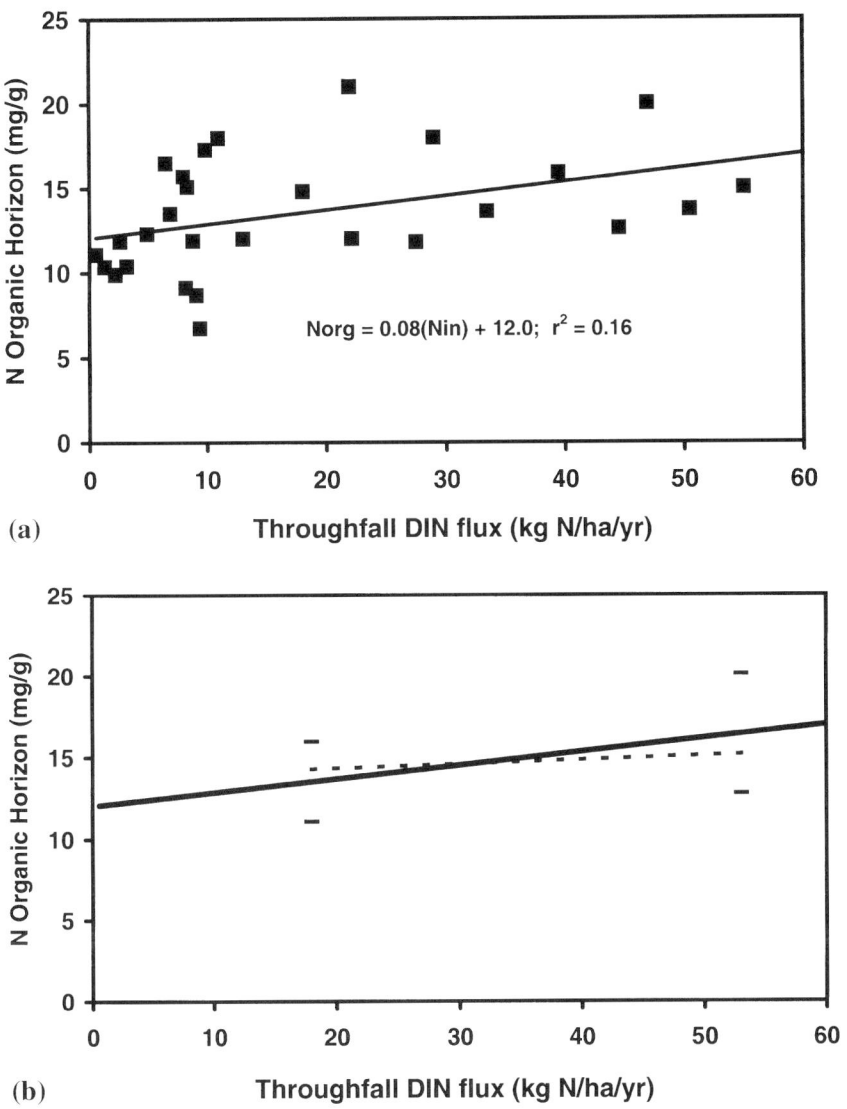

Figure 3. N concentration in the organic soil horizon versus N input in throughfall. For details see Figure 1 legend.

3. Results and Discussion

3.1. COMPARATIVE RESPONSE

N_{in} explained between 16 and 63% of the variability in the response variables. The best models are those for vegetation and water chemistry; the poorest are those for soil chemistry (Figures 1a–5a). Variability that cannot be explained by N deposition

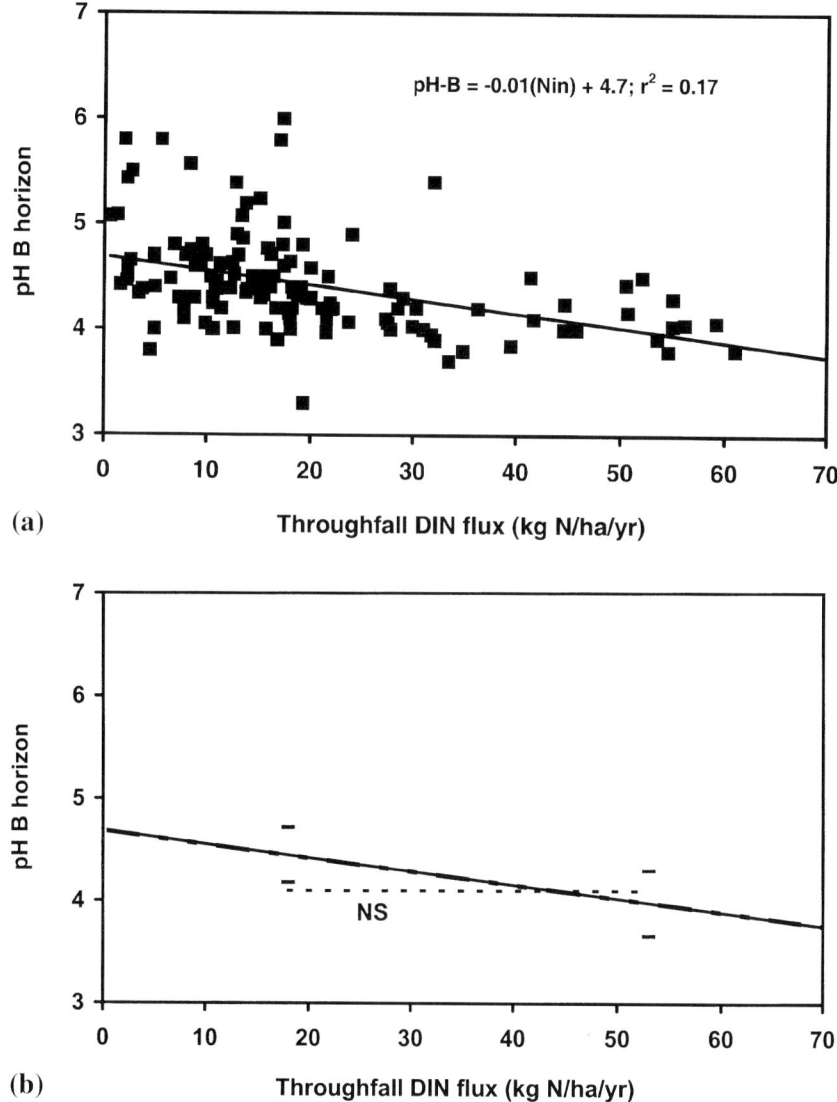

Figure 4. pH of the B horizon versus N input in throughfall. For details see Figure 1 legend.

may include factors such as bedrock/soil type, climate, tree age and species, soil chemistry, land use history, and site management.

3.2. EXPERIMENTAL RESPONSE

The overall response of the forest at Klosterhede to experimental changes in N deposition was an increase in the N status of the site (Table I; Figures 1b–5b). Current-year needles, fresh litter and seepage water N responded rapidly, the con-

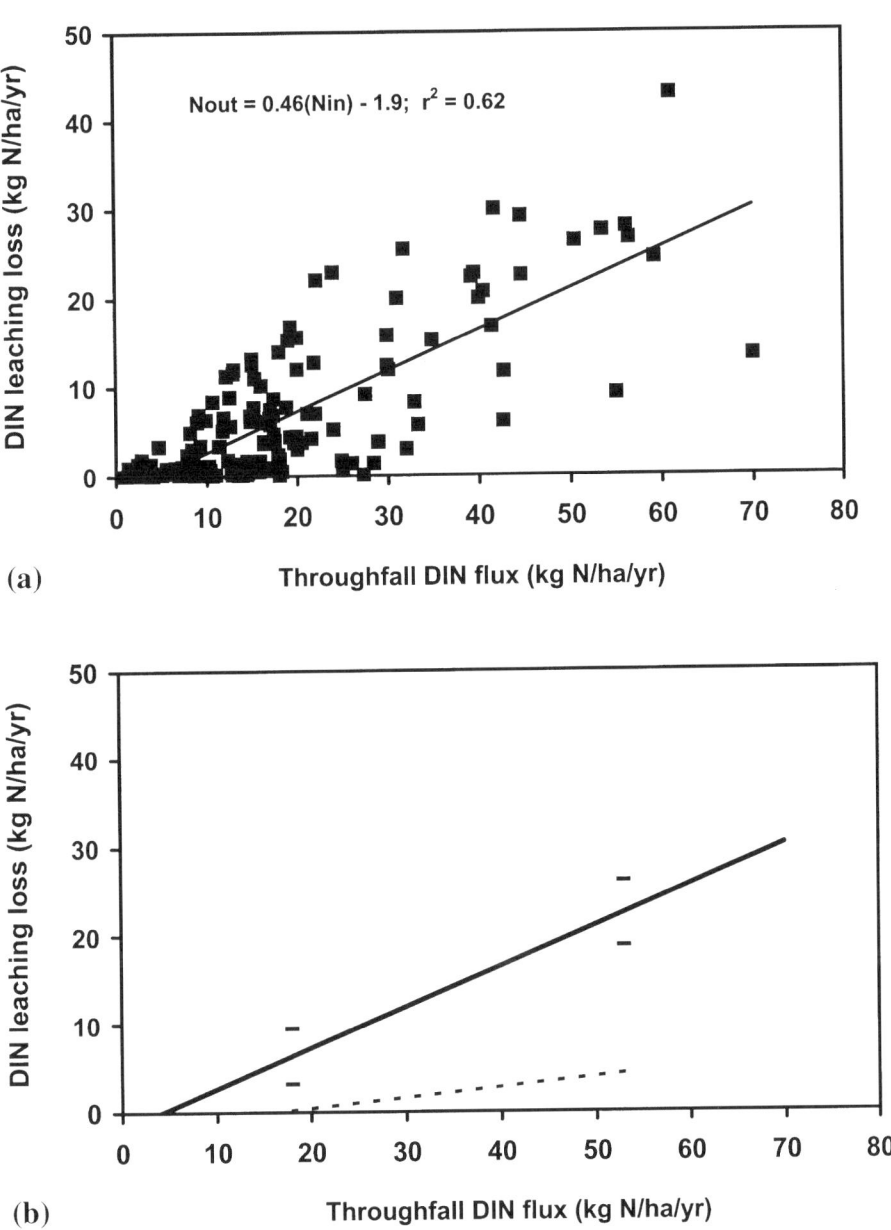

(a)

(b)

Figure 5. N leaching loss versus N input in throughfall. For details see Figure 1 legend.

centration of N in the organic layer significantly increased only after 7 yr, and there was no significant pH change over the course of the study.

TABLE I

Ecosystem responses to changes in N input at Klosterhede. The first column shows the concentration or flux of an ecosystem component at the start of the manipulation experiment; the second column shows the value at the end of the experiment (4 yr for all except organic layer and pH, which is 7 yr, but with 2 yr in 1997 and 1998 of no treatment), and the third column shows the first year in which a significant effect was detected. An asterisk by the end value shows whether or not the ending flux or concentration is significantly ($p < 0.05$) different from the starting flux or concentration (Gundersen, 1998). Fluxes in kg N ha^{-1} a^{-1}; concentrations in mg N g^{-1} dry mass

	Starting level	Ending level	Significant change in year
Current yr needle N	13.1	16.3*[a]	3
Litter N	12.0	14.4*	2
N in organic layer	14.3	15.2*	7
pH	4.1	4.1	NS
Seepage water N	0.2	4.3*	1

[a] In 1999, after 2 yr of no treatment, current-year needle N declined to 14.0 and was not significantly different from controls at 13.3.

3.3. LINKED RESPONSE

3.3.1. *Concentration of Nitrogen in Needles*

About half the variability in the N concentration of current year needles in the IFEF database can be explained by variability in the throughfall flux of DIN (Figure 1a). The regression line suggests that the N concentration of needles increases from about 12 mg g^{-1} dry mass at negligible levels of N deposition to 18.0 mg g^{-1} dry mass (95% CI 15.5–20.5) at levels of N deposition of 53 kg N ha^{-1} a^{-1}. Figure 1b shows this relationship as a solid black line, as well as the 95% CIs on the predicted value of needle N for a site receiving either 18 or 53 kg N ha^{-1} a^{-1}.

The dotted line (Figure 1b) shows the trajectory demarcated by the concentration of N in needles at the beginning of the experiment at Klosterhede (at N input of 18 kg N ha^{-1} a^{-1}) and after 4 yr of experimentally increased N input to 53 kg N ha^{-1} a^{-1} (Table I). The increase in N input changed N concentration in needles from 13.1 to 16.3 mg N g^{-1} dry mass, and the increase was first significant in year 3 (Table I). In addition, the final value of needle N was within the 95% CI of the expected value from the comparative regression. Thus, both the comparative and the experimental changes were significant and the experimental response was within the expected range of the long-term response.

We conclude from this comparison that: (1) needle N at Klosterhede responds rapidly to a tripling of N deposition, (2) the response reaches the expected long-term state within a few years, and (3) needle N would not be expected to significantly increase further, given continued N deposition at the level of 53 kg N ha^{-1} a^{-1}. The hypothesis of a rapid and dynamic response is strengthened by the observation that 2 yr after treatments stopped, needle N reversed back to a concentration not significantly different from the controls. However, it should be noted that the rapid return of needle-N to pre-treatment levels does not necessarily mean it will recover rapidly, or at all, after an extended interval of elevated N input (decades rather than years).

3.3.2. *Concentration of N in Needle Litter*

About 60% of the variability in the N concentration of current-year litter can be explained by variability in the throughfall flux of DIN (Figure 2a). The regression suggests that the N concentration of litter increases from about 7 mg g^{-1} dry weight at negligible levels of N deposition to 16.6 mg g^{-1} dry mass at 53 kg N ha^{-1} a^{-1}.

Klosterhede responded to an increase in N input with a change in needle litter N from 12 to 14.4 mg N kg^{-1} dry mass (Figure 2b, Table I). A significant response was detected after only 2 yr of treatment (Table I). It is interesting that a significant litter N response was detected sooner than a significant needle N response. This may indicate that translocation of N from senescent needles decreases soon after the increase in N input. The final concentration of needle litter N is within the 95% CI of the predicted long-term value (13.8–19.4). Thus, both lines of evidence suggest that litter N has also reached its long-term state.

3.3.3. *Concentration of N in the Organic Layer*

About 16% of the variability in the N concentration of the soil organic horizon can be explained by variability in the throughfall flux of DIN (Figure 3a). The slope of the regression line suggests that the N concentration of the organic horizon significantly increases from about 12 mg g^{-1} dry mass at negligible levels of N deposition to 16.4 mg g^{-1} dry mass (12.8–20.1) at 53 kg N ha^{-1} a^{-1}.

In the Klosterhede experiment, there was no significant response in organic horizon N concentrations over the first 4 yr of the study. However, the re-sampling of soils in 1999 revealed a significant response after 7 yr (5.5 yr treatment). This suggests that the N concentration in the organic horizon approaches its long-term state within 5–10 yr, although the wide variability in the data means that further change cannot be excluded.

3.3.4. *pH*

There was no significant change in soil pH during the 7 yr of the Klosterhede experiment (Table I). In contrast, the comparative dataset suggests a significant decline in pH with N deposition, from a mean of 4.83 at negligible values of DIN input to 3.98 (3.66–4.30) at N deposition levels of 53 kg N ha^{-1} a^{-1} (Figure 4).

In comparison to the IFEF sites, soil pH at Klosterhede (4.1) was already low for the N deposition it was receiving at the start of the manipulation experiment

TABLE II

Conclusions on direction of response, rate of change, time to long-term state, and evidence for these conclusions, from comparing the comparative (CO) with the experimental (EX) evidence for the response of Klosterhede forest to an increase in N deposition from 18 to 53 kg N ha^{-1} a^{-1}. 'CO, EX' = either comparative or experimental data can be used independently to draw conclusion. 'CO + EX'= both are needed together to draw conclusion

Variable	Response	Evidence for response	Rate	Evidence for rate of response	Time to long-term state	Evidence for time to long-term state
Needle N	Increase	CO, EX	Moderate (3–5 yr)	EX	Fast	CO + EX
Litter N	Increase	CO, EX	Fast (1–2 yr)	EX	Moderate	CO + EX
Organic layer N	Increase	CO, EX	Slow (5–10 yr)	EX	Slow	CO + EX
pH B-hor.	Decrease	CO	Slow (>7 yr)	EX	Slow	CO + EX
N leaching	Increase	CO, EX	Fast (1–2 yr)	EX	Slow	CO + EX

(expected 4.45, range 4.18–4.72). The fact that there is no change yet suggests that changes in pH, if they occur, will be slow and probably minor.

3.3.5. *Flux of DIN in Seepage Water*

About 60% of the variability in the leaching of DIN in runoff or seepage water in the IFEF database can be explained by variability in the throughfall flux of DIN (Figure 5a). The regression suggests that the expected value of DIN leaching increases from very low values at negligible levels of N deposition to a mean of 22.4 kg N ha^{-1} a^{-1} (18.7–26.1) at 53 kg N ha^{-1} a^{-1}.

The change in N leaching from the Klosterhede experiment, to 4.3 kg N ha^{-1} a^{-1} after 4 yr (Table I) is significant. However, this value lies well out of the 95% CI for the expected value (Figure 5b). We conclude that output fluxes of DIN at Klosterhede respond rapidly to a tripling of N deposition, but the response does not reach the expected long-term state within 4 yr. Thus, DIN output flux will be expected to continue to increase over time given continued elevated N deposition.

4. Conclusion

Table II summarises the interpretation of the data from linking the comparative with the experimental evidence. Both the comparative and the (4–7 yr) experimental evidence suggest that N deposition increases needle N, litter N, organic horizon N and N leaching fluxes. Only the comparative evidence indicates that N deposition also decreases soil pH.

The experimental data are necessary to determine the rates of change for those ecosystem components that change relatively quickly, and the comparative data are necessary to identify all of the components than change (short and long-term) and the value of the parameter at the long-term state. We conclude that needle and new litter chemistry reach their long-term state within a few years, initiation of changes in soil chemistry lag by about a decade and continue to change slowly, and, although N appears in runoff water quickly (1–2 yr), it will take many years (5–50, and from the low level of N_{out} after 4 yr, closer to 50) before the long-term state is reached. This suggests that a long-term 'filling up' of other ecosystem components will delay the runoff response. As N is gradually increases in the system, however, a larger and larger proportion of that input N will come out in runoff.

The rapid response in leachate DIN fluxes may imply that there are some parts of the forest ecosystem that are highly responsive to input chemistry and can rapidly 'switch off' or 'switch on' processes depending on that chemistry. Alternatively, it suggests that there may be some proportion of N input that simply bypasses vegetation and microbial processing. The rapid detection of enhanced NO_3^- in soil water down the soil profile (Gundersen and Rasmussen, 1995) and evidence from ^{15}N isotope enrichment studies at the site suggest the latter (Gundersen, unpublished data).

Our results support overall the Aber model of nitrogen saturation (Aber et al., 1998), although suggest somewhat different rates of change for various ecosystem components. For instance, the Aber model postulates continuing increasing foliar N with increasing stages of N saturation; our results suggest a rapid increase to a plateau well before saturation is reached. The Aber model suggests N leaching in runoff or seepage water is delayed until stage 2; our study suggests some leaching begins immediately. Finally the Aber model suggests a steady decline in soil Ca:Al ratios occurs from the onset of N deposition; our study (using pH as a proxy) suggests this does not occur until other elements of the system (vegetation, soil N) are N-saturated. Using this approach, we are also able to estimate the absolute time scale for the rate of change of different components of the ecosystem.

We conclude from this exercise that combining comparative with experimental studies yields a much more thorough understanding of the nature, direction, magnitude and rate of environmental change than is achievable by focusing on one approach only. These results, of course, must be combined with long-term studies and laboratory experimentation to refine our understanding. In addition, we recognise that ecosystem responses to chronic N additions are often non-linear and may only partly be reversible, and there are feedbacks and linkages across the N cycle beyond those suggested by this simplified approach. Further refinements of this work will attempt to reduce the variability of the comparative data by a number of filters, while still endeavouring to keep enough sites over a wide enough range in deposition chemistry to achieve good empirical models.

Acknowledgements

Financial support came from the European Union (EU) through article 254 contract No 2000.60.NL, 3B (the DYNAMIC project) and contract No. QLK5-2001-00596 (the CNTER project) as well as SamNordisk Skogsforskning, contract SNS-72.

References

Aber, J., McDowell, W., Nadelhoffer, K., Magill, A., Berntson, G., Kamakea, M., McNulty, S., Currie, W., Rustad, L. and Fernandez, I.: 1998, 'Nitrogen saturation in temperate forest ecosystems', *Bioscience* **48**, 921–34.
Armbruster, M., MacDonald, J., Dise, N. B. and Matzner, E.: 2002, 'Throughfall and output fluxes of Mg in European forest ecosystems: A regional assessment', *For. Ecol. Manage.* **164**, 137–147.
Carpenter, S. R., Chisholm, S. W., Krebs, C. J., Schindler, D. W. and Wright, R. F.: 1995, *Science* **269**, 324–227.
Dise, N. B., Matzner, E., Armbruster, M. and MacDonald, J.: 2001, 'Aluminum output fluxes from forest ecosystems in Europe: A regional assessment', *J. Environ. Qual.* **30**, 1747–1756.
Dise, N. B., Matzner, E. and Forsius, M.: 1998, 'Evaluation of organic horizon C:N ratio as an indicator of nitrate leaching in conifer forests across Europe', *Environ. Pollut.* **102**, 453–456.
Dise, N. B., Matzner, E. and Gundersen, P.: 1998, 'Synthesis of nitrogen pools and fluxes from European forest ecosystems. *Water, Air, Soil Pollut.* **105**, 143–154.
Gundersen, P., Boxman, A. W., Lamersdorf, N., Moldan, F. and Andersen, B. R.: 1998, 'Experimental manipulation of forest ecosystems: Lessons from large roof experiments', *For. Ecol. Manage.* **101**, 339–352.
Gundersen, P.: 1998, 'Effects of enhanced nitrogen deposition in a spruce forest at Klosterhede, Denmark, examined by moderate NH_4NO_3 addition', *For. Ecol. Manage.* **101**, 251–268.
Gundersen, P. and Rasmussen, L.: 1995, 'Nitrogen mobility in a nitrogen limited forest at Klosterhede, Denmark, examined by NH_4NO_3 addition', *For. Ecol. Manage.* **71**, 75–88.
Gundersen, P.: 1989, 'Air pollution with nitrogen compounds: Effects in coniferous forest', *Ph.D. Thesis*, Laboratory of Environmental Sciences and Ecology, Technical University of Denmark, pp. 292 (in Danish).
Hovmand, M. F. and Bille-Hansen, J.: 1999, 'Atmospheric input to Danish spruce forests and effects on soil acidification and forest growth based on 12 yr measurements', *Water, Air, Soil Pollut.* **116**, 75–88.
MacDonald, J. A., Dise, N. B., Matzner, E., Armbruster, M., Gundersen, P. and Forsius, M.: 2002, 'Nitrogen input together with ecosystem nitrogen enrichment predict nitrate leaching from European forests', *Global Change Biol.* **8**, 1028–1033.
Wright, R. F. and van Breemen, N.: 1995, 'The NITREX project – An introduction', *For. Ecol. Manage.* **71**, 1–5.

SPECTRAL ANALYSIS OF CHEMICAL TIME SERIES FROM LONG-TERM CATCHMENT MONITORING STUDIES: HYDROCHEMICAL INSIGHTS AND DATA REQUIREMENTS

XIAHONG FENG[1], JAMES W. KIRCHNER[2] and COLIN NEAL[3]

[1] Department of Earth Sciences, Dartmouth College, Hanover, New Hampshire 03755, U.S.A.;
[2] Department of Earth and Planetary Science, 307 McCone Hall, University of California, Berkeley, California 94720-4767, U.S.A.; [3] Center for Ecology and Hydrology, McLean Building, Wallingford, Oxon OX10 8BB, U.K.
(* author for correspondence, e-mail: xiahong.feng@dartmouth.edu; phone: 603-646-1712; fax: 603-646-3922)

(Received 20 August 2002; accepted 18 April 2003)

Abstract. Hydrological and hydrochemical data from long-term monitoring stations are vital for inferring travel times and flowpaths of water, and transport of contaminants through catchments. Spectral analysis is particularly powerful for studying the hydrological and chemical dynamics of catchments across a wide range of time scales. Here, recent work is reviewed that illustrates how spectral analyses of long-term monitoring data can be used to infer the travel-time distribution of water through catchments, and to measure the chemical retardation of reactive solutes at the catchment scale. For spectral analysis, it is desirable to have data sets with high sampling frequency and long periods of coverage. Using two data sets, a 3-yr daily data series and a 17-yr weekly data series from the Hafren catchment at Plynlimon, Wales, we demonstrate that high-frequency sampling (e.g., daily or more frequent) is particularly useful for revealing the short-term chemical dynamics that most clearly reflect the interplay of subsurface chemical and hydrological processes. However, data sets that combine high-frequency sampling during storm events with low-frequency sampling between storms can cause spectral artifacts and must be treated with special care.

Keywords: catchments, hydrochemistry, Plynlimon, solute transport, spectral analysis, time series analysis, tracers

1. Introduction

Catchment studies are important for understanding water quality, contaminant transport, biogeochemical processes, and ecosystem responses to natural and anthropogenic disturbances (Černy et al., 1995). It has been increasingly recognized that many catchment-level processes, such as soil responses to acid deposition and ecosystem responses to deforestation and forest fires, operate on timescales of decades or longer, and that observing these processes requires catchment monitoring programs spanning similar lengths of time. As a result, many monitoring programs have been set up, and their long-term time series of water and chemical fluxes not only record the history of ecosystem responses to disturbance, but also provide insight into the structure and function of ecosystem processes at the landscape scale (e.g., Church, 1997). For example, long-term data have enabled landscape-scale input-output budget calculations for various chemical species. Such calculations are essential for constraining rates of chemical weathering, biological uptake and release, and nutrient cycling at the catchment scale (e.g., Likens and Bormann, 1995). Further, these long-term data sets have made it possible to quantify temporal trends in major and trace element budgets. They also have provided understanding of catchment responses to various ecosystem disturbances, both natural and anthropogenic, including acid deposition, climate change, land use change (deforestation, agriculture), hurricanes and forest fires (e.g., Britton, 1991; Neal *et al.*, 1992; Kirchner and Lydersen, 1995; Wesselink *et al.*, 1995; Schaefer *et al.*, 2000). Long-term monitoring data also havebeen used to clarify subsurface flowpaths and reaction mechanisms, and to calibrate and test mathematical models of hydrological and geochemical processes (e.g., Hooper *et al.*, 1988; Kirchner, 1992; Kirchner *et al.*, 1992; Ferrier *et al.*, 1995).

In this contribution, we discuss spectral analysis as a little-explored use of long-term catchment monitoring data. We examine the utility of spectral methods in the context of long-term catchment studies at Plynlimon, mid-Wales (e.g., Reynolds *et al.*, 1986; Durand *et al.*, 1994; Neal *et al.*, 1997). We illustrate the importance of long-term data sets from research catchments like Plynlimon, and emphasize, using the Plynlimon data, the importance of both long-term coverage and high sampling frequency in catchment monitoring data sets.

2. Significance of Spectral Analysis for Catchment Studies

In a series of papers (Kirchner *et al.*, 2000a, 2001; Feng *et al.*, in review), we recently have used long-term monitoring data from Plynlimon to show how spectral analysis of naturally-occurring chemical tracers can be used to measure the travel-time distribution of water moving through a catchment, as well as the catchment-scale retardation factor for reactive solutes. In this section, we review the important contributions, emphasizing the value of time-series data sets and the utility of spectral methods. In the following section, we discuss desirable qualities of data sets for such analyses.

2.1. CATCHMENT-SCALE TRAVEL-TIME DISTRIBUTIONS

Some fraction of the rain that falls on a catchment today will reach the stream today; some fraction will reach the stream tomorrow, some fraction the day after, and so forth. These timescales over which a catchment transmits precipitation to streamflow are quantified by its travel-time distribution, which is the probability distribution of the relative amounts of water reaching the stream after a given travel time through the catchment. The travel-time distribution is an important characteristic of a catchment because it determines how long it takes for the catchment to be flushed and, therefore, how long it would take for soluble contaminants to be cleaned up. Recently, Kirchner *et al.* (2000a) used long-term records of Cl$^-$ in precipitation and stream water at Plynlimon to demonstrate how one can empirically determine a catchment's travel-time distribution from spectral analyses of passive tracer concentrations.

The Plynlimon catchments generally are covered with thin acid soils (podzols and gleys) that overlie fractured bedrock of slates and shales. Rainfall is typically about 2500 mm a^{-1} and evaporation plus transpiration amounts to 25 to 50% of the input, depending upon the type of vegetation cover. The available Cl$^-$ data at Plynlimon include weekly measurements of precipitation and streamwater from mid-1983 to the present and daily measurements for three years (1994–1997) (Neal and Kirchner, 2000). This long sampling period, combined with three years of high-frequency data, was particularly useful for determining the travel-time distribution of the Plynlimon catchment (Kirchner *et al.*, 2000a).

For a chemical tracer that is supplied to the catchment entirely by rainfall, the concentration in the stream $c_S(t)$ at any time t will be the convolution of the travel-time distribution $h(\tau)$ and the rainfall concentration at all previous times $c_R(t-\tau)$, where τ is the lag time between rainfall and runoff:

$$c_s(t) = \int_0^\infty h(\tau) c_R(t-\tau) d\tau \tag{1}$$

Because the flow rate varies through time, Equation (1) is strictly valid when t and τ are expressed in terms of the cumulative flow through the catchment, rather than calendar time (Neimi, 1977; Rodhe *et al.*, 1996), but the mathematics are the same in either case (Neimi, 1977). The rainfall and stream Cl-time series can be used to constrain the travel-time distribution h(τ) by employing the convolution theorem, which states that the convolution in Equation (1) is equivalent to multiplying the Fourier transforms of each of its terms:

$$C_S(f) = H(f) \, C_R(f) \quad \text{and} \quad |C_S(f)|^2 = |H(f)|^2 \, |C_R(f)|^2 \tag{2}$$

where f is frequency (cycles/time); $C_S(f)$, $H(f)$, and $C_R(f)$ are the Fourier transforms of $c_S(t)$, $h(\tau)$, and $c_R(t-\tau)$; and $|C_S(f)|^2$, $|H(f)|^2$, and $|C_R(f)|^2$ are their power spectra (Gelhar, 1993). This equation allows one to test alternative travel-time models $h(\tau)$ by calculating their power spectra $|H(f)|^2$, and testing whether they are consistent with the relationship between the input and output power spectra $|C_R(f)|^2$ and $|C_S(f)|^2$. Kirchner *et al.* (2000a) computed the power spectra of Cl$^-$ in rainfall ($|C_R(f)|^2$) and Plynlimon streams ($|C_S(f)|^2$), and found that the spectral power of rainfall Cl$^-$ scales roughly as white noise and the spectral power of stream Cl$^-$ scales roughly as fractal $1/f$ noise (Figure 1), with spectral power increasing proportionally to wavelength. In addition, they showed that the

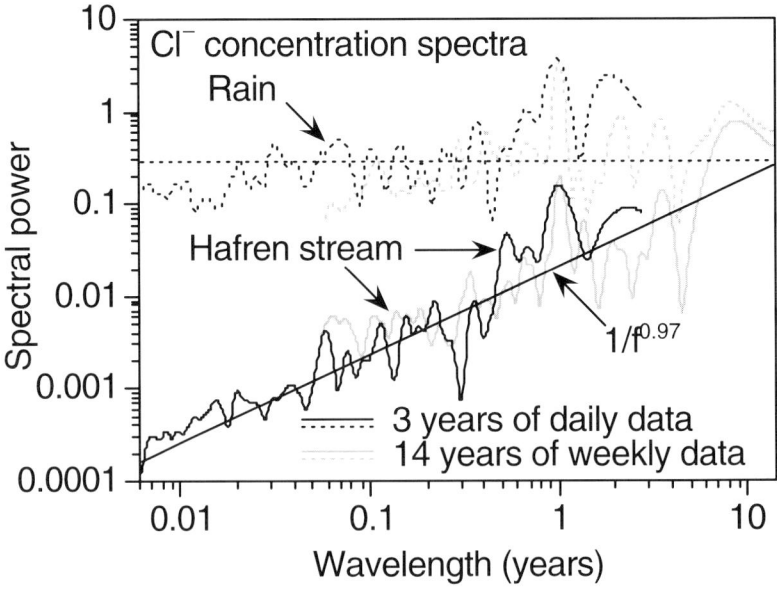

Figure 1. Power spectra of Cl⁻ variations in rainfall and Hafren stream water at Plynlimon, Wales.

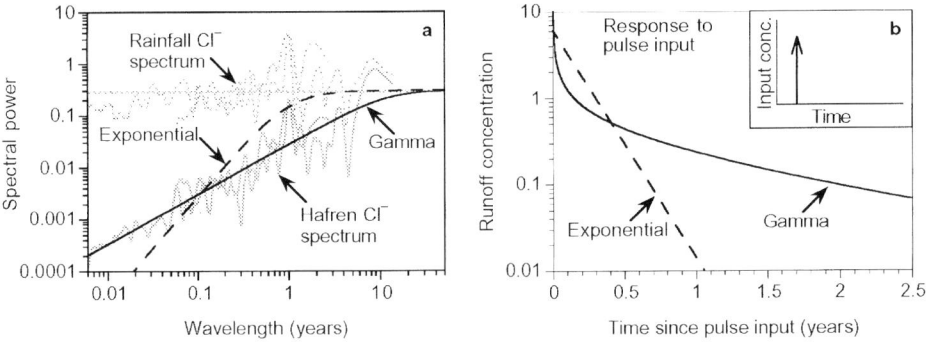

Figure 2. Comparison of two alternative travel-time distributions, their power spectra, and their consequences for contaminant transport. (a) Best-fit power spectra of the two distributions, superimposed on the rainfall and Hafren chloride concentration spectra. The gamma distribution is consistent with the observed power spectrum of Cl⁻ in Hafren streamflow, but the exponential distribution is not, showing that the catchment does not function as a homogeneous mixing tank. (b) Response of streamflow concentrations to a delta-function pulse input of contaminants. Compared to the exponential travel-time distribution, the gamma distribution would seem to remove the contaminant rapidly at the beginning. However, since the gamma distribution has a much longer tail than the exponential distribution, it sustains substantial contaminant concentrations for much longer time spans. The inset depicts the delta-function contaminant input.

Plynlimon Cl⁻ power spectra were consistent with a travel-time distribution that can be empirically approximated by the gamma distribution,

$$h(\tau) = \frac{\tau^{\alpha-1}}{\beta^\alpha \, \Gamma(\alpha)} \, e^{-\tau/\beta} \qquad (3)$$

where β is a scale parameter and α (≈ 0.5) is a shape parameter. Figure 2a shows that this gamma function closely reproduces the scaling of the Cl⁻ power spectra in stream water. In contrast, a conventional catchment 'box' model would predict an exponential distribution of travel times, which is inconsistent with the spectral scaling observed at Plynlimon (Figure 2a).

The inferred travel-time distribution has significant implications for contaminant transport through catchments. Figure 2b shows how a pulse input of a soluble contaminant is removed by natural flushing if the travel-time distribution is a gamma function versus an exponential function (see Catchment Models, below). Compared to the exponential distribution, the catchment having a gamma travel-time distribution would appear to flush out the contaminant quickly at first but then very slowly thereafter, delivering low-level contamination to the stream for a long time.

The $1/f$ scaling behavior of stream Cl⁻ is not unique to the Plynlimon catchments. A wide array of catchments, with substrates ranging from deeply fractured shales to glacially scoured gneisses, and with drainage areas ranging over three orders of magnitude, exhibit fractal tracer scaling similar to that shown in Figure 1 (Kirchner *et al.*, 2000b). By comparing the power spectra, and thus the travel-time distributions, from different catchments, it may be possible to determine how (or whether) the residence time of water is related to characteristics such as catchment geometry, hillslope gradient, soil depth, and substrate properties.

2.2. CATCHMENT MODELS

Empirical travel-time distributions, like those presented above, are even more useful for studying catchment transport processes if we know the mechanism(s) that generate them. This requires creating physical models that are consistent with the observed scaling in tracer fluctuations. The simplest catchment model is a 'box' or 'mixing tank' model, in which the catchment is viewed as a well-mixed water reservoir. In such a model, the travel-time distribution is an exponential function. However, as indicated in Figure 2a, this travel-time distribution is fundamentally inconsistent with the observed scaling of stream Cl⁻ at Plynlimon.

Kirchner *et al.* (2001) built a simple one-dimensional model in which spatially distributed rainfall tracer inputs advect and disperse with water flowing downhill. They showed that the power spectrum of the model tracer scales roughly as $1/f$ noise, as long as the Peclet number of the advection-dispersion system is of order 1 or smaller; that is, as long as the subsurface flow system is highly dispersive, with characteristic dispersion length scales on the order of the average hillslope

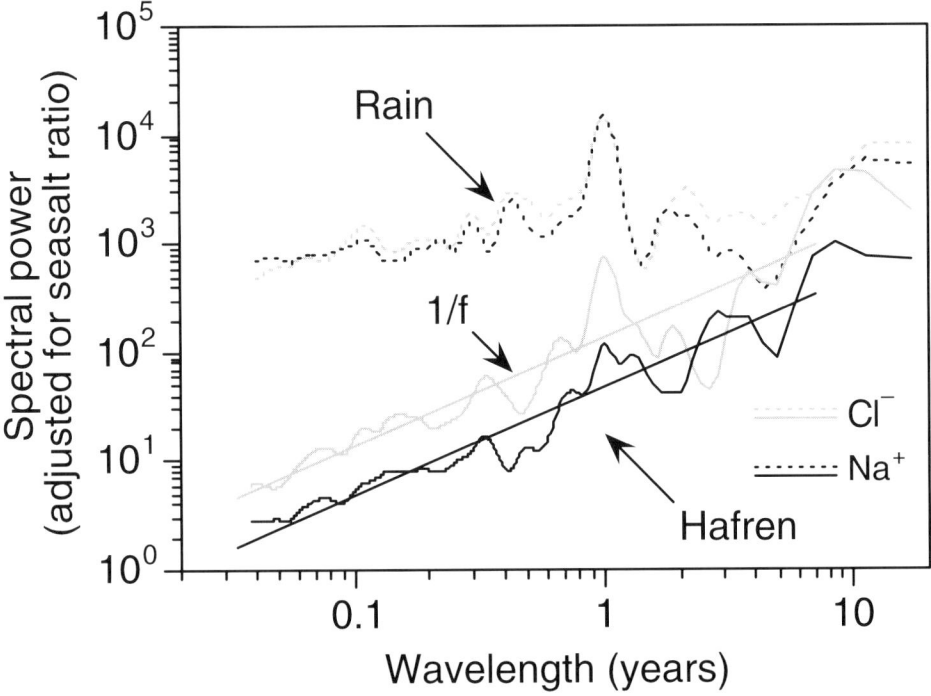

Figure 3. Power spectra of Na^+ and Cl^- in rainfall and Hafren stream at Plynlimon, Wales. The average retardation factor can be obtained from the vertical offset of the two spectra. In this case, the Na^+ retardation is 2.7 (Feng et al. (in press)).

length. This in turn implies that the subsurface flow is affected by large variations in conductivity, on all scales up to the hillslope scale itself, results that are in keeping with general field observations at Plynlimon (Neal, 1997).

Scher et al. (2002) have shown that a continuous-time random-walk (CTRW) model can also produce the $1/f$ scaling of the Cl^- spectrum and the corresponding travel-time distribution. We believe that there may be still other physical models that are consistent with the spectral tracer scaling that we have observed. Searching for these models will improve our understanding of subsurface flow routing and its consequences for physical, chemical and biological processes in catchments.

2.3. QUANTIFYING REACTIVE TRACER TRANSPORT USING SPECTRAL ANALYSIS

Recently, we compared Na^+ and Cl^- time series in rainfall and stream water at Plynlimon (Neal and Kirchner, 2000), and used them to derive whole-catchment chemical retardation factors for Na^+ transport at four Plynlimon catchments (Feng et al. (in press)). At Plynlimon, both Na^+ and Cl^- are almost entirely derived from precipitation, and the flow weighted Na^+/Cl^- molar ratio in stream water is close to that of seasalt (0.86). When compared with stream Cl^-, stream Na^+ has a longer mean travel time and its fluctuations are more strongly damped in the stream relative to precipitation. This additional damping of Na^+ compared to Cl^- can be attributed to adsorption/desorption of Na^+ in the subsurface, most likely by cation exchange. The spectral power of both Na^+ and Cl^- scale as white noise in rainfall and as $1/f$ noise in streamwater. However, streamwater Na^+ has consistently lower spectral power than streamwater Cl^- across the range of wavelengths studied (e.g., Figure 3). From the vertical offset between the Cl^- and Na^+ spectra, we can calculate the whole-catchment chemical retardation factor for Na^+ (which equals 2.7 in this example). To our knowledge, whole-catchment retardation factors have never been reported before. This spectral method opens up new opportunities for studying and quantifying transport properties of reactive tracers.

These are a few examples showing the usefulness of spectral analyses for catchment studies. The long-term monitoring data at Plynlimon have made these studies possible. More studies at Plynlimon and similar studies for other catchments are yet to come. The spectral analysis methods reviewed here allow us to quantify hydrologically and geochemically important properties of catchments at catchment scale (such as their whole-catchment travel-time distributions and whole-catchment retardation factors). They thus provide new opportunities for determining whether these properties are shared among catchments generally, or whether they are specific to individual catchments with particular characteristics (substrates, geometries, soil types, climates, vegetation covers, etc.).

3. Data Sets for Spectral Analysis

For spectral analysis of time-series data, it is always desirable to have both long-term coverage and high sampling frequency. The importance of long-term catchment monitoring has been widely recognized in the scientific community (Church, 1997). Here, while acknowledging the importance of long-term data sets, we emphasize the utility of high-frequency sampling for revealing catchment behavior. This is because streams not only have a long chemical memory of precipitation, but also exhibit prompt responses to rainfall inputs.

Figure 4 shows three pairs of diagrams plotting the time series and power spectra of Cl^- concentrations in rainfall and stream water in the Hafren catchment at Plynlimon. From the top down, the figures show monthly, weekly and daily sampling frequencies. For all three sampling frequencies, the spectral power of stream Cl^- is consistently lower than the corresponding spectral power of rainfall Cl^-. In addition, these figures all suggest that high frequency rainfall variations are damped more in the stream water than low frequency variations are. With each increase in sampling frequency, one can see that this trend of greater damping at shorter wavelengths extends from monthly to weekly and to daily time scales.

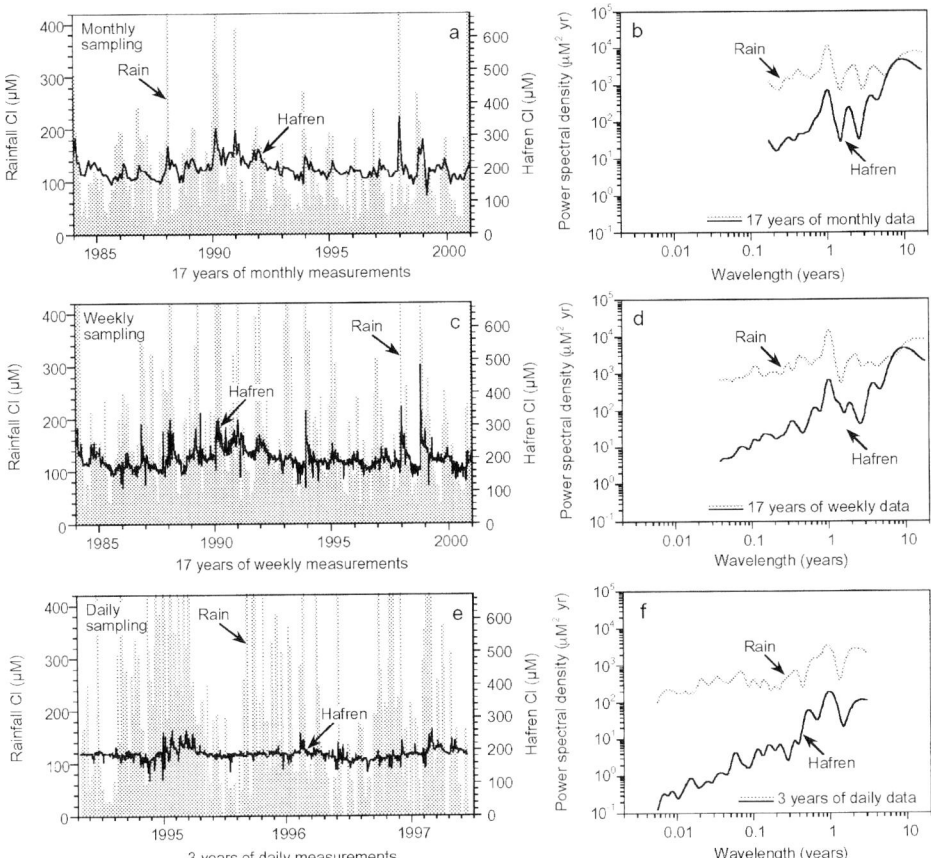

Figure 4. Time series and power spectra at monthly, weekly, and daily sampling frequencies for rainfall and stream Cl^- at the Hafren catchment. (a,b) Monthly measurements of Cl^- in rainfall and stream water, subsampled from a 17-yr data set of weekly measurements, and their corresponding power spectra. (c,d) Weekly measurements for 17 yr, and their corresponding power spectra. (e,f) Daily measurements for three years, and their corresponding power spectra. High-frequency sampling (e.g., daily data) more clearly shows short-wavelength features, better defining the $\sim 1/f$ scaling of stream spectra. Such high-frequency information is intrinsically missing in low-frequency data sets (e.g., monthly sampling, in this example).

This example demonstrates that the daily data set is particularly valuable for clarifying the $1/f$ spectral scaling of stream Cl^- at short wavelengths. One reason for this is that seasonal variations in Cl^- create an annual cycle with a strong spectral peak that dominates the spectrum at wavelengths near 1 yr. This strong annual peak makes it difficult to see the underlying $1/f$ scaling behavior unless the spectrum extends to wavelengths significantly shorter than 1 yr. The higher the sampling frequency, the farther the spectrum can be extended into the short-wavelength domain.

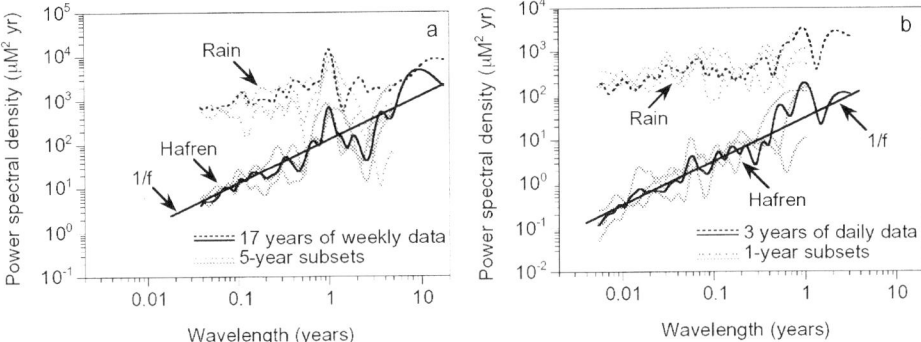

Figure 5. Comparison of power spectra of Cl$^-$ data having different temporal lengths. (a) Comparison of the 17-yr weekly data set (black) with three 5-yr subsets (grey). (b) Comparison of the 3-yr daily data set (black) with three 1-yr subsets (grey). In both cases, the longer data set gives information at longer wavelengths than the short data sets. However, the daily data better define the $1/f$ scaling behavior of the spectra than weekly data regardless of the record length, because the daily data contain high-frequency information that is not present in the weekly data sets.

For a given sampling interval, one inevitably gets more information from a long data set than from a short one. Figure 5 illustrates two comparisons of spectra generated from long versus short data sets. Figure 5a shows the power spectra (solid lines) for 17 yr of weekly data in comparison to three 5-yr subsets (grey lines) from the same 17-yr data set. Of course, the 5-yr data sets cannot provide information for wavelengths longer than five years. Their spectra are also more variable than those from the 17-yr data set (particularly in the rainfall spectra), especially in the long-wavelength range near the annual peak and beyond it. This variability results from the fact that there are fewer cycles at these long wavelengths in the shorter data sets. For example, there are five annual cycles in the 5-yr data sets, but 17 in the 17-yr data set, so the longer data set can more accurately constrain the average spectral signature of the annual cycles in Cl$^-$.

Similar observations can be made from Figure 5b, in which the power spectra of the 3-yr daily time series are compared with those of three one-year subsets of the same data. Note, however, that even though the one-year data sets are three times shorter than the 3-yr data set, they are nevertheless helpful in constraining the spectral behavior at the short-wavelength end. The damping of stream Cl$^-$ spectral power relative to that of rainfall Cl$^-$ is more clearly shown from these data sets than from the weekly data sets (Figure 5a), and the $1/f$ scaling of stream Cl$^-$ is better defined as well.

While stressing the importance of high-frequency sampling, we caution that if high-frequency data are only available for a relatively short window of time (as will usually be the case), the analysis may be biased if the catchment's behavior during that time window is unrepresentative. Hydrological conditions fluctuate significantly from year to year. Daily samples from a stormy year may lead to different

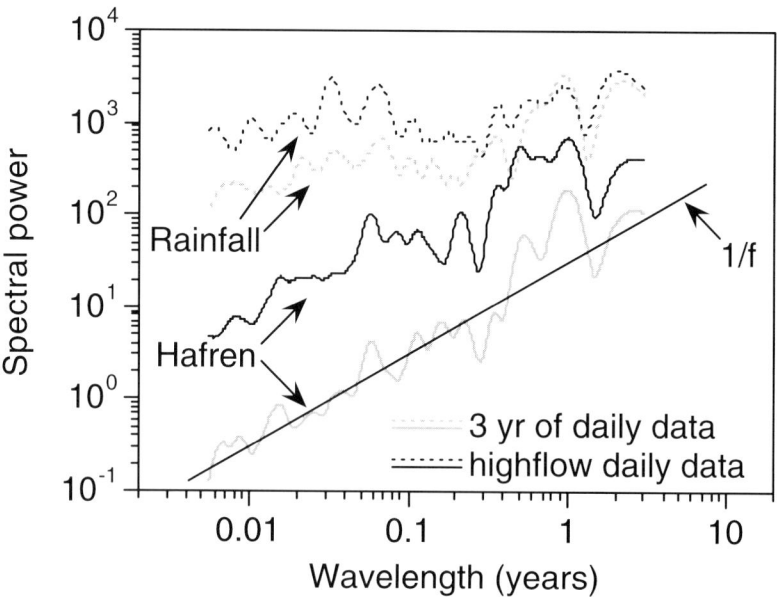

Figure 6. Spectral effects of sampling bias. A high flow data series is generated by subsampling the 20% of days in the 3-yr daily data with the highest stream flows. The power spectra from this highflow data set are compared with that of the complete 3-yr daily data set. The spectral power of stream Cl^- in the high-flow data is significantly higher than that in the complete data set, but both exhibit approximate $1/f$ scaling.

results than data sampled from a hydrologically calm year. Disturbances that affect catchment chemistry may contribute to this bias; examples are hurricanes, droughts, fire, ENSO events, etc. What this means in practice is that one should be careful when combining spectra from different data sets with different sampling frequencies, particularly when they cover different spans of time.

Many catchment data sets include long-term measurements taken regularly each week, with higher-frequency sampling during storm events. The recently-developed spectral analysis methods for unevenly-sampled data (Scargle, 1982; Foster, 1996) make it tempting to analyze such records, but caution is needed in interpreting the resulting spectra. The high-frequency sampling (and thus the short-wavelength characteristics of the spectrum) will be inherently biased toward the catchment's behavior during storm events, while the low-frequency regular sampling (and thus the long-wavelength part of the spectrum) will reflect the catchment's average behavior, of which storm events are just one component. Thus, the high-frequency data, and the short-wavelength end of the spectrum, will be unrepresentative of the average catchment behavior in the regular weekly data. The weekly monitoring data and the event data both contain useful information, but they cannot be uncritically combined.

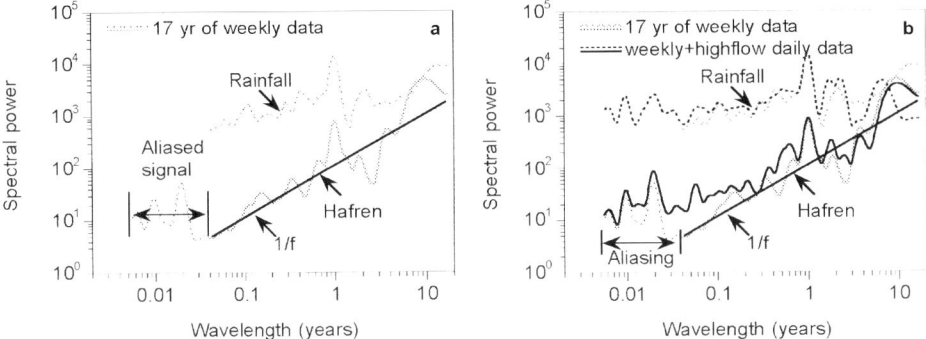

Figure 7. Spectral aliasing arising from analysis of time series combining regular weekly sampling with daily sampling during highflow periods. (a) Analyzing the weekly data at wavelengths shorter than the Nyquist limit (~ two weeks) generates marked spectral aliases. (b) These spectral aliases dominate the short-wavelength behavior of the combined weekly-plus-highflow-daily time series, obscuring the $1/f$ scaling that is observed in both the weekly and highflow daily time series when they are analyzed separately (see Figures 5a and 6).

To demonstrate this point, we generated a new data set by subsampling the existing 3-yr daily data set at Hafren, retaining only those points for days with the highest 20% of stream flows over the 3-yr period. This yields a data set covering only the high-flow periods (which are of course unevenly distributed through the three-year record). As Figure 6 shows, the spectral power of rainfall and streamflow Cl^- during these high-flow periods is substantially higher than in the continuous three-year data set. Thus, although event data contain useful information about catchment hydrochemical properties at high flow, it should not be misinterpreted as the average catchment behavior.

In addition to this sampling bias, spectral aliasing can substantially distort the power spectra of data sets that combine long periods of regular (or nearly regular) sampling with more frequent sampling during short episodes. Figure 7 demonstrates this aliasing effect. In Figure 7a, the weekly data from Hafren are intentionally analyzed down to wavelengths of only two days, corresponding to the conventional Nyquist limit for daily sampling. At wavelengths shorter than roughly 14 days (corresponding to the Nyquist limit for weekly sampling), the $1/f$ spectrum disappears and is replaced by a flat spectrum (white noise) punctuated by two peaks at wavelengths of 7 days and 3.5 days. These wavelengths correspond to the sampling frequency and its first harmonic; the peaks are aliases of the large low-frequency power in the signal. This spectral aliasing can persist, even if the time series contains higher-frequency sampling during brief episodes. We can demonstrate this aliasing effect by combining the long-term weekly data from Hafren with the high-flow subset of the daily sampling data. Figure 7b shows the spectrum of this combined time series. Note that the spectral aliasing of the weekly data dominates the short-wavelength end of the spectrum. Remember that the weekly data set and the highflow daily data set both exhibit $1/f$ scaling when they are analyzed

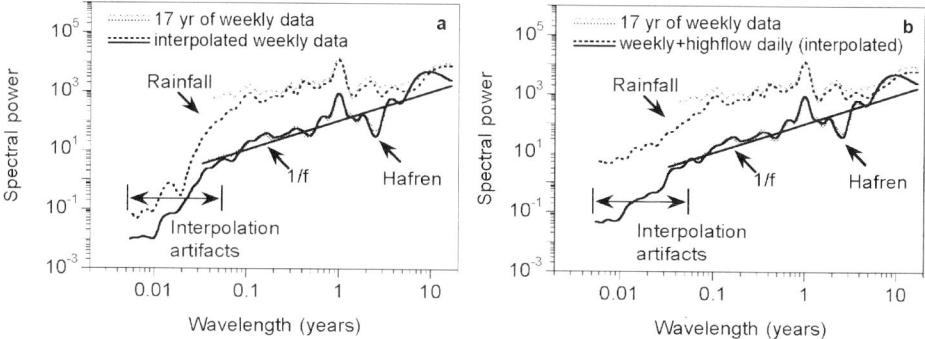

Figure 8. Spectral distortions arising from linear interpolation of time series combining regular weekly sampling with daily sampling during highflow periods. (a) Linear interpolation generates marked damping of spectral power at wavelengths shorter than the Nyquist limit for the original, un-interpolated data. (b) This interpolation artifact dominates the short-wavelength end of the spectrum even for time series that include high-frequency sampling during episodes.

separately, each over their appropriate ranges of wavelengths (see Figures 5a and 6). However, when the two time series are combined and analyzed together, the weekly data are analyzed beyond their Nyquist limit, resulting in spectral aliasing.

It is often assumed that chemical concentrations in stream water vary little between storms. It is therefore tempting to analyze records that mix long-term regular sampling with short-term episode sampling, by creating an evenly spaced high-frequency time series, using linear interpolation to bridge the gaps in the long-term regularly sampled data. We can test this approach using our combined weekly-plus-highflow data set, by interpolating between each of the real data to construct an evenly-sampled daily time series. At wavelengths shorter than the true sampling interval, linear interpolation will artificially reduce the spectral power, because the interpolated time series will be inherently less variable than the real, unsampled time series would be. As Figure 8a shows, interpolating between the weekly data to create a daily data set distorts the spectrum, artificially steepening it at short wavelengths. Including the highflow daily data in the interpolated time series reduces the interpolation artifact somewhat (see Figure 8b), but the short-wavelength end of the spectrum is still artificially steepened.

Great care must be taken when analyzing catchment data sets that combine long, regularly sampled records with shorter periods of high-frequency sampling during episodes. Here we have identified three types of distortions that can arise. First, the behavior during episodes may be atypical, generating bias (Figure 6). Second, the spectrum may be extended to wavelengths shorter than the Nyquist limit for the regularly sampled data, generating aliases (Figure 7). Finally, interpolating such records can produce artifactually low spectral power at short wavelengths (Figure 8). These distortions appear to be intrinsic to the time series themselves, as they appear with both of the widely used spectral analysis methods that are designed for unevenly sampled data (Scargle, 1982; Foster, 1996). It may be possible that at some specific combination of high-frequency episode sampling and low-frequency regular sampling, these three artifacts might offset one another. An undistorted spectrum might thus be obtained, but only through the sheer coincidence of the various distortions canceling each other out. A wiser approach, in our view, is to not merge the data sets in the first place, but instead analyze each of the different types of data separately, using spectral methods and wavelength ranges that are appropriate to each (as in Figures 1 and 4).

4. Conclusions

Spectral analyses of time series from long-term catchment monitoring stations are valuable for studying the hydrological and chemical dynamics of catchments across many time scales. Our recent work has demonstrated that travel-time distributions of water moving through catchments can be determined using spectral analysis of passive tracer concentrations in precipitation and in stream flow. By comparing the power spectra of passive and reactive tracers, it is possible to estimate chemical retardation factors for reactive solutes at the whole-catchment scale.

The ideal data set for spectral analysis would have a long temporal span (a decade or longer) and a relatively high sampling frequency (e.g., daily sampling); no such 'ideal' data set currently exists. High-frequency variations in rainfall and stream chemistry are particularly useful for understanding catchments' chemical response to precipitation at short time scales. At the Plynlimon catchments in Mid-Wales, three years of daily measurements reveal the spectral scaling behavior of stream Cl^- more clearly than 17 yr of weekly measurements. Higher-frequency sampling during storm events is common at many monitoring stations. However, these data should not be uncritically combined with regular monitoring data reflecting average catchment behavior, as spectral biases and aliasing can result.

Acknowledgements

Our collaboration was supported by National Science Foundation Grants EAR-9903281, EAR-0125338 and EAR-0125550. Sample collection and analysis were supported by the Natural Environment Research Council, the Environment Agency of England and Wales, and the Forestry Commission. We thank the Plynlimon field staff for sample collection and M. Neal for sample analysis. The manuscript was improved by comments from the editor and an anonymous reviewer.

References

Černý, J., Novák, M., Paăes, T. and Wieder, R. K. (eds): 1995, *Biogeochemical Monitoring in Small Catchments*, Kluwer Academic Publishers, Dordrecht, 432 pp.

Britton, D. L.: 1991, 'Fire and the chemistry of a south-African mountain stream', *Hydrobiologia* **218**, 177–192.

Church, M. R.: 1997, 'Hydrochemistry of forested catchments', *Ann. Rev. Earth Planet. Sci.* **25**, 23–59.

Durand, P., Neal, C., Jeffery, H. A., Ryland, G. P. and Neal, M.: 1994, 'Major, minor and trace element budgets in the Plynlimon afforested catchments (Wales): General trends, and effects of felling and climate variations', *J. Hydrol.* **157**, 139–156.

Feng, X., Kirchner, J. W. and Neal, C.: 'Measuring catchment-scale chemical retardation using spectral analysis of reactive and passive chemical tracer time series', *J. Hydrol.* (in press).

Ferrier, R. C., Wright, R. F., Cosby, B. J. and Jenkins, A.: 1995, 'Application of the MAGIC model to the Norway spruce stand at Solling, Germany', *Ecol. Model.* **83**, 77–84.

Foster, G.: 1996, 'Time series analysis by projection. 1. Statistical properties of Fourier analysis', *Astron. J.* **111**, 541–554.

Gelhar, L. W.: 1993, *Stochastic Subsurface Hydrology*, Prentice-Hall, Englewood Cliffs, New Jersey, 390 pp.

Hooper, R. P., Stone, A., Christophersen, N., de Grosbois, E. and Seip, H. M.: 1988, 'Assessing the Birkenes model of stream acidification using a multisignal calibration methodology', *Water Resour. Res.* **24**, 1308–1316.

Kirchner, J. W.: 1992, 'Heterogeneous geochemistry of catchment acidification', *Geochim. Cosmochim. Acta* **56**, 2311–2327.

Kirchner, J. W., Dillon, P. J. and LaZerte, B. D.: 1992, 'Predicted response of stream chemistry to acid loading tested in Canadian catchments', *Nature* **358**, 478-482.

Kirchner, J. W., Feng, X. and Neal, C.: 2000a, 'Fractal stream chemistry and its implication for contaminant transport in catchments', *Nature* **403**, 524–527.

Kirchner, J. W., Feng, X. and Neal, C.: 2001, 'Catchment-scale advection and dispersion as a mechanism for fractal scaling in stream tracer concentrations', *J. Hydrol.* **254**, 82–101.

Kirchner, J. W., Feng, X., Neal, C., Skjelkvaale, B. L., Clair, T. A., Langan, S., Soulsby, C., Kahl, J. S. and Norton, S. A.: 2000b, 'Generality of – and a proposed mechanism for – fractal fluctuations in stream tracer chemistry', *EOS, Trans. Am. Geophys. Union* **81**, F554.

Kirchner, J. W. and Lydersen, E.: 1995, 'Base cation depletion and potential long-term acidification of Norwegian catchments', *Environ. Sci. Technol.* **29**, 1953–1960.

Likens, G. E. and Bormann, F. H.: 1995, *Biogeochemistry of a Forested Ecosystem*, Springer-Verlag.

Neal, C.: 1997, 'A view of water quality at the Plynlimon catchment', *Hydrol. Earth Syst. Sci.* **1**, 743–754.

Neal, C., Forti, M. C. and Jenkins, A.: 1992, 'Towards modeling the impact of climate change and deforestation on stream water-quality in Amazonia – a perspective based on the MAGIC model', *Sci. Total Environ.* **127**, 225–241.

Neal, C. and Kirchner, J. W.: 2000, 'Sodium and chloride levels in rainfall, mist, streamwater and groundwater at the Plynlimon catchments, mid-Wales: Inferences on hydrological and chemical controls', *Hydrol. Earth Syst. Sci.* **4**, 295–310.

Neal, C., Wilkinson, J., Neal, M., Harrow, M., Wickham, H., Hill, S. and Morfitt, C.: 1997, 'The hydrochemistry of the headwater of the river Severn Plynlimon', *Hydrol. Earth Syst. Sci.* **1**, 583–617.

Niemi, A. J.: 1977, 'Residence time distributions of variable flow processes', *Int. J. Appl. Radiat. Is.* **28**, 855–860.

Reynolds, B., Neal, C., Hornung, M. and Stevens, P. A.: 1986, 'Baseflow buffering of streamwater acidity in five mid-Wales catchments', *J. Hydrol.* **87**, 167–185.

Rodhe, A., Nyberg, L. and Bishop, K.: 1996, 'Transit times for water in a small till catchment from a step shift in the oxygen 18 content of the water input', *Water Resour. Res.* **32**, 3497–3511.

Scargle, J. D.: 1982, 'Studies in astronomical time series analysis. II. Statistical aspects of spectral analysis of unevenly spaced data', *Astrophys. J.* **263**, 835–853.

Schaefer, D. A., McDowell, W. H., Scatena, F. N. and Asbury, C. E.: 2000, 'Effects of hurricane disturbance on stream water concentrations and fluxes in eight tropical forest catchments of the Luquillo Experimental Forest, Puerto Rico', *J. Trop. Ecol.* **16**(Part 2), 189–207.

Scher, H., Margolin, G., Metzler, R., Klafter, J. and Berkowitz, B.: 2002, 'The dynamical foundation of fractal stream chemistry: The origin of extremely long retention times', *Geophys. Res. Lett.* **29**, 10.1029/2001GL014123.

Wesselink, L. G., Meiwes, K.-J., Matzner, E. and Stein, A.: 1995, 'Long-term changes in water and soil chemistry in spruce and beech forests, Solling, Germany', *Environ. Sci. Technol.* **29**, 51–58.

NEUTRALISATION OF SULPHUR DIOXIDE DEPOSITION IN A CONIFEROUS CANOPY

MARTIN FERM* and HANS HULTBERG

*IVL Swedish Environmental Research Institute, P.O. Box 47086, SE-402 58 Gothenburg, Sweden
(* author for correspondence, e-mail: Martin.ferm@ivl.se; phone: +46 31 7256224;
fax: +46 31 7256290)*

Abstract. Previously, it has been observed that the internal circulation (ion leakage) of calcium from a coniferous forest is caused by uptake of sulphur dioxide (SO_2). Here we show that this correlation was not changed when the forest floor is covered with a roof. The reaction takes place in the canopy and is not influenced by deposition and root uptake of calcium and sulphate. The ion leakage of calcium is linked to the loss of acidity in throughfall. The process can, for one of the catchments, schematically be written: $SO_2 + H_2O + 0.5\,O_2 + 0.58\,CaA_2 \rightarrow SO_4^{2-} + 0.94\,H^+ + 0.58\,Ca^{2+} + 1.16\,HA$, in which A denotes the anion to a weak acid. This reaction also takes place today when the SO_2 concentration is very low, but when the precipitation is still acidic. The ion leakage of manganese also is caused by the uptake of SO_2, but only 0.12 manganese ions are released per SO_2 molecule.

Keywords: atmospheric deposition, calcium, coniferous forest canopy, internal circulation, ion leakage, manganese, throughfall

1. Introduction

Swedish rivers, lakes, groundwater, and soils have been acidified significantly during the 20th century. In forest soils, the acidification causes the flux of calcium in runoff to exceed the weathering rate of the soil plus litterfall and atmospheric deposition minus the uptake by biota. When most of the exchangeable Ca^{2+} in the soil has been lost, aluminium starts to leach from the soil. The acidification in Sweden is caused by deposition of acidifying compounds, most of them long-range transported (Ferm and Hultberg, 1998) and in the latter part of the 20th century the decreased deposition of base cations (Hedin *et al.*, 1994) has worsened the acidification. Acidifying pollutants can also damage the needles of conifers.

A very simple method to estimate the SO_2 uptake in conifer needles has previously been developed (Ferm and Hultberg, 1995). It is based on the assumption that a negligible fraction of the S amount being deposited or taken up by the tree through stomata is retained in the tree (Hultberg *et al.*, 1983; Lindberg and Garten, 1988; Lindberg and Lovett, 1992; Hultberg and Grennfelt, 1992). The throughfall of SO_4^{2-} represents the wet deposition of SO_4^{2-} plus the dry deposition of particulate SO_4^{2-} and the SO_2 uptake. The throughfall of Ca^{2+} represents the wet deposition of Ca^{2+} plus the dry deposition of particulate Ca^{2+} and the internal circulation. The dry deposition of particulate SO_4^{2-} and Ca^{2+} are measured using a surrogate surface that resembles the needles. In several studies, it has been found that the internal circulation of Ca^{2+} in spruce is well correlated with the uptake of SO_2 (Hultberg and Ferm, 1995; Skeffington and Sutherland, 1995; Ferm and Hultberg, 1999; Ferm et al., 2000). Retention of S has, however, been observed in deciduous trees (Lindberg and Garten, 1988; Garten et al., 1988). The correlation between internal circulation of Ca^{2+} and the uptake of SO_2 has been studied for over two decades during which the SO_2 concentration has decreased by a factor 20 (Hultberg and Ferm, in press). The mechanism for this interaction is, however, not understood. However, Fink (1991) showed that calcium oxalate crystals crystals occur extracellularly on the outside of the walls of mesophyll cells facing the intercellular spaces. The uptake of SO_2 is highly correlated with the SO_2 concentration in air. The annual average acidity and SO_4^{2-} concentration of precipitation in Gårdsjön are correlated to the annual average SO_2 concentration in air ($r^2 = 0.83$ and $r^2 = 0.89$ respectively). The concentration of other acidifying pollutants (except for particulate SO_4^{2-}) has been rather constant during these decades. The net acidity of throughfall (throughfall minus wet deposition) has also decreased. Igawa et al. (2002) has shown that acidic fog also can leach out Ca^{2+} from needles. The measurements presented here were not designed for investigating exchanges in the canopy, but were part of monitoring within the UN ECE programme on Integrated Monitoring and the ROOF project (Hultberg and Skeffington, 1998).

2. Experimental

The measurements were carried out at two forested catchments at the Gårdsjön Experimental Area about 15 km from the open Swedish west coast (58.07 °N, 12.02 °E, and altitude 120 m). The roof-covered catchment was 0.6 ha and consisted of a coniferous forest dominated by Norway spruce [*Picea abies* (L.) Karst.] (83% with 587 stems ha^{-1} and about 80 to 100 years old and 17% Scots pine). Both the spruce and pine are about 18 m tall. Deionised water containing some KNO_3 (40 mmol m^{-2} a^{-1}) and seawater (corresponding to the natural input in the reference area which was 210 mmol m^{-2} a^{-1} of Na^+ and 13 mmol m^{-2} a^{-1} of SO_4^{2-}) was sprinkled on the soil under the roof from April 1991 to summer 2001.

The forest in the reference catchment consists of ca. 13 m tall and 100-yr-old Norway spruce. It is 3.7 ha and has 653 stems ha^{-1}. Continuous measurements in this catchment started in 1981. Manganese was, however, not measured in precipitation and throughfall until 1989. Throughfall collectors were placed on top of the roof and inside the reference forest. Net throughfall (NTF), dry deposition (DD), and internal circulation (leaching, denoted IC) of Ca^{2+} and S were calculated using throughfall, wet deposition, and a surrogate surface (Ferm and Hultberg, 1995, 1999; Hultberg and Ferm, 1995).

NEUTRALISATION OF SO₂ DEPOSITION IN A CONIFEROUS CANOPY   239

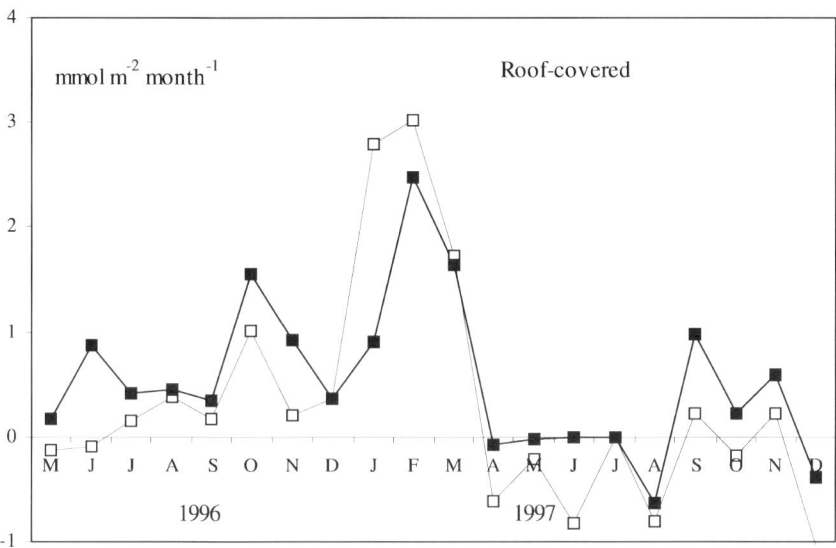

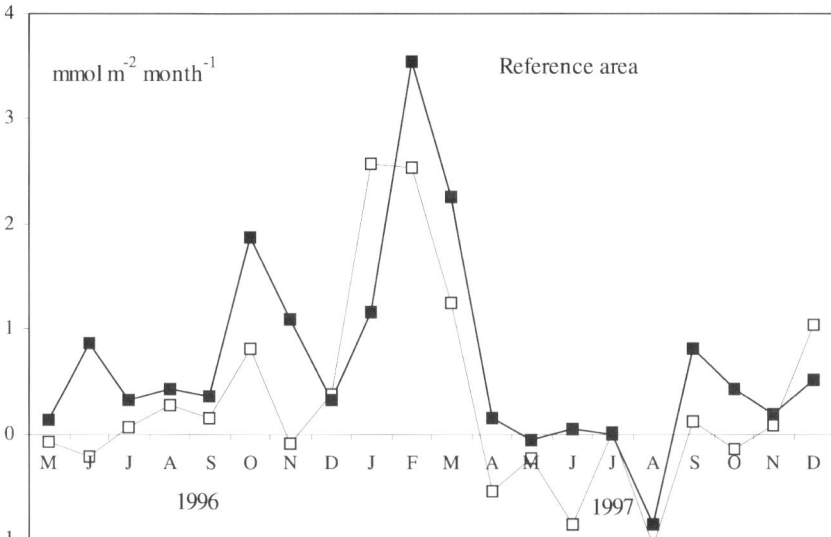

Figure 1. SO$_2$ uptake (open squares) and calcium leakage (filled squares) by Norway spruce as a function of month. The upper plot represents the roof-covered part and the lower the reference area.

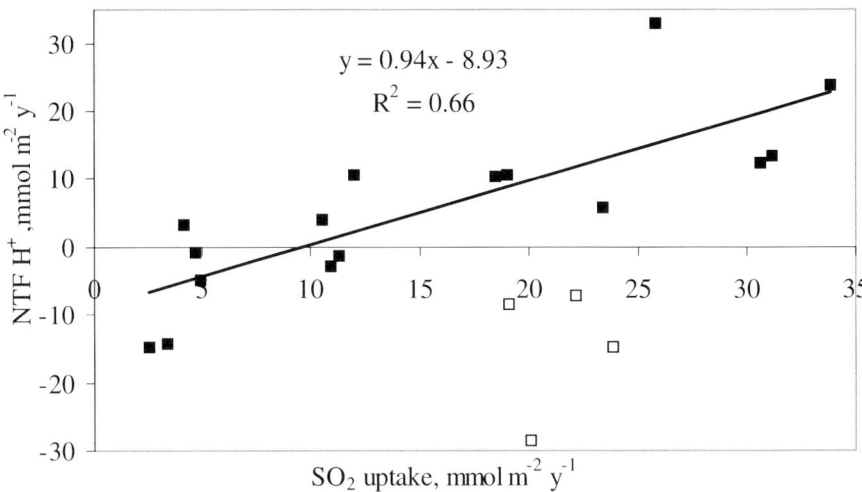

Figure 2. Annual net throughfall of hydrogen ions as a function of annual SO_2 uptake from 1981 to 2000. Four outliers (open squares) have been excluded from the regression line due to extreme hot summers in Europe affecting the pH of samples in the field.

3. Results and Discussion

The uptake of SO_2 and the ion leakage of Ca^{2+} as a function of time in the roof-covered forest and the reference forest are shown in Figure 1. The ion leakage of Ca^{2+} follows the uptake of SO_2 at both catchments. Maximum fluxes occur in the wintertime when the SO_2 concentration has a maximum. This has been observed previously during several years of observations (Ferm and Hultberg, 1999). The inputs of Ca^{2+}, SO_4^{2-}, and H^+ ions to the soils are very different at the two catchments (*cf.* Table I). The roof-covered soil receives only 19% of the SO_4^{2-} deposition, 14% of the Ca^{2+} deposition, and 1 % of the H^+ deposition compared with the reference area. The average concentration of SO_2 was 14 nmol m^{-3} (S.D. = ± 8 nmol m^{-3}, n = 20) and the average particulate SO_4^{2-} concentration was 21 nmol m^{-3} (S.D. = ± 10 nmol m^{-3}, n = 20) in air during the period of May 1996 to December 1997. The relationship between SO_2 uptake and ion leakage of Ca^{2+} is not affected by the deposition of SO_4^{2-} or Ca^{2+} on the ground. It seems to be a process taking place in the canopy.

The annual net acidity in throughfall as a function of SO_2 uptake in the reference area is shown in Figure 2. There is a wide scatter in the data. It is, however, evident that the acidity increases with increasing SO_2 deposition. If some of the early years are excluded in the regression (indicated by open squares in Figure 2), a slope of 0.94 is obtained. This implies that the formed H_2SO_4 is partly neutralised before reaching the ground. If it were not neutralised the slope would be 2.0. The annual internal circulation of Ca^{2+} increases as a function of the annual SO_2 uptake during the same period with a slope of 0.58 (Hultberg and Ferm, in press). This implies that for each SO_2 we get one SO_4^{2-}, 0.94 H^+, and 0.58 Ca^{2+}. The charge of the

TABLE I

Integrated fluxes (mmol m^{-2}) of sulphur, calcium, hydrogen ions, and manganese at the two forest plots during 20 months

	Roof	Reference
Sulphur		
WD	37	37
DD	21	22
SO$_2$ uptake	+6.4	+6.0
Sum deposition	64	65
reaching ground	13^a	65
runoff	47	76
Calcium		
WD	8.4	8.4
DD	9.3	9.6
Ion leakage	+11	+14
Sum deposition	28	32
reaching ground	4.5*	32
runoff	7.8	18
Hydrogen ions		
WD	47	47
DD + interaction	+18	+(-3.4)
Sum deposition	65	43
reaching ground	3.5*	43
runoff	28	51
Manganese		
WD	0.8	0.8
DD	0.1	0.1
Ion leakage	+1.1	+0.9
Sum deposition	1.9	1.8
reaching ground	0.0*	1.8
runoff	0.2	0.2

^{a} = Added by sprinkling under the roof

leached Ca^{2+} almost balances the SO$_4^{2-}$, implying that Ca^{2+} must be accompanied by an anion that is a rather strong base (e.g., HCO$_3^-$) or an ion forming a biological decomposable acid (e.g., oxalic acid). The reaction for the reference area schematically can be written: SO$_2$ + H$_2$O + 0.5 O$_2$ + 0.58 CaA$_2$ → SO$_4^{2-}$ + 0.94 H$^+$ + 0.58Ca^{2+} + 1.16 HA, in which A denotes the unidentified anion. The coefficients in this reaction represent this reference area only. The neutralisation

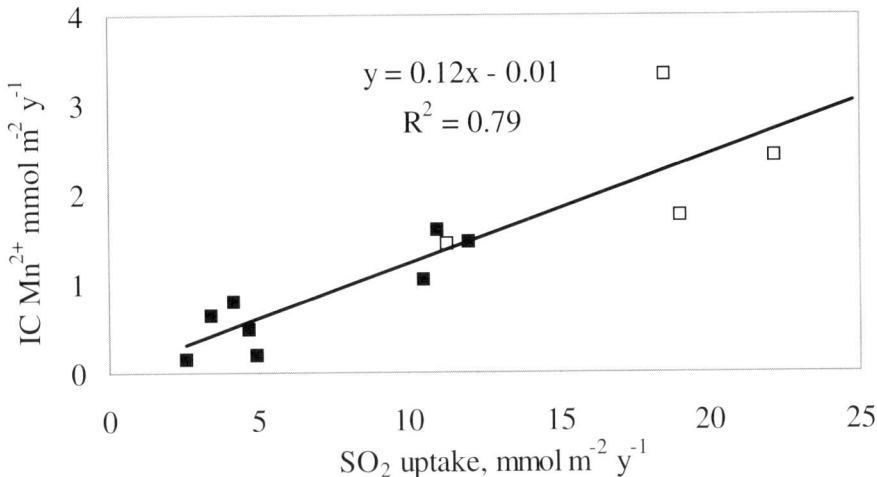

Figure 3. Annual internal circulation of manganese as a function of annual SO_2 uptake from 1989 to 2000. The open squares represent measurements from 1989 to 1992 when the dry deposition of manganese was not measured.

was smaller in the roof-covered area. Lindberg *et al.* (1986) have earlier observed that H^+ is retained in the canopy of deciduous trees (*Quercus prinus, Quercus alba*) by exchange with Ca^{2+} and K^+.

The leakage of Ca^{2+} is not solely caused by uptake of SO_2. When the ion leakage of Ca^{2+} was plotted as a function of SO_2 uptake during 20 years, the intercept (no SO_2 uptake) was 4.6 mmol Ca^{2+} m^{-2} a^{-1} (Hultberg and Ferm, in press). The intercept for net throughfall of H^+ during the same 20 years (Figure 2) is negative, indicating that the unknown anion also is extruded from the needles when there is no SO_2 in the air. The intercept is -8.9 mmol m^{-2} a^{-1}, about twice the intercept for ion leakage of calcium, indicating a release of CaA_2. The uncertainty of this intercept is, however, large. The present ion leakage of calcium (2000) is 8 mmol m^{-2} y^{-1} and has been as high as 28 mmol m^{-2} y^{-1} in 1988 when the SO_2 concentration was concentration was 70 nmol m^{-3}. The present (2000) SO_2 concentration is 9 nmol m^{-3}. The ion budget of throughfall generally shows a cation excess of 5 % and presumably the leaching of Ca^{2+} and K^+ is accompanied by an organic ion.

The ion leakage of Mn^{2+} also is correlated with the uptake of SO_2. The slope is not as steep as for Ca^{2+} and the intercept is very small. The relationship is shown in Figure 3. Between 1993 and 2000, the dry deposition of Mn^{2+} was 0.05 ($\pm$ 0.01) mmol m^{-2} a^{-1}. This figure was used to estimate the ion leakage between 1989 and 1992 when the surrogate surface was not installed. The highest fluxes of Mn^{2+} are, as for Ca^{2+}, also observed during the winter. If the ion leakage of Mn^{2+} and the uptake of SO_2 are plotted as a function of month (Figure 4), there are no large differences between the roof-covered area and the reference area. The ion leakage of Mn^{2+} is therefore not liked to the flux of Mn^{2+} reaching the ground (Table I).

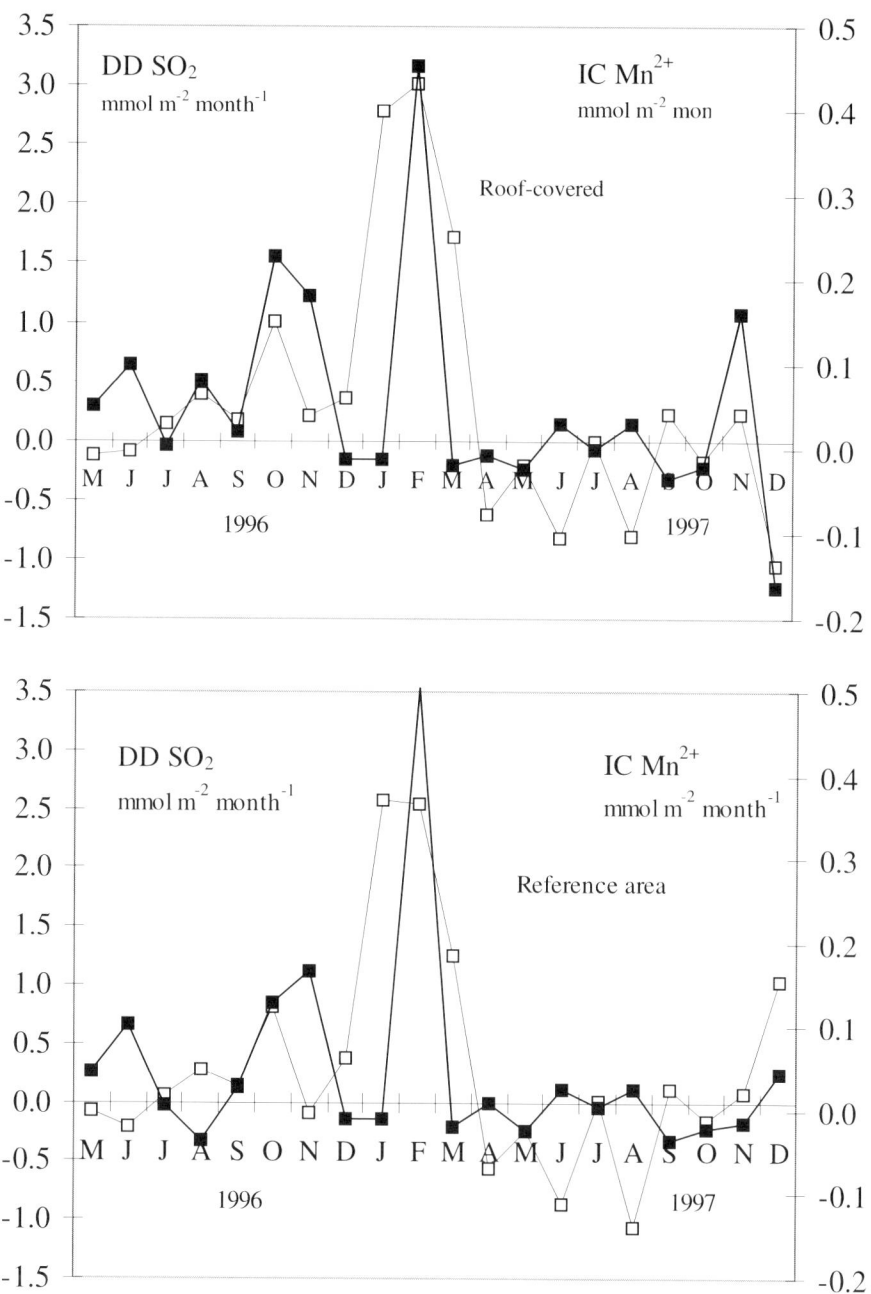

Figure 4. SO$_2$ uptake (open squares) and manganese leakage (filled squares) by Norway spruce as a function of month. The upper plot represents the roof-covered part and the lower the reference area.

The ion leakage of K^+ and Mg^{2+} was also plotted as a function of the uptake of SO_2 in a similar way. The correlations were however very poor.

4. Conclusions

The SO_2 that has been dry deposited by the needles is found as SO_4^{2-} in throughfall. The throughfall should thus be more acidic than the precipitation. A spruce canopy can, however, partly neutralise the acidic precipitation before it reaches the soil. This neutralisation as well as the ion leakage of Ca^{2+} is strongly linked to the SO_2 uptake. The ion leakage of Mn^{2+} also is strongly linked to the SO_2 uptake. The composition of throughfall that reaches the ground does not seem to affect this process. The fluxes have been measured using simple sampling techniques, such as bulk collectors and a rain-shielded surrogate surface of inert material (Ferm and Hultberg, 1999).

Acknowledgements

Funding of the research projects and monitoring at Lake Gårdsjön has come from many different sources over the last 20 years. Some of the most important are the Swedish Environmental Protection Agency who fund the UN ECE programme Integrated Monitoring and Swedish and British electric power industry together with IVL Swedish Environmental Research Institute who funded the Roof-project. Some additional funding was received from the ASTA (International and National Abatement Strategies for Transboundary Air Pollution) programme.

References

Ferm, M. and Hultberg, H.: 1995, 'Method to estimate atmospheric deposition of base cations in coniferous throughfall', *Water, Air, Soil Pollut.* **85**, 2229–2234.

Ferm, M. and Hultberg, H.: 1998, 'Atmospheric Deposition to the Gårdsjön Research Area', in H. Hultberg and R. A. Skeffington (eds), *Experimental Reversal of Acid Rain Effects: The Gårdsjön Roof Project*, John Wiley & Sons Ltd., London, UK, pp. 71–84.

Ferm, M. and Hultberg, H.: 1999, 'Dry deposition and internal circulation of nitrogen, sulphur and base cations to a coniferous forest', *Atmos. Environ.* **33**, 4421–4430.

Ferm, M., Westling, O. and Hultberg, H.: 2000, 'Atmospheric deposition of base cations, nitrogen and sulphur in coniferous forests in Sweden – a test of a new surrogate surface', *Boreal Environ. Res.* 5, 197–207.

Fink, S.: 1991, 'The micromorhological distribution of bound calcium in needles of Norway spruce [*Picea abies* (L.) Karst.]', *New Phytol.* **119**, 33–40.

Garten, T. C. Jr,, Bondietti, E. A. and Lomax, R. D.: 1988, 'Contribution of foliar leaching and dry deposition to sulfate in net throughfall below deciduous trees', *Atmos. Environ.* **22**, 1425–1432.

Hedin, L. O., Granat, L., Likens, G. E., Buishand, T. A., Galloway, J. N., Butler, T. J. and Rodhe, H.: 1994, 'Steep declines in atmospheric base cations in regions of Europe and North America', *Nature* **367**, 351–354.

Hultberg, H. and Ferm M.: 1995, 'Measurements of atmospheric deposition and internal circulation of base cations to a forested catchment area', *Water, Air, Soil Pollut.* **85**, 2235–2240.

Hultberg, H. and Grennfelt, P.: 1992, 'Sulphur and seasalt deposition as reflected by throughfall and runoff chemistry in forested catchments', *Environ. Pollut.* **75**, 215–222.

Hultberg, H. and Skeffington, R. A. (eds): 1998, *Experimental Reversal of Acid Rain Effects: The Gårdsjön Roof Project*, John Wiley & Sons Ltd., London, UK, 484 pp.

Hultberg, H. and Ferm, M.: 2004, 'Temporal changes and fluxes in wet and dry deposition, internal circulation as well as in runoff in a coniferous forest in Sweden during two decades', *Biogeochemistry*, in press.

Hultberg, H., Grennfelt, P. and Olsson, B.: 1983, 'Sulphur and chloride deposition and ecosystem transport in a strongly acidified lake watershed', *Water Sci. Technol.* 15, 81–103.

Igawa, M., Kase, T., Satake, K. and Okochi, H.: 2002, 'Severe leaching of calcium ions from fir needles caused by acid fog', *Environ. Pollut.* **119**, 375–382.

Linberg, S. E., Lovett, G. M., Richter, D. D. and Johnson, D.W.: 1986, 'Atmospheric deposition and canopy interactions of major ions in a forest', *Science* **231**, 141–145.

Lindberg, S. E. and Garten, C. T. Jr.: 1988, 'Sources of sulphur in forest canopy throughfall', *Nature* **336**, 148–151.

Linberg, S. E. and Lovett, G. M.: 1992, 'Deposition and forest canopy interactions of airborne sulfur: Results from the integrated forest study', *Atmos. Environ.* **26A**, 1477–1492.

Skeffington, R. A. and Sutherland, P. M.: 1995, 'The effects of SO_2 and O_3 fumigation on acid deposition and foliar leaching in the Liphook forest fumigation experiment', *Plant, Cell Environ.* **18**, 247–261.

RECENT RECOVERY OF LAKE WATER QUALITY IN SOUTHERN QUÉBEC FOLLOWING REDUCTIONS IN SULFUR EMISSIONS

DANIEL HOULE[1,2], CHRISTIAN GAGNON[1], SUZANNE COUTURE[1] and ALAIN KEMP[1]

[1] *Centre Saint-Laurent, Environnement Canada, 105 rue McGill, Montréal, Québec, Canada H2Y 2E7;* [2] *Ministère des Ressources Naturelles du Québec, Direction de la recherche forestière, 2700 Einstein St., Sainte-Foy, Québec, Canada G1P 3W8*
(* author for correspondence, e-mail: daniel.houle@mrn.gouv.qc.ca; phone: 418-643-7994, ext: 6543; fax: 418-643-2165)

(Received 20 August 2002; accepted 11 April 2003)

Abstract. Since 1985, monitoring activities have been conducted in a network of 43 lakes comprising the Québec portion of the Long-Range Transport of Airborne Pollutants (LRTAP) program. The results to date indicate that Québec lakes generally are responding positively to the generalized decline in precipitation sulfate (SO_4^{2-}), with 40 of the 43 lakes now showing steep declines in SO_4^{2-} concentrations. The drop in SO_4^{2-} was associated with a significant decrease in Ca^{2+} concentrations in 77% of the lakes (67% for Mg^{2+} concentrations). Overall, the acid-neutralizing capacity was increasing in 19 lakes and decreasing only in three, while 21 lakes showed no temporal trends. Compared with previous trend studies of the LRTAP-Québec network for the period of 1985–1993, the longer period (1985–1999) shows a clear improvement, with the proportion of lakes that were acidifying changing from 24 to 7% and with the proportion of lakes that were recovering changing from 16 to 35%. These observations suggest that the recent drop in SO_4^{2-} deposition in the northeastern U.S. and eastern Canada was significant enough to allow chemical recovery for a significant proportion of Québec lakes.

Keywords: acid deposition, acidification reversibility, alkalinity, basic cations, lake chemistry, lake recovery, Québec, sulfate, surface water recovery

1. Introduction

In recent decades, major reductions in SO_2 emissions have resulted in a generalized decrease in SO_4^{2-} concentrations in surface waters in many regions of the northeastern United States (Stoddard *et al.*, 1999) and eastern Canada (Clair *et al.*, 1995; Houle *et al.*, 1997; Kemp, 1999). In some lakes, the reduced SO_4^{2-} concentrations led to decreased acidity, while producing no improvement in the acid-base status of the water column in others (Bouchard, 1997; Stoddard *et al.*, 1999). A lack of improvement in water chemistry may have many causes. Some authors have posited that NO_3^- may contribute more to precipitation acidity (Driscoll *et al.*, 1989; Hedin *et al.*, 1994). Others note that atmospheric deposition of base cations has decreased considerably in the last 20 yr (Hedin *et al.*, 1994). Another possibility is the increase in dissolved organic carbon (DOC) concentrations in lake waters (Bouchard, 1997), and therefore in the contribution of organic acids (OA^-) to the acid-base chemistry of surface water. All these possibilities could potentially reduce any expected benefits of the SO_4^{2-} reduction in precipitation and in lakes. However, the main concern emerging from the literature and now getting widespread recognition is the reduction in reserves of exchangeable basic cations in the soil of watersheds (Likens *et al.*, 1996; Houle *et al.*, 1997; Lawrence *et al.*, 1997) and therefore in their capacity to supply well-buffered water to lakes.

The objectives of this study were to determine recent temporal trends in the main parameters related to the acidification of aquatic ecosystems in the lakes of the LRTAP-Québec network ($n = 43$) and to test the hypothesis that lake recovery (or the lack thereof) is linked to the base cations status of the lake water chemistry.

2. Methods

2.1. ATMOSPHERIC DEPOSITION AND PRECIPITATIONS

The precipitation data used in this study were taken from the Chalk River and Montmorency stations of the Canadian Air and Precipitation Monitoring Network (CAPMoN). The Chalk River station is located west of the study area, near the Ontario-Québec border, 150 km northwest of Ottawa, while the Montmorency station is located 80 km north of Québec City. Monthly data for precipitation quantity, average pH, concentration and deposition of H^+, SO_4^{2-}, NO_3^-, Ca^{2+}, and Mg^{2+} were used. Detailed information on the sampling protocols and analytical methods are provided in Vet *et al.* (1989). The data available cover the period of September 1983 to December 1996 for the Chalk River station, and the period January 1981 to December 1996 for the Montmorency station.

2.2. NETWORK DESCRIPTION AND FIELD SAMPLING

The 43 lakes of the LRTAP-Québec network are located in an area defined as a strip 150-km wide, parallel to the St. Lawrence River, between the Ottawa and Saguenay rivers (Figure 1). They were selected based on a presampling of 158 headwater lakes chosen at random (Bobée *et al.*, 1983). As one of the criteria, the lake watersheds had to be free of major human activity and peat bogs to minimize the influence of organic acids. The selected lakes were then grouped into six statistically homogeneous regions in terms of pH, alkalinity, mineralization (the sum of the Ca^{2+} and Mg^{2+} concentrations), and SO_4^{2-} concentration (Bobée *et al.*, 1983).

With the exception of the lakes in Region 5 and some lakes in Region 4, which are considered relatively insensitive to acidification due to the nature of the underlying bedrock and unconsolidated deposits relatively rich in carbonate materials (marble and metamorphic limestone), most lakes are particularly susceptible to

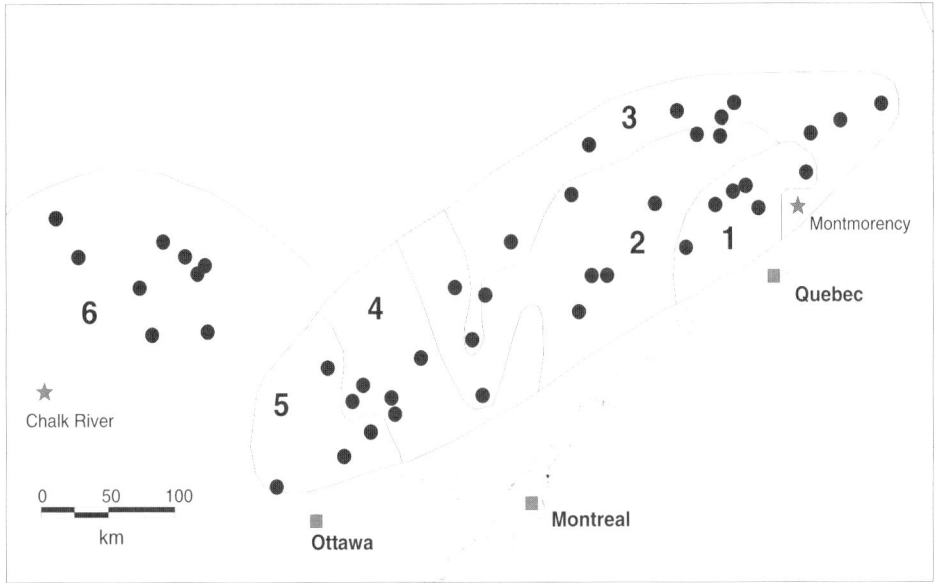

Figure 1. Location of the network lakes and boundaries of regions 1 to 6. The two sampling stations for precipitation (Montmorency and Chalk River) are also indicated.

acidification because their watersheds are composed of soils and rocks with a low capacity for neutralizing acid inputs.

The network lakes were sampled twice a year at the end of May and at the end of November. Of these 43 lakes, 15 were sampled six times a year (every two months) from 1985 to 1997, and twice a year thereafter. The lakes previously selected for this purpose were representative of average regional values, as determined from the presampling. Integrated water samples (5 m) were collected by helicopter at the geometric center of the lake. When the lake was less than 5 m deep, integration was done 1 m from the bottom of the lake to the surface. All samples were kept cool and sent to the laboratory within 24 hr. Prior to shipping to the laboratory, samples for NH_4^+ determination were fixed with sulfuric acid (H_2SO_4 30%).

2.3. CHEMICAL ANALYSIS

Sample pH and conductivity were measured potentiometrically. Alkalinity was measured by potentiometric titration. Sulfate (Cronan, 1979) and NO_3^- (Stainton *et al.*, 1974) were measured by colorimetry. Dissolved organic carbon (DOC) was measured by infrared spectrometry after ultraviolet oxidation in the presence of persulfate. Before 1991, the cations (Ca^{2+}, Mg^{2+}, Na^+, K^+) were measured by atomic absorption; thereafter, inductively coupled plasma emission was employed (Environment Canada, 1994). An overlap study comparing base cation concentrations revealed that the two methods produced similar results.

2.4. DATA PROCESSING AND STATISTICAL ANALYSIS

Differences in water chemistry among regions were evaluated using the non-parametric Kruskal–Wallis test, at a significance level of 5%. Nonparametric tests (Hirsch and Slack, 1984; Berryman *et al.*, 1988) were used to check for the existence of temporal trends using the DETECT software (Cluis *et al.*, 1989). Trends in NO_3^- were not tested because too many values were below the detection limit (2.8 μeq L^{-1}).

3. Results

3.1. PRECIPITATION AT BOTH STATIONS

Concentrations of H^+, SO_4^{2-}, NO_3^-, Ca^{2+} and Mg^{2+} in precipitation were lower at the Montmorency station than at the Chalk River station (Table I). These results are consistent with the locations of the two stations, as Chalk River is further west and closer to industrial areas and pollution sources. However, because of the greater volume of precipitation at the Montmorency station (38% on average), average deposition at the two sites is comparable for the entire study period.

Between the early 1980s and 1996, monthly SO_4^{2-} concentrations and deposition at the Chalk River and Montmorency stations fell by 36% and 28%, and by 45 and 34%, respectively. Concurrent with the drop in SO_4^{2-} concentrations, precipitation H^+ concentrations declined at both stations, resulting in a significant rise in pH. No significant trend in NO_3^- could be detected at either station.

The concentration and deposition of Ca^{2+} showed strong downward trends at both stations. They declined by nearly half from levels recorded in the early 1980s. At Chalk River, Mg^{2+} concentrations and deposition have also dropped significantly, whereas at Montmorency, there is no significant trend for this cation.

3.2. SURFACE WATER COMPARISON BETWEEN REGION

The spatial distribution of Ca^{2+} and Mg^{2+} concentration means (Table II) in each region shows a decreasing gradient from the southwest to the northeast (*i.e.*, regions 5 to 1). The lakes of the Outaouais region (region 5) have the highest values, with averages of 350.0 and 78.1 μeq L^{-1} for Ca^{2+} and Mg^{2+} respectively, while the lakes of region 1 have the lowest, at 54.0 and 18.0 μeq L^{-1} for Ca^{2+} and Mg^{2+}, respectively. The same gradient is observed for pH and alkalinity, which changed respectively from 7.1 to 5.5 and from 282.0 to 7.8 μeq L^{-1} from regions 5 to 1.

The SO_4^{2-} concentration gradient in network lakes (Table II) corresponds closely to the distance from SO_2 emission sources. Region 5, the one closest to emission sources, has the highest average: 145.6 μeq L^{-1}. Regions 1, 2 and 3, the easternmost and farthest away, have average SO_4^{2-} concentrations of 75.2, 91.5 and 81.1 μeq L^{-1}, respectively. No regional differences can be clearly demonstrated

TABLE I

Trends in monthly precipitation, concentration and wet deposition measured at the Chalk River and Montmorency stations of the CAPMoN network

	Trend	Start	End	% Variation	Mean	Test	Significance level
Chalk River, 1983–1996 ($n = 148$)							
Precipitation (mm)	–	66.6	72.5		69.6	KS	
pH	↑	4.21	4.44	5	4.28	S-L	($p = 0.025$)
H^+ conc. ($\times 10^{-5}$ µeq L^{-1})	↓	6.1	3.6	–41	5.2	S-L	($p = 0.051$)
H^+ dep. (kg ha^{-1} mo^{-1})	↓	0.04	0.03	–25	0.03	KS	($p = 0.004$)
SO_4^{2-} conc. (µeq L^{-1})	↑	52.8	33.9	–35	43.0	H/S	($p = 0.003$)
SO_4^{2-} dep. (kg ha^{-1} mo^{-1})	↓	1.7	1.2	–29	1.4	KS	($p = 0.006$)
NO_3^- conc. (µeq L^{-1})	–	37.8	30.0		33.5	KS	
NO_3^- dep. (kg ha^{-1} mo^{-1})	–	0.3	0.2		0.3	K	
Ca^{2+} conc. (µeq L^{-1})	↓	10.0	5.5	–45	7.6	H/S	($p = 0.038$)
Ca^{2+} dep. (kg ha^{-1} mo^{-1})	↓	0.14	0.07	–50	0.1	KS	($p = 0.007$)
Mg^{2+} conc. (µeq L^{-1})	↓	2.4	1.6	–33	2.0	KS	($p = 0.072$)
Mg^{2+} dep. (kg ha^{-1} mo^{-1})	↓	0.02	0.01	–50	0.01	KS	($p = 0.020$)
Montmorency, 1981–1996 ($n = 192$)							
Precipitation (mm)	–	106.1	121.8		114.0	KS	
pH	↑	4.43	4.54	3	4.45	K	($p = 0.002$)
H^+ conc. ($\times 10^{-5}$ µeq L^{-1})	↓	3.7	2.8	–24	3.5	K	($p = 0.002$)
H^+ dep. (kg ha^{-1} mo^{-1})	–	0.04	0.04		0.04	KS	
SO_4^{2-} conc. (µeq L^{-1})	↓	36.4	19.9	–45	28.2	KS	($p = 0.000$)
SO_4^{2-} dep. (kg ha^{-1} mo^{-1})	↓	1.87	1.23	–34	1.55	KS	($p = 0.000$)
NO_3^- conc. (µeq L^{-1})	–	20.0	20.0		20.0	KS	
NO_3^- dep. (kg ha^{-1} mo^{-1})	–	0.27	0.33		0.30	KS	
Ca^{2+} conc. (µeq L^{-1})	↓	5.5	2.5	–55	4.1	H/S	($p = 0.012$)
Ca^{2+} dep. (kg ha^{-1} mo^{-1})	↓	0.11	0.06	–45	0.09	H/S	($p = 0.078$)
Mg^{2+} conc. (µeq L^{-1})	–	1.6	1.6		1.4	S/L	
Mg^{2+} dep. (kg ha^{-1} mo^{-1})	–	0.02	0.02		0.02	H/S	

for NO_3^- concentrations, mainly because 60% of the measurements were below the detection limit (2.8 µeq L^{-1}).

The values for DOC concentration exhibit no longitudinal gradient. Although there are significant differences among regions (Kruskal–Wallis, $p < 0.0000$), they are nonetheless relatively weak. Region 5 has the lowest average, at 3.2 mg L^{-1}, and region 3 the highest, at 4.4 mg L^{-1}. The relatively low DOC levels in the

TABLE II
Regional averages (± standard deviation) of water quality and morphometric characteristics

Parameter	Region (number of lakes)					
	1 (6) $\bar{x} \pm \sigma$	2 (10) $\bar{x} \pm \sigma$	3 (9) $\bar{x} \pm \sigma$	4 (5) $\bar{x} \pm \sigma$	5 (4) $\bar{x} \pm \sigma$	6 (9) $\bar{x} \pm \sigma$
pH	5.50 ± 0.39	6.05 ± 0.39	6.56 ± 0.35	6.87 ± 0.25	7.11 ± 0.31	5.94 ± 0.39
Alkalinity (μeq L^{-1})	7.8 ± 10.6	31.6 ± 14.4	110.6 ± 52.4	157.2 ± 40.2	282.0 ± 77.2	22.6 ± 17.0
Ca^{2+} (μeq L^{-1})	54.0 ± 10.0	86.5 ± 18.0	138.0 ± 45.5	207.5 ± 30.5	350.0 ± 23.5	83.0 ± 15.5
Mg^{2+} (μeq L^{-1})	18.1 ± 3.2	33.7 ± 8.2	46.9 ± 18.9	64.1 ± 19.7	78.1 ± 31.2	36.2 ± 7.4
SO$_4^{2-}$ (μeq L^{-1})	75.2 ± 20.3	91.5 ± 22.8	81.1 ± 25.1	116.4 ± 23.7	145.6 ± 20.3	110.2 ± 22.0
NO$_3^-$ (μeq L^{-1})[1]	2.8 ± 3.0	2.6 ± 2.9	3.6 ± 4.8	4.0 ± 5.7	2.5 ± 3.6	1.9 ± 2.4
DOC (mg L^{-1})	3.4 ± 1.4	4.4 ± 1.1	4.4 ± 1.2	3.6 ± 1.1	3.3 ± 0.9	3.9 ± 1.9
Basin area (ha)	85.7 ± 23.2	131.8 ± 84.3	137.2 ± 88.5	108.6 ± 66.6	106.7 ± 53.8	191.4 ± 202.8
Lake area (ha)	12.9 ± 6.3	21.0 ± 14.6	20.2 ± 15.7	19.8 ± 13.2	25.2 ± 18.0	42.3 ± 61.8
Ave. maximum depth (m)	12.2 ± 5.7	15.8 ± 12.8	10.8 ± 6.2	21.4 ± 12.9	28.0 ± 15.7	10.4 ± 7.8

1 = Between 37% and 46% of the data under detection limit.

network lakes are not surprising, in view of the deliberate selection of lakes with a color value below 75 Hazen units at the time of network design.

3.3. COMPARISON OF ANNUAL AVERAGE WATER CHEMISTRY BASED ON TWO SAMPLING SCENARIOS

A number of chemical parameters have seasonal cycles that can affect the calculation of annual averages. In order to make sure that the lake water chemistry measured only twice a year was comparable to an average of six measurements a year, which is more representative of annual variability, we compared the annual averages of 15 lakes that were sampled 6 times a year to the value obtained when only the November and June samplings were included. The annual averages (Table III) using both scenarios were not significantly different for Ca^{2+}, Mg^{2+}, SO$_4^{2-}$, and DOC (Mann–Whitney, $p > 0.05$). Alkalinity (Lake Thomas) and pH (Lake 88188) were different ($p < 0.05$) only on one occasion, although with very small absolute differences. We therefore concluded that the sampling scenario has no effect on the statistical detection of temporal trends.

3.4. TEMPORAL TRENDS

3.4.1. *Sulfate (SO$_4^{2-}$)*

An individual analysis of trends in the network lakes revealed that the SO$_4^{2-}$ concentrations had declined significantly ($p = 0.05$) in 40 of the 43 lakes studied between 1985–1999 and between 1992–1999 (Table IV). The reductions vary from 23.9 to 69.8 μeq L^{-1} between lakes, with an average 47.8 μeq L^{-1} decrease. On a

TABLE III

Comparison of annual average water chemistry based on two sampling scenarios. The difference is calculated by subtracting the mean obtained with two samples per year from the mean obtained with six samples per year. The values in bold print indicate a significant difference with $p < 0.05$. The comparison was done using a non-parametric Mann–Whitney test

	Parameters					
	pH (unit)	Alkalinity (μeq L^{-1})	Ca^{2+} (μeq L^{-1})	Mg^{2+} (μeq L^{-1})	SO$_4^{2-}$ (μeq L^{-1})	DOC (mg L^{-1})
Region 1						
Bonneville	0.07 (1%)	0.6 (13%)	2.0 (4%)	1.4 (6%)	2.4 (3%)	−0.2 (−6%)
Lagou	0.03 (0.5%)	−0.2 (−3%)	0.4 (0.8%)	0,08 (0.5%)	1.6 (2%)	−0.05 (−2%)
Region 2						
Éclair	−0.04 (−0.6%)	0.0 (0%)	−0.3 (−0.4%)	0.5 (2%)	1.6 (2%)	0.07 (3%)
Lemaine	−0.02 (−0.3%)	2.0 (8%)	2.3 (3%)	0.5 (2%)	2.7 (3%)	−0.1 (−2%)
Truite-Rouge	0.00 (0%)	3.2 (14%)	2.6 (3%)	0.9 (3.2%)	2.2 (2%)	−0.02 (−0.5%)
Francina	−0.10 (2%)	2.0 (5%)	3.6 (3%)	1.5 (3%)	3.7 (3%)	0.2 (3%)
Region 3						
Chômeur	−0.04 (−0.6%)	9.8 (10%)	5.5 (4%)	1.4 (5%)	2.2 (3%)	−0.3 (−7%)
Thomas	0.01 (1%)	**15.6 (9%)**	12.4 (7%)	2.8 (6%)	3.7 (3%)	−0.4 (−11%)
Nolette	−0.03 (−0.5%)	9.6 (11%)	3.9 (4%)	2.2 (3%)	5.4 (5%)	0.07 (1%)
Region 4						
Chevreuil	−0.05 (1%)	3.0 (2%)	1.7 (0.8%)	1.8 (4%)	4.9 (5%)	0.04 (2%)
Kidney	−0.06 (−1%)	3.0 (2%)	4.0 (2%)	3.0 (4%)	1.0 (1%)	0.2 (5%)
Region 5						
Blais	−0.06 (−1%)	0.2 (0.07%)	2.6 (0.7%)	1.1 (2%)	4.5 (3%)	0.1 (4%)
Region 6						
6827	−0.04 (−0.5%)	1.0 (2%)	0.9 (1%)	0.3 (0.8%)	2.2 (2%)	−0.2 (−6%)
88188	**−0.12 (−2%)**	0.4 (1%)	1.6 (2%)	2.0 (6%)	4.9 (5%)	−0.1 (5%)
Poirier	−0.06 −1%)	0.4 (6%)	1.3 (2%)	0.3 (1%)	2.7 (3%)	−0.2 (5%)

regional basis, the observed decreasing rates were quite similar, and vary between −2.9 μeq L^{-1} yr^{-1} (region 1) and −3.7 μeq L^{-1} yr^{-1} (region 5).

3.4.2. *pH and Alkalinity*

In the western part of the study area (regions 4, 5 and 6), most lakes are stable (12/18) or improving (6/18) with respect to pH and/or alkalinity, with no sign of acidification being recorded. The greatest chemical recovery was observed in region 6. Overall S deposition reductions in the northeastern U.S., combined with the steep decline in local S emissions from the Noranda smelter (70%, Dupont, 1997), have resulted in a clear improvement. From the 25 eastern lakes (regions 1,

TABLE IV

Number of lakes exhibiting the various trend types and overall regional changes detected from 1985 to 1999

Region	Trend type	SO_4^{2-} (μeq L^{-1})	Ca^{2+} (μeq L^{-1})	Mg^{2+} (μeq L^{-1})	DOC (mg L^{-1})	pH (units)	Alkalinity (μeq L^{-1})
1	↑	0	0	0	5	1	3
	↓	6	5	3	0	1	0
	0	0	1	3	1	4	3
Mean regional change		−42.6 (−46%)	−10.0 (−16%)	−2.3 (−11%)	+0.775 (23%)	−0.003 (−0.04%)	+5.7 (45%)
2	↑	0	0	0	4	3	4
	↓	9	9	9	1	2	2
	0	1	1	1	5	5	4
Mean regional change		−45.9 (−39%)	−18.0 (−18%)	−7.3 (−20%)	+0.258 (11%)	+0.031 (0.5%)	+2.0 (6%)
3	↑	0	0	0	2	1	2
	↓	9	4	4	0	1	1
	0	0	5	5	7	7	6
Mean regional change		−39.8 (41%)	−8.9 (−8%)	−3.0 (−7%)	+0.195 (4%)	−0.008 (-0.1%)	+0.2 (0.2%)
4	↑	0	0	0	0	3	0
	↓	4	4	4	1	0	0
	0	1	1	1	4	2	5
Mean regional change		−43.6 (−31%)	−11.8 (−6%)	−5.4 (−8%)	+0.156 (5%)	+0.149 (2.2%)	0
5	↑	0	0	0	1	1	2
	↓	4	3	3	0	0	0
	0	0	1	1	3	3	2
Mean regional change		−58.0 (−35%)	−13.9 (−5%)	−5.2 (−7%)	+0.393 (14%)	+0.052 (0.7%)	+10.6 (3%)
6	↑	0	0	0	5	4	4
	↓	8	8	6	2	0	0
	0	1	1	3	2	5	5
Mean regional change		−38.2 (−31%)	−12.3 (−13%)	−5.0 (−12%)	+0.44 (13%)	+0.097 (1.6%)	+3.8 (33%)

2 and 3), 13 lakes have remained stable, 3 were acidifying and 9 were improving in term of alkalinity.

3.4.3. Ca^{2+} and Mg^{2+}

Thirty-three of the 43 lakes (*i.e.*, 77%) in the network have undergone significant reductions ($p < 0.05$) in Ca^{2+} concentrations (29 for Mg^{2+} concentrations) (Table IV). For the study period, the reductions vary between 2.0 and 42.2 μeq L^{-1} for Ca^{2+} and between 2.2 and 15.1 μeq L^{-1} for Mg^{2+}. For both Ca^{2+} and Mg^{2+} concentrations, the highest rates of reduction were found in regions 2 (4.4 μeq L^{-1} yr^{-1} for Ca^{2+} and 2.0 μeq L^{-1} yr^{-1} for Mg^{2+}) and region 6 (2.7 μeq L^{-1} yr^{-1} for Ca^{2+} and 1.8 μeq L^{-1} yr^{-1} for Mg $^{2+}$), and the lowest in region 1 (0.2 μeq L^{-1} yr^{-1} for Ca^{2+} and 0.1 μeq L^{-1} yr^{-1} for Mg^{2+}). No significant increasing trend was recorded.

3.4.4. *Dissolved organic carbon (DOC)*

Between 1985 and 1999, 18 of the 43 lakes showed a significant increasing trend in DOC concentrations. These 18 lakes, distributed mainly in regions 1, 2 and 6, increased an average of 0.07 mg L^{-1} yr^{-1}. Only three lakes show significant downward trends ($p < 0.05$). DOC increases could have been caused by variations in temperature and precipitation, which are known to affect soil leaching and the export of DOC to lakes (Molot and Dillon, 1997).

4. Discussion

The most obvious observation emerging from these results is the widespread reduction in SO_4^{2-} concentrations in almost all the network lakes (Table IV). Moreover, the average rate of concentration decrease was quite similar in all regions (from -2.9 μeq L^{-1} yr^{-1} in region 1 to -3.7 μeq L^{-1} yr^{-1} in region 5), showing that lakes and watershed soils are responding in a uniform fashion throughout the study area. This overall reduction was likely caused by the significant decrease in SO_4^{2-} depositions, and it suggests that the lake-watershed systems are rapidly reaching a new steady state with regard to the changing S depositions. This situation is probably caused by the SO_4^{2-} adsorption-desorption reactions in watershed soils. These reactions have been shown to control mid-term SO_4^{2-} fluxes and to reach a new steady state within four years in response to various scenarios of changing S depositions (Houle and Carignan, 1995) at the Laflamme Lake watershed, one of the network lakes. The soil in this watershed is of the type (humo-ferric podzols) found throughout most of the study. The reduction in SO_4^{2-} concentrations in surface water also has been reported in Ontario (Clair *et al.*, 1995; Mallory *et al.*, 1998) and in several regions of the northeastern United States (Driscoll *et al.*, 1995; Stoddard *et al.*, 1998).

The widespread and fairly uniform decline in SO_4^{2-} levels was not uniformly reflected in the acidity or the acid-neutralizing capacity of the network lakes. There was a strong spatial pattern in trend responses: lake status in the western part of the study area (regions 4, 5 and 6) was clearly improving or remaining stable, with no single lake showing negative trends in pH or alkalinity and a particularly strong recovery in region 6. However, certain lakes, mainly in the eastern part of the study area (regions 1, 2 and 3), have remained stable ($n = 13$) or are becoming more acidified ($n = 3$) despite noticeable improvements in alkalinity in nine of a total of 25 lakes in regions 1, 2 and 3, respectively.

Compared with previous results for the LRTAP-Quebec network, including trend studies for the period 1985–1993 (Bouchard, 1997), the present situation shows an overall clear improvement indicating that recovery is a recent phenomenon for many lakes. It shows that in the last 6 yr only, the proportion of lakes that were acidifying changed from 24 to 7% for the time series that ended in 1993, and 1999 respectively, while the proportion of lakes that were recovering shifted from 16 to 35%. The situation suggests that the recent decrease in SO_4^{2-} deposition in the northeastern U.S. and eastern Canada was significant enough to allow recovery in moderately sensitive regions and in some lakes (9/25) in more sensitive regions (1, 2 and 3).

4.1. Recovery mechanisms

Lake acidification and recovery involve a series of complex, interrelated processes that operate not only in lakes, but also, and primarily, in the soils of watersheds. Recovery (or acidification) is measured in term of alkalinity, which is generally described using the following equation:

$$[\text{Alkalinity}] = ([Ca^{2+}] + [Mg^{2+}] + [Na^+] + [K^+] + [H^+]) -$$

$$([SO_4^{2-}] + [NO_3^-] + [Cl^-] + [OA^-])$$

Alkalinity thus basically reflects the difference between the cations and anions of strong acids (SO_4^{2-} and NO_3^-). According to the above equation, the decrease in lake SO_4^{2-} concentration would lead to an increase in lake alkalinity if the concentrations of the other ions remain constant. For lakes that are not showing alkalinity gain, the three more likely hypotheses that may prevent or delay recovery are an increase in NO_3^-, an increase in DOC, and a decrease in Ca^{2+} and Mg^{2+}. The first two possibilities appear unlikely, however.

First, although temporal trends in NO_3^- concentration cannot be determined because most of the data were below the detection limit (2.8 μeq L^{-1}), it is doubtful that NO_3^- could make an increasing or important contribution to the acidity of network lakes. The NO_3^- concentrations in lake water are generally considerably lower (below 2.8 μeq L^{-1}) than concentrations in precipitation (32.1 and 19.2 μeq L^{-1} at the Chalk River and Montmorency stations, respectively). This observation indicates that NO_3^- is intensively retained in watersheds and that there are no signs of N saturation in the soils surrounding the network of lakes.

Also, the significant DOC increase observed in 18 of the network lakes (Table IV) may suggest that the contribution of organic acids to the acid-base chemistry is increasing. However, there is no general link between variations in DOC concentration and lake acidity and the absence of recovery cannot be attributed solely to increased DOC.

Consequently, the overall decline in Ca^{2+} and Mg^{2+} concentration (Table IV) remains the most likely explanation for the mitigated recovery of some lakes, particularly in the eastern part of the network, despite clear decreases in SO_4^{2-} concentration. Such a declining trend in Ca^{2+} and Mg^{2+} often has been reported in areas where concomitant declining trends have been observed in SO_4^{2-} concentration (Driscoll *et al.*, 1995; Likens *et al.*, 1996; Mallory *et al.*, 1998; Stoddard *et al.*, 1998). Because the SO_4^{2-} anion acts as a carrier for cations (Galloway *et al.*, 1983), less SO_4^{2-} leaching from soil means less cation leaching and therefore decreasing Ca^{2+} and Mg^{2+} concentrations in surface waters.

According to the conceptual model of Galloway *et al.* (1983), the decline in surface water cation concentrations is the first sign of recovery after a reduction in SO_4^{2-} deposition. However, if the decline rate of basic cations were similar to the decline rate of SO_4^{2-}, then alkalinity would remain unchanged. On the other hand, a more rapid reduction in SO_4^{2-} than in cations would increase alkalinity. To better illustrate how lake recovery is linked to the temporal behavior of both SO_4^{2-} and basic cation concentrations, selected data from those network lakes initially sampled six times a year till 1997 were used. Because Ca^{2+} and Mg^{2+} are strongly correlated and Ca^{2+} makes the highest contribution to positive charges, only Ca^{2+} was included in the figures.

A good example of recovery is Lake Éclair (Figure 2), for which alkalinity significantly increased (data not shown). From 1984 to about 1991, temporal variations in SO_4^{2-} and Ca^{2+} concentrations were quite synchronous, after which the relationship between these two variables changed as SO_4^{2-} began to decrease faster than did Ca^{2+}. This suggests that the weathering rate of basic cations in the watershed's soil and the supply from the exchangeable reservoir were sufficient to overcome the loss due to co-leaching of SO_4^{2-} and Ca^{2+}, thus allowing recovery. For the overall period there was a significant, but weak, relationship between SO_4^{2-} and Ca^{2+} ($r^2 = 0.27$).

By contrast, variations in SO_4^{2-} and Ca^{2+} concentrations in Lake Truite-Rouge, a lake for which alkalinity is decreasing, were highly synchronized (Figure 2) and both variables were strongly correlated ($r^2 = 0.79$). Such behavior suggests that supply from the exchangeable Ca^{2+} reservoir and weathering is too low to allow recovery. A reduction in the exchangeable Ca^{2+} reservoir probably occurred in this watershed over the last few decades, and the degree to which the soil's exchangeable cation reserve has been depleted is impacting present recovery. Although the trends observed in some of the network lakes tend to confirm this hypothesis, valid-

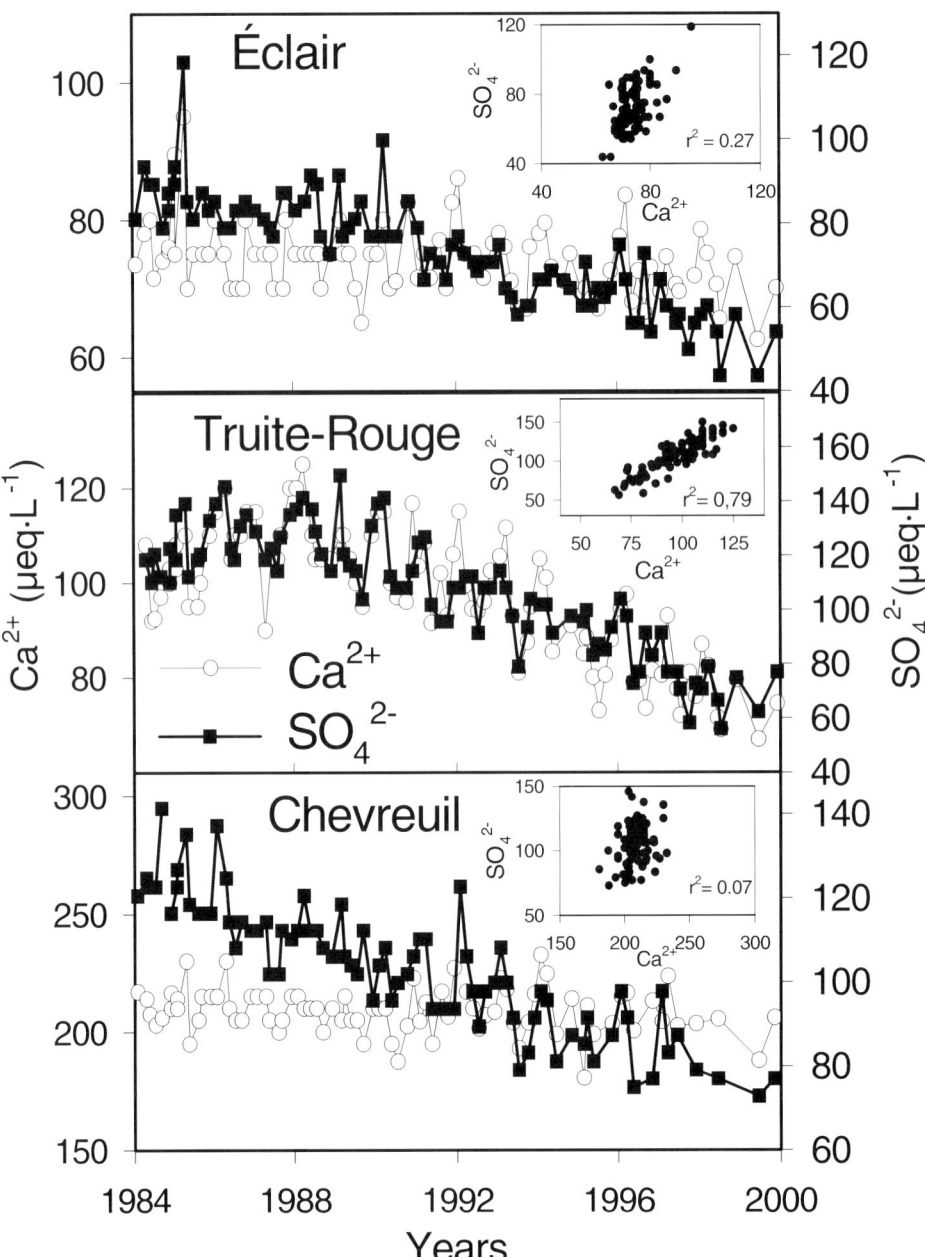

Figure 2. Temporal variations of SO_4^{2-} and Ca^{2+} concentrations between 1984 and 1999 in three selected lakes of the network.

ation thereof would require an assessment of the exchangeable cation reserves and mass balances. Work in the Hubbard Brook Experimental Forest in New Hampshire has demonstrated, in fact, that the decline in soil cation concentration, and especially Ca^{2+}, was due to leaching caused by acid precipitation over the past 50 yr (Likens et al., 1996, 1998). Mitchell et al. (1995) also found a net export of cations associated with that of SO_4^{2-} in the Arbutus Lake watershed in the Adirondacks. At Clair Lake, Québec, 50 km northwest of Québec City, Houle et al. (1997) observed that atmospheric and rock weathering inputs were not enough to offset losses due to cation leaching and accumulation in vegetation.

The decline in cation deposition (Table I) also may play a role in the perceived delay in lake recovery further to the reduction in acidifying deposition (Driscoll et al., 1989; Hedin et al., 1994). The Ca^{2+} and Mg^{2+} deposition levels and rates of decrease in deposition are comparable for the two precipitation analysis stations, one of which is in the western part of the study area and the other in the eastern part (Table I). Nevertheless, their magnitude relative to the reserve of exchangeable cations may vary from one region to another. The reductions of nearly 50% found at the Chalk River and Montmorency stations represent an overall decline of about 1 kg ha^{-1} for the last 15 yr. At the Clair Lake watershed, the net Ca^{2+} loss from the exchangeable cations reservoir was 5 kg ha^{-1} yr^{-1} and the decrease in deposition may thus represent 20% of the exchangeable reservoir annual loss (Houle et al., 1997). Similarly, Likens et al. (1996) determined that 20% of the drop in Ca^{2+} reserves at Hubbard Brook could be explained by the reduction in atmospheric inputs from 1940 to 1995. These observations suggest that even though the decline in Ca^{2+} deposition is relatively low and clearly not sufficient on its own to explain the reduction in Ca^{2+} concentration in surface water, it is still a factor that can delay lake recovery in regions where cation reserves are already low.

4.2. Assessment of lake status

The overall interpretation of the lake status in the network greatly depends on the meaning given to the term 'stability'. In regions 4 and 5, for instance, stability cannot be associated with absence of recovery, in most cases, because the lakes have high pH and alkalinity values and several probably were never acidified in the past. A good example of this situation is Lake Chevreuil (Figure 2), located in region 5, an area rich in carbonate deposits: there is no correlation between SO_4^{2-} and Ca^{2+} concentrations variations, suggesting that most of the lake Ca^{2+} comes from weathering reactions taking place in soils and till deposits. In regions 1, 2 and 3, however, stability for some lakes, may signify no recovery, but the length of time for the recent period series limits our interpretation. We decided to use the pH threshold of 6.0 (Jeffries et al., 2000) and to arbitrarily determine that the lakes showing stability with a pH lower than 6.0 are not recovering, while those showing a higher pH were not acidified significantly in the past. Using this assumption, we estimate that, for the sensitive regions 1, 2 and 3, which are more representative of most of the Québec lakes not covered by the network, 12% of the lakes are still acidifying, 20% are not recovering from past acidification, 36% are recovering and, lastly, that 32% did not experience significant acidification in the past.

5. Conclusion

There was a widespread decline in SO_4^{2-}, Ca^{2+} and Mg^{2+} concentrations throughout the LRTAP-Quebec network lakes. In the western part of the network (regions 4, 5 and 6), an increase in alkalinity levels indicates recovering aquatic ecosystems. Farther east (regions 1, 2 and 3), however, the response is more mixed. Some lakes show signs of recovery, some are stable and others continue to become acidified.

It seems that soil processes in the watersheds are largely responsible for the apparently differing lake responses. Recovery is not governed solely by the rate of SO_4^{2-} reduction, but rather by the difference between the rate of SO_4^{2-} and cation reduction, which is in turn related to the reserves of exchangeable cations in the watershed soil. Our observations also suggest that even though the decline in Ca^{2+} deposition is relatively low and clearly not sufficient on its own to explain the reduction in Ca^{2+} concentration in surface water, it is still a factor that can delay lake recovery, particularly in regions where cation reserves are already low. Consequently, although the reduction in cation deposition is similar in both the western and eastern parts of the study area, the impact of cation reduction on lake recovery is more significant in the eastern regions.

References

Berryman, D., Bobée, B., Cluis, D. and Haemmerli, J.: 1988, 'Nonparametric tests for trend detection in water quality time series', *Water Res. Bull.* **24**, 545–556.

Bobée, B., Lachance, M., Haemmerli, J., Tessier, A., Charrette, J. Y. and Kramer, J.: 1983, 'Évaluation de la sensibilité à l'acidification des lacs du sud du Québec et incidences sur le réseau d'acquisition de données', Report No. 157, INRS-Eau, Sainte-Foy, Québec.

Bouchard, A.: 1997, 'Recent lake acidification and recovery trends in southern Québec, Canada', *Water, Air, Soil Pollut.* **94**, 225–245.

Clair, T. A., Dillon, P. J., Ion, J., Jeffries, D. S., Papineau, M. and Vet, R. J.: 1995, 'Regional precipitation and surface water chemistry trends in southeastern Canada (1983–1991)', *Can. J. Fish. Aquat. Sci.* **52**, 197–212.

Cluis, D., Langlois, C., van Coillie, R. and Laberge, C.: 1989, 'Development of a software package for trend detection in temporal series: Application to water and industrial effluent quality data for the St. Lawrence River', *Environ. Monit. Assess.* **12**, 429–441.

Cronan, C. S.: 1979, Determination of sulfate in organically colored water samples', *Anal. Chem.* **51**, 1333–1335.

Driscoll, C. T., Likens, G. E., Hedin, L. O., Eaton, J. S. and Bormann, F. H.: 1989, 'Changes in the chemistry of surface waters', *Environ. Sci. Technol.* **23**, 137–142.

Driscoll, C. T., Postek, K. M., Kretser, W. and Raynal, D. J.: 1995, 'Long-term trends in the chemistry of precipitation and lake water in the Adirondack region of New York, U.S.A.', *Water, Air, Soil Pollut.* **85**, 583–588.

Dupont, J.: 1997, 'Projet Noranda Phase III – Effet des réductions d'émissions de SO_2 sur la qualité de l'eau des lacs de l'Ouest québécois', Envirodoq EN980066, Report No. PA-53/1. Ministère de l'Environnement et de la Faune du Québec, Direction des écosystèmes aquatiques, Québec.

Environment Canada: 1994, *Manual of Analytical Methods. Volume 2: Trace Metals*. Canada Centre for Inland Waters, National Laboratory of Environmental Testing, Burlington, Ontario.

Galloway, J. N., Norton, S. A. and Church, M. R.: 1983, 'Freshwater acidification from atmospheric deposition of sulfuric acid: A conceptual model', *Environ. Sci. Technol.* **17**, 541–545.

Hedin, L. O., Granat, L., Likens, G. E., Buishand, T. A., Galloway, J. N., Butler, T. J. and Rodhe, H.: 1994, 'Steep declines in atmospheric base cations in regions of Europe and North America', *Nature* **367**, 351–354.

Hirsch, R. M. and Slack, J. M.: 1984, 'A nonparametric trend test for seasonal data with serial dependence', *Water Resour. Res.* **20**, 727–732.

Houle, D. and Carignan, R.: 1995, 'Role of SO_4^{2-} adsorption and desorption in the long term S budget of a coniferous catchment on the Canadian Shield', *Biogeochemistry* **28**, 161–182.

Houle, D., Paquin, R., Camiré, Ouimet, C. R. and Duchesne, L.: 1997, 'Response of the Lake Clair watershed (Duchesnay, Québec) to changes in precipitation chemistry (1988–1994)', *Can. J. For. Res.* **27**, 1813–1821.

Jeffries, D. S., Lam, D. C. L., Wong, I. and Moran, M. D.: 2000, 'Assessment of changes in lake pH in southeastern Canada arising from present levels and expected reductions in acidic deposition', *Can. J. Fish. Aquat. Sci.* **57**(suppl. 2), 40–49.

Kemp, A.: 1999, 'Trends in lake water quality in southern Québec following reductions in sulphur emissions', Scientific and Technical Report STE-212. Environment Canada – Québec Region, Environmental Conservation, St. Lawrence Centre, Montreal, Québec.

Lawrence, G. B., David, M. B., Bailey, S. W. and Shortle, W. C.: 1997, 'Assessment of soil calcium status in red spruce forests in the northeastern United States', *Biogeochemistry* **38**, 19–39.

Likens, G. E., Driscoll, C. T. and Buso, D. C.: 1996, 'Long-term effects of acid rain: Response and recovery of a forest ecosystem', *Science* **272**, 244–246.

Likens, G. E., Driscoll, C. T., Buso, D. C., Siccama, T. G., Johnson, C. E., Lovett, G. M., Fahey, T. J., Reiners, W. A., Ryan, D. F., Martin, C. W. and Bailey, S. W.: 1998, 'The biogeochemistry of calcium at Hubbard Brook', *Biogeochemistry* **41**, 89–173.

Mallory, M. L., McNicol, D. K., Cluis, D. A. and Laberge, C.: 1998, 'Chemical trends and status of small lakes near Sudbury Ontario, 1983–1995: Evidence of continued chemical recovery', *Can. J. Fish. Aquat. Sci.* **55**, 63–75.

Mitchell, M. J., Raynal, D. J. and Driscoll, C. T.: 1995, 'Biogeochemistry of a forested watershed in the central Adirondack Mountains: Temporal changes and mass balances', *Water, Air, Soil Pollut.* **88**, 355–369.

Molot, L. A. and Dillon, P. J.: 1997, 'Colour mass balances and colour-dissolved organic carbon relationships in lakes and streams in central Ontario', *Can. J. Fish. Aquat. Sci.* **54**, 2789–2795.

Stainton, M. P., Capel, M. J. and Armstrong, F. A. J.: 1974, 'The Chemical Analysis of Freshwater', Fisheries Research Board of Canada, Miscellaneous Special Publication 25.

Stoddard, J. L., Driscoll, C. T., Kahl, J. S. and Kellogg, J. P.: 1998, 'A regional analysis of lake acidification trends for the Northeastern U.S., 1982–1994', *Environ. Monit. Assess.* **51**, 399–413.

Stoddard, J. L., Jeffries, D. S., Lukewille, A., Clair, T. A., Dillon, P. J., Driscoll, C. T., Forsius, M., Johannessen, M., Kahl, J. S. Kellogg, J. H., Kemp, A., Mannio, J., Monteith, D. T., Murdoch, P. S., Patrick, S., Rebsdorf, A., Skjelkvale, B. L., Stainton, M. P., Traaen, T., van Dam, H., Webster, K. E., Wieting, J. and Wilander, A.: 1999, 'Regional trends in aquatic recovery from acidification in North America and Europe', *Nature* **401**, 575–578.

Vet, R. J., Sukloff, W. B., Still, M. E., McNair, C. S., Martin, J. B., Kobelka, W. F. and Gaudenzi, A. J.: 1989, 'Canadian Air and Precipitation Monitoring Network (CAPMoN): Precipitation Data Summary - 1987', Environment Canada, Atmospheric Environment Service, Air Quality and Interenvironmental Research Branch, Measurements and Analysis Research Division, Downsview, Ontario.

FIRE EFFECTS ON CARBON AND NITROGEN BUDGETS IN FORESTS

D. W. JOHNSON[1]*, R. B. SUSFALK[2], T. G. CALDWELL[2], J. D. MURPHY[1], W. W. MILLER[1] and R. F. WALKER[1]

[1] *Environmental and Resource Sciences, University of Nevada, Reno, Nevada, U.S.A. 89557;*
[2] *Division of Hydrologic Sciences, Desert Research Institute, Reno, Nevada, U.S.A. 89512*
(* author for correspondence, e-mail: dwj@cabnr.unr.edu; phone: (775) 784-4511; fax: (775) 784-4789)

(Received 20 August 2002; accepted 1 April 2003)

Abstract. Estimates of C and N loss by gasification during a wildfire in a Jeffrey pine (*Pinus Jeffreyii* [Grev. and Balf.]) forest in Little Valley, Nevada are compared to potential losses in more mesic forests in the Integrated Forest Study (IFS). In Little Valley, the fire consumed the forest floor, foliage, and an unknown amount of soil organic matter, but little standing large woody material. On an ecosystem level, the fire consumed approximately equal percentages of C and N (12 and 9%, respectively), but a considerably greater proportion of aboveground N (71%) than C (21%). Salvage logging was the major factor in loss, and C lost from the site will not be replenished until forest vegetation is established and succeeds the current shrub vegetation. N_2 fixation by *Ceanothus velutinus* [Dougl.] in the post-fire shrub vegetation appears to have more than made up for N lost by gasification in the fire over the first 16 yr, and may result in long-term increases in C stocks once forest vegetation takes over the site. N loss from the fire equaled > 1,000 years of atmospheric N deposition and > 10,000 years of N leaching at current rates. Calculations of C and N losses from theoretical wildfires in the IFS sites show similar patterns to those in Little Valley. Calculated losses of N in most of the IFS sites would equal many centuries of leaching. Conceptual models of biogeochemical cycling in forests need to include episodic events such as fire.

Keywords: atmospheric deposition, carbon, fire, leaching, N-fixation, nitrogen

1. Introduction

The conceptual model of biogeochemical cycling in forest ecosystems that has dominated the literature for the last few decades considers nutrient pool changes to be result of slow, steady processes such as atmospheric deposition, leaching, and uptake (e.g., Likens *et al.*, 1977; Swank and Crossley, 1988; Johnson and Van Hook, 1989). The effects of harvesting and prescribed fire sometimes are considered in managed forests (e.g., Likens *et al.*, 1978; Richter *et al.*, 1982; Johnson *et al.*, 1988), but the role of wildfire largely is ignored, presumably because of its low frequency. Most forests of the world experience fires at some interval ranging from decades to centuries (Auclair, 1985; Fisher and Binkley, 2000), however, and fires can have major effects on long-term carbon (C) and nitrogen (N) budgets (e.g., Gessel *et al.*, 1973). In Mediterranean-type forest ecosystems, wildfire is commonplace and is known to have a major effect on long-term N budgets (Gessel et al., 1973; Grier, 1975; Raison et al., 1985; Trabaud, 1994; Carrieria et al., 1996; Johnson et al., 1998; Baird et al., 1999; Neary et al., 1999). Furthermore, fire often stimulates the establishment of N_2-fixing vegetation that can dominate the N inputs to these ecosystems for prolonged periods (Gessel et al., 1973; Zavitovski and Newton, 1968; Youngberg and Wollum, 1978; Binkley et al, 1982; Cole et al., 1995; Johnson, 1995). Thus, the biogeochemical cycling paradigm developed for mesic forest ecosystems needs to be modified in order to be usefully applied to Mediterranean-type forest ecosystems (Johnson et al., 1998).

The contributions of fire to global C and N cycles are also significant. Carbon dioxide emissions from fire have been estimated to rival those of fossil fuel emissions (Olson, 1981; Crutzen and Andreae, 1990; Mack et al., 1996), and N emissions from fire also make a substantial contribution to the global N budget (Crutzen and Andreae, 1990; Galloway et al., 1995). The pulse-like CO_2 emissions resulting directly from fires often are not particularly large because only a fraction of total organic C in forest ecosystems (typically < 5–25%) actually burns (Auclair 1985; Dixon and Krankina, 1993). It has been assumed that post-fire decomposition can result in much larger emissions of CO_2 over time (Dixon and Krankina, 1993), but the data sets upon which this assumption are based are sparse. Totally lacking in any regional or global analyses of fire effects upon C budgets is the effect of post-fire N-fixing vegetation on soil C and N pools, which can be very substantial (Johnson and Curtis, 2001).

In this paper, we present a case study of the C and N budgets with wildfire and post-fire N fixation in a Sierran forest, and attempt to place the case study in perspective with an theoretical assessment of the potential role of fire on N losses in more mesic, temperate forest ecosystems from the Integrated Forest Study (IFS) (Johnson and Lindberg, 1992).

2. Little Valley, Nevada: A Case Study

2.1. SITE AND METHODS

The Little Valley site is located approximately 30 km southwest of Reno, Nevada in the eastern Sierra Nevada Mountains. Elevation in the valley ranges from 2010 to 2380 m, and elevation at the site is 2010 m. The climate is characterized by warm, dry summers and cold winters; the major hydrologic event is snowmelt. Mean annual air temperature near the valley floor is 5 °C and mean annual precipitation is 550 mm, approximately 50% of which falls as snow. Vegetation at the study site is dominated by 110–130 year old Jeffrey pine (*Pinus* Jeffrey*ii* [Grev. and Balf.]) and lodgepole pine (*Pinus contorta* Dougl.). Soils are the Corbett series, mixed, typic, frigid Xeropsamments derived from colluvium of decomposed granite.

The entire valley, as well as much of the eastern Sierra Nevada Mountains, was logged and probably burned during the Comstock mining era between 1870 and

1890. An area just east of the valley, which had consisted of 100-yr-old Jeffrey pine forest, burned in a stand-replacing wildfire in 1981. As is usually the case, the wildfire did not consume much of the large standing woody tissues (tree boles and large branches); and thus the area was salvage logged for merchantable timber (snags) the following year. Since that time, the burned area has been dominated by snowbush (*Ceanothus velutinus* Dougl.), a species that often invades after fire and fixes N (Zavitovski and Newton, 1968; Youngberg and Wollum, 1976; Binkley *et al.*, 1982) with lesser amounts of manzanita (*Arcostaphylos patula* [Greene]) and spotty regeneration of Jeffrey pine planted in 1985.

In a previous paper, we estimated that the wildfire caused the loss of 300–600 kg ha^{-1} of N based on the N contents of nearby forests (Johnson *et al.*, 1998) and the conservative assumption that all forest floor and foliage was consumed in the fire (no soil or woody tissue). In 1997, we established three 0.1 ha plots in the former fire site and one in an adjacent forested site with the same soils, slope, and aspect in order to check on the original estimates. The forested site had received underburning (forest floor burning, but not crown fire), as evidenced by the presence of fire scars on trees, but had experienced little mature tree mortality. Within the forested plot, all trees were measured at dbh and six randomly assigned points were established for litter and soil sampling. At each sampling point, all litter within a 15-cm diameter ring was removed by horizon (Oi, Oe, and Oa) and transported to the laboratory for drying, weighing, and analyses. After litter was removed, the soil was sampled at three depths corresponding to the A (0–7 cm) and AB (7–20 cm) and BC (20–40 cm) horizons. Bulk density was determined by the core method for each horizon at three of the sampling points in each plot. Coarse fragment contents were negligible. In the former fire site, diameters of all stumps were measured at 40 cm above ground level and used to estimate pre-fire dbh based on a regression established from live trees in the forested plot. In some cases, bark had sloughed off, and this was corrected for using regressions of bark thickness vs. diameter. Tree biomass in the forested plot and pre-fire tree biomass in the former fire were then estimated from regressions provided by Gholz *et al.* (1978) (checked earlier for accuracy in Little Valley trees), and N contents were calculated from N concentrations measured on foliage, branch, and boles of live trees in Little Valley (Stark, 1973; Johnson *et al.*, 1998). In two of the plots in the former fire site, six randomly established 1 m^2 subplots were established for measurement of current (1997) aboveground shrub, litter and soil C and N pools. Within each subplot, all aboveground vegetation was removed, sorted by species and component (foliage and woody), weighed, and subsampled for moisture content. After the vegetation was removed, all litter and soils were sampled as described above. All samples taken from the site were analyzed for total C and N using a Perkin-Elmer 2400 CHN analyzer at the Desert Research Institute. Statistical analyses, where possible (i.e., Student's t-tests for differences in soils) were performed using DataDesk®software.

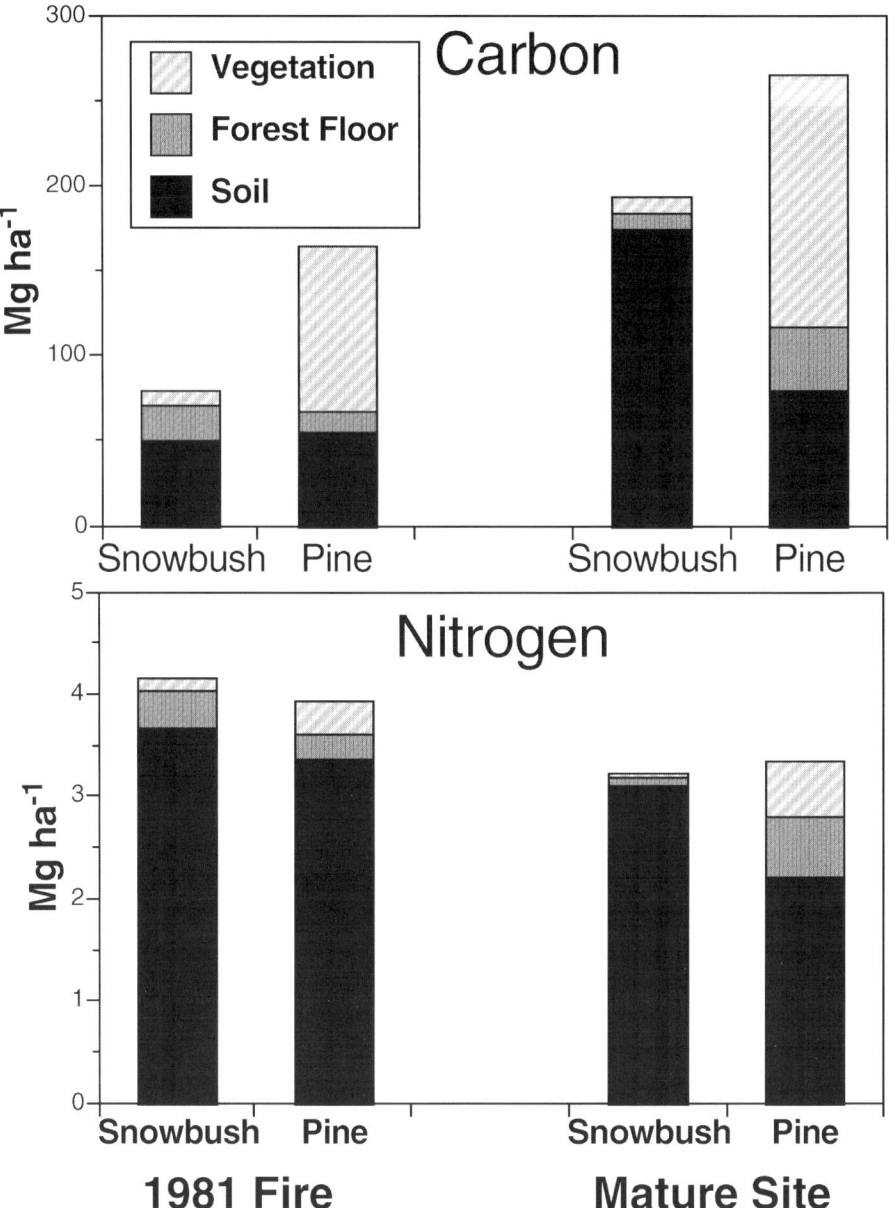

Figure 1. Carbon and nitrogen contents of adjacent snowbush and Jeffrey pine stands in the 1981 fire and in a mature ecosystem in Little Valley, Nevada.

2.2. RESULTS AND DISCUSSION

The distribution of C and N in the adjacent forest and shrub systems in the 1981 fire are shown in the left hand side of Figure 1, and data from a previous study in a part of Little Valley not burned recently (Johnson, 1995) are shown on the right. The data for snowbush in the 1981 fire represent the average of two plots; the other plots were unreplicated. Despite the lack of replication at any given site, it is clear from these two data sets that the Jeffrey pine ecosystem in each case contains considerably greater C because of greater aboveground biomass. In the mature site, soils beneath snowbush contained more C than the soils beneath pine, a pattern also noted in four other mature snowbush-pine plots (Johnson, 1995). In the 1981 fire site, however, soil C contents in the two vegetation types were similar. Ecosystem N contents were similar in the pine and snowbush ecosystems at each site because soil N in the snowbush ecosystems was greater than soil N in the pine ecosystems.

The estimates of C and N lost during the 1981 fire, subsequent removals in salvage logging, and restoration of C and N to the ecosystem by the shrub vegetation are depicted in Figure 2. Estimated losses during the fire are minimal estimates: gasification of all C and N in foliage and litter was assumed, but no losses from either woody biomass or soil. Post-fire gains are assumed to be limited to those in vegetation and litter, excluding soils, and are thus minimum estimates as well. According to these estimates, post-fire salvage logging had a greater impact on C loss than gasification during the fire. Auclair (1985) found that C losses due to decomposition of woody materials exceeded those due to the fire itself in a boreal forest ecosystem where no salvage logging occurred. Had salvage logging not taken place after the fire, C losses from woody decomposition may have dominated the long-term C budget in the Little Valley fire as well. Salvage logging propably reduced net CO_2 release to the atmosphere compared to leaving woody debris on site to decay. Vegetation and litter layer gains in C in the 16 years following the fire more than made up for the estimated losses during the fire, but fell far short of making up for losses associated with salvage logging. For N, on the other hand, losses due to salvage logging were far less than those resulting from gasification during the fire, and post-fire gains in litter and vegetation (presumably including substantial inputs from N fixation) equaled total estimated losses due to the fire and salvage logging combined. Collectively, these data indicate that post-fire salvage logging had a substantially greater impact on C loss from the Little Valley fire site than the fire did itself whereas the reverse was true for N.

Estimated N losses due to gasification of the foliage and litter layers in the Little Valley fire (370 kg ha^{-1}) fall at the low end of the range of our previous estimations (300 to 600 kg ha^{-1}) and within the low range of values reported in the literature for wildfires and prescribed fires in general (300 to 855 kg ha^{-1} and 10 to 1505 kg ha^{-1}, respectively; Belillas and Feller, 1998; Johnson *et al.*, 1998). The use of the forest floor data from the partially burned forest as an estimate of pre-fire values may bias this estimate downward in that part of the forest floor in this stand

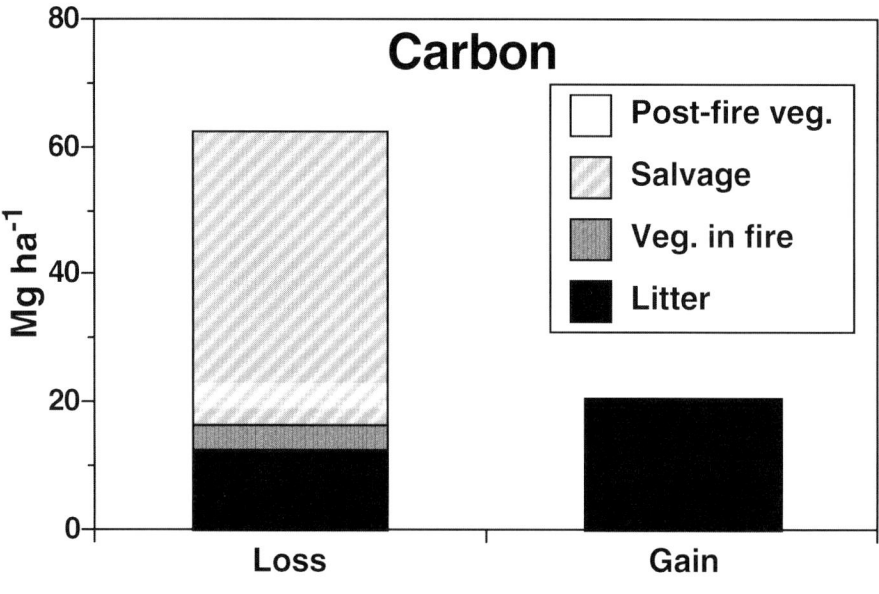

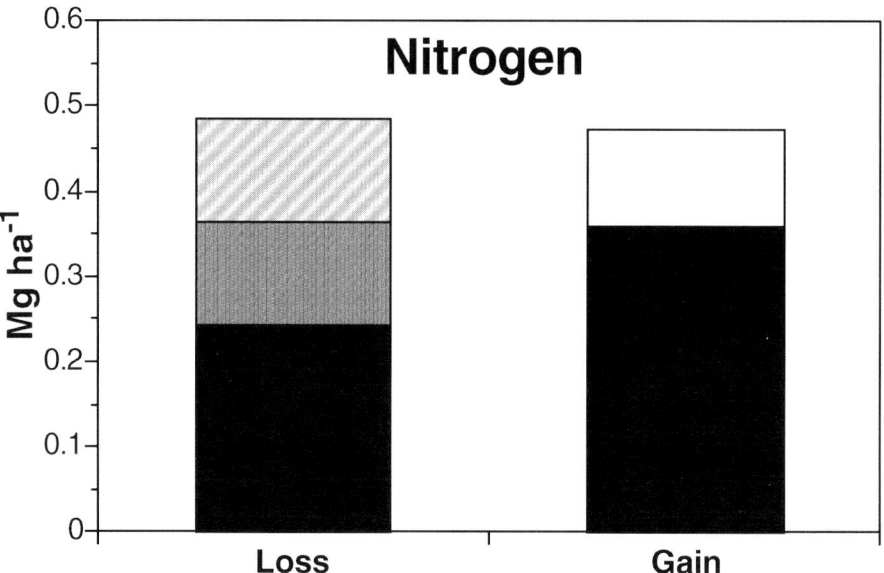

Figure 2. Estimated carbon and nitrogen losses due to gasification during the 1981 fire (veg. in fire and litter), and post-fire salvage logging (salvage). Gains are associated with revegation and litter 16 yr after the fire.

was burned in 1981. In the mature pine stand depicted in Figure 1, the forest floor contained 37,000 kg C ha^{-1} and 580 kg N ha^{-1} as compared to 12,700 kg C ha^{-1} and 240 kg N ha^{-1} in the partially burned forest near the fire. The estimates of C and N loss during the fire are also biased low because of the implicit assumption of no loss of soil C or N during the fire. We did not estimate soil N loss in this study because we did not have a suitable, reasonable means for doing so and also because the 1997 sampling showed no indication of lower soil N in burned plots. It is probable that there was some soil N loss during the Little Valley fire and that this N was replaced by fixation during the ensuing 16 yr.

3. Comparison of Little Valley to Temperate Forests of the Integrated Forest Study

The C and N contents of forests from the Integrated Forest Study (IFS; Johnson and Lindberg, 1992) are compared to those in the Little Valley site (LV) in Figure 3. When compared to the more mesic IFS sites, the Little Valley site is on the low end of the scale for ecosystem C and N content, and theoretical C and N losses by fire in the IFS sites are generally higher than for Little Valley if equivalent ecosystem components (forest floor, understory, and foliage) are presumed to be consumed. If we presume for the moment that a fire passes through these ecosystems and consumes the forest floor, the potential C and N losses vary by an order of magnitude, from 6.8 to 70.8 Mg C ha^{-1} and from 240 to 2,600 kg N ha^{-1}. The largest potential forest floor losses are in the red spruce sites in Maine, New York, and North Carolina and in the red alder site in Washington. Stand-replacing fires are relatively infrequent in these areas, but can nevertheless occur. If we presume that the hypothetical fire consumes all understory and tree foliage biomass, the potential C and N losses increase only slightly in most cases (from 16.1 to 79.6 Mg C ha^{-1} and from 340 to 2,700 kg N ha^{-1}). As a percentage of total ecosystem capital, the potential losses of C and N similar, varying from 6 to 30% for C and from 4 to 29% for N. As a percent of aboveground capital, however, N is preferentially lost (54 to 89%) compared to C (14 to 58%), reflecting the assumption that high C:N ratio woody material remains unburned. The Little Valley site does not differ remarkably from the other IFS sites in any of the above respects.

When expressed on an annualized basis (using stand age as the fire interval), the potential loss of N with fire is far greater than that due to leaching in all but the red alder and Smokies red spruce site (Figure 4). In the red alder site, N leaching is very high due to excessive N fixation by red alder (*Alnus rubra* Bong) (Van Miegroet and Cole, 1984), and in the Smokies red spruce site, N leaching is high because of high rates of atmospheric N deposition and low vegetation N uptake (i.e., the site is 'N-saturated'; Johnson *et al.*, 1991). Atmospheric deposition rates for N are greater than or equal to potential losses with fire in some sites (Oak Ridge loblolly

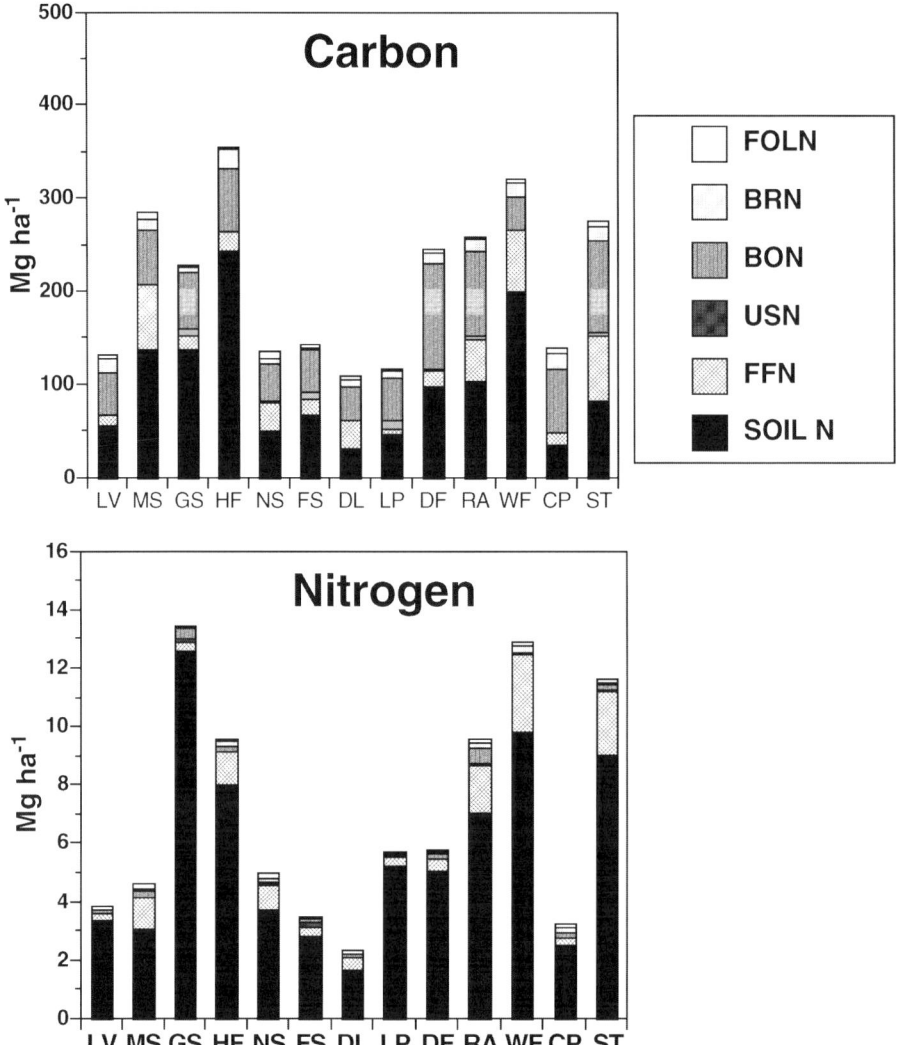

Figure 3. Ecosystem contents of carbon and nitrogen from the Integrated Forest Study sites (from Johnson and Lindberg, 1992) and the Little Valley, Nevada (LV) sites. Legend: CP = *Pinus strobus* stand at Coweeta, NC; DL = *Pinus taeda* stand at Duke, NC; GS = *Pinus taeda* stand at B.F. Grant Forest, GA; LP = *Pinus taeda* stand at Oak Ridge, TN; FS = *Pinus eliottii* at Bradford Forest, FL; DF = *Pseudotsuga menziesii* stand at Thompson, WA; RA = *Alnus rubra* stand at Thompson, WA; NS = *Picea abies* stand at Nordmoen, Norway; HF = northern hardwood stand at Huntington Forest, NY; MS = *Picea rubens* stand at Howland, ME; WF = *Picea rubens* at Whiteface, NY; ST = *Picea rubens* stand at Clingman's Dome, NC; LV = *Pinus contorta/P.* Jeffreyii stand at Little Valley, NV. The sites are arranged in order of increasing annual precipitation.

Potential Nitrogen Losses with Fire Compared to Deposition and Leaching

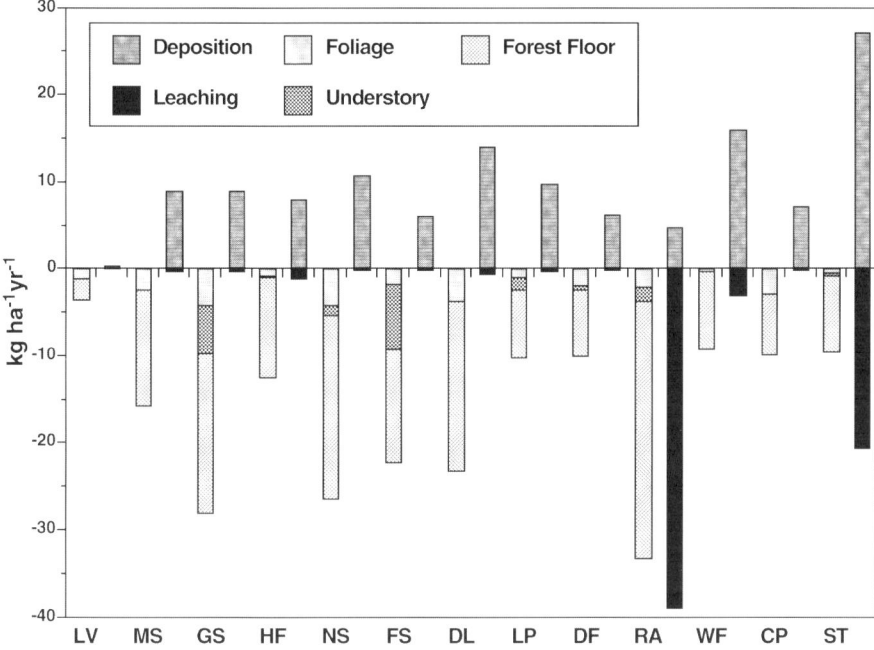

Figure 4. Potential N losses with fire, expressed on an annualized basis, compared to atmospheric N deposition and leaching in the Integrated Forest Study sites. See Figure 3 for legend (after Johnson and Lindberg, 1992).

pine, Coweeta white pine, Whiteface red spruce, Smokies red spruce), but less than potential fire loss in most cases (Figure 4).

A convenient way to put potential fire N losses into perspective is to divide the calculated total N loss with fire by either atmospheric deposition rate (giving an index of the number of years needed for atmospheric deposition to replenish the potential losses of N with fire) or by N leaching rate (giving the number of years of leaching that would equal losses in fire under different scenarios). These values, along with the ages of the various ecosystems, are plotted in Figure 5. Because of the very low atmospheric deposition rate, calculated N replenishment time for Little Valley is by far the highest: over 1100 yr (Figure 5A). These calculations suggest that fire in the Little Valley ecosystem will result in long-term declines in N status unless N is replenished by fixation. A fixation rate of 40 kg N ha^{-1} yr^{-1}, the average value currently estimated for snowbush in the regenerating fire at present, would result in the replenishment of N lost in the fire within only nine years. Similarly, the calculated N replenishment time for the red alder site (380 yr) is far greater than the stand age (50 yr), but average rates of N fixation at this

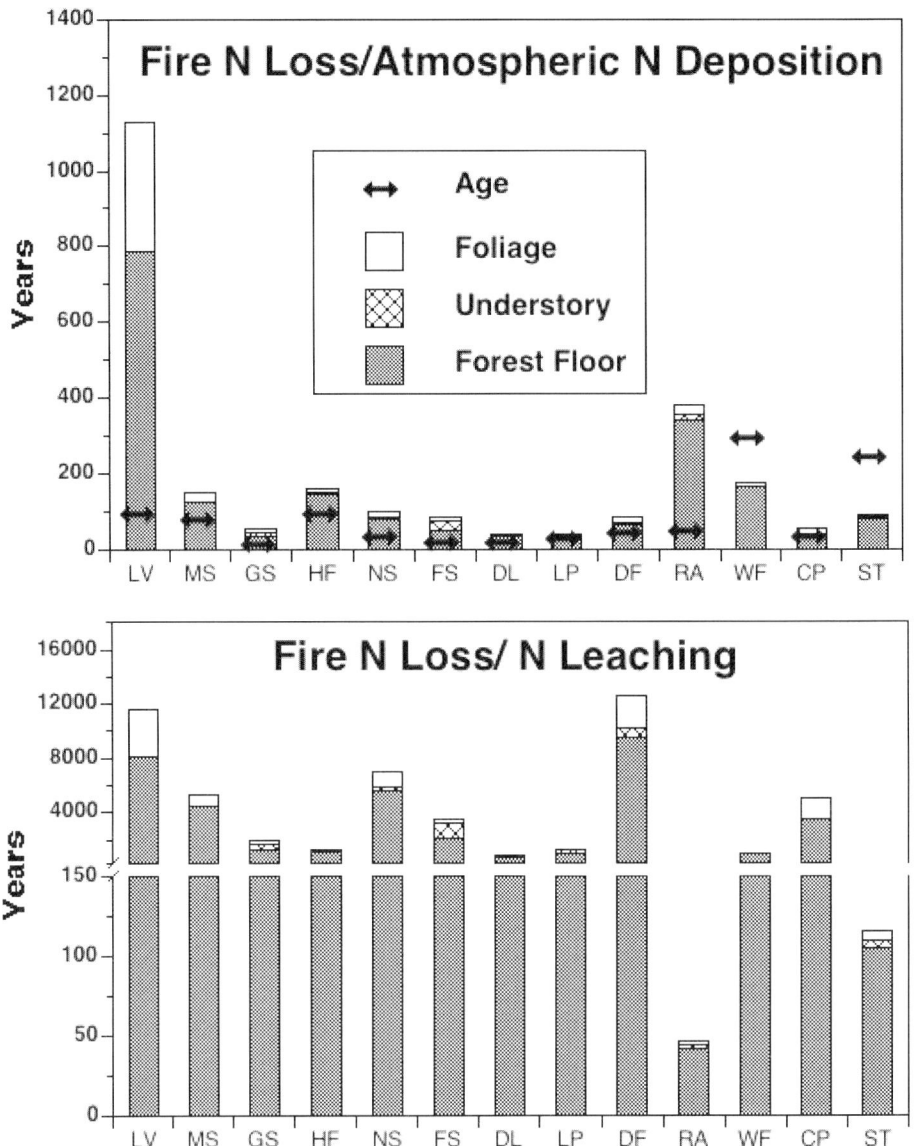

Figure 5. Comparison of estimated nitrogen losses due to fire and atmospheric deposition (A), and comparison of estimated nitrogen losses due to fire with nitrogen leaching (B) in the Integrated Forest Study sites. See Figure 3 for legend (after Johnson and Lindberg, 1992).

site (100 kg N ha^{-1} yr^{-1}) would make up the N lost in a mere eight years. For the other IFS sites, the calculated N replenishment times are lower (from 30 to 170 years) and closer to the stand age at the time of sampling. Aside from the two sites with high rates of N leaching (Red Alder and Smokies Spruce), the calculations of the number of years of N leaching to equal N losses from a hypothetical stand-replacing wildfire (where foliage, understory and forest floor are consumed) are on the order of centuries. This is true even if only half the forest floor is assumed to be consumed. Given the normally low rate of leaching in most temperate forests, even a moderate ground fire every century will cause more N loss on an annual basis than would be lost to leaching.

These budget calculations are crude – we do not know the exact fire interval in these stands and our assumptions as to how much forest floor, understory, and foliage would be consumed cannot be verified. Fires of different intensities will burn either more or less organic matter than we have assumed here. However, these calculations serve to illustrate the potential role of fire in C and N balances in more mesic forest ecosystems. They show that potential N losses by fire could be substantially greater than those due to N leaching over the long term. The calculations also show that atmospheric N deposition in all but the most polluted sites may not readily replenish N losses by a catastrophic fire.

4. Conclusions

1. Fire and post-fire N_2 fixation dominate the long-term N budgets of the Little Valley site. N_2 fixation more than made of for N losses during the fire and could conceivably enrich the site in N such that long-term productivity and C sequestration are higher than they would have been without the fire.
2. Salvage logging was the major factor in C loss from the Little Valley site, and such losses will not be replenished until forest vegetation is re-established on the site.
3. Calculated losses of N from a hypothetical fire would equal many centuries of leaching in all but two of the Integrated Forest Study sites.

Better numbers on fire frequencies and the extent of organic matter consumption during fire in these systems are needed to refine these calculations. In theory, however, even infrequent fire could substantially deplete the N reserves of many of these systems. While fire is infrequent in many of these systems at present, climate warming or even an unusually dry year may cause an abrupt change in this situation, as was the case in the 1998 fire season in Florida (NOAA, 1998). Thus, the potential role of fire in ecosystem C and N budgets of temperate, mesic forest ecosystems merits more consideration than it has been given to date.

Acknowledgements

Research supported in part by the Nevada Agricultural Experiment Station, Publication 52 031327, and by the USDA Joint Fire Science Program.

References

Auclair, A. N.: 1985, 'Postfire regeneration of plant and soil organic pools in a *Picea mariana-Cladonia stellaris* ecosystem', *Can. J. For. Res.* **15**, 297–291.

Baird, M., Zabowski, D. and Everett, R. L.: 1999, 'Wildfire effects on carbon and nitrogen in inland coniferous forests', *Plant Soil* **209**, 233–243.

Belillas, C. M. and Feller, M. C.: 1998, 'Relationships between fire severity and atmospheric and leaching nutrient losses in British Columbia's coastal western hemlock zone forests', *Int. J. Wildland Fire* **8**, 87–101.

Binkley, D., Cromack, K. and Fredriksen, R. L.: 1982, 'Nitrogen accretion and availability in some snowbrush ecosystems', *For. Sci.* **28**, 720–724.

Carriera, J. A., Arvevalo, J. R. and Neill, F. X.: 1996, 'Soil degradation and nutrient availability in fire-prone Mediterranean shrublands of southeastern Spain', *Arid Soil Res. Rehab.* **10**, 53–.

Cole, D. W., Compton, J. E., Edmonds, R. S., Homann, P. S. and Van Miegroet, H.: 1995, 'Comparison of carbon accumulation in Douglas-fir and red alder forests', in W. W. McFee and J. M. Kelly (eds), *Carbon Forms and Functions in Forest Soils*, Soil Science Society of America, Madison, Wisconsin, U.S.A., pp. 527–547.

Crutzen, P. J. and Andreae, M. O.: 1990, 'Biomass burning in the tropics: Impact on atmospheric chemistry and biogeochemical cycles', *Science* **250**, 1669–1678.

Dixon, R. K., and Krankina, O. N.: 1993, 'Forest fires in Russia: carbon dioxide emissions to the atmosphere', *Can. J. For. Res.* **23**, 700–705.

Fisher, R. F. and Binkley, D.: 2000, *Ecology and Management of Forest Soils*, John Wiley & Sons, New York, New York, U.S.A.

Galloway, J. N., Schlesinger, W. H., Levy II, H., Michaels, A. and Schnoor, J. L.: 1995, 'Nitrogen fixation: Anthropogenic enhancement-environmental response', *Global Biogeochem. Cycles* **9**, 235–252.

Gessel, S. P., Cole, D. W. and Steinbrenner, E. C.: 1973, 'Nitrogen balances in forest ecosystems of the Pacific Northwest', *Soil Biol. Biochem.* **5**, 19–34.

Gholz, H., Grier, C. C., Campbell, A., and Brown, A:. 1979, 'Equations for estimating biomass and leaf area of plants in the Pacific Northwest', Research Paper 41, Forest Research Laboratory, Oregon State University, Corvallis, Oregon, U.S.A.

Grier, C. C.: 1975, 'Wildfire effects on nutrient distribution and leaching in a coniferous ecosystem', *Can. J. For. Res.* **5**, 599–607.

Johnson, D. W.: 1995, 'Soil properties beneath *Ceanothus* and pine stands in the Eastern Sierra Nevada', *Soil Sci. Soc. Amer. J.* **59**, 918–924.

Johnson, D. W. and Curtis, P. S.: 2001, 'Effects of forest management on soil carbon and nitrogen storage: Meta Analysis', *For. Ecol. Manage.* **140**, 227–238.

Johnson, D. W., Kelly, J. M., Swank, W. T., Cole, D. W., Van Miegroet, H., Hornbeck, J. W., Pierce, R. S. and van Lear, D.: 1988, 'Effects of whole-tree and stem-only clear-cutting on postharvest hydrologic losses, nutrient capital, and regrowth', *J. Environ. Qual.* **17**, 418–424.

Johnson, D. W. and Lindberg, S. E.: 1992, *Atmospheric Deposition and Forest Nutient Cycling: A Synthesis of the Integrated Forest Study*, Ecological Studies 91, Springer-Verlag, New York, New York, U.S.A.

Johnson, D. W., Susfalk, R. B., Dahlgren, R. A. and Klopatek, J. M.: 1998, 'Fire is more important than water for nitrogen fluxes in semi-arid forests', *Environ. Sci. Policy* **1**, 79–86.

Johnson, D. W. and Van Hook, R. I.: 1989, *Analysis of Biogeochemical Cycling Processes in Walker Branch Watershed*, Springer-Verlag, New York, New York, U.S.A., 401 pp.

Johnson, D. W., van Miegroet, H., Lindberg, S. E., Harrison, R. B. and Todd, D. E.: 1991, 'Nutrient cycling in red spruce forests of the Great Smoky Mountains', *Can. J. For. Res.* **21**, 769–787.

Likens, G. E., Bormann, F. H., Pierce, R. S. and Reiners, W. A.: 1978, 'Recovery of a deforested ecosystem', *Science* **199**, 492–496.

Likens, G. E., Bormann, F. H., Pierce, R. S., Eaton, J. S. and Johnson, N. M.: 1977, *Biogeochemistry of a Forested Ecosystem*, Springer-Verlag, New York, New York, U.S.A.

Mack, F., Hoffstadt, J., Esser, G. and Goldammer, J. G.: 1996, 'Modeling the influence of vegetation fires on the global carbon cycle', in J. S. Levine (ed.), *Biomass Burning and Global Change*, The MIT Press, Cambridge, Massachusetts, U.S.A., pp. 149–159.

National Oceanic and Atmospheric Administration: 1998, 'Recent abnormal weather in Florida set the stage for Wild Fires', http://www.ncdc.noaa.gov/oa/climate/research/1998/fla/florida.html#recent

Neary, D. G., Klopatek, C. C., DeBano, L. F. and Ffolliot, P. F.: 1999, 'Fire effects on belowground sustainability: A review and synthesis', *For. Ecol. Manage.* **122**, 51–71.

Olson, J. S.: 1981, 'Carbon balance in relation to fire regimes', in H. A. Mooney, T. M. Bonnicksen, N. L. Christensen, J. E. Lotan and W. A. Reiners (eds), *Fire Regimes and Ecosystem Properties*, USDA Forest Service Gen. Tech. Rep. WO-26. Washington, DC, pp. 327–378.

Raison, R. J., Khanna, P. K. and Woods, P. V.: 1985, 'Mechanisms of element transfer to the atmosphere during vegetation fires', *Can. J. For. Res.* **15**, 132–140.

Richter, D. E., Ralston, C. W. and Harms, W. R.: 1982, 'Prescribed fire: Effects on water quality and forest nutrient cycling', *Science* **215**, 661–663.

Stark, N. M.: 1973, *Nutrient Cycling in a Jeffrey Pine Ecosystem*, University of Montana Press, Missoula, Montana, U.S.A.

Swank, W. T. and Crossley, D. A. (eds): 1988, *Forest Hydrology and Ecology of Coweeta*, Springer-Verlag, New York, 469 pp.

Trabaud, L.: 1994, 'The effect of fire on nutrient losses and cycling in a *Quercus coccifera* garrigue (southern France)', *Oecologia* **99**, 379–386.

Van Miegroet, H. and Cole, D. W.: 1984, 'The impact of nitrification on soil acidification and cation leaching in red alder ecosystem', *J. Environ. Qual.* **13**, 586–590.

Wan, S., Hui, D. and Liu, Y.: 2001. 'Fire effects on nitrogen pools and dynamics in terrestrial ecosystems: A meta analysis', *Ecol. Appl.* **11**, 1349–1365.

Youngberg, C. T. and Wollum, A. G.: 1976, 'Nitrogen accretion in developing *Ceanothus velutinus* stands', *Soil Sci. Soc. Amer. J.* **40**, 109–112.

Zavitovski, J. and Newton, M.: 1968, 'Ecological importance of snowbrush *Ceanothus velutinus* in the Oregon Cascades', *Ecology* **49**, 1134–1145.

GROWTH AND NITROGEN AVAILABILITY OF RED PINE SEEDLINGS UNDER HIGH NITROGEN LOAD AND ELEVATED OZONE

TATSURO NAKAJI[1], TAKUYA KOBAYASHI[2], MIHOKO KUROHA[3], KUMIKO OMORI[3], YUKO MATSUMOTO[3], TETSUSHI YONEKURA[3], KATSUHIKO WATANABE[3], JARKKO UTRIAINEN[3] and TAKESHI IZUTA[3]*

[1] *Center for Global Environmental Research, National Institute for Environmental Studies, Tsukuba, Ibaraki 305–8506, Japan;* [2] *Central Research Institute of Electric Power Industry, Abiko, Chiba 270–1194, Japan;* [3] *Tokyo University of Agriculture and Technology, Fuchu, Tokyo 183–8509, Japan*
(* author for correspondence, e-mail: izuta@cc.tuat.ac.jp; phone: +81-42-3675728; fax: +81-42-3675728)

(Received 20 August 2002; accepted 11 April 2003)

Abstract. To evaluate the effect of increasing nitrogen (N) deposition and tropospheric ozone (O_3) concentrations on N-saturated forest ecosystems, we investigated the response of Japanese red pine (*Pinus densiflora*), an N-saturation sensitive tree species, to increasing N load under elevated O_3 concentrations. One-year-old seedlings of red pine were treated with three levels of N supply (0, 50 and 100 mg N L^{-1} fresh soil volume) under two levels of atmospheric O_3 concentration (< 5 and 60 ppb) for two growing seasons. Nitrogen treatment did not stimulate dry matter production of the seedlings. Growth inhibition was observed in the highest N treatment under low O_3 and in the two higher N treatments under elevated O_3. Irrespective of the O_3 concentration, increasing N supply negatively affected root growth and mycorrhizal development in fine roots, resulting in a reduction in P and Mg uptake from the soil. Net photosynthetic rate was significantly reduced by both the highest N treatment under low O_3 and the two higher N treatments under elevated O_3, together with decreased N-availability to Rubisco. Nitrogen assimilated from NO_3^- to amino acid in the needles was not affected by the treatments. However, needle protein concentration was reduced by the highest N-treatment under low O_3 and by the two higher N-treatments under elevated O_3. These results suggest that elevated O_3 potentially disturbs the N-availability in the form of protein including Rubisco, and may advance the negative effects of excessive N-deposition on N-sensitive plant species in N-saturated forests.

Keywords: nitrogen, nitrogen availability, ozone, photosynthesis, red pine

1. Introduction

Nitrogen (N) availability is an important constraint on plant growth, and atmospheric N deposition acts as a fertilizer in many temperate forests. However, in some terrestrial ecosystems, increasing anthropogenic N deposition induces over-nutrition with N, described as 'nitrogen saturation', resulting in reduced tree health by nutrient imbalance and soil acidification (Schulze, 1989). The attendant increase of atmospheric NO_x concentration also increases tropospheric ozone (O_3) concentration (Penkett, 1988). In Japanese suburban forests, it has been reported that some regions around Tokyo metropolis may exhibit N saturation with relatively high atmospheric N deposition between 10 and 40 kg N ha^{-1} yr^{-1} (Baba et al., 1995; Ohrui and Mitchell, 1997). Furthermore, relatively high tropospheric O_3 concentrations above 60 nmol mol^{-1} (ppb) are commonly observed in mountainous areas around Tokyo metropolis from spring to autumn (Wakamatsu et al., 1999). Therefore, it is important to investigate the tree responses to increasing N deposition with elevating O_3 concentration to evaluate the health condition of suburban forests. Based on the results of our previous study, Japanese red pine (Pinus densiflora), a major Japanese coniferous species, is relatively sensitive to high N load as compared to the other coniferous species (Nakaji et al., 2001). In this study, we investigated the responses of red pine seedlings to increasing N load under elevated O_3 concentration to discuss the potential hazards of O_3 to the health condition of N-sensitive tree species grown in N-saturated forest.

2. Materials and Methods

In December 1999, brown forest soils were collected from the University Forest (Kusaki, Gunma Prefecture, Japan) where N saturation has been demonstrated previously (Ohrui and Mitchell, 1997). The N input near the sampling site ranged from 10.5 to 13.9 kg N ha^{-1} yr^{-1} as bulk precipitation (Ohrui and Mitchell, 1997, 1998). Total N content and C/N ratio of the soil were 8.3 mg g^{-1} and 16.5 g g^{-1}, respectively. The collected soil was sieved through a 5 mm mesh immediately after the collection.

One-yr-old seedlings of Japanese red pine (Pinus densiflora Sieb. et Zucc.), which were naturally colonized by ectomycorrhizal fungi, were obtained from an experimental field of the Forestry and Forest Products Research Institute (Ibaraki Prefecture, Japan). The whole-plant dry mass at the beginning of the experiment was 1.52 ± 0.30 g. On 10 March 2000, 48 seedlings were individually planted into 5.3 L plastic pots that contained 5 L of the collected soil. Then, all the seedlings were grown at the experimental field of Tokyo University of Agriculture and Technology (Fuchu, Tokyo, Japan) under field conditions. From 25 April to 25 May 2000, N was added as 25 mM NH_4NO_3 solution to the potted soil at rates of 0 (N0), 50 (N50) and 100 (N100) mg N L^{-1} fresh soil volume (16 pots per N treatment). The amounts of N added to the soil were equivalent to 0, 90 and 180 kg N ha^{-1} yr^{-1}, respectively. The seedlings were irrigated using deionized water as necessary. The drainage from the bottom of each pot was collected after irrigation, and re-added to the potted soil. From 10 May to 22 November 2000 and 14 May to 5 November 2001, all the seedlings were grown in four naturally-lit growth chambers (S-180-SP, Koito Co.), and exposed to two levels of O_3. In the growth chambers, air temperature, atmospheric CO_2 concentration and relative air humidity were maintained at 20.0 ± 1.0/15.0 ± 1.0 °C (6:00–18:00/18:00–6:00), 350 ± 10/420 ± 10 μmol mol^{-1} and 65 ± 5/75 ± 5%, respectively. Half of the seedlings were exposed daily to charcoal-filtered air with O_3 at < 5 nmol mol^{-1}. The remaining seedlings were exposed daily to 60 ± 10 nmol mol^{-1} O_3 for 7 hr a day from 11:00 to 18:00. The O_3 concentration and exposure duration were determined based on the typical pollution scenarios in surrounding suburban forests of Tokyo metropolis. Two replicated chambers were randomly assigned to each gas treatment.

Determinations of net photosynthesis and N metabolites in the current-year needles were carried out on 22 October 2001. Light-saturated net photosynthetic rate (A_{sat}) was measured under photosynthetic photon flux density (PPFD) of 1600 ± 5 μmol m^{-2} s^{-1}, by using an infrared gas analyzer system (LCA-4, ADC Co. Ltd., UK).

Immediately after the measurements of net photosynthetic rate, the needles were collected, rinsed with deionized water and used for the determination of N metabolism. Nitrate reductase (NR) activity was determined by a modified method of Akama (1992). The fresh needles (200 mg) were cut into 1.5 mm pieces and individually transferred to 10 mL glass tubes that contained 8 mL of 100 mM KH_2PO_4 buffer (pH 7.5, 3 °C) with 200 mM KNO_3 and 2% n-propanol. After three subsequent vacuums (–0.06 MPa), the tubes were incubated in the dark for 1 hr at 30 °C. Aliquots of incubation buffer were boiled for 3 min to stop enzyme activity, and centrifuged at 20,000 g for 20 min at 3 °C. Nitrite concentrations in 1 mL supernatant were analyzed spectrophotometrically (540 nm) after adding 1 ml of 1% sulfanilamide in 1 M HCl and 0.02% N-(1-napthyl) ethylenediamine dihydrochloride. Nitrate reductase activity was described as an increase in the rate of NO_2^- concentration. For the determination of nitrite reductase (NiR) activity and soluble protein concentrations, needle samples (300 mg fw.) were frozen by liquid N_2, homogenized with 30 mg PVPP and 3 mL of extraction buffer containing 100 mM HEPES (pH 8.0, 3 °C), 5 mM EDTA, 0.7% (w/v) PEG 20,000, 1 mM phenylmethylsulfonyl fluoride, 1% Tween-80 and 24 mM 2-mercaptoethanol. The homogenate was centrifuged at 9,000 g for 1 min. Nitrite reductase activity in the supernatant was determined by using methyl viologen as the electron carrier (Ida et al., 1974). Concentrations of total soluble protein and ribulose-1,5-bisphosphate carboxylase/oxygenase (Rubisco) were assayed by the method described by Nakaji et al. (2001). To analyze inorganic-N (NO_3^-, NO_2^- and NH_4^+) concentrations, 100 mg of fresh needles were frozen, homogenized with 10 mg PVPP and 4 ml H_2O at 3 °C. After centrifugation at 20,000 g for 20 min, the supernatant was filtered by centrifugal UF membrane (Microcon YM-3, Millipore Co.) to remove high molecular compounds. Concentrations of NO_3^- and NH_4^+ in the filtered samples were determined by ion chromatography (IC200, Yokogawa Co., Japan). Nitrite concentrations were spectrophotometrically analyzed by sulfanilamide-napthyl ethylenediamine method (Klute, 1982). Free amino acid was extracted from 100 mg fresh needles by 4 ml of 80% ethanol. The needles were frozen, homogenized with solvent, and centrifuged at 20,000 g for 20 min. Amino acid concentrations in the super-

TABLE I

Chemical properties of soil solutions. Analyses were conducted prior to ('Initial') and four times during the experiment in May and August of every year. Each value represents the mean of four determinations. Because O_3 did not affect soil chemistry significantly, the data were pooled. Different lowercase letters indicate significant effects of N treatment on the average values (Tukey's HSD test; $p < 0.05$), and asterisks indicate significant differences between initial soil and N0 ($p < 0.05$)

	Initial	N0	N50	N100
pH	4.95	5.02	4.79	4.72
NO_3^- (mM)	4.9	6.0 b	11.3 a	13.6 a
NH_4^+ (mM)	0.001	0.002 b	0.261 ab	0.817 a
P (μM)	0.57	0.93	1.16	1.39
K (μM)	0.04	0.07 b	0.09 ab	0.13 a
Mg (mM)	0.14	0.42*	0.48	0.50

natant were determined colorimetrically by ninhydrin method (Oyama, 1990) using glutamic acid as a standard.

After exposing seedlings to two growing seasons, all the seedlings were harvested, and the formation ratio of ectomycorrhizal roots was determined (Nakaji et al., 2002). Seedlings were separated into plant organs, dried at 60 °C for 1 week, and weighed. For determination of needle nutrients, dried needles from each seedling were homogenized. Needle N concentrations were analyzed on a C/N analyzer (MT-500, Yanagimoto Co.). Concentrations of P, K, Ca, Mg, Mn, Mo and Al were determined with an ICP-AES (JY48P, Seiko Inst. Inc.) after the digestion by HNO_3 and H_2O_2 (Nakaji et al., 2001).

Soil solution was taken from the potted soil with a soil moisture sampler (Eijkelkamp Co.). The pH, ion concentrations and element concentrations in the soil solution were determined with a pH meter (M-12, Horiba Co.), ion chromatography and ICP-AES, respectively.

Analysis of variance (ANOVA) was used to test the effects of N and O_3 treatment on soil and red pine seedlings. Variation from chamber to chamber not observed, the data of replicated growth chambers were pooled. All statistical analyses were performed using SPSS® software (11.0J, SPSS Inc.).

3. Results and Discussion

By adding N to soil as NH_4NO_3, mean concentrations of NO_3^- and NH_4^+ in the soil solutions were significantly increased (Table I). Total inorganic N concentration in the soil solution decreased by approximately 28% during the experiment, possibly

TABLE II

Dry mass (g) of plant organ and litter in red pine seedlings grown under varied N load and O_3 concentrations during two growing seasons. Each value represents the mean ($\pm$ s.d.) of six determinations. Different Letters indicate significant effects of soil N on the dry mass within each gas treatment (Tukey's HSD test; $p < 0.05$), and asterisks indicate significant O_3 effects on dry mass within each soil treatment (ANOVA; $p < 0.05$)

	Low O_3			Elevated O_3		
	N0	N50	N100	N0	N50	N100
Current-year needles	10.5	10.8	10.3	10.7	10.2	11.1
	(0.7)	(0.9)	(1.5)	(1.1)	(1.4)	(1.7)
Previous-year needles	2.7 a	2.2 a	0.9 b	2.4 a	1.3 b	0.3 c
	(0.7)	(1.1)	(1.1)	(0.7)	(0.7)	(0.6)
Stem	8.6	8.7	7.6	8.4 a	7.2 ab	6.5 b*
	(1.3)	(1.3)	(1.0)	(1.3)	(0.4)	(0.6)
Coarse roots	9.0	8.1	8.8	9.0	8.3	7.5
	(0.7)	(0.8)	(1.2)	(1.0)	(0.9)	(1.2)
Fine roots	2.6 ab	3.0 a	1.8 b	2.3 a	2.5 ab	1.6 b
	(1.1)	(0.8)	(0.4)	(0.3)	(0.9)	(0.4)
Whole-plant	33.3 a	32.9 a	29.4 b	32.8 a	29.4b*	26.9 b*
	(1.6)	(1.2)	(3.0)	(1.8)	(0.6)	(2.2)
Litter	0.2 b	0.7 ab	1.0 a	0.5 b	0.9 ab	1.2 a
	(0.2)	(0.4)	(0.5)	(0.4)	(0.4)	(0.5)

due to plant uptake of both forms of N and consumption of NH_4^+ in nitrification (data not shown). Although N treatment tended to reduce soil pH possibly due to accelerated nitrification, it was not less than 4.0 which could induce growth reduction in red pine seedlings growing in acidified brown forest soil (Lee et al., 1998). Concentration increases of P, K, and Mg in the soil solution were caused by soil acidification (Table I).

Adding N to soil did not stimulate dry mass production in the red pine seedlings (Table II). Conversely, whole-plant dry mass was significantly reduced by the N100 treatment under low O_3 and by N50 and N100 treatments under elevated O_3. Irrespective of O_3 concentrations, the highest N treatment significantly reduced the dry masses of fine roots and previous-year needles, and accelerated defoliation. Stem dry mass of the seedlings exposed to elevated O_3 was reduced significantly by the highest N treatment. No significant treatment effects were found for the dry mass of current-year needles. These results indicate that O_3 itself reduces growth and secondly, decreases the threshold level of N-induced growth suppression in red pine seedlings. As a result, O_3-induced increase in the sensitivity of the seedlings

TABLE III

Needle concentrations (mg g dw^{-1}) and ratios (g g^{-1}) of red pine seedlings grown under varied N loadings and O$_3$ concentrations. Each value represents the mean ($\pm$ s.d.) of four determinations. Different letters indicate significant effects of soil N on the element concentrations within each gas treatment (ANOVA; $p < 0.05$). Ozone did not have a significant effect on needle concentrations

	Low O$_3$			Elevated O$_3$		
	N0	N50	N100	N0	N50	N100
N	22(1)	20(1)	21(2)	19(4)	22(3)	21(4)
P	0.75 a	0.54 b	0.47 b	0.68 a	0.51 ab	0.48 b
	(0.03)	(0.16)	(0.06)	(1.1)	(0.15)	(0.09)
K	3.1	3.0	2.7	3.3	3.0	2.7
	(0.2)	(0.3)	(0.3)	(0.6)	(0.4)	(0.4)
Ca	4.3	4.4	4.0	4.8	4.1	3.6
	(0.3)	(0.2)	(0.2)	(0.6)	(1.1)	(0.7)
Mg	1.2 a	1.1 ab	1.0 b	1.2 a	1.2 ab	1.0 b
	(0.1)	(0.1)	(0.2)	(0.1)	(0.2)	(0.1)
Mn	1.0	1.1	1.0	0.9	1.1	1.0
	(0.2)	(0.3)	(0.2)	(0.3)	(0.3)	(0.3)
Mo	0.84	0.63	0.59	0.61	0.47	0.54
	(0.16)	(0.03)	(0.25)	(0.11)	(0.12)	(0.20)
Al	0.18	0.15	0.14	0.18	0.16	0.13
	(0.02)	(0.01)	(0.01)	(0.04)	(0.04)	(0.03)
N/P	29(1) b	43(8) a	45(9) a	29(5) b	44(9) a	45(11) a
N/Ca	5.1(0.3)	4.5(0.4)	5.2(0.4)	3.9(0.8)	5.5(1.9)	6.0(2.1)
N/Mg	19(2)	18(3)	21(4)	16(4)	19(5)	20(3)

to N is mainly expressed as lowered stem growth. Significant growth reduction, induced by elevated O$_3$, was found in the N-treated seedlings.

In several studies, excess N-induced growth reduction has been observed with nutrient imbalances such as high needle N/P ratio with low needle P concentration or excessive needle accumulation of Mn (Mohren et al., 1986; Nakaji et al., 2002). In the present study, needle concentrations of P and Mg were reduced, and the N/P ratio was significantly increased by the highest N treatment (Table III). There were no significant changes in the needle concentrations of other nutrient elements and Al, and O$_3$ did not affect the elemental responses of the needles to increasing N (Table III). Therefore, although N-induced growth reduction of the seedlings can be closely related to nutrient imbalances of P rather than toxicity of excessive

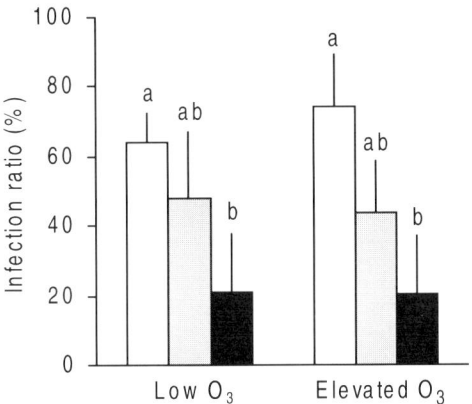

Figure 1. The ratio of mycorrhiza-infected root tips to total root tips (infection ratio) of red pine seedlings grown under varied N load and atmospheric O_3 concentration. Each bar is the mean of eight determinations ± the standard deviation. Different letters indicate significant differences among the N treatments within each gas treatment (Tukey's HSD test; $p < 0.05$). N treatments; square = N0, greysquare = N50, blacksquare = N100.

Mn and Al, O_3-induced increase in the sensitivity of the seedlings to N cannot be explained by a P imbalance.

Nitrogen-induced reduction in P and Mg concentrations in needles has been attributed to dilution effects caused by stimulation in growth (Flückiger and Braun 1998; Sogn and Abrahamsen 1998) and by inhibition of root uptake (Wilson and Skeffington, 1994). In the present study, because the dry mass of plant organs did not increase with N amendments (Table II), a dilution effect was not the cause of the observed nutrient imbalances. Meanwhile, fine root and ectomycorrhizal infection generally contributed to P and Mg uptake in the roots of coniferous species (Wallander and Nylund, 1992); and in our study, we demonstrated that the fine root biomass and infection ratio of ectomycorrhiza were significantly reduced by the N treatment (Table II, Figure 1). Therefore, the reduction of fine root biomass and poor mycorrhizal development might reduce the concentrations of P and Mg in red pine seedlings under high N load.

As for the seedlings grown under low O_3 concentration, A_{sat} and Rubisco concentration in the current-year needles were significantly reduced by the highest N treatment (Figure 2). Furthermore, A_{sat} and Rubisco concentration were significantly reduced by two of the N treatments under the elevated O_3 concentration (Figure 2). Ozone-induced reductions in A_{sat} and Rubisco concentration were found only in the highest N treatment (Figure 2). Nakaji et al. (2002) reported that continuous reduction, which occurred in A_{sat} under high N load (> 100 mg N L^{-1} fresh soil volume) over two growing seasons, was the main cause for the high N load-induced growth reduction of red pine seedlings. This is mainly due to the decline of CO_2 fixation in the chloroplasts linked with lowered Rubisco

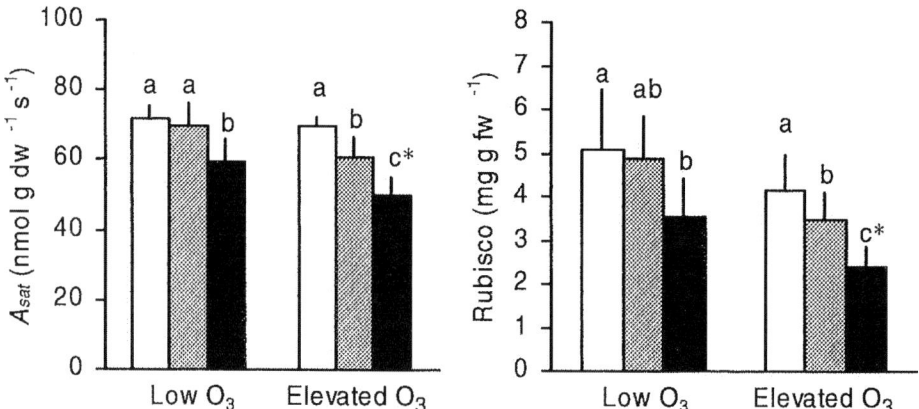

Figure 2. Light-saturated net photosynthetic rate (A_{sat}) and rubisco concentrations in current-year needles of red pine seedlings. Each bar is the mean of four determinations ± the standard deviation. Different letters indicate significant differences of A_{sat} and Rubisco concentrations among the N treatments within each gas treatment (Tukey's HSD test; $p < 0.05$); an asterisk indicates significant effect of O_3 on A_{sat} and Rubisco concentration in each N treatment (ANOVA, $p < 0.05$). N treatments; square = N0, greysquare = N50, blacksquare = N100.

concentration. In this study, decreased Rubisco concentration is considered to be the main cause of reduced A_{sat}.

Excessive N-induced reduction of plant growth has also been explained by toxicity of accumulated NO_2^- in leaves, especially in crop plants exposed to relatively high doses of NO_2 (Wellburn, 1990; Shimazaki *et al.*, 1992). In this study, to clarify the contribution of internal NO_2^- and cause of reduced Rubisco in the needles of red pine seedlings, inorganic N assimilation and the fraction of soluble N were investigated. As for the seedlings under low O_3 concentration, needle NO_3^- concentrations were increased with a stimulation of NR activity when N was supplied to the soil (Table IV). Alternatively, there was no clear response of needle NO_3^- concentration and NR activity to the N supply in the O_3 treated seedlings (Table IV). Negative effects of N treatment and elevated O_3 concentration were neither found in concentrations of NO_2^- and NH_4^+, nor in NiR activity in the current-year needles (Table IV). These results indicate that inorganic-N assimilation was not inhibited by excessive N supply and/or elevated O_3, and accumulation of NO_2^- was not a cause of growth reduction in the seedlings under the highest N treatment.

Nitrogen availability in the form of amino acids was significantly increased by the N treatment, despite O_3 concentrations (Table IV). However, protein-N was significantly reduced by the N100 treatment under low O_3 and by N50 and N100 treatments under elevated O_3 (Table IV). The ratio of Rubisco-N to protein-N ranged from 50 to 54% with no significant treatment effects. Therefore, high N-induced reduction in Rubisco concentration in the needles was mainly due to depression of the biosynthesis of soluble protein from amino acid.

TABLE IV

Nitrate reductase (NR) and nitrite reductase (NiR) activity, and soluble N concentrations in current year needles of red pine seedlings. Enzyme activities are expressed as μmol g fw^{-1} h^{-1}. All concentrations of soluble N are expressed as μmol g fw^{-1} except for NO_2^-, which has units of nmol g fw^{-1}. Each value represents the mean ($\pm$ s.d.) of six determinations. Different letters indicate significant effects of soil N on the enzyme activities and N concentrations within each gas treatment (Tukey's HSD test; $p < 0.05$). No significant O_3 effects were found for enzyme activities and N concentrations (ANOVA; $p < 0.05$)

	Low O_3			Elevated O_3		
	N0	N50	N100	N0	N50	N100
NR activity	0.10 b	0.15 a	0.15 a	0.10 b	0.14 a	0.11 ab
	(0.03)	(0.04)	(0.05)	(0.03)	(0.03)	(0.02)
NiR activity	9.7(2.2)	8.5(1.5)	10.0 (1.6)	10.2(2.3)	11.6(2.1)	9.3(1.1)
NO_3^-	0.39 b	0.47 ab	0.52 a	0.48	0.49	0.49
	(0.05)	(0.11)	(0.09)	(0.08)	(0.10)	(0.09)
NO_2^-	4.9(0.9)	3.5(1.3)	4.3(1.8)	4.5(2.2)	5.8(2.3)	5.7(2.1)
NH_4^+	2.3(0.7)	2.4(0.8)	2.2(0.8)	2.1(0.6)	2.1(1.0)	1.9(0.7)
Amino acid-N	46(6) b	46 (9) ab	53(8) a	45(7) b	57(5) a	59(10) a
Protein-N*	95(13) a	94(16) a	75(11) b	92(13) a	73(12)b	65(8)b

* Protein-N was calculated from the amount of total soluble protein including rubisco.

4. Conclusion

Adding N to brown forest soils collected from N-saturated forests did not increase the growth of Japanese red pine seedlings, but high N load over 90 kg ha^{-1} yr^{-1} induced growth inhibition. The N load threshold of growth reduction was lowered by O_3. Mechanisms for N-induced growth reduction could not be explained by the toxicity of accumulated NO_2^- in the needles. Excessive N supply to the rhizosphere induced needle nutrient imbalances of P and Mg, and reduced net photosynthesis. Reductions in fine root biomass and poor mycorrhizal development might be the cause of excessive-N induced reduction in needle concentrations of P and Mg. The main cause of excessive N-induced reduction in net photosynthesis was the reduced concentration of Rubisco in the needles. Although elevated O_3 did not affect needle nutrient status, it accelerated high N-induced depression of N availability in the form of needle protein including Rubisco. Elevating tropospheric O_3 stimulated negative N effects for Japanese red pine seedlings. Consequently, we conclude that tropospheric O_3 should be considered as an important factor when evaluating the effects of increasing N deposition on N-saturated forest ecosystems.

Acknowledgments

This study was financially supported in part by Global Environment Research Fund (Ministry of the Environment, Japan) and OMC Card Inc. The authors are greatly indebted to Dr. A. Akama (Forestry and Forest Products Research Institute) and Prof. Y. Dokiya (Edogawa University) for their invaluable advice.

References

Akama, A.: 1992, 'Influences of nitrogen sources and an aeration on the seedlings of *Pinus densiflora* Sieb. et Zucc. and *Cryptomeria japonica* D. Don in a hydroponics', *Proc. 133th Mtg. Jpn. For. Soc.*, 279 (in Japanese).

Baba, M., Okazaki, M. and Hashitani, T.: 1995, 'Effect of acidic deposition on forested Andisols in the Tama Hill region of Japan', *Environ. Pollut.* **89**, 97–106.

Flückiger, W. and Braun, S.: 1998, 'Nitrogen deposition in Swiss forests and its possible relevance for leaf nutrient status, parasite attacks and soil acidification', *Environ. Pollut.* **102**, 69–76.

Ida, S., Mori, E. and Morita, Y.: 1974, 'Purification, stabilization and characterization of nitrite reductase from barley roots', *Planta* **121**, 213–224.

Klute, A.: 1982, *Methods of Soil Analysis (2nd edit.)*, American Society for *Agronomy*, pp. 682–687.

Lee, C. H., Izuta, T., Aoki, M., Totsuka, T. and Kato, H.: 1998, 'Growth and photosynthetic responses of red pine seedlings grown in brown forest soil acidified by adding H_2SO_4 solution', *Jpn. J. Soil Sci. Plant Nut.* **69**, 53–61 (in Japanese with English summary).

Mohren, G. M. J., van den Burg, J. and Burger, F. W.: 1986, 'Phosphorus deficiency induced by nitrogen input in Douglas fir in the Netherlands', *Plant Soil* **95**, 191–200.

Nakaji, T., Fukami, M., Dokiya, Y. and Izuta, T.: 2001, 'Effects of high nitrogen load on growth, photosynthesis and nutrient status of *Cryptomeria japonica* and *Pinus densiflora* seedlings, *Trees* **15**, 453–461.

Nakaji, T., Takenaga, S., Kuroha, M. and Izuta, T.: 2002, 'Photosynthetic responses of *Pinus densiflora* seedlings to high nitrogen load', *Environ. Sciences* **9**, 269–282.

Ohrui, K. and Mitchell, M. J.: 1997, 'Nitrogen saturation in Japanese forested watersheds', *Ecol. Appl.* **7**, 391–401.

Ohrui, K. and Mitchell, M. J.: 1998, 'Effects of nitrogen fertilization on stream chemistry of Japanese forested watersheds', *Water, Air, Soil Pollut.* **107**, 219–235.

Oyama, T.: 1990, 'Amino acid and ureide', in *Method of plant nutrition analysis*, Hakuyusya, Tokyo, pp. 181–182 (in Japanese).

Penkett, S. A.: 1988, 'Indications and causes of ozone increase in the troposphere', in F.S. Rowland and I. S. A. Isaksen (eds), *The Changing Atmosphere*, John Wiley and Sons, London, pp. 91–103.

Schulze, E.-D: 1989, 'Air pollution and forest decline in a spruce (*Picea abies*) forest', *Science* **244**, 776–783.

Shimazaki, K., Yu, S.-W., Sasaki, T. and Tanaka, K.: 1992, 'Differences between spinach and kidney bean plants in terms of sensitivity to fumigation with NO_2, *Plant Cell Physiol.* **33**, 267–273.

Sogn, T. A. and Abrahamsen, G.: 1998, 'Effects of N and S deposition on leaching from an acid forest soil and growth of Scots pine (*Pinus sylvestris* L.) after 5 yr of treatment', *Forest Ecol. Manage.* **103**, 177–190.

Wakamatsu, S., Uno, I., Ohara, T. and Schere, K. L.: 1999, 'A study of the relationship between photochemical ozone and its precursor emissions of nitrogen oxides and hydrocarbons in Tokyo and surrounding areas', *Atmos. Environ.* **33**, 3097–3108.

Wallander, H. and Nylund, J.-E.: 1992, 'Effects of excess nitrogen and phosphorus starvation on the extramatrical mycelium of ectomycorrhizas of *Pinus sylvestris* L.', *New Phytol.* **120**, 495–503.

Wellburn, A. R.: 1990, 'Why are atmospheric oxides of nitrogen usually phytotoxic and not alternative fertilizers?', *New Phytol.* **115**, 395–429.

Wilson, E. J. and Skeffington, R. A.: 1994, 'The effects of excess nitrogen deposition on young Norway spruce trees. Part II The vegetation', *Environ. Pollut.* **86**, 153–160.

ACIDIFICATION TRENDS AND THE EVOLUTION OF NEUTRALIZATION MECHANISMS THROUGH TIME AT THE BEAR BROOK WATERSHED IN MAINE (BBWM), U.S.A.

STEPHEN A. NORTON[1*], IVAN J. FERNANDEZ[2], JEFFREY S. KAHL[3] and RAQUEL L. REINHARDT[4]

[1] *Earth Sciences, Bryand Global Sciences Center, University of Maine, Orono, Maine, 04469-5790, U.S.A.;* [2] *Plant, Soil, and Environmental Sciences, University of Maine, Orono, Maine 04469, U.S.A.;* [3] *George Mitchell Center, University of Maine, Orono, Maine 04469, U.S.A.;* [4] *Ecology and Environmental Sciences, University of Maine, Orono, Maine 04469-5790, U.S.A.*
(* author for correspondence, e-mail: norton@maine.edu, phone: 207 581 2156, fax: 207 581 2202)

(Received 20 August 2002; accepted 9 April 2003)

Abstract. The paired catchment study at the forested Bear Brook Watershed in Maine (BBWM) U.S.A. documents interactions among short- to long-term processes of acidification. In 1987–1989, runoff from the two catchments was nearly identical in quality and quantity. Ammonium sulfate has been added bi-monthly since 1989 to the West Bear catchment at 1800 eq ha^{-1} a^{-1}; the East Bear reference catchment is responding to ambient conditions. Initially, the two catchments had nearly identical chemistry (e.g., Ca^{2+}, Mg^{2+}, SO_4^{2-}, and alkalinity $\approx$82, 32, 100, and 5 μeq L^{-1}, respectively). The manipulated catchment responded initially with increased export of base cations, lower pH and alkalinity, and increased dissolved Al, NO_3^- and SO_4^{2-}. Dissolved organic carbon and Si have remained relatively constant. After 7 yr of treatment, the chemical response of runoff switched to declining base cations, with the other analytes continuing their trends; the exports of dissolved and particulate Al, Fe, and P increased substantially as base cations declined. The reference catchment has slowly acidified under ambient conditions, caused by the base cation supply decreasing faster than the decrease of SO_4^{2-}, as pollution abates. Export of Al, Fe and, P is mimicking that of the manipulated watershed, but is lower in magnitude and lags in time. Probable increasing SO_4^{2-} adsorption caused by acidification has moderated the longer-term trends of acidification of both watersheds. The trends of decreasing base cations were interrupted by the effects of several short-term events, including severe ice storm damage to the canopy, unusual snow pack conditions, snow melt and rain storms, and episodic input of marine aerosols. These episodic events alter alkalinity by 5 to 15 μeq L^{-1} and make it more difficult to determine recovery from pollution abatement.

Keywords: acid rain, acidification, aluminum, episodic acidification, iron, models, phosphorus, soil, stream water, sulfate

1. Introduction

Acidification in ecosystems results in chemical changes to soil and to surface and groundwater as a consequence of depletion of base cations in soils and a decrease of base cations relative to strong acid anions in runoff. The most fundamental changes during naturally occurring chronic acidification are an increase in exchangeable

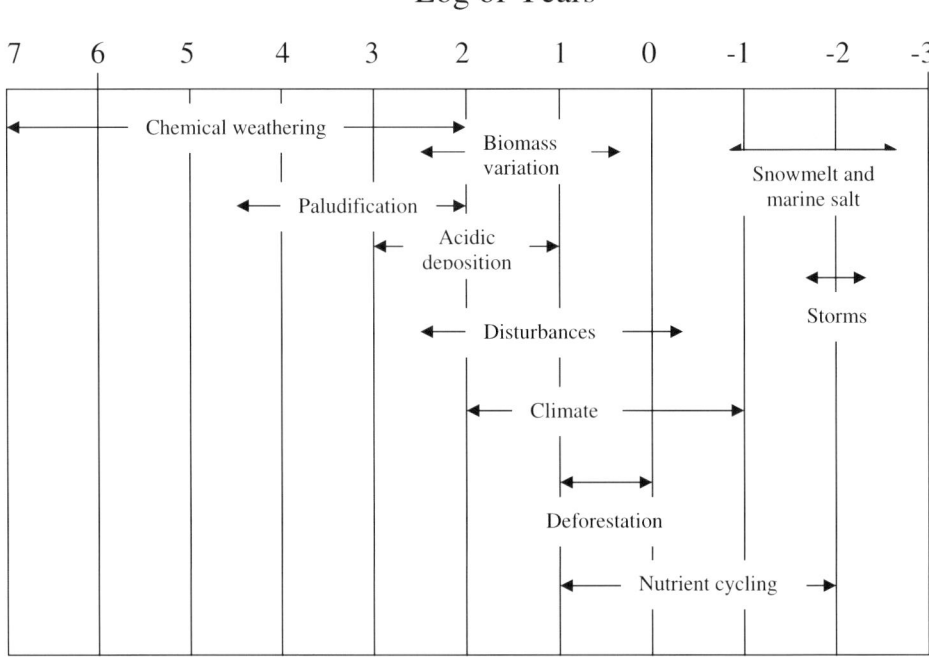

Figure 1. Conceptual model of time scales for mechanisms of surface water acidification.

H^+ or Al^{3+} ions in soils, an increase in H^+ activity in soil water, and a decrease in alkalinity in waters draining from the catchment. As systems acidify, the biotic community changes composition. Poorly drained terrestrial systems acidify and may develop into ombrotrophic bogs or blanket peat systems – where atmospheric inputs of water and nutrients and mineralization of organic matter control water chemistry and vegetation. Acidified systems may include dystrophic to clear-water oligotrophic streams and lakes.

There is no rigorous definition of when a system has become 'acidic', nor is there sharp definition between chronic and episodic acidification. Acidification processes operate at different time scales (Figure 1). Natural chronic acidification occurs over time frames of 10^2 to 10^6 yr as base cation-rich soluble minerals are depleted from the soil profile. During the process, concentrations of base cations decrease in runoff and, with no change in the supply of acidity from any source, the pH and bicarbonate alkalinity decline. Even SO_4^{2-} concentrations, if derived from the weathering of sulfide minerals, may decline. Such acidification trajectories have been demonstrated by inference using fossil diatom and chrysophyte assemblages in sediments (Whitehead *et al.*, 1989) and soil chronosequences, and using contemporaneous water chemistry from spatially distributed lakes of varying age (Engstrom *et al.*, 2000). The pH in lakes in glaciated terrain typically declines about 1.5 to 2 pH units over periods ranging from a few hundred to a few thousand years after deglaciation. With no change in other processes that may cause acidification, runoff chemistry approaches a steady state value. If paludification of the catchment occurs, base cation supply may be decreased through hydrologic isolation of mineral soil, production and export of organic acidity (dissolved organic carbon) may increase, and alkalinity and pH decline further. Paludification may occur over hundreds to thousands of years and result in new steady state chemistry for runoff.

Acidification processes operating at intermediate time frames (10^1 to 10^3 yr) include acidic deposition and biomass variation. Biomass increases in an aggrading forest, resulting in a sink for base cations (Nilsson *et al.*, 1982). The effect of net accumulation of biomass is acidification. Decreasing biomass may be gradual, driven by climate or soil evolution. The result, not well documented and hidden in the noise of other processes, would be an increase in mineralization and release of base cations, thereby reducing or reversing acidification. Loss of biomass may be catastrophic (short-time disturbance) due to fire, wind-throw, insect invasion, or ice storms, thereby producing unpredictable consequences for pH and alkalinity. Effects on acid-base status in the range of 10^1 to 10^{-1} yr include afforestation, climate, and deforestation. These processes affect acid-base status through variations in input of aerosols, production of organic acidity, storage of base cations in biomass, and mineralization. Lastly, a number of processes operate on a time scale of <1 yr: nutrient cycling, snowmelt, marine aerosol incidents, and rain storms. As each of these vary in strength from year to year, their impact on the acid-base status of a watershed will vary. Isolating their individual effects requires high frequency data over relatively long periods of time.

We report here on the 'natural' acidification and artificially induced acidification of two adjacent low alkalinity systems, as well as several natural short-term acidification mechanisms at Bear Brook Watershed in Maine, U.S.A. (BBWM) during the period 1987–2002.

2. Methods

The Bear Brook Watershed in Maine, U.S.A. (BBWM) is a paired watershed study designed to test models of surface water acidification caused by atmospheric deposition of acidic or acidifying substances (Norton and Fernandez, 1999). The two contiguous, forested watersheds are drained by low alkalinity headwater streams. The watersheds have very similar aspect, topography, vegetation (Eckhoff and Wiersma, 2002), soils (Fernandez *et al.*, 1999; Swoboda-Colberg and Drever, 1993), bedrock lithology (Norton *et al.*, 1999), and stream hydrology (Chen and Beschta, 1999). Continuous monitoring of precipitation quantity and quality (weekly), discharge volume (5 min intervals), and stream chemistry (weekly or more frequently at high flow) from the watersheds started in 1987 (Norton *et al.*, 1999). The West Bear catchment has been treated bi-monthly since November 1989 with 1800 eq ha^{-1} a^{-1} of (NH$_4$)$_2$SO$_4$. Additional details about the watersheds and analytical methods are in Norton and Fernandez (1999). Analytical methods have changed for cations and trace metals since 2000. They are determined by ICP analysis on single solutions. Phosphorus was determined by standard colorimetric procedures and ICP. The two methods gave comparable results.

Unique to the induced stream water chemistry in West Bear Brook are the extremely high concentrations of dissolved Al and particulate Al and Fe. Our protocol for analysis of various analytes consisted of removal of aliquots from field samples within 48 hr for anions (ion chromatography), dissolved organic carbon (DOC) (IR detection), alkalinity (Gran titration), pH (ISE), and P (colorimetric and ICP) and processing appropriately for each of the respective analytical methods. For ISCO^{TM} samples that remain in the field for as long as one week, we decanted and filtered 60 mL of sample for determination of 'dissolved Al' and we removed 45 mL of sample and acidified it to pH = 0 to 1 for the determination of 'total laboratory Al'. We then acidified the remaining contents of the sample in the field bottle to 0.5% (v/v) acid. The Al in this sample is termed 'total field Al'. Repeated samples, repeated speciation steps, and repeated reanalysis of the various solutions reveal that 'total field Al' > 'total laboratory Al' » 'dissolved Al' (Table I). Table I shows the partial chemistry of a sequence of samples taken during a two-day snowmelt event in February of 2002 that affected West Bear Brook. Samples were collected into polypropylene bottles by an automated ISCO^{TM} collector. The samples were retrieved after several days and then processed as described above. Dissolved Al shows a typical pattern for an acidic episode, with concentrations of Al increasing as pH decreases. The 'laboratory total Al' also increases. The difference between them is in particulate form. The 'field total Al' is substantially greater than the 'laboratory total Al' and follows the same trend of increasing as pH declines. We interpret these data to indicate that Al had precipitated within the field sample bottle as a continuous process during warming and as excess CO$_2$ degassed from them (Norton and Henriksen, 1983). The existence of an overpressure of CO$_2$ is indicated by the increase in pH induced during the measurement of the air-equilibrated pH compared to the pH at the beginning of the Gran titration, even after several days of shelf time. Apparently the precipitated Al (and Fe) must plate out on the container wall as well as some of it remaining in suspension, the latter to be determined as part of the 'laboratory total Al'. As a consequence, the concentration reported as 'laboratory total Al' underestimates the true Al in the original sample, some Al having been lost to the container surface. The 'field total Al' overestimates the true Al in the original sample, some Al having been added to the residual field sample in the ISCO^{TM} bottle (because of removal of aliquots) by acidifying it and dissolving any particulate Al left on the container wall. The true total Al lies between the 'field' and 'laboratory' totals. In summary, dissolved Al and particulate acid-soluble Al are operationally defined and may not be reproducible or comparable among studies.

TABLE I

Partial chemistry of sequential samples from West Bear Brook, Maine during a snow melt event in February 2002

ID	Date	Time	ANC pH	Equil. pH	Total P	Diss. P	Lab. total Al	Field Total Al	Field Diss. Al	Lab. total Fe	Field Total Fe	Field Diss. Fe
							(μg L^{-1})					
WB-1	02/26/2002	1600	4.72	5	5.1	0.6	609	997	545	440	1480	12
WB-4	02/27/2002	800	4.64	4.9	6	0.4	1720	2080	730	200	1020	4.27
WB-7	02/27/2002	1400	4.53	4.8	34.2	0.7	2630	4600	1060	585	2870	5.07
WB-10	02/27/2002	2000	4.49	4.7	93.4	1.2	3400	13900	1080	1130	11900	6.10
WB-12	02/28/2002	0	4.48	4.7	18.4	1.5	1420	8910	1130	248	5620	2.39
WB-14	02/28/2002	400	4.52	4.7	9.8	0.7	1060	7590	1100	135	6000	3.68
WB-18	02/28/2002	1200	4.55	4.8	4.6	0.3	1020	4290	1080	161	4050	2.65
WB-24	02/01/2002	0	4.60	4.8	1.8	0.7	1090	1360	1000	86	679	0.93

TABLE II

Concentrations of Al, Ca and Fe in solution decanted from sampling bottle and acidified, and from the field collection bottle filled with distilled water and then acidified

ISCO ID	Al (μg L^{-1})	Ca (μg L^{-1})	Fe (μg L^{-1})
Collection bottle filled with DI and acidified			
E2	21.8	0	0
E5	20.9	0.001	0
E6	14.6	0.006	0
E7	24.3	0.001	4.68
E8	31.8	0.008	0
E10	11.8	0.005	0
Decanted solution, acidified			
E2	135	0.989	0
E5	120	1.05	2.03
E6	117	1.07	0
E7	119	1.09	12.7
E8	123	1.07	0

To test this hypothesis, we collected daily ISCO™ samples from East Bear (reference) for five days during a period of low and relatively constant flow. Stream pH remained near 6. Each 1 L sample was decanted into a clean container and acidified. The resulting solution was analyzed for major analytes, total Al (Table II), and trace metals. The field ISCO™ container was refilled with deionized water and acidified, and the resulting solution was analyzed to determine the amount of Al and other elements on the wall of the collection bottle. The decanted solution typically had total Al concentrations between 117 and 135 μg L^{-1}, while the acidified container plus deionized water had total Al concentrations between 12 and 32 μg L^{-1}. These data from relatively non-acidic and severely acidified surface waters suggest that it is extremely difficult to determine dissolved Al in waters where Al concentrations are elevated because of the input of groundwater charged with CO_2. The concentration of dissolved Al is probably decreasing with time, and the precipitated Al is partitioning between the container and the solution. If our data are representative of low alkalinity streams undergoing episodic or chronic acidification, it is likely that the flux of both dissolved Al and particulate acid-soluble Al has not been well characterized in many studies. The results for Fe are parallel with respect to the particulate matter. Dissolved Fe was relatively low in both the acidic and non-acidic conditions in both streams with a small increase during periods of depressed pH. The partitioning of Fe to the container walls is equally as dramatic as for Al, and apparently faster. The numerical value of the concentration of dissolved Al and Fe may be a moving target, changing rapidly as a result of degassing of CO_2 from, and introduction of oxygen to, emerging groundwater. The dramatic amount of P associated with the particulate Al and Fe (Table I) suggests that the concentration of total and dissolved P in acidic solutions may be equally difficult to characterize.

In November of 1989, bi-monthly applications (1800 eq ha^{-1} a^{-1}) of pelletized $(NH_4)_2SO_4$ to West Bear (10.2 ha) were initiated and continue to the present. These applications were designed to be a realistic way of accelerating the processes related to acid rain-driven acidification. In 1989, the dose represented a 200% increase in the flux of S to the watershed and an increase of 300% for N. East Bear (10.7 ha) serves as a reference watershed. During the period of study (1987–2002) we assume that the long-term chemical weathering acidification trajectory was essentially at steady state, paludification is nearly absent and at steady state, living biomass is relatively constant in quality and quantity, and climate (temperature and moisture) has not changed systematically or appreciably. However, there was one defoliation episode (both catchments) in the mid-1990s that preferentially attacked the beech component of the forest (and the beech is in decline), and a major ice storm in January of 1998 destroyed approximately 1/3 of the hardwood canopies. Canopy closure was essentially complete after four yr. The paired catchment design allows us to isolate the effects of the chemical manipulation from these other variable processes.

3. Results

Prior to the start of the chemical manipulation (November 1989), the chemistry of both streams was highly variable as a consequence of the seasonality of nutrient uptake and mineralization, variable discharge due to rain storms and snowmelt, and variable input of marine aerosols (Figures 2a and b). However, volume-weighted means of major analytes were nearly identical. The volume-weighted means for Ca and Mg in the two streams for 1989 and subsequent years are shown on Figure 3. The temporal variability of the chemistry of the two catchments during the period 1987–1989 was virtually identical with respect to all major analytes, pH and ANC. The similarity of behavior for the pre-manipulation period is emphasized by inspection of a 'difference' diagram (Figure 4) on which the differences between concentrations of simultaneous samples from both streams are plotted. Even compounds with seasonally highly variable values (e.g., NO_3^-) or elements that may be considerably diluted during high discharge events (e.g., Ca and Mg) behaved very similarly in the two watersheds, prior to the manipulation.

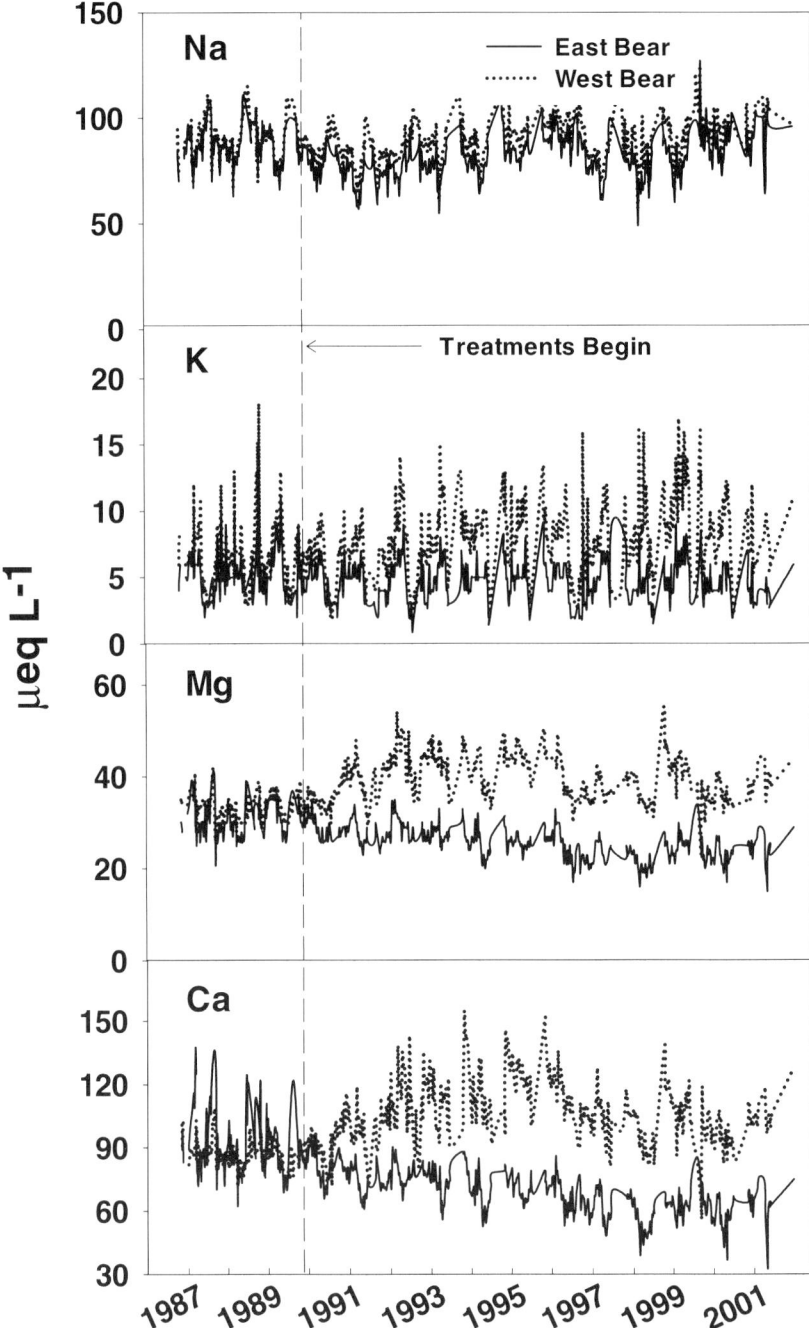

Figure 2. Concentrations of cations (a) and anions (b) at East and West Bear Brooks, Maine for 1989–2000. The vertical line on both graphs at November 1989 is the onset of the chemical manipulation. All data are in μeq L^{-1}. West Bear is shown by the dotted line, East Bear by the solid line.

ACIDIFICATION TRENDS AND NEUTRALIZATION MECHANISMS 297

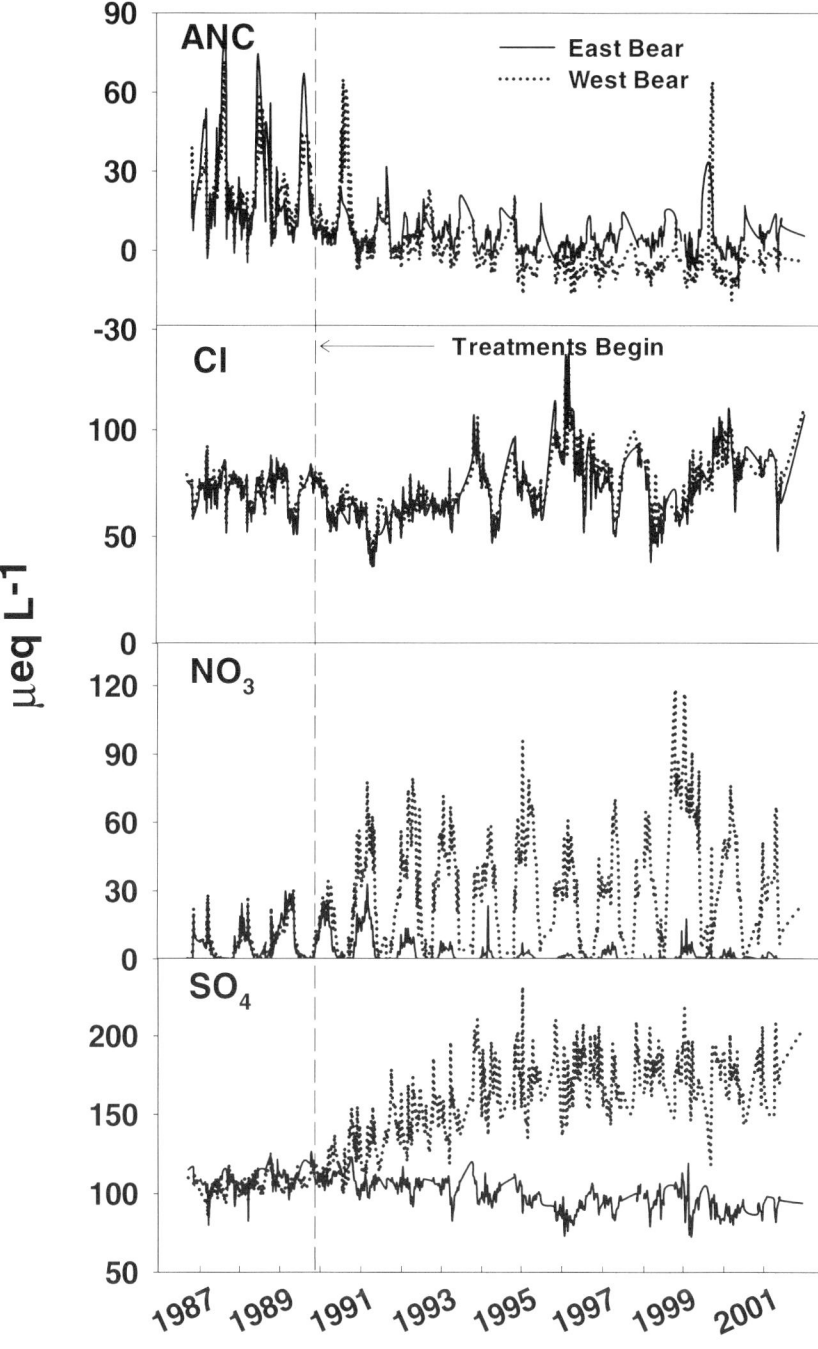

Figure 2. (Continued).

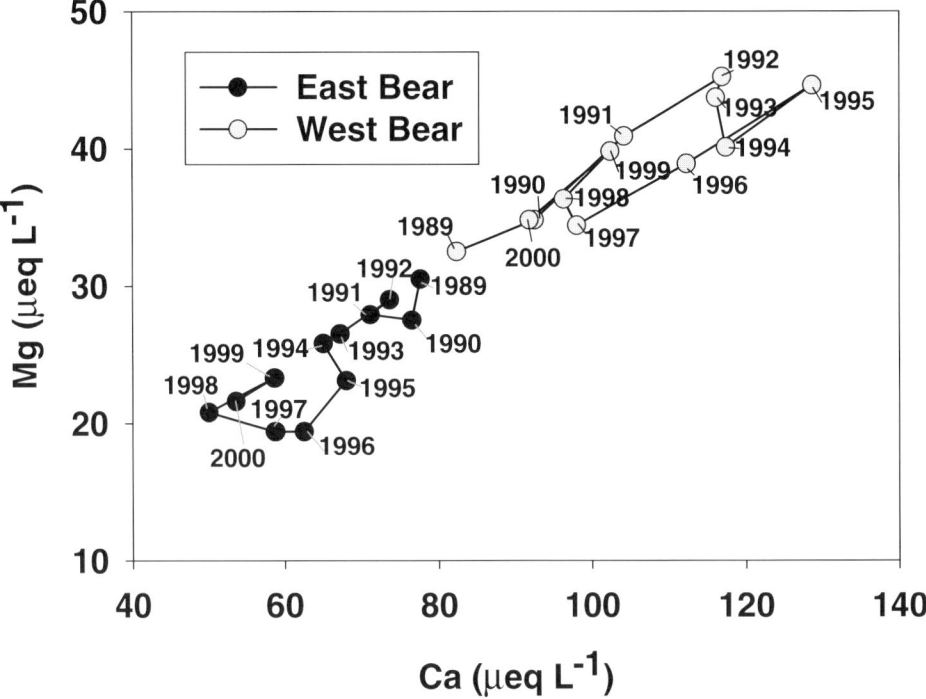

Figure 3. Mean annual volume-weighted concentrations of Ca and Mg for West (grey dots) and East Bear (black dots) Brooks, Maine for 1989 to 2000. All data are in μeq L^{-1}.

3.1. EAST BEAR

Runoff from East Bear Brook, the reference catchment, slowly acidified between 1989 to 2001, as SO_4^{2-} and NO_3^- concentrations declined, and base cations declined even more on an equivalent basis (Figures 2 and 3). The mean annual volume-weighted concentrations of Ca and Mg decreased irregularly from 1989 (78 and 30 μeq L^{-1}, respectively) to 1998. Starting in 1998; the mean values increased slightly for 2 yr and then declined again in 2000 to 52 and 21 μeq L^{-1}, respectively. The DOC (ca. 2 mg L^{-1}) and Si (ca. 2 mg L^{-1}) have remained nearly unchanged. The concentration of SO_4^{2-} in stream water declined from about 105 to 85 μeq L^{-1}, as SO_4^{2-} declined in precipitation. The concentration of NO_3 was strongly seasonal, with minimum values during the growing season, and spikes of high concentration during periods of high flow in the spring and fall. Annual export of NO_3^- from East Bear, the reference watershed, has declined since the early 1990s, a pattern repeated at many localities in the eastern United States. Mitchell *et al.* (1996) attributed the period of high NO_3^- export in the late 1980s and early 1990s to short term climate fluctuations, particularly cold and dry winters. The acidification of East Bear, in the context of declining NO_3^- and SO_4^{2-} in stream water is enigmatic. It is clearly driven by base cations declining more rapidly than the acid anions but the cause of

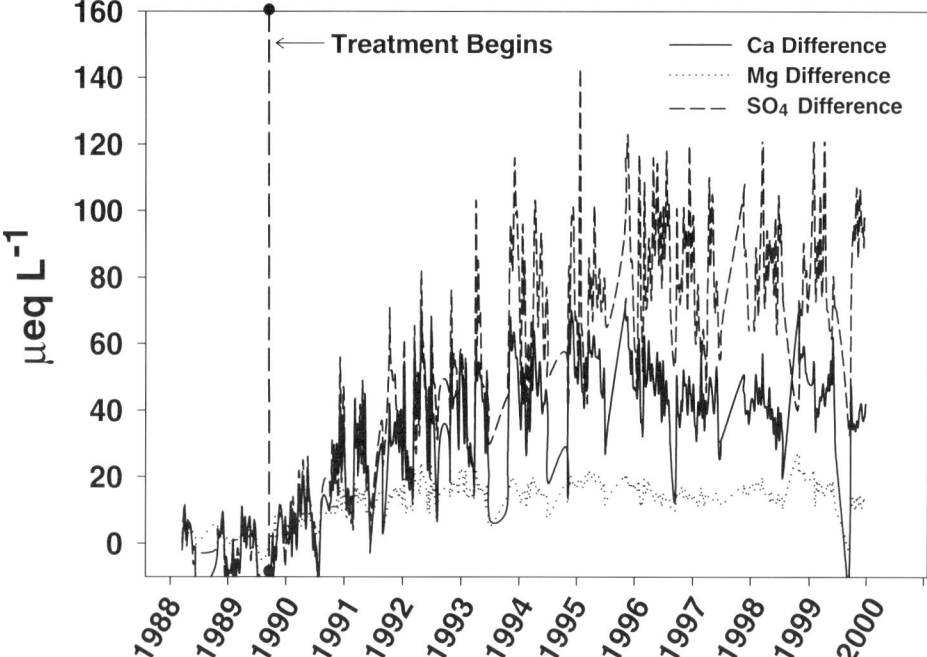

Figure 4. (Concentration of analytes in West Bear Brook, Maine) minus (concentration of analytes in East Bear Brook, Maine) for 1987–2000. All data are in μeq L^{-1}. The vertical dashed line indicates the beginning of chemical manipulation. Dashed line = SO_4^{2-}. Solid line = Ca. Dotted line = Mg.

the decline in base cations is unclear. Such a declining pH and alkalinity in spite of decreased SO_4^{2-} is common in the northeastern U.S. (Stoddard *et al.*, 1998, 2003). The declines are not caused at BBWM by changes in atmospheric deposition of non-marine base cations (Norton *et al.*, 1999). The decline of stream concentrations of base cations may relate in part to recovery of base saturation by the soils as SO_4^{2-} declined. Such a decline in response to decreased SO_4^{2-} is predicted by ion exchange theory. However, the decline in stream base cations should not exceed that of the strong acid anions, as it did. It may be that watershed soils have not yet reached equilibrium even with reduced SO_4^{2-} loading from the atmosphere.

A short-term de-acidification climate-related event not caused by atmospheric pollution occurred in 1989–1991. Then, we had concurrent measurement of soil PCO_2 and stream chemistry. High soil PCO_2 during one winter exceeded summer values (Fernandez *et al.*, 1993) and coincided with elevated export of base cations (and thus alkalinity). The high PCO_2 was coincident with a substantial and continuous snow pack. The tendency for pH of soil solutions to be depressed by the high PCO_2 was apparently offset by desorption of more base cations in the soil, producing surplus bicarbonate alkalinity of 10 to 15 μeq L^{-1}. Lower than normal PCO_2 resulted in slightly lower pH and alkalinity in runoff. Modeling with MAGIC (Cosby *et al.*, 1985) indicated that the variations of soil PCO_2 were sufficient to

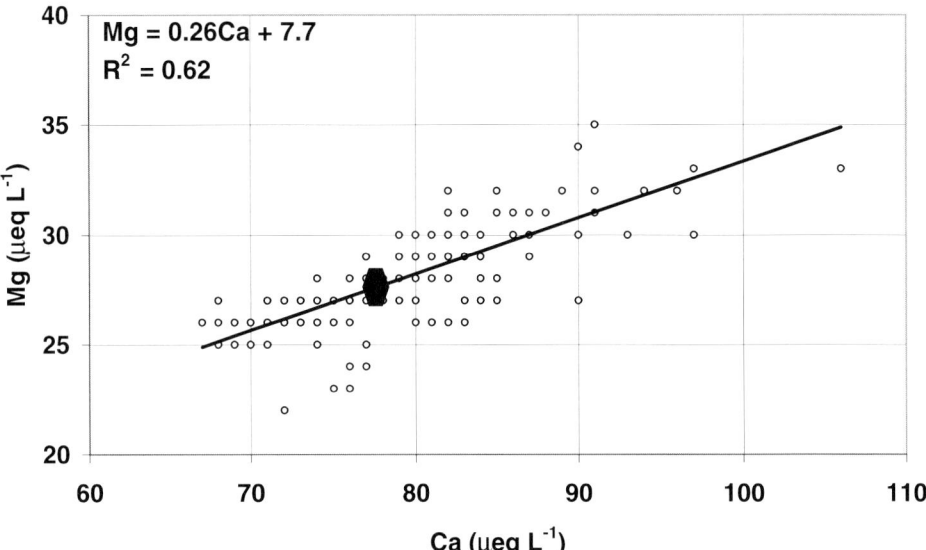

Figure 5. Variation of Ca and Mg for East Bear Brook, Maine for all samples in 1990. The volume-weighted annual mean is indicated by the filled symbol. All data are in μeq L^{-1}.

explain the variation in alkalinity from winter to winter (Norton et al., 2001). Nearly half of the runoff in any year at BBWM is associated with the melting of the snowpack, making this mechanism important in the alkalinity budget. Systematic variations in alkalinity production are not sustainable over long periods of time. The variability makes it more difficult to statistically identify the acidification due to the chemical treatment in West Bear, and the longer term decline in pH in East Bear that is possibly driven by base cation depletion.

A second meteorological event appears to have caused a 2 yr (1998–1999) reversal of the decline of base cation concentrations in stream water. In January of 1998, both Bear Brook watersheds were subjected to a freezing rainstorm that lasted nearly one week. The storm deposited up to 10 cm of ice on horizontal surfaces and destroyed as much as one third of the hardwood canopy. Subsequently, base cation concentrations and annual volume-weighted means increased for 2 yr (Figures 2a and 3), after which the decline in base cations continued. Nitrate was also higher in both watersheds than in the bracketing years (Figure 2b).

Periods of higher discharge during snowmelt or rainstorms are accompanied by slightly lower concentrations for base cations, reflecting dilution. The concentrations of cations, particularly Ca and Mg (Figure 5), are statistically highly related. This relationship is most likely caused by ion exchange equilibria. These periods of higher discharge also have slightly higher concentrations of DOC (Δ = +1 to 2 mg DOC L^{-1} = 5 to 10 μeq organic acidic anion L^{-1}) and NO$_3^-$ (Δ = +10 to

20 μeq L^{-1}), and relatively unchanged SO$_4^{2-}$. Consequently pH and alkalinity decline during higher discharge episodes. Only rarely is precipitation rich enough in marine aerosols to cause measurable episodic acidification at BBWM (e.g., winter 1995/1996), driven by ion exchange of Na$^+$ and Mg^{2+} for H$^+$, although the effects are persistent (Norton and Kahl, 2000). The increasing Cl$^-$ in runoff in 1998–2000 lags behind the sharp increase in Ca and Mg, suggesting that marine aerosols are not the dominant process involved in the sharply increased concentrations and fluxes of base cations starting early in 1998.

The annual mean volume-weighted concentration of dissolved Al varied between 4 and 8 μmol L^{-1} from 1989 to 2002. Higher concentrations occurred during higher discharge and lower pH. The increases in Al were all in the inorganic form (Postek et al., 1996). In high discharge events in 1995, dissolved Al reached 10 μmol L^{-1} and acid-soluble particulate Al increased slightly to as much as 5 or 6 μmol L^{-1} (Roy et al., 1999). The increased particulate Al was dissolved at pH = 0 in H$_2$SO$_4$, but there were no comparable increases in base cations, indicating that the particulate Al was not associated with alumino-silicate minerals. In snowmelt episodes in the winter of 2001/2002, dissolved Al reached as high as 11 μmol L^{-1} (Reinhardt et al., 2004) and particulate Al reached 185 μmol L^{-1}. Dissolved and particulate Fe both were below 0.5 μmol L^{-1} in the 1995 high discharge episodes. In 2001/2002 episodes, dissolved Fe remained <0.5 μmol L^{-1} while particulate, acid-soluble Fe increased to as much as 275 μmol L^{-1}, a 500-fold increase over 1995. As indicated above, these concentrations may slightly overestimate the original concentration of particulate Al and Fe, but it is clear that enormous amounts of Al and Fe hydroxide are leaving the catchment as a consequence of acidification. The dissolved P typically remains below 2 μg L^{-1}, regardless of pH. However, particulate acid-soluble P has increased from a maximum of about 50 μg L^{-1} in 1995 to nearly 100 μg L^{-1} in 2002. Particulate P is about 50 times dissolved P.

3.1.1. *West Bear*

In the first 12 yr of treatment (1989–2001), as a result of the (NH$_4$)$_2$SO$_4$ addition, West Bear acidified more than East Bear. Within a few years, the response by most major analytes (Figure 4) was statistically significant (Uddameri et al., 1995). Alkalinity decreased by about 20 μeq L^{-1} and and pH declined from ca. 5.5 to 4.7.

Concentrations of dissolved base cations increased by up to 75 μeq L^{-1} by 1995 (Figures 2a, 3 and 4). West Bear has responded to the (NH$_4$)$_2$SO$_4$ addition by exporting more base cations at higher flow, the reverse of East Bear. The excess export is Ca > Mg > Na > K. Consequently, the volume-weighted annual means for Ca and Mg generally increased for the first 6 yr and then started to decline, apparently as base cations were depleted in the exchangeable soil pool (Fernandez et al., 2003). During the decline, runoff was depleted in Mg relative to Ca, suggesting that the soils had been preferentially depleted of Mg. The decline was interrupted in 1998, concurrent with the ice storm and for 2 yr, Ca and Mg *increased* in runoff. This brief period of accelerated loss of base cations in both catchments was

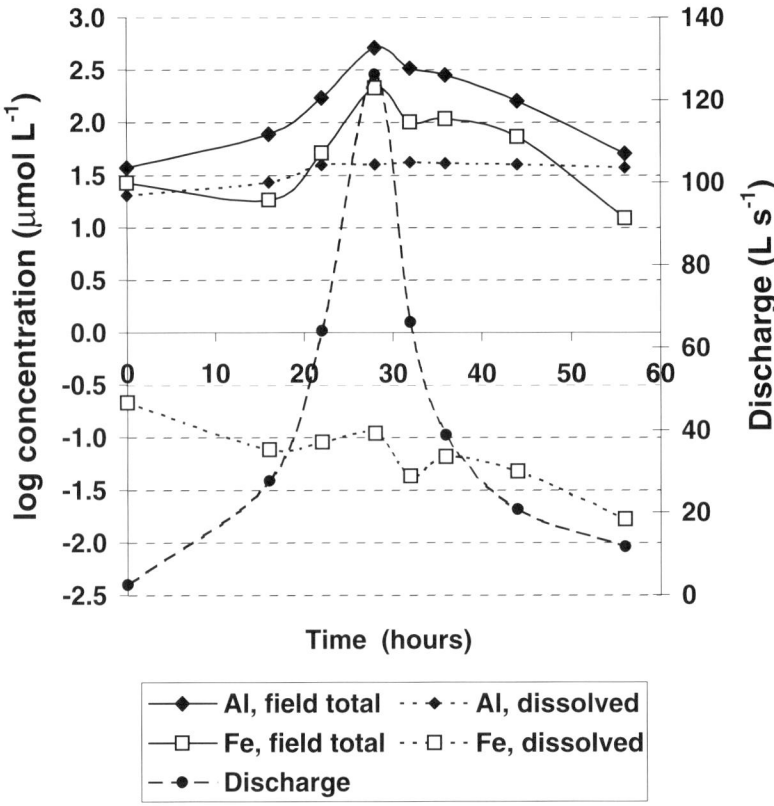

Figure 6. Variation of Al and Fe during an acidic episode at West Bear Brook, Maine in February 2002. All data are in μmol L^{-1}. Diamonds = Al. Squares = Fe. Solid lines = acid soluble particulate. Dotted line = dissolved. Filled circles with dashed line = discharge.

most likely caused by a combination of: mineralization of destroyed biomass lying on the forest floor, increased mineralization rates caused by higher moisture and temperature in the forest floor, reduced base cation uptake due to the reduction in canopy area, and increased marine aerosols displacing some base cations from soil. By 2001, the decrease of Ca and Mg resumed, with the mean volume-weighted value approaching that prior to the chemical manipulation, but at a substantially lower stream pH. The synchronicity of base cation trends in both watersheds from 1998 to 2000 indicates that the ice storm was the cause, rather than the chemical manipulation.

Mean annual volume-weighted dissolved Al has increased irregularly from about 5 μmol L^{-1} in 1989 to annual means greater than 20 μmol L^{-1}, with concentrations commonly well over 40 μmol L^{-1}. In 2000, the increase of dissolved Al exceeded the sum of all the excess base cations. The concentration of acid-soluble particulate Al in the 2001/2002 winter increased episodically up to 520 μmol L^{-1} (Figure 6) with associated particulate P (Δ = up to nearly 100 μg L^{-1}) (see Reinhardt *et al.*, 2004). The particulate Al is likely to have been derived from the B-horizon of mineral soil by either a one-stage or a two-stage stage process. As soil waters have become progressively more acidic, more Al is mobilized in dissolved form and transported to the stream in shallow flow. Upon emergence into the stream and mixing of soil water with stream water with higher pH, combined with degassing of excess CO_2, much of the dissolved Al in the water column is precipitated (Norton and Henriksen, 1983). At low flow, some Al hydroxide may adhere to the stream substrate and some dissolved Al may adsorb on exchange surfaces. Subsequent higher stream flow with lower pH may physically and chemically mobilize the Al from the stream channel (Tipping and Hopwood, 1988; Norton *et al.*, 1992) and add to the Al transported directly from the soil. Mobilization of Al from the soil consumes H^+ in the soil; precipitation of Al in the stream produces H^+. Effectively there is a translocation of Al due largely to carbonate equilibria with no net change in alkalinity of the system. However, the long-term effect is to mobilize the more labile $Al(OH)_3$ from the soil, with the consequence that more lower pH water will be transported to the stream. We caution that collection of samples and preserving their integrity to distinguish between dissolved and particulate Al is extremely difficult, due to changes in temperature, PCO_2, and other reactions (see the Methods section).

In 1995 high discharge episodes in West Bear Brook, dissolved Fe never exceeded 0.5 μmol L^{-1} while particulate Fe reached 11 μmol L^{-1}. In the winter of 2001/2002, concentrations of dissolved Fe reached 0.5 μmol L^{-1}, while Mn concentrations reached 0.1 μmol L^{-1}. Concurrently, high concentrations of particulate acid-soluble Fe (up to 255 μmol L^{-1}) occurred during periods of high discharge caused by rain and snowmelt (Figure 6). High concentrations of particulate P (up to 100 μg L^{-1}) accompanied the high Fe (Table I). We have too few samples to assess whether the P is more strongly associated with Al or Fe in samples when both Al and Fe are high. The origin of Fe and Mn is also most likely from the Fe-, Mn-, and P-rich B-horizon of the forest spodosols (Fernandez, unpub.). The mechanism of mobilization should be the same as for Al. As soil solutions have become progressively more acidic, particularly along high flow paths, more Fe, Mn, and P are mobilized and transported toward the stream. Precipitation of mobilized Fe in the stream should be caused by increased pH in the stream related to mixing, degassing of excess CO_2, and exposure to higher PO_2. Historically, Fe and Mn were low in concentration in East and West Bear. Increases of dissolved and particulate Fe were documented for West Bear by Roy *et al.* (1999) in 1995 for high discharge events. However, in 1995 both dissolved and particulate Fe remained lower than dissolved and particulate Al. In 2002, particulate Fe is higher than particulate Al. Fe is emerging as an important indicator of lowering pH.

The pre-manipulation concentration of SO_4^{2-} was approximately 100 μeq L^{-1} with only slight variation during periods of variable discharge (Figures 2b and 4). Soils apparently buffered the value quite well. The chemical amendment of $(NH_4)_2SO_4$, in the absence of other influences, should cause the runoff SO_4^{2-} to increase to approximately 300 μeq L^{-1} when steady state has been reached. However, by 2001 the volume-weighted annual mean SO_4^{2-} concentration was approximately 200 μeq L^{-1}. Retention of added SO_4^{2-} in 2001 was about 50% and over the 12 yr of manipulation averages about 75%. The rate of increase in retention is decreasing, even as anion adsorption capacity may be increasing as the soil pH decreases (Nodvin et al., 1988) or possibly Al-SO_4^{2-} phases may be forming in the soil.

Although NO_3^- concentrations have increased up to 100 μeq L^{-1} during high discharge, there is still strong apparent retention of added N and a strong seasonal signal (Figures 2b and 4). We do not know if N retention has resulted in increased biomass production. Uptake of NO_3^- offsets acidification, but uptake and storage of base cations in biomass would be an additional source of acidification. The extra N might stimulate mineralization of dead biomass, with the reverse effects. Nitrate was higher in 1998 than in the year before or after suggesting an effect from the ice storm.

4. Discussion

Both East and West Bear Brooks have acidified during the period 1987–2001. East Bear has acidified more slowly than West Bear and under ambient atmospheric deposition of excess SO_4^{2-}, NO_3^-, and NH_4^+. The East Bear acidification was caused by decreasing export of base cations, apparently because of continued (but reduced) loading of excess SO_4^{2-} in excess of the rate of supply of base cations from weathering and other less significant sources. Superimposed on this slow acidification are variations linked with more episodic processes including: ice damage to the canopy (base cation mineralization and reduced uptake), unusual snow cover conditions (elevated soil PCO_2 producing more alkalinity), episodic inputs of marine aerosols (ion exchange of Na^+ and Mg^{2+} for H^+, and the reversal of this exchange), highly variable hydrology (dilution at high discharge for East Bear and the increased desorption of base cations for West Bear), and climate-driven variation in nitrogen retention (released as NO_3^-). Initially, resistance to the acidification (short- and long-term) was provided by a combination of titration of bicarbonate alkalinity, cation desorption, Al desorption and dissolution, and possibly protonation of dissolved organic carbon. As acidification of the East Bear watershed has progressed, base cation concentrations, pH, and alkalinity in runoff have decreased and dissolved Al concentrations have increased. Progressively more Al and Fe are leaving the watershed in particulate form, probably as amorphous hydroxide. This particulate material does not represent a loss of alkalinity to the system. The translocation of Al and Fe requires H^+ to mobilize the metals from the soil and releases H^+ in the stream, with no net change in the acid-base status of the water. However, as the sources of labile Al and Fe become depleted in the soil, flow paths will progressively have lower pH, dissolved Al and Fe will increase further along in soil hydrologic pathways, and the acidification of stream water (particularly as measured by pH) will accelerate.

The addition of $(NH_4)_2SO_4$ to West Bear has accelerated the acidification processes documented for the East Bear catchment. Because of the additional strong acid anions SO_4^{2-} and NO_3^-, added and derived from increased nitrification, respectively, base cation desorption accelerated for about 6 to 7 yr until both of the dominant base cations (Ca^{2+} and Mg^{2+}) became depleted on the soil exchange complex. Base cation depletion at BBWM (Fernandez et al., 2003) has the potential (under current investigation) to impact the chemistry of vegetation as well as to disrupt normal forest processes, including regeneration and seedling growth. Despite increasing SO_4^{2-}, NO_3^-, and H^+ in soil water and runoff, desorption of base cations from soils has decreased since 1996. The dominant processes for neutralization of acidity shifted to desorption and/or dissolution of Al after about 6 to 7 yr of treatment. A similar switch from Ca release to Al release was demonstrated experimentally by Carnol et al. (1997) using undisturbed soil cores also treated with $(NH_4)_2SO_4$. A shift in the Ca/Al ratio in soils has the potential for direct impact on root physiology and tree health (Cronan and Grigal, 1995). Most recently, Fe has started to mobilize as pH has declined, providing yet another line of defense against acidification. Mn, in short supply in these soils, is being mobilized, but is insignificant in its capacity to neutralize.

The expectation for chemical behavior of West Bear Brook is shown on the schematic plot of chemistry through time (Figure 7). Galloway et al. (1983) proposed a 7-stage conceptual model for base cation behavior during acidification and recovery. Prior to acidification (stage I), base cation concentrations in runoff are in steady state with chemical weathering, input from marine aerosols, and net biomass. Under an imposed stress of elevated acid input, there is a period of increased export of base cations during which base saturation of soils is reduced. This period (stage II) corresponds to years 1990 to 1995 in Figure 3 for West Bear. Stage III corresponds to a period of declining base cations and is the model for East Bear for the entire study period of 1987 to 2001 and in the context of ambient acid rain. It is also the model for West Bear since 1996 with its increased acid deposition due to the manipulation. Stage IV corresponds to a new steady state where base cation export is essentially controlled again by the rate of chemical weathering and input from marine aerosols. However, soil base saturation is lower. Stage V corresponds to decreased base cation concentrations in runoff, after the acid stress is removed from the system and while soils are resorbing base cations. Stage VI corresponds to the period of increasing concentrations of base cations in runoff, as soils approach chemical equilibrium (Stage VII) with inputs from weathering and the atmosphere.

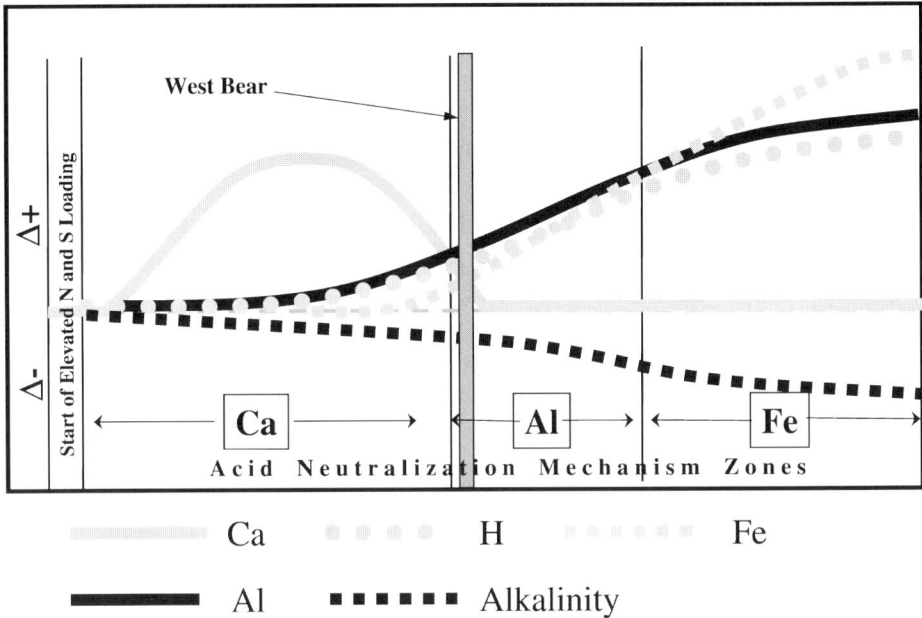

Figure 7. Conceptual model of mechanisms of neutralization through time, based partly on the behavior of West Bear Brook, Maine.

This simple trajectory of acidification and recovery can be confounded by many of the short-lived transient effects noted in Figure 1.

We suggest that a more complete conceptual model of watershed acidification should include the dynamics of Al and Fe. Based on the dynamics of West Bear, manipulated over the last 13 yr, it is clear that as desorption of base cations declines as the major acid neutralization mechanism, it is replaced by Al. This has been the impetus for research on the effects of acidic deposition on surface waters, *viz.* acid rain lowered the pH of soil and surface waters. Lowered pH mobilized Al. Inorganic (unbound) Al is toxic to fish. These acidification relationships and recovery have been demonstrated empirically, using spatial studies (e.g. Henriksen *et al.*, 1998) and elegantly in experiments (e.g., Wright *et al.*, 1993). In turn, it appears that Fe may replace Al, either at lower pH and after much acidification, or in Al-deficient soils (e.g., Borg, 1986). The runoff event in February of 2002 was sampled every 4 hours. The concurrent sampling of the two streams found that export of particulate Fe was greater from East Bear at pH = 5 than from West Bear at pH = 4.6. It is possible that we missed a brief period when the more acidic stream exported more Fe. It is also possible that the threshold for increased mobilization of particulate Fe from these forest soils is reached at a stream pH of about 5.0 and that the most labile soil Fe is quickly depleted. The moblized labile Fe may be precipitated as groundwater emerges into streams and lakes and be subsequently deposited downstream as sediment. Many studies indicate that currently acidic lakes have a period during their recent acidification when deposition of Fe in lake sediment was substantially increased (Norton, 1989). Not surprisingly, these catchment-scale responses to acidification are parallel to the buffer soil classes (base cations followed by Al followed by Fe) defined by Ulrich (1981).

There are many studies of watersheds at various stages of acidification, including stages I–IV, but it is extremely difficult to establish in which stage a system lies because very long-term data are necessary to determine this. We are fortunate to have created a long-term paired watershed study where one system (East Bear) was acidifying through a critical stage under ambient acidic deposition conditions, and we were able to force the other system (West Bear) to accelerate the process of acidification. The volume-weighted chemistry of West Bear Brook has evolved through at least three stage of acidification (the end of I, II, III, and into IV). However, shallow rapid flow paths heavily bias the volume-weighted chemistry. Base flow at West Bear has chemical characteristics of stage III, while some high flow paths may be dominated by Fe mobilization.

Lastly, the increased movement of dissolved Al and Fe through the groundwater system and into surface waters as particulates may have profound implications for the availability of dissolved P in systems that are acidifying. At the Bear Brooks, P export is 25 to 50 times greater in acid-soluble particulate form than dissolved. The presence of the two Al- and Fe-hydroxide phases in suspension may effectively decrease the dissolved P in streams and lakes through adsorption, causing oligotrophication (Dickson, 1978), thereby altering the trophic status of receiving waters (Kopáček et al., 2000). The release of P from sediments even during development of anoxia in the hypolimnia of lakes may be prevented if $Al(OH)_3$ is present in the sediment.

5. Summary

Acidification of the paired watersheds at Bear Brook Watershed in Maine over a 14 yr period has occurred as a consequence of progressive depletion of several acid-neutralizing processes. The East Bear watershed has served as a reference and is acidifying largely as a consequence of the decreased base cation export exceeding the decline in strong acid anions (SO_4^{2-} plus NO_3^-). The West Bear watershed has been treated with $(NH_4)_2SO_4$ to accelerate acidification. The treatment accelerated the export of dissolved base cations and Al with a concurrent depression of alkalinity and pH. Base cation export then decreased, especially Mg^{2+}, even as pH and alkalinity continued to decline.

Short-term acidification processes include: a *decrease* of base cation in runoff at higher discharge because of shallow flow paths in the soil being dominated by soils with low base saturation (East Bear), an *increase* of base cation in runoff at higher discharge because of a substantially increased flux of SO_4^{2-} and NO_3^- from the treatment (West Bear), an increased in dissolved organic acid concentrations during higher discharge (both catchments), and an episodic input of marine aerosols producing a salt effect (ion exchange for H^+, both catchments). The consequences of these episodic events of varying duration is a decrease in alkalinity, pH, and base cations (East Bear only) and an increase in dissolved Al, DOC, and NO_3^- concentrations.

Short-term alkalization has been caused by an increased export of base cations caused by mineralization of biomass related to a severe ice storm and reduced photosynthesis (biomass production), and by a higher-than-normal soil PCO_2 in the winter that causes excess desorption of base cations into spring runoff.

In both catchments, substantial loss of Al, Fe, and P occurs because of acidification of soil flow paths and elevated soil PCO_2. The Al and Fe, occurring dominantly as amorphous hydroxide in the soil, are transported from the soil in solution to the streams where lower PCO_2 and higher pH and PO_2 cause precipitation. The P, probably adsorbed and occluded in soil phases, is mobilized with the Al and Fe, and then is resorbed by the Al and Fe particulate material in the stream. Primary productivity of downstream systems may be reduced by the sequestration of P by Al, in particular.

Acknowledgements

Funding for this research has been provided by the U.S. Environmental Protection Agency, U.S. National Science Foundation, U.S. Geological Survey, and U.S. Department of Agriculture. This is the Maine Agricultural Experiment Station Contribution #2625. We thank the staff of the Environmental Chemistry Laboratory at the University of Maine for years of terrific assistance in the field and laboratory. Many graduate and undergraduate students have assisted us in practical ways, by contributing research, and by asking good questions!

References

Borg, H.: 1986, 'Metal speciation in acidified mountain streams in central Sweden', *Water, Air, Soil Pollut.* **30**, 1007–1014.

Carnol, M., Ineson, P. and Dickinson, A. L.: 1997, 'Soil solution nitrogen and cations influenced by $(NH_4)_2SO_4$ deposition in a coniferous forest', *Environ. Pollut.* **97**, 1–10.

Chen, H. and Beschta, R.: 1999, 'Dynamic hydrologic simulation of the Bear Brook Watershed in Maine (BBWM)', *Environ. Monit. Assess.* **55**, 53–96.

Cosby, B. J., Hornberger, G. M., Galloway, J. N. and Wright, R. F.: 1985, 'Modelling the effects of acid deposition: Assessment of a lumped-parameter model of soil water and stream water chemistry', *Water Resour. Res.* **21**, 51–63.

Cronan, C. S. and Grigal, D. F.: 1995, 'Use of calcium/aluminum ratios as indicators of stress in forest ecosystems', *J. Environ. Qual.* **24**, 209–226.

Dickson, W.: 1978, 'Some effects of the acidification of acidification of Swedish lakes', *Verh. int. Ver. Limnol.* **20**, 851–856.

Eckhoff, J. D. and Wiersma, G. B.: 2002, 'Baseline data for long-term forest vegetation monitoring at Bear Brook Watershed in Maine', *Tech. Bull. 180*, Maine Agricultural and Forest Exp. Station, Orono, Maine, U.S.A., pp. 202.

Engstrom, D. R., Fritz, S. C., Almendinger, J. E. and Juggins, S.: 2000, 'Chemical and biological trends during lake evolution in recently deglaciated terrain', *Nature* **408**, 161–166.

Fernandez, I., Son, Y., Kraske, C. R., Rustad, L. E. and David, M. B.: 1993, 'Soil carbon dioxide characteristics under different forest types and after harvest', *Soil Sci. Soc. Am. J.* **57**, 1115–1121.

Fernandez, I. J., Rustad, L., David, M., Nadelhoffer, K. and Mitchell, M.: 1999, 'Mineral soil and solution responses to experimental N and S enrichment at the Bear Brook Watershed in Maine (BBWM)', *Environ. Monit. Assess.* **55**, 165–185.

Fernandez, I. J., Rustad, L. E., Norton, S. A., Kahl, J. S. and Cosby, B. J.: 2003, 'Experimental acidification causes soil base cation depletion in a New England forested watershed', *Soil Sci. J.* (in press).

Galloway, J. N., Norton, S. A. and Church, M. R.: 1983, 'Freshwater acidification from atmospheric deposition of H_2SO_4 – A conceptual model', *Environ. Sci. Technol.* **17**, 541A–545A.

Henriksen, A., Skjelkvaale, B.-L., Manio, J., Wilander, A., Harriman, R., Curtis, C., Jensen, J. P., Fjeld, E. and Moiseenko, T.: 1998, 'Northern European Lake Survey, 1995', *Ambio* **27**, 80–91.

Kopáček, J., Hejzlar, J., Borovec, J., Porcal, P. and Kotorová, I.: 2000, 'Phosphorus inactivation by aluminum in the water column and sediments: Lowering of in-lake phosphorus availability in an acidified watershed-lake ecosystem', *Limnol. Oceanogr.* **45**, 212–225.

Mitchell, M. J., Driscoll, C. T., Kahl, J. S., Likens, G. E., Murdoch, P. S. and Pardo, L. H.: 1996, 'Climatic control of nitrate loss from forested watersheds in northeast United States', *Environ. Sci. Technol.* **30**, 2609–2612.

Nilsson, S. I., Miller, H. G. and Miller, J. D.: 1982, 'Forest growth as a possible cause of soil and water acidification: An examination of the concepts', *Oikos* **39**, 40–49.

Nodvin, S. C., Driscoll, C. T. and Likens, G. E.: 1988, 'Soil processes and sulfate loss at the Hubbard Brook Experimental Forest', *Biogeochemistry* **5**, 185–200.

Norton. S. A. and Henriksen, A.: 1983, 'The importance of CO_2 in evaluation of effects of acidic deposition', *Vatten* **39**, 346–354.

Norton, S.A.: 1989, 'Watershed acidification – A chromatographic process' in J. Kamari *et al.* (eds), *Models to Describe the Geographic Extent and Time Evolution of Acidification and Air Pollution Damage*, Springer-Verlag, New York, pp. 89–102.

Norton, S. A., Brownlee, J. C. and Kahl, J. S.: 1992, 'Artificial acidification of a non-acidic and an acidic headwater stream in Maine, U.S.A.', *Environ. Pollut.* **77**, 123–128.

Norton, S. A. and Fernandez, I. J. (eds): 1999, 'The Bear Brook Watershed in Maine', *Environ. Monit. Assess.* **55**, pp. 250.

Norton, S., Kahl, J., Fernandez, I., Haines, T., Rustad, L., Nodvin, S., Scofield, J., Strickland, T., Erickson, H., Wigington, P. and Lee, J.: 1999, 'The Bear Brook Watershed, Maine U.S.A. (BBWM)', *Environ. Monit. Assess.* **55**, 7–51.

Norton, S. A. and Kahl, J. S.: 2000, 'Impacts of marine aerosols on surface water chemistry at Bear Brook Watershed, M aine U.S.A.', *Verhandlungen, 27th Congress, Internat. Assoc. Theor. and Appl. Limnology* **27**, pp. 1280–1284.

Norton, S. A., Cosby, B. J., Fernandez, I. J., Kahl, J. S. and Church, M. R.: 2001, 'Long-term and seasonal variations in CO_2: Linkages to catchment alkalinity generation', *Hydrol. Earth Syst. Sci.* **5**, 83–91.

Postek, K. M., Driscoll, C. T., Kahl, J. S. and Norton, S. A.: 1996, 'Changes in the concentrations and speciation of aluminum in response to an experimental addition of ammonium sulfate to the Bear Brook Watershed, Maine, U.S.A.', *Water, Air, Soil Pollut.* **85**, 1733–1738.

Reinhardt, R. L., Norton, S. A., Handley, M. and Amirbahman, A.: 2004, 'Dynamics of P, Al, and Fe during high discharge episodic acidification at the Bear Brook Watershed in Maine, U.S.A.', *Water, Air, Soil Pollut.: Focus* (this volume).

Roy, S., Norton, S., Fernandez, I. and Kahl, J.: 1999, 'Linkages of P and Al export at high discharge at the Bear Brook Watershed in Maine', *Environ. Monit. Assess.* **55**, 133–147.

Stoddard, J. L., Driscoll, C. T., Kahl, J. S. and Kellogg, J. H.: 1998, 'A regional analysis of lake acidification trends for the northeastern U.S., 1982–1994', *Environ. Monit. Assess.* **51**, 399–413.

Stoddard, J., Kahl, J. S., Deviney, F., DeWalle, D., Driscoll, C., Herlihy, A., Kellogg, J., Murdoch, P., Webb, J. and Webster, K.: 2003, *Response of Surface Water Chemistry to the Clean Air Act Amendments of 1990*, EPA/620/R-03/001, U.S. Environmental Protection Agency, Washington, D.C., pp. 78.

Swoboda-Colberg, N. G. and Drever, J. L.: 1993, 'Mineral dissolution rates in plot-scale field and laboratory experiments', *Chem. Geol.* **105**, 51–69.

Tipping, E. and Hopwood, J. E.: 1988, 'Estimating stream water concentrations of aluminum released from stream beds during 'acid episodes', *Environ. Tech. Lett.* **9**, 703–712.

Uddameri, V., Norton, S. A., Kahl, J. S., and Schofield, J. P.: 1995, 'Randomized intervention analysis of the response of the West Bear Watershed, Maine to chemical manipulation', *Water, Air, Soil Pollut.* **79**, 131–146.

Ulrich, B.: 1981, 'Ökologische Gruppierung von Böden nach ihrem chemischen Bodenkunde', *Z. Pflanzenernähr. Bodenkd.* **144**, 289–305.

Whitehead, D. R., Charles, D. F., Jackson, S. T., Smol, J. P. and Engstrom, D. R.: 1989, 'The developmental history of Adirondack (N.Y.) lakes', *J. Paleolimnol.* **2**, 185–206.

Wright, R. F., Lotse, E., and Semb, A.: 1993, 'Rain project: Results after 8 years of experimentally reduced acid deposition to a whole catchment', *Can. J. Fish. Aquat. Sci.* **50**, 258–268.

DYNAMICS OF P, Al, AND Fe DURING HIGH DISCHARGE EPISODIC ACIDIFICATION AT THE BEAR BROOK WATERSHED IN MAINE, U.S.A.

RAQUEL L. REINHARDT[1], STEPHEN A. NORTON[2*], MICHAEL HANDLEY[3] and ARIA AMIRBAHMAN[4]

[1] *Ecology and Environmental Sciences, University of Maine, Orono, Maine 04469, U.S.A.;*
[2] *Department of Earth Sciences, University of Maine, Orono, Maine 04469, U.S.A.;*
[3] *Environmental Chemistry Laboratory, University of Maine, Orono, Maine 04469, U.S.A.;*
[4] *Department of Civil and Environmental Engineering, University of Maine, Orono, Maine 04469, U.S.A.*

(* *author for correspondence, e-mail: norton@maine.edu, phone: (207) 581 2156, fax: (207) 581 2202)*

(Received 20 August 2002; accepted 8 April 2003)

Abstract. Phosphorus (P), aluminum (Al), and iron (Fe) stream chemistry were assessed for high discharge snowmelt events at the Bear Brook Watershed, Maine (BBWM) during December 2001 and February 2002 and compared with results from a January 1995 study of the same streams. The West Bear catchment has been subjected to artificial acidification since 1989. The East Bear catchment is the untreated reference. Total (acid soluble) Al, Fe, and P were positively correlated with discharge during the 2001–2002 events. However, dissolved P concentrations remained low (≤ 0.1 μmol L^{-1}) during high discharge events as pH decreased in both streams. For example, in 2001, total P concentration increased to 1.7 μmol L^{-1} during the rising limb of the hydrograph in West Bear, approximately five times the value in East Bear. During the same event, in West Bear and East Bear dissolved Al concentrations increased to 21 and 6.3 μmol L^{-1}, respectively, while total Al concentrations increased to 166 and 30 μmol L^{-1}, respectively. Dissolved Fe concentrations remained ≤ 0.9 μmol L^{-1} in both streams during all study events. However, total Fe concentrations in 2001 increased to 239 and 4.1 μmol L^{-1} for West Bear and East Bear, respectively. Total Al and Fe declined parallel to total P after peaking during all study periods. Nearly all of the base cations were in dissolved form during the three events, indicating that total Al in West and East Bear Brooks is not associated with primary minerals such as feldspars. We conclude that particulate Al, Fe, and P are chemically linked during transport at high discharge in these episodically and chronically acidified streams.

Keywords: acidification, aluminum, base cations, episodic acidification, iron, phosphorus, watershed manipulation

1. Introduction

Understanding the relationship between acid rain and phosphorus (P) mobilization in headwater systems is essential in understanding the trophic status of surface waters. Lake eutrophication from excess P is a significant water quality problem. Particulate P in stream water increases with increasing discharge due to surface runoff and entrainment of material eroded from the stream bed or banks (Likens *et al.*, 1977). Therefore, the bulk of annual particulate P export from northeastern U.S. forested watersheds typically coincides with brief, high discharge events (i.e., fall rains, snowmelt). Dissolved P, however, is highly variable in relation to stream discharge among watersheds, and fluctuations may be correlated with, but not caused by, declining pH (Likens *et al.*, 1977; Meyer and Likens, 1979; Munn and Prepas, 1986; Prairie and Kalff, 1988). Phosphorus adsorbed to secondary Al and Fe phases on soil particle surfaces may dissolve as acidified groundwater rises into the solum during high discharge (Fernandez and Struchtemeyer, 1985).

During catchment acidification, the accumulation of $Al(OH)_3$ and P varies episodically and reversibly in stream sediments (Norton *et al*, 1992; Roy *et al.*, 1999). Aluminum (Al) mobilization is attributed primarily to the acid-catalyzed dissolution of Al-bearing secondary phases in the soil. Aluminum mobilization from soil and subsequent precipitation downstream may reduce the concentration of dissolved P by adsorption or co-precipitation (Kopáček *et al.*, 2000).

Mobilization of iron (Fe) and other metals may occur in watersheds during high flow events and is promoted by acidic and anoxic soil solutions with a high PCO_2 and high DOC (Borg, 1986). Metal mobilization is enhanced in catchments subject to acidic precipitation, as documented at the Bear Brook Watershed in Maine (BBWM), U.S.A. Roy *et al.* (1999) compared the chemistry of artificially acidified West Bear Brook and untreated East Bear Brook catchments during a January 1995 high discharge event, when pH declined. They hypothesized that particulate Al, Fe, and P are chemically associated during transport, with the Al phase dominating. The pH decline was typical of episodic acidification of Maine streams (Kahl *et al.*, 1992; Roy *et al.*, 1999). Particulate Al and Fe covaried with total particulate P during the January 1995 event. Roy *et al.* (1999) suggested that P- and Al- or Fe-bearing acid-soluble materials were eroded from the catchment, streambed, or stream banks, or that P was chemically associated with particulate Al and Fe derived from the soil.

In this study, total and dissolved Al, Fe, and P in stream water during high discharge events at BBWM in December 2001 and February 2002 were determined and compared to those reported by Roy *et al.* (1999) for January 1995. Continuous artificial acidification of the West Bear catchment, from 1995 to 2002, has increased the mobilization rates of Al, Fe, and P. We infer mechanisms controlling P export in surface waters undergoing chronic and episodic acidification.

2. Methods

BBWM is comprised of the adjacent catchments of West and East Bear Brooks with areas of 10.7 and 10.2 ha, respectively. Soils are primarily Spodosols, Inceptisols, and Folists with subsoil accumulations of sesquioxides (Fernandez and Struchtemeyer, 1985). Bimonthly treatments of 1800 eq $(NH_4)_2SO_4$ ha^{-1} a^{-1} have been applied to the West Bear catchment since 1989. The East Bear catchment is an untreated reference.

Stream water samples were collected simultaneously at short intervals from West and East Bear Brooks during a high discharge event in December 2001 and two high discharge events in February 2002. We used ISCO™ automated samplers fitted with 1 L acid-washed, plastic containers. Equilibrated pH was determined on 30 mL unfiltered aliquots by equilibrating the sample with air containing 300 ppm CO_2 at room temperature. We used a Radiometer combination pH electrode (model GK273920B). Alkalinity (ANC) was determined on 50 mL aliquots via Gran Titration using a Radiometer Titration Manager with the same electrode. Sixty mL aliquots were filtered through Fisher 0.45 μm nylon filters into HDPE bottles, preserved with 2 drops of concentrated H_2SO_4, and stored at 4 °C until analysis for dissolved organic carbon (DOC) with an OI Corporation model 1010 TOC analyzer.

Total P and dissolved P were each determined on 50 mL aliquots by colorimetry. Samples for dissolved P were filtered through 0.45 μm nylon filters prior to digestion. Total and dissolved P samples were digested by ammonium persulfate oxidation, autoclaved, and analyzed using the ascorbic acid method (Standard Method #4500 P-E) on a Thermospectronics Genesys5 Spectrophotometer. The detection limit was 1 μg L^{-1} with a precision of ± 1 μg L^{-1}.

Samples for determination of dissolved fractions of base cations and other metals were filtered through Fisher 0.45 μm nylon filters and acidified to pH < 2 with HNO_3. Samples for total base cations and other metals were microwave-digested with 5 mL HNO_3 at 180 °C for 15 min prior to analysis. Base cations (Ca, K, Mg, Na) and Al, Fe, and Si were determined using a Perkin-Elmer Optima 3300XL inductively coupled plasma optical emission spectrometer (ICP-OES) with axially-viewed plasma and a CETAC International, Inc. ultrasonic nebulizer.

3. Results

Stream discharge for the three events at West and East Bear Brooks are reported in Figures 1–3, panels a and b, respectively. Maximum discharge increased in each successive event. Discharge ranged from 0.7 to 24.9 L s^{-1} (28 L s^{-1} = 1 mm hr^{-1}) in East Bear during the 23–26 December 2001 event (Figure 1b). The hydrograph data in West Bear were not collected due to equipment failure during the same storm event. Discharge ranges in West and East Bear Brooks were 0.7 to 35.8 and 0.7 to 47.0 L s^{-1} for the 10–12 February 2002 event (Figures 2a and b, respectively), and 2.5 to 126.2 and 3.5 to 132.6 L s^{-1} for the 26–28 February 2002 event (Figures 3a and b, respectively).

The average equilibrated pH decreased in each successive event. Equilibrated pH decreased from 6.20 to 4.99 and from 6.10 to 5.60 in December 2001 in West Bear and East Bear, respectively (Figures 1a and b). Equilibrated pH decreased

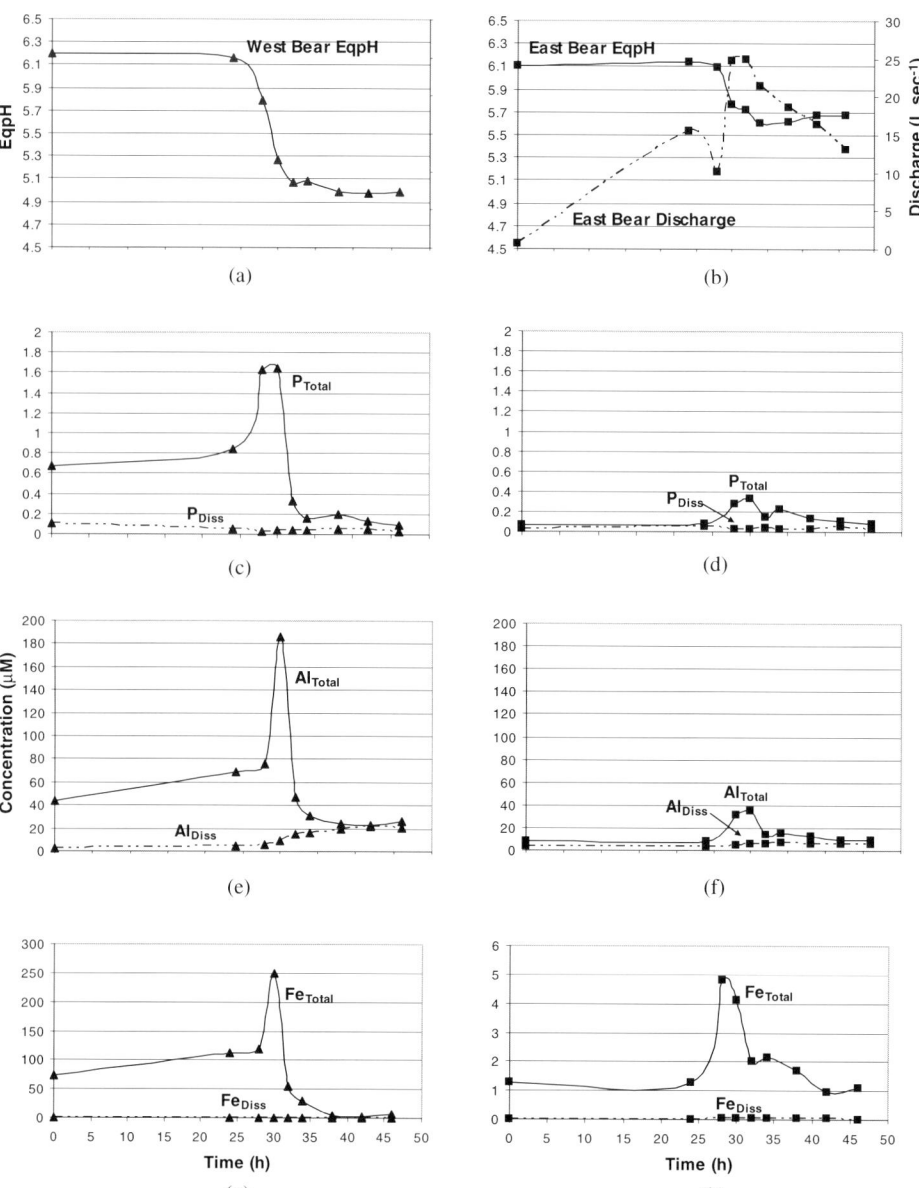

Figure 1. December 23–26, 2001 event: Time versus (a) equilibrated pH (EqpH) in West Bear Brook (WB), (b) EqpH and discharge (Q, L s^{-1}) in East Bear Brook (EB), (c) total and dissolved P in WB, (d) total and dissolved P in EB, (e) total and dissolved Al in WB, (f) total and dissolved Al in EB, (g) total and dissolved Fe in WB, (h) total and dissolved Fe in EB. Time zero equals 1600 hr on 12/23/2001. The subscripts 'Total' and 'Diss' refer to total and dissolved concentrations, respectively. All concentrations are μmol L^{-1}. Note that the scales for concentrations of Fe in (g) and (h) are not the same.

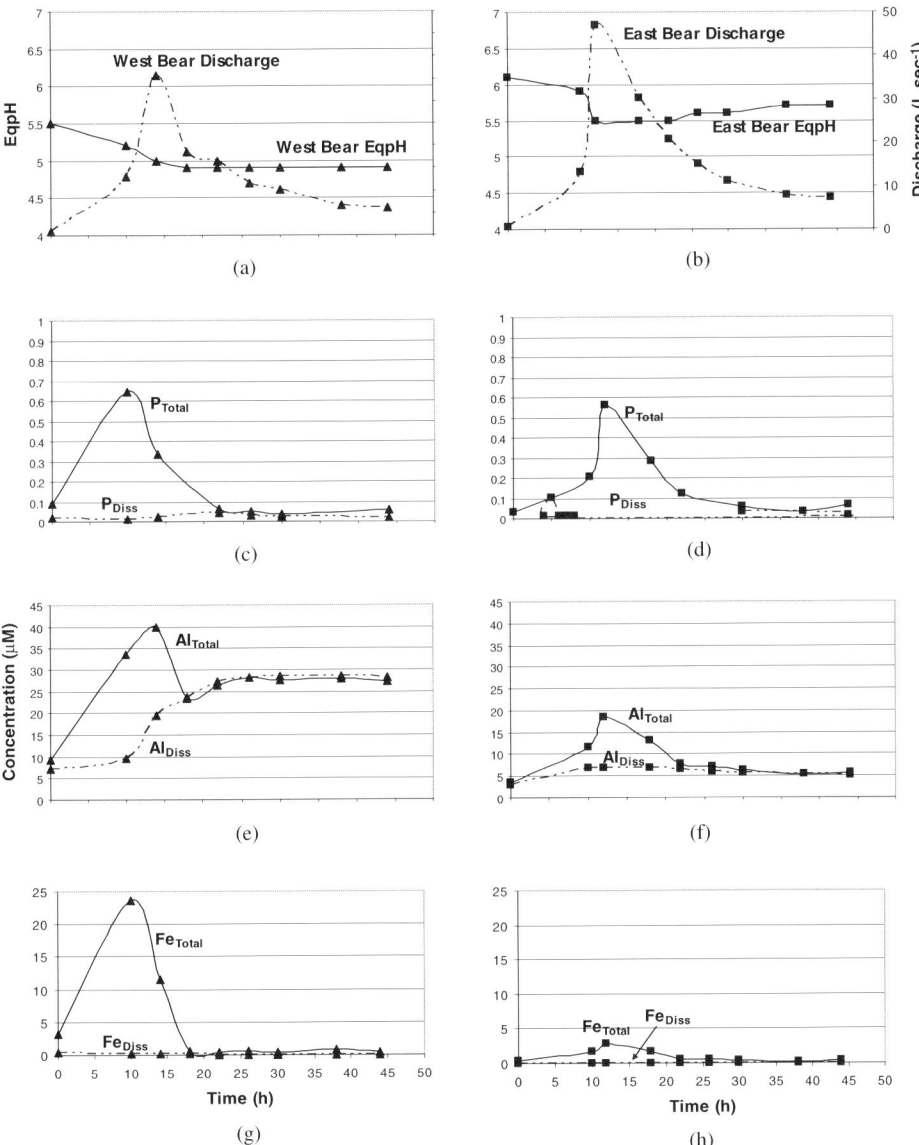

Figure 2. February 10–12, 2002 event: Time versus (a) equilibrated pH (EqpH) and discharge(Q, L s^{-1}) in West Bear Brook (WB), (b) EqpH and discharge (Q, L s^{-1}) in East Bear Brook (EB), (c) total and dissolved P in WB, (d) total and dissolved P in EB, (e) total and dissolved Al in WB, (f) total and dissolved Al in EB, (g) total and dissolved Fe in WB, (h) total and dissolved Fe in EB. Time zero equals 1600 hr on 2/10/2002. The subscripts 'Total' and 'Diss' refer to total and dissolved concentrations, respectively. All concentrations are μmol L^{-1}.

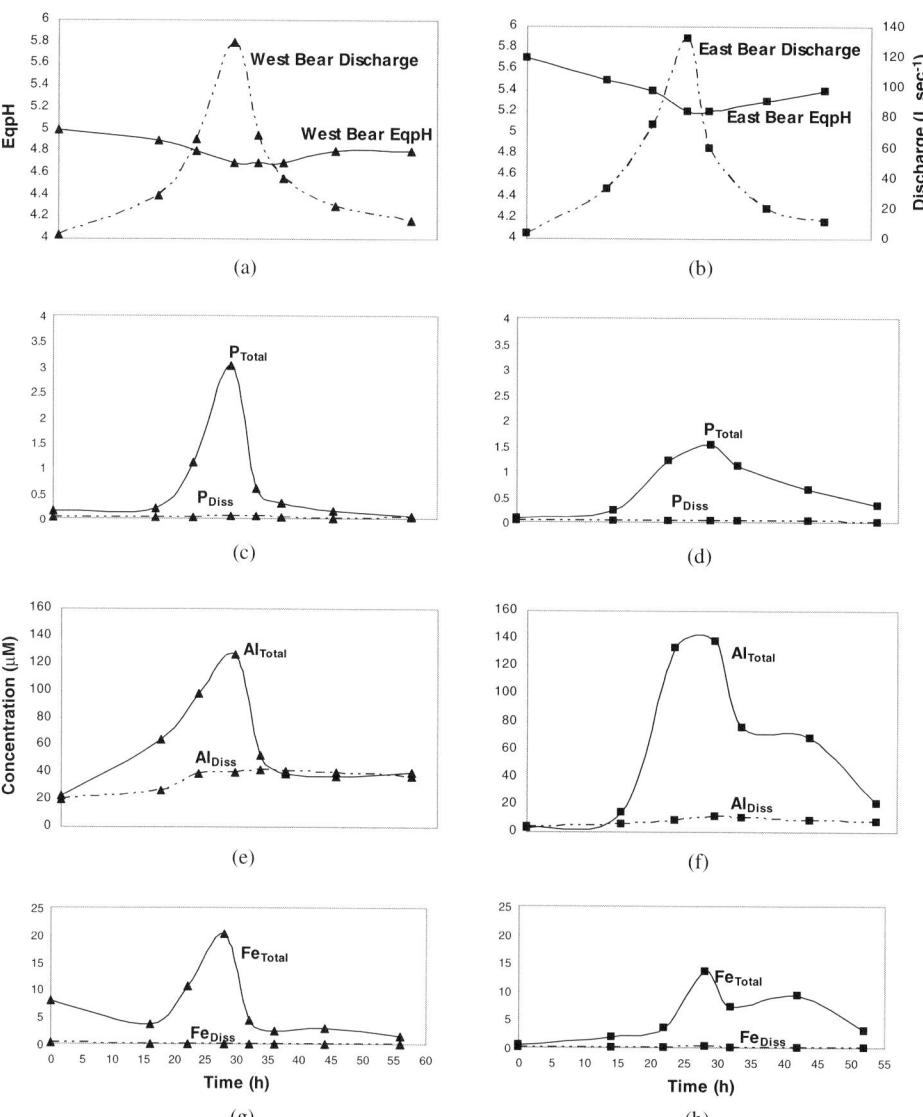

Figure 3. February 26–28, 2002 event: Time versus (a) equilibrated pH (EqpH) and discharge(Q, L s^{-1}) in West Bear Brook (WB), (b) EqpH and discharge (Q, L s^{-1}) in East Bear Brook (EB), (c) total and dissolved P in WB, (d) total and dissolved P in EB, (e) total and dissolved Al in WB, (f) total and dissolved Al in EB, (g) total and dissolved Fe in WB, (h) total and dissolved Fe in EB. Time zero equals 1600 hr on 2/26/2002. The subscripts 'Total' and 'Diss' refer to total and dissolved concentrations, respectively. All concentrations are μmol L^{-1}.

from 5.5 to 4.9 and 6.1 to 5.5 in West Bear and East Bear, respectively, during the early February event (Figures 2a and b) and from 5.0 to 4.7 and 5.7 to 5.2 during the late February event (Figures 3a and b).

Dissolved P concentrations remained low (≤ 0.10 μmol L^{-1}) in both streams during high discharge events during all three study periods (Figures 1–3c and d). In December 2001, total (acid-soluble) P concentration increased to 1.7 μmol L^{-1} during the rising limb of the hydrograph in West Bear compared to 0.3 μmol L^{-1} in East Bear (Figures 1c and d, respectively). Total P concentrations reached 0.6 μmol L^{-1} in both streams during the early February 2002 event (Figures 2c and d) and reached 3 and 1.5 μmol L^{-1} during the late February event in West Bear and East Bear Brook, respectively (Figures 3c and d).

Maximum concentrations of dissolved Al increased with each successive event, consistent with the decreasing pH. Dissolved Al concentrations increased to maxima of 20.0 and 6.30 μmol L^{-1} in West Bear and East Bear, respectively, during the December 2001 event (Figures 1e and f). Dissolved Al concentrations reached 28.5 and 7.04 μmol L^{-1} in West and East Bear, respectively during early February 2002 (Figures 2e and f), and 40.0 and 11.5 μmol L^{-1} during late February 2002 (Figures 3e and f). Total (acid-soluble) Al concentrations in 2001 increased to 186 and 35.6 μmol L^{-1} in West Bear and East Bear, respectively (Figures 1e and f). Total Al concentrations reached maxima of 40.0 and 18.5 μmol L^{-1} during the early February 2002 event (Figures 2e and f), and 138 and 126 μmol L^{-1} during the late February event (Figures 3e and f), in West and East Bear Brooks, respectively.

Dissolved Fe concentrations remained low (≤ 0.9 μmol L^{-1}) in both streams during the three events (Figures 1–3g and h). However, maximum total (acid-soluble) Fe concentrations in December 2001 increased to 251 and 4.1 μmol L^{-1} for West Bear and East Bear, respectively (Figures 1g and h). Concentrations of total Fe in West and East Bear Brooks reached maxima of 23.6 and 6.9 μmol L^{-1} during the early February 2002 event (Figures 2g and h), and 13.4 and 20.2 μmol L^{-1} during the late February event (Figures 3g and h). From peak values, total Al and total Fe declined parallel to total P during all events.

Dissolved base cations comprised nearly all of the total base cations and remained relatively constant or decreased slightly as discharge increased during the three events (Table I). The difference between total and dissolved base cation concentrations were typically one to three μmols L^{-1}. Differences in ranges of base cation concentration between West and East Bear Brook are attributed to cation desorption caused by the chemical manipulation of West Bear Brook (Norton *et al.*, 2004). Particulate acid-soluble Al concentration also substantially exceeded concentrations of particulate base cations in both streams during each of the events.

TABLE I

West and East Bear Brook chemistry for 26–28 February, 2002

Date	Time (hr)	Discharge (L s^{-1})	DOC (mg L^{-1})	Ca (mM) Total	Ca (mM) Diss.[a]	K (mM) Total	K (mM) Diss.	Mg (mM) Total	Mg (mM) Diss.	Na (mM) Total	Na (mM) Diss.	Si (mM) Total	Si (mM) Diss.
West Bear Brook													
26 February 2002	1600	2.46	1.75	0.054	0.055	0.007	0.007	0.020	0.020	0.095	0.100	0.083	0.074
27 February 2002	800	27.58	2.24	0.050	0.051	0.008	0.008	0.019	0.020	0.089	0.090	0.071	0.065
27 February 2002	1400	64.00	2.66	0.050	0.048	0.013	0.012	0.021	0.020	0.096	0.094	0.079	0.070
27 February 2002	2000	126.15	3.47	0.064	0.050	0.020	0.016	0.022	0.021	0.087	0.088	0.091	0.070
28 February 2002	0	66.09	3.08	0.052	0.051	0.014	0.013	0.021	0.022	0.084	0.089	0.078	0.071
28 February 2002	400	38.74	2.89	0.050	0.051	0.012	0.012	0.021	0.021	0.095	0.091	0.077	0.072
28 February 2002	1200	20.67	2.52	0.050	0.051	0.011	0.011	0.021	0.021	0.090	0.094	0.077	0.073
01 March 2002	0	11.67	2.22	0.051	0.051	0.009	0.009	0.021	0.021	0.084	0.094	0.076	0.073
East Bear Brook													
26 February 2002	1600	3.45	2.09	0.031	0.032	0.004	0.003	0.012	0.012	0.077	0.091	0.061	0.058
27 February 2002	600	32.90	2.54	0.030	0.031	0.005	0.005	0.012	0.012	0.079	0.089	0.058	0.051
27 February 2002	1400	74.95	3.41	0.039	0.029	0.008	0.007	0.013	0.012	0.076	0.082	0.066	0.046
27 February 2002	2000	132.55	4.17	0.031	0.025	0.010	0.008	0.011	0.011	0.066	0.072	0.072	0.046
28 February 2002	0	59.95	3.49	0.030	0.025	0.008	0.007	0.011	0.011	0.070	0.072	0.075	0.048
28 February 2002	1000	19.62	2.96	0.030	0.026	0.006	0.007	0.011	0.011	0.074	0.077	0.068	0.050
28 February 2002	2000	11.67	2.72	0.028	0.026	0.006	0.005	0.011	0.011	0.069	0.074	0.062	0.052

[a] Diss. = Dissolved.

4. Discussion

The particulate $Al(OH)_3$ and $Fe(OH)_3$ in stream water during high flow may originate from erosion of the streambed or from precipitated colloids from solutions leached directly from B horizon soil. In the latter case, the particulate $Al(OH)_3$ and $Fe(OH)_3$ are likely formed as a consequence of increased pH caused by CO_2 degassing from soil solutions emerging into the stream and mixing with higher pH and oxygen-rich water. The generalized reactions are, within the soil:

$$\text{Amorphous } Al(OH)_3 + 3H^+ = Al^{+3} + 3H_2O \tag{1a}$$

$$\text{Amorphous } Fe(OH)_3 + 3H^+ + 1e^- = Fe^{+2} + 3H_2O. \tag{1b}$$

The anion accompanying the release of the Al and Fe is HCO_3^-, which is relatively abundant even at low pH because of the high PCO_2 in the soil (Stumm and Morgan, 1996). As the pH of the stream water rises because of CO_2 degassing and mixing with higher pH water, and as oxygen is introduced, the reactions reverse.

Phosphorus sequestered (adsorbed and occluded) in the sesquioxide-rich forest soils is released as the Al and Fe are mobilized (Equations (2a) and (2b)), and is recaptured as the Al and Fe are precipitated in the stream water.

$$[Al\text{-}O\text{-}AlOPO_3]^- + H^+ \Leftrightarrow Al\text{-}OH + Al^{3+} + PO_4^{3-} \tag{2a}$$

$$[Fe^{III}\text{-}O\text{-}Fe^{III}OPO_3]^- + H^+ + e^- \Leftrightarrow Fe^{III}\text{-}OH + Fe^{2+} + PO_4^{3-}. \tag{2b}$$

Induced acidification of the treated catchment resulted in increased transport of Al, Fe, and P from the treated catchment compared to the reference catchment. Both total Al and total Fe are strongly correlated to total P (Table II). Discharge relates more strongly with total Al and Fe as the pH of the system decreases in the three successive events. Dissolved P, however, does not correlate with the discharge. Similar significant relationships were observed during the 1995 discharge events (Roy et al., 1999).

Concentrations of total Al and Fe were comparable in West Bear Brook during the 2001–2002 events. Total P tracks total Fe more closely in time during a given discharge event. Total Al and P concentrations in West Bear in 2001–2002 were comparable to those in 1995. However, 2001–2002 discharge events had higher total Fe concentrations than the 1995 event, perhaps due to earlier selective dissolution of more labile Al phases and lower pH during the 1995 event. Then, concentrations of total Al (160 μmol L^{-1}) were substantially higher than total Fe (10 μmol L^{-1}) in West Bear during the 1995 discharge event. As a result, Roy et al. (1999) concluded that most of the particulate P was associated then with $Al(OH)_3$, and to a lesser extent with $Fe(OH)_3$.

Two lines of evidence indicate that the total P is not in primary minerals. Nearly all base cations are in dissolved form (Table I). First, the presence of Ca primarily in a dissolved form precludes P associated with phosphate minerals such as

TABLE II

Correlation matrix for the chemistry and discharge of West and East Bear Brooks, Maine for three high discharge events, December 2001–February 2002. The subscripts 'Total' and 'Diss' refer to total and dissolved fractions, respectively. R-square values greater than 0.500 are in bold print

		P_{Total}		P_{Diss}		Al_{Total}		Al_{Diss}		Fe_{Total}		Fe_{Diss}		Discharge	
		WB	EB	WB	EB	WB	EB	WB	EB	WB	EB	WB	EB	WB	EB
(a)	P_{Total}	1.000	1.000												
	P_{Diss}	0.000	0.146	1.000	1.000										
	Al_{Total}	**0.699**	**0.904**	0.000	0.176	1.000	1.000								
	Al_{Diss}	0.500	0.213	0.229	0.034	0.213	0.086	1.000	1.000						
	Fe_{Total}	**0.811**	**0.808**	0.004	0.163	**0.951**	**0.938**	0.413	0.031	1.000	1.000				
	Fe_{Diss}	**0.641**	**0.676**	0.001	0.018	0.209	**0.529**	**0.679**	0.396	0.380	0.494	1.000	1.000		
	Discharge	HF	0.221	HF	0.075	HF	0.089	HF	**0.581**	HF	0.032	HF	0.350	1.000	1.000
	EqpH	0.260	0.013	0.311	0.006	0.038	0.005	**0.864**	**0.837**	0.154	0.047	**0.593**	0.117	HF	0.407
(b)	P_{Total}	1.000	1.000												
	P_{Diss}	0.414	0.026	1.000	1.000										
	Al_{Total}	0.261	**0.952**	0.022	0.022	1.000	1.000								
	Al_{Diss}	0.384	0.398	0.335	0.011	0.080	**0.550**	1.000	1.000						
	Fe_{Total}	**0.996**	**0.943**	0.441	0.046	0.218	**0.963**	0.455	0.391	1.000	1.000				
	Fe_{Diss}	0.091	**0.637**	0.145	0.009	0.199	**0.806**	**0.836**	**0.587**	0.151	**0.813**	1.000	1.000		
	Discharge	0.145	**0.902**	0.006	0.001	**0.528**	**0.873**	0.003	**0.506**	0.108	**0.763**	0.001	**0.506**	1.000	1.000
	EqpH	0.096	0.220	0.199	0.009	0.293	0.224	**0.863**	**0.513**	0.170	0.093	**0.886**	0.064	0.101	0.470
(c)	P_{Total}	1.000	1.000												
	P_{Diss}	0.299	0.418	1.000	1.000										
	Al_{Total}	**0.813**	**0.928**	0.139	0.464	1.000	1.000								
	Al_{Diss}	0.113	**0.817**	0.163	0.141	0.131	**0.639**	1.000	1.000						
	Fe_{Total}	**0.890**	**0.569**	0.196	0.008	**0.692**	0.484	0.004	**0.759**	1.000	1.000				
	Fe_{Diss}	0.027	**0.699**	0.005	**0.591**	0.000	**0.592**	**0.539**	**0.552**	0.204	0.379	1.000	1.000		
	Discharge	**0.880**	**0.834**	0.461	0.436	**0.800**	**0.762**	0.286	**0.609**	**0.655**	**0.507**	0.003	**0.801**	1.000	1.000
	EqpH	0.210	**0.637**	0.324	0.081	0.174	0.476	**0.884**	**0.945**	0.027	**0.720**	0.409	0.452	0.438	0.422

apatite ($Ca_5(PO_4)_3(OH)$) in the streams. All stream waters are also undersaturated with respect to apatite. Total P ranges up to 100 μmol L^{-1} whereas particulate Ca is in the range of a few μmols L^{-1} and dissolved Ca is less than 100 μmol L^{-1}, at all discharges. Second, there is no stoichiometric relationship between P and Al or Fe or Si. In both the January 1995 event (Roy et al., 1999) and the 2001–2002 events, molar ratios of total Al to total P ranged from 50:1 to 100:1, suggesting no stoichiometric precipitation of P as an essential element of some mineral phase, but rather that P was adsorbed to or co-precipitated with $Al(OH)_3$ (Norton and Henriksen, 1983; Roy et al., 1999). Similarly, the total Al in both streams is not associated with primary minerals such as feldspars ($KAlSi_3O_8$ or $NaAlSi_3O_8$) because base cations, in contrast to Al, are almost all in the dissolved phase (Table I).

Total Al was high in each of the three closely-spaced events, suggesting that much of the Al was derived directly from the soil as discharge increased, with relatively little intervening precipitation on the stream bed and subsequent erosion of Al during increasing discharge. The maximum total Fe decreased in each successive event, perhaps suggesting a different mechanism for mobilization, or differences in the dissolution kinetics due to the degree of crystallinity of Al and Fe hydroxide phases.

Unlike dissolved Al that increased in both streams with decreasing pH, dissolved Fe and P did not relate to pH. Dependence of dissolved Al on pH (Table II) suggests relatively rapid dissolution kinetics of $Al(OH)_3$ in the soil leading to equilibrium with respect to this phase. Lack of dependence of dissolved Fe concentrations on pH may relate to the fact that a changing redox environment, in addition to pH, lead to the mobilization of Fe. Lack of dependence of dissolved P concentrations in the stream may be due to adsorption of P onto colloidal Al and Fe hydroxides.

5. Summary

Aluminum- and Fe-linked export of P at the Bear Brook Watershed in Maine appears to be episodic, occurring concurrently with increasing discharge and depressed pH. Difficulties in accurately determining the dissolved and particulate species arise due to time-dependent processes that are causing dissolved Al and Fe to precipitate, including degassing of excess PCO_2, mixing of acidic groundwater with higher pH stream water, and increasing the PO_2 as groundwater emerges into the stream. At BBWM, export of Al, Fe, and P in particulate acid-soluble form dominates the stream budget for these three elements. Total P is strongly associated with particulate Al and Fe, with P in particulate form exceeding dissolved P by a factor of approximately 50. Dissolved Al, in contrast to dissolved Fe and P, is strongly controlled by the pH. Mobilization of these elements has been accelerated by progressive chronic acidification of the catchments and is particularly enhanced at high flow. Given that particulate Al and Fe concentrations reached high values at peak flow of three closely spaced events, most mobilization to the stream appears to be a one-stage process from the soil to the stream water. If the Al and Fe had been derived from the streambed, Al and Fe concentrations would have decreased noticeably as a result of the successive high discharge events. However, temporary storage of precipitated Al and Fe, with adsorbed P, may occur in the streambed at lower flow. The mobilization of Al and Fe in this fashion does not provide any net alkalinity change to the system, only a removal of the solid phase from the soil. However, downstream P budgets may be substantially affected by the presence of the Al and Fe solids, both of which strongly adsorb P, thereby affecting trophic status of water (Kopáček et al., 2000).

Acknowledgements

This research was funded by the United States Geological Survey Grant 2001ME1418G, as well as the U.S. Department of Agriculture and the U.S. National Science Foundation. We thank the staff of the Environmental Chemistry Laboratory at the University of Maine, and in particular, John Cangelosi, Kate Mahaffey, and Tiffany Wilson, for their help and cooperation on this project. We also appreciate the thoughtful, helpful comments of our reviewers. This is the Maine Agricultural Experiment Station Contribution #2626.

References

Borg, H.: 1986, 'Metal speciation in acidified mountain streams in central Sweden', *Water, Air, Soil Pollut.* **30**, 1007–1014.

Fernandez, I. J. and Struchtemeyer, R. A.: 1985, 'Chemical characteristics of soils under spruce-fir forests in eastern Maine', *Can. J. Soil Sci.* **65**, 61–69.

Kahl, J. S., Norton, S. A., Haines, T. A., Rochette, E. A., Heath, R. H. and Nodvin, S. C.: 1992, 'Factors controlling episodic acidification in low-order streams in Maine, U.S.A.', *Environ. Pollut.* **78**, 37–44.

Kopáček, J., Hejzlar, J., Borovec, J., Porcal, P. and Kotorová, I.: 2000, 'Phosphorus inactivation by aluminum in the water column and sediments: Lowering of in-lake phosphorus availability in an acidified watershed-lake ecosystem', *Limnol. Oceanogr.* **45**, 212–225.

Likens, G. E., Bormann, F. H., Pierce, R. S., Eaton, J. S. and Johnson, N. M.: 1977, *Biogeochemistry of a Forested Ecosystem*, Springer-Verlag, New York, pp. 146.

Meyer, J. L. and Likens, G. E.: 1979, 'Transport and transformation of phosphorus in a forest stream ecosystem', *Ecology* **60**, 1255–1269.

Munn, N. and Prepas, E.: 1986, 'Seasonal dynamics of phosphorus partitioning and export in two streams in Alberta, Canada', *Can. J. Fish. Aquat. Sci.* **43**, 2464–2471.

Norton, S. A. and Henriksen, A.: 1983, 'The importance of CO_2 in evaluation and effects of acidic deposition', *Vatten* **39**, 346–354.

Norton, S. A., Brownlee, J. C., Kahl, J. S.: 1992, 'Artificial acidification of a non-acidic and an acidic headwater stream in Maine, U.S.A.', *Environ. Pollut.* **77**, 123–128.

Norton, S. A., Fernandez, I. J., Kahl, J. S. and Reinhardt, R. L.: 2004, 'Acidification trends and the evolution of neutralization mechanisms through time at the Bear Brook Watershed in Maine (BBWM), U.S.A.', *Water, Air, Soil Pollut.: Focus* (2004).

Prairie, Y. T. and Kalff, J.: 1988, 'Dissolved phosphorus dynamics in headwater streams', *Can. J. Fish. Aquat. Sci.* **45**, 200–209.

Roy, S., Norton, S., Fernandez, I. and Kahl, J.: 1999, 'Linkages of P and Al export at Bear Brook Watershed in Maine', *Environ. Monit. Assess.* **55**, 133–147.

Stumm, W. and Morgan, J.: 1996, *Aquatic Chemistry Chemical Equilibria and Rates in Natural Waters*, John Wiley & Sons, Inc., New York, pp. 1022.

A BIOGEOCHEMICAL COMPARISON OF TWO WELL-BUFFERED CATCHMENTS WITH CONTRASTING HISTORIES OF ACID DEPOSITION

JAMES B. SHANLEY[1]*, PAVEL KRÁM[2], JAKUB HRUŠKA[2] and THOMAS D. BULLEN[3]

[1] *U.S. Geological Survey, P.O. Box 628, Montpelier, Vermont 05601, U.S.A.;* [2] *Czech Geological Survey, Klárov 3, CZ 118 21 Prague 1, Czech Republic;* [3] *U.S. Geological Survey, MS 420, 345 Middlefield Rd., Menlo Park, California 94025, U.S.A.*
(* author for correspondence, e-mail: jshanley@usgs.gov, phone: (802) 828 4466, fax: (802) 828 4465)

(Received 20 August 2002; accepted 10 April 2003)

Abstract. Much of the biogeochemical cycling research in catchments in the past 25 years has been driven by acid deposition research funding. This research has focused on vulnerable base-poor systems; catchments on alkaline lithologies have received little attention. In regions of high acid loadings, however, even well-buffered catchments are susceptible to forest decline and episodes of low alkalinity in streamwater. As part of a collaboration between the Czech and U.S. Geological Surveys, we compared biogeochemical patterns in two well-studied, well-buffered catchments: Pluhuv Bor in the western Czech Republic, which has received high loading of atmospheric acidity, and Sleepers River Research Watershed in Vermont, U.S.A., where acid loading has been considerably less. Despite differences in lithology, wetness, forest type, and glacial history, the catchments displayed similar patterns of solute concentrations and flow. At both catchments, base cation and alkalinity diluted with increasing flow, whereas nitrate and dissolved organic carbon increased with increasing flow. Sulfate diluted with increasing flow at Sleepers River, while at Pluhuv Bor the sulfate-flow relation shifted from positive to negative as atmospheric sulfur (S) loadings decreased and soil S pools were depleted during the 1990s. At high flow, alkalinity decreased to near 100 μeq L^{-1} at Pluhuv Bor compared to 400 μeq L^{-1} at Sleepers River. Despite the large amounts of S flushed from Pluhuv Bor soils, these alkalinity declines were caused solely by dilution, which was greater at Pluhuv Bor relative to Sleepers River due to greater contributions from shallow flow paths at high flow. Although the historical high S loading at Pluhuv Bor has caused soil acidification and possible forest damage, it has had little effect on the acid/base status of streamwater in this well-buffered catchment.

Keywords: acidification, alkalinity, base cations, buffering, catchments, Czech Republic, nitrate, sulfate, Vermont

1. Introduction

A rapid increase in coal burning in the present-day Czech Republic after World War II caused extremely high acid loadings (Kopáček *et al.*, 2001; Hruška *et al.*, 2002), which by the 1980s led to large areas of forest dieback and mortality. High-S emissions in close proximity to mountain ranges with little innate buffering capacity

The U.S. Government's right to retain a non-exclusive, royalty free licence in and to any copyright is acknowledged.

exacerbated the problem. In the U.S.A., damage has been more subtle, but over a much broader region. Forest decline and limited mortality have been linked to acid deposition in some areas (Driscoll *et al.*, 2001). As in the U.S.A., some areas of the Czech Republic have lithologies that are resistant to chemical weathering and have limited ability to buffer soils and waters.

In this paper, we compare the biogeochemistry of two well-buffered catchments. Sleepers River Research Watershed (Sleepers River) in the U.S.A. is buffered primarily by calcite within the phyllite bedrock. The Pluhuv Bor catchment (Pluhuv Bor) in the Czech Republic is buffered primarily by the weathering of antigorite in the serpentinite bedrock (Krám and Hruška, 1994). Present-day precipitation at both sites has a pH of 4.6. Pluhuv Bor is not in the area of the most serious forest damage, but it received much higher loadings of atmospheric acidity than Sleepers River during the last two decades. Our central question is how stream chemical patterns and trends and forest health may differ in these two well-buffered catchments as a result of their differing histories of atmospheric deposition.

Catchment biogeochemistry research has proliferated in the past few decades, largely due to funding for study of acid-deposition effects. This research has understandably focused on base-poor ecosystems, which are most susceptible to acidification and adverse ecological effects. Well-buffered sites, thought to be relatively immune from acidification, have received less study. The two well-buffered catchments discussed here are exceptions, having been the focus of intensive hydrologic and biogeochemical investigation for more than a decade.

Pluhuv Bor was established as a research site in 1991 by the Czech Geological Survey. It is the well-buffered catchment in a paired catchment study with the acidic Lysina catchment nearby (Krám and Hruška, 1994; Krám *et al.*, 1997; Hruška *et al.*, 2002). These catchments became part of the GEOMON network of 14 monitoring catchments in the Czech Republic (Novák *et al.*, 1996; Fottová and Skořepová, 1998). Sleepers River Research Watershed was established as a research site in 1957 by the Agricultural Research Service. In 1991, it was selected as one of five sites in the U.S. Geological Survey's Water, Energy, and Biogeochemical Budgets (WEBB) Program (Shanley and Chalmers, 1999; Shanley *et al.*, 2002a, b). At both sites, regular monitoring of water and solute fluxes in atmospheric deposition and streamflow has been ongoing for more than a decade. This paper is part of a collaboration between the U.S. and Czech Geological Surveys.

The objectives of this paper are (1) to compare trends in stream chemistry and solute fluxes in response to contrasting trends in atmospheric deposition at two similarly well-buffered catchments, (2) to compare biogeochemical processes and solute behavior at different flow regimes and to evaluate whether high loadings of atmospheric acidity may compromise the buffering capacity of well-buffered catchments, and (3) to compare forest health at the two catchments.

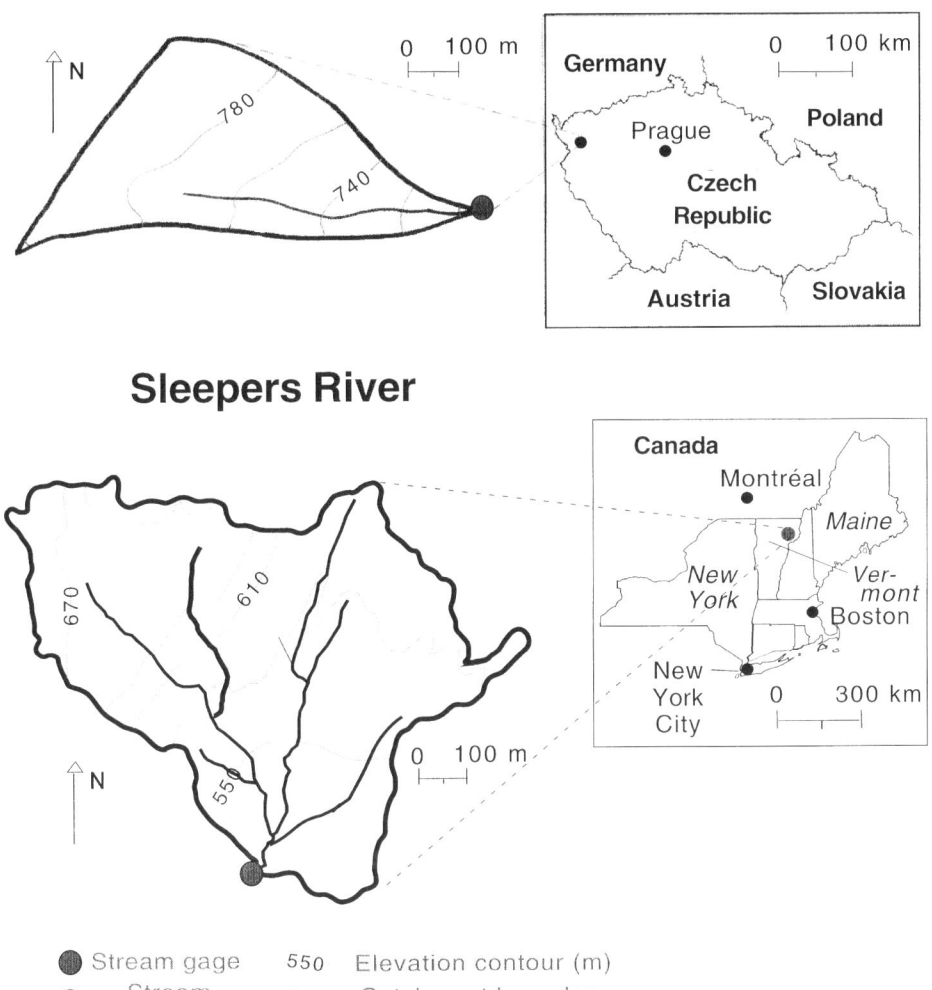

Figure 1. Location maps of (A) Pluhuv Bor, Czech Republic, and (B) Sleepers River W-9, Vermont, U.S.A.

2. Site Descriptions

Pluhuv Bor is a 22 ha catchment in the Slavkov Forest, western Bohemia, Czech Republic (Figure 1), ranging in elevation from 690 to 804 m (Table I). It is underlain by serpentinite, dominated by antigorite, an easily weathered magnesium (Mg)-silicate mineral that imparts high Mg and bicarbonate concentrations to drainage waters. Drainage waters are also high in sulfate (SO_4^{2-}) as a result of recent high sulfur (S) deposition. Apart from Mg, streamwater is very low in base cations,

TABLE I
Catchment comparison

	Pluhuv Bor	Sleepers River
Catchment area (ha)	22.0	40.5
Outlet elevation (m)	690	524
Relief (m)	114	155
Mean slope (%)	13	22
Aspect	SE	S
Annual average precipitation (mm)	910	1323
Annual average runoff (mm)	252	738
Annual average temperature (°C)	6.0	4.6
Vegetation	Norway spruce, with minor Scots pine	Sugar maple, yellow birch, white ash, American beech, red spruce, balsam fir
Bedrock	Serpentinite	Phyllite, calcareous schist
Soils	Inceptisols	Inceptisols, spodosols, histosols
Glacial deposits	None	1–4 m

reflecting the chemistry of the serpentinite. Pluhuv Bor was not glaciated during the last Ice Age. Brown earths and gleyey peats have developed to a depth of 0.5 to 4 m. The catchment is 94% forested by a 100 yr old plantation of mainly Norway spruce with a minor amount of Scots pine. The nonforested area is in grasses. Pluhuv Bor has a mean annual air temperature of 6 °C.

The U.S.A. site is the W-9 catchment at Sleepers River Research Watershed in northeastern Vermont (Figure 1). W-9 is a 40.5 ha catchment underlain by a quartz-mica phyllite with beds of calcareous granulite (Table I). Elevation ranges from 524 to 672 m. The Sleepers River watershed was glaciated and is mantled by 1–4 m of fairly dense basal till with a large percentage of fine silt. Inceptisols and spodosols have developed on the till in upland settings to an average depth of 70 cm. Histosols prevail in riparian zones and are characterized by as much as several tens of centimeters of black muck overlying the dense till. Swampy areas in headwater settings have as much as 2 m of peat. Weathering of calcite in the till leads to a calcium-bicarbonate-sulfate water. The SO_4^{2-} derives primarily from sulfides within the till and bedrock (Bailey *et al.*, in press). The forest cover is northern hardwoods dominated by sugar maple, yellow birch, white ash, and American beech. Softwoods make up about 15% of the basin and include balsam fir and red spruce. The forest was partially logged in 1929, and some additional yellow birch was removed in 1960. The mean annual temperature is 4.6 °C.

Several differences in the physical and chemical characteristics of these two catchments confound the effort to compare response to differential acid loading. The prominent differences, in order of descending importance, are (1) primary buffering by weathering of magnesium silicate (Pluhuv Bor) vs. calcium carbonate (Sleepers River); (2) hydrologic setting – 50% more precipitation and three times more runoff at Sleepers River, and evidence for greater importance of shallow flow paths at Pluhuv Bor; (3) glaciation at Sleepers River vs. none at Pluhuv Bor; and (4) planted coniferous forest at Pluhuv Bor vs. natural regeneration deciduous forest at Sleepers River. Despite these differences, these two sites are among the best-studied well-buffered catchments in the world and offer an excellent opportunity for comparison.

3. Results

3.1. Hydrology

Pluhuv Bor is a considerably drier environment than Sleepers River. Sleepers River W-9 catchment receives an annual average of 1320 mm of precipitation (25% as snow) compared to 950 mm at Pluhuv Bor (15% as snow) (Figure 2). Annual runoff averages 760 mm at Sleepers River and 270 mm at Pluhuv Bor. Evapotranspiration, calculated as the difference between precipitation and runoff, is 560 ± 40 mm at Sleepers River and 680 ± 150 mm at Pluhuv Bor. The greater evapotranspiration at Pluhuv Bor compared to Sleepers River reflects a longer growing season, with slightly warmer mean temperature and much shorter duration of snow cover. Runoff at Pluhuv Bor is greatest during the winter months, when precipitation is least, and least during the summer months, when precipitation is greatest. This pattern is driven by seasonal evapotranspiration demand, with some modification by storage and release from the snowpack. High runoff is sustained through the winter because of limited snowpack storage, and a modest runoff peak occurs in March from snowmelt. Runoff at Pluhuv Bor decreases to less than 10 mm mo^{-1} during the summer months. Compared to the nearby Lysina catchment, Pluhuv Bor has similar runoff in winter but less runoff in summer, consistent with its greater rooting depth.

The Sleepers River W-9 catchment has greater month-to-month variation in streamflow compared to Pluhuv Bor, despite a more uniform distribution of monthly precipitation (coefficient of variation for monthly precipitation was 18% compared to 33% at Pluhuv Bor). The greater hydrologic variation is due to a prominent influence of the seasonal snowpack at Sleepers River (Figure 2). Snow typically covers the landscape from mid-November until mid-April, and flow gradually recedes through the winter. Water stored in the seasonal snowpack is released during spring snowmelt, when approximately half of the annual runoff occurs during a 6 wk period from late March to early May. Streamflow recedes through the summer,

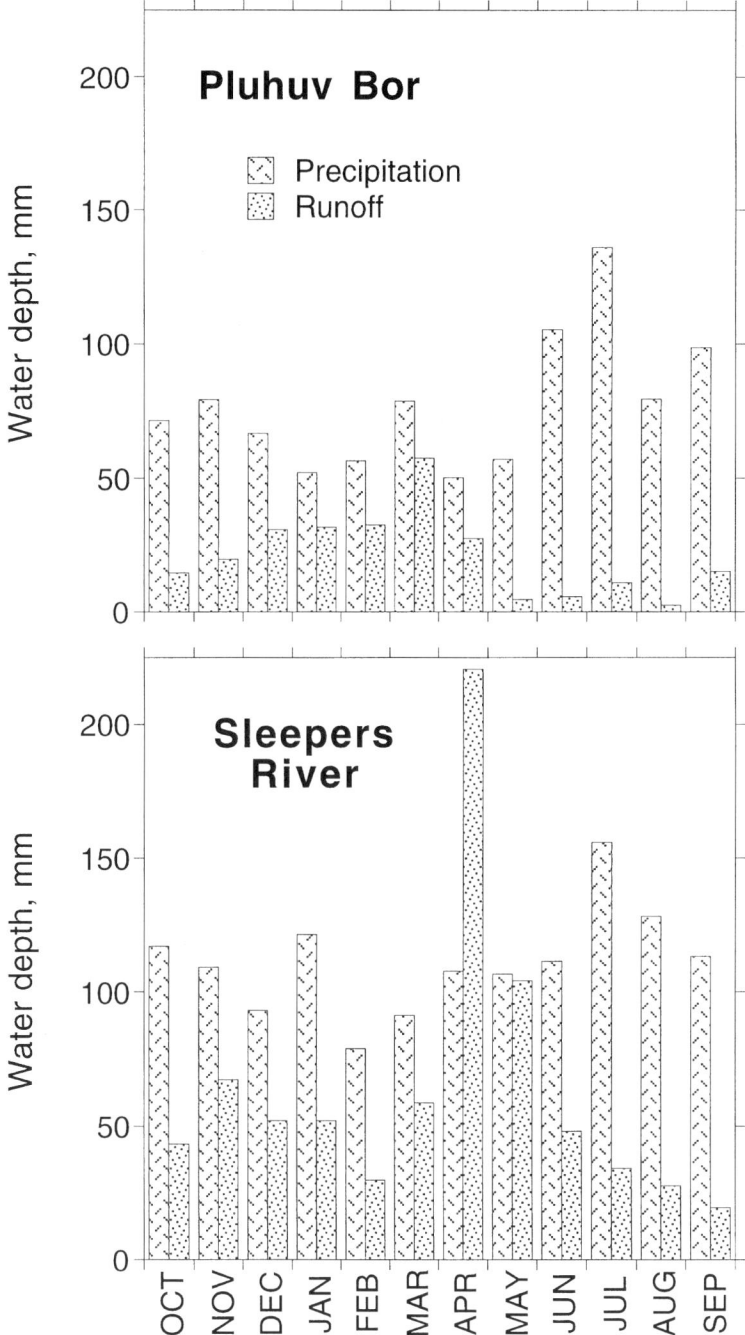

Figure 2. Mean monthly precipitation and runoff at Pluhuv Bor and Sleepers River W-9, 1991–1999.

punctuated by storms that may cause high peak discharges but that have limited duration. Low-intensity fall storms cause more significant runoff as the aquifer recharges. At Pluhuv Bor, the snowpack is less persistent and of shorter duration than at Sleepers River. Flow increases in winter due to reduced evapotranspiration and periodic rain and melt events. Event-driven flow increases occur occasionally at any time of year, but high flows are sustained for significant periods only during snowmelt. The smaller interannual variability of evapotranspiration at Sleepers River reflects the more consistent moisture supply relative to Pluhuv Bor.

At Sleepers River, groundwater makes the greatest contribution to streamflow, even during most high-flow events. Dunne and Black (1970) demonstrated that saturation overland flow is the main streamflow generation mechanism at Sleepers River, but isotopic data demonstrate that most of the overland flow is caused not by the rain or snowmelt triggering the event, but by discharge of displaced groundwater (Shanley *et al.*, 2002b). Melting snow on the expanded saturated areas in a fully recharged condition causes high and broad hydrograph peaks. During summer, high-intensity storms may also cause high discharge peaks, but they are short-lived because saturated contributing areas are small. At Pluhuv Bor, base flow is sustained by persistent groundwater discharge. Snowmelt is less important compared to Sleepers River. The overall drier and slightly warmer climate at Pluhuv Bor appears to limit the extent of saturated areas. However, response to rainfall is flashy; during events, chemical evidence suggests a large dilution by water following a shallow flow path. This water probably is some combination of channel interception, near-channel saturated overland flow, and shallow subsurface flow, possibly through macropores.

3.2. ACID DEPOSITION

Acid deposition in central Europe peaked later and with greater severity than in eastern North America. In the present-day Czech Republic, sulfur dioxide (SO_2) emissions peaked in 1982 at 2.4 Tg a^{-1}. This represents only 7.5% of the peak SO_2 emission in the U.S.A. in 1973. On a per capita basis, however, the peak annual Czech emissions were more than 50% greater than the peak in the U.S.A. in 1969 (Figure 3). On an areal basis, the peak of Czech SO_2 emissions per unit area of land surface was 10 times that of the U.S.A. peak. Since the peak years, SO_2 emissions have declined sharply in the Czech Republic and gradually in the U.S.A. By 1999, SO_2 emissions in the Czech Republic were at 10% of the 1982 peak, and were only 1.7 times greater per unit area than in the U.S.A. Current S loadings range widely in the Czech Republic, but they average 5–15 kg ha^{-1} a^{-1} in the area of Pluhuv Bor (Czech Hydrometeorological Institute, 2001), similar to those in the eastern U.S.A.

Modern precipitation acidity (from bulk collectors near each site) is similar at Pluhuv Bor and Sleepers River. In particular, sulfate and hydrogen ion concentrations are nearly the same (Figure 4a). Pluhuv Bor has considerably higher

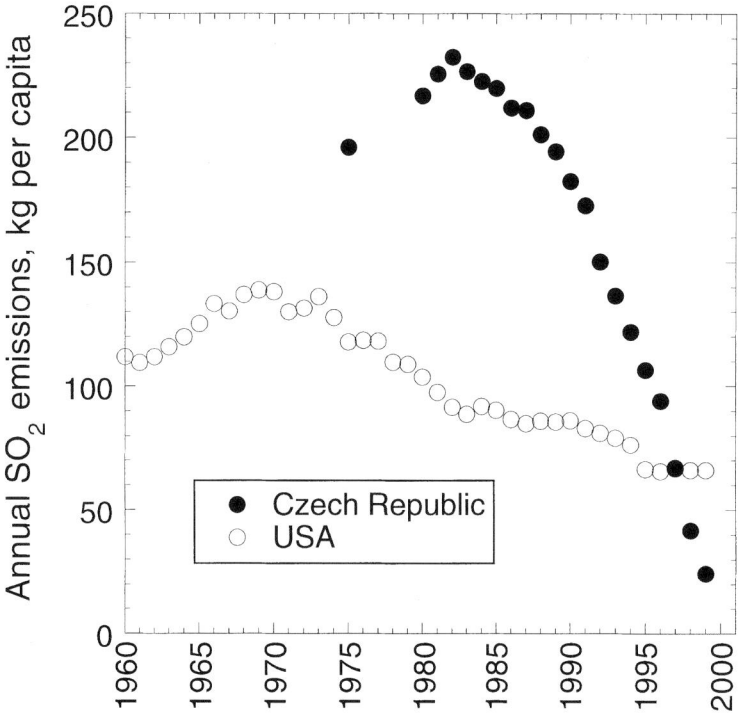

Figure 3. Plot of annual emissions of SO_2 per capita in the Czech Republic and U.S.A. Modified from Hruška *et al.* (2002); U.S.A. emissions data from U.S. EPA (2000).

chloride (Cl^-) and nitrate (NO_3^-) concentrations, balanced by greater base cation and ammonium (NH_4^+) concentrations. However, given the 50% greater precipitation amount at Sleepers River, SO_4^{2-} and H^+ loadings are greater at Sleepers River whereas NH_4^+ and NO_3^- loadings are similar at the two sites. Note that there is likely significant dry deposition input of both S and nitrogen (N) to these sites that is not captured in bulk deposition. Dry deposition of S may have a greater relative importance at Pluhuv Bor because of its proximity to point sources.

3.3. STREAM CHEMISTRY OVERVIEW

Acidic deposition is thoroughly neutralized at each site (Figure 4). Weathering of antigorite in the serpentinite at Pluhuv Bor gives rise to a Mg-bicarbonate water, whereas calcite weathering at Sleepers River leads to calcium bicarbonate streamwater. At both sites, base cations other than the dominant cation are minor contributors. In both catchments, SO_4^{2-} is the second most abundant anion. At Pluhuv Bor, high SO_4^{2-} concentrations reflect the legacy of high atmospheric S loadings, and a geologic source also may exist. At Sleepers River, where historical S deposition was considerably lower, SO_4^{2-} concentrations were lower than Pluhuv Bor despite an important geologic source of S in sulfide minerals.

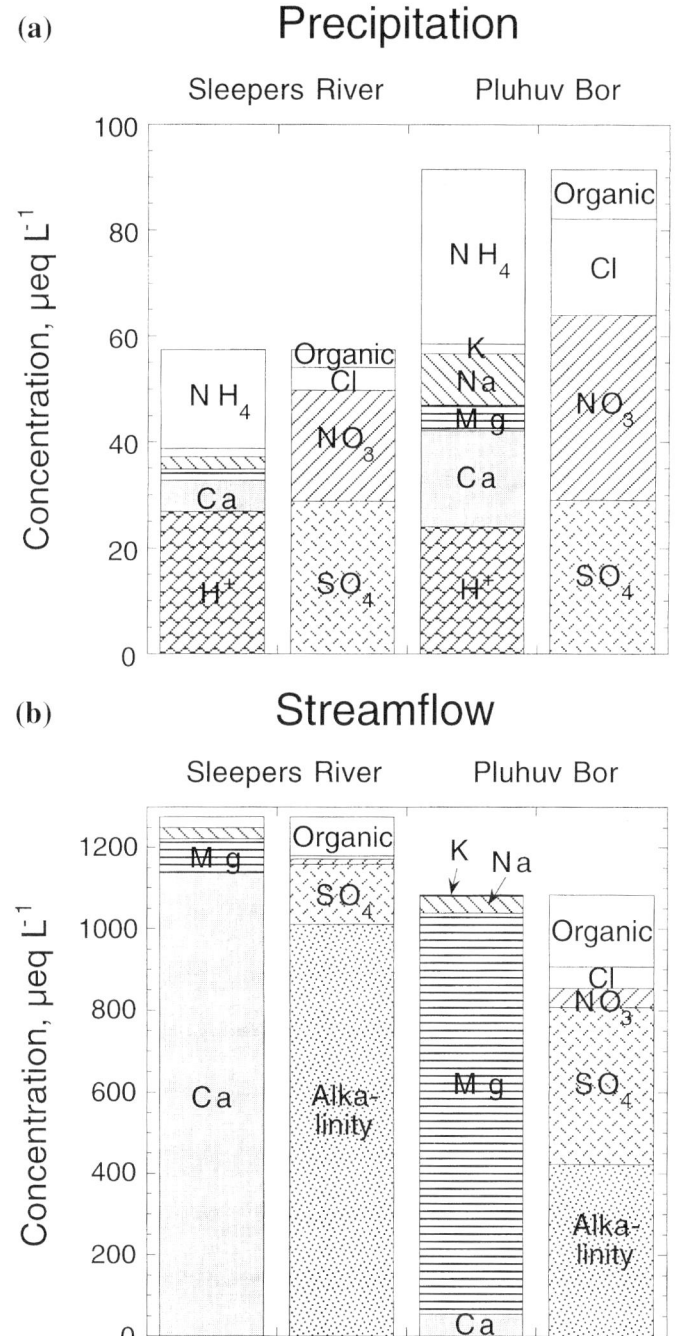

Figure 4. Major-ion chemistry of (A) precipitation and (B) streamflow of both catchments; mean annual concentrations based on water year 1999.

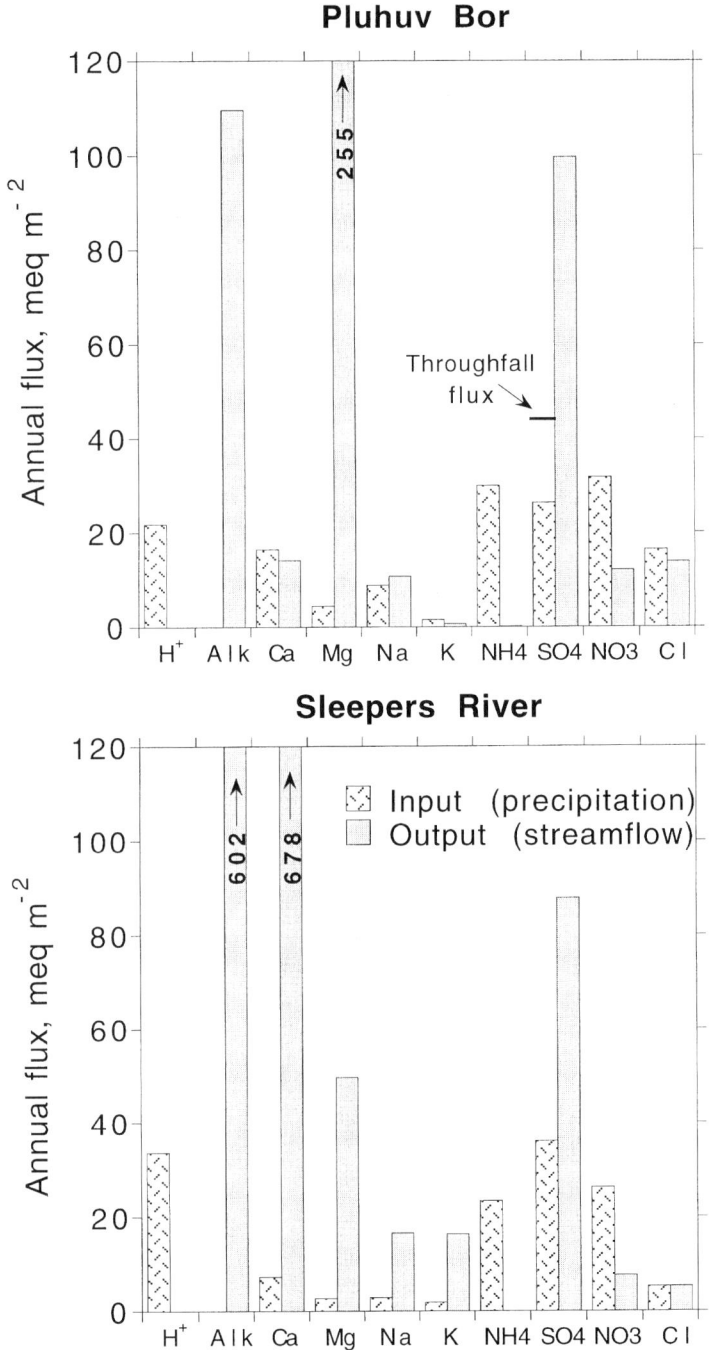

Figure 5. Chemical mass balance: Inputs in precipitation and outputs in streamflow at each site, based on water year 1999. Inputs based on bulk deposition, which may underestimate dry deposition, especially of sulfate. Sulfate deposition in throughfall is indicated for Pluhuv Bor. 'Alk' is alkalinity.

3.4. INPUT-OUTPUT BUDGETS

Because of the trends in atmospheric S deposition at Pluhuv Bor, we compare budgets for the two catchments during the most recent year with a complete data set, water year 1999. Input-output budgets (Figure 5) attest to the overwhelming ability of these catchments to buffer acidic deposition. On an annual basis, each catchment exports its dominant cation in amounts nearly two orders of magnitude greater than that cation in deposition and about one order of magnitude greater than the H^+ in deposition (charge basis). Secondary base cations at Pluhuv Bor are in approximate balance, suggesting little input from weathering and tight biogeochemical cycling of calcium (Ca) and potassium (K). At Sleepers River, in contrast, where Ca^{2+} is the dominant cation in streamwater, export of Mg is about 20 times the input, and sodium (Na^+) and K^+ also have a high net export. Net export of these secondary base cations is attributed primarily to the weathering of silicates, especially plagioclase.

The two catchments have similar mass-balance patterns for anions. In each catchment, Cl^- is in approximate mass balance, NO_3^- export is about half of NO_3^- deposition, and SO_4^{2-} export is about two times SO_4^{2-} deposition (assumed to be the sulfate flux in throughfall). Except for Cl^-, which has nearly 3 times higher flux at Pluhuv Bor, the magnitudes of the anion input/output fluxes are similar at the two sites. Contemporary S deposition is similar at the two catchments (Figure 5) despite differing trajectories of recent trends. Sulfate export exceeds input at Pluhuv Bor because past elevated S deposition is stored in catchment soil and biomass and is slowly released to surface water. In addition, there is evidence of a geologic S source at Pluhuv Bor (Krám *et al.*, 1997). At Sleepers River, young glaciated soils should have limited SO_4^{2-} adsorption capacity (Rochelle *et al.*, 1987), but most atmospheric S input appears to be incorporated in soil organic matter (J. Shanley *et al.*, unpublished data). In the absence of a strong atmospheric S deposition trend, soil incorporation of atmospheric S is probably balanced by mineralization of soil organic S. With some allowance for unmeasured dry deposition, the amount of net S export represents geologic S.

3.5. TRENDS IN SOLUTE CONCENTRATIONS

Since atmospheric monitoring began in 1991, Pluhuv Bor has shown a modest declining trend in NO_3^- flux and a sharp declining trend in SO_4^{2-} flux in bulk deposition. Nitrate flux has decreased from about 45 to about 35 meq m^{-2} a^{-1}, while SO_4^{2-} flux has decreased from near 80 to about 30 meq m^{-2} a^{-1} during the past decade (Figure 6). For throughfall, SO_4^{2-} flux decreased from 220 to 50 meq m^{-2} a^{-1}, suggesting an even more rapid decrease in dry deposition of S. At Sleepers River, no trends are apparent in SO_4^{2-} or NO_3^- deposition during the same period.

In streamflow, there is no trend in NO_3^- concentration at either site. At Pluhuv Bor, the decline in mean streamwater SO_4^{2-} concentration, from greater than 1000

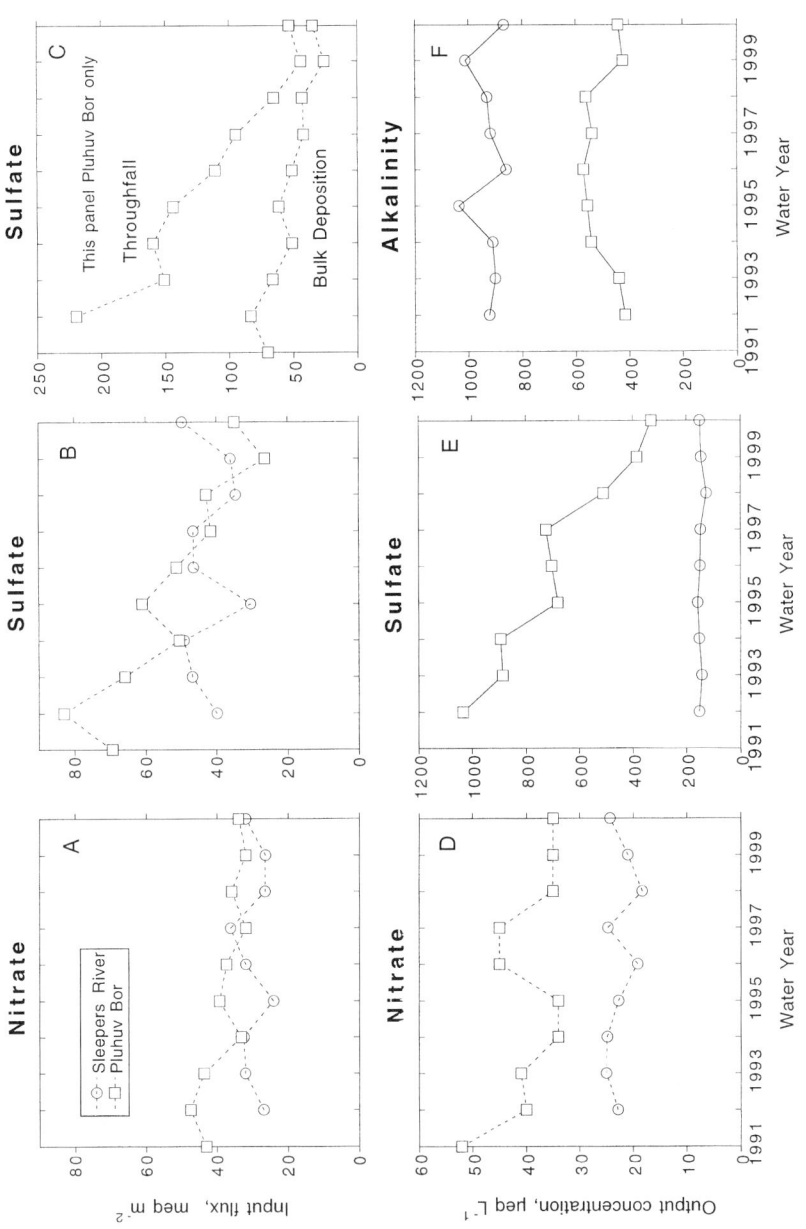

Figure 6. Chemical trends in the two catchments. Top row shows input fluxes in bulk deposition. Bottom row shows output concentrations in streamwater (concentrations rather than fluxes used for comparison purposes, to effectively normalize for interannual variation in flow). (A) Nitrate input. (B) Sulfate input. (C) Sulfate input in bulk deposition and throughfall at Pluhuv Bor. Difference attributed to dry deposition of S. (D) Nitrate output. (E) Sulfate output. (F) Alkalinity output.

μeq L^{-1} to less than 400 μeq L^{-1}, reflects the decrease in deposition. Mean annual SO$_4^{2-}$ concentration in Sleepers River streamwater is remarkably constant near 170 μeq L^{-1}. Despite the declining trend in SO$_4^{2-}$ concentration in Pluhuv Bor streamwater, there was no trend in alkalinity during the decade at either site (Figure 6).

3.6. SOLUTE CONCENTRATION VARIATIONS WITH STREAMFLOW

Base cations show the characteristic dilution with increasing streamflow in both catchments. The range of concentrations, however, is greater at Pluhuv Bor (Figure 7). Absolute concentrations are a function of weathering rates and water residence times, but greater evapotranspiration at Pluhuv Bor contributes to the higher concentrations at low flow relative to Sleepers River. A greater dilution effect from precipitation may cause the occasional lower concentrations relative to Sleepers River. The surficial aquifer is presumably less areally extensive at Pluhuv Bor than at Sleepers River, so contributions to streamflow from direct channel precipitation, near-channel saturated overland flow, and shallow subsurface flow may assume greater importance.

As mentioned previously, these dilution episodes at Pluhuv Bor occasionally nearly deplete the alkalinity in this normally well-buffered catchment. The consistent input of groundwater at Sleepers River limits alkalinity depressions to about one-third of concentrations at low flow. The nondominant base cations at each site (Ca^{2+}, Na$^+$, and K$^+$ at Pluhuv Bor and Mg^{2+}, Na$^+$, and K$^+$ at Sleepers River) dilute with increasing flow similarly to the dominant cations and alkalinity.

In contrast to base cations and alkalinity, NO$_3^-$ and DOC concentrations increase with increasing flow at each site. This is common behavior for these two solutes, which have a primary source in the forest floor from which they are flushed to streamwater during rain and snowmelt. As mentioned previously, the export of NO$_3^-$ is about the same in the two catchments. DOC export averages 44 kg ha^{-1} a^{-1} at Pluhuv Bor compared to 13 kg ha^{-1} a^{-1} at Sleepers River. The spruce forest at Pluhuv Bor likely produces more DOC than the deciduous forest at Sleepers River.

Sulfate behavior displays the greatest contrast between the two catchments. At Sleepers River, SO$_4^{2-}$ dilutes with increasing flow much like base cations and alkalinity (Figure 7). Sulfate and base cation concentrations co-vary at Sleepers River because most of the SO$_4^{2-}$ in Sleepers River streamwater derives from weathering of sulfide minerals in the bedrock and till (Bailey *et al.*, in press). Sulfate does not behave as a strong acid anion because it is balanced by base cations. At Pluhuv Bor, by contrast, a large amount of anthropogenic SO$_4^{2-}$ has been stored in the surface soil horizons, probably by adsorption. During high-flow events, this soil SO$_4^{2-}$ is flushed to the stream. As at Sleepers River, there is also a deeper groundwater source of S at Pluhuv Bor. Although there is regional evidence of anthropogenic S leaking to the groundwater system (Lischeid *et al.*, 2002), the convergence of

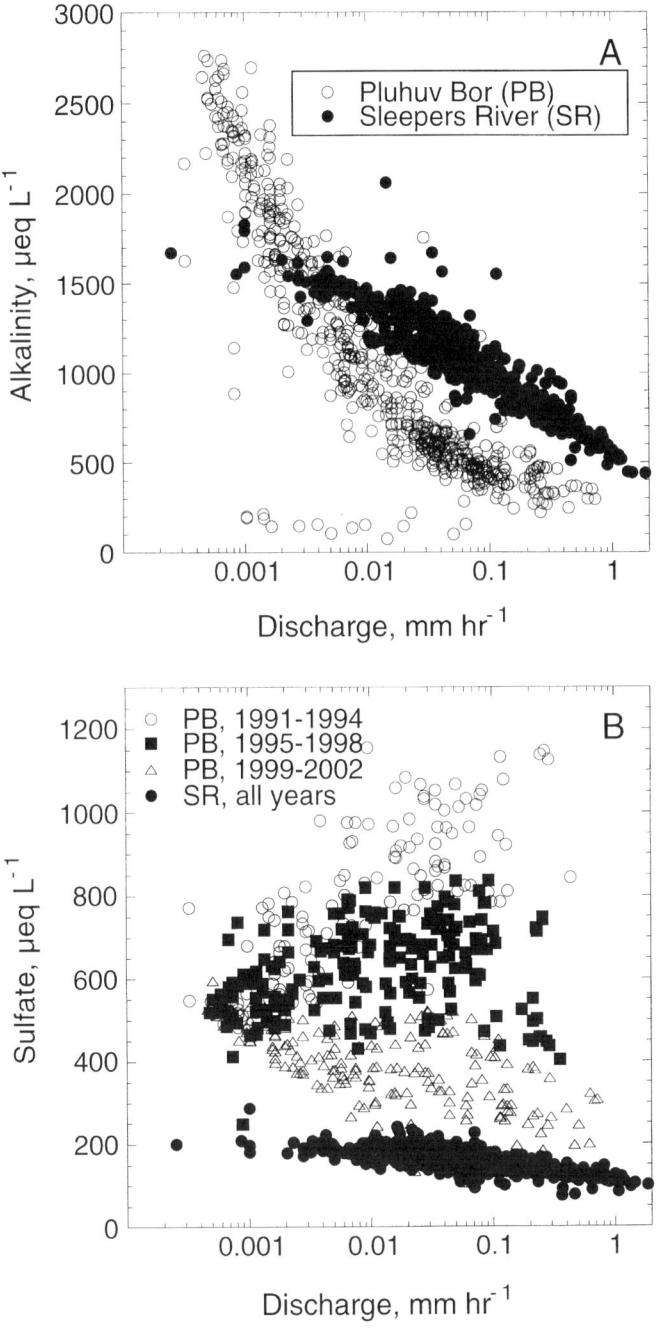

Figure 7. Concentration vs. discharge for (A) alkalinity, and (B) sulfate. Because of the strong declining trend in sulfate with time at Pluhuv Bor, sulfate data were split into three time periods, showing a shifting relation.

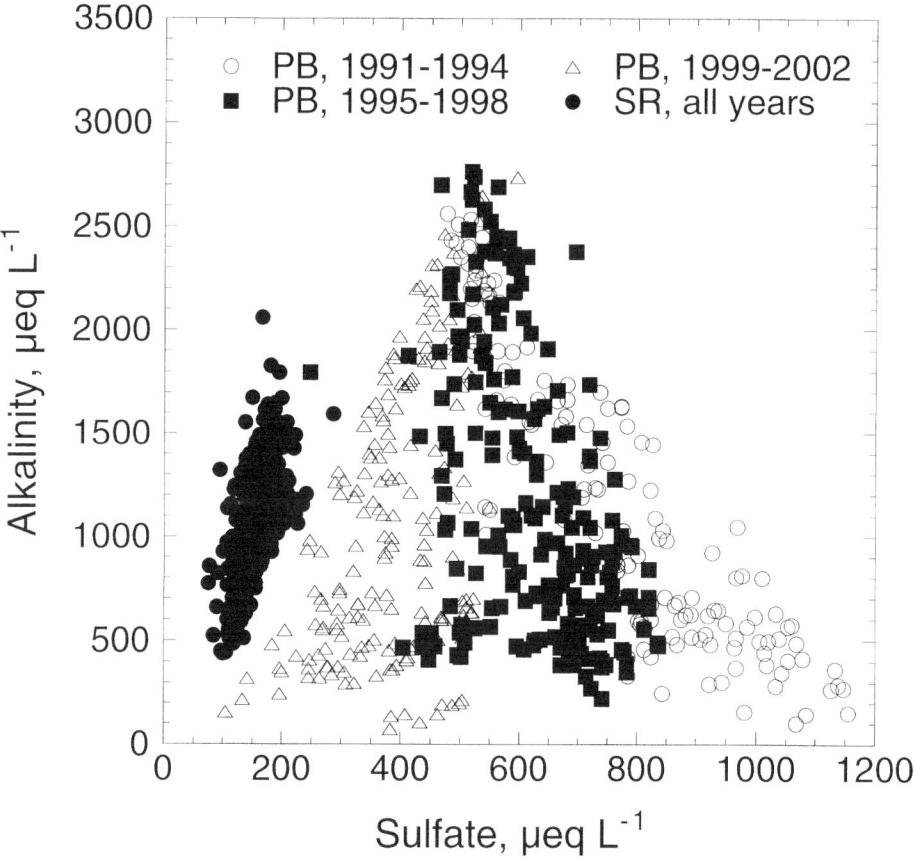

Figure 8. Sulfate vs. alkalinity for both sites. For Pluhuv Bor, data are split into three time periods.

low-flow SO_4^{2-} concentrations to a single value throughout a decade of sharply decreasing atmospheric S deposition argues for a geologic source. There is also independent evidence that this S may be geologic (Krám *et al.*, 1997). Streamwater SO_4^{2-} at Pluhuv Bor probably is a mixture of the deep and shallow sources, but because of the strong declining trend in SO_4^{2-} deposition over time, the soil SO_4^{2-} pool has become progressively depleted. The relation of SO_4^{2-} to flow therefore has shifted from positive in the early 1990s to negative at present, as soil-water SO_4^{2-} has shifted from higher to lower concentrations than SO_4^{2-} in groundwater, which remained constant through the period (Figure 7).

Similar to the relation of SO_4^{2-} to discharge at Pluhuv Bor, the relation between SO_4^{2-} and alkalinity has shifted from a positive to an inverse correlation during the same period (Figure 8). This unusual triangular distribution is formed by a series of mixing lines that move clockwise with time as soil S is depleted. The apex of the triangle represents the fixed endpoint SO_4^{2-} and alkalinity concentrations at base flow. Interestingly, the base of the triangle is horizontal, indicating that alkalinity depressions to near 100 μeq L^{-1} have occurred regularly throughout the period. These excursions are caused by dilution at high flows and are independent of the SO_4^{2-} dynamics. Although SO_4^{2-} was the acidifying compound in shallow soil water (e.g., decreasing soil base saturation), all SO_4^{2-} was neutralized completely within the soils and affected only the ionic strength of streamwater, not its acid/base character; otherwise, alkalinity depressions would be less severe at lower SO_4^{2-} concentrations. Alkalinity values at a given discharge were almost identical in the 'high SO_4^{2-}' period in the early 1990s and the 'low SO_4^{2-}' period in the early 2000s.

3.7. FOREST HEALTH

If geology and forest type were similar at Sleepers River and Pluhuv Bor, it would be ideal to compare forest health under two different S loading regimes in similarly well-buffered catchments. However, outright comparison of forest health is confounded by differences in forest type, forest management practice, and geology. The method of forest health assessment also differed at each site, making comparison difficult. Despite the abundance of total cations from the weathering of serpentinite at Pluhuv Bor, the paucity of calcium is known to cause forest nutritional deficiencies on serpentinite landscapes (Roberts and Proctor, 1992).

The Norway spruce plantation at Pluhuv Bor exhibited opposing trends in two different indicators of forest health, as measured in 1994, 1996, and 2002 (P. Moravčík, Institute for Forest Ecosystem Research, Davle, unpublished data). The percentage of trees that exhibited some browning of needles decreased from 96% in 1994 to 50% in 2002. In contrast, average whole crown defoliation increased monotonically from 28 to 37% (19 to 27% if only the upper one-third of the crown is considered) during the period. As mentioned previously, Pluhuv Bor is outside the area of heaviest S deposition, and even forests on poorly-buffered lithologies in the Slavkov Forest have not experienced significant dieback. It is not clear why defoliation increased during a period when air quality was improving and SO_4^{2-} concentrations in soil water were decreasing.

Health of the northern hardwood forest at Sleepers River was assessed once in 2001 (R. Hallett, U.S. Forest Service, unpublished data). Percent branch dieback of dominant and codominant trees averaged from 5 to 15% in three separate stands. Crown vigor classification averaged 'healthy' in one stand, 'slight decline' in another, and intermediate between these two classes in the third stand. At both of these sites, weathering in the soil zone has depleted the buffering capacity, leading to surficial soil and soil solution pH less than 5 (Sleepers River: David and Lawrence, 1996; Pluhuv Bor: Krám *et al.*, 1997). Acidic soil conditions, possibly exacerbated by elevated S deposition, may be contributing to subpar forest health at both sites.

4. Discussion

Catchments with alkaline-weathering lithologies have received less study than their acid-sensitive counterparts. This study presents an opportunity to compare two of the best-studied well-buffered catchments. Because the Czech catchment has received higher acid loadings, we can evaluate the common assumption that easily weathered alkaline substrates provide 'infinite' buffering capacity. By comparing these two catchments, we have attempted to evaluate whether differing S loading has affected natural biogeochemical cycles in different ways.

Although present-day SO_4^{2-} export is comparable from the two catchments, the S source is primarily atmospheric at Pluhuv Bor (Krám et al., 1997), whereas half or more of SO_4^{2-} comes from sulfide weathering at Sleepers River (J. Shanley et al., unpublished data). However, because of the lower water flux, SO_4^{2-} concentrations average 3 times higher at Pluhuv Bor, and the range of SO_4^{2-} concentrations, even relative to the mean, is far greater. The high SO_4^{2-} loadings have acidified the soils and may be detrimental to forest health through both soil and atmospheric pathways. However, the deep alkalinity depressions at Pluhuv Bor result from hydrologic factors – namely, dilution by rain and snowmelt following shallow flow paths at high flow. The downward trend in SO_4^{2-} concentrations through time has been largely incidental to the acid/base status of the stream.

5. Conclusions

Pluhuv Bor in the Czech Republic and Sleepers River in Vermont, U.S.A. are both well-buffered catchments that currently receive mildly acidic deposition, but Pluhuv Bor has received much higher acid loadings in the recent past. Pluhuv Bor has a higher base flow alkalinity than Sleepers River but experiences episodic decreases in alkalinity during high flows, sometimes to 100 μeq L^{-1}, compared to 400 μeq L^{-1} at Sleepers River. Surprisingly, these decreases are not related to the S deposition, but rather to simple dilution. At Sleepers River, alkalinity depressions also are attributed to dilution, but they are less severe because of the sizeable groundwater contribution to streamflow. In well-buffered catchments such as these, episodic acidification will occur only by dilution of streamflow with a large volume of event water following shallow pathways where buffering is negligible.

Acknowledgements

This study was funded by the Grant Agency of the Czech Republic (Grant No. 205/01/1426) and by the U.S Geological Survey Water, Energy, and Biogeochemical Budgets (WEBB) Program. We appreciate the valuable reviews of Owen Bricker, Doug Burns, and an anonymous reviewer. Rich Hallett kindly provided forest health data.

References

Bailey, S. W., Mayer, B. and Mitchell, M. J.: 'The influence of mineral weathering on drainage water sulfate in Vermont and New Hampshire', *Hydrological Processes* (in press).

Czech Hydrometeorological Institute: 2001, *Air pollution in the Czech Republic in 2000*, Czech, Hydrometeorological Institute, Prague, Czech Republic, pp. 213.

David, M. B. and Lawrence, G. B.: 1996, 'Soil and soil solution chemistry under red spruce stands across the northeastern U.S.A.', *Soil Sci.* **161**, 314–328.

Driscoll, C. T., Lawrence, G. B., Bulger, A. J., Butler, T. J., Cronan, C. S., Eagar, C., Lambert, K. F., Likens, G. E., Stoddard, J. L. and Weathers, K. C.: 2001, 'Acidic deposition in the northeastern United States: Sources and inputs, ecosystem effects, and management strategies', *BioScience* **51**, 180–198.

Dunne, T. and Black, R. D.: 1970, 'An experimental investigation of runoff production in permeable soils', *Water Resour. Res.* **6**, 478–490.

Fottová, D. and Skořepová, I.: 1998, 'Changes in mass element fluxes and their importance for critical loads: GEOMON network, Czech Republic', *Water, Air, Soil Pollut.* **105**, 365–376.

Hruška, J., Moldan, F. and Krám, P.: 2002, 'Recovery from acidification in central Europe – Observed and predicted changes of soil and streamwater chemistry in the Lysina catchment, Czech Republic', *Environ. Pollut.* **120**, 261–274.

Kopáček, J., Veselý, J. and Stuchlík, E.: 2001, 'Sulphur and nitrogen fluxes and budgets in the Bohemian Forest and Tatra Mountains during the Industrial Revolution (1850–2000)', *Hydrol. Earth Sys. Sci.* **5**, 391–405.

Krám, P. and Hruška, J.: 1994, 'Influence of bedrock geology on elemental fluxes in two forested catchments affected by high acidic deposition', *Appl. Hydrogeol.* **2**, 50–58.

Krám, P., Hruška, J., Wenner, B. S., Driscoll, C. T. and Johnson, C. E.: 1997, 'The biogeochemistry of basic cations in two acid-impacted forest catchments with contrasting lithology', *Biogeochemistry* **37**, 173–202.

Lischeid, G., Böttcher, H., Krám, P. and Hruška, J.: 2002, 'Comparative analysis of hydrochemical time series of adjacent catchments by process-based and data-oriented modelling', in G. H. Schmitz (ed.), *Water Resources and Environmental Research, Vol. 2, Matter and Particle Transport in Surface and Subsurface Flow*, Technical University Dresden, Germany, pp. 237–241.

Novák, M., Bottrell, S. H., Fottová, D., Buzek, F., Groscheová, H. and Žák, K.: 1996, 'Sulfur isotope signals in forest soils in Central Europe along an air pollution gradient', *Environ. Sci. Technol.* **30**, 3473–3476.

Roberts, B. A. and Proctor, J. (eds): 1992, *The Ecology of Areas with Serpentinized Rocks: A World View*, Geobotany 17, Kluwer Academic Publishers, Dordrecht, pp. 427.

Rochelle, B. P., Church, M. R., and David, M. B.: 1987, 'Sulfur retention at intensively-studied watersheds in the U.S. and Canada', *Water, Air, Soil Pollut.* **33**, 73–83.

Shanley, J. B. and Chalmers, A. T.: 1999, 'The effect of frozen soil on snowmelt runoff at Sleepers River, Vermont', *Hydrol. Process.* **13**, 1843–1857.

Shanley, J. B., Schuster, P. F., Reddy, M. M., Roth, D. A., Taylor, H. E. and Aiken, G. R.: 2002a, 'Mercury on the move during snowmelt in Vermont', *EOS, Trans. Am. Geophys. Union* **83**, 45–48.

Shanley, J. B., Kendall, C., Smith, T. E., Wolock, D. M. and McDonnell, J. J.: 2002b, 'Controls on old and new water contributions to streamflow in some nested catchments in Vermont, U.S.A.', *Hydrol. Process.* **16**, 589–609.

U.S. EPA: 2000, *National Air Pollutant Emission Trends, 1900–1998*, U.S. Environmental Protection Agency, EPA-454/R-00-002, pp. 238.

VALIDATING A NEW MODEL FOR N SEQUESTRATION IN FOREST SOIL ORGANIC MATTER

BJÖRN BERG[1]* and NANCY DISE[2]**

[1] Lehrstuhl für Bodenökologie, BITÖK, Postfach 101251, Universität Bayreuth, Dr-Hans-Frisch-Strasse 1-3, DE-954 48, Bayreuth, Germany; [2] Department of Earth Sciences, The Open University, Milton Keynes, MK7 6AA, U.K.; ** Present address: Department of Biology, Villanova University, Villanova, Pennsylvania 19085, U.S.A.
(* author for correspondence, e-mail: bjoern.berg@bitoek.uni-bayreuth.de, phone: +49 (171) 858 8212, fax: +49 (921) 555799)

(Received 20 August 2002; accepted 7 April 2003)

Abstract. A conceptual model for N sequestration into the terrestrial nitrogen (N) sink is presented. The model uses foliar litter-fall data, limit values for litter decomposition, and calculated N concentration at the limit value (N_{limit}), giving the N concentration in the hypothesized stable remains. The N_{limit} values were determined extrapolating a linear relationship between accumulated litter mass loss and the increasing litter N concentration to the limit value. The sequestration rates for N in boreal forest humus were calculated and validated for a Scots pine (*Pinus sylvestris* L.) monocultural stand and mixed stands with Scots pine, Norway spruce (*Picea abies* L.), and silver birch (*Betula pendula* L.). The calculated stable N fraction was compared to actually measured amounts of N in humus layers that started to accumulate 2984, 2081, 1106, and 120 yr BP. Sequestration rates of N were measured to be 0.255, 0.221, 0.147, and 0.168 g m^{-2} yr^{-1} and modeled to be 0.204, 0.207, 0.190, and 0.190 g m^{-2} yr^{-1}, respectively, with missing fractions being 11.0, 1.5, 30.8, and 13.3%, respectively. The more N-rich the litter, the larger was the N fraction sequestered. This was found for experimental Scots pine needle litter ($n = 6$) and for 53 decomposition studies, encompassing seven litter species. The amounts of N sequestered annually ranged from ca. 1–2 kg ha^{-1} yr^{-1} under nutrient-poor boreal conditions to about 30 kg ha^{-1} yr^{-1} in temperate, more nutrient-rich forests.

Keywords: C sequestration, C sink, humus, model, N sequestration, N sink, soil organic matter

1. Introduction

The quantification of terrestrial sources and sinks for carbon dioxide and N-based greenhouse gases is one of the most important tasks facing environmental scientists today. Central to this is the determination of mechanisms for carbon (C) and nitrogen (N) sequestration in the soil organic matter (SOM). The accumulation rates of SOM layers depend not only on the amount of litterfall and its quality but also on the completeness of its decomposition. Furthermore, the accumulation of SOM is a slow process that spans generations of scientists, thereby causing continuity problems in studying the buildup as well as the mechanisms controlling it.

Numerous attempts have been made to use mathematical models to describe organic matter decomposition and thus SOM accumulation rates. Field studies, using litterbags have confirmed that the decay rate, which instantaneously is high, decreases over time (e.g., Meentemeyer, 1978). If the decay rate approaches zero before all of the organic matter has been decomposed, the accumulated mass loss will asymptotically approach a limit value. The limit-value concept is based on the degradation of lignin in late decomposition stages, when its degradation regulates the decomposition of the whole litter. Nitrogen has a suppressing effect on microbial lignin degradation, and the higher its concentration the stronger the effect (Berg and McClaugherty, 2003). Low-molecular weight N compounds may react with lignin degradation products chemically to form more recalcitrant N compounds (Nömmik and Vahtras, 1982), an hypothesis that has recently been confirmed (Spaccini et al., 2002). These effects on a higher resolution level may be seen as a negative relationship between litter N concentration and the estimated extent of litter decomposition (Berg and McClaugherty, 2003).

We define the term 'limit value' as the accumulated mass of decomposing litter that has been lost (in %) at the time when the decomposition rate is zero. It can be calculated as (Berg and Ekbohm, 1991) as

$$m.l. = m(1 - e^{-kt/m}), \tag{1}$$

where $m.l.$ is litter mass loss, t is time in days, m is the asymptotic level that the accumulated mass loss will ultimately reach, and the k is the decomposition rate at the beginning of the decay. The recalcitrant 'remaining fraction' (Berg et al., 2001) is calculated as

$$\text{Remaining fraction} = (100 - \text{limit value})/100. \tag{2}$$

Asymptotic models for mass loss during decay have been proposed by several scientists. Such models have the advantage that the recalcitrant or extremely slowly decomposing part may be quantified. Using data for litter incubated in the field, Berg and Johansson (1998) calculated about 130 limit values for several litter types. Limit values have been significantly related to the nutrients N, Mn, and Ca (Berg, 2000), which influence both the activity and composition of the lignolytic microbial community (Eriksson et al., 1990; Hatakka, 2001).

The remaining fraction for a particular forest litter type, multiplied by the foliar litterfall (in unit of kg ha^{-1} yr^{-1}) allows the annual accumulation rate of SOM to be calculated. Given an estimate for the accumulated litterfall for a particular forest since the time it colonized a new area, the total accumulation of SOM over that time can be calculated. Finally, it is possible to estimate the total accumulation of selected nutrients in SOM, such as C, N, P, or trace metals, if the concentration of that component in the remaining fraction is known. This may be validated by comparing the estimate with actually measured amounts in the SOM.

In decomposing litter, the N concentration increases linearly with litter mass loss, with R^2 values normally well above 0.9 (Berg et al., 1999; Figure 1). Different litter types show different relationships, with the slopes depending on such factors

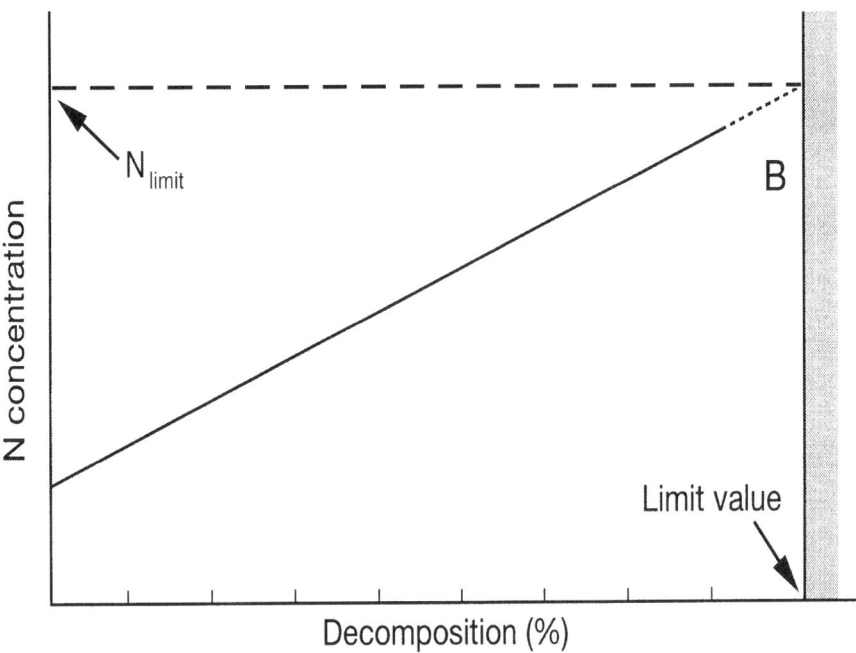

Figure 1. The increasing N concentration in decomposing litter is linear to litter mass loss and the N concentration at the limit value (N_{limit}) can be estimated through extrapolation. The R^2 value for the relationship is normally > 0.9. From Berg *et al.* (1999).

as species and its initial N concentration (Berg *et al.*, 1999). The increase continues until decomposition reaches a stage at which it is extremely slow or – at the limit value – appears to cease completely (Berg *et al.*, 1999).

By extrapolation to the limit value, the N concentration at the limit value (N_{limit}) is calculated as

$$N_{limit} = k * m + N_{init}, \quad (3)$$

where k, the slope of the line, is a unique coefficient for each set of data, and N_{init} is the initial litter N concentration (cf. Figure 1).

We define the capacity of a given litter type to accumulate N (N_{capac}; in mg g^{-1}) as the amount of N remaining at the limit value after decomposing an initial mass of 1 g of litter;

$$N_{capac} = N_{limit}(100 - m)/100 . \quad (4)$$

Using the above equations, data sets from decomposing litter from boreal forests were used to calculate the expected amount of N bound in SOM accumulated for four different periods. The calculated values were compared to the measured amounts of N in SOM accumulated over the same periods ranging from 120 to 2984 yr.

The aims of this synthesis paper are to validate a model for N sequestration on a long-term basis and to show that decomposing foliar litter may store greater quantities of N in humus as the initial N concentration increases. To reach these aims, we have combined a conceptual model for the accumulation of SOM with a model for calculating humus-N concentrations. The model for SOM accumulation has been developed and validated in previous studies (Berg *et al.*, 1999; 2001).

2. Methods

2.1. THE MODEL FOR QUANTIFYING N SEQUESTRATION – OVERVIEW

In a first step the limit value (m) for decomposing litter is estimated using Equation (1). The decomposition continues until this limit value is reached, at which time it comes to a stop or continues at an extremely low rate, estimated to be less than 1% in 30–300 yr (cf. Couteaux *et al.*, 1998). The causal relationships underlying this model are described in detail by Berg *et al.* (1996) and Berg and McClaugherty (2003).

The concentration of N increases with decomposition until the limit value is reached when the increase stops and the N concentration becomes the same as that in humus (i.e., F and H layers formed from the very same litter) (Berg *et al.*, 1999). This increase in N concentration can be described empirically as a straight line (Aber and Melillo, 1982; Staaf and Berg, 1982) with R^2 values normally well above 0.9 (Berg *et al.*, 1999). For calculating the N concentration at the limit value, we used Equation (3). The absolute amount of N that remained in the litter was estimated using Equation (4), thus giving the remaining amount of N from initially 1 g of litter. The sequestration of N is then estimated simply by multiplying by litterfall. The recalcitrance of the C and N in the remains at the limit value have been discussed in earlier papers (Berg *et al.*, 1999, 2001) and by Berg and McClaugherty (2003).

2.2. THE SITES USED FOR VALIDATION

2.2.1. *Site Jädraås*
Site Jädraås in Central Sweden, about 200 km north to north-west of Stockholm has a Scots pine (*Pinus sylvestris* L.) monoculture, about 150 yr old (in 2000) located at 60°49′N; 16°30′E at an elevation of 185 m a.s.l. The forest is situated on a very nutrient-poor sediment soil. Mean annual precipitation is 609 mm and the long-term mean annual temperature is 3.8 °C. The ground vegetation is composed mainly of blueberry (*Vaccinium myrtillus*), lingonberry (*Vaccinium vitis-idaea*), heather (*Calluna vulgaris*), mosses, and lichens. The humus form is mor, the soil profile is a podzol, and the soil texture is fine sand.

TABLE I

Estimated and observed accumulation of soil organic matter and N in the forest floor of known age in boreal forests in Swedish Lapland and central Sweden. The stands in north Sweden were located on islands and values are averages. Data are from Wardle et al. (1997) and Berg et al. (1995b, 2001). Standard errors are given within parentheses. The site in north Sweden has today no N deposition and data for the Jädraås site originated from 1974, when N deposition was negligible

	North Sweden – islands			Jädraås
	$n = 14$	$n = 24$	$n = 12$	$n = 1$
Mean stand age (yr)	2984 ($\pm$340)	2081 ($\pm$424)	1106($\pm$495)	120
Estimated litterfall[c] (kg m^{-2} yr^{-1})	0.081	0.081	0.081	15.2
Average limit values[a] (%)				
Scots pine (SE)	86.4 (3.85)	86.4 (3.85)	86.4 (3.85)	89.0[b](3.63)
Norway spruce(SE)	70.3 (13.1)	70.3 (13.1)	70.3 (13.1)	–
Silver birch(SE)	84.7 (7.14)	84.7 (7.14)	84.7 (7.14)	–
Humus				
Measured forest floor mass (kg m^{-2})[e]	49.08	34.62	14.33	1.54
Modeled forest floor mass (kg m^{-2})	47.2	31.0	14.56	1.67
Missing fraction (%)[d]	3.8	10.5	1.6	8.4
Nitrogen				
Estimated N conc at limit value (mg g^{-1})				
Scots pine	12.4	12.4	12.4	12.8
Norway spruce	12.76	12.76	12.76	–
Silver birch	22.71	22.71	22.71	–
Measured N in SOM (g m^{-2})	761	460	163	18.8
Estimated N in SOM (g m^{-2})	677.3	453.2	213.2	21.4
Missing fraction (%)	11.0	1.5	30.8	13.3

[a] For the north-Swedish sites average limit values were estimated from existing limit values for Scots pine ($n = 12$), Norway spruce ($n = 5$), at sites in Sweden north of ca. 59°N and available Scandinavian ones for silver birch leaf litter ($n = 3$; data from Berg and Johansson, 1998).
[b] Average limit value for needle litter decomposition at site Jädraås (Berg et al., 1995b).
[c] Litterfall for the Scots pine stand at Jädraås was estimated over a stand age and calculated using Equation (5) as described by Berg et al. (1995b). Litterfall at Hornavan – Uddjaur was interpolated vs. latitude using available Scandinavian data for Scots pine and Norway spruce forests between 59°N (about north of the line Oslo-Stockholm-Helsinki) and 67°N (cf. Berg et al., 2001).
[d] Missing fractions as calculated by Berg et al. (2001) to be 16, 17 and 6% for the 2984, 2081 and 1106 yr old stands. The present calculation considered an equal occurrence of all three species and took into account the fact that Norway spruce invaded the area about 2000 years ago.
[e] Humus mass is here calculated from Wardle et al. (1997). The total C per meter square on the '2984 yr' islands was 26.8 ± 8.8 kg, from which biomass with 2.25 kg C m^{-2} was subtracted. For the '2081 yr' islands, the figures were 20.3 ± 8.2 from which 2.99 kg C m^{-2} was subtracted as biomass-C, and for the '1106 yr' islands, the numbers were 10.7 ± 4.1 kg C m^{-2} from which 3.99 kg C m^{-2} were subtracted.

2.2.2. *Site Hornavan*

Site Hornavan, in Swedish Lapland. This site is made up of about 50 small islands in the two lakes of Uddjaure and Hornavan, a lake system located in the area defined by 65°55′–66°09′N and 17°43′–17°53′E, north-west and south-east of the village of Arjeplog, and about 280 km north-west of the city of Umeå, in northern Sweden. The lake system has an elevation of 425 m a.s.l. and the area is in a part of Swedish Lapland that has remained remote up to the present time. The location on small islands ranging in area from 0.2 to 15 ha in the relatively large lakes has protected the forests and the humus on each island from both forest practice and from wildfires (Wardle *et al.*, 1997). In their investigation Wardle *et al.* (1997) divided the islands into three groups; *viz.* those smaller than 0.1 ha, those between 0.1 and 1.0 ha and those larger than 1.0 ha. The average time since the last wildfire was 2984, 2081, and 1106 yr respectively, the smallest islands thus being more protected from fire (Wardle *et al.*, 1997). All islands were located on till, the smallest ones with the oldest systems were characterized by Norway spruce (*Picea abies* L.) and some silver birch, relatively well developed communities of *Empetrum nigrum* and very thick (up to 1.4 m) humus layers. The medium-sized islands had more mixed cultures of Norway spruce and Scots pine, less developed communities of *Empetrum nigrum* and, on average, less thick humus layers (Table I). Finally, the youngest systems (on the largest islands) were dominated by Scots pine and had thinner humus layers than the other two groups. Site data are from G. Hörnberg (pers. comm.) and from Wardle *et al.* (1997).

2.3. Validation of the model

The model was validated using the stands above with a known site history since the last wildfire. For the Jädraås site the data set used originated from 1974 when N deposition was negligible and no nitrification was found. In the remote part of Swedish Lapland where site Hornavan is located, N deposition is negligible.

For the Hornavan site, average limit values were estimated by using all available limit values for decomposing Norway spruce and Scots pine needle litter from northern Sweden (north of about 59°N) and all available data for silver birch (Berg and Johansson, 1998). The mean (± standard error) values are: for Scots pine, $86.4 \pm 3.9\%$ ($n = 12$); for Norway spruce, $70.3 \pm 13.1\%$ ($n = 5$); and for silver birch, $84.7 \pm 7.1\%$ ($n = 3$) (Table I). The values were weighted for the species composition of the three groups of stands with different ages at Hornavan. For the oldest stands the weighted limit value is 80.5, for the intermediate-aged stands it is 81.7, and for the youngest stands it is 83.9%. These different values reflect the fact that the oldest stands (on the 14 smallest islands) are dominated by Norway spruce, with a low limit value, and the 12 youngest stands (on the largest islands) are dominated by Scots pine with a high limit value. Norway spruce invaded this area about 2000 yr ago, and the weighted values were used for the period after that year. For the period before, the species-specific limit values were used.

The N_{limit} value was estimated using Equation (3). The value for N_{init} used was the average initial N concentration for stands north of 59°'N, including site Jädraås. Mean (± standard error) N_{init} for Scots pine was 3.94 ± 0.70 mg g^{-1} ($n = 12$), for Norway spruce was 5.02 ± 0.80 mg g^{-1} ($n = 5$), and for silver birch was 10.17 ± 2.43 mg g^{-1} ($n = 3$). The coefficients (k) are species-specific and were calculated as averages for each species and taken from an earlier study (Berg et al., 1999); for Scots pine, $k = 0.0979$, for Norway spruce, $k = 0.1101$, and for silver birch, $k = 0.1476$. These values were used (in Equation (3)) to calculate average N_{limit} values which are 12.40 mg g^{-1} N, 12.76 mg g^{-1} N, and 22.71 mg g^{-1} N, for Scots pine, Norway spruce, and silver birch, respectively (Table I). The N_{limit} values were weighted relative to occurrence of the three species. For the oldest stands at Hornavan, the weighted N_{limit} value was 14.34 mg g^{-1}, for the intermediate-aged stands, it was 14.62 mg g^{-1}, and for the youngest stands, it was 14.64 mg g^{-1}, reflecting a slightly higher relative frequency of silver birch in the two younger stands.

For the Hornavan islands, litterfall was estimated using a function relating litterfall to latitude. Because these stands are very old (individual trees would be replaced over the years, but not the entire stand at once), we can ignore the fact that litterfall in the first 20 yr (approximately) of a forest's existence is much lower than litterfall at stand maturity, which we could not ignore in the Jädraås calculations (below). We thus only need to estimate annual litterfall at maturity to calculate mean annual litterfall. To do this, we use a relationship between annual litterfall at maturity and latitude, based on data from 39 sites in Scandinavia (6 for Norway spruce and 33 for Scots pine) located between 67 and 59°N (Berg et al., 1999b, c):

$$M.a.l.f. = 10782.3(558.1) - 150.8(34.9) \times \text{Latitude}, \quad (5)$$

where $M.a.l.f.$ stands for mean annual foliar litterfall (kg ha^{-1}) at stand maturity and latitude is given in decimal degrees (standard errors in parentheses; $R^2 = 0.336$). We assumed that the mean annual litterfall for silver birch was similar to that of the two conifers (B. Berg, unpubl. data), and thus we used the same relationship for all three major tree species found on the islands. For the Hornavan forest stands, the average latitude of 66°10'N gives a mean annual litterfall of 813 kg ha^{-1} (Table I). This equation gives an estimated $M.a.l.f.$ for Jädraås of 1635 kg ha^{-1} a^{-1}, in good agreement with the value of 1620 kg ha^{-1} a^{-1} at maturity determined by the serial approximation method described by Berg et al. (1995b).

The annual litterfall values were multiplied by the number of years for the different time periods, giving the accumulated litterfall. In the 2984 yr old stands (Hornavan) the accumulated foliar litterfall is 242 kg m^{-2} (0.081 kg m^{-2} a^{-1} × 2984 yr), for the 2081 yr stands it is 169 kg m^{-2}, for the 1106 yr old stands it is 89.6 kg m^{-2} (Table I).

For site Jädraås, limit values were taken from Berg et al. (1995b) and were the average from nine sets of Jädraås Scots pine litter with the average limit value of 89.0% (standard error = 3.63) (Table I). This value was multiplied by the estimated

TABLE II

Modeled increasing C and N sequestration rates for pine species (mainly Scots pine) under different climate conditions as indexed by actual evapotranspiration (AET). Equations for calculating the different steps are given in the Methods Section

AET (mm)	Litterfall (kg ha^{-1})	Init N conc. (mg g^{-1})	Limit value (%)	Recalcitrant fraction (–)	N conc. at limit value (mg g^{-1})	Sequestration rate C (kg ha^{-1} yr^{-1})	Sequestration rate N (kg ha^{-1} yr^{-1})
400	396	3.5	83.0	0.170	13.3	34	0.89
450	1422	4.4	81.4	0.186	13.9	132	3.7
500	2448	5.3	79.8	0.202	14.6	247	7.2
550	3475	6.2	78.2	0.218	15.4	379	11.7
600	4501	7.1	76.5	0.235	16.1	529	17.0
650	5527	8.0	74.9	0.251	16.8	694	23.3

accumulated litterfall (15.2 kg m^{-2}) over 120 yr to give an estimated forest floor mass of 1.67 kg m^{-2} (Berg et al., 1995b) (Table II). The N_{limit} value was calculated using Equation (3) giving a value of 12.8 mg g^{-1} (Table I) (Berg et al., 1999).

2.4. PREDICTION OF C AND N SEQUESTRATION IN SCOTS PINE FORESTS UNDER DIFFERENT CLIMATES

For this prediction, we used the climate index actual evapotranspiration (AET) to calculate litterfall, initial litter N concentration, and limit values (Table II). Annual pine needle litterfall has been related to climate as indexed by actual evapotranspiration (AET) (Berg and Meentemeyer, 2001). We used a linear relationship developed by them for the region of northern Europe, with an AET range covered by the function;

$$M.a.l.f. = -7814.2 + 20.525 \times AET, \qquad (6)$$

where $M.a.l.f.$ is in kg ha^{-1} and AET in mm.

Nitrogen concentration in foliar litter has been positively related to climate. For Scots pine needle litter, Berg et al. (1995a) developed a relationship covering the region from the Barents Sea to the Carpatians with increasing N concentrations from about 2.4 to about 8 mg g^{-1} that were related to *AET*. Berg and Meentemeyer (2002) included several pine species and both deciduous and coniferous litter. For Scots pine needle litter combined with litter from other pine species, they developed the linear function

$$N_{init} = -3.6479 + 0.01797 \times AET, \qquad (7)$$

($R^2 = 0.548$, $n = 40$), which we used to calculate the initial litter N concentration N_{init}.

Limit values have been related to litter chemical composition (Berg et al., 1996; Berg and McClaugherty, 2003) and by knowing the litter chemical composition we can estimate limit values (m) for decomposing Scots pine needle litter e.g. by using the N_{init} values (Equation (7)) and the function

$$m = 89.2942 - 1.7948 \times N_{init}, \quad (8)$$

with $R^2 = 0.404$ ($n = 53$; B. Berg, unpubl.) where m is in % and N_{init} is in mg g^{-1}.

To calculate the sequestration rate of C or humus, we used the relationship from Berg et al. (2001), in which the remaining fraction was multiplied by average annual foliar litterfall ($M.a.l.f.$) to give the annual accumulation ($Ann. acc.$) (kg ha^{-1});

$$Ann.\ acc. = M.a.l.f. \times (100 - m)/100. \quad (9)$$

Nitrogen concentration at the limit value (N_{limit}) was calculated with Equation (3), using a coefficient (k) that was specific for each data set. Within one species and type of stand, the variation in k is rather small, so for modeling the N_{limit} for pine, we used an average value for pine species (Berg et al., 1999). As the intercept, we used N_{init} as calculated from Equation (7).

$$N_{limit} = 0.117169 \times m + N_{init}, \quad (10)$$

where N_{init} (mg g^{-1}) was the calculated initial N concentration for different climate situations (Table II).

3. Results and Discussion

3.1. Accumulation of humus (C)

A humus accumulation model was improved (Table I) for boreal stands ranging in age from 120 to 2984 yr. Using calculated litterfall based on several sets of measurements, Berg et al. (2001) applied an average limit value (Equations (1) and (2)) for the three species and recalculated the accumulation of humus. We used the same litterfall information as Berg et al. (2001), but used the new information that Norway spruce invaded this area about 2000 yr ago (G. Hörnberg, pers. comm.). This means that for the oldest stand, we used data for just pine and birch when calculating C and N sequestration for the first 984 yr. After that, we assumed an equal occurrence of the three species Scots pine, Norway spruce, and silver birch in terms of biomass and their different limit values. We then compared our calculated accumulation to the measured values for stored humus (Table I). This approach resulted in rather small missing fractions. We define a missing fraction as the modeled mass subtracted from that actually measured divided by the measured mass. The fraction is expressed in percent. Thus the missing fraction for the humus accumulated over 2984, 2081, 1106, and 120 yr was 3.8, 10.5, 1.6, and 8.6%, respectively.

3.2. COMPARISONS OF MEASURED AND CALCULATED AMOUNTS OF SEQUESTERED N – VALIDATION OF A MODEL

The amount of N sequestered in litter at the limit value was calculated in four main steps: (i) N_{limit} values were calculated separately for Scots pine, Norway spruce, and silver birch (Equation (3); Table I); (ii) the amount of recalcitrant matter formed annually was calculated as [litterfall×(100 – limit value)/100] (in kg ha^{-1}), extrapolated to the three different periods for the three groups of stands and summarized for the youngest one (cf. Berg et al., 1995b); (iii) using the results of (i) and (ii) in Equation (4), the amount of sequestered N (N_{capac}) was calculated; and (iv) N_{capac} was compared to the measured amounts of sequestered N (Table I).

The modeled amounts of N were compared to those measured in the field during the periods 2984, 2081, 1106, and 120 yr (Table I). The modeled annual accumulation rates were 0.204, 0.207, 0.190, and 0.190 g N m^{-2} respectively, and the measured rates were 0.255, 0.221, 0.147, and 0.168 g m^{-2}, respectively. The humus accumulated over 2984 yr was modeled to have stored, in all, 677 g N m^{-2}, which gave a missing fraction of 11% of the measured amount (761 g m^{-2}). Over 2081 yr, the calculated 453 g N m^{-2} gave a missing fraction of 1.5% of the amount actually stored (460 g). Over 1106 yr, the modeled amount was 213 g m^{-2}, and the amount stored 163 g m^{-2}, giving an excess fraction of 30.8%. For the 120 yr old Scots pine stand, we obtained 21 g N m^{-2}, a fraction 13.3% too high as compared to the measured value.

3.3. IS THE MODEL VALID FOR OTHER TREE SPECIES?

Berg et al. (1999) examined eight species using 48 decomposition studies and found a highly significant correlation between N in the organic part of humus (N_{humus}) and calculated N_{limit} values for local litter decomposing in the same stands ($R^2 = 0.632$; $n = 48$; $p < 0.0001$). The investigated stands encompassed Scots pine, lodgepole pine, silver birch, Norway spruce, common oak, common alder, silver fir, and common beech, and the geographical distribution of the plots ranged from the Arctic Circle in Sweden to southern Italy. The highly significant relationship between modeled and measured humus N for eight species suggests a general relationship for litter initial N concentrations ranging from 4.0 to 12.8 mg g^{-1} (Berg et al., 1999).

3.4. HOW STABLE IS THE LONG-TERM STORED N IN HUMUS?

That the amount of N stored in humus increased with time (Table I) indicates a certain stability of the compounds holding N. The fact that there was a long-term predictability based on the limit-value concept further supports this.

The stability of stored humus and humus N is in part dependent on the composition and activity of the microbial community and factors controlling their activity. A given humus that has accumulated for a century, for example, may be decomposed in a relatively short time if the limiting conditions for the microbial community change. Possibly, nutrient stress for the trees opens a mechanism for a high fungal activity (e.g., Hintikka and Näyki, 1967). Still, we have reconstructed the amounts of mor humus C and N stored over almost three millennia, indicating that the stored N has a long-term stability. In all cases, the SOM was located under growing forest stands, a factor that may influence the stability.

Wardle et al. (1997) concluded that the N sequestered in the oldest humus of the Hornavan stands was less available than that of the younger ones based on experiments on the availability of N to plants. They also found that humus N concentration (range of about 10–15 mg g^{-1}) was related to the age of the humus. This may be interpreted that there had been a certain turnover of C, but that N had been kept in the system, for example by fixation of NH_3 (Nömmik and Vahtras, 1982). Another interpretation is that the oldest islands had a dominant vegetation of birch with higher N levels (cf. Table III) for some time before the conifers started to dominate.

3.5. CAN DIFFERENT CAPACITIES TO STORE N BE RELATED TO SPECIES OR TO THE INITIAL LITTER N CONCENTRATION?

The capacity to sequester N (N_{capac}; Equation (4)) is related to initial litter N concentrations. Using all available data with N_{capac} estimated from each of 53 litter decomposition studies encompassing 7 litter species, we obtained a highly significant, positive linear relationship ($R^2 = 0.700$) over the range in initial N concentrations from 2.9 to 15.1 mg g^{-1} with litter from unpolluted forest systems (Figure 2a). For six litter types, we calculated average values for both litter N concentration and N_{capac} (Figure 2b). In this relationship, the species-determined N levels from unpolluted systems were related to N_{capac} (Figure 2b) ($R^2 = 0.982$; $p < 0.001$). A set of experimental Scots pine needle litter showed the same trend within the species, in which case N availability in the soil influenced the litter N level ($R^2 = 0.913$; Figure 2c).

When relating N_{capac} to species (Table III; Figure 2b) the lowest storage was found for lodgepole pine litter with 0.68 mg N sequestered per gram of initial litter. For Scots pine litter the storage was higher (2.39 mg N g^{-1}) and Norway spruce litter had an even higher capacity with 3.74 mg N g^{-1}, which may be ascribed to a higher amount remaining and a high N_{limit} value. For silver birch the capacity was considerably higher (7.34 mg N g^{-1}), mainly due to a high N_{limit} value (22.71 mg N

TABLE III

The capacity of some different litter types to sequester N as based on calculations using limit values and N concentration at the limit value. Data in part from Berg and Johansson (1998) (cf. Figure 2b)

Litter type	Average values for init N conc. ($mg\ g^{-1}$)		Average limit value ($mg\ g^{-1}$)			Average N conc. at limit value (N_{limit}) ($mg\ g^{-1}$)			Sequestered N (N_{capac}) ($mg\ g^{-1}$)	Sequestered fraction (%)
	N	SD	Limit value	SD	n	N	SD	n		
Lodgepole pine	4.0	0.51	94.91	5.14	7	13.6	1.16	5	0.68	17
Scots pine	4.19	0.57	81.3	6.11	23	12.76	1.63	20	2.39	57
Norway spruce	5.44	1.42	74.07	13.9	15	14.46	2.14	14	3.74	69
Silver birch	9.55	2.74	77.7	15.6	4	22.71	1.19	3	7.34	77
Common beech	11.9	4.85	59.12	8.51	5	24.05	2.45	2	9.84	83
Silver fir	12.85	0.66	51.5	2.52	4	21.93	1.36	3	10.86	85

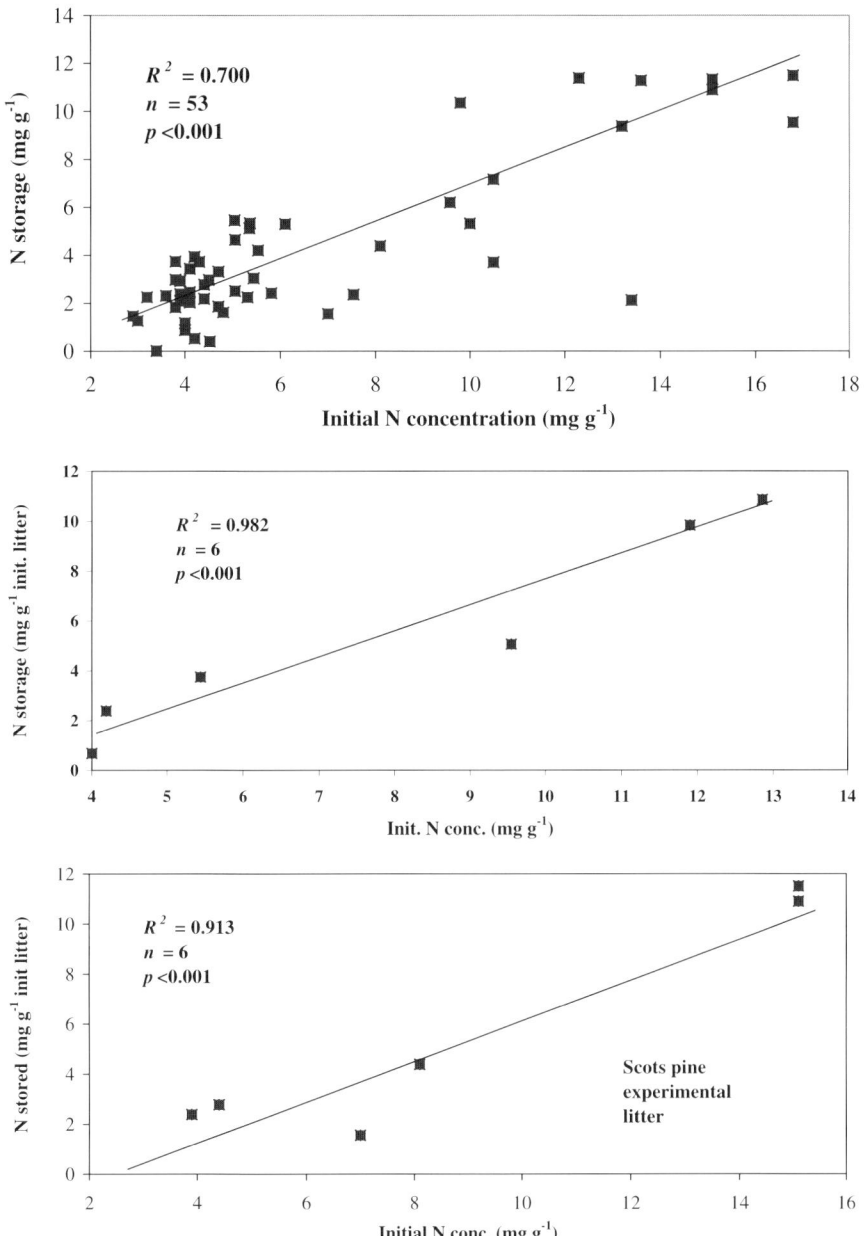

Figure 2. Initial litter N concentrations compared to the capacity to sequester N in the humus as related to litter N concentration. The estimated accumulation is defined as $N_{capac} = N_{limit} \times (100\text{-limit value})/100$ (Equation (4)). (a) All available data with each point representing a decomposition study. (b) Average values for N_{capac} and N_{init} for six litter types, the different N levels being species-related. Lodgepole pine ($n = 5$), Scots pine ($n = 20$), Norway spruce ($n = 14$), silver birch ($n = 3$), common beech ($n = 2$), silver fir ($n = 3$). (c) Scots pine needle litter. The different N levels were obtained by using needle litter from N-fertilized trees and green needles. Data are from Berg *et al*. (1999) and from Berg and Johansson (1998).

g^{-1}). Common beech and silver fir had even higher capacities (9.84 and 10.89 mg N g^{-1}, respectively), due to both rather low limit values and high values for N_{limit}.

The fraction of N sequestered [N_{capac}/N_{init}] expressed as percent increased with increasing initial N levels in a nonlinear way (Table III) from 17% for lodgepole pine to 83 and 85% for common beech and silver fir, respectively.

Nitrogen pollution will produce litter with higher initial N concentrations, which will lead to a higher storage of N (cf. Figure 2c). That litter in N-fertilized stands takes up more N and has a higher N concentration increase rate (cf. Equation (3)) was observed by Berg and Tamm (1994). That the higher uptake also results in a higher N concentration at the limit value (cf. Figure 2c) supports the idea that the initial N concentration is a regulating factor.

3.6. THERE IS A VARIATION IN N SEQUESTERED IN SOM OVER A PINE CLIMATE TRANSECT

The concentrations of N in fresh Scots pine needle litter are positively related to the climate index *AET* (Berg et al., 1995a). Berg and Meentemeyer (2002) found a more general relationship for pine species (Equation (7)) and also related available European data for N concentration in foliar litter to *AET* with a geographical range from the Barents Sea to the Mediterranean. Equation (6) gives us litterfall along a pine transect with AET from 400 to 650 mm, corresponding to Europe north of the Alps and the Carpatians, and Equation (7) gives the initial N concentrations in the same transect (Table II). Annual litterfall for pine would thus increase from almost 400 to ca. 5500 kg ha^{-1} within the *AET* interval 400 to 650 mm and initial N concentrations would increase from 3.5 to 8.0 mg g^{-1}. The limit values, which were calculated with respect to litter N concentration (Equation (8)) decreased from 83% at the site with the lowest *AET* to about 75% at the site with the highest *AET*. The recalcitrant fraction of litter/humus thus increased from 0.17 to 0.251 (Table II). With increasing initial N concentration from 13.3 to 16.8 mg g^{-1} the N_{limit} increased (Table II). This emphasized the N sequestration rate that increased from 0.89 kg ha^{-1} yr^{-1} to 23.3 kg ha^{-1} yr^{-1}, spanning a factor of 26 over this range in *AET*.

Acknowledgement

The study was carried out within the framework of the European Union project CN-ter (contract number QLK5-2001-00596).

References

Aber, J. D. and Melillo, J. M.: 1982, 'Nitrogen immobilization in decaying hardwood leaf litter as a function of initial nitrogen and lignin content', *Can. J. Bot.* **60**, 2263–2269.

Berg, B.: 2000, 'Litter decomposition and organic matter turnover in northern forest soils', *For. Ecol. Manage.* **133**, 12–22.

Berg, B. and Ekbohm, G.: 1991, 'Litter mass-loss rates and decomposition patterns in some needle and leaf litter types. Long-term decomposition in a Scots pine forest VII', *Can. J. Bot.* **69**, 1449–1456.

Berg, B. and Johansson, M.-B.: 1998, 'Maximum limit for foliar litter decomposition – A synthesis of data from forest systems, Part 1', in B. Berg (ed.), *A Maximum Limit for Foliar Litter Decomposition – A Synthesis of Data from Forest Systems*, Reports from the Departments of Forest Ecology and Forest Soils, Swedish University of Agricultural Sciences, Report 77, pp. 158.

Berg, B. and McClaugherty, C.: 2003, *Plant Litter. Decomposition. Humus Formation. Carbon Sequestration*, Springer-Verlag, Heidelberg, Berlin, pp. 286.

Berg, B. and Meentemeyer, V.: 2001, 'Litterfall in some European coniferous forests as dependent on climate – A synthesis', *Can. J. For. Res.* **31**, 292–301.

Berg, B. and Meentemeyer, V.: 2002, 'Litter quality in a north European transect versus carbon storage potential', *Plant Soil* **242**, 83–92.

Berg, B. and Tamm, C. O.: 1994, 'Decomposition and nutrient dynamics of litter in long-term optimum nutrition experiments. II. Nutrient concentration changes in decomposing Norway spruce (*Picea abies*) needle litter', *Scand. J. For. Res.* **9**, 99–105.

Berg, B., Johansson, M., Tjarve, I., Gaitnieks, T., Rokjanis, B., Beier, C., Rothe, A., Bolger, T., Göttlein, A. and Gerstberger, P.: 1999b, 'Needle Litterfall in a North European Spruce Forest Transect', Reports from the Departments of Forest Ecology and Forest Soils, Swedish University of Agricultural Sciences, *Report 80*, pp. 54.

Berg, B., Albrektson, A., Berg, M., Cortina, J., Johansson, M.-B., Gallardo, A., Madeira, M., Pausas, J., Kratz, W., Vallejo, R. and McClaugherty, C.: 1999c, 'Amounts of litter-fall in pine forests in the northern hemisphere, especially Scots pine', *Ann. For. Sci.* **56**, 625–639.

Berg, B., Calvo de Anta, R., Escudero, A., Johansson, M. B., Laskowski, R., Madeira, M., McClaugherty, C., Meentemeyer, V., Reurslag, A. and Virzo De Santo, A.: 1995a, 'The chemical composition of newly shed needle litter of different pine species and Scots pine in a climatic transect. Long-term decomposition in a Scots pine forest X', *Can. J. Bot.* **73**, 1423–1435.

Berg, B., Ekbohm, G., Johansson, M.-B., McClaugherty, C., Rutigliano, F. and Virzo De Santo, A.: 1996, 'Some foliar litter types have a maximum limit for decomposition – A synthesis of data from forest systems', *Can. J. Bot.* **74**, 659–672.

Berg, B., Laskowski, R. and Virzo De Santo, A.: 1999, 'Estimated N concentration in humus as based on initial N concentration in foliar litter – A synthesis', *Can. J. Bot.* **77**, 1712–1722.

Berg, B., McClaugherty, C., Virzo De Santo, A. and Johnson, D.: 2001, 'Humus buildup in boreal forests – Effects of litterfall and its N concentration', *Can. J. For. Res.* **31**, 988–998.

Berg, B., McClaugherty, C., Virzo De Santo, A., Johansson, M.-B. and Ekbohm, G.: 1995b, 'Decomposition of forest litter and soil organic matter – A mechanism for soil organic matter buildup?', *Scand. J. For. Res.* **10**, 108–119.

Couteaux, M. M., McTiernan, K. B., Berg, B., Szuberla, D., Dardenne, P. and Bottner, P.: 1998, 'Chemical composition and carbon mineralisation potential of Scots pine needles at different stages of decomposition', *Soil. Biol. Biochem.* **30**, 597–610.

Eriksson, K.-E., Blanchette, R. A. and Ander, P.: 1990, *Microbial and Enzymatic Degradation of Wood and Wood Components*', Springer Series in Wood Science, Springer-Verlag, Berlin, pp. 407.

Hatakka, A.: 2001, 'Biodegradation of lignin', in M. Hofman and A. Stein (eds), *Biopolymers, Vol. 1. Lignin, Humic substances and Coal*, Wiley, Weinheim, pp. 129–180.

Hintikka, V. and Näykki, O.: 1967, 'Notes on the effects of the fungus *Hydnellum ferrugineum* on forest soil and vegetation', *Comm. Inst. For. Fenn.* **62**, 1–22.

Meentemeyer, V.: 1978, 'Macroclimate and lignin control of litter decomposition rates', *Ecology* **59**, 465–472.

Nömmik, H. and Vahtras, K.: 1982, 'Retention and fixation of ammonium and ammonia in soils', in F. J. Stevenson (ed.), *Nitrogen in Agricultural Soils*. Agronomy Monographs, No. 22, American Society of Agronomy, Madison, Wisconsin, pp. 123–171.

Spaccini, R., Piccolo, A., Conte, P., Haberhauer, G. and Gerzabek, M. H.: 2002, 'Increased soil organic carbon sequestration through hydrophobic protection by humic substances. *Soil Biol. Biochem.* **34**, 1839–1851.

Staaf, H. and Berg, B.: 1982, 'Accumulation and release of plant nutrients in decomposing Scots pine needle litter. Long-term decomposition in a Scots pine forest II', *Can. J. Bot.* **60**, 1561–1568.

Wardle, D. A., Zachrisson, O., Hörnberg, G. and Gallet, C.: 1997, 'The influence of island area on ecosystem properties', *Science* **277**, 1296–1299.

NITRATE LEACHING FROM MOORLAND SOILS: CAN SOIL C:N RATIOS INDICATE N SATURATION?

C. J. CURTIS[1]*, B. A. EMMETT[2], B. REYNOLDS[2] and J. SHILLAND[1]

[1] *Environmental Change Research Centre, University College London, 26 Bedford Way, London WC1H 0AP, U.K.;* [2] *CEH Bangor, Orton Building, Deiniol Road, Bangor, Gwynedd LL57 2UP, U.K.*
(* author for correspondence, e-mail: ccurtis@geog.ucl.ac.uk; phone: +44 (0)20 7679 7553; fax: +44 (0)20 7679 7565)

(Received 20 August 2002; accepted 4 April 2003)

Abstract. Links between forest floor carbon:nitrogen (C:N) ratios, atmospheric N deposition and nitrate leaching into surface waters have been reported for forest ecosystems, but similar studies have not been reported previously for the equivalent compartments of moorland ecosystems in Great Britain, despite the importance of nitrate in contributing to the acidification of moorland streams and lakes in British uplands. In this paper, the relationships between the C:N ratio of moorland soil surface organic matter, N deposition, and nitrate leaching are explored for 13 soils in four moorland catchments. Although there is spatial variability in the C:N ratio of soils, major differences are apparent between soils and especially between catchments. The C:N ratio appears to be inversely related to modelled inorganic N deposition and, to a lesser degree, measured nitrate leaching, for three of the four catchments studied (Allt a'Mharcaidh, Afon Gwy, and Scoat Tarn). Nitrification may make an important contribution to nitrate leaching at the two higher deposition sites. At the fourth site, the heavily acidified River Etherow catchment, extremely high rates of nitrate leaching are not accompanied by low C:N ratios or high nitrification potentials in the upper soil horizons. Hence the C:N ratio of surface soil organic matter may have potential as an indicator of nitrogen saturation and leaching in some systems, but it is not universally applicable.

Keywords: C:N ratio, moorland, nitrate leaching, nitrification, nitrogen saturation

1. Introduction

Forest floor carbon:nitrogen (C:N) ratios have been proposed in recent years as indicators of nitrate leaching from forest ecosystems in Europe, in response to chronic deposition inputs of N (Dise *et al.*, 1998; Emmett *et al.*, 1998; Gundersen *et al.*, 1998). In Great Britain, the significant contribution of nitrate leaching to the acidification of the lakes and streams of many upland areas has been recognised (Allott *et al.*, 1995), but few of these upland catchments are forested. Attempts to model the process of N saturation, by which the accumulation of N in biomass and soils could potentially lead to enhanced nitrate leaching, have been hampered by the paucity of data from non-forest, moorland systems (NEGTAP, 2001).

The effects of N deposition on C:N ratio, microbial N cycling, and nitrate leaching in forests have been summarised by Gundersen (1991) and Aber (1992). In strongly N limited systems, most of the inorganic N cycled is in the form of NH_4^+, with plant uptake and microbial immobilisation outcompeting nitrifiers for available NH_4^+. Gross nitrification rates are therefore low because of a lack of NH_4^+, and any NO_3^- produced is rapidly immobilised to result in zero net nitrification. High rates of gross immobilisation are supported by large pools of labile organic C produced by the decomposition of N-poor organic matter, maintaining the tightly closed cycling of N and high C:N ratios in plants, litter, soil microbial biomass and soil organic matter.

Increased N supply through atmospheric deposition of reduced or oxidised N species results in greater plant and microbial uptake of inorganic N, increasing the N content (and reducing the C:N ratio) of living tissues, which in turn provide N-rich litter and soil organic matter after litterfall or plant death. Abiotic immobilisation processes, described by Ågren and Bosatta (1996) as condensation reactions between decomposition products such as phenols with either amino acids or ammonia, may also be favoured by large, pulsed N inputs (Aber, 1992). Biological demand for inorganic N increasingly is met by deposition inputs and gross mineralisation of N enriched organic matter. Furthermore, N deposition may create a positive feedback in forest N cycling because N-rich litter with a low C:N ratio is more readily mineralised, further increasing N availability to trees (Gundersen, 1991, 1992; Emmett et al., 1998). Hence for plants adapted to low N availability, an indirect effect of N deposition is the positive feedback of stimulated mineralisation and nitrification, which increases the potential N supply rate within the soil (NEGTAP, 2001).

While NH_4^+ generally is limiting in N-poor systems, its availability is increased by deposition inputs and accelerated N cycling. Excess NH_4^+ availability has the dual effect of promoting nitrification (for which it is the substrate) and inhibiting NO_3^- immobilisation, both directly (Bradley, 2001; Rennenberg and Gessler, 1999) and indirectly, because NH_4^+ is preferentially immobilised by microbes (van den Dreissche, 1971). Therefore, NO_3^- immobilisation is reduced despite increased NO_3^- availability from deposition and/or net nitrification, while NH_4^+ assimilation reduces tissue C:N ratios. With sufficiently high inputs of inorganic N, NO_3^- will accumulate in the soil and the potential for leaching will increase.

Although theories on the mechanisms by which N deposition could reduce C:N ratios and induce NO_3^- leaching have been proposed for forest systems, little has been published on the equivalent processes in moorland systems, where the N cycle is influenced by very different management practises (e.g., burning on grouse moors) and land use (e.g., grazing). Here we present an attempt to link the C:N ratio of moorland soil surface organic material to NO_3^- leaching fluxes in order to test its potential as an index of N saturation for use in modelling NO_3^- leaching and surface water acidification in response to N deposition.

2. Methods

Four sites were selected from the UK Acid Waters Monitoring Network (Monteith

TABLE I

Experimental N budget sites

Site	O.S. Grid Ref.	Precipitation (mm)	NO_3^- μeq L^{-1}	N_{dep} kg N ha^{-1} a^{-1}	N_{leach}	% N
Allt a' Mharcaidh	NH 881 045	1210	1	7.3	0.1	1.1
Afon Gwy	SN 824 854	2258	6	27.0	1.8	6.7
Scoat Tarn	NY 159 104	2217	15	33.6	4.9	14.5
River Etherow	SK 116 996	1272	42	33.8	11.0	32.5

and Evans, 2000) across gradients of total N deposition and leaching losses of inorganic N (Table I). Site locations are provided on the UK Ordnance Survey (O.S.) grid. Precipitation and total inorganic N deposition flux (N_{dep}) are based on 5 km grid modelled data for 1995–97, comprising wet (orographically enhanced) plus dry deposition of oxidised and reduced N, but uncertainty in modelled data is large, especially for reduced N (NEGTAP, 2001). Nitrate (NO_3^-) concentrations are mean values for the period of October 1999-October 2000, with leaching fluxes (N_{leach}) calculated from site-specific flow data.

Within each site, sample plots were established on the main soil types, giving a total of 13 soils from within the four catchments. The Allt a'Mharcaidh site comprises wet and dry heath with grasses, shrubs, mosses, and *Sphagnum* species. The Afon Gwy and Scoat Tarn catchments are dominated by acid grasslands, although shrubs also occur on the peaty hilltops at the Afon Gwy. The Gwy and Scoat Tarn catchments are subject to low intensity grazing by sheep whilst deer graze in the Allt a'Mharcaidh. The study area at the River Etherow is dominated by *Calluna vulgaris* and is managed as a grouse moor by burning.

Retention of deposited N in catchment soils and vegetation ranges from almost 99% at the Allt a'Mharcaidh to 67.5% at the River Etherow (Table I). Nitrate leaching increases with total inorganic N deposition across the four catchments, but leaching from the two highest deposition sites (Scoat Tarn and River Etherow), for which modelled deposition figures are very similar, differs by more than a factor of 2 in both absolute and percentage terms.

The C:N ratios of moorland soil organic matter were measured in each of the 13 soils (Table II) that were collected as baseline samples for a ^{15}N tracer study into N immobilisation processes (Curtis, 2003). Three replicate samples were obtained from each soil type by cutting two 10 × 5 cm rectangles in each of 3 plots to a depth of approximately 5 cm into the soil surface with a knife and removing the distinct surface horizon of organic material, including roots. The two samples from each plot were bulked to form a composite sample, which was analysed for C and N content. This horizon, although variable in thickness, is assumed to be analogous to the forest floor compartment of forest ecosystems. Samples were analysed at

TABLE II
Experimental areas at N budget sites ($n = 13$)

Site	Soil code	Soil type	Mean soil temp. (°C)	Mean soil water pH
Allt a'Mharcaidh	M1	Peaty ranker	4.4	3.4
Allt a'Mharcaidh	M2	Valley peat	5.5	3.7
Allt a'Mharcaidh	M3	Peaty podsol	5.8	4.0
Allt a'Mharcaidh	M4	Shallow peat	6.2	3.9
Afon Gwy	G1	Hilltop peat	7.3	3.4
Afon Gwy	G2	Peaty gley	–	3.6
Afon Gwy	G3	Podsol	8.4	4.1
Afon Gwy	G4	Valley peat	8.3	3.8
Scoat Tarn	S1	Podsol	6.5	3.9
Scoat Tarn	S2	Peaty gley	6.6	3.8
Scoat Tarn	S3	Deep peat	6.9	3.9
River Etherow	E1	Deep peat (burnt *Calluna*)	7.2	3.0
River Etherow	E2	Deep peat (unburnt *Calluna*)	7.1	3.3

the University of Wales, Bangor, using a LECO CHN-2000 total element analyser. Mean annual soil temperature was measured using temperature dataloggers (Gemini Tiny Talk) buried at the 5–10 cm depth in 100 mL water-tight plastic bottles adjacent to experimental soil plots. One datalogger was located at each set of three replicated plots, and took readings at 2 hr intervals. Soil water was sampled every two weeks in porous cup lysimeters at about 20 cm depth for pH measurement.

Mineralisation and nitrification potentials also were measured for separate samples of the same soil horizons within the 3×1 m study plots as part of a wider study (Curtis, 2003). Composite samples, made up of subsamples from four 5 cm diameter soil cores in each plot, were homogenised after removal of surface vegetation, roots and stones by hand, after which available NH_4^+-N and NO_3^--N were measured by 1 M KCl extraction and analysis by Skalar SA-40 autoanalyser using the modified Berthelot (indophenol) reaction and sulphanilamide/NEDA/Cd/Cu reduction methods, respectively. Net mineralisation and nitrification potentials were measured by aerobic incubation of the remaining intact, field moist, halved soil cores in gas permeable polythene bags at 15 °C for 33 days (Page *et al.*, 1982). After measurement of available NH_4^+-N and NO_3^--N using the above methods, mineralisation and nitrification potentials were calculated as the difference in total inorganic N ($NH_4^+ + NO_3^-$) or NO_3^-, respectively, between pre- and post-incubation

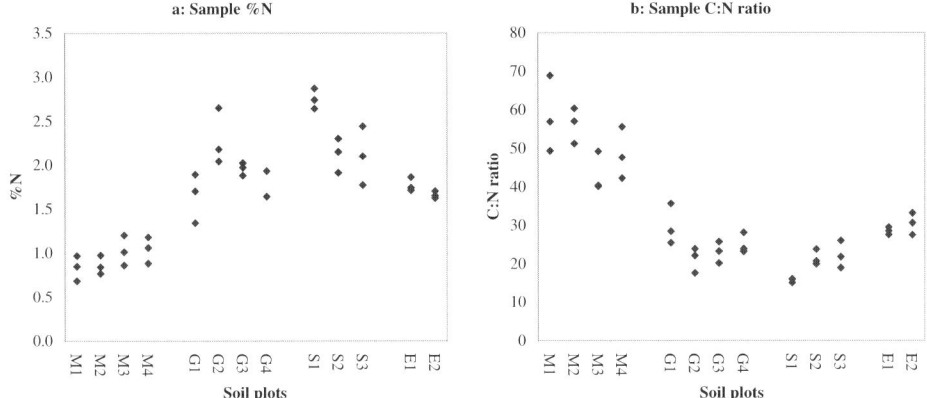

Figure 1. %N and C:N ratio of soil surface organic matter. Sites are ordered along the *x*-axis in order of increasing deposition.

samples, divided by the incubation period. Results were expressed per gram organic matter of soil, with loss on ignition at 375 °C used to determine organic matter content.

3. Results

Mean soil temperature and soil water pH are presented in Table II. The N content (% N) and C:N ratios of sampled soils are illustrated in Figure 1, with sites in order of increasing N deposition. There is an increase in N content from the Mharcaidh to the Gwy and Scoat Tarn, but then a decrease at the Etherow. Of the mineral soils at Scoat Tarn, only the podsol (S1) has a higher N content than the mineral Gwy soils (G2 and G3), while the peaty gley (S2) shows a similar range to these soils. The peat from Scoat Tarn (S3) has a slightly higher N content than the peats from the Gwy (G1 and G4), but is very variable. The Etherow peats are comparable in N content to those from the Gwy.

Figure 1 shows that although there is spatial variability in C:N ratio, there are major differences between soils and especially between sites. Maximum C:N ratios of about 43–58 in the Mharcaidh soils compare with a minimum value of 16 in the podsol at Scoat Tarn (S1). A decline from the Mharcaidh to the Gwy and Scoat Tarn is evident, while the higher mean C:N ratio in the surface of the Etherow peats is similar to that in the hilltop peat from the Gwy (G1), at about 30.

A measure of total inorganic N supply is the mineralisation potential, while nitrification potential provides an indication of the level of NO_3^- supply. These potentials are plotted against C:N ratio of the surface organic horizon in Figure 2. There is an apparent threshold C:N ratio of around 30 below which mineralisation potentials are above zero. Mineralisation potential broadly increases with

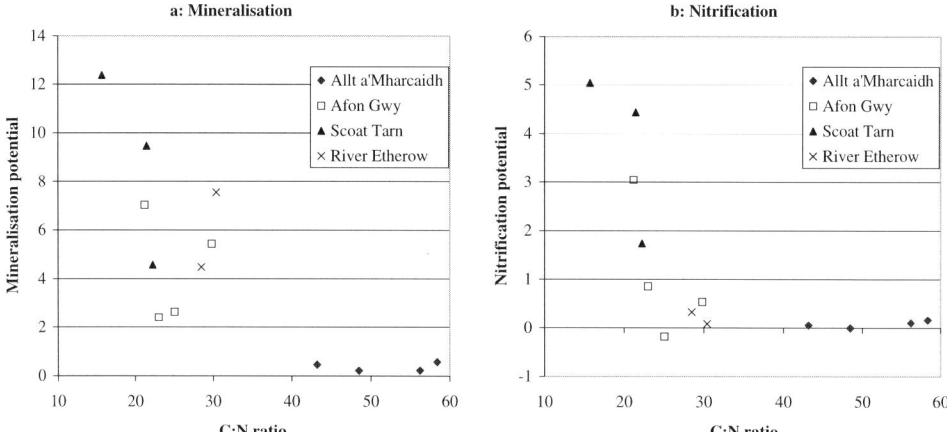

Figure 2. Mineralisation and nitrification potentials (μg N g^{-1} organic matter d^{-1}) and C:N ratio.

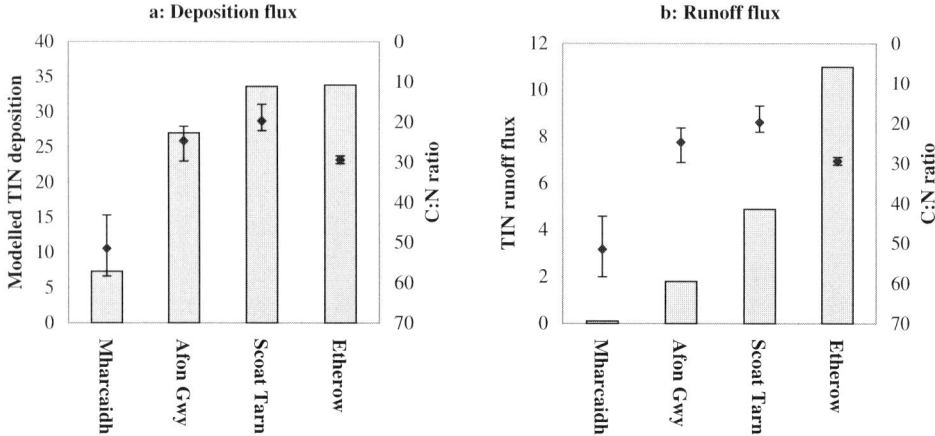

Figure 3. Deposition input and leaching output fluxes of total inorganic N (TIN: $NH_4^+ + NO_3^-$, columns: kg N ha^{-1} a^{-1}) against C:N ratio (points) averaged for all sampled soils within the catchment (bars indicate min. and max. C:N).

decreasing C:N ratio below this threshold, but there is much scatter in the data. For nitrification, the threshold appears to lie between C:N values of 25 to 30, i.e. slightly lower than for mineralisation. Furthermore, there is a much sharper increase in nitrification potential with decreasing C:N ratio below this threshold. Hence it appears that high rates of nitrification are associated with C:N ratios below a value of around 25.

To examine the relationship between C:N ratio and deposition inputs or catchment leaching outputs in streamwaters, the soil C:N data have to be aggregated at the catchment scale. In the absence of data on soil distribution within each catchment to facilitate weighting of the C:N data by soil type, an arithmetic mean value for all sampled soils is used here, assuming that the soil types are representative for each catchment. Overall mean, minimum and maximum values (using means for each soil type) are plotted against N deposition data in Figure 3. Figure 3a shows modelled deposition data from the 5 km grid 1995–97 average dataset, while Figure 3b shows measured total inorganic N fluxes in surface waters. Note that the C:N scale is reversed on the right hand side of the charts.

While the number of data points is too small for statistical analysis, there is a clear relationship between the ranges in C:N ratio observed at the Mharcaidh, Gwy, and Scoat Tarn catchments and deposition inputs of inorganic N. At the Etherow, however, the C:N ratio is greater than at Scoat Tarn despite the very similar deposition load. N leaching increases at each site along the N deposition gradient, showing a weaker relationship with C:N ratio than for total inorganic N deposition. In particular, the very high leaching flux at the Etherow is not reflected by a low C:N ratio in the surface soils there.

4. Discussion

For three of the four study catchments, the C:N ratio shows an apparent relationship with N deposition and, to a lesser extent, NO_3^- leaching. At the fourth site, the River Etherow, the relationship does not hold, as extremely high NO_3^- leaching rates are accompanied by relatively high soil C:N ratios. Furthermore, nitrification is suppressed in the surface soils at the Etherow, even though climatic conditions may be more favourable for nitrification there than elsewhere, with relatively high mean soil temperatures and lower precipitation compared with other sites (Tables I and II). A possible reason for this discrepancy is the disruption of the N cycle there in response to severe acidification and/or the presence of potentially toxic levels of heavy metals (Smith *et al.*, in press) caused by very high contemporary and historical atmospheric deposition loadings.

The nitrification data support the hypothesis that nitrifying activity is suppressed in the uppermost horizons of the Etherow peats. Nitrification potentials are negligible in the Etherow soils, despite high mineralisation potentials compared with peat soils from the other study catchments (Curtis, 2003). Only the Mharcaidh soils and the valley peat at the Afon Gwy (G4) show similar, low nitrification potentials (Figure 2). Hence, nitrification in the surface soil horizon is not causing the high NO_3^- leaching observed at the Etherow. Nitrification can be one of the most acid-sensitive microbial processes in soils (Paul and Clark, 1996), and may be suppressed by very low soil pH (3.0–3.3) at the Etherow.

However, work on Dutch forest and moorland soils has demonstrated that nitrification can be significant even in very acid conditions, through the activity of acid tolerant autotrophs as well as heterotrophic nitrifiers (van Breemen *et al.*, 1987; de Boer *et al.*, 1989, 1990; Tietema *et al.*, 1992). Studies in Belgian forest soils have also shown that very high inputs of NH_4^+ may bypass the upper organic layers, where nitrification is negligible, to reach deeper mineral horizons with lower

C:N ratios and higher pH where significant nitrification occurs (de Schrijver *et al.*, 2000). In the Etherow catchment, nitrification potentials in deeper soils (5–20 cm) beneath unburnt *Calluna* are close to zero, but at similar depths beneath burnt *Calluna*, nitrification potentials are relatively high (0.63 μg N g^{-1} organic matter d^{-1}) compared to equivalent peat horizons at other sites (maximum 0.24 μg N g^{-1} organic matter d^{-1}; Curtis, 2003). However, ratios of C:N increase with depth at the Etherow, up to 44.2 (burnt) and 46.9 (unburnt), while soil pH is lower under the burnt *Calluna* than anywhere else (Table II). Hence the process described by de Schrijver *et al.* (2000) cannot be responsible for NO_3^- leaching at the Etherow. While the potential role of nitrification at depth in contributing to NO_3^- leaching there, stimulated by burning of surface vegetation, cannot be ruled out, the mechanisms are unclear and the flux of water at this depth in the peat is likely to be small.

Another contributory factor to the high NO_3^- leaching at the Etherow may be the relatively low proportional vegetation uptake of N from atmospheric sources (about 40%) compared with the other catchments (43–100%), as shown by ^{15}N tracer additions experiments (Curtis, 2003). A major source of streamwater NO_3^- at this site may therefore be direct leaching of deposition inputs that do not participate in the terrestrial N cycle, rather than biologically mediated NO_3^- production in the upper soil profile. Hence a large proportion of oxidised N deposition may not be contributing to a decrease in soil C:N ratio at the Etherow.

At the three catchments where a relationship between C:N ratio and N deposition is observed, other factors must also be considered. For example, the very high C:N ratios at the Allt a'Mharcaidh correspond to almost complete N retention in soils and vegetation, but it cannot be assumed that elevated N deposition at the site would necessarily lead to reduced soil C:N ratios. Instead, the accumulation of soil organic matter may increase, effectively providing a permanent store for atmospheric inputs of N (*cf*. Berg and Matzner, 1997). White *et al.* (1996) found that at low levels of N inputs (<9 kg N ha^{-1} a^{-1}) to *Calluna* moorland on peaty podsol soils, total soil N content and C:N ratio *increased* almost linearly with N deposition, presumably due to the effect of acidification on organic matter accumulation rate, such that C accumulated faster than N. At higher total N deposition rates, the C:N ratio declined with increasing deposition. Hence there may be a lower threshold of N deposition for certain soils, below which C:N ratio does not decline and no leaching of NO_3^- occurs (e.g., Dise and Wright, 1995).

Furthermore, recent studies have contradicted the long held belief that increased N content in organic residues always leads to accelerated decomposition (INDITE, 1994). N deposition may lead to changes in the quantity and quality of plant litter, which may accelerate the initial stages of decomposition, but slow down later stages (Berg *et al.*, 1998; NEGTAP, 2001). Fog (1988) suggested a general pattern linked to both C:N ratio and 'quality' of organic matter. If the C pool is not readily metabolised by soil microbes, then mineralisation (and hence nitrification) is slow regardless of the C:N ratio of the substrate (Smith, 1994). In poor quality substrates with high C:N ratios (> about 50), decomposition is largely controlled by lignin content, and inorganic N inputs may inhibit the synthesis of enzymes required for lignin degradation, thus slowing the later stages of decomposition (Berg and Matzner, 1997). A change in the balance of microbial populations from lignin to cellulose decomposers may occur, causing reduced rates of decomposition over a timescale of many years (INDITE, 1994). Litter quality (defined by lignin or polyphenol content), as well as C:N ratio, may therefore be an important control on decomposition and nitrification rates, so that there is no universal value of C:N ratio at which the switch from immobilisation to net mineralisation and NO_3^- leaching occurs (Richards, 1987; Gundersen, 1992; Vinten and Smith, 1993). Organic matter quality, possibly linked to the shrub vegetation that dominates at the Etherow catchment, may therefore explain the very low nitrification rates observed there, even though climatic conditions and NH_4^+ availability, though perhaps not pH, are otherwise favourable.

A final consideration is the representativeness of the soil samples used. The sample size is very small given the attempts to link soil properties to leaching at the catchment scale, but this limitation was recognised from the outset and the sampling strategy was designed accordingly. Sampling methods were stratified to focus on up to four dominant soils within each catchment, with three replicated plots on each soil type. The wider project aimed to characterise N dynamics and C:N ratios of each soil type, on the assumption that if characterised correctly, the soil properties would show some relationship to stream chemistry. The major differences in C:N ratio between sites despite within soil variability (Figure 1) vindicate, to a large extent, the sampling strategy adopted.

5. Conclusions

For three of the four study catchments, the C:N ratio appears to be related to N deposition and NO_3^- leaching, suggesting that it has potential as an index for representing these processes in acidification models such as MAGIC7 (Cosby *et al.*, 2001). While our results are consistent with observations from forest systems, our study contains too few sites, both within catchments and along the N deposition gradient, to thoroughly investigate links between moorland soil C:N ratio, N saturation, and NO_3^- leaching into surface waters. However, data from the Etherow catchment suggest that there are limitations to the concept as the hypothesised relationships between surface soil C:N ratio and N input/output fluxes are not observed. The reasons for this are not clear, but may relate to the influence of long-term chronic acid deposition and accumulation of metals or other factors which influence the N cycle, such as litter quality.

Future work should focus on the differences in soil organic matter quality between soil and vegetation types, testing the relationship between soil C:N ratio and leaching on a soil specific basis to better understand whether certain soil types might be driving NO_3^- leaching fluxes. Finally, it could prove useful in terms of understanding processes and for improving models to investigate the possibility (and assumption in MAGIC7; Cosby et al., 2001) that there are both upper and lower thresholds of N deposition beyond which the assumed relationship between N deposition, C:N ratio, and NO_3^- leaching does not apply.

Acknowledgements

This work was funded under the DETR contract 'Acidification of freshwaters: the role of nitrogen and the prospects for recovery' (EPG1/3/117). Many colleagues from CEH Bangor, the ECRC at University College London, CEH Merlewood, MLURI, Aberdeen, and the Freshwater Fisheries Laboratory, Pitlochry, contributed to the collection of N budget data. Special thanks are due to Andy Owen at the University of Wales, Bangor, Ewan Shilland at the ECRC, and Jo Porter. Deposition data were provided by CEH Edinburgh.

References

Aber, J. D.: 1992, 'Nitrogen cycling and nitrogen saturation in temperate forest ecosystems', *Trends Ecol. Evol.* **7**, 220–224.

Ågren, G. I. and Bosatta, E.: 1996, *Theoretical Ecosystem Ecology: Understanding Element Cycles*, Cambridge University Press, Cambridge, U.K., 234 pp.

Allott, T. E. H., Curtis, C. J., Hall, J., Harriman, R. and Battarbee, R. W.: 1995, 'The impact of nitrogen deposition on upland surface waters in Great Britain: A regional assessment of nitrate leaching', *Water, Air, Soil Pollut.* **85**, 297–302.

Berg, B. and Matzner, E.: 1997, 'Effect of N deposition on decomposition of plant litter and soil organic matter in forest systems', *Environ. Rev.* **5**, 1–25.

Berg, M. P., Kniese, J. P., Zoomer, R. and Verhoef, H. A.: 1998, 'Long-term decomposition of successive organic strata in a nitrogen saturated Scots pine forest soil', *For. Ecol. Manage.* **107**, 159–172.

Bradley, R. L.: 2001, 'An alternative explanation for the post-disturbance NO_3^- flush in some forest ecosystems', *Ecol. Lett.* **4**, 412–416.

Cosby, B. J., Ferrier, R. C., Jenkins, A. and Wright, R. F.: 2001, 'Modelling the effects of acid deposition: Refinements, adjustments and inclusion of nitrogen dynamics in the MAGIC model', *Hydrol. Earth Syst. Sci.* **5**, 499–517.

Curtis, C. J.: 2003, 'An assessment of the representation of moorland nitrogen sinks in static critical load models for freshwater acidity', *Unpublished Ph.D. Thesis*, University of London, U.K.

de Boer, W., Klein Gunnewiek, P. J. A., Troelstra, S. R. and Laanbroek, H. J.: 1989, 'Two types of chemolithotrophic nitrification in acid heathland humus', *Plant and Soil* **119**, 229–235.

de Boer, W., Klein Gunnewiek, P. J. A. and Troelstra, S. R.: 1990, 'Nitrification in Dutch heathland soils. II. Characteristics of nitrate production', *Plant Soil* **127**, 193–200.

Dise, N. B. and Wright, R. F.: 1995, 'Nitrogen leaching from European forests in relation to nitrogen deposition', *For. Ecol. Manage.* **71**, 153–161.

Dise, N. B., Matzner, E. and Forsius, M.: 1998, 'Evaluation of organic horizon C:N ratio as an indicator of nitrate leaching in conifer forests across Europe', *Environm. Pollut.* **102**, 453–456

Emmett, B. A., Boxman, D., Bredemeier, M., Gundersen, P., Kjønaas, O. J., Moldan, F., Schleppi, P., Tietema, A. and Wright, R. F.: 1998, 'Predicting the effects of atmospheric nitrogen deposition in conifer stands: Evidence from the NITREX ecosystem-scale experiments', *Ecosystems* **1**, 352–360.

Fog, K.: 1988, 'The effect of added nitrogen on the rate of decomposition of organic matter', *Biol. Rev.* **63**, 433–462.

Gundersen, P.: 1991, Nitrogen deposition and the forest nitrogen cycle: Role of denitrification', *For. Ecol. Manage.* **44**, 15–28.

Gundersen, P.: 1992, 'Mass balance approaches for establishing critical loads for nitrogen in terrestrial ecosystems', in P. Grennfelt and E. Thörnelöf (eds), *Critical Loads for Nitrogen – A Workshop Report*, Nord 1992:41, Nordic Council of Ministers, Copenhagen, Denmark, pp. 55–109.

Gundersen, P., Callesen, I. and de Vries, W.: 1998, 'Nitrate leaching in forest ecosystems is controlled by forest floor C/N ratio', *Environ. Pollut.* **102**, 403–407.

INDITE: 1994, *Impacts of nitrogen deposition in terrestrial ecosystems.* UK DOE, London, 110 pp.

Monteith, D. T. and Evans, C. D.: 2000, 'UK acid waters monitoring network: 10 yr report. Analysis and interpretation of results, April 1988 – March 1998', ENSIS Publishing, London, U.K., 364 pp.

NEGTAP: 2001, 'Transboundary air pollution: Acidification, eutrophication and ground-level ozone in the UK', CEH Edinburgh, Penicuik, U.K., 314 pp.

Page, A. L., Miller, R. H. and Keeney, D. R. (eds): 1982, *Methods of Soil Analysis, Part 2: Chemical and Microbiological Properties*, Agronomy Monograph No. 9, 2nd edition, American Society of Agronomy and Soil Science Society of America, Madison, Wisconsin, U.S.A., pp. 163–165.

Paul, E. A. and Clark, F. E.: 1996, *Soil Microbiology and Biochemistry*, Academic Press, San Diego, California, U.S.A., 340 pp.

Rennenberg, H. and Gessler, A.: 1999, 'Consequences of N deposition to forest ecosystems – Recent results and future research needs', *Water, Air, Soil Pollut.* **116**, 47–64.

Richards, B. N.: 1987, *The Microbiology of Terrestrial Ecosystems*, Longman, Harlow, 399 pp.

de Schrijver, A., van Hoydonck, G., Nachtergale, L., de Keersmaeker, L., Mussche, S. and Lust, N.: 2000, 'Comparison of nitrate leaching under silver birch (*Betula pendula*) and Corsican pine (*Pinus nigra* ssp. *laricio*) in Flanders (Belgium)', *Water, Air, Soil Pollut.* **122**, 77–91.

Smith, E. J., Hughes, S., Lawlor, A. J., Lofts, S., Simon, B. M., Stevens, P. A., Stidson, R. T., Tipping, E. and Vincent, C. D.: 'Potentially toxic metals in ombrotrophic peat along a 400 km English-Scottish transect', *Water, Air, Soil Pollut.* (in press).

Smith, J. L.: 1994, 'Cycling of nitrogen through microbial activity', in J. L. Hatfield and B. A. Stewart (eds), *Soil Biology: Effects on Soil Quality*, Lewis, Boca Raton, Florida, U.S.A., pp. 91–120.

Tietema, A., de Boer, W., Riemer, L. and Verstraten, J. M.: 1992, 'Nitrate production in nitrogen-saturated acid forest soils-vertical-distribution and characteristics', *Soil Biol. Biochem.* **24**, 235–240.

van Breemen, N., Mulder, J. and van Grinsven, J. J. M.: 1987, 'Impacts of acid atmospheric deposition on woodland soils in the Netherlands: II. Nitrogen transformations', *Soil Sci. Soc. Am. J.* **51**, 1634–1640.

van den Dreissche, R.: 1971, 'Response of conifer seedlings to nitrate and ammonium sources of nitrogen', *Plant Soil* **34**, 421–439.

Vinten, A. J. A. and Smith, K. A.: 1993, 'Nitrogen cycling in agricultural soils', in T. P. Burt, A. L. Heathwaite and S. T. Trudgill (eds), *Nitrate: Processes, Patterns and Management*, Wiley, Chichester, U.K., pp. 39–73.

White, C. C., Dawod, A. M. and Cresser, M. S.: 1996, 'Nitrogen accumulation in surface horizons of moorland podzols: Evidence from a Scottish survey', *Sci. Total Environ.* **184**, 229–237.

NUTRIENTS ON ASPHALT PARKING SURFACES IN AN URBAN ENVIRONMENT

DIANE HOPE[1,2*], MARKUS W. NAEGELI[2§], ANDY H. CHAN[1§§] and NANCY B. GRIMM[1,2]

[1] *Center for Environmental Studies, P.O. Box 873211, Arizona State University, Tempe, Arizona 85287, U.S.A.;* [2] *School of Life Sciences, Arizona State University, Box 874501, Tempe, Arizona 85287, U.S.A.;* [§] *Current address: Canon (Schweiz) AG, Industriestrasse 12, CH-8305 Dietlikon, Switzerland;* [§§] *Current address: SmartTool Technologies, 1717 Grant Street, Santa Clara, California 95050, U.S.A.*
(* *author for correspondence, e-mail: di.hope@asu.edu; phone: 480-965-2887; fax: 480-965-8087*)

(Received 20 August 2002; accepted 4 April 2003)

Abstract. Amounts of readily soluble nutrients on asphalt parking lot surfaces were measured at four locations in metropolitan Phoenix, Arizona, U.S.A. Using a rainfall simulator, short intense rainfall events were generated to simulate 'first flush' runoff. Samples were collected from 0.3 m^2 sections of asphalt at 8 to 10 sites on each of four parking lots, during the pre-monsoon season in June-July 1998 and analyzed for dissolved NO_3^--N, NH_4^+-N, soluble reactive phosphate (SRP), and dissolved organic carbon (DOC). Runoff concentrations varied considerably for NO_3^--N and NH_4^+-N (between 0.1 and 115.8 mg L^{-1}) and DOC (26.1 to 295.7 mg L^{-1}), but less so for SRP (0.1 to 1.0 mg L^{-1}), representing average surface loadings of 191.3, 532.2, and 1.8 mg m^{-2} respectively. Compared with similar data collected from undeveloped desert soil surfaces outside the city, loadings of NO_3^--N and NH_4^+-N on asphalt surfaces were greater by factors of 91 and 13, respectively. In contrast, SRP loads showed little difference between asphalt and desert surfaces. Nutrient fluxes in runoff from a storm that occurred shortly after the experiments were used to estimate input-output budgets for 3 of the lots under study. Measured outputs of DOC and SRP were similar to those predicted using rainfall and experimentally determined surface loadings, but for NH_4^+-N and particularly for NO_3^--N, estimated rainfall inputs and surface runoff were significantly higher than exports in runoff. This suggests that parking lots may be important sites for nutrient accumulation and temporary storage in arid urban catchments.

Keywords: ammonium, arid urban environments, asphalt parking lot surfaces, DOC, nitrate, phosphorus, storm runoff

1. Introduction

A significant proportion of contaminant loads to recipient systems originate from atmospheric deposition (Burian *et al.*, 2001; Grennfeld and Hultberg, 1986; Howarth *et al.*, 1996; Hicks, 1998). Such deposition can be significantly enhanced in and around urban areas, particularly for nutrients (Bytnerowicz and Fenn, 1996; Russell *et al.*, 1993; Lovett *et al.*, 2000; Smith *et al.*, 2000) and may make a measurable contribution to the nutrient cycling of urban ecosystems (Baker *et al.*, 2001).

Water, Air, and Soil Pollution: Focus **4:** 371–390, 2004.
© 2004 *Kluwer Academic Publishers. Printed in the Netherlands.*

However, little has been done to determine direct measurements of nutrient dry deposition to inert urban surfaces, and modeled estimates are hampered by a lack of knowledge of deposition velocities for typical urban surface types such as asphalt (Sehmel, 1980). Urbanization not only enhances pollutant deposition, but also modifies natural drainage patterns and hydrologic pathways. Replacement of natural ground cover by paved surfaces increases the volume and rate of storm runoff (Leopold, 1968), prevents natural infiltration of stormwater to the subsurface and increases soil erosion and pollutant runoff (Burian et al., 2001). This increase in impervious surface area coupled with enhanced pollutant deposition has meant that most cities are net exporters of nutrients and contaminants. Urban drainage from paved areas transports dissolved, colloidal, and solid constituents in a heterogeneous mixture, of which heavy metals are typically the most prevalent pollutant constituents, along with organic and inorganic compounds (Sansalone et al., 1998). Such urban stormwater can degrade the quality of streamflow with oil and grease, pesticides, and trace metals (e.g., Lopes et al., 1995).

Materials accumulating on street surfaces and contributing to pollutant loads in urban storm runoff have been studied by the National Urban Runoff Program or NURP (Sartor and Boyd, 1972; EPA, 1983). The focus of such work has been largely on the contaminant chemistry of runoff, in particular heavy metals (e.g., Pitt and Amy, 1973), petroleum hydrocarbons (e.g., Latimer et al., 1990; Lopes et al., 2000), and particulates (e.g., Sansalone et al., 1998). Furthermore, most studies have concentrated on street surfaces where nutrient concentrations in runoff typically are not very high (Sartor and Boyd, 1972; EPA, 1983; Lopes et al., 1995). However, little attention has been paid to nutrient accumulation and storage on parking lot surfaces, which not only have significant extent in cities, but also potentially accumulate larger surface loads than streets, due to slower vehicle speeds, leakage of petroleum hydrocarbons from stationary vehicles, and less frequent surface cleaning.

In the arid desert ecosystems of the southwestern US, the pools of major nutrients stored in plants and soils is low compared to other terrestrial ecosystems (Peterjohn and Schlesinger, 1990), while dry deposition of nutrients typically comprises a significant fraction of total deposition (Bytnerowicz and Fenn, 1996). Hence parking lot surfaces in urban environments could constitute important nonpoint sources of nutrients, potentially contributing to chronic, low-level degradation of stormwater quality.

The aims of this study were to: i) measure the maximum amounts of readily soluble nutrients likely to have accumulated on asphalt parking lot surfaces in Phoenix after an extended period of dry weather, ii) examine variations in surface loads with respect to site/land use, traffic type and density, surface slope, pavement condition and distance to the nearest curb, iii) compare nutrient loadings on parking lot surfaces with those for an adjacent undeveloped desert soil surface, iv) use a mass balance approach to compare nutrient inputs in rainfall and from asphalt surfaces, with losses in runoff during an actual storm event in the mini-watersheds

TABLE I

Site descriptions

Site name	U.S.G.S. station number	Land use & surface type	Land use (%)	Basin area (ha)	Impervious area (%)	Number of plots
Box Culvert	09512185	Light industrial & asphalt	84	0.159	80	9
Peoria Avenue	09513885	Commercial & asphalt	97	1.4	96	9
Olive Avenue	09513925	Residential & asphalt	100	7.2	60	10
Papago Park	no station	Desert park & soil	99	n. a.	3	10
Sycamore Creek	09510200	Desert & soil	100	n. a.	<2	12

studied, and v) use these findings to clarify the likely role of parking lot surfaces in urban nutrient cycling.

1.1. STUDY SITES

The Central Arizona-Phoenix Long Term Ecological Research (CAP LTER) study site encompasses an extensive, rapidly expanding metropolitan area of 3 million people, located on a large alluvial floodplain in the Sonoran desert of the southwestern US. The climate is hot (average annual temperature 30 °C) and dry, with a mean annual rainfall of 180 mm, approximately half of which occurs during low-intensity winter rains (Feb-April) that span the entire basin. The remaining rainfall is produced by summer (July-September) convective storms, which are intense and highly-localized (Ricci, 1984). Rapid development of the Phoenix metropolitan area over the past 50 yr has led to an extensive urban area characterized by heterogeneous landscape consisting of residential, commercial, industrial, transportation, and remnant desert land uses, with agriculture on the periphery of the urban development in the west and southeast parts of the valley (Stefanov *et al.*, 2001). The developed urban core of metropolitan Phoenix consists of a mosaic of land use types (residential, transportation corridors, commercial, and institutional) that include significant (around 50%) impervious surface cover.

Parking lots were selected in areas surrounded by four different land uses: desert, commercial, light industrial, and residential (Table I, Figure 1). At three of these sites (considered as urban 'mini-catchments') a suite of chemical constituents in storm runoff have been monitored by the United States Geological Survey (USGS) since the early 1990s as part of the National Pollutant Discharge Elimination System permitting requirements of the Environmental Protection Agency (Lopes *et al.*, 1995). The commercial parking lot site is surrounded by businesses including a restaurant and retail stores; about half of the pervious area is undeveloped and half has desert landscaping with some irrigation. At the light in-

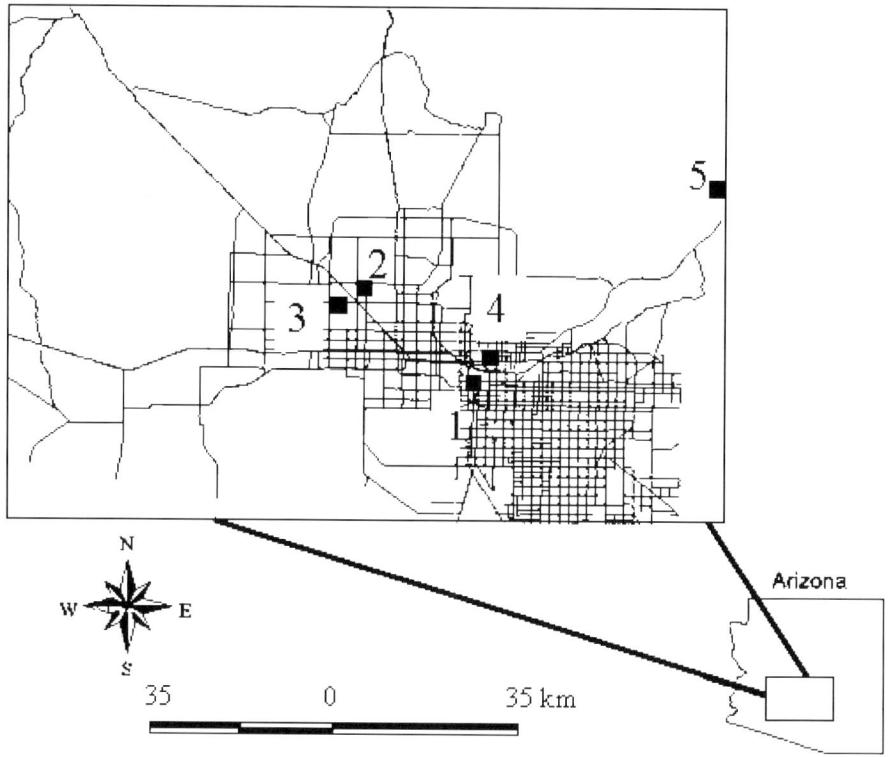

Figure 1. Location of the research sites within the Central Arizona-Phoenix study area. The sites are: 1 – Box Culvert at 48th St., Tempe; 2 – Peoria Avenue and 43rd Ave., Phoenix; 3 – Olive Ave. and 67th Ave., Glendale; 4 – Papago Park, Tempe; 5 – Sycamore Creek.

dustrial site, businesses include offices, warehouses, small manufacturing, heavy equipment rental, a hotel, and a restaurant; pervious areas consist of irrigated landscaping strips with a surface cover of decomposing granite adjacent to parking lots, along with a number of long grassy swales that line several of the channels and route storm runoff to the USGS-gauged outlet draining the site. The residential site consists mostly of homes, of which half have desert landscaping and half irrigated lawns. At the desert site, asphalt parking surfaces were surrounded by an undeveloped desert remnant (Papago Park) within the urban matrix. In addition to the parking lot sites, comparable experiments to determine the nutrient content of runoff from undisturbed desert soil surfaces were carried out in Sycamore Creek ($33°45'N$, $111°30'W$), a 505 km^2 undeveloped Sonoran desert watershed 35 km northeast of the Phoenix metropolitan area. The watershed is underlain by pre-Cambrian granite and some Quaternary or Tertiary basalt. Annual rainfall at Sycamore Creek averages 295 mm divided roughly in half between summer and winter rains (Thomsen and Shumann, 1968). Soils are compact with abundant eroded mineral grains scattered over the surface; plant litter is largely confined to the base of vegetation clumps except where annual grass growth is locally abundant.

2. Methods

The runoff experiments were performed in early June 1998 shortly before the onset of summer 'monsoon' storms. This is typically the driest season in the northern Sonoran Desert; experiments were preceded by several weeks of dry weather during which dry-deposited material could accumulate.

2.1. SAMPLE PLOT LOCATION AND CHARACTERIZATION

Rainfall-runoff experiments were conducted on between 8 and 10 asphalt plots at each of the four sites, for a total of 38 plots overall (Table I). Plots for runoff simulations were stratified within each site by positioning a similar number in each of two categories: 'byways' (areas where vehicles move) and 'stationary' (areas where vehicles park). Within these general areas sites were chosen by walking a random number of paces from an arbitrary point. The following characteristics of each plot were recorded: pavement condition, plot type (either byway or stationary), parking density, slope of the asphalt surface, and distance from the nearest curb. Pavement condition was described according to the method used by the NURP study (Sartor and Boyd, 1972), which involved assigning a score between 1 and 4 corresponding to excellent (score = 1: smooth surface, no cracks, essentially new condition), good (2: few cracks, near new condition), fair (2: cracks, some pavement deterioration), and poor (4: many cracks, moderate to extensive deterioration). Parking density was also described and scored following Sartor and Boyd (1972) as light (1: very few vehicles parked), moderate (2: around half available spaces filled), or heavy (3: parking mostly continuous). Simulated rainfall experiments also were carried out on 12 plots at the undisturbed desert site in the lower portion of the Sycamore Creek watershed.

2.2. SIMULATED RAINFALL-RUNOFF EXPERIMENTS ON ASPHALT

The rainfall simulator consisted of a water-filled 0.6×0.6 m tank punctured by metal needles arranged in a regular grid (spacing 25 mm). This was designed specifically for rainfall-runoff of studies, with the needles sized to deliver a continuous stream of discrete droplets from a tank containing deionized water, suspended on an aluminum frame at a height 1.64 m above the ground surface. This arrangement produced droplets of realistic hydrometeor size and velocity, while flow rate was controlled by a regulator valve. Water was applied at a rate of 0.82 L min^{-1} for 5 min, corresponding to a rainfall intensity of 170 mm hr^{-1}. This rate was chosen to simulate the upper end of the range seen in rainfall events in Phoenix; rain of this intensity typically falls once or twice every 6 yr (Miller *et al.*, 1973). Due to the large volumes of water required, along with the variable chemistry of monsoon rainfall, it was not considered practical to attempt to simulate natural rainfall chemistry.

All the runoff was captured using a 0.3 m^2 circle of 15 cm wide aluminum soldered to form a circular hoop. This size of 'capture area' was the largest that could be rained on effectively by the simulator; it was also such that a 5 min simulated rain storm could be generated with a practical, portable amount of applied water (5 L). Leakage from beneath the mini-watershed was minimized using a foam lining that formed a temporary seal when pressed lightly onto the asphalt surface. All the surface runoff was collected via aspiration using rubber tubing and a peristaltic pump into a 3 L pre-washed Nalgene bottle. Samples of rainfall from the simulator also were collected for analysis during each experiment and blanks of deionised water stored in sample bottles for a similar length of time also were collected and analyzed. All samples were stored on ice during transport back to the laboratory.

2.3. DESERT SOIL SURFACE RAINFALL-RUNOFF EXPERIMENTS

Similar artificial runoff experiments were performed on 12 plots in the lower part of the Sycamore Creek watershed. Both sites were on open desert slopes with minimal annual plant growth and litter accumulation. As in the parking lots, the experiments used the same mini-watersheds constructed of aluminum flashing, in this case compressed at one end to form a tear-drop-shape to funnel the runoff. A hole was drilled at the compressed ends, covered with nylon mesh and a polyethylene tube was fitted to the outside opening from which to collect the runoff, which flowed via gravity into the sample bottle. Leakage from beneath the mini-watershed was minimized using a foam lining that sealed it temporarily to the soil surface. Water was added using a garden watering can with sprinkler head, since the rainfall simulator was insufficiently portable to use at the desert sites. Runoff volumes (on average 47% of the amount applied) were recorded and subsamples collected in 60 mL acid-washed bottles, which were refrigerated prior to analysis.

2.4. SAMPLE HANDLING (SPLITTING, FILTRATION) AND LAB ANALYSES

For the parking lot samples, the total volume of each sample was measured on return to the laboratory and then split into 5 to 10 subsamples using a Teflon cone splitter to produce identical subsamples in terms of both volume and particle size distribution. Subsamples were filtered through ashed Whatman GF/F (0.7 μm pore size) filters to ensure no organic carbon contamination (Likens and Wetzel, 1991) and analyzed within 48 hr for nitrate (NO_3^--N) and soluble reactive phosphorous (SRP) on a Bran Lubbe TRAACS 800 flow injection analyzer. NH_4^+-N was analyzed colorimetrically (Solorzano, 1969) using the phenol-hypochlorite method on a Shimadzu-UV160U spectrometer. Dissolved organic carbon (DOC)

TABLE II
Storm event duration, intensity, total precipitation and estimated nutrient inputs in rainfall to the mini-urban watersheds during the storm on July 6th 1998

Site	Duration	Intensity	Total precip	Average precipitation composition[a]			
				NO_3^--N	NH_4^+-N	SRP	DOC
	mins	mm min^{-1}	mm	mg L^{-1}	mg L^{-1}	mg L^{-1}	mg L^{-1}
Storm event duration, intensity, total precipitation amount and composition[a]							
Light industrial	45	0.66	7.6	1.0	1.2	0.022	4.2
Commercial	52	3.96	17.0	1.1	1.3	0.047	3.7
Residential	72	6.40	6.6	1.1	1.3	0.047	3.7
Estimated contribution of nutrients input in rainfall to exports measured in storm runoff at the watershed outlets (%)							
Light industrial				119	117	8	6
Commercial				66	28	14	3
Residential				34	38	6	3

[a] Taken from average concentrations in June & July rain events from 1999 and 2000 at CAP monitoring sites (light industrial from Brooks Rd, Mesa; commercial & residential both from Central Phoenix).

was analyzed on Shimadzu TOC-5000 analyzer. Duplicate analyses were run in all cases.

2.5. WHOLE WATERSHED STORM RUNOFF SAMPLING AND ANALYSIS

Shortly after the simulation experiments, on July 6th 1998, a monsoon rain storm produced significant surface runoff from all 3 of the USGS gauged watersheds. Rainfall lasted for 45, 52 and 72 min, with total precipitation amounts of 7.6, 17.0 and 6.6 mm, at the light industrial, commercial and residential sites respectively (Table II). Samples of the resulting runoff were collected by the USGS with automatic-pumping samplers (ISCO, Inc., Model 3700) set to trigger when storm runoff began. Samples were collected in 1 L teflon-lined polyethylene bottles and combined to form a flow-weighted composite sample, then split into subsamples using a Teflon-lined churn splitter, filtered and analyzed for nitrate (NO_3^--N), ammonium (NH_4^+-N), ortho-phosphorus (corresponds closely to SRP), and DOC by the USGS National Water-Quality Laboratory.

2.6. CALCULATION OF SURFACE LOADS AND MASS BALANCES

Data from the simulated rainfall experiments were used to calculate parking lot surface nutrient loadings (*i.e.*, the amount of accumulated material available to be exported in storm runoff) during the first 5 min of an intense summer monsoon rainfall event. Because the experiments were done after the prolonged dry season, which typically occurs prior to the monsoon, these surface loadings are likely to represent maximum figures, rather than typify accumulations throughout the entire year. For each plot, the total mass of each nutrient was determined from sample volume and concentration, then divided by the area washed (0.3 m^2) to convert the concentration into a nutrient amount (loading) per m^2 of asphalt. Similarly, the mass of each nutrient in surface runoff from the desert plots was normalized to the surface area of desert surface from which the sample was taken. To calculate the available nutrient input from the entire parking lot surface ($N_{surface}$) the area covered by asphalt in the whole watershed was determined from aerial photographs and multiplied by the mean surface loadings for each nutrient. Since there were also areas of roof and pervious surfaces potentially contributing some water to the overall storm runoff at the catchment outlets, these estimates are likely to represent minimum values for inputs from surfaces in the contributing drainages.

Precipitation chemistry of the rainfall at the three monitored urban catchments was not available, so rainfall inputs (N_{precip}) were estimated using average concentrations of NO_3^--N, NH_4^+-N, SRP, and DOC for the monsoon rainfall period, collected in adjacent wet deposition collectors as part of CAP LTER routine sampling. Catchment outflows were sampled during the first natural storm runoff after the simulated rainfall-runoff experiments were carried out. Nutrient exports (N_{export}) were calculated by multiplying the total volume of storm runoff by the concentration of each constituent in the flow-weighted composite samples (Lopes *et al.*, 1995). Inputs were then compared with exports and the inferred change in nutrient storage determined as follows:

$$\Delta N_{storage} = N_{precip} + N_{surface} - N_{export}.$$

2.7. STATISTICAL ANALYSES

To look for differences between sites, pavement conditions, and plot types, one and two-way ANOVAs were performed on the entire data set (which was log transformed to adjust for a slight positive skew). Pairwise differences between individual sites and pavement condition categories were examined using Bonferroni multiple comparison tests.

3. Results

Asphalt runoff samples generated by the rainfall experiments were highly colored, varying from yellow-brown at the desert site to dark brown-black at the commercial, residential, and light industrial sites. Of the 4–5 L of water rained onto the plots, on average a sample volume of 2.7 L (66%) was recovered, with the remaining water having been lost via evaporation (air temperatures were typically in excess of 35 °C when the experiments were conducted) and infiltration into the asphalt. On the desert soil plots, slightly more (just over half) water was lost to infiltration and evaporation, with 47% of the added rainfall collected in runoff. Concentrations of nutrients in blank samples collected directly from the rainfall simulator were usually very low (below detection limit); where they were above this level, the average concentration in the blank were used to correct sample data collected for the particular nutrient and site concerned.

3.1. SIMULATED 'FIRST FLUSH' RUNOFF CONCENTRATIONS

Concentrations of dissolved NO_3^--N, NH_4^+-N, and DOC in runoff samples ranged widely both between plots within individual sites and across all plots, while SRP concentrations showed much less variation (Table III). Despite a slight positive skew, means were mostly very similar to medians. Inorganic N in runoff from the asphalt plots was dominated by NO_3^--N (Table III) with a NO_3^--N:NH_4^+-N ratio of 14 overall, although this varied widely between sites. Nitrate-N concentrations varied by two orders of magnitude from 1.1 to 115.9 mg L^{-1} (Table III) and were highest at the light industrial and desert parking lot sites and lowest at the residential site. Ammonium-N was a significant component of the inorganic N in runoff from the commercial and residential plots (approximately one third and two-thirds, respectively); the highest NH_4^+-N concentrations were found at the residential and commercial sites (both in the NW valley) and the lowest concentrations were obtained at the light industrial and desert sites (both in the east-central part of the valley). Inorganic N concentrations in runoff from the undeveloped desert site were much lower than from any of the urban sites, averaging 0.7 mg L^{-1} and 0.4 mg L^{-1} for NH_4^+-N and NO_3^--N, respectively.

DOC in simulated runoff showed the highest concentrations of all nutrients, ranging from 26.1 to 295.9 mg L^{-1}, with more variation within sites than between them (Table III). Mean DOC concentrations were highest at the commercial site, this being largely due to one exceptionally high value. In contrast to the other nutrients, SRP concentrations were consistently low with a maximum concentration of 1.02 mg L^{-1} and differed by a factor of only 15 across the entire data set. The variance in SRP concentrations at individual sites was much lower than for the other nutrients (Table III). SRP concentrations in desert runoff (mean of 0.19 mg L^{-1}) were similar to those from asphalt.

3.2. ESTIMATED SURFACE LOADINGS

Concentrations in the 'first flush' runoff samples from asphalt represented mean surface loadings of 151.1 mg m^{-2} for NO_3^--N, 40.2 mg m^{-2} for NH_4^+-N, 532.2 mg m^{-2} for DOC, and 1.8 mg m^{-2} for SRP (Table IV). The mean surface load for SRP in the artificially-generated runoff from desert soil surfaces at Sycamore Creek was 2.2 mg m^{-2}, with surface loads of 1.6 mg m^{-2} for NO_3^--N and 3.1 mg

TABLE III

Concentrations of dissolved nutrients in simulated storm runoff from asphalt parking lot surfaces

Site	NO_3^--N	NH_4^+-N	SRP	DOC
Mean, range and variance (in mg L^{-1})				
Box Culvert (light industrial)				
Mean	26.6	0.8	0.18	47.6
Variance	1009.9	0.2	> 0.01	146.4
Range	6.2–115.8	0.2–1.7	0.13–0.26	32.3–73.1
Peoria & 43rd Ave (commercial)				
Mean	14.2	9.6	0.30	81.2
Variance	21.9	10.2	0.08	6876.2
Range	6.3–21.0	6.5–17.0	0.14–1.02	32.8–295.9
Olive & 67th Ave (residential)				
Mean	3.4	6.7	0.13	59.1
Variance	1.1	2.7	0.00	133.3
Range	1.9–5.7	4.4–9.1	0.07–0.23	41.1–81.5
Papago Park (remnant desert)				
Mean	16.0	1.2	0.15	42.2
Variance	8.0	0.1	0.01	107.2
Range	11.6–19.8	0.6–1.8	0.09–0.39	26.1–57.4
All parking lot sites				
Mean	15.4	4.6	0.19	57.7
Variance	339.8	17.4	> 0.1	1888.3
Range	1.9–115.8	0.1–17.0	0.07–1.02	26.1–295.9
Sycamore Creek (undeveloped desert)				
Mean	0.4	0.7	0.19	n.a.

m^{-2} for NH_4^+-N. These latter two values were lower those for the asphalt parking lot surfaces by factors of 13 and 91 respectively (Table IV).

An ANOVA on asphalt surface loads (Table V) showed significant differences between the different parking lot sites for all nutrients except DOC, which was uniformly high, irrespective of site. NH_4^+-N loads differed for most pairwise site comparisons; differences in NO_3^--N loads were between the residential site and the

TABLE IV

Nutrients on surfaces calculated from simulated storm runoff

Site	NO_3^--N	NH_4^+-N	SRP	DOC
Surface loadings (in mg m^{-2})				
Box Culvert (light industrial)				
Mean	156.3	11.5	1.5	408.9
Std. error	10.9	1.2	0.4	36.2
Peoria & 43rd Ave (commercial)				
Mean	127.1	84.3	2.7	724.2
Std. error	17.3	10.2	0.9	255.7
Olive & 67th Ave (residential)				
Mean	28.7	57.1	1.1	504.1
Std. error	3.3	5.7	0.2	50.5
All parking lot sites				
Mean	151.1	40.2	1.8	532.2
Std. error	33.7	6.1	0.3	66.6
Desert sites				
Mean	1.6	3.1	2.2	n.a.

TABLE V

Differences in asphalt plot surface loadings due to site, pavement condition and plot type from one and two-way ANOVA

Dependent variable	Site	Pavement condition	Plot type	Site* pavement condition	Site* plot type	Pavement condition *plot type
Significance levels (log-transformed data, $n = 36$ samples)[§]						
$L_n NH_4^+$	0.000[a,b,c,d,e]	0.003[C,E]	0.806	0.041	0.153	0.380
L_n DOC	0.557	0.108	0.241	0.866	0.008	0.004
L_n SRP	0.015[f]	0.069	0.041	0.167	0.323	0.055
$L_n NO_3^-$	0.000[c,e,f]	0.057[F]	0.998	0.332	0.229	0.182

[§] Superscripts denotes where Bonferroni multiple comparison test showed significant differences between individual pavement condition categories ([A] poor:fair; [B] poor:good; [C] poor:excellent; [D] fair:good; [E] fair:excellent; [F] good:excellent) and between different sites ([a] desert:industrial; [b] desert:commercial; [c] desert:residential; [d] industrial:commercial; [e] industrial:residential; [f] commercial:residential).* Denotes test for interaction.

TABLE VI

Nutrient mass balances for three urban parking lots in metropolitan Phoenix during a monsoon storm on July 6th 1998

NH_4^+-N and NO_3^--N (mg m^{-2}) Site	NO_3-N				NH_4^+-N			
	Inputs		Export	Δ Storage	Inputs		Export	Δ Storage
	Surface runoff	Precipitation			Surface runoff	Precipitation		
Light industrial	75.3	3.0	2.6	75.8	3.3	3.7	3.2	3.8
Commercial	6.7	11.8	17.9	60.9	43.4	13.5	47.9	9.1
Residential	3.9	1.2	3.4	1.7	8.9	1.3	3.5	6.7

SRP (mg m^{-2}) and DOC (g m^{-2}) Site	SRP				DOC			
	Inputs		Export	Δ Storage	Inputs		Export	Δ Storage
	Surface runoff	Precipitation			Surface runoff	Precipitation		
Light industrial	0.7	0.1	0.9	$-$ 0.1	182.4	12.6	226.4	$-$ 31.4
Commercial	1.0	0.5	3.6	$-$ 2.1	214.3	35.7	1285.7	$-$1000.0
Residential	0.1	0.04	0.7	$-$ 0.5	69.4	4.2	111.1	$-$ 27.8

Inputs from rainfall and nutrients washed off the asphalt surfaces (calculated using the data from the simulated rainfall experiments) were compared with nutrient exports in storm runoff measured at the parking lot outlets. Total storm input and export estimates are expressed per square meter of parking lot, to allow comparison between the different sites.

others; for SRP there was only a significant difference between the commercial and residential sites (Table V). These differences in asphalt surface loads of NH_4^+-N and NO_3^--N were related to pavement condition. This relationship was an inverse one; loads were highest on the poorest pavement conditions. The commercial site had the highest frequency of asphalt plots rated to be in 'poor' condition and these plots had particularly high NH_4^+-N loads. There also was a significant interaction between site and pavement condition for NH_4^+-N loads (Table V). SRP and DOC asphalt surface loads showed no differences according to pavement condition, although pavement condition and site together did produce a significant difference in DOC load (Table V). SRP load also differed according to whether the plot was in classified as 'stationary' or 'byway', with higher loads measured on stationary plots at all but the residential site. Other measured variables (traffic type, distance from curb and slope) did not appear to influence nutrient surface loads and these variables are not discussed further.

3.3. PARKING LOT NUTRIENT MASS BALANCES

The natural rainfall event on July 6th 1998 at the three USGS-monitored watersheds lasted from between 45 to 72 min and deposited from 6.6 to 17.0 mm of rain (Table II). Total runoff exports for the entire storm, calculated from measured flow and analyses of composite runoff samples at the parking lot outlets, were divided by the total surface area of each site to allow comparison between sites (Table VI). These exports ranged from 2.6 to 17.9 mg m^{-2} for NO_3^--N and from 3.2 to 47.9 mg m^{-2} for NH_4^+-N; measured SRP losses in storm runoff almost an order of magnitude lower at 0.7 to 3.6 mg m^{-2}, while DOC exports were over an order of magnitude higher at between 111.1 and 1285.7 mg m^{-2}.

The degree to which storm exports at the parking lot outlets corresponded with predicted inputs from the readily soluble nutrient loads on the asphalt surfaces (as determined by the simulation experiments) varied with constituent and site. When the contribution from the asphalt surfaces was estimated for the entire parking lot surface, it generally was sufficient to account for export at catchment outlets. Indeed for NO_3^--N and NH_4^+-N, rainfall inputs alone were sufficient to account for between 30 to 120% of inorganic N exiting the catchments in storm runoff (Table II) and more inorganic N appeared to be entering than was exiting. In the case of NO_3^--N at the light industrial site, predicted inputs from the asphalt surface were 43.5 times greater than exports at the outlet. Conversely for SRP and DOC the amounts predicted to be contributed by asphalt washoff and estimated rainfall inputs were similar to, or very slightly lower than, total losses from the parking lots in storm runoff. For these two nutrients, rainfall inputs accounted for only a small proportion of storm exports from the basins (6–14% in the case of SRP and 3–6% for DOC).

4. Discussion

A major aim of the study was to quantify the maximum likely amount of accumulated nutrients that could be readily dissolved and transported in the first flush conditions produced by an intense monsoon rainfall following an extended dry period. Advantages of using the rainfall simulator were that the conditions generated would correspond closely to natural rainfall:runoff conditions, while at the same time standardizing conditions for all the plots, rather than sampling natural storms where the amount of surface material solubilized and transported is partly influenced by the size (precipitation and runoff volume) of the event. In addition, natural runoff integrates sources from the whole catchment surface area, giving no information on the contribution from individual surface types and locations within the drainage area, whereas a rainfall simulator permits the investigator to determine the degree of spatial variability across the contributing area.

Our findings differ from the findings of the NURP (Sartor and Boyd, 1972; EPA, 1983) in that the amount of exportable nutrients measured on asphalt parking lot surfaces – particularly for NO_3^--N, are a good deal higher than the mean value of 4.6 mg N m^{-2} reported for street surfaces in the NURP study. One explanation of this difference is that traffic turbulence on street surfaces limits the accumulation of nutrients (specifically those in particulate form) compared to parking lots, where turbulence is likely to be a lot lower due to much slower vehicle speeds. However, our measurements do correspond closely with more recently published findings – for example the estimated loads of 120 mg m^{-2} NO_3^--N and 2 mg m^{-2} NH_4^+-N for urban surfaces in Los Angeles (Burian et al., 2001) and also of 239 mg m^{-2} NO_3^--N and 41 mg m^{-2} NH_4^+-N for French highway surfaces (Pagotto et al., 2000). Conversely for phosphorus, our mean parking lot loading of 1.8 mg m^{-2} is an order of magnitude lower than the mean for city streets nationwide of 19.1 mg P m^{-2} reported by NURP (Sartor and Boyd, 1972). Moreover we found SRP surface loads to be consistently low, showing little within-site variation. DOC concentrations and loadings (not measured during NURP) were high, with an average concentration of 57.7 mg L^{-1} in experiment runoff across all the asphalt sites, compared to typical concentrations in treated wastewater in the region (e.g., 17 mg L^{-1} found by Westerhoff and Pinney, 2000).

4.1. SITE FACTORS RELATED TO NUTRIENT LOADS ON ASPHALT SURFACES

Previous work has shown that the quantity of contaminant material on city street surfaces varies widely and depends principally on surrounding land use, duration since last rainfall or street-cleaning, and street surface type and condition (Sartor and Boyd, 1972). In our study, we only found a direct relationship between nutrient loads on asphalt surfaces and pavement condition in the case of NH_4^+-N. There was no clear relationship between asphalt surface loadings of NO_3^--N, SRP, DOC and pavement condition, although there was sometimes an interactive effect of site and pavement condition combined. Because we had only one site for each dominant surrounding land use type, it was not possible to discern a clear land use effect, but our data do at least indicate that location is a factor.

4.2. SOURCES OF ENHANCED NUTRIENT LOADS ON PARKING LOT SURFACES

Runoff from asphalt at urban sites contained considerably higher soluble inorganic N loads than from natural desert soil surfaces – by a factor of 91 for NO_3^--N and 13 for NH_4^+-N. There are a number of potential sources for elevated levels of leachable inorganic N species in the urban environment, of which the primary one is atmospheric. Deposition rates of oxidized N species to soil and artificial urban surfaces when dry are similar (Grossman-Clarke et al., submitted) and are likely to be low after an extended dry period (Steinberger and Sarig, 1993; Vishnevetsky and Steinberger, 1997), so the low values in soil are unlikely to reflect recent depletion due to plant or microbial uptake in the desert soils. It is more likely that our findings result from enhanced atmospheric N deposition and accumulation rates on the urban asphalt surfaces.

Atmospheric N deposition (particularly dryfall) is significantly enhanced in and around urban areas such as Phoenix (Russell *et al.*, 1993; Bytnerowitz and Fenn, 1996; Lovett *et al.*, 2000; Baker *et al.*, 2001; Grossman-Clarke *et al.*, submitted) due to elevated ambient air concentrations of fine particulate and gaseous nitrogen oxides from fossil fuel combustion (especially vehicle exhausts), as well as emissions from irrigated croplands and landscapes, manure, and wastewater treatment operations (Dignon and Hameed, 1989; Goulding *et al.*, 1998; Baker *et al.*, 2001). While inert urban surfaces have low deposition velocities for many N species relative to vegetation surfaces and open water (Sehmel, 1980), during prolonged periods between rainfall events, significant accumulation may occur on these surfaces. Additional sources of inorganic N to the parking lot surface are from people spilling liquids, animals urinating, and drift of fertilizer applied to landscape areas adjacent to the asphalt surfaces.

The predominance of inorganic N as NO_3^- rather than NH_4^+ on parking lot surfaces (ratio 14:1 overall, but varied significantly between sites) is likely to reflect the relative stability of those compounds and the proximity of the individual parking lot study sites to sources of NH_3 versus NO_x derivatives. Nitric acid and other nitrates are terminal compounds that are relatively long-lived, so deposition can occur over large areas, some distance from source (Singh, 1987). In this study NO_3^--N was dominant at the light industrial and desert sites, located in the central part of the urban area known to be under the plume of pollutants that develops each day (Fernando *et al.*, 2001). In the case of NH_4^+-N, the highest loadings and worst pavement condition were both at the industrial and commercial parking lots, located in the same general (NW) part of the metropolitan area. Ammonia is very reactive with a much shorter atmospheric lifetime and would therefore be expected to show high concentrations near sources, decreasing rapidly with distance from those sources (Asman *et al.*, 1998). We found NH_4^+-N to be highest at the commercial and residential sites, both located in the NW of Phoenix-Glendale (Figure 1) closer to agricultural areas which are likely to emit NH_3 (Baker *et al.*, 2001).

In contrast to inorganic N, phosphate loads did not differ much between the parking lot and desert soil plots. Mean SRP loads were 2.2 mg m^{-2} in the desert compared to 1.8 mg m^{-2} urban plots. Phosphate concentrations in desert soils of the region are typically not limiting to plant growth, as phosphate is readily derived from weathering of P-containing minerals in bedrock. In urban environments, many urban building materials, including asphalt, have moderately high SRP contents; another major source of dry deposition in urban areas such as Phoenix is soil-derived dust (Kleeman and Cass, 1998; ADEQ, 1999; Artaxo *et al.*, 1999). The finding that SRP loads tended to be higher on asphalt plots classified as 'stationary' rather than 'byways' supports the importance of soil-derived dust that is likely to accumulate more on less disturbed surfaces. Overall we conclude that urbanization appears to have little effect on SRP accumulations to terrestrial surfaces in arid systems such as this.

The DOC loadings on the parking lot surfaces could represent a significant impact to recipient systems from storm runoff. This DOC has three likely major sources – leakage from vehicles (e.g., dripping oil, gasoline, and other hydrocarbon fluids; Lopes et al., 2000), leaching of surface particulates derived from the breakdown of the asphalt surface itself (Sartor and Boyd, 1972; EPA, 1983), and atmospheric deposition (Lovett et al., 2000).

4.3. NUTRIENT MASS BALANCES FOR PARKING LOT WATERSHEDS

From the urban mini-watershed mass balance calculations for the July 6th storm event we found that measured NO_3^--N and NH_4^+-N exports in storm runoff at the parking lot 'catchment' outlets tended to exceed estimated inputs in rainfall and soluble surface nutrient accumulations combined. Indeed at the commercial and light industrial sites, only 27 and 3%, respectively, of the potential surface-accumulated NO_3^--N appeared to be exported, while only 40% of the estimated surface NH_4^+-N loading was exported from the residential site. There are a number of possible explanations for this. We may have overestimated the total amount of NO_3^--N on the asphalt surface due to the high spatial variation from plot to plot at the light industrial site. However, in the case of DOC at the commercial site, where the variance of the DOC data from plot to plot is also very high, measured exports were far greater than estimated inputs to the whole parking lot. Because the difference is in the opposite direction to that observed for NO_3^--N, we suggest that the data are representative of a real difference and not solely a result of errors involved in estimating the total surface nutrient amounts on the study sites. Another possible explanation is that we only simulated the first flush of runoff, which may represent a relatively small proportion of the total storm runoff load (e.g., Deletic, 1998). However, limited data from an ancilliary experiment where the same asphalt plot was rewashed repeatedly five times indicate that the majority of soluble material (on average 70%) is lost in the first 5 min of washing. A recent modeling study by Burian et al. (2001) suggested that only a small percentage of the NO_3^--N (1.5%) and NH_4^+-N (1.3%) dry deposited onto an urban catchment in Los Angeles, California was actually discharged in runoff, indicating that there may be other processes that act to remove or store atmospherically deposited inorganic N before it can be transported from the catchment.

Given the above evidence, along with the fact that the contribution of nutrients in runoff from other impervious surfaces (e.g., roofs) was not measured, we conclude it likely that N is accumulated on these parking lot catchments converted to another form or was lost from the catchment by another route (volatilization, denitrification, or microbial uptake). We suggest that significant retention and/or transformation of NO_3^--N and NH_4^+-N must occur either on the parking lot surface or at sites along the flowpath to the catchment outlet. The latter explanation seems to be supported by the fact that the difference in input:output was largest at the light industrial site, the only watershed which included a number of grassy swales through which much of the urban runoff is channeled prior to reaching the outlet.

For SRP differences in the estimated input:output balances for the study catchments are small. Hence, while there may be additional sources of this nutrient in the watersheds (e.g., from other impervious surfaces not measured, such as roofs), they are probably not that significant. For DOC, the mismatches may be due, at least in part, to the extremely patchy distribution of oil and grease deposits on the parking lots, causing the amount of DOC available for export to have been significantly underestimated by the simulation experiments.

Finally, we very much consider the work reported here to form a useful precursor study to further work. Future expansion of the approach outlined here could usefully include: i) simulations of natural rainfall chemistry from the simulator, to test the effect of this on surface wash off chemistry; ii) repeatedly washing off the same section of parking lot until the surface load is exhausted, to enable a 'wash off' function to be computed for the various ions; iii) measure a wider suite of nutrients and other urban contaminants, e.g. total N and organic N, heavy metals, and polyaromatic hydrocarbons; iv) repeat the study on a more widespread basis with more replicate plots, in order to investigate the effect of variables such as different land use types, distance from strong sources of NO_x emissions, e.g. major roadways and from the urban center, as well as studying the variation in N and DOC across individual parking lots and other impervious urban surfaces in more detail.

5. Conclusions

Nitrogen continues to be ubiquitous and poorly controlled, largely due to numerous non-point sources (Smith *et al.*, 1994; Howarth *et al.*, 1996; Vitousek *et al.*, 1997). Atmospheric deposition may constitute a significant input of macronutrients in urban environments (Russell *et al.*, 1993; Lovett *et al.*, 2000; Smith *et al.*, 2000). A recently completed mass balance of N for Phoenix found that NO_x emissions from fossil fuel combustion may contribute around 30% of the total annual N inputs to the ecosystem (Baker *et al.*, 2001). Russell *et al.* (1993) estimated that in the Los Angeles basin, 52% of NO_x emissions and 53% of NH_3 emissions were deposited within the modeled airshed. The methodology used in this study allowed examination of point-to-point variability of the type of area, which in previous work has been treated as a uniform non-point source. The results show that impervious urban surfaces can constitute important sources of nutrients in urban storm runoff, but that the amount of such material exported can vary significantly for different nutrients and with the characteristics of the urban catchment area. Input-output budgets for an actual rain event show that much of the available load on a parking lot surface does not appear to reach surface waters during floods. In contrast, for SRP and especially DOC, parking lot surfaces would appear to be significant source areas, the readily leachable loadings on asphalt accounting for the majority of the runoff export measured during a storm event.

Acknowledgments

Thanks to Tom Colella, Jacqueline Walters, Damon Bradbury, Mike Myers, Sarah Quinlivan, Roy Erickson for field and laboratory assistance, to Erik Wenninger, Lisa Dent, Aaron McDade, Ayoola Folarin, and Mark Compton for field assistance, and John Brock for the use of his rainfall simulator. Also thanks to Ken Fossum of the U.S. Geological Survey, Water Resources Division, for providing the storm runoff data and information about urban watersheds. This research was carried out as part of the Central Arizona-Phoenix LTER research project, which is supported by the National Science Foundation's Long-Term Studies Program, grant number DEB 9714833.

References

ADEQ: 1999, 'Air Quality Report' (A.R.S. #49–424.10) Appendix I. Arizona Department of Environmental Quality, Phoenix, Arizona.

Artaxo, P., Oyola, P. and Martinez, R.: 1999, 'Aerosol composition and source apportionment in Santiago de Chile', *Nucl. Instrum. Methods Phys. Res. B* **150**, 409–416.

Asman, W. A. H., Sutton, M. A. and Schjorring, J. K.: 1998, 'Ammonia: emission, atmospheric transport and deposition', *New Phytol.* **139**, 27–48.

Baker, L. A., Hope, D., Xu, Y., Lauver, L. and Edmonds, J.: 2001, 'Nitrogen balance for the Central Arizona-Phoenix ecosystem', *Ecosystems* **4**, 582–602.

Burian, S. J., Streit, G. E., McPherson, T. N., Brown, M. J. and Turin, H. J.: 2001, 'Modeling the atmospheric deposition and stormwater washoff of nitrogen compounds', *Environ. Model. Software* **16**, 467–479

Bytnerowicz, A. and Fenn, M. E.: 1996, 'Nitrogen deposition in California forests: A review', *Environ. Pollut.* **92**, 127–146.

Deletic, A.: 1998, 'The first flush load of urban surface runoff', *Water Res.* **32**, 2462–2470.

Dignon, J. and Hameed, S.: 1989, 'Global emissions of nitrogen and sulfur-oxides from 1860 to 1980', *J. Air Water Manage. Assoc.* **39**, 180–186.

EPA: 1983, 'Results of the Nationwide Urban Runoff Program – Volume 1 – Final Report. Water Planning Division, US Environmental Protection Agency', Report # PB84-185552, Washington D.C.

Fernando, H. J. S., Lee, S. M., Anderson, J., Princevac, M. Pardyjak, E. and Grossman-Clarke, S.: 2001, 'Urban fluid mechanics: Air circulation and contaminant dispersion in cities', *Environ. Fluid Mech.* **1**, 107–164.

Goulding, K. W. T., Bailey, N. J., Bradbury, N. J, Hargreaves, P., Howe, M., Murphy, D. V., Poulton, P. R. and Willison, T. W.: 1998, 'Nitrogen deposition and its contribution to nitrogen cycling and associated soil processes', *New Phytol.* **139**, 49–58.

Grennfeld, P. and Hultberg, H.: 1986, 'Effects of nitrogen deposition on the acidification of terrestrial and aquatic ecosystems', *Water, Air, Soil Pollut.* **30**, 945–963.

Grossman-Clarke, S. G., Hope, D., Lee, S. M., Fernando, H. J. S., Hyde, P. G., Stefanov, W. L. and Grimm, N. B.: 'Modeling temporal and spatial characteristics of nitrogen dry deposition in the Phoenix metropolitan area', *Environ. Sci. Technol.*, submitted.

Hicks, B. B.: 1998, 'Wind, water, earth, and fire – a return to an Aristotelian environment', *Bull. Am. Meteorol. Soc.* **79**, 1925–1933.

Howarth, R. W., Billen, G., Swaney, D., Townsend, A., Jaworski, N., Lajtha, K., Downing, J. A., Elmgren, R., Caraco, N., Jordan, T., Berendse, F., Freney, J., Kudeyarov, V., Murdoch, P. and Zhu, Z.-L.: 1996, 'Regional nitrogen budgets and riverine N and P fluxes for the drainages to the North Atlantic Ocean: Natural and human influences', *Biogeochemistry* **35**, 75–139.

Kleeman, M. J. and Cass, G. R.: 1998, 'Source contributions to the size and composition of urban particulate air pollution', *Atmos. Environ.* **32**, 2803–2816.

Latimer, J. S., Hoffman, E. J., Hoffman, G., Fasching, J. L. and Quinn, J. G.: 1990, 'Sources of petroleum hydrocarbons in urban runoff', *Water, Air, Soil Pollut.* **52**, 1–21.

Leopold, L. B.: 1968, 'Hydrology for Urban Land Planning – A Guidebook on the Hydrologic Effects of Urban Land Use', Geological Survey Circular 554, U.S. Geological Survey, Washington, D.C.

Likens, G. E. and Wetzel, R. G.: 1991, *Limnological Analysis*, 2nd edition, Springer-Verlag, New York.

Lopes, T. J., Fossum, K. D., Phillips, J. V. and Monical, J. E.: 1995, 'Statistical summary of selected physical, chemical, and microbial characteristics, and estimates of constituents loads in urban stormwater', Maricopa County, Arizona, U.S. Geological Survey, Water Resources Investigations Report 94-4240, U.S. Geological Survey, Tucson, Arizona, U.S.A.

Lopes, T. J., Fallon, J. D., Rutherford, D. W. and Hiatt, M. H.: 2000, 'Volatile organic compounds in storm water from a parking lot', *J. Environ. Eng.* **126**, 1137–1143.

Lovett, G. M., Traynor, M. M., Pouyat, R. V., Carreiro, M. M., Zhu, W. X. and Baxter, J. W.: 2000, 'Atmospheric deposition to oak forests along an urban-rural gradient', *Environ. Sci. Technol.* **34**, 4294–4300.

Miller, J. F., Frederick, R. H. and Tracy, R. J.: 1973, 'NOAA atlas 2: Precipitation frequency atlas of the western United States, Volume VIII – Arizona', U.S. Department of Commerce, Silver Spring, Maryland, U.S.A.

Pagotto, C., Legret, M. and Le Cloirec, P.: 2000, 'Comparison of the hydraulic behaviour and the quality of highway runoff water according to the type of pavement', *Water Res.* **34**, 4446–4454.

Peterjohn, W. T. and Schlesinger, W. H.: 1990, 'Nitrogen loss from deserts in the southwestern United States', *Biogeochemistry* **10**, 67–79.

Pitt, R. E. and Amy, G.: 1973, 'Toxic materials analysis of street surface contaminants', EPA Report # EPA-R2-73-283, U.S. Environmental Protection Agency, Washington, D.C.

Ricci, E. D.: 1984, 'Final Report Urban Stormwater Runoff: A Review Submitted to the Environmental Services Department Salt River Project', Center for Environmental Studies, Arizona State University, Tempe, Arizona.

Russell, A. G., McRae, G. J. and Cass, G. R.: 1993, 'Mathematical modeling of the formation and transport of ammonium nitrate aerosol', *Atmos. Environ.* **17**, 949–964.

Sansalone, J. J., Koran, J. M., Smithson, J. M. and Buchberger, S. G.: 1998, 'Physical characteristics of urban roadway solids transported during rain events', ASCE *J. Environ. Eng.* **124**, 427–440.

Sartor, J. D. and Boyd, G. B.: 1972, 'Water Pollution Aspects of Street Surface Contaminants', EPA-R2-72-081, U.S. Environmental Protection Agency, Washington, DC.

Schlesinger, W. H.: 1997, *Biogeochemistry*, second edition, Academic Press, London.

Sehmel, G. A.: 1980, 'Particle and gas dry deposition: A review', *Atmos. Environ.* **14**, 983–1011.

Singh, H. B.: 1987, 'Reactive nitrogen in the troposphere', *Environ. Sci. Technol.* **21**, 320–326.

Smith, R. A., Alexander, R. B. and Lanfear, K. J.: 1994, 'Stream Water Quality in the Conterminous United States – Status and Trends of Selected Indicators During the 1980's', U.S. Geological Survey, Washington, DC.

Smith, R. I., Fowler, D., Sutton, M. A., Flechard, C. and Coyle, M.: 2000, 'Regional estimation of pollutant gas dry deposition in the UK: Model description, sensitivity analysis and outputs', *Atmos. Environ.* **34**, 3757–3777.

Solorzano, L.: 1969, 'Determination of ammonia in natural waters by the phenolhypochlorite method', *Limnol. Oceanogr.* **14**, 799–801.

Stefanov, W. L., Ramsey, M. S. and Christensen, P. R.: 2001, 'Monitoring urban land cover change: An expert system approach to land cover classification of semiarid to arid urban centers', *Remote Sens. Environ.* **77**, 173–185.

Steinberger, Y. and Sarig, S.: 1993, 'Response by soil nematode populations and the soil microbial biomass to a rain episode in the hot, dry Negev desert', *Biol. Fertility Soils* **16**, 188–192.

Thomsen, B. W. and Shuman, H. H.: 1968, 'Water resources of the Sycamore Creek watershed, Maricopa County, Arizona, U.S.A.', U.S. Geological Survey, Water-Supply Paper 1861, Tempe, Arizona.

Vishnevetsky, S. and Steinberger, Y.: 1997, 'Bacterial and fungal dynamics and their contribution to microbial biomass in desert soil', *J. Arid Environ.* **37**, 83–90.

Vitousek, P. M., Aber, J. D., Howarth, R. W., Likens, G. E., Matson, P. A., Schindler, D. W., Schlesinger, C. A. and Tilman, G. D.: 1997, 'Human alterations of the global nitrogen cycle: Causes and consequences', Issues in Ecology 1, Ecological Society of America, Washington, D. C.

Westerhoff, P. and Pinney, M.: 2000, 'Dissolved organic carbon transformations during laboratory-scale groundwater recharge using lagoon-treated wastewater', *Waste Manage.* **20**, 75–83.

NITROGEN TURNOVER AND NITRATE LEACHING AFTER BARK BEETLE ATTACK IN MOUNTAINOUS SPRUCE STANDS OF THE BAVARIAN FOREST NATIONAL PARK

CHRISTIAN HUBER[1]*, MANUELA BAUMGARTEN[2], AXEL GÖTTLEIN[1] and VERENA ROTTER[1]

[1] Fachgebiet für Waldernährung und Wasserhaushalt, Department für Ökologie, Wissenschaftszentrum Weihenstephan, TU-München, Am Hochanger 13, D-85354 Freising, Germany; [2] Forstliche Versuchs- und Forschungsanstalt Baden Württemberg, Abteilung Biometrie und Informatik, Wonnhaldestr. 4, D-79100 Freiburg i. Br., Germany
(* author for correspondence, e-mail: Huber@forst.tu-muenchen.de, phone: +49 8161 71 4789, fax: +49 8161 71 4738)

(Received 20 August 2002; accepted 6 April 2003)

Abstract. More than 85% of the mountainous spruce forest of the Bavarian Forest National Park died after bark beetle attack during the last decade. The elemental budget of intact stands and of different stages after the dieback was investigated. N-fluxes in throughfall of intact stands were lower (12–16 kg ha^{-1} a^{-1}) than in an earlier study in an intact mountainous spruce stand in the Bavarian Forest National Park and were reduced in the first years after the dieback (3–5 kg N ha^{-1} a^{-1}). Nitrate-N fluxes by seepage water of intact stands at 40 cm depth, which is below the main rooting zone, were moderate (5–9 kg ha^{-1} a^{-1}). After the dieback of the stands, NH_4^+ concentrations were increased in humus efflux as were NO_3^- concentrations in mineral soil. Due to the relatively high precipitation, dilution of the elemental concentrations in seepage was considerable. Therefore, NO_3^- concentrations were usually below the level of drinking water (806 μmol NO_3^- L^{-1}), with lowest concentrations after the snowmelt and highest in autumn. Nitrate concentrations were elevated from the first year until the 7th year after the dieback. Total NO_3^--N losses by seepage until the 7th year after the dieback equalled 543 kg N ha^{-1}. Aluminium fluxes after the dieback were enhanced in the mineral soil from 55 to 503 mmol$_c$ m^{-2} a^{-1} (average of 8 yr), K$^+$ fluxes from 8 to 37 mmol$_c$ m^{-2} a^{-1}, and Mg^{2+} fluxes from 13 to 35 mmol$_c$ m^{-2} a^{-1}. The consequences for the nutritional status of the ecosystem, the hydrosphere, and forest management are discussed in the paper.

Keywords: aluminium, ammonium, bark-beetle, clear-cutting, magnesium, nitrate, nitrogen cycle, potassium, spruce, water quality

1. Introduction

The Bavarian Forest National Park, founded 1970, covers an area of 24 250 ha. Mild winters, warm summers, and storm-thrown trees promoted the development of bark beetles (*Ips typographus*) in the park during the last decade, especially in the mountainous spruce ecosystem (1100–1450 m altitude). According to the 'leaving nature alone' directive of the World Nature Protection Organisation, IUCN, natural processes are given priority in the National Park (Rall, 1999). Therefore, no countermeasures against the bark beetle attack were undertaken except for the creation of a border zone of 500 to 1000 m to prevent further spread. In 1988, the National Park started to document the spread of bark beetles with an aerial photo survey. Since 1992, the rate of dieback of the spruce stands increased. In the highlands of the Bavarian Forest National Park, over 85% of the trees died, along with 18% of the trees in the mixed woodlands of the slopes and lowlands (through 2001).

The investigation of effects of disturbed ecosystems on the water and elemental budgets of forest ecosystems was one of the first main tasks in ecosystem research (e.g., Likens *et al.*, 1969, 1977; Vitousek *et al.*, 1979; Swank 1988; Hauhs, 1989; Frazer *et al.*, 1990; Führer and Hüser, 1991; Bauhus, 1994). After processes like clear-felling or other harvesting methods, there is an abrupt increase in dead biomass (roots and aboveground parts), a strong decrease in water and nutrient uptake, an increase in soil moisture, and higher energy fluxes to the ground (elevated radiation, higher mean air temperatures). Transformation processes like mineralisation and nitrification usually are enhanced. If there are no sufficient sinks for N in the ecosystem, 'excess nitrification' leads to further acidification in the soil, NO_3^- and Al leaching, and to losses of nutrient-cations like Mg^{2+}, Ca^{2+}, and K^+.

However, investigations about processes like clear cutting, storm damages, or thinning are not completely comparable to a natural forest dieback. For example, after the dieback the whole biomass remains on the site, and the stems remain for up to 3 to 5 yr until they break down, some even longer (Heurich *et al.*, 2001). Therefore radiation, air temperature, and interception in such stands is, relatively to a clear-cut, unchanged for the first several years (Huber and Göttlein, 2001).

In the slopes of the Bavarian National Park, changes in the elemental turnover after bark beetle attack already are under investigation in the 'UN/ECE Monitoring Project Forellenbach', situated at 820 m above sea level (Beudert, 1999). However, N transformation processes may be different in the mountainous region because of different site conditions, including lower air temperatures, higher precipitation, longer lasting snow cover, shorter vegetation period, and slower growth of the regeneration as compared to site conditions at lower elevations.

Our main questions in the investigation of the elemental turnover after bark beetle attack in the high elevations of the Bavarian Forest National Park are: (1) does the input of N with throughfall change after the dieback of a stand?, (2) how high are the N fluxes in different compartments of an intact spruce stand? (do we have excess nitrification?), (3) what are the changes in seepage water chemistry? (how high are the NO_3^- and Al^{3+} concentrations in seepage water? how long does a potential NO_3^- peak last?), and (4) how high are the N fluxes after the dieback? (are the nutrient losses, especially of N, Mg^{2+}, and K^+, a potential risk for healthy regeneration of the ecosystem?) The study is also of political interest, because of an intensive discussion about effects and further management of the bark beetle attack in the National Park. Therefore, results should be achieved in a relatively short time period.

TABLE I

Yield and growth data at stands close to our main investigation plots (investigations and inventories mentioned in Jehl, 2001)

Age	80–250
Number of trees per ha	Up to 920
Growing stock (m^3 ha^{-1})	App. 500 to 800
Mean height (m)	25.7

2. Site Description

The Bavarian National Park covers an area of 24 250 ha and is situated in the Inner Bavarian Forest in eastern Bavaria, around 150 km east of Munich along the border with the Czech Republic. The altitudes of the highlands are from 1050 to 1450 m above sea level. The highland area is stocked uniformly by a mountain spruce forest (*Soldanello-Picetum barbilophoietosum*). Trees are 80 to 300 yr old and sporadically mixed with rowan trees (*Sorbus aucuparia*). Table I shows some yield and growth data according to the investigations and inventories mentioned in Jehl (2001) at stands near our main investigation plots. The ground flora at the plots is dominated by *Calamagrostis villosa* and *Deschampsia flexuosa*. Other species present were *Vaccinium myrtillus, Luzula sylvatica*, and *Oxalis acetosella*, in intact stands as well *Sphagnum* spp., *Dicranum scoparium*, and *Polytrichum attenuatum*, and on older wind thrown stands *Rubus fructicosa*.

The mean annual air temperature is 3.0 to 4.5 °C, mean annual precipitation is around 1600 mm with a maximum in some years of more than 2000 mm. About 50% of precipitation is snow. Water supply for the trees is usually sufficient; the vegetation period is short. On average snow cover lasts 7 months, from October until May. The load of SO_2 (1.4 μg m^{-3}) and NO_2 (4.5 μg m^{-3}) is low with reference to Central European conditions. The O_3 concentrations are elevated in the range of 60–80 μg m^{-3}, with daily peak concentrations of 190 μg m^{-3} (Beudert, 1999).

The typical soil in the highlands is an acidic brown earth (FAO: Dystric Cambisol), rich in humus with indications of podzolisation, with a strongly acidified topsoil (Elling *et al.*, 1987). The soils are derived from periglacial weathering material of granite and gneiss.

The mineral soil is covered by a 10 to 18 cm thick organic layer subdivided into LOf1 (4.5 cm), Of2 (5.0 cm), and Oh (6.0 cm). The humus form is raw humus. Some important characteristics about nutrient supply are shown in Table II. pH values are low with a minimum in the Oh horizon. In contrast to the humus layer, the mineral soil exhibits a very low base saturation of the cation exchange capacity

TABLE II

Cation exchange capacity and pH of a typical profile of an intact stand. The cation exchange capacity was determined by extracting 3 g material from the organic layer and 5 g from the mineral soil with 1 M NH_4Cl, one hour shaking, centrifuging, filtering, and analysis by ICP OES. CEC – cation exchange capacity, BS – base saturation

Horizon	Layer thickness	pH ($CaCl_2$)	Ca	Mg	K	Fe	Mn	H	Al	CEC	BS
	(cm)				($mmol_c$ 100 g^{-1})						(%)
LOf1	4.5	3.4	7.2	2.7	4.0	0.1	0.7	1.3	0.5	17	84
Of2	5.0	2.6	6.1	2.0	1.5	0.8	0.3	6.4	3.9	21	45
Oh	6.0	2.6	2.5	1.1	0.8	1.8	0.1	5.3	13.3	25	20
0–5 cm		2.7	0.4	0.2	0.2	0.8	0.0	1.6	6.0	9	10
5–10 cm		2.8	0.0	0.1	0.1	0.7	0.0	1.1	6.9	9	2
10–20 cm		3.3	0.0	0.0	0.1	0.2	0.0	0.0	4.9	5	2
20–40 cm		3.9	0.0	0.0	0.1	0.2	0.0	0.0	4.6	5	3

TABLE III

Concentration and amount of total N in different compartments in the soil of an intact spruce stand

Horizon	N_{tot} (%)	N_{tot} (kg ha^{-1})	(C/N)
Lof1	1.45	149	35
Of2	1.85	911	26
Oh	1.76	1170	23
0–5 cm	0.48	979	22
5–10 cm	0.47	940	18
10–20 cm	0.2	1986	22
20–40 cm	0.08	2107	24

of only 2 to 10%, with Al-saturation of 90%. Therefore, nearly all of the plant-available cations are stored in the biomass of the vegetation or in the humus layer. The supply of exchangeable K^+ and especially Mg^{2+} is low and is restricted more or less exclusively to the humus horizons. Therefore, excess nitrification in the humus horizons can cause both further acidification and a depletion of the cations (K^+ and Mg^{2+}) accompanying the mobile anion NO_3^- in humus efflux and seepage water. Considering the very low pH values, there are noticeably high N contents, or

TABLE IV

Short characterisation of the investigated plots

Plot	Description
00	1999 alive, dieback 2000, dead wood standing
99	Dieback 1999, dead wood standing
96	Dieback 1996, dead wood still standing
94	Dieback 1994, storm thrown in winter 1999/2000
83	Wind thrown 1983
E	Spruce stand in the expansion zone of the national park, dead trees were removed

low C/N ratios, in the organic layer (Table III). This means that the microbial decay of organic matter and release of nutrients for recycling is functioning well in the uppermost soil horizons, despite strong acidification and limitation by cold temperature and long snow cover. Therefore, Elling (1987) found higher N-concentrations in needles in mountainous spruce trees than in the slopes. The supply of N in the soil is high. Approx. 8000 kg N ha^{-1} are stored in the humus until 40 cm depth (Table III).

The plots listed in Table IV have been under investigation since May 1999. Five of six plots are located on a plateau in the area of the Reschbachklause close to Finsterau (48.933°N latitude × 13.583°E longitude; altitude 1150 m above sea level) in the old part of the National Park. Additionally, the plot Buchenau (49.033°N latitude × 13.333°E longitude; 1100 m above sea level) is situated in the 'Expansion Area' (approx. 10 000 of ha added to the Bavarian National Park 1997), where bark beetle prevention is planned to be conducted until 2017.

The criteria for selecting the different sites were: (1) stands and sites had to be representative of a large area in the mountainous region of the Bavarian National Park, (2) soil surface had to be flat with prevailing vertical seepage, (3) stands had to represent a mosaic of dead wood stands of different years of the dieback, (4) intact old mountain spruce stands had to be situated close to the investigated dead wood stands (at the beginning of investigation), and (5) a spruce stand in Buchenau was taken to be representative for of 'intact' managed mountainous spruce forest in the 'Expansion Area'.

The date of the dieback was determined on the site for the plots '00' and '99' or with aerial photo interpretation of the yearly flights usually made in early summer by order of the administration of the Bavarian National Park for the plots '96', '94', and '83' (Heurich et al., 2001).

3. Materials and Methods

Throughfall was sampled with five open polyethylene collectors (LÖLF) on each plot at 1 m above the ground. Samples were taken during beginning of May until end of November every 2–3 weeks. From December until end of April, precipitation was collected with 5 polyethylene drums. Water samples were taken after the snow melt (April, May) on a 2 week to monthly basis with 10 to 20 subsamples per plot with zero tension lysimeters below the humus layer (0 cm) and tension ceramic lysimeters (SKF 100, Haldenwanger GmbH, Berlin, applied tension 60 kPa) at 40 cm depth.

pH was measured with an Ingold glass-electrode, HCO_3^- by titration to pH 4.2 on unfiltered samples. Water samples were filtered using membrane filters with a pore size of 0.45 μm (Schleicher & Schuell, NC45) and stored until they were analysed at 4 °C. Aluminium (Al), calcium (Ca), iron (Fe), magnesium (Mg), manganese (Mn), sodium (Na), and potassium (K) were analysed with an ICP (Perkin Elmer, Optima 3000). Chloride (Cl^-), nitrate (NO_3^-), and sulphate (SO_4^{2-}) were determined with ion chromatography (Dionex, IC2020I). Ammonium (NH_4^+) was measured photometrically with a segmented flow analyser (Skalar San plus System, Skalar analytic, GmbH). Dissolved organic carbon (DOC) was measured with a TOC-analyser (Shimadzu 5050).

Meteorological standard data were available from the weather station Waldhäuser (945 m a.s.l.) operated by the German Weather Service and the National Park Administration. Additionally, air-temperature, relative humidity, and photosynthetically active radiation were measured on each plot at hourly intervals. Measured and derived data for global radiation, maximum and minimum air-temperature, water-vapour pressure, and wind speed were used for the calculation of water fluxes with the model BROOK90 (Federer, 1997).

Deposition fluxes were calculated by multiplying of measured water quantity and the concentration of each sampling event. Soil water fluxes were calculated by multiplying the average concentration on a distinct sampling date with the sum of the daily modelled water fluxes since the prior sampling date. Flow-weighted yearly elemental concentrations were calculated by division of the yearly elemental fluxes by yearly water fluxes.

Within the investigation period, a chronosequence from the beginning of the dieback (year 0) to the 7th year after the dieback and the 16th to 18th year after the dieback was established. The scheme of the chronosequence is presented in Table V. The spreading of dieback was monitored using aerial photograph surveys, which were taken once a year in early summer. In 1999, there was an additional survey in October (Heurich *et al.*, 2001). Data were provided by the Administration of the Bavarian National Park.

TABLE V

Scheme of the chronosequence with beginning of the dieback (year 0) till year 18 after the dieback, the year of the investigations with the corresponding plot

		'Plot': Year of investigation
Intact stands	Year – 1	'E': 1999, 2000, and 2001
(control)	Year – 1	'00': 1999
Dieback	Year 0	'00': 2000 and '99': 1999
	Year 1	'00': 2001 and '99': 2000
	Year 2	'99': 2001
	Year 3	'96': 1999
	Year 4	'96': 2000
	Year 5	'96': 2001 and '94': 1999
	Year 6	'94': 2000
	Year 7	'94': 2001
	Year 16	'83': 1999
	Year 17	'83': 2000
	Year 18	'83': 2001

4. Results

4.1. WATER AND ELEMENTAL FLOW IN THROUGHFALL

Water and elemental fluxes in throughfall are presented in Table VI. In the investigated years, precipitation was 1880 mm in the first period (November 1999–October 2000, plot '83'), 1820 mm in the second period (November 2000–October 2001) on the Finsterau site, and 1800 mm in the second period on the Buchenau site. Interception losses were between 300 to 370 mm on the Finsterau plots. No differences were detected for the interception loss between dead wood stands and intact spruce stands while the boles remained standing. Interception loss was lower at the Buchenau site (200 mm).

The healthy stand, not infested by bark beetle, plot 'E' (spruce stand in the expansion area) showed the highest N-input to the soil with throughfall (87 $mmol_c$ m^{-2} a^{-2} N in the first period, respectively, 111 $mmol_c$ m^{-2} a^{-1} in the second period). Bulk precipitation, measured only in the second period, was in the same range (116 $mmol_c$ m^{-2} a^{-1}). The storm-thrown dead wood stands with 'bulk precipitation character' plots '83' and '94' showed nearly the same N-input, 80–99 $mmol_c$ m^{-2} a^{-1}. In contrast, N-fluxes in throughfall in dead wood stands (boles

TABLE VI

Elemental deposition by bulk precipitation and throughfall in the first (November 1999–October 2000) and second period (November 2000–October 2001) of the investigation at the different plots. 'E' is a spruce stand in a still managed expansion zone of the National Park. The other plots are situated in the unmanaged old part of the National Park. '00' was 1999 alive, the 'dieback' happened 2000. On '99' the dieback was 1999, on '96' in 1996, on '94' in 1994. The plot '83' was wind thrown 1983. 'E' – throughfall at plot 'E', EBulk – bulk precipitation at plot 'E'

Plot	mm	H^+	Al^{3+}	Mn^{2+}	Fe^{3+}	NH_4^+	Ca^{2+}	Mg^{2+}	K^+	Na^+	NO_3^-	SO_4^{2-}	Cl^-	HCO_3^-	DOC
	(L m^{-2})						(mmol$_c$ m^{-2} yr^{-1})								(g m^{-2} yr^{-1})

November 1999–October 2000

Plot	mm	H^+	Al^{3+}	Mn^{2+}	Fe^{3+}	NH_4^+	Ca^{2+}	Mg^{2+}	K^+	Na^+	NO_3^-	SO_4^{2-}	Cl^-	HCO_3^-	DOC
'00'	1553	11	1	7	4	43	235	6	78	12	16	53	45	243	28
'99'	1597	5	0	10	4	20	225	6	78	11	14	54	33	213	28
'96'	1610	6	1	4	4	20	224	1	65	8	18	52	31	247	22
'94'	1913	13	0	3	5	54	189	1	57	9	28	61	27	220	16
'83'	1878	13	0	0	5	51	212	1	10	10	48	56	18	207	10
'E'	1690	12	1	2	4	42	256	1	39	10	45	67	28	212	17
'E' Bulk															

November 2000–October 2001

Plot	mm	H^+	Al^{3+}	Mn^{2+}	Fe^{3+}	NH_4^+	Ca^{2+}	Mg^{2+}	K^+	Na^+	NO_3^-	SO_4^{2-}	Cl^-	HCO_3^-	DOC
'00'	1444	7	1	9	7	17	147	6	70	21	8	34	33	189	24
'99'	1519	15	1	7	7	36	161	7	47	24	11	38	30	199	28
'96'	1482	13	1	4	7	50	130	1	55	20	15	35	33	178	21
'94'	1855	15	0	1	8	56	137	0	19	25	32	41	26	166	17
'83'	1824	16	0	0	8	46	147	0	18	20	34	39	21	167	11
'E'	1609	14	3	2	6	65	137	1	52	23	46	54	27	160	21
'E' Bulk	1803	15	2	1	7	72	126	0	18	21	44	54	21	189	12

remained standing) were significantly lower. The lowest N input was measured for the plot '00', which died in 2000, with 25 mmol$_c$ m^{-2} a^{-1} in the second period.

Sulphate deposition decreased on average from the first period (99/00) to the second (00/01), and was nearly equal on all plots in 99/00, but higher in Buchenau compared with Finsterau in 00/01. Slightly higher SO_4^{2-} fluxes were determined in 00/01 on the plots with 'bulk precipitation character' than in plots with stand character, mostly because of the higher water fluxes, while SO_4^{2-} concentrations were equal. DOC, Mn^{2+} and K^+ fluxes were highest on stands that had died recently (Plots '99' and '00').

4.2. Seepage water chemistry in humus efflux

Ammonium, which is taken up by spruce preferentially to NO_3^- (Gessler *et al.*, 1996), is one of the first chemical markers after the dieback (Figure 1). In summer of 2000, single trees died at the plot '00'. At the same time enhanced NH_4^+ concentrations in humus efflux could be detected. All of the bark beetle attacked stands show higher NH_4^+ (Figure 1) and DOC (Figure 2) concentrations in humus efflux than intact stands. Nitrate concentrations in seepage water were only significantly enhanced from summer until autumn on bark beetle attacked stands that had died more than 4 yr ago (Figure 1). The highest NO_3^- concentrations were detected on plot '94' and '96'. The lowest NO_3^- and NH_4^+ concentrations were determined on plot '83'. Concentrations of Ca and Mg were elevated in humus efflux. K^+-concentrations were more than doubled after the bark beetle attack (Figure 3). Nitrate concentrations were rather low in humus efflux compared to seepage water concentrations (compare Figure 1 with Figure 4).

4.3. Seepage water chemistry in mineral soil

Nitrate concentrations in seepage water at 40 cm depth were low in the intact spruce stands 'E' and plot '00', yr 1999 (Figure 4). Plot '00' showed NO_3^- concentrations in 1999, before the dieback in 2000, of 26 μmol$_c$ L^{-1} NO_3^-, the stand in Buchenau of 30 μmol$_c$ L^{-1} NO_3^- on average in the period 1999–2001. The lowest NO_3^- concentrations in the mineral soil were measured for the 1983 storm-thrown stand '83' with 11 μmol$_c$ L^{-1} NO_3^-, independent if influenced either by regeneration or grassland.

However, all of the investigated bark beetle stands, except plot '83', showed elevated NO_3^- concentrations in the mineral soil. Peak concentrations of up to 1980 μmol$_c$ L^{-1} NO_3^- were reached for a single suction cup (on plot '96', August 2000). A marked seasonality of the NO_3^- concentration in all of the bark beetle attacked stands could be determined. Highest average NO_3^- concentrations were measured usually at the end of the vegetation period, between September and November, when ground vegetation declined. Lowest NO_3^- concentrations were usually measured after the snow melt. Highest average NO_3^- concentrations were determined in 1999 on plot '94' with 1101 μmol$_c$ L^{-1} NO_3^-, in the 5th year after

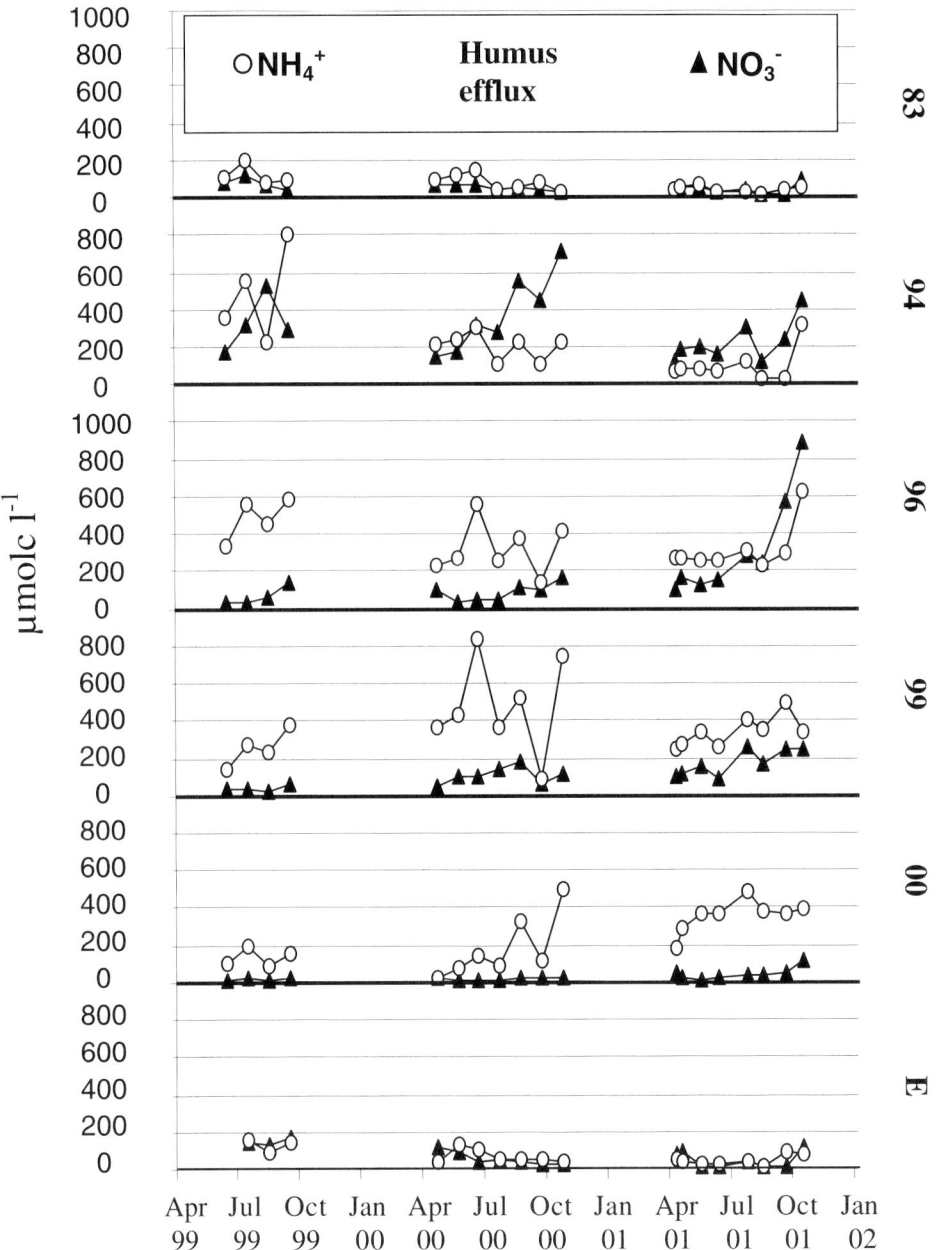

Figure 1. NH_4^+- and NO_3^--concentrations (μmol_c L^{-1}) in humus efflux at the plots 'E', '00', '99', '96' '94', and '83'. 'E' is a intact spruce stand in a still managed expansion zone of the National Park. The other plots are situated in the unmanaged old part of the National Park. '00' was 1999 alive the 'dieback' occurred in the year 2000. On '99' the dieback was 1999, on '96' in 1996, on '94' in 1994. The plot '83' was wind thrown 1983.

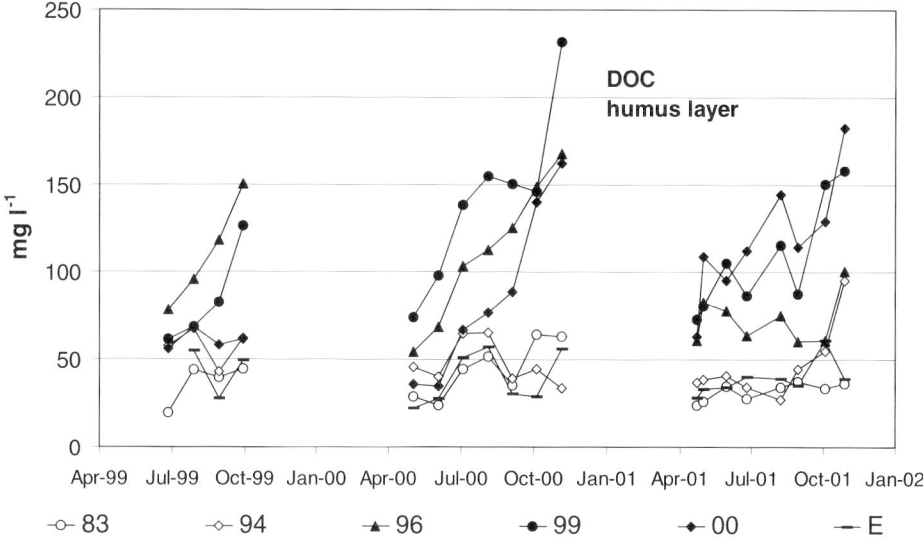

Figure 2. DOC concentrations (mg L^{-1}) in the humus efflux at the different plots 'E', '00', '99','96' '94', and '83'. 'E' is an intact spruce stand in a still managed expansion zone of the National Park. The other plots are situated in the unmanaged old part of the National Park. '00' was 1999 alive the 'dieback' occurred in the year 2000. On '99' the dieback was 1999, on '96' in 1996, on '94' in 1994. The plot '83' was wind thrown 1983.

the dieback and at the end of the year. Average NO_3^- concentrations above the limit for drinking water (50 mg L^{-1} or 806 μmol_c L^{-1} NO_3^-) occurred rarely on the bark beetle attacked stands of the highland area because of a dilution effect due to frequent high amounts of precipitation there. Elevated NO_3^- concentrations in seepage water were correlated very closely to Al^{3+} concentrations (Figure 4). Al^{3+} in the bark beetle attacked stands reached maximum concentration of 2016 μmol_c L^{-1}. Lowest Al^{3+}-concentrations were observed on plot '83' (average concentration from 1999–2001 was 50 μmol_c L^{-1}).

Figure 5 shows NO_3^- concentrations in a chronosequence from the beginning to seven years after the dieback of a stand and in the later period after 16 to 18 yr. From the second year until year 7 after the dieback, NO_3^- concentrations were elevated. In the later period, NO_3^- concentrations were below 80 μmol_c L^{-1} NO_3^-. Six years after the dieback, the maximum NO_3^- concentration was reached. In Figure 5, the flow-weighted yearly NO_3^- concentrations of the chronosequence are shown, where the effect of snow melting has a high influence. At this time, NO_3^- concentrations were usually the lowest of the year. Because of this dilution effect, average yearly flow-weighted NO_3^- concentrations never exceeded 806 μmol_c L^{-1} NO_3^-. The Figure shows two peaks – a first peak 3 yr and a second 6 yr after the dieback.

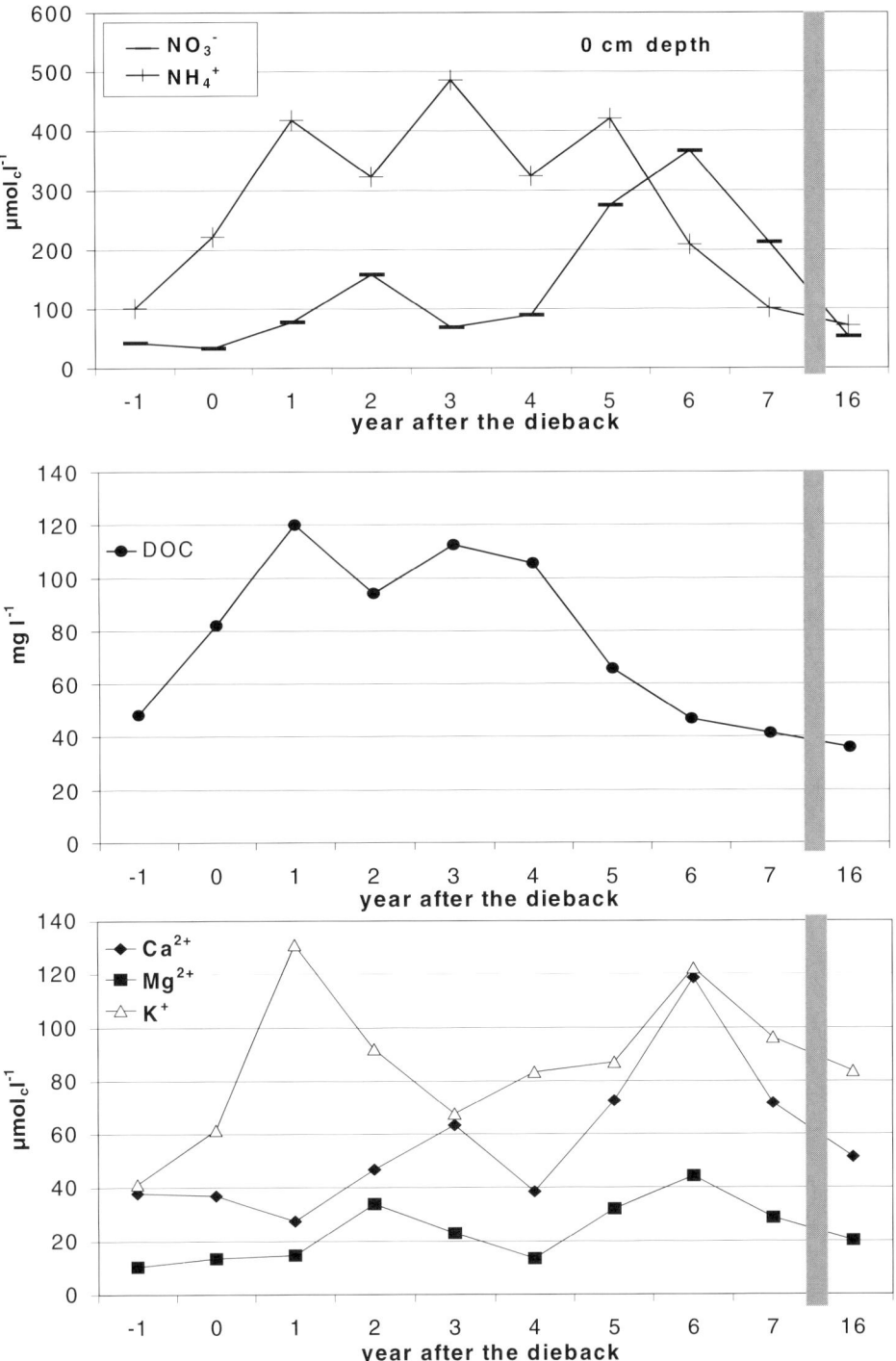

Figure 3. Yearly average NO_3^-, NH_4^+, DOC, Ca^{2+}, Mg^{2+}, and K^+ concentrations in humus efflux (0 cm depth) before (−1) and after the dieback (year 0–8). Year 0 is the year of the dieback.

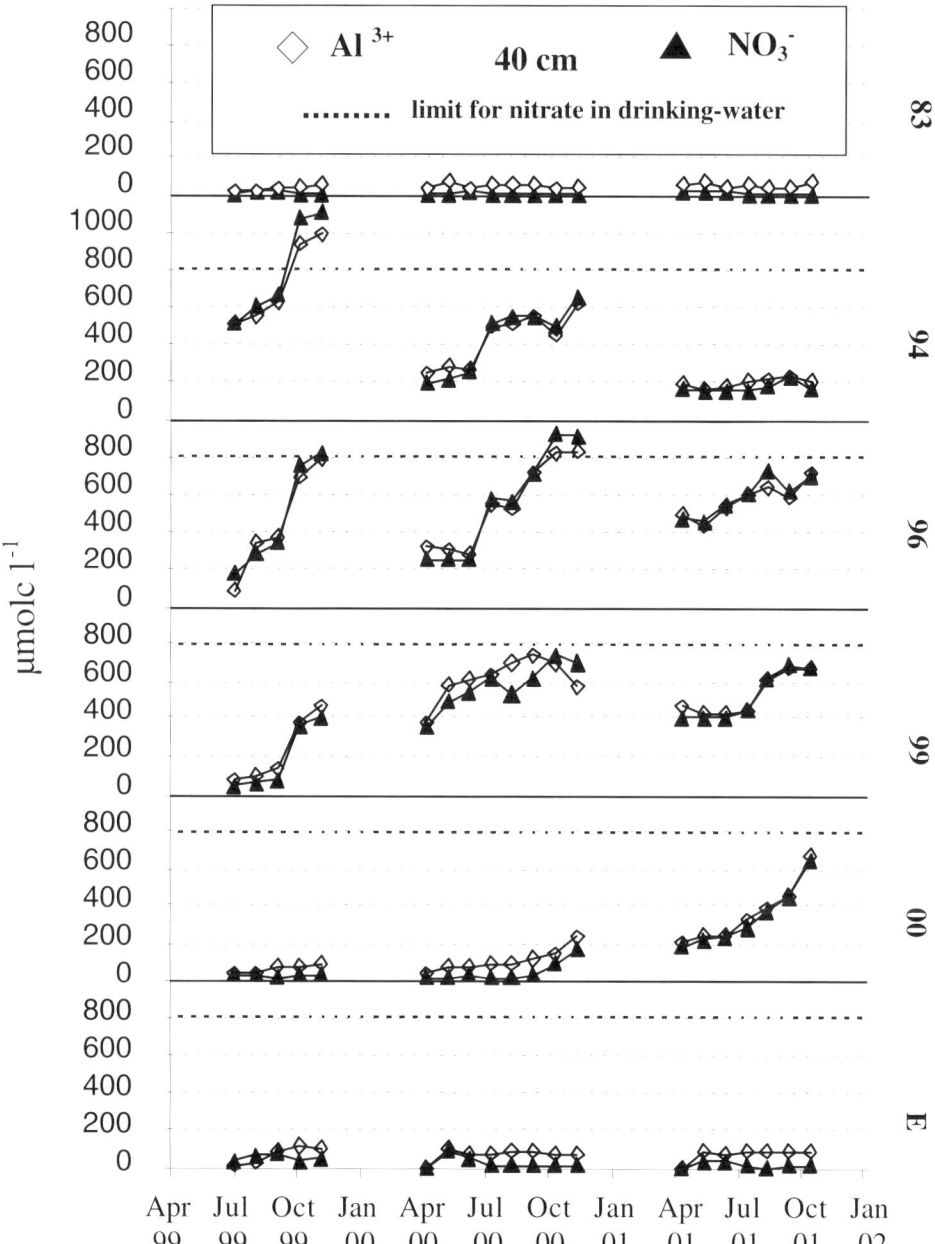

Figure 4. NO_3^- and Al^{3+} concentrations below the main rooting zone (40 cm depth) at the plots 'E', '00', '99', '96' '94', and '83'. 'E' is a intact spruce stand in a still managed expansion zone of the National Park. The other plots are situated in the unmanaged old part of the National Park. '00' was 1999 alive the 'dieback' occurred in the year 2000. On '99' the dieback was 1999, on '96' in 1996, on '94' in 1994. The plot '83' was wind thrown 1983.

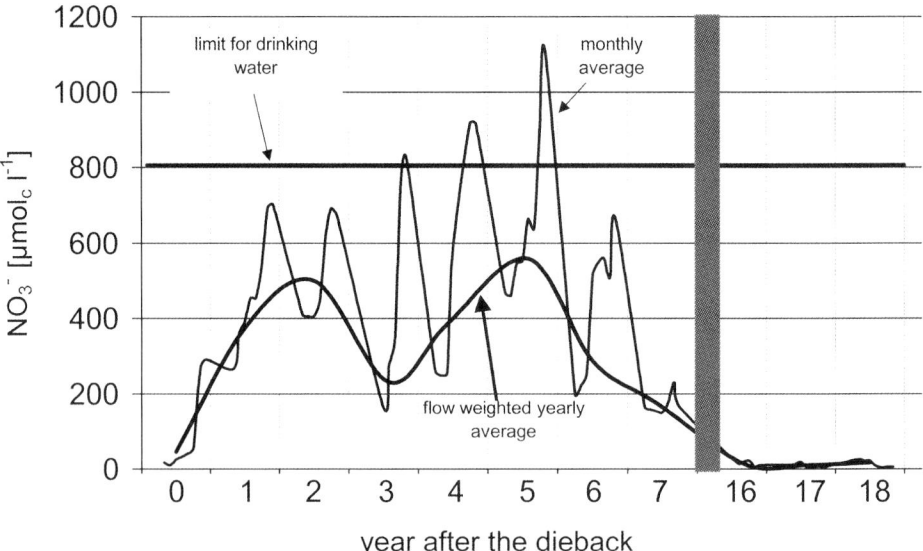

Figure 5. Monthly average and flow weighted yearly average NO_3^- concentration in a chronosequence from the beginning to 7 yr after the 'dieback' and later (16 to 18 yr after the 'dieback'). The European limit for NO_3^- in drinking water (806 μmol_c L^{-1}) is marked.

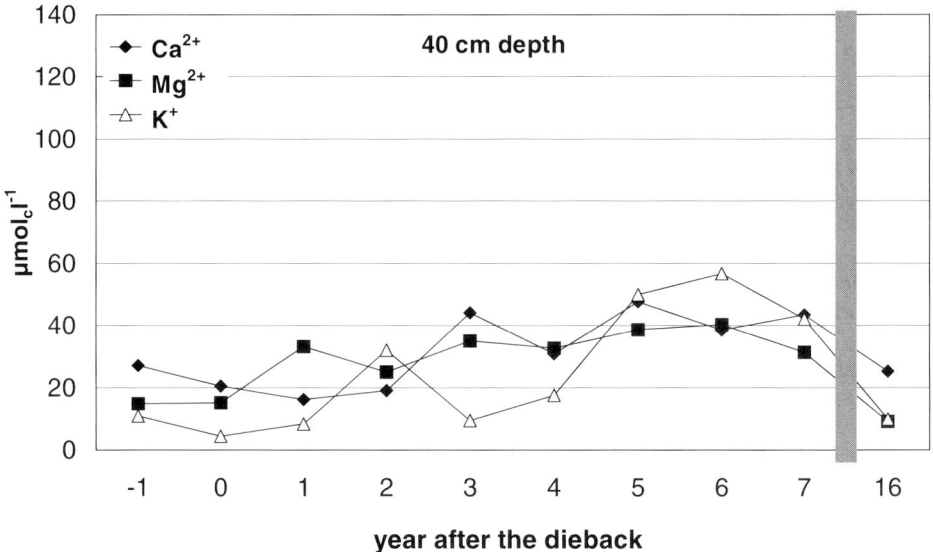

Figure 6. Yearly average concentrations of Ca^{2+}, Mg^{2+}, and K^+ in mineral soil (40 cm depth) before (−1) and after the 'dieback' (year 0–8). Year 0 is the year of the 'dieback'.

Concentrations of K^+, Mg^{2+}, and Ca^{2+} were lower in the mineral soil (Figure 6) than in the humus efflux (Figure 3). However, concentrations of K^+ and Mg^{2+} were elevated in the mineral soil after the dieback as compared to intact stands.

4.4. ELEMENTAL FLUXES IN HUMUS EFFLUX AND MINERAL SOIL

Approximately 90% of the inorganic N in humus efflux in an undisturbed spruce forest in Finsterau was NH_4^+. For the intact spruce stand of Finsterau, we calculated a humus efflux of 149 $mmol_c$ m^{-2} a^{-1} (129 $mmol_c$ m^{-2} a^{-1} of NH_4^+ and 20 $mmol_c$ m^{-2} a^{-1} of NO_3^-). In the mineral soil (40 cm depth) N-fluxes were low, NH_4^+ disappeared. Nitrate fluxes were slightly higher in the mineral soil than in humus efflux (36 $mmol_c$ m^{-2} a^{-1} NO_3^-).

N-fluxes in the mountain spruce stand of Buchenau were 244 $mmol_c$ m^{-2} a^{-1} N in the humus efflux (106 $mmol_c$ m^{-2} a^{-1} of NH_4^+ and 138 $mmol_c$ m^{-2} a^{-1} of NO_3^-). Nitrate losses with seepage were 64 $mmol_c$ m^{-2} a^{-1}.

The dieback of the trees enhanced the N-flux in the humus layer and in the mineral soil. In the first year of the dieback, N-fluxes remained nearly unchanged in the humus layer and mineral soil. N-fluxes were enhanced from year 2 until year 7 after the dieback. From year 0 to year 5 after the dieback, humus efflux was dominated by NH_4^+ (70–90% of inorganic N). Maximum fluxes in humus were reached in the 5th year with 768 $mmol_c$ m^{-2} a^{-1} of N (476 $mmol_c$ m^{-2} a^{-1} of NH_4^+ and 293 $mmol_c$ m^{-2} a^{-1} NO_3^-). N-fluxes for the first 8 yr (year 0 to 7 after the dieback) in humus efflux added up to total 4185 $mmol_c$ m^{-2} (2875 $mmol_c$ m^{-2} of NH_4^+ and 1311 $mmol_c$ m^{-2} of NO_3^-).

In the mineral soil (Table VII), maximum N-fluxes were also reached in the 5th year after the dieback (872 $mmol_c$ m^{-2} a^{-1}). The NO_3^--N-losses for the first 8 yr after the dieback totaled 3880 $mmol_c$ m^{-2} (543 kg ha^{-1} NO_3^--N). This is a yearly average N-loss of 68 kg ha^{-1}. In the intact spruce stand, the N-loss was around 5 kg N ha^{-1} (data from plot '00' in 1999). The intact stand 'E' showed an average yearly NO_3^--N-loss of 9 kg ha^{-1}.

Average SO_4^{2-} fluxes in the mineral soil were slightly higher in the bark beetle attacked stands (123 $mmol_c$ m^{-2} a^{-1} in the first 8 yr) than in intact stands (103 $mmol_c$ m^{-2} a^{-1} in Buchenau,; 110 $mmol_c$ m^{-2} a^{-1} at the Finsterau plot).

Al^{3+}-fluxes were in the same order of magnitude as NO_3^- fluxes (503 $mmol_c$ m^{-2} a^{-1}). Magnesium fluxes were 13–21 $mmol_c$ m^{-2} a^{-1} in intact stands and nearly doubled after the dieback (average of 8 yr: 35 $mmol_c$ m^{-2} a^{-1}). Ca^{2+} fluxes were slightly lower in dead stands (40 $mmol_c$ m^{-2} a^{-1}) than in intact stands (46 $mmol_c$ m^{-2} a^{-1} in Buchenau and 56 $mmol_c$ m^{-2} a^{-1} in Finsterau). K^+ fluxes were elevated in the bark beetle attacked stands (37 $mmol_c$ m^{-2} a^{-1} on average in the first 8 yr) compared with intact stands (17 $mmol_c$ m^{-2} a^{-1} in Buchenau and 8 $mmol_c$ m^{-2} a^{-1}). The lowest NO_3^-, SO_4^{2-}, and Al^{3+} fluxes were determined for plot '83'.

TABLE VII

Chronosequence of the elemental fluxes after the dieback in 40 cm depth. 'Control 1' is the intact spruce stand 'E' in the managed expansion area, 'Control 2' is the intact spruce stand '00' in the unmanaged part of the National Park in the year 1999 (stand was attacked 2000 by bark beetle)

Year	H^+	Al^{3+}	Mn^{2+}	Fe^{3+}	NH_4^+	Ca^{2+}	Mg^{2+}	K^+	Na^+	NO_3^-	SO_4^{2-}	Cl^-	HCO_3^-	DOC
						($mmol_c$ m^{-2} yr^{-1})								(g m^{-2} yr^{-1})
0	35	116	0	8	0	32	14	8	34	61	99	23	48	11
1	53	616	0	10	0	21	36	12	52	573	93	63	50	10
2	72	832	1	17	0	21	36	44	40	791	124	54	33	16
3	63	241	3	20	0	64	36	27	50	354	94	33	20	9
4	64	633	0	9	0	42	34	21	45	594	120	39	37	13
5	61	868	4	16	0	57	55	71	39	872	155	32	43	15
6	44	440	2	5	0	42	36	66	18	395	161	18	42	8
7	50	276	4	11	0	42	36	51	22	239	134	22	50	16
Sum	443	4022	14	98	0	321	281	299	300	3880	985	285	312	98
Average	55	503	2	12	0	40	35	37	37	485	123	36	40	12
15–17	10	56	0	14	0	42	12	15	29	19	48	17	75	13
Control 1	50	93	1	16	0	46	21	17	31	64	103	27	40	14
Control 2	35	55	0	7	0	56	13	8	33	36	110	26	57	11

5. Discussion

5.1. N-BUDGET OF INTACT SPRUCE STANDS IN THE NATIONAL PARK

N-fluxes in throughfall of intact stands were lower in our study (12–16 kg N ha^{-1} a^{-1}) than in an earlier study in the highlands of the Bavarian National Park (23 kg N ha^{-1} a^{-1} at the Böhmweg site) carried out in 1987/88 (Göttlein and Kreutzer, 1991; Heil, 1996), and lower than the average N-input in 86 investigated level II forest ecosystems in Germany (Anonymous, 2002). Needle analysis showed that the stands in the highlands are not N-limited (Bosch, 1986; Elling *et al.*, 1987; Heil, 1996). The N-demand of an adult spruce stand in the mountain region is low. Heil (1996) calculated for the 'Böhmweg' site approximately 5 kg N ha^{-1} annual incorporation of the growing stand. The amount of N-lost with litter fall is approx. 25–30 kg N ha^{-1} a^{-1} at the Böhmweg site (Huber, 1999). Therefore, the yearly N-uptake of a stand should be around 30 to 35 kg N ha^{-1} a^{-1}.

There is a high risk of NO_3^- leaching, when the N-input in throughfall is raised. In the former study at the Böhmweg, a higher N-input with throughfall caused increased NO_3^- losses in seepage water (21 kg NO_3^--N ha^{-1} a^{-1}). Excess nitrification already occurred in the humus efflux (Heil, 1996), causing significant K and Mg losses with seepage water. In our study, the ecosystem lost NO_3^--N with seepage water, however in small amounts (5–9 kg N ha^{-1} a^{-1}), nearly the same amount of NO_3^- as determined in throughfall input.

In the slopes of the Bavarian National Park, fluxes similar to those determined in our study in the highlands were measured in an intact spruce stand. Beudert (1999) found 10 kg N ha^{-1} a^{-1} input in throughfall and 6 kg NO_3^--N loss with seepage water.

5.2. CHANGES IN N-INPUT AFTER THE DIEBACK OF THE STAND

In the first years after the dieback, N-input decreased significantly (3 to 5 kg N ha^{-1} a^{-1}). Beudert (1999) also reported a large drop in N-input, from about 10 kg N ha^{-1} a^{-1} to 4 kg N ha^{-1} a^{-1}, in the first years after the dieback of a stand in the slopes of the Bavarian National Park. Some effects are possibly responsible for the decrease, including N-consumption by microbes during the decay of the standing dead wood (low C/N ratio) and gaseous losses (N_2, NO, and N_2O) in the standing dead wood stand.

After the dead trees fall to the ground, the supposed process of N-consumption of microbes decomposing bark and wood goes on. Though the decay of wood is slow, in the long run significant amounts of N may be accumulated by microbes.

5.3. NITRATE IN SEEPAGE WATER AFTER BARK BEETLE ATTACK – COMPARISON WITH OTHER INVESTIGATIONS OF 'DISTURBED ECOSYSTEMS'

Nitrate leaching, despite clear-cutting and previous immense N-fertilisation, can be very low on N-limited forests (Ring, 1995). In several case studies of clear-felling in U.S. forests, relatively low N-losses were determined. However, at Hubbard Brook, clear-cutting caused a NO_3^--N loss of 97 kg ha^{-1} in the first year after felling (Vitousek *et al.*, 1979). As mentioned above, the spruce stands in the highlands of the Bavarian Forest National Park are obviously not N-limited. Nitrate-N losses with seepage water of bark beetle attacked stands in our study were on the upper level of 'disturbed ecosystem'. 543 kg NO_3^--N ha^{-1} a^{-1} were lost with seepage water in the first 7 yr after the dieback; a maximum NO_3^--N loss of 122 kg N ha^{-1} was determined in the 5th year after the dieback. According to our flux calculations the main source for NO_3^- in seepage water is NH_4^+ from the forest floor, which is then nitrified in the upper mineral horizons. This was also the case in the study of Dahlgren and Driscoll (1994).

High NO_3^- losses frequently are found in German case studies of disturbed forests. Beudert (1999) determined for a stand in the lower regions of the Bavarian Forest National Park high NO_3^- losses after bark beetle attack: 40 kg NO_3^--N ha^{-1} in the year of the dieback, and as much as 180 kg NO_3^--N ha^{-1} in the first year after the dieback. Beudert (1999) mentioned that further losses of N can be taken for granted until there is a new demand of N by the regeneration of beech and spruce. Mellert *et al.* (1996) investigated 13 different wind-thrown spruce forests in Bavaria, Germany. Excess nitrification with higher NO_3^- concentrations in seepage water was detected on nearly all sites. Nitrate-N leaching during 5 yr after the windthrow caused a N-loss of 70 to 310 kg ha^{-1} NO_3^--N. Weis *et al.* (2001) investigated three clear-cut stands in Bavaria. 140 kg ha^{-1} NO_3^--N was lost in the first year on a clear-cut at the N-saturated Höglwald site, 110 kg ha^{-1} in Ebersberg, and 34 kg ha^{-1} in Flossenbürg.

On sites with high risk of N-losses, a 'good forest practice' should reduce NO_3^- losses during the regeneration of a stand. Weis *et al.* (2001) reported significant lower NO_3^- fluxes in a group shelterwood system, where 20 to 30% of the basal area of a stand was removed. Losses were smaller than in the clear-cut area and not significantly higher than in the intact stand. However, it's unclear if the losses add-up in the long run. Some further thinning and felling, until the regeneration is established and the adult stand is finally removed, are still under investigation in these studies and therefore no final conclusions can be made.

Despite high N-fluxes, NO_3^- concentrations in seepage water in our study were mostly below the European limit for drinking water (50 mg L^{-1}). This is due to the immense dilution effect of the relatively high amount of precipitation and snow melt. In stands with lower precipitation, NO_3^- concentrations can be significantly higher. Mellert *et al*, (1996) found maximum NO_3^--N concentrations of more than 300 mg L^{-1} and average concentrations up to 200 mg L^{-1} in windthrown spruce stands. In a clear-cut at the Höglwald site average peak concentrations of 140 mg L^{-1} NO$_3^-$ were observed. Also, in a clear-cut with additional herbicide treatment made at the Hubbard Brook Experimental Forest average NO$_3^-$ concentrations raised from 0.9 mg L^{-1} to 38 mg L^{-1} in the first year and to 52 mg L^{-1} NO$_3^-$ in the second year after the treatment (Likens et al., 1970, 1977).

Most of the disturbed stands show a seasonal NO$_3^-$ peak around September to November (Mellert et al., 1996; De Keerksmaeker et al., 2000; Weis et al., 2001). Increased NO$_3^-$ concentrations were observed in Hubbard Brook one year following the whole tree harvest, reached their maximum in the next year, and declined to near reference concentrations after 4–5 yr (Dahlgren and Driscoll, 1994).

In our case, later occurring peak NO$_3^-$ concentrations are most probably due to a shading effect of standing dead trees in the first years. However, seasonal effects were similar. In the bark beetle attacked stands also NO$_3^-$ concentrations were higher in autumn, due to lower precipitation amounts and the end of the main growing season of the ground vegetation. Also, in the Hubbard Brook experiment a distinct delay in the release of NO$_3^-$ to the soil solutions could be detected. Borman and Likens (1979) explained the delay by high C:N and other element ratios of organic material after felling. Therefore, dissolved nutrients are assimilated by increasing populations of micro-organisms. As the C:N ratio decreases and demand by microorganisms for N declines, excess nutrient ions are released by decomposition and can be leached.

5.4. Nutrients added to the soil in tree biomass versus loss of NO$_3^-$-N

Table VIII describes the amounts of N, K, Ca, and Mg stored in aboveground and underground parts of comparable stands in South Germany (Dietrich et al., 2002; Raspe, 2002), the amount of N and exchangeable nutrients stored in the upper 40 cm depth of a typical soil profile, and the loss with seepage water in the first 8 yr after the dieback in our investigation. Storage of N in aboveground and underground parts of a comparable stand in Flossenbürg is 4350 mmol$_c$ m^{-2}. According to this data, the N-loss with seepage water of 3880 mmol$_c$ m^{-2} (543 kg N ha^{-1}) in the first seven years will be more or less compensated by the addition of around 600 kg of organic N added to the soil. Most critical are the losses of K$^+$ and Mg^{2+} after the dieback (300 and 282 mmol$_c$ m^{-2}, respectively), which are in the range of the amount of K$^+$ stored in easily decomposable material like needles, bark, and twigs. Losses of Mg^{2+} and K$^+$ with seepage water are critical for healthy regeneration of the stand, because these elements are in a low supply. Magnesium deficiency symptoms were widespread in the highlands in the 1980's (Bosch, 1986). However, the elemental content of these elements in the organic material left on the site exceeded leaching with seepage.

TABLE VIII

Elemental storage in aboveground parts and roots of comparable spruce stands (Dietrich *et al.*, 2002; Raspe, 2002) and in the upper 40 cm of the soil of an intact spruce stand in the Bavarian Forest National Park (plot '00', time of investigation 1999) compared with seepage losses below the main rooting zone in the first eight years after the dieback of a spruce stand. Note that N is the total supply, while K^+, Ca^{2+} and Mg^{2+} were calculated from the exchangeable cations

	N	K	Ca	Mg
	($mmol_c$ m^{-2})			
Wood	790	160	530	150
Bark	490	90	520	110
Branches	290	40	190	50
Twigs	460	60	130	40
Needles	1430	200	190	80
Roots	890	150	500	150
Stand (Total)	4350	700	2050	580
Soil (upper 40 cm)	58880	450	780	320
Loss with seepage water (year 0 to 7)	3880	300	321	282
Loss per year	485	37	45	35
Intact stand	36–64	8–17	50–56	13–21

5.5. GROUND VEGETATION AS LIMITED SINK FOR N

Vitousek *et al.* (1979) mentioned eight processes that could delay or prevent solution losses of NO_3^- in disturbed forests. According to the authors, all processes except uptake of regrowing vegetation are insufficient on fertile sites. Mellert *et al.* (1996) estimated the N-content of the ground vegetation 3 yr after a windthrow of 13 spruce stands in Bavaria. Average storage in aboveground parts was approx. 40 kg N ha^{-1} (minimum of 19 kg N ha^{-1} and maximum of 70 kg N ha^{-1}). In their study the ground vegetation was supposed as an effective sink to reduce NO_3^- concentrations in seepage water. This was mentioned also in other investigations (for example Stevens and Hornung, 1990; Emmet *et al.*, 1991; Bauhus, 1994). Weis *et al.* (2001) could show that a 'high ground vegetation' at a clear-cut could reduce NO_3^- concentrations from 1 to 0.1 mmol L^{-1}.

However, in our investigation ground vegetation was obviously not as effective in reducing NO_3^- losses in the first 7 yr after the dieback of the stand because the process of nutrient-uptake and mineralisation of dead parts of the vegetation is in a steady state, when the ground vegetation is established. Then the net uptake of the ground vegetation is very limited. However, a deep rooting-system is able to work as a pump for the nutrients lost with seepage from the humus layer. In addition, coverage of the ground vegetation was sometimes decreased on dead wood stands, because the loss of needles, bark, and fallen trees caused a mulch effect. Finally, on the long run, more nutrients are stored in ground vegetation of attacked than in intact stands. However, the difference from N-storage of attacked to intact stands is small compared to the fluxes with seepage water (data will be presented in a separate publication).

6. Conclusions

In the first seven years after the dieback of the stand obviously more sources than sinks of N are in the system. The main sources for N are: (1) enhanced litter fall, (2) accumulated large amounts of N with low C:N ratio, and (3) accelerated mineralisation and nitrification. The main sinks for N are not sufficient to prevent NO_3^- losses because of: (1) lack of N-uptake by the roots of the 'died-back' spruce stand, (2) regeneration is slow, (3) a low C:N ratio in the humus layer, indicating a limited potential for immobilisation of N by decomposers, microorganisms etc., (4) N-accumulation of microorganisms during the decay of wood with high C:N ratio is likely to occur, but is effective to prevent NO_3^- leaching only on the long run, after a significant decay of the wood, and (5) net N-uptake of the ground vegetation is limited.

N-loss does most probably not cause actual or future N-deficiency for the healthy regeneration in the Bavarian National Park. The losses are in the range of the added N-fluxes of the dead organic substrates left on the sites. Furthermore, the N-input is already higher than the regeneration or an adult spruce stand can store. The pool of Mg^{2+} and K^+ is small. These losses are more severe, comparable to a whole tree harvest, and cannot be compensated by throughfall input.

Average yearly NO_3^- concentrations in seepage water do not exceed the limit for drinking water quality (806 μmol_c L^{-1} NO_3^-) because a relatively high amount of precipitation causes a dilution of NO_3^- fluxes in seepage water. Due to the shading effect of standing dead trees, mineralisation and nitrification is spread over a longer time, with a lower NO_3^- amplitude in the first years. Elevated NO_3^- concentrations will last for a minimum of around seven years after the dieback. Afterwards, NO_3^- concentration is low, independent of the extent of regeneration.

The bark beetle outbreak occurring in the Bavarian Forest National Park during the last decade is restricted to a distinct remote area. Some negative side effects are 'compensated' in a way that a managed forest can hardly achieve. Remaining dead wood is important to realise a site specific biodiversity, and a natural regeneration of the stands. The negative side effects may most probably also occur after processes like harvesting, especially after clear-cutting.

Acknowledgements

We would like to thank the Bayerische Staatsministerium für Landwirtschaft und Forsten for their financial support of the Projekt V50b. Our thanks belong also to the Deutsche Wetterdienst for the prompt providing of the meteorological data and the Nationalparkverwaltung Bayerischer Wald for their help. We also thank the technical assistants for their skilful analytical work in the laboratory and the time consuming and exhausting work in the field, and Jared David May for valuable comments and the proof-reading of the manuscript.

References

Anonymous: 2002, 'Berichte über den Zustand des Waldes 2001. Ergebnisse des forstlichen Umweltmonitorings', Bundesministerium für Verbraucherschutz, Ernährung und Landwirtschaft (BMVEL) Referat Öffentlichkeitsarbeit, Bonn.

Bauhus, J.: 1994, 'Stoffumsätze in Lochhieben', Berichte des Forschungszentrums Waldökosysteme der Univ. Göttingen, Reihe A, Bd.113.

Beudert, B.: 1999, 'Veränderungen im Stoffhaushalt eines abgestorbenen Fichtenökosystems im Forellenbachgebiet des Nationalpark Bayerischer Wald', in Einzugsgebiet Große Ohe – 20 Jahre Hydrologische Forschung im Nationalpark Bayerischer Wald, Grafenau, 11.05.1999, Wasserhaushalt und Stoffbilanzen im naturnahen Einzugsgebiet Große Ohe Bd.7, pp. 93–106.

Borman, F. H. and Likens, G. E.: 1979, *Pattern and Process in a Forested Ecosystem: Disturbance, Development and Steady State based on the Hubbard Brook Ecosystem Study*, Springer, New York.

Bosch, C.: 1986, 'Standorts- und ernährungskundliche Untersuchungen zu den Erkrankungen der Fichte (*Picea abies* (L.) Karst.) in höheren Gebirgslagen', Forstliche Forschungsberichte München 75.

Dahlgren, R. A. and Driscoll, C. T.: 1994, 'The effects of whole-tree clear-cutting on soil processes at the Hubbard Brook Experimental Forest, New Hampshire, U.S.A.', *Plant Soil* **158**, 239–262.

De Keersmaeker, L., Neirynck, J., Maddelein, D., De Schrijver, A. and Lust, N.: 2000, 'Soil water chemistry and revegetation of a limed clearcut in a N saturated forest', *Water, Air, Soil Pollut.* **122**, 49–62.

Dietrich, H.-P., Raspe, S., Schwarzmeier, M. and Ilg, S.: 2002, 'Biomasse- und Nährstoffinventuren zur Ermittlung von Ernteentzügen an drei bayerischen Fichtenstandorten', in *Inventur von Biomasse- und Nährstoffvorräten in Waldbeständen*, Freising, Seminar der Bayer. Landesanstalt für Wald und Forstwirtschaft, Forstliche Forschungsberichte München 186, pp. 59–72.

Elling, W., Bauer, E., Klemm, G. and Koch, H.: 1987, *Klima und Böden*, Wissenschaftliche Schriftenreihe Nationalpark Bayerischer Wald, 1. Bayerisches Staatsministerium für Ernährung Landwirtschaft und Forsten, München, pp. 255.

Emmett, B. A., Anderson, J. M. and Hornung, M.: 1991, 'The control on N-losses following two intensities of harvesting in a Sitka spruce forest (N. Wales)', *Forest Ecol. Manag.* **41**, 81–93.

Federer, C. A.: 1997, 'Brook90: A simulation model for evaporation, soil water and streamflow. Version 3.2', Computer share ware and documentation, USDA Forest Service, Durham, NH, U.S.A., pp. 84.

Frazer, D. W., McColl, J. G. and Powers, R. F.: 1990, 'Soil nitrogen mineralisation in a clearcutting chronosequence in a northern California conifer forest', *Soil Sci. Soc. Am. J.* **54**, 1145–1152.

Führer, H.-W. and Hüser, R.: 1991, *Bioelementausträge aus mit Buche bestockten Wassereinzugsgebieten im Krofdorfer Forst: Zeittrends und Effekte von Verjüngungseingriffen*, Forstw. Cbl. 110, pp. 240–247.

Gessler, A., Schneider, S., Von Sengbusch, D., Weber, P., Hanemann, U., Huber, C., Rothe, A., Kreutzer, K. and Rennenberg, H.: 1996, 'Field and laboratory experiments on net uptake of nitrate and ammonium by the roots of spruce (*Picea abies*) and beech (*Fagus sylvatica*) trees', *The New Phytologist* **138**(2), 275–285.

Göttlein, A. and Kreutzer, K.: 1991, 'Der Standort Höglwald im Vergleich zu anderen ökologischen Fallstudien.' in K. Kreutzer and A. Göttlein (eds), *Ökosystemforschung Höglwald, Forstwissenschaftliche Forschungen 39*, Paul Parey, Hamburg, pp. 22–29.

Hauhs, M.: 1989, 'Lange Bramke: An Ecosystem study of a forested catchment', in D. C. Adriano and M. Hauhs (eds), *Acid Precipitation*, Springer, New York, NY, pp. 275–305.

Heil, K.: 1996, 'Wasserhaushalt und Stoffumsatz in Fichten- (*Picea abies* (L.) Karst.) und Buchenökosystemen (*Fagus sylvatica* L.) der höheren Lagen des Bayer. Waldes', *Dissertation* zur Erlangung des Doktorgrades der Forstwissenschaftlichen Fakultät der Ludwig-Maximilians-Universität München.

Heurich, M., Reinelt, A. and Fahse, L.: 2001, 'Die Buchdruckermassenvermehrung im Nationalpark Bayerischer Wald', in *Waldentwicklung im Bergwald nach Windwurf und Börkenkäferbefall*, Nationalpark Bayerischer Wald 14, pp. 10–47 (with English abstract: 'Bark beetle outbreak in the Bavarian Forest National Park').

Huber, C. and Göttlein, A.: 2001, 'Auswirkungen des Borkenkäferbefalls auf Wasserqualität und Waldernährung in den Hochlagen des Nationalpark Bayerischer Wald', in *Chemische und Physikalische Schlüsselprozesse der Speicher-, Regler- und Reaktorfunktionen von Waldböden*, Freiburg, 5–6 April 2001, Berichte Freiburger Forstliche Forschung 33, pp. 99–108.

Huber, C.: 1999, 'Auswirkungen des Borkenkäferbefalls auf Wasserqualität und Waldernährung in den Hochlagen im Nationalpark Bayerischer Wald: Forschungskonzept', in *Einzugsgebiet Große Ohe – 20 Jahre hydrologische Forschung im Nationalpark Bayerischer Wald*, Grafenau, 11-05-1999, Wasserhaushalt und Stoffbilanzen im naturnahen Einzugsgebiet Große Ohe Bd.7, pp. 137–146.

Jehl, H.: 2001, 'Die Waldentwicklung nach Windwurf in den Hochlagen des Nationalparks Bayerischer Wald', in *Waldentwicklung im Bergwald nach Windwurf und Borkenkäferbefall*, Nationalpark Bayerischer Wald 14, pp. 49–97 (with English abstract: 'Forest development in high elevations of the Bavarian Forest National Park').

Likens, G. E., Bormann, F. H. and Johnson, N. M.: 1969, 'Nitrification: Importance of nutrient losses from a cutover forested ecosystem', *Science* **163**, 1205–1206.

Likens, G. E., Borman, F. H., Johnson, N. M., Fisher, D. W. and Pierce, R.: 1970, 'Effects of forest cutting and herbicide treatment on nutrient budgets in Hubbard Brook watershed – Ecosystem', *Ecol. Monogr.* **40**(1), 23–47.

Likens, G. E., Bormann, F. H., Pierce, R. S., Eaton, J. S. and Johnson, N. M.: 1977, *Biogeochemistry of a Forested Ecosystem*, Springer, New York, pp. 146.

Mellert, K.-H., Kölling, C. and Rehfuess, K. E.: 1996, 'Bioelement leaching from Norway spruce ecosystems in Bavaria after windthrow', *Forstw. Cbl.* **115**, 363–377.

Rall, H.: 1999, 'Ziele und Aufgabenschwerpunkte der Forschung im Nationalpark Bayerischer Wald', in *Einzugsgebiet Große Ohe – 20 Jahre Hydrologische Forschung im Nationalpark Bayerischer Wald*, Grafenau, 11-05-1999, Wasserhaushalt und Stoffbilanzen im naturnahen Einzugsgebiet Große Ohe Bd.7, pp. 1–5 (with English abstract).

Raspe, S.: 2002, 'Beispiele zur Erhebung der Biomasse- und Mineralstoffvorräte von Fichtenwurzeln', in *Inventur von Biomasse- und Nährstoffvorräten in Waldbeständen*, Freising, Seminar der Bayer. Landesanstalt für Wald und Forstwirtschaft, Forstliche Forschungsberichte München 186, pp. 73–82.

Ring, E.: 1995, 'Nitrogen leaching before and after clear-felling of fertilised experimental plots in a Pinus sylvestris stand in central Sweden', *Forest Ecol. Manage.* **72**, 151–166.

Stevens, P. A. and Hornung, M.: 1990, 'Effect of harvest intensity and ground flora establishment on inorganic N-leaching from a Sitka spruce plantation in north Wales', *Biogeochemistry* **10**, 53–65.

Swank, W. T.: 1988, 'Stream chemistry responses to disturbance', in W. T. Swank and D. A. Crossley (eds), *Forest Hydrology and Ecology at Coweeta*, Springer, New York.

Vitousek, P. M., Gosz, J. R., Grier, C. C., Melillo, J. M., Reiners, W. A. and Todd, R. L.: 1979, 'Nitrate losses from disturbed ecosystems', *Science* **204**, 469–474.

Weis, W., Huber, C. and Göttlein, A.: 2001, 'Regeneration of mature Norway spruce stands: Early effects of selective cutting on seepage water quality and soil fertility', in *Optimizing Nitrogen Management in Food and Energy Production and Environmental Protection, Proceedings of the 2nd International Nitrogen Conference on Science and Policy*, The ScientificWorld 1(S2), pp. 493–499.

EFFECTS OF LONG-TERM APPLICATION OF AMMONIUM SULPHATE ON NITROGEN FLUXES IN A BEECH ECOSYSTEM AT SOLLING, GERMANY

HENNING MEESENBURG[1]*, AGUSTÍN MERINO[2], KARL J. MEIWES[1] and FRIEDRICH O. BEESE[3]

[1] *Forest Research Institute of Lower Saxony, Grätzelstr. 2, D-37079 Göttingen, Germany;*
[2] *Department of Soil Science and Agriculture Chemistry, E.P.S., Universidad de Santiago de Compostela, 27002 Lugo, Spain;* [3] *Institute of Soil Science and Forest Nutrition, University of Göttingen, Büsgenweg 2, D-37085 Göttingen, Germany*
(* author for correspondence, e-mail: henning.meesenburg@nfv.gwdg.de)

(Received 20 August 2002; accepted 11 April 2003)

Abstract. To study the effects of elevated inputs of acidity and nitrogen (N), 1000 mmol m^{-2} a^{-1} of ammonium sulphate (NH$_4$NO$_3$) equivalent to an input of potential acidity of 2000 mmol m^{-2} a^{-1} was applied annually for 11 yr between 1983 and 1993 in a beech forest at Solling, Germany. Most of the applied NH$_4^+$ was nitrified in the litter layer and in the upper mineral soil. N in soil leachate quickly responded to the elevated input, but most of the applied N was stored in the soil or left the ecosystem via pathways other than soil output. Leaching of N from the soil increased until the last year of N addition. After the last N application, N fluxes decreased rapidly to low values. The buffering of acidity produced by the nitrification of the applied NH$_4^+$ was caused mainly by three different processes: (i) sulphur (S) retention, (ii) release of aluminium, (iii) release of base cations. Retention of S took place mostly in the subsoil. 72% of the S input was recovered in output after 14 years of the experiment. Due to the increased fluxes of mobile anions with soil solution, outputs of cations increased drastically.

Keywords: acid/base budget, beech forest, exchangeable cations, nitrogen saturation, soil solution

1. Introduction

Nitrogen (N) saturation and soil acidification caused by atmospheric deposition are the most severe problems for the stability and sustainability of forest ecosystems of the northern hemisphere. Ammonium sulphate ((NH$_4$)$_2$SO$_4$) actually is the major component of atmospheric deposition in large regions of Europe (Ulrich, 1994). Aber *et al.* (1989) defined N saturation as a state where the availability of inorganic N is in excess of plant and microbial demand, resulting in increased leaching of N from the soil. Leaching of nitrate (NO$_3^-$) intensifies the depletion of other nutrients in the soil, may impair the suitability of groundwater for drinking water purposes, and may lead to eutrophication of surface waters.

A delay in the response of N leaching from the soil to changing input of N and other acidifying substances is related to the retention characteristics of the ecosystem. Whereas N is incorporated mainly in living or dead organic material (Aber *et al.*, 1998), sulphate (SO_4^{2-}) is retained mainly by abiotic mechanisms (Prenzel, 1994; Alewell, 2001).

To study the effects of elevated inputs of acidity and N, $(NH_4)_2SO_4$ was applied annually for 11 yr between 1983 and 1993 in a beech forest at Solling, Germany. This paper focuses on the fate of N in the ecosystem and the relative importance of buffering processes in the soil. Similar experiments have been conducted elsewhere at several locations, both at the catchment scale (Norton *et al.*, 1992; Feger, 1992; Edwards *et al.*, 2002) and at the plot scale (Andersson *et al.*, 2001; Bergholm and Majdi, 2001).

2. Study Site and Methods

2.1. STUDY SITE

The study site is located at the center of the Solling plateau at 504 m a.s.l. in an 153 yr old (2002) European beech (*Fagus sylvatica* L.) stand. Mean annual air temperature is 7.3 °C (1975–2000) and annual precipitation is 1100 mm. The soil is a Dystric Cambisol (FAO classification) or a Typic Dystrochrept (USDA) with a typical Moder as litter layer (Benecke and Mayer, 1971). The parent material of the soil is a loess-derived solifluction layer (to 60 cm depth), which covers a transient layer (to 95 cm) and solifluction layer derived from Triassic claystone (to 150 cm). Quarz, feldspars, and illite are the major minerals in the fine earth, whereas haematite, kaolinite, vermiculite, smectite, and some mixed-layer clay minerals occur in traces. The soils are acidic (Al buffer range) and low in exchangeable base cations. The main rooting zone is restricted to the litter layer and the upper mineral soil.

The $(NH_4)_2SO_4$ addition experiment took place at a plot of 25×50 m adjacent to a control plot within the same stand. $(NH_4)_2SO_4$ was applied for 11 yr between 1983 and 1993 with a dose of 1000 mmol m^{-2} a^{-1} (140 kg N ha^{-1} a^{-1} and 160 kg S ha^{-1} a^{-1}). The dose corresponds to a N input of 3–7 times and a S input of 3–8 times the annual atmospheric deposition. Because deposition changed during the treatment, a constant relation between deposition and addition could not be followed. The addition is equivalent to an input of potential acidity of 2000 mmol m^{-2} a^{-1}. The treatment took place manually by one single operation each year in spring (1983–1986), summer (1987–1988), and autumn (1989–1993). Details of the experimental setup are given by Meiwes *et al.* (1998).

2.2. METHODS

Open field precipitation and throughfall were collected from 1982 to 1997 during May to October in 15 funnel type bulk samplers and during November to April in 9 buckets (Meiwes *et al.*, 1984). At the treated plot, throughfall was collected from 1983 to 1984 and from 1986 to 1992. Stemflow was collected with polyurethane collars around the stems. Stemflow was measured from 1983 until 1987 at both plots, whereas it was measured from 1991 afterwards only at the control plot. Weekly samples were bulked to a monthly sample for subsequent analysis. Total deposition was estimated according to Ulrich (1994).

Soil water samples were collected at the treated plot from 1982 to 1997 (with an interruption in 1985) by ceramic tension lysimeters and analyzed monthly to quarterly. The lysimeters were installed with 2 to 5 replicates in depths of 10, 20, 40, and 100 (or 90) cm.

Analysis of pH was carried out using a glass electrode. Sodium (Na^+), potassium (K^+), magnesium (Mg^{2+}), calcium (Ca^{2+}), and manganese (Mn^{2+}) were determined by atomic absorption spectrometry (AAS) until 1989, and subsequently by inductively coupled plasma – atomic emission spectrometry (ICP-AES). Aluminium (Al) was determined colorimetrically until 1982, by AAS until 1990, and subsequently by ICP-AES. All Al in soil solution was assumed to be Al^{3+}. NH_4^+, NO_3^- and Cl^- were analysed colorimetrically with a continuous flow system. Sulphate was measured by precipitation with Ba^{2+} and by potentiometric titration of excess Ba^{2+} with EDTA until 1982, from 1983 to 1992 by the methyl-thymol-blue method, and subsequently by ICP-AES. Details of the chemical analysis are given by König and Fortmann (1996).

We used the hydrologic model SIMPEL for the estimation of water fluxes beneath the rooting zone. SIMPEL uses a capacitive approach for the simulation of water fluxes (Hörmann, 1997). The soil water storage was defined by field capacity, permanent wilting point, and maximum rooting depth. The reduction from potential to actual evapotranspiration was implemented as a linear reduction function dependent on available soil water content. For the present simulations, stand precipitation was used as infiltration into the soil storage. For the estimation of potential evapotranspiration the Penman/Monteith formula was used. Modelled water fluxes in soil output were very similar for the control and treated plot (Table I).

Element fluxes with soil output at 100 cm depth were obtained by multiplying the monthly element concentrations in soil solution with the respective water flux within that period. For periods with no soil solution, a linear interpolation between the preceding and the following period with soil solution data was applied.

Element and acid/base budgets were calculated using the approach of Ulrich (1994). The budget of any solute is the difference between input and output. Input was defined as the sum of total deposition, addition, and weathering. The release of basic cations due to weathering was estimated using the PROFILE model (Sverdrup and Warfvinge, 1993a). N release by weathering was assumed to be negligible. Soil output is the only output flux for most solutes. No export of biomass took place during the experiment. If the budget of the soil is calculated, a net accumulation in the biomass also has to be regarded as an output flux. Net accumulation of nutrients in the biomass was estimated by Rademacher (2004).

According to Ulrich (1994) the acid load of the ecosystem is the sum of acid input (positive budgets (input > output) of major acid cations (H^+, Mn^{2+}, Al^{3+},

TABLE I

Mean input and output fluxes (at 100 cm depth) of water and elements at the control and at the treated plot at Solling during the period 1982–1996

Flux	Water	Na^+	K^+	Mg^{2+}	Ca^{2+}	H^+	Mn^{2+}	Al^{3+}	SO_4^{2-}	Cl^-	NH_4^+	NO_3^-	$\Sigma^+ - \Sigma^-$	N_{tot}	DON
	(mm a^{-1})	(mmol$_c$ m^{-2} a^{-1})												(mol m^{-2} a^{-1})	
Control															
Total depositon[a]	1112	62	12	15	44	136	2	11	195	77	122	96	35	250	32
Soil leachate	551	47	5	18	18	24	9	171	210	57	3	9	18	12	0
Silicate weathering[b]	–	26	52	27	8	–113	–	–	–	–	–	–	–	–	–
Biomass accumulation[c]	–	–	13	9	29	–	–	–	–	–	–	–	–	63	–
Treated															
Total depositon + addition	1112	62	12	15	44	136	2	11	928	77	855	96	35	983	32
Soil leachate	551	108	7	45	74	38	30	892	665	93	11	401	47	426	14
Silicate weathering	–	26	52	27	8	–113	–	–	–	–	–	–	–	–	–

[a] Estimated according to Ulrich (1994).
[b] Estimated with the PROFILE model.
[c] From Rademacher (2004).

Fe^{3+}), N transformation (NH_4^+ input – NH_4^+ output – NO_3^- input + NO_3^- output), S release (negative budgets of SO_4^{2-} and weak acids), and base cation retention. Weak acids have been calculated as the charge balance difference. Cl^- is taken into account only if a negative budget cannot be attributed to the loss of neutral salts, i.e. if its budget is not balanced by base cations. Base cation retention is calculated as positive budgets of base cations (Na^+, K^+, Mg^{2+}, Ca^{2+}), corrected for neutral salt accumulation. The acid load is balanced by acid/base (buffering) reactions: acid cation release (negative budgets of acid cations, S retention (positive budgets of SO_4^{2-} and weak acids), and base cation release (negative budgets of base cations, corrected for the release of neutral salts).

In contrast to Ulrich (1994), proton generating and consuming N transformations were calculated using one single equation. If the value of N transformation was negative, it was considered as an acid/base reaction.

3. Results

No significant differences between the treated and the control plot were found in throughfall and stemflow fluxes (Meiwes et al., 1998). Hence, for the calculation of element budgets, total deposition at the control was used for the treated plot. The atmospheric deposition of SO_4^{2-}, NO_3^-, NH_4^+, H^+, Ca^{2+}, Mg^{2+}, and K^+ decreased significantly between 1981 and 1994 at the Solling beech stand (Meesenburg et al., 1995). N input was, on average, 49% as NH_4^+, 38% as NO_3^-, and 13% as organic N (DON).

N in soil solution quickly responded to the elevated input at the treated plot. Most of the applied NH_4^+ was nitrified in the organic layer and in the upper 10 cm of the mineral soil (Meiwes et al., 1998). Below 20 cm depth there was almost no NH_4^+ in soil solution. During the first years of treatment, the decrease of N fluxes down the soil profile was most pronounced between the 40 and 100 cm depths. In later years, there were no significant differences in NO_3^- concentrations over the soil profile. NO_3^- concentrations at 20 and 40 cm depth peaked during the first half of the treatment, whereas at the 100 cm depth, there was an increase until the last N application. After the last N application (1993), N concentrations decreased rapidly to low values.

Sulphate concentrations in the upper soil also showed an immediate response to the treatment (Meiwes et al., 1998). At the 40 and 100 cm depths, there was a delay in response of one year and three years, respectively. Sulphate concentrations at the 100 cm depth decreased only slightly after the last application, whereas at the 10 to 40 cm depths, concentrations decreased to pre-treatment levels.

Concentrations of major cations responded to elevated concentrations of mobile anions. The increase was most pronounced for Al, but was also visible for Ca^{2+}, Mg^{2+}, Na^+, K^+, and H^+. With the exception of H^+, concentrations decreased after the last application to pre-treatment values or even lower at the 10 to 40 cm depths.

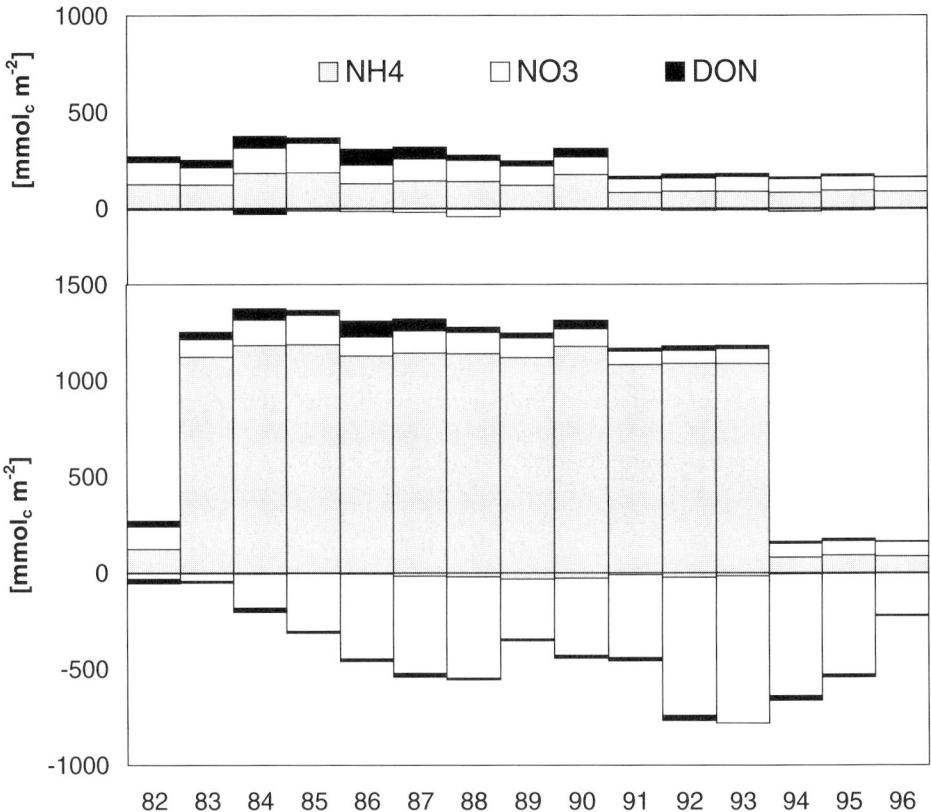

Figure 1. N input (positive values) and N output (negative values) fluxes at the control (upper panel) and at the treated plot (lower panel) from 1982 to 1996.

Whereas N output at the control plot was almost negligible over the course of the experiment, the treated plot showed a strong response to the elevated input (Figure 1). More than 90% of the N output was in the form of NO_3^-. About 46% of the N input (deposition plus addition) between 1983 and 1996 left the ecosystem via soil leachate.

At the control plot, the acid load was mainly caused by an accumulation of base cations (Figure 2). The input of acids (H^+) was relatively important during the mid-1980s, whereas the release of SO_4^{2-} contributed to the acid load at the beginning of the 1990s. N transformation processes always caused an additional input of acids. However, the contribution to the total acid load was not very substantial. The acid load was buffered mostly by the release of acid cations, i.e. Al. During the mid-1908s, S retention also contributed to the buffering of acids.

The treatment had a distinct effect on the acid/base budget of the ecosystem (Figure 2). The acid load at the plot was mainly caused by N transformation processes. Since 1993, the release of SO_4^{2-} was also a major proton generating process. The buffering of acids was caused initially mainly by S retention processes. During

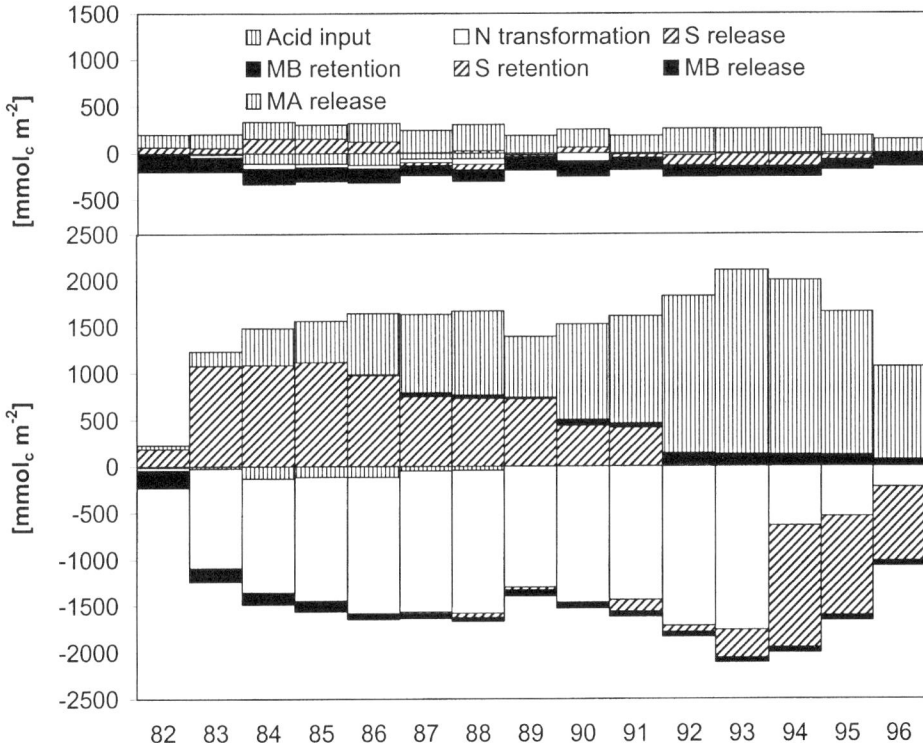

Figure 2. Components of acid load (negative values) and acid/base (buffering) reactions (positive values) at the control (upper panel) and at the treated plot (lower panel) from 1982 to 1996.

the course of the experiment, the S retention decreased continuously and turned into a release of SO_4^{2-}. Retention of S took place mainly in the subsoil below the 40 cm depth. 72% of the S input from 1982 to 1996 was found in soil leachate. The release of acid cations became the most important acid consuming process during the latter phase of the experiment. Al comprised more than 90% of the acid cations in soil leachate. The release of base cations played a minor role especially at the end of the period.

4. Discussion

Both the control and the treated plots had strongly positive budgets of N during the observation period. At the control, there was almost no N leaching from the soil. Dise and Wright (1995) showed for European forest ecosystems with N input rates (inorganic N in stand precipitation) between 65 and 180 $mmol_c$ m^{-2} a^{-1}, some had significant N leaching but some did not. Above the upper threshold, all ecosystems had significant N leaching rates. The control plot with a mean inorganic N flux in stand precipitation of 183 $mmol_c$ m^{-2} a^{-1} is at the upper threshold indicating a state close to N saturation. The immediate response of the treated plot to the N addition is another indication of a state close to N saturation. Moreover, Corre *et al.* (2003) found a closed microbial N cycle, with gross N mineralization and gross nitrification rates similar to NH_4^+ and NO_3^- immobilization rates by microbes at the control plot. In contrast, the treated plot showed lower NH_4^+ and NO_3^- immobilization rates than gross N mineralization and gross nitrification rates, indicating an open microbial N cycle. Such characteristics, together with high N leaching losses, are indicative of N saturated or approaching to N saturated conditions. Several other N manipulation studies, which also found a rapid response of soil solution to increases or decreases in N input, confirmed this observation (Gundersen *et al.*, 1998). Nevertheless, about 54% of the N input (deposition plus addition) at the treated plot was retained or left the ecosystem via pathways other than soil leachate between 1983 and 1996.

Possible sink mechanisms for N are accumulation in tree biomass, accumulation in the soil, and gaseous emission. Rademacher (2004) estimated a mean annual N accumulation of 63 mmol m^{-2} a^{-1} in the biomass of the control plot for the period 1982–1996 (Table I). Rapp (1991) found 10 to 20% higher N contents in fine roots of the treated plot. Due to a lower fine root production rate, the N demand for fine root production was lower at the treated plot than at the control plot. Contents of fine roots at the treated plot were lower in the litter layer, but higher at the 20 to 40 cm depths of the mineral soil compared to the control after five years of treatment (Rapp, 1991). The N content of green leaves was elevated by 20% compared to the control (Meiwes *et al.*, 1998), leading to a surplus N retention of 10 mmol m^{-2} a^{-1}. For other biomass compartments, no data are available. Even though some studies revealed a substantial N accumulation in biomass after N addition (Bergholm and Majdi, 2001), it does not seem very likely that N accumulation in biomass is much higher than at the control. Biomasses of fine roots and leaves are more or less at steady state for old growth forests such as the Solling stand. N contents in other compartments are generally much lower and are not affected by N addition (Seibt and Wittich, 1965; Schleppi *et al.*, 1999). Tree ring measurements showed that the growth rate at the treated plot did not increase compared to the control (Makowka *et al.*, 1991). Thus, it is concluded that N accumulation in the biomass of the treated plot is similar to the control.

Aber *et al.* (1998) discussed the mechanisms by which retention of N, which can be as high as 1000 mmol$_c$ m^{-2} a^{-1}, takes place in forest soils. They favoured the hypothesis of mycorhizal assimilation over abiotic incorporation and assimilation by free-living microorganisms. Long-term monitoring of the organic layer at the control revealed N accumulation in the litter layer between 1966 and 2001 at an average rate of 150 mmol m^{-2} a^{-1} (Meiwes *et al.*, 2002). This is in good agreement with the retention rate calculated from the N fluxes, which is 129 mmol m^{-2} a^{-1} for the period 1976–2000 and 160 mmol m^{-2} a^{-1} for the period 1982–1996 (Table I). However, for the duration of the experiment, there is no indication of an accumulation of the applied N in the organic layer of the treated plot.

Monitoring of N_2O emissions from the soil from 1990 to 2000 revealed emission rates of 15 mmol m^{-2} a^{-1} at the control (Brumme and Borken, 2004). Episodic measurements at the treated plot showed no significant differences from the control. Compared to the beech stand at Solling, most other temperate forest ecosystems studied had lower N_2O emissions (Brumme et al., 1999). As N_2 fluxes were not measured, it remains unclear whether the emission of N_2 is of importance for the N budget of the stand. According to Brumme et al. (1999), N_2 losses from forest soils are higher than N_2O losses only under anaerobic conditions.

Although the fate of the applied N is still unknown, is seems to be most likely that about 500 mmol$_c$ m^{-2} yr^{-1} of the applied N was retained in the mineral soil. However, because the C and N pools in the mineral soil are very large, the change in storage cannot be measured with sufficient accuracy to show the change during the treatment.

In addition to the N status of the ecosystem, the acid/base status was also changed drastically by the treatment (Figure 2). Not only did the total acid load increase strongly, but also the relative importance of the various sources of acid load and acid/base reactions changed drastically. The mean acid load of 1610 mmol$_c$ m^{-2} yr^{-1} during the 11 yr of treatment was close to the potential acid load of mmol$_c$ m^{-2} yr^{-1} introduced by the application. Whereas the retention of base cations was the most important component at the control, N transformations determined most of the acid load at the treated plot. Acid produced by N transformations increased until the last year of application. This leads to the conclusion that steady state conditions were not achieved within 11 yr.

The retention of SO_4^{2-} buffered the acid load during the first nine years of the treatment. Afterwards, the release of SO_4^{2-} generated an acid load for the ecosystem (Figure 2). The SO_4^{2-} retention took place almost exclusively in the subsoil (40–100 cm). Sulphate retention in Solling soils was attributed to the formation of Al hydroxy sulphate (Prenzel, 1994) or to an adsorption at the soil matrix (Meiwes et al., 1980). As most of the SO_4^{2-} was exported from the ecosystem during the application and the three years after the last application, the retention seems to be quite reversible. Meiwes et al. (1980) showed for the control, that desorption of newly retained SO_4^{2-} followed the adsorption characteristics with almost no hysteresis. Initially present SO_4^{2-} behaved different suggesting the existence of two fractions of weakly and tightly bound SO_4^{2-}. A deposition exclosure experiment at Solling showed a strong delay of SO_4^{2-} in soil solution after decreased SO_4^{2-} input (Alewell et al., 1997). Alewell (2001) claimed that under high input conditions S retention is predominantly a reversible sorption of SO_4^{2-}, whereas organic S pools with low turnover rates become more important when S input is low.

Whereas the N leaching seems to be completely reversible after the end of the treatment, there was only a partial recovery of the induced acidification. The pH and the ratio of nutrient cations (base cations: Ca^{2+}, Mg^{2+}, K^+) to Al (base cations/Al ratio) in soil solution (an indicator of possibly toxic conditions for tree roots; Sverdrup and Warfinge, 1993b) decreased strongly at all sampling depths of the treated plot to far below the critical level for beech of 0.6 mol mol^{-1}. Rapp (1991) observed a lower mycorrhiza frequency at the treated plot compared to the control. Other biotic effects of the decreased base cations/Al ratio were not observed. The pH showed only a slight recovery after the last application. Except at 10 cm depth there was no recovery at all of the base cations/Al ratio, suggesting an irreversible deterioration of the acid/base status of the treated soil. Presumably the leaching of base cations with mobile anions introduced by the treatment lead to an exhaustion of the exchangeable fraction of these ions. The export of base cations from the treated plot was >2000 mmol$_c$ m^{-2} higher than from the control during the period 1983–1996, which is about 60% of the actual pool of exchangeable Mb cations. As Ca^{2+}, Mg^{2+}, and K$^+$ are important nutrients, the sustainability of forestry also may be affected by a depletion of these ions from forest soils.

5. Conclusions

Addition of (NH$_4$)$_2$SO$_4$ to the beech stand at Solling strongly increased element fluxes. Leaching of NO$_3^-$ accounted for only 50% of the N input. The fate of the retained N is unclear, but accumulation in the soil is the most likely sink. In addition to N transformations, S retention was an important process for the acid/base budget of the ecosystem. S retention seems to be almost completely reversible within relatively short periods. In contrast, the induced acidification caused a partly irreversible depletion of base cations from the soil. The nutrient supply of the forest stand may be hampered by this nutrient depletion for a long time span.

Acknowledgements

The study was done with support of the Ministry of Nutrition, Agriculture and Forestry of Lower Saxony and from the Soil Monitoring Programme of Lower Saxony.

References

Aber, J. D., McDowell, W., Nadelhoffer, K. J., Magill, A., Berntson, G., Kamakea, M., McNulty, S., Currie, W., Rustad, L. and Fernandez, I.: 1998, 'Nitrogen saturation in temperate forest ecosystems: Hypotheses revisited', *BioScience* **48**, 921–934.

Aber, J. D., Nadelhoffer, K. J., Steudler, P. and Melillo, J. M.: 1989, 'Nitrogen saturation in northern forest ecosystems: Hypotheses and implications', *BioScience* **39**, 378–386.

Alewell, C.: 2001, 'Predicting reversibility of acidification: The European sulfur story', *Water, Air, Soil Pollut.* **130**, 1271–1276.

Alewell, C., Bredemeier, M., Matzner, E. and Blanck, K.: 1997, 'Soil solution response to experimentally reduced acid deposition in a forest ecosystem', *J. Environ. Qual.* **26**, 658–665.

Andersson, P., Berggren, D. and Johnsson, L.: 2001, '30 years of N fertilisation in a forest ecosystem: The fate of added N and effects on N fluxes', *Water, Air, Soil Pollut.* **130**, 637–642.

Benecke, P. and Mayer, R.: 1971, 'Aspects of soil water behaviour as related to beech and spruce stands: Some results of the water balance investigations', *Ecol. Studies* **2**, 153–163.

Bergholm, J. and Majdi, H.: 2001, 'Accumulation of nutrients in above and below ground biomass in response to ammonium sulphate addition in a Norway spruce stand in southwest Sweden', *Water, Air, Soil Pollut.* **130**, 1049–1054.

Brumme, R. and Borken, W.: 2004, 'Gaseous nitrogen and carbon fluxes', in R. Brumme (ed.), *Functioning and Management of European Beech Ecosystems*, Ecol. Studies (in prep).

Brumme, R., Borken, W. and Finke, S.: 1999, 'Hierarchical control of nitrous oxide emission in forest ecosystems', *Glob. Biogeochem. Cycles* **13**, 1137–1148.

Corre, M. D., Beese, F. O. and Brumme, R.: 2003, 'Internal N cycle in high nitrogen deposition forest soil: Changes under nitrogen saturation and liming', *Ecol. Appl.* **13**, 287–298.

Dise, N. B. and Wright, R. F.: 1995, 'Nitrogen leaching from European forests in relation to nitrogen deposition', *Forest Ecol. Manage.* **71**, 153–161.

Edwards, P. J., Kochenderfer, J. N., Coble, D. W. and Adams, M. B.: 2002, 'Soil leachate responses during 10 years of induced whole-watershed acidification', *Water, Air, Soil Pollut.* **140**, 99–118.

Feger, K.-H.: 1992, 'Nitrogen cycling in two Norway spruce (*Picea abies*) ecosystems and effects of a $(NH_4)_2SO_4$ addition', *Water, Air, Soil Pollut.* **61**, 295–307.

Gundersen, P., Emmet, B. A., Kjonaas, O. J., Koopmanns, C. J. and Tietema, A.: 1998, 'Impact of nitrogen deposition on nitrogen cycling in forests: A synthesis of NITREX data', *Forest Ecol. Manage.* **101**, 37–56.

Hörmann, G.: 1997, 'SIMPEL – Ein einfaches, benutzerfreundliches Bodenwassermodell zum Einsatz in der Ausbildung', *Dtsch. Gewässerkundl. Mitt.* **41**, 67–72.

König, N. and Fortmann, H.: 1996, *Probenvorbereitungs-, Untersuchungs- und Elementbestimmungs-Methoden des Umweltanalytik-Labors der Niedersächsischen Forstlichen Versuchsanstalt und des Zentrallabor II des Forschungszentrums Waldökosysteme*, Teil 1–4, Ber. Forschungszentrum Waldökosysteme **B46–B49**.

Makowka, I., Stickan, W. and Worbes, M.: 1991, 'Jahrringbreitenmessung an Buchen (*Fagus sylvatica* L.) im Solling', *Ber. Forschungszentrum Waldökosysteme* **B18**, 83–159.

Meesenburg, H., Meiwes, K. J. and Rademacher, P.: 1995, 'Long-term trends in atmospheric deposition and seepage output in northwest German forest ecosystems', *Water, Air, Soil Pollut.* **85**, 611–616.

Meiwes, K. J., Hauhs, M., Gerke, H., Asche, N., Matzner, E. and Lamersdorf, N.: 1984, 'Die Erfassung des Stoffkreislaufes in Waldökosystemen: Konzept und Methodik', *Ber. Forschungszentrum Waldökosysteme/Waldsterben* **7**, 70–142.

Meiwes, K. J., Khanna, P. K. and Ulrich, B.: 1980, 'Retention of sulphate by an acid brown earth and its relationship with the atmospheric input of sulphur to forest vegetation', *Z. Pflanzenernaehr. Bodenk.* **143**, 402–411.

Meiwes, K. J., Meesenburg, H., Bartens, H., Rademacher, P. and Khanna, P. K.: 2002, 'Akkumulation von Auflagehumus im Solling: Mögliche Ursachen und Bedeutung für den Nährstoffkreislauf', *Forst und Holz* **57**, 428–433.

Meiwes, K. J., Merino, A. and Beese, F. O.: 1998, 'Chemical composition of throughfall, soil water, leaves and leaf litter in a beech forest receiving long term application of ammonium sulphate', *Plant Soil* **201**, 217–230.

Norton, S. A., Wright, R. F., Kahl, J. S. and Scofield, J. P.: 1992, 'The MAGIC simulation of surface water acidification at, and first year results from, the Bear Brook Watershed, Maine, U.S.A.', *Environ. Pollut.* **77**, 279–286.

Prenzel, J.: 1994, 'Sulfate sorption in soils under acid deposition: Comparison of two modeling approaches', *J. Environ. Qual.* **23**, 188–194.

Rademacher, P.: 2004, 'Ermittlung der Ernährungssituation, der Biomasseproduktion und der Nährelementakkumulation mit Hilfe von Inventurverfahren sowie Quantifizierung der Entzugsgrößen auf Umtriebsebene in forstlich genutzten Beständen', *Habilitation Thesis*, Univ. of Göttingen (in prep.).

Rapp, C.: 1991, 'Untersuchungen zum Einfluss von Kalkung und Ammoniumsulfat-Düngung auf Feinwurzeln und Ektomykorrhizen eines Buchenaltbestandes im Solling', *Ber. Forschungszentrum Waldökosysteme*, A72.

Schleppi, P., Bucher-Wallin, I., Siegwolf, R., Saurer, M., Muller, N. and Bucher, J. B.: 1999, 'Simulation of increased nitrogen deposition to a montane forest ecosystem: Partitioning of the added ^{15}N', *Water, Air, Soil Pollut.* **116**, 129–134.

Seibt, G. and Wittich, W.: 1965, *Ergebnisse langfristiger Düngungsversuche im Gebiet des nordwestdeutschen Dilluviums und ihre Folgerungen für die Praxis*, Schr. Forstl. Fak. Univ. Göttingen u. Nieders. Forstl. Versuchsanst. 27/28.

Sverdrup, H. and Warfvinge, P.: 1993a, 'Calculating field weathering rates using a mechanistic geochemical model PROFILE', *Appl. Geochem.* **8**, 273–283.

Sverdrup, H. and Warfvinge, P.: 1993b, 'The effect of soil acidification on the growth of trees, grass and herbs as expressed by the (Ca + Mg + K)/Al ratio', *Report of Ecology and Environmental Engineering 2*, 1993.

Ulrich, B.: 1994, 'Nutrient and acid-base budget of Central European forest ecosystems', in A. Hüttermann and D. L. Godbold (eds), *Effects of Acid Rain on Forest Processes*, Wiley Publishers, New York, NY, U.S.A., pp. 1–50.

NITROUS OXIDE EMISSIONS FROM TWO RIPARIAN ECOSYSTEMS: KEY CONTROLLING VARIABLES

S. E. MACHEFERT[1*], N. B. DISE[1,4], K. W. T. GOULDING[2] and P. G. WHITEHEAD[3]

[1] *Department of Earth Sciences, The Open University, Milton Keynes, MK7 6AA, U.K.;*
[2] *Agriculture and the Environment Division, Rothamsted, Harpenden, AL5 2JQ, U.K.;* [3] *Aquatic Environments Research Centre, Department of Geography, University of Reading, Reading, RG6 6AB, U.K.;* [4] *Department of Biology, Villanova University, Villanova, Pennsylvania 19085, U.S.A.*
(* *author for correspondence, e-mail: S.E.Machefert@open.ac.uk, phone: 01908 655975, fax: 01908 655151)*

(Received 20 August 2002; accepted 6 April 2003)

Abstract. Nitrous oxide (N_2O) emissions were measured weekly to fortnightly between April 2001 and March 2002 from two riparian ecosystems draining different agricultural fields. The fields differed in the nature of the crop grown and the amount of fertiliser applied. Soil water content and soil temperature were very important controls of N_2O emission rates, with a 'threshold' response at 24% moisture content (by volume) and 8 °C, below which N_2O emission was very low. N_2O fluxes were higher at the site that had received the most fertiliser N, but NO_3^- was not a limiting factor at either site. There was also a 'threshold' effect of rainfall, in which major rainfall events (≥ 10 mm) triggered a pulse of high N_2O emission if none of the other environmental factors were limiting. These results suggest the existence of multiple controls on N_2O emissions operating at a range of spatial and temporal scales and that non-linear relationships, perhaps with a hierarchical structure, are needed to model these emissions from riparian ecosystems.

Keywords: buffer strips, mineral N, nitrous oxide, rainfall, riparian, soil temperature, soil water content

1. Introduction

Nitrous oxide (N_2O) is one of the most important anthropogenically-enhanced greenhouse gases, behind carbon dioxide (CO_2) and methane (CH_4). It contributes about 6% to global warming (Denmead, 1991) and is involved in the destruction of stratospheric ozone (Crutzen, 1970). About 70% of the total globally emitted N_2O is derived from soils (Bouwman, 1990) and agriculture as a whole (i.e., animal excreta, denitrification of leached nitrate (NO_3^-), etc.) contributes about 81% of the anthropogenic N_2O emissions (Brown *et al.*, 2001).

Stream riparian zones form an important transition between land and freshwater systems (Gregory *et al.*, 1991), with a significant potential to reduce diffuse pollution, especially NO_3^-, phosphate and pesticides, from agriculture and other human activities. For example, forested and grass buffer strips can reduce N in subsurface waters by 40–100% and 10–60%, respectively (Osborne and Kovacic, 1993). The principal process that removes NO_3^- from water moving through riparian zones is denitrification, in which the NO_3^- is reduced to N_2O and N_2 (e.g., Burt et al., 1999). Complete reduction to N_2 effectively closes the N cycle and benefits the environment as NO_3^- is removed from water without release of N_2O to the atmosphere. Partial reduction to N_2O, however, swaps one pollutant for another. Riparian areas are known to be 'hotspots' of N_2O production in the landscape especially when they receive and process large amounts of excess N from agricultural fields (Groffman et al., 2000).

Amounts of N_2O emitted from riparian zones depend on the physical, chemical and biological attributes of soil, on climate and weather conditions, and on complex interactions among these factors (Teira-Esmatges et al., 1998). Factors known to influence soil N_2O emission include available N (NO_3^- and NH_4^+), temperature, soil moisture content, carbon availability (Conrad, 1996), climate (Ambus and Christensen, 1995), and hydrologic flow. Very few studies of N_2O emission from riparian ecosystems have considered both N_2O fluxes and all of the major environmental controls measured at regular intervals over a full year, however. The high variability of the fluxes, complex interaction of the controls and lack of understanding of the processes involved make such studies difficult. Models of N_2O emission exist, but there is a need for more integrated approaches to tackle the problem.

This paper presents the results obtained from measurements of N_2O fluxes from two riparian ecosystems differing in their hydrology and the nature of the agricultural fields they are draining between April 2001 and March 2002. The study aims to determine the factors controlling the N_2O emissions at these sites, whether the site hydrological differences have any influence on emissions and controlling factors, and the most appropriate way for modelling these fluxes.

2. Site and Methods

Two experimental plots were chosen within the Great Ouse River catchment (U.K.). They are located near the town of Chicheley, north east of Milton Keynes, and will be referred to as Chicheley North and Chicheley East in the rest of the paper. Both sites are riparian ecosystems situated in an active farm and drain different agricultural fields into the same small stream: the Chicheley Brook. Chicheley East plot (7.6 × 7.6 m) is characterised by a gradual slope (17%) down to the stream and a long runoff from the field. From April to September 2001, it drained an oilseed rape field receiving 302 kg N ha^{-1} a^{-1}. Chicheley North plot (2.5 × 19.7 m) is representative of the more common riparian ecosystem found in the U.K., with a steep drop (40% slope) from the field to the stream. From April to September 2001, it drained a wheat field that receives 226 kg N ha^{-1} a^{-1}. From September 2001, both crops were reversed. The fertiliser was applied as urea and Liquid N37 on two different occasions at both sites: 6 March 2001 and 12 April 2001. The soils are from the Fladbury Series, a grey clayey pelo-alluvial gley with >50% clay in the plough layer (0–25 cm), and the soil pH (in H_2O, 1:2.5) is 7.6.

Twenty-two 'static' chambers were installed *in situ* at the two sites: 10 at Chicheley North and 12 at Chicheley East. The chambers were made of 30 cm diameter PVC rings inserted 5 cm deep in the topsoil at two and three levels above the stream surface at Chicheley North and Chicheley East, respectively. The number of replicate chambers at each level was 5 at Chicheley North and 4 at Chicheley East. N_2O fluxes were determined using the closed chamber technique (Hutchinson and Mosier, 1981). Gas samples (20 mL) were withdrawn from the headspace using a 60 mL syringe immediately after closing the chamber, and 30 and 60 min later, and after the atmosphere in the chamber had been mixed by pumping the syringe plunger 6 times. Each sample was injected into an evacuated container (Labco Exetainer, 10 mL). The change in the N_2O concentration as a function of time gave the flux rate for N_2O emission or, in the case of a decrease in concentration over time, absorption by the soil. After each 60 min sampling period, the chambers were left open until the following sampling. Ambient air samples were also taken at each visit. The Exetainers were transported to the laboratory and N_2O concentrations were determined using a gas chromatograph fitted with an electron capture detector and equipped with a Porapak Q, 50–80, 6 ft column. The carrier gas (N_2) flow rate was 58 mL min^{-1}, the detector temperature was 320 °C, and the injector and the oven were at 45 and 60 °C, respectively.

Air temperature using a digital thermometer, soil temperature at 10 cm depth and soil moisture content (% vol. TDR, 6 cm probe) were monitored on a weekly to fortnightly basis at each sampling date. Total daily rainfall was obtained from a nearby meteorological station about 6 km south of the sites. Nitrate (NO_3^-), ammonium (NH_4^+), and DON were determined every two weeks in the soil solutions obtained from 13 and 8 ceramic cup (Fairey Industrial Ceramics Limited) samplers inserted 35 cm deep around the gas sampling points at Chicheley East and Chicheley North, respectively, using a Skalar SAN^{PLUS} System.

3. Results and Discussion

3.1. FLUXES

Nitrous oxide fluxes measured throughout the year displayed wide temporal and spatial variation at each sampling site within each riparian zone, and between them (Figure 1). Similar high variability has been found for N_2O fluxes from other temperate climate riparian zones (Groffman and Tiedje, 1989; Hanson *et al.*, 1994). Fluxes were generally higher in spring/summer than in winter. Some small negative fluxes were observed at Chicheley North, which indicates that the soil was able to take up atmospheric N_2O (Granli and Bockman, 1994). At Chicheley North, the maximum mean N_2O flux observed was 0.109 mg N_2O-N m^{-2} hr^{-1} (0.026 kg N

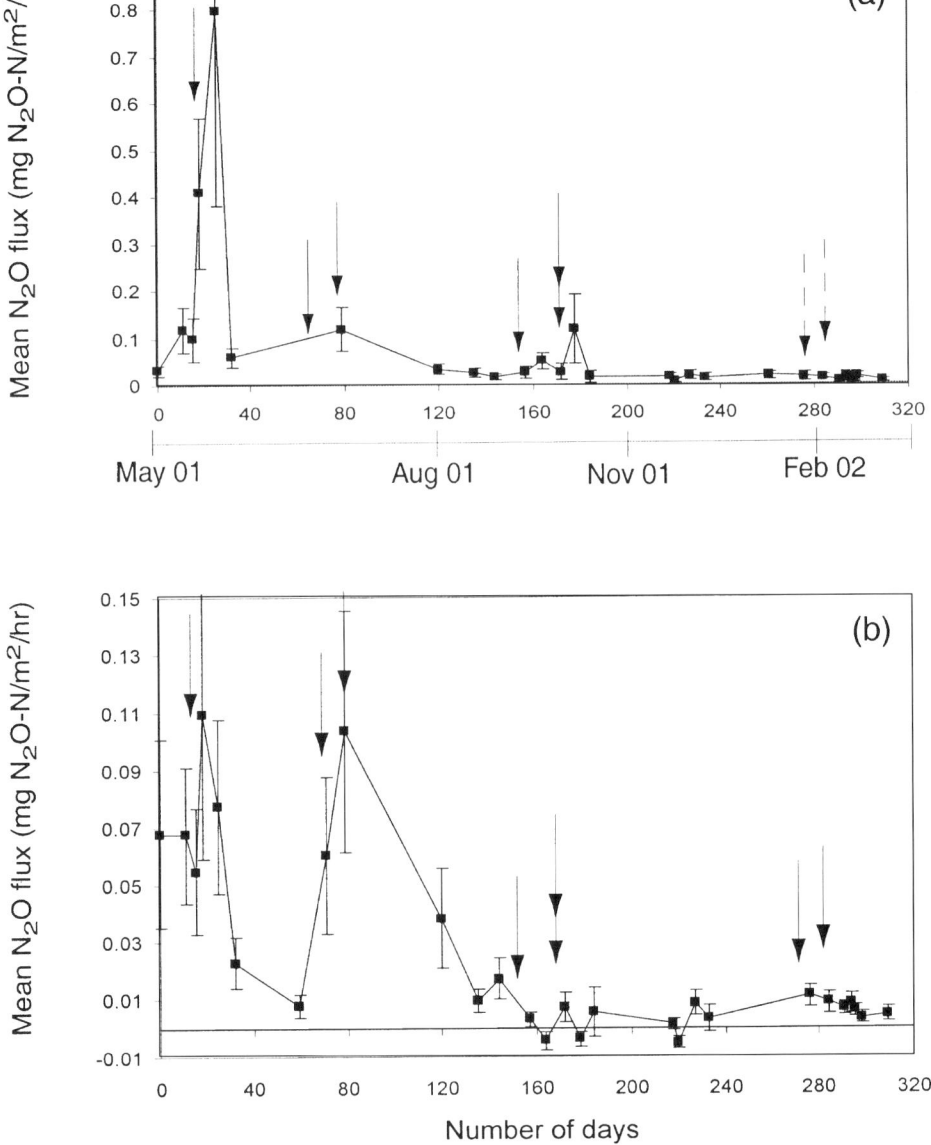

Figure 1. Mean N₂O fluxes (with standard errors) and principal rainfall events at (a) Chicheley East and (b) Chicheley North sites. The black arrows represent total daily rainfall events of 10 mm or more when moisture and temperature are above the thresholds. The dotted arrows represent total daily rainfall events of 10 mm or more when at least one other factor was limiting. Number of days are from 30 April 2001 (day 0) to 05 March 2002 (day 309).

ha^{-1} d^{-1}) and the maximum single value was 0.344 mg N$_2$O-N m^{-2} hr^{-1} (0.083 kg N ha^{-1} d^{-1}) whereas the maximum mean value observed at Chicheley East was 0.798 mg N$_2$O-N m^{-2} hr^{-1} (0.191 kg N ha^{-1} d^{-1}) and the maximum single value was 4.54 mg N$_2$O-N m^{-2} hr^{-1} (1.09 kg N ha^{-1} d^{-1}). At both locations, the highest values were observed at the downslope positions where the water table fluctuated closest to the surface, especially at CH East downslope, which sometimes was flooded. The total N$_2$O fluxes at CH East and CH North were 5.502 and 0.625 kg N ha^{-1}, corresponding to 1.82 and 0.28% of the N applied, respectively. These values are within the range of 'emission factors' inventoried by Brown et al. (2001) for direct emission from soil (0.25–2.25). Total fluxes from both riparian ecosystems studied are also characteristic of emissions from N-enriched ecosystems (Machefert et al., 2002).

3.2. SOIL TEMPERATURE AND MOISTURE

The individual N$_2$O fluxes for each of the 22 chambers on each sampling date were plotted against the environmental factors monitored at both experimental sites (Figure 2). Despite the observation of higher water table at the lowest slope locations soil moisture as well as water filled pore space (WFPS) differed little from one plot to the other (data not shown). This suggests that weekly measurements may be insufficient to capture the dynamics of rapid water level changes, at the lowest locations on the slope.

Throughout the measurement period, the overall soil moisture increased, as spring 2001 was dry and the rest of the year was relatively wet. The soil temperature at both sites increased from spring to summer 2001 and then decreased through autumn and winter. No linear relationships were found between the fluxes and these two factors. However, a clear non-linear pattern was observed in these relationships. N$_2$O fluxes were very low or negligible until a 'threshold' was reached for soil temperature (8 °C) above which fluxes were observed on some occasions (Figure 2a). A non-parametric T-test (Wilcoxon test) showed that fluxes were significantly higher for temperatures above 8 °C than below ($p < 0.001$). For soil moisture, we also observed a 'threshold' at 24% vol., (Figure 2b) although not statistically significant. A similar threshold (25% vol.) was found by Granli and Bockman (1994). We hypothesize that these thresholds are necessary for the potential emission of N$_2$O to occur.

Maximum N$_2$O emissions were measured at soil temperatures of 14–15 °C and soil moisture of 32–36% vol. Above these levels, fluxes started to decline. However, temperature will have an effect on emissions only when the other main soil parameters are not limiting (Dobbie et al., 1999). The decrease in flux rates, observed here when soil temperatures reached 18 °C, occurred in 80% of the cases at soil moistures below the threshold of 24% vol. and corresponds to the summer when conditions were dryer and soil moisture was limiting. The decrease in fluxes observed when soil moisture reached levels >40% vol. corresponded, in 84% of the

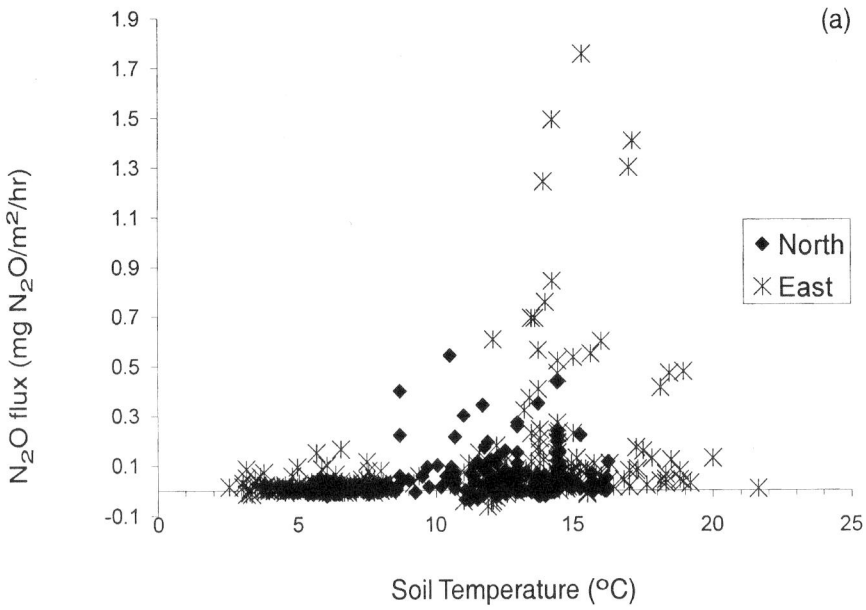
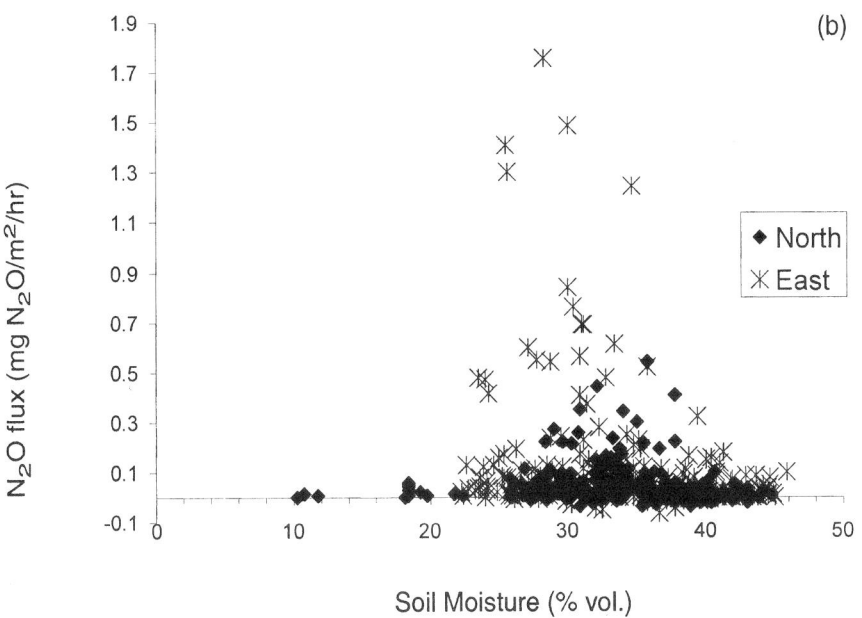

Figure 2. Relationship between N_2O flux at both experimental plots and (a) the soil temperature measured at a depth of 10 cm, and (b) the soil moisture.

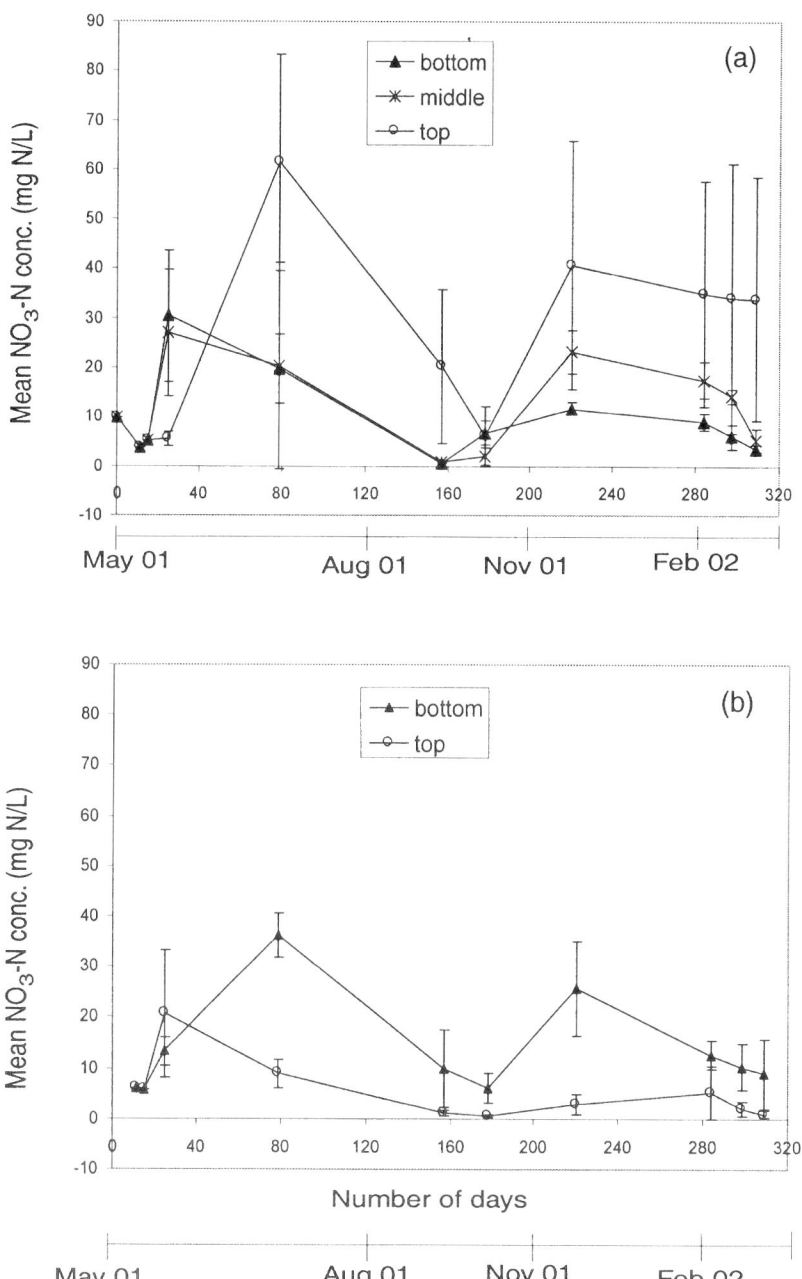

Figure 3. Nitrate-N mean concentrations in soil solution (with standard errors) measured for (a) the top, middle and bottom position of the slope at Chicheley East and (b) the top and bottom position of the slope at Chicheley North. The time of fertiliser application was 06 March–12 April 2001. Number of days are from 30 April 2001 (day 0) to 05 March 2002 (day 309).

cases, to sampling dates when temperature was limiting. It is also expected because N_2O is further reduced to N_2 in soils at higher moisture contents.

Previous studies have shown that N_2O fluxes were related to soil temperatures with a steep increase in N_2O emissions for soil temperatures between 5 and 11 °C (Smith et al., 1998). Other studies showed that substantial N_2O emission occurred when temperatures increased for many soils in temperate climates (Skiba et al., 1998). Most of these studies applied linear regressions which we suggest are not appropriate for modelling N_2O. In contrast, Butterbach-Bahl et al. (2002a), in a study of forest soils, could not demonstrate any trend of N_2O in relation to soil moisture or temperature. However they observed a weak tendency for higher $N_2O:N_2$ ratios with increasing temperature. They also found that at soil temperatures >6.5 °C the correlation between N input and the trace gas fluxes was more pronounced (Butterbach-Bahl et al., 2002b), suggesting that when N deposition was higher a type of temperature threshold existed for the soils they studied. Their work was also done on soil cores in the lab and at a lower frequency of measurement than this study.

3.3. Soil nitrate content

Figure 3 shows the changes in NO_3^- concentration in the soil solution at both locations and along the gradient of the slope. Ammonium content of the soil solution over the study period was constantly close to the detection limit and so the NO_3^- data reflect the concentrations of dissolved inorganic N. Nitrate in soil was generally plentiful so that it had no limiting effect on the N_2O fluxes. More NO_3^- was measured in the soil solution at Chicheley East, potentially reflecting the greater application rates of fertilisers. There was some evidence of a decrease in NO_3^- over the season, with some short-term increases following fertiliser application. At Chicheley East, the NO_3^- concentration in the soil solution decreased from top to bottom, suggesting that NO_3^- was removed from percolating water as it moved downslope before reaching the stream. This is in agreement with findings from previous studies of riparian ecosystems (Smith and Duff, 1988; Ambus and Christensen, 1993). However, this pattern was not observed at Chicheley North.

3.4. Rainfall

At both sites, we observed large pulses of N_2O following total daily rainfall events ≥10 mm (Figure 1). This effect only occurred when other factors controlling denitrification were above the thresholds identified above. Others (e.g., Davidson and Swank, 1986; Ashby et al., 1998) have observed a dramatic increase in N_2O emissions immediately after precipitation. The reason for this may be that soil moisture directly stimulates microbial activity. Nitrate can also accumulate in dry soil (Davidson et al., 1990) as mineralisation occurs and be released with readily available carbon (Davidson et al., 1987). Subsequent precipitation then stimulates denitrification. However, there can also be a piston effect, with rainwater pushing out N_2O trapped in the soil.

4. Conclusion

The results showed that N_2O fluxes were highly variable both temporally and spatially. Fluxes were controlled by soil moisture and soil temperature, but not simply; no linear relationships existed. However, a significant threshold response of the fluxes was observed at 8 °C soil temperature and a similar threshold at 24% moisture content by volume is inferred. NO_3^- content in the soil solution was not a limiting factor at the sites. High pulses of N_2O were generated following main rainfall events providing the other factors influencing the fluxes were not limiting. These results suggest that multiple controls exist on N_2O emissions and non-linear relationships, perhaps with a hierarchical structure, are needed to model these emissions from riparian ecosystems.

Acknowledgements

This research is supported by the European Commission (Project EVK1-1999-00011), and the results are derived from collaborations between all project partners, who are listed on the project website: http://www.reading.ac.uk/INCA. Rothamsted receives grant-aided support from the U.K. Biotechnology and Biological Sciences Research Council. The authors would also like to thank the owner of the land on which measurements were made, Mr. Lewton, for his very helpful collaboration.

References

Ambus, P. and Christensen, S.: 1993, 'Denitrification variability and control in a riparian fen irrigated with agricultural drainage water', *Soil Biol. Biochem.* **25**, 915–923.

Ambus, P. and Christensen, S.: 1995, 'Spatial and seasonal nitrous oxide and methane fluxes in Danish forest-, grassland-, and agroecosystems', *J. Environ. Qual.* **24**, 993–1001.

Ashby, J. A., Bowden, W. B. and Murdoch, P. S.: 1998, 'Controls on denitrification in riparian soils in headwater catchments of a hardwood forest in the Catskill mountains, U.S.A.', *Soil. Biol. Biochem.* **30**, 853–864.

Bouwman, A. F.: 1990, *Soils and the Greenhouse Effect*, Springer, New York, pp. 103–148.

Brown, L., Armstrong Brown, S., Jarvis, S. C., Syed, B., Goulding, K. W. T., Phillips, V. R., Sneath, R. W. and Pain, B. F.: 2001, 'An inventory of nitrous oxide emissions from agriculture in the U.K. using IPCC methodology: emission estimate, uncertainty and sensitivity analysis', *Atmos. Environ.* **35**, 1439–1449.

Burt, T. P., Matchett, L. S., Goulding, K. W. T., Webster, C. P. and Haycock, N. E.: 1999, 'Denitrification in riparian buffer zones: The role of floodplain hydrology', *Hydrol. Process.* **13**, 1451–1463.

Butterbach-Bahl, K., Willibald, G. and Papen, H.: 2002a, 'Soil core method for direct simultaneous determination of N_2 and N_2O emissions from forest soils', *Plant Soil* **240**, 105–116.

Butterbach-Bahl, K, Gasche, R., Huber, CH., Kreutzer, K. and Papen, H.: 2002b, 'Impact of N-input by wet deposition on N-trace gas fluxes and CH_4-oxidation in spruce forest ecosystems of the temperate zone in Europe', *Atmos. Environ.* **32**, 559–564.

Conrad, R.: 1996, 'Soil microorganisms as controllers of atmospheric trace gases', *Microbiol. Rev.* **60**, 609–640.

Crutzen, P. J.: 1970, 'The influence of nitrogen oxides on the atmospheric ozone content', *Q. J. R. Meteorol. Soc.* **96**, 320–325.

Davidson, E. A. and Swank, W. T.: 1986, 'Environmental parameters regulating gaseous nitrogen losses from two forested ecosystems via nitrification and denitrification', *Appl. Environ. Microbiol.* **52**, 1287–1292.

Davidson, E. A., Galloway, L. F. and Strand, M. K.: 1987, 'Assessing available carbon: Comparison of techniques across selected forest soils', *Commun. Soil Sci. Plant Anal.* **18**, 45–64.

Davidson, E. A., Stark, J. M. and Firestone, M. K.: 1990, 'Microbial production and consumption of nitrate in an annual grassland', *Ecology* **71**, 1968–1975.

Denmead, O. T.: 1991, 'Sources and sinks of greenhouse gases in the soil-plant environment', *Vegetatio* **91**, 73–86.

Dobbie, K. E., McTaggart, I. P. and Smith, K. A.: 1999, 'Nitrous oxide emissions from intensive agricultural systems: Variations between crops and seasons, key driving variables, and mean emission factors', *J. Geophys. Res. (D. Atmos.)* **104**, 26,891–26,899.

Granli, T. and Bockman, O. C.: 1994, 'Nitrous oxide from agriculture', *Norw. J. Agri. Sci.* **12**(Suppl), 1–125.

Gregory, S. V., Swanson, F. J., McKee, W. A. and Cummins, K. W.: 1991, 'An ecosystem perspective of riparian zones', *BioScience* **41**, 540–551.

Groffman, P. M. and Tiedje, J. M.: 1989, 'Denitrification in North temperate forest soils: Spatial and temporal patterns at the landscape and seasonal scales', *Soil Biol. Biochem.* **21**, 613–620.

Groffman, P. M., Gold, A. J. and Addy, K.: 2000, 'Nitrous oxide production in riparian zones and its importance to national emission inventories', *Chemos.: Global Change Sci.* **2**, 291–299.

Hanson, G. C., Groffman, P. M. and Gold, A. J.: 1994, 'Denitrification in riparian wetlands receiving high and low groundwater nitrate inputs', *J. Environ. Qual.* **23**, 917–922.

Hutchinson, G. L. and Mosier, A. R.: 1981, 'Improved soil cover method for field measurements of nitrous oxide fluxes', *Soil Sci. Soc. Amer. J.* **45**, 311–316.

Machefert, S. E., Dise, N. B., Goulding, K. W. T. and Whitehead, P. G.: 2002, 'Nitrous oxide emission from a range of land uses across Europe', *Hydrol. Earth Syst. Sci.* **6**, 325–337.

Osborne, L. L. and Kovacic, D. A.: 1993, 'Riparian vegetated buffer strips in water-quality restoration and stream management', *Freshwat. Biol.* **29**, 243–258.

Skiba, U. M., Sheppard, L. J., MacDonald, J. and Fowler, D.: 1998, 'Some key environmental variables controlling nitrous oxide emissions from agricultural and semi-natural soils in Scotland', *Atmos. Environ.* **32**, 3311–3320.

Smith, K. A., Thomson, P. E., Clayton, H., McTaggart, I. P. and Conen, F.: 1998, 'Effects of temperature, water content and nitrogen fertilisation on emissions of nitrous oxide by soils', *Atmos. Environ.* **32**, 3301–3309.

Smith, R. L. and Duff, J. H.: 1988, 'Denitrification in a sand and gravel aquifer', *Appl. Environ. Microbiol.* **54**, 1071–1078.

Teira-Esmatges, M. R., Van Cleemput, O. and Porta-Casanellas, J.: 1998, 'Fluxes of nitrous oxide and molecular nitrogen from irrigated soils of Catalonia (Spain)', *J. Environ. Qual.* **27**, 687–697.

NITROUS OXIDE IN AGRICULTURAL DRAINAGE WATERS FOLLOWING FIELD FERTILISATION

DAVID S. REAY[1]*, KEITH A. SMITH[1] and ANTHONY C. EDWARDS[2]

[1] *School of GeoSciences, University of Edinburgh, Darwin Building, Mayfield Road, Edinburgh EH9 3JU, U.K.;* [2] *Macaulay Institute, Craigiebuckler, Aberdeen AB15 8QH, U.K.*
(* author for correspondence, e-mail: David.Reay@ed.ac.uk, phone: +44(0)131 6507723, fax: +44 (0)131 662 0478)

(Received 20 August 2002; accepted 6 April 2003)

Abstract. Dissolved nitrous oxide (N_2O), nitrate (NO_3^-), and ammonium (NH_4^+) concentrations in an agricultural field drain were intensively measured over the period of field nitrogen (N) fertilisation and for several weeks thereafter. Supersaturations of dissolved N_2O were observed in field drain waters throughout the study. On entry to an open drainage ditch, concentrations of dissolved N_2O rapidly decreased and a total N_2O-N emission via this pathway of 13.2 g over the period of study (45 days) was calculated. This compared with a predicted emission of the order of 300 g, based on measured losses of NO_3^- and NH_4^+ in the field drainage water, and the default IPCC emission factor of 0.01 kg N_2O-N per kg N entering rivers and estuaries. In contrast to widespread evidence of a clear relationship between the amount of N applied to agricultural land and subsequent direct N_2O emission from the soil surface, the relationship between the amount of N_2O in soil drainage waters and the amount of N applied was poor. We conclude that the complexity, both spatially and temporally, of the processes ultimately responsible for the amount of N_2O in agricultural drainage waters make a straightforward relationship between N_2O concentration and N application rate unlikely in all but the simplest of systems.

Keywords: ammonium, degassing, denitrification, fertiliser, leachate, nitrate, nitrification, nitrogen

1. Introduction

Nitrous oxide (N_2O) is a powerful greenhouse gas and one of its largest global sources is agriculture (Nevison, 1999). Emissions of N_2O arise both directly from cultivated and/or fertilised agricultural soils (e.g., Bouwman, 1996; Hénault *et al.*, 1998; MacKenzie *et al.*, 1998; Smith *et al.*, 1998), and indirectly from drainage streams, groundwaters, rivers and estuaries (Dowdell *et al.*, 1979; Law *et al.*, 1992; Groffman *et al.*, 1998; McMahon and Dennehy, 1999; Hiscock *et al.*, 2002). It is estimated that around 30% of the nitrogen (N) applied to agricultural soils subsequently is lost in water runoff and soil leachate (IPCC, 1997). The amounts of N_2O produced, and the factors that control N_2O emissions from these indirect routes are poorly defined (Nevison, 2000).

Nitrous oxide in agricultural drainage waters can originate from formation in the soil, from which it is then leached, and also in the drainage waters themselves (Figure 1). Additionally, the loss of inorganic N from the soil to drainage waters,

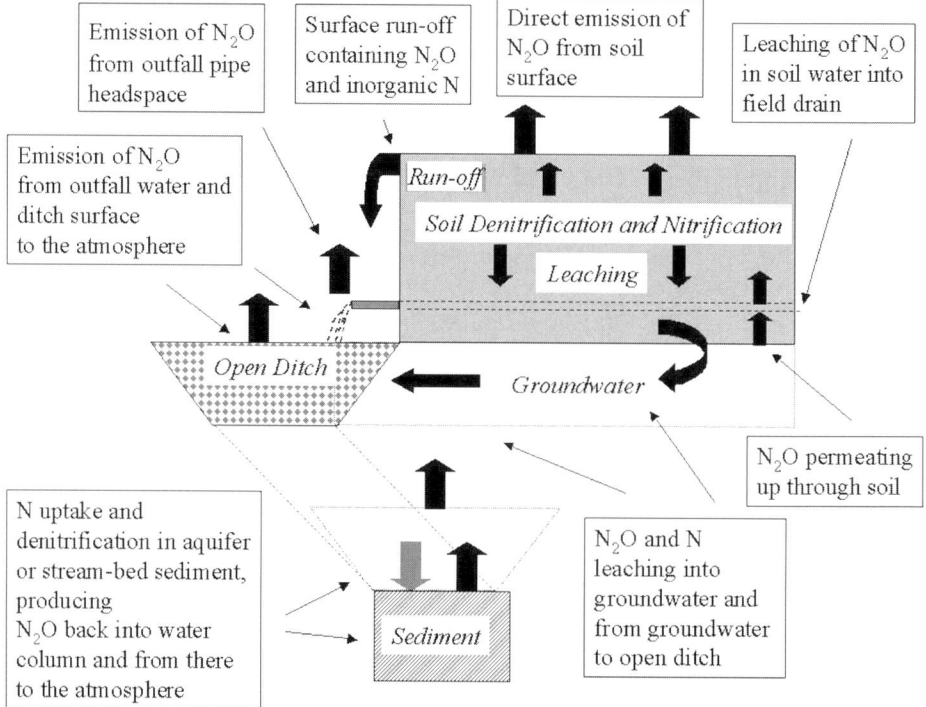

Figure 1. Schematic showing N_2O production and loss from a generalised agricultural system. Solid black arrows denote movement of N_2O through system. Note that N_2O consumption via denitrification may at times be important. Arrow sizes do not reflect relative magnitude of flux.

via leaching and run off, can promote N_2O formation in stream, river and estuarine sediments (Mosier *et al.*, 1998). Nitrous oxide formation in soils is predominantly via denitrification in anaerobic microsites, brought about by the inhibition of aeration at high water contents (Linn and Doran, 1984; Smith *et al.*, 1998; Davidson and Verchot, 2000). However, in well-aerated and warm soils nitrification also can result in the production of N_2O (Yoshida and Alexander, 1970; Poth and Focht, 1985). In moist climates, such as that of the U.K., the former pathway is regarded as being the dominant one (Davidson, 1991; Dobbie *et al.*, 1999).

The application of N fertilisers provides additional substrate for N_2O production by these processes (e.g., Velthof and Oenema, 1995; Hénault *et al.*, 1998; Kaiser *et al.*, 1998; Dobbie *et al.*, 1999), and is associated with elevated concentrations of dissolved N_2O and inorganic N in soil water runoff and leachate (Hasegawa *et al.*, 2000).

Figure 1 demonstrates the numerous stages at which N_2O can be both formed and emitted to the atmosphere in a generalised arable agricultural system. The inorganic N load in drainage waters, arising from leaching and runoff, may not encounter conditions suitable for sediment denitrification and/or nitrification until many kilometres downstream from the point of source (Garcia-Ruiz et al., 1998, 1999). Consequently, the total amount of N_2O emission attributable to a given amount of N fertiliser may be spread over many kilometres and its quantification is difficult.

The N_2O emission factor adopted by the Intergovernmental Panel on Climate Change, IPCC, for leaching and runoff (EF_5) includes a component for groundwater and drainage ditches (EF_5-g), rivers (EF_5-r), and estuaries (EF_5-e) (IPCC, 1996). Each of these three components may act as a source of N_2O through degassing of N_2O dissolved in soil leachate and through *in situ* production of N_2O via nitrification and/or denitrification. However, the balance is likely to be towards degassing of supersaturated leachate for drainage ditches, swinging to a predominance of *in situ* production in rivers and estuaries.

Studies of direct N_2O emission from the soil surface, following N fertiliser application on agricultural land in the U.K., have shown a strong positive relationship between N application and subsequent peaks in N_2O emission rates, with soil water content also being a key determining factor (Clayton et al., 1997; Dobbie et al., 1999). However, to date, no intensive measurements of indirect N_2O emissions, via the appearance of dissolved N_2O in field drainage waters, after fertiliser application, have been reported.

Here we have examined such indirect N_2O emissions, over the 45 days following spring fertilisation, in a field drain outfall fed from a large, and easily definable, arable field catchment in south east Scotland. This has been done to establish whether the relationship between N fertiliser application and direct emission of N_2O from the soil surface also holds for indirect emissions via soil water leaching. Additionally, we have examined the relationship between N application rate and the concentration of inorganic N, particularly nitrate (NO_3^-), in the drainage water, and the fate of dissolved N_2O entering an open drainage ditch.

2. Methods

2.1. EXPERIMENTAL SITE

The experimental site was located on Bush Estate in Midlothian, Scotland, U.K. (Figure 2). The study focused on one particular field drain outfall entering an open agricultural drainage ditch approximately 50 cm in width. The outfall was fed from a large arable field (Kimming Hill) planted to spring barley in both 2001 and 2002. Outfall flow rates were consistently high (1.5 to 2 L sec^{-1}) throughout the period of study, with water temperatures ranging from 9 °C at the start of the study (April 2002) to 13 °C by the end (June 2002). The outfall pipe itself was 12 cm in diameter, with a headspace of about 8 cm. The normal depth of the drains in this and adjacent fields is 50–75 cm. The outfall entered an open ditch, with a gradient of approximately 1:30, between the Kimming Hill and Hay Knowes fields

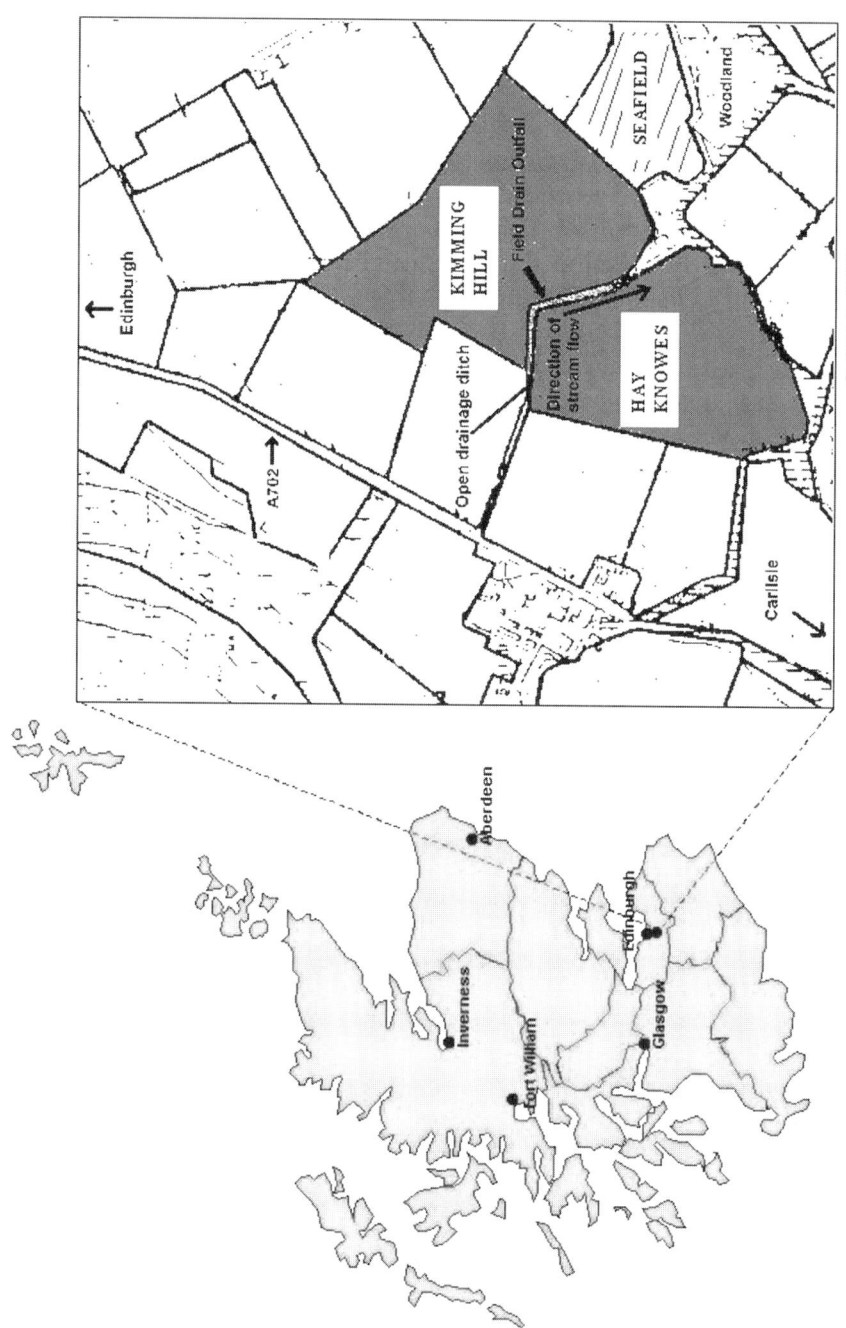

Figure 2. Sample site located in Mid Lothian, Scotland, U.K. The arable fields dominating the open drainage ditch catchment are highlighted. The hatched field (Seafield) is that studied in 1997 by Dobbie *et al.* (1999) during assessment of direct N_2O flux. Scale 1:10 000.

TABLE I

Fertiliser applications for the year preceding (2001) and year of sampling (2002) on the field catchment of the agricultural drainage ditch studied

Field name	2001		2002	
	Date	Rate, kg N ha^{-1}	Date	Rate, kg N ha^{-1}
Kimming Hill	04 April	54 (S)	10 April	70 (S)
	24 May	56 (L)	03 May	88 (L)
Hay Knowes	07 April	56 (S)	25 March	60 (S)
	25 May	56 (L)	22 April	120 (L)

Kimming Hill was planted with Spring Barley in both 2001 and 2002, Hayknowes was planted to Spring Wheat in 2001 and to Winter Wheat in 2002.
S Fertiliser applied as top dressing in granular form (N:P:K 16:16:16), N as NH_4NO_3.
L Fertiliser applied in liquid form (37% N), N as NH_4NO_3.

(Figure 2). Flow rates in the ditch upstream of the outfall were very similar (1.5–2 L sec^{-1}) to those from the outfall itself, and thus the combined flow downstream of the confluence ranged between 3 and 4 L sec^{-1} over the course of the study.

The Kimming Hill (11.5 ha) and Hay Knowes (10.9 ha) fields received several applications of N fertiliser (as NH_4NO_3) in both granular and liquid form during both the year of sampling and in 2001 (see Table I). The soil in Hay Knowes was predominantly alluvium of medium to poor drainage, while that of Kimming Hill was an imperfectly drained Macmerry Series soil. The field drain from Kimming Hill accounted for all observable water input to the open drain over the stretch studied (no increase in ditch flow was observed in this stretch, downstream of the drain entry point; the bulk of the drainage water from Hay Knowes field entered the ditch further downstream, beyond this stretch).

Data obtained during this study are compared to those of a study of direct N_2O flux from the surface of a neighbouring field, Seafield (4.1 ha) planted to winter wheat, in 1997 (Dobbie et al., 1999). This field had comparable N applications, soil, and drainage properties to those of the fields investigated in this study.

2.2. SAMPLING METHODS

Sampling began on the day of seed sowing and the first N fertiliser application of the year at Kimming Hill field (9 April 2002). Daily sampling of drainage waters was then continued for about 1 week, after which time regular, but less frequent (every 2–3 d), sampling continued. The total period of sampling covered a 45 d period and incorporated two separate N fertiliser applications.

Water samples were taken, in triplicate, at the point of entry of the Kimming Hill field drain into the ditch. Additional water samples were taken at regular intervals (approx. 20 m) downstream at six sites along a 150 m length of the ditch. Three samples also were collected from a 20 m section upstream of the field drain outfall. All water samples were collected in 250 mL HDPE containers. Containers were filled completely and sealed with a gas tight inner seal, held in place by a screw-top lid. Water samples were stored in an insulated box containing ice-packs immediately after collection and during transfer to the laboratory ($\sim$20 min), whereupon samples were stored at 4 °C until analysis. Samples of air in the headspace of the field drain outfall pipe were also taken on several occasions. These samples (40 mL) were collected at 10 cm inside the mouth of the outfall pipe using a 50 mL syringe and a three-way gas tight tap. Headspace gas samples were analysed for N_2O, as described below, within 2 hr of collection.

2.3. SAMPLE ANALYSES

Concentrations of both dissolved N_2O and inorganic N were measured in water samples collected from the field drain outfall, with N_2O concentrations also being analysed in the water samples collected from the open drain above and below the field drain outfall. All samples were analysed within 48 hr of collection, control samples preserved with mercuric chloride (Kattner, 1999) showed that no observable changes in dissolved N_2O and inorganic N concentrations occurred within this period (data not shown).

Nitrous oxide concentrations were assessed in the laboratory by analysis of duplicate 5 mL subsamples from each initial sample. Each subsample was injected with a syringe into a 22 mL vial sealed with a septum and shaken vigorously for 2 min, followed by a 30 min standing period. Preliminary experiments showed this technique to give full equilibration between N_2O in the gas and aqueous phases. Nitrous oxide concentrations in the headspaces were determined by gas chromatography using an Agilent 6890 GC fitted with a 1.8 m Porapak-N column and electron capture detector. *In situ* dissolved N_2O concentrations were then calculated, based on N_2O solubility at laboratory temperature and pressure versus *in situ* temperature and pressure (Weiss and Price, 1980), and allowing for the atmospheric N_2O concentration. Nitrous oxide concentrations in outfall pipe headspace samples were measured in the same way, but without the need for water-gas phase equilibration. Inorganic N concentrations were measured by colorimetry using a segmented flow auto-analyser (Bran and Luebbe, Norderstedt, Germany); results are presented as NO_3^--N and NH_4^+-N; nitrite (NO_2^--N) concentrations were negligible (<0.01 mg L^{-1}) in all samples. Data were analysed using linear regression analysis and ANOVA carried out using the data analysis package Systat (version 5.04, Systat Inc.).

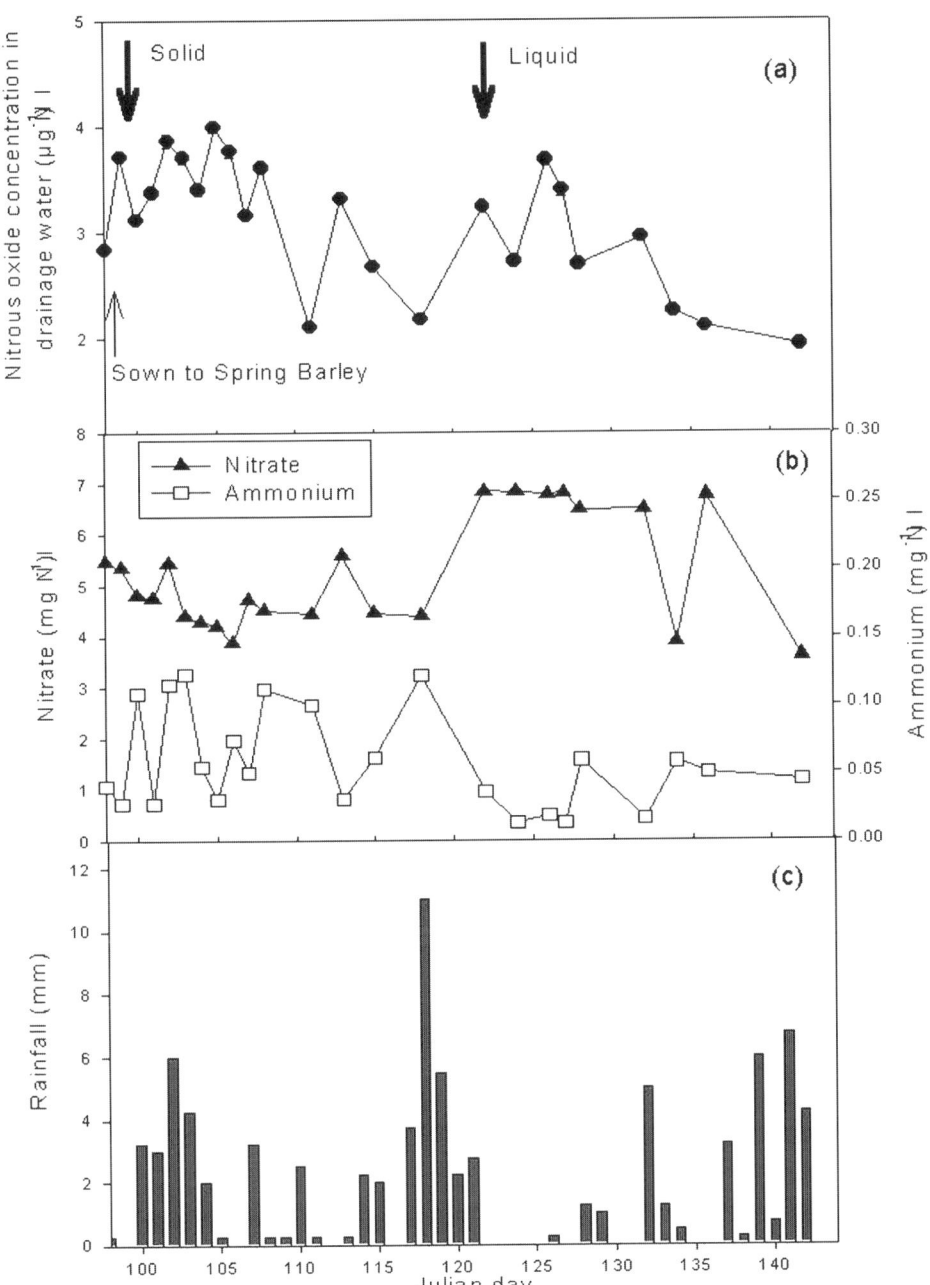

Figure 3. (a) Time-course of dissolved N_2O concentration in water emerging from Kimming Hill outfall. Arrows indicate times of fertilizer applications, see Table I for details. Error bars represent standard error about the mean ($n = 3$). (b) Time-course of dissolved NO_3^- and NH_4^+ concentration in water emerging from the Kimming Hill outfall. (c) Time-course of daily rainfall variation at the study site.

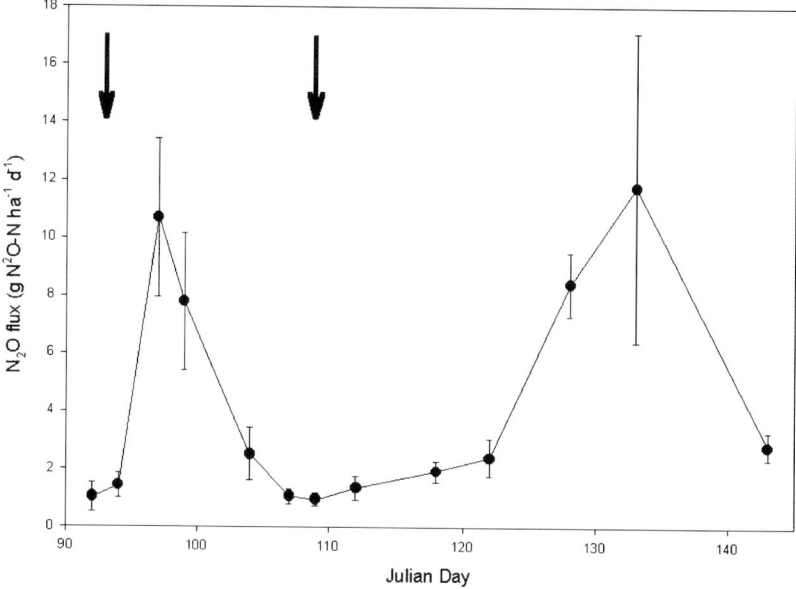

Figure 4. Direct nitrous oxide flux from soil surface following N fertilizer application (arrows) on Winter Wheat in Seafield, at Boghall Farm, SE Scotland in 1997. N applications were 60 and 140 kg N ha^{-1} respectively. Data were obtained from Dobbie *et al.*, 1999.

3. Results and Discussion

3.1. Field drain output

Nitrous oxide concentrations in the Kimming Hill field drain output were consistently greater (5 to 10 times) than that of air equilibrium (~0.35 μg N L^{-1}). Such elevated N$_2$O concentrations are, though, towards the bottom of the range reported previously for agricultural field drains (e.g., Dowdell *et al.*, 1979; Hasegawa *et al.*, 2000; Hack and Kaupenjohann, 2002). Concentrations fluctuated from day to day (Figure 3a). Some effect of N application was apparent, with a generally positive response for several days after each application event, followed by an eventual falling away of dissolved N$_2$O concentrations around two weeks after application. However, the often large and easily discernible effect of such applications on direct agricultural N$_2$O flux reported previously (e.g., Dobbie *et al.*, 1999; Figure 4) and in studies like those of Hénault *et al.* (1998) and Ruser *et al.* (1998), was absent from the indirect N$_2$O flux via drainage waters measured here. The N$_2$O appearing in outfall water is likely to be derived, at least in part, from both nitrification and denitrification at the soil surface, in deeper soil, and even in the field drain itself. The heterogeneous nature of these various potential N$_2$O production zones means that a large surface input of N may induce N$_2$O production, and subsequent appearance in soil drainage waters, over varying time scales and to various intensities. In contrast, direct emission of N$_2$O from the soil surface is likely to be dominated by soil mediated nitrification and/or denitrification in the upper soil layers, and so a larger and more direct response of N_2O flux to N application is apparent.

Nitrate showed no observable response to the first N application on April 10th. However, on the day of the second N application a slight increase in NO_3^--N concentrations ($\sim$1.5 mg L^{-1}) was observed in drainage waters. This increase, though, also coincided with a period of little or no rainfall. It may be that decreased dilution of soil leachate, rather than increased leaching of soil NO_3^-, was responsible for this change in NO_3^- concentration. Certainly, where significant rainfall ($\sim$5 mm d^{-1}) fell after this dry period, NO_3^- concentrations in the drainage waters rapidly decreased, rather than increased (Julian day 118). However, the lag-time between precipitation falling on the field and its appearance in the drainage water was a key unknown.

Overall, the relatively low rainfalls, following application of the fertiliser on both occasions, are likely to have favoured the retention of applied N within the surface soil. Fertiliser granules from the first application were observable on the soil surface for more than a week.

Ammonium concentrations in the field drain water also were variable and failed to show a clear response to fertiliser application, with concentrations generally being almost two orders of magnitude smaller than NO_3^-, even though the N applied was in the form of liquid and granular NH_4NO_3. Nitrate has a much greater mobility through the soil and into drainage waters via leaching and run-off (Valiela and Bowen, 2002), and any nitrification occurring would add to the predominance of the NO_3^- form.

3.2. Cumulative N_2O and N emissions

From measurement of water flow rates at the Kimming Hill field drain outfall, we have calculated a total emission of N_2O to the atmosphere from this single source over the period of study. We have assumed that all of the dissolved N_2O, at concentrations above that of air saturation was subsequently lost to the atmosphere. Where data were not available for a particular day, a concentration was estimated by integration of the closest concentrations before and after it. A total N_2O-N emission of 13.2 g was obtained for the 45 days of study. Extrapolated to an annual basis, such an emission rate would equate to about 107 g of N_2O-N a^{-1} from this single field drain, though of course variations in rainfall, temperature, flow rate, and the absence of additional N fertilisation greatly increase the uncertainty of such an annual emission estimate. Compared with direct N_2O emissions from agricultural soils in Scotland with similar rainfall and N application rates, such as Seafield 1997 (Dobbie et al., 1999; Figure 4), such indirect emission is slight. Furthermore, for much of agriculture elsewhere in western Europe, where soil moisture contents and leaching rates are generally lower than those in Scotland, the relative significance of indirect N_2O emission via drainage waters is likely to be even smaller.

Total losses of inorganic N, in the form of dissolved NO_3^--N and NH_4^+-N in drainage waters, amounted to 29.5 and 0.6 kg, respectively, over 45 days. Previous studies of N application and leaching in intensive agricultural systems have reported total NO_3^- leaching equivalent to between 20 and 30% of applied N (Ramos *et al.*, 2002). With a total N application rate on Kimming Hill of 160 kg N ha^{-1} a^{-1}, and a total field area of 11.5 ha, around 450 kg of N might be expected to be lost via leaching. The dissolved NO_3^- and NH_4^+ measured in the outfall water over the 45 day study was equivalent to nearly 7% of this predicted annual loss.

Using the IPCC default value of 0.01 for N_2O emission arising from leached N in rivers and estuaries, a further emission of 300 g of N_2O-N would be predicted, 98% of this resulting from leached NO_3^--N. Combined with the emission of leached N_2O calculated above, a total N_2O-N emission from the Kimming Hill field drain equivalent to 313 g, over the 45 day study period, was predicted. On this basis the N_2O arising from in-stream, river, and estuarine processing of leached inorganic N, particularly NO_3^-, vastly outweighs that emitted to the atmosphere from field drainage water supersaturated with N_2O. However, as noted earlier, dissolved N_2O concentrations measured during this study are towards the bottom of the range previously reported for such agricultural drainage systems (e.g., Dowdell *et al.*, 1979; Hack and Kaupenjohann, 2002).

3.3. N_2O AND RAINFALL

The occasional strong relationship observed between rainfall and direct N_2O emission from the soil surface in the days following N fertiliser application may be less obvious for indirect fluxes of N_2O in leachate. The effect of individual rainfall events will be made even more complex due to time lag effects, and the simultaneous stimulation of N_2O production rates in the soil, increased leaching rates, and dilution of N_2O in some drainage waters. In this study, neither dissolved N_2O, NO_3^- nor NH_4^+ concentrations in the drainage waters showed a clear relationship with rainfall (Figure 3c). It is likely that there is a lag time between variations in rainfall and their effect on N_2O, NO_3^- and NH_4^+ processing in the soil. However, this lag time is itself likely to be extremely variable due to the spatial heterogeneity of soil N processing and clear relationships between N application and N_2O emission under such conditions will inevitably be difficult to detect.

Across the study period, no significant relationship ($p > 0.1$) was observed between dissolved N_2O and NO_3^- or NH_4^+ concentrations in the field drainage water samples. However, given the complexity of N processing in the soil, and the differential effects rainfall and N applications might be expected to have on N_2O, NO_3^-, and NH_4^+ processing, such a relationship is unlikely to be evident over the relatively short time scale of this study. In their year-long study of N_2O and NO_3^- discharge from tile drains in Southern Germany, Hack and Kaupenjohann (2002) reported that higher concentrations of N_2O did indeed coincide with higher concentrations of NO_3^-, though the stoichiometric relationship was highly variable.

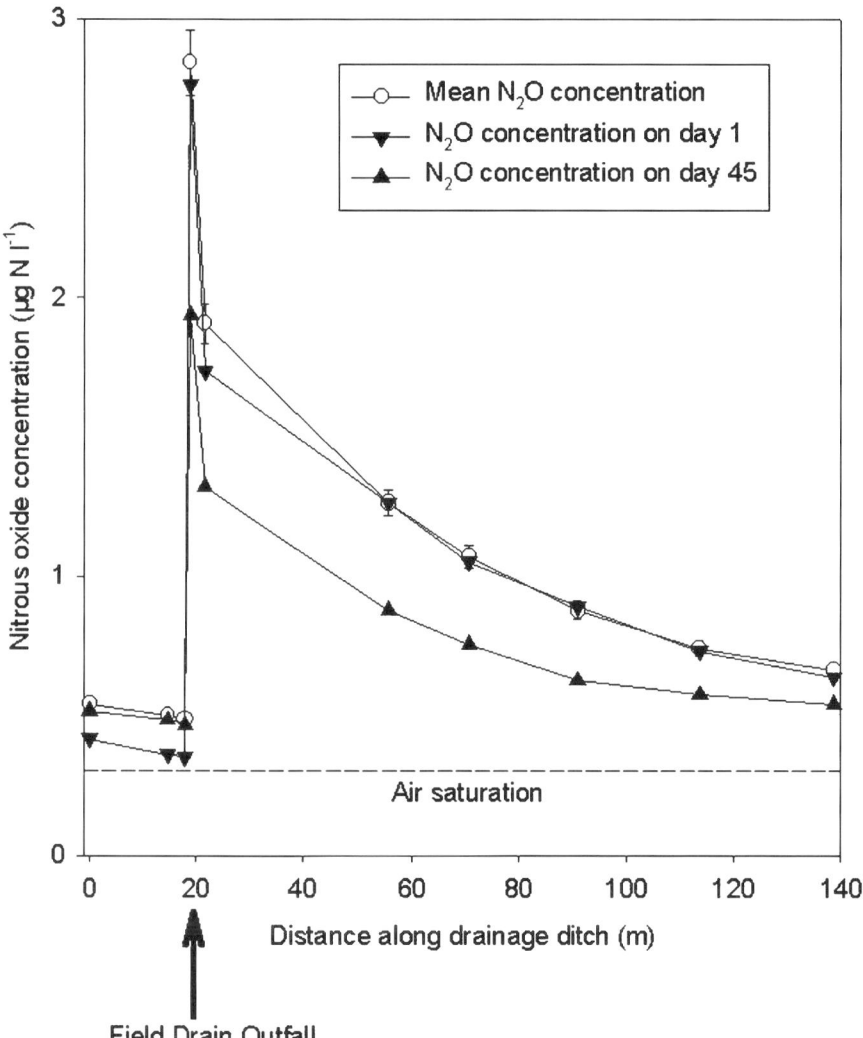

Figure 5. Dissolved N_2O concentration in drainage water with distance along the ditch on the first and last sampling day, compared to the mean, over the 45 day sampling period. Error bars represent standard error about the mean ($n = 24$). Note that the Kimming Hill outfall enters the ditch at $\sim$20 m. Background dissolved N_2O concentration ($\sim$0.3 μg L^{-1}) denoted by dashed line.

3.4. OPEN DITCH

In the open ditch, into which the Kimming Hill field drain flowed, dissolved N_2O concentrations upstream from the outfall were close to that of air equilibrium throughout the course of the study. The sharp increase in dissolved N_2O concentrations in Figure 5 represents that of the Kimming Hill field drain outfall, before mixing. Subsequent dilution of the Kimming Hill field drain water by that of the

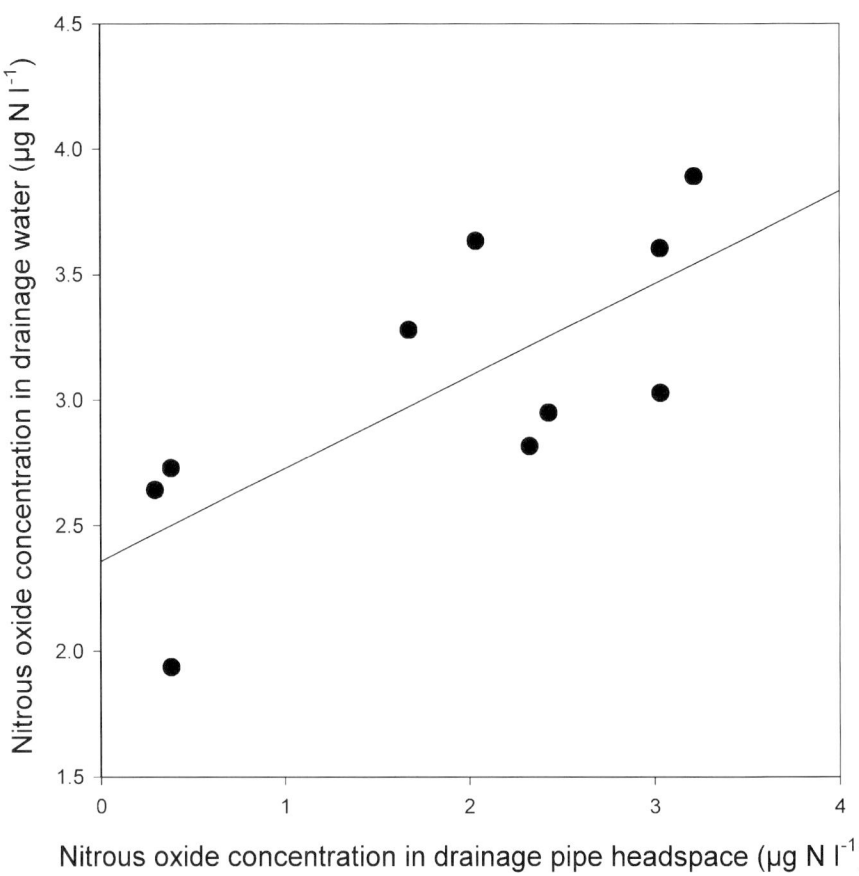

Figure 6. Relationship between dissolved N_2O in outfall water and N_2O in pipe headspace on 10 different sampling occasions during the 45 day study period.

open field drain was apparent in the rapid drop in dissolved N_2O concentrations downstream of the confluence. Despite this dilution effect, dissolved N_2O concentrations observed immediately downstream from the confluence of the Kimming Hill field drain outfall and the open ditch (21 m) remained several times those of air equilibrium. These elevated concentrations of dissolved N_2O decreased over the subsequent 120 m of open drainage ditch, with concentrations having fallen back to less than 1 μg N L^{-1} by the bottom of the study transect. This trend was remarkably consistent across the period of study, despite variations in the initial N_2O loading. Dissolved NO_3^- and NH_4^+ concentrations remained relatively constant downstream of the confluence, as described in Reay *et al.* (2003), with no evidence for *in situ* N_2O production in the drainage ditch itself. Increasing water temperatures over the course of the study are likely to have slightly increased the rate of loss of dissolved N_2O in the open ditch, due to a decrease in N_2O solubility at higher water temperatures (Weiss and Price, 1980).

3.5. N_2O IN FIELD DRAIN HEADSPACE

Concentrations of N_2O in the air headspace of the outfall pipe showed a significant positive relationship ($p < 0.05$) with those of N_2O dissolved in the drainage water itself (Figure 6). The variations in this relationship are likely to have resulted from variable wind speeds, and so headspace flushing times, over the study period. Rates of direct N_2O emission to the atmosphere via this pathway were not determined, but on windy days, where air movement in the mouth of the outfall pipe is rapid, emissions by this route may be important. However, Hack and Kaupenjohann (2002), in their study, calculated that such emissions via diffusion from the pipe headspace accounted for less than 0.1% of total N_2O discharge in drainage water. In general, the perforations in field drainage pipes required to allow water infiltration presumably also allow a certain amount of N_2O flux from the pipe headspace back into the surrounding soil, and eventually to emission at the soil surface (Figure 1) via diffusion and convection.

4. Conclusions

Of the total N measured in field drainage waters over a 45 day period, no more than 0.07% occurred as N_2O. This is equivalent to less than 5% of that predicted by the IPCC default N_2O emission factor (EF_5-g) for such drainage waters. Downstream processing of dissolved N, predominantly as NO_3^-, appeared to be of much greater importance to total N_2O emissions arising from the leaching of agricultural N, assuming the IPCC emission factor is correct in this instance. Supersaturations of N_2O in agricultural drainage water emerging from a field drain outfall varied from between 5 and 10 times that at air equilibrium. Within only 150 m downstream of entry of into an open ditch, dissolved N_2O concentrations had consistently fallen back to a level close to that at air equilibrium. The complexity of soil N processing common to most agricultural systems means that a simple relationship between field N application and the amount of dissolved N_2O in field drainage waters is unlikely.

Acknowledgements

We thank Karen Dobbie of the University of Edinburgh and Ken Hargreaves of CEH Edinburgh for their provision of direct N_2O field flux data and meteorological data, respectively. We also thank Donald Dunbar and Robin Carey of SAC Edinburgh (Boghall Farm) for their help with fertiliser application history and field information. This work was funded by a grant provided as part of the NERC Global Nitrogen Enrichment (GANE) programme.

References

Bouwman, A. F.: 1996, 'Direct emissions of nitrous oxide from agricultural soils', *Nutr. Cycling Agroecosyst.* **46**, 53–70.

Clayton, H., McTaggart, I. P., Parker, J., Swan, L. and Smith, K. A.: 1997, 'Nitrous oxide emissions from fertilised grassland: A two-year study of the effects of N fertiliser form and environmental conditions', *Biol. Fertil. Soils*, **25**, 252–260.

Davidson, E. A.: 1991, 'Fluxes of nitrous oxide and nitric oxide from terrestrial ecosystems', in J. E. Rogers and W. B. Whitman (eds), *Microbial Production and Consumption of Greenhouse Gases: Methane, Nitrogen Oxides, and Halo-methanes*, Am. Soc. for Microbiol., Washington, D.C.

Davidson, E. A. and Verchot, L. V.: 2000, 'Testing the hole-in-the-pipe model of nitric and nitrous oxide emissions from soils using the TRAGNET database', *Global Biogeochem. Cycles*, **14**, 1035–1043.

Dobbie, K. E., McTaggart, I. P. and Smith, K. A.: 1999, 'Nitrous oxide form intensive agricultural systems: Variations between crops and seasons, key driving variables, and mean emission factors', *J. Geophys. Res.* **104**, 26,891–26,899.

Dowdell, R. J., Burford, J. R. and Crees, R.: 1979, 'Losses of nitrous oxide dissolved in drainage water from agricultural land', *Nature* **278**, 342–343.

Garcia-Ruiz, R., Pattinson, S. N. and Whitton, B. A.: 1998, 'Denitrification in river sediments: Relationship between process rate and properties of water and sediment', *Freshwat. Biol.* **39**, 467–476.

Garcia-Ruiz, R., Pattinson, S. N. and Whitton, B. A.: 1999, 'Nitrous oxide production in the River Swale-Ouse, North-East England', *Water Res.* **33**, 1231–1237.

Groffman, P. M., Gold, A. J. and Jacinthe, P. A.: 1998, 'Nitrous oxide production in riparian zones and groundwater', *Nutr. Cycling Agroecosyst.* **52**, 179–186.

Hack, J. and Kaupenjohann: 2002, 'N_2O discharge with drain water from agricultural soils of the upper Neckar region in Southern Germany', in J. Van Ham, A. P. M. Baede, R. Guicherit and J. G. F. M. Williams-Jacobse (eds), *Proc. Third Int. Symp. Non-CO_2 Greenhouse Gases*, Maastricht, The Netherlands, Millpress, Rotterdam, *Symp. Non-CO_2 Greenhouse Gases*, Maastricht, The Netherlands, pp. 185–190.

Hasegawa, K., Hanaki, K., Matsuo, T. and Hidaka S.: 2000, 'Nitrous oxide from the agricultural water system contaminated with high nitrogen', *Chemosphere – Global Change Sci.* **2**, 335–345.

Hénault, C., Devis, X., Kucas, J. L. and Germon, J. C.: 1998, 'Influence of different agricultural practices (type of crop, form of N-fertiliser) on soil nitrous oxide emissions', *Biol. Fertil. Soils*, **27**, 299–306.

Hiscock, K. M., Bateman, A. S., Fukada, T., Dennis, P. F.: 2002, 'The concentration and distribution of groundwater N_2O in the Chalk aquifer of eastern England', in J. Van Ham, A. P. M. Baede, R. Guicherit and J. G. F. M. Williams-Jacobse (eds), *Non-CO_2 Greenhouse Gases: Scientific Understanding, Control Options and Policy Aspects, Proc. Third Int. Symp. Non-CO_2 Greenhouse Gases*, Maastricht, The Netherlands, Millpress, Rotterdam, pp. 179–184.

IPCC Intergovernmental Panel on Climate Change: 1996, Houghton, J. T., Meira, L., Filho, G., Lim, B., Treanton, K., Mamaty, I., Bonduki, Y., Griggs, D. J., Callender, B. A. (eds) *Revised 1996 IPCC Guidelines for National Greenhouse Gas Inventories*. IPCC/OECD/IEA. U.K. Meteorological Office, Bracknell, U.K.

IPCC Intergovernmental Panel on Climate Change: 1997, *Guidelines for National Greenhouse Gas Inventories*, OECD, Paris.

Kaiser, E. A., Kohrs, K., Kücke, M., Schnug, E., Heinemeyer, O. and Munch, J. C.: 1998, 'Nitrous oxide release from arable soil: Importance of N fertilization, crops and temporal variation', *Soil. Biol. Biochem.* **30**, 1553–1563.

Kattner, G.: 1999, 'Storage of dissolved inorganic nutrients in seawater: Poisoning with mercuric chloride', *Mar. Chem.*, **67**, 61–66.

Law, C. S., Rees, A. P. and Owens, N. J. P.: 1992, 'Nitrous oxide: Estuarine sources and atmospheric flux', *Estuar. Coast. Shelf Sci.* **35**, 301–314.

Linn, D. M. and Doran, J. W.: 1984, 'Effect of water filled pre space on carbon dioxide and nitrous oxide production in tilled and nontilled soils' *Soil Sci. Soc. Am. J.* **48**, 1267–1272.

MacKenzie, A. F., Fan, M. X. and Cadrin, F.: 1998, 'Nitrous oxide emission in three years as affected by tillage, corn-soybean-alfalfa rotations, and nitrogen fertilization', *J. Environ. Qual.* **27**, 698–703.

McMahon, P. B. and Dennehy, K. F.: 1999, 'N_2O emissions from a nitrogen-enriched river', *Environ. Sci. Technol.* **22**, 21–25.

Mosier, A. R., Kroeze, C., Nevison, C., Oenema, O., Seitzinger, S. and van Kleemput, O.: 1998, 'Closing the global atmospheric N_2O budget: Nitrous oxide emissions through the agricultural nitrogen cycle', *Nutr. Cycling Agroecosyst.* **52**, 225–248.

Nevison, C.: 1999, Indirect nitrous oxide emissions from agriculture. Background paper for IPCC expert group meeting on Good Practice in Inventory Preparation – *Agricultural Sources of Methane and Nitrous Oxide*, IPCC, Wageningen, The Netherlands.

Nevison, C.: 2000, 'Review of the IPCC methodology for estimating nitrous oxide emissions associated with agricultural leaching and runoff', *Chemosphere – Global Change Sci.* **2**, 493–500.

Poth, M. and Focht, D. D.: 1985, '^{15}N kinetic analysis of N_2O production by *Nitrosomonas europea*: An examination of nitrifier denitrification', *Appl. Environ. Microbiol.* **49**, 1134–1141.

Ramos, C., Agut, A. and Lidon, A. L.: 2002, 'Nitrate leaching in important crops of the Valencian Community region (Spain)', *Environ. Pollut.* **118**, 215–223.

Reay, D. S., Smith, K. A. and Edwards, A. C.: 2003, 'Nitrous oxide in agricultural drainage waters', *Global Change Biol.* **9**, 195–203.

Ruser, R., Flessa, H., Schilling, R., Steindl, H. and Beese, F.: 1998, 'Soil compaction and fertilization effects on nitrous oxide and methane fluxes in potato fields', *Soil Sci. Soc. Am. J.* **62**, 1587–1595.

Smith, K. A., Thomson, P. E., Clayton, H., McTaggart, I. P. and Conen, F.: 1998, 'Effects of temperature, water content and nitrogen fertilization on emissions of nitrous oxide by soils', *Atmos. Environ.* **32**, 3301–3309.

Valiela, I. and Bowen, J. L.: 2002, 'Nitrogen sources to watersheds and estuaries: role of land cover mosaics and losses within watersheds', *Environ. Pollut.* **118**, 239–248.

Velthof, G. L. and Oenema, O.: 1995, 'Nitrous oxide fluxes form grassland in the Netherlands: II. Effects of soil type, nitrogen fertilizer application and grazing', *Eur. J. Soil Sci.* **46**, 541–549.

Weiss, R. F. and Price, B. A.: 1980, 'Nitrous oxide solubility in water and seawater', *Mar. Chem.* **8**, 347–359.

Yoshida, T. and Alexander, M.: 1970, 'Nitrous oxide formation by *Nitrosomonas europea* and heterotrophic microorganisms', *Soil. Sci. Soc. Am. Proc.* **34**, 880–882.

NITRATE LEACHING FROM A MOUNTAIN FOREST ECOSYSTEM WITH GLEYSOLS SUBJECTED TO EXPERIMENTALLY INCREASED N DEPOSITION

PATRICK SCHLEPPI*, FRANK HAGEDORN and ISABELLE PROVIDOLI

Swiss Federal Institute for Forest, Snow and Landscape Research (WSL), CH-8903 Birmensdorf, Switzerland
(* author for correspondence, e-mail: schleppi@wsl.ch; phone: +41 1 739 24 22; fax: +41 1 739 24 88)

(Received 20 August 2002; accepted 10 May 2003)

Abstract. Nitrate leaching was measured over seven years of nitrogen (N) addition in a paired-catchment experiment in Alptal, central Switzerland (altitude: 1200 m, bulk N deposition: 12 kg ha^{-1} a^{-1}). Two forested catchments (1500 m^2 each) dominated by *Picea abies* were delimited by trenches in the Gleysols. NH$_4$NO$_3$ was added to one of the catchments using sprinklers. During the first year, the N addition was labelled with ^{15}N. Additionally, soil N transformations were studied in replicated plots. Pre-treatment NO$_3^-$-N leaching was 4 kg ha^{-1} a^{-1} from both catchments, and remained between 2.5 and 4.8 kg ha^{-1} a^{-1} in the control catchment. The first year of treatment induced an additional leaching of 3.1 kg ha^{-1}, almost 90% of which was labelled with ^{15}N, indicating that it did not cycle through the large N pools of the ecosystem (soil organic matter and plants). These losses partly correspond to NO$_3^-$ from precipitation bypassing the soil due to preferential flow. During rain or snowmelt events, NO$_3^-$ concentration peaks as the water table is rising, indicating flushing from the soil. Nitrification occurs temporarily along the water flow paths in the soil and can be the source of NO$_3^-$ flushing. Its isotopic signature however, shows that this release mainly affects recently applied N, stored only between runoff events or up to a few weeks. At first, the ecosystem retained 90% of the added N (2/3 in the soil), but NO$_3^-$ losses increased from 10 to 30% within 7 yr, indicating that the ecosystem became progressively N saturated.

Keywords: forest ecosystem, Gleysols, nitrate leaching, nitrogen deposition, paired-catchment experiment

1. Introduction

Deposition of inorganic nitrogen (N) to terrestrial ecosystems has increased over the past decades due to emissions associated with human activities. Increased inputs of ammonium (NH$_4^+$) and nitrate (NO$_3^-$) can induce eutrophication of previously N-limited systems, including forests, and finally cause their 'nitrogen saturation' (Ågren and Bosatta, 1988; Aber *et al.*, 1989). According to these authors, increases in plant N uptake and N mineralisation occur before saturation is reached. Then, at saturation, there is a sharp increase in nitrification. Nitrate leaching, finally, increases as the N saturation continues. Nitrate leaching is thus usually regarded as the main symptom of N saturation in forests (Gundersen et al., 1998).

As part of the European research project NITREX (Wright and Rasmussen, 1998; Emmett et al., 1998), we studied the effects of increased N deposition on a mountain forest in Alptal, Switzerland (Schleppi et al., 1998). Nitrogen was added underneath the tree crowns both in a paired-catchment experiment and in a replicated plot design. We found that under ambient conditions as well as with increased N deposition, NO_3^- leaching from the Gleysols of this site did not fit into the concept of N saturation by Aber et al. (1989). Specifically, NO_3^- leaching occurred even if the trees were still slightly deficient in N (1.1% N in needles) and if no effects of the N addition were observed on the vegetation (Schleppi et al., 1999b). Further, neither the inorganic N pools in the soil (extractable NO_3^- and NH_4^+) nor the net nitrification were affected by two years of N addition, but NO_3^- leaching clearly increased (Hagedorn et al., 2001).

In the model of N saturation by Stoddard (1994), some of the deposited NO_3^- may bypass the ecosystem during winter and spring before any N saturation. Concentrations measured at Alptal during the dormant season would fit to this 'stage 0'. Nitrate leaching, however, also happens regularly during the summer, which would point to advanced (stage 2) saturation. It is therefore not possible to classify our site on Stoddard's scale.

In most cases, NO_3^- leaching from forest ecosystems is studied on upland soils without waterlogging. Leaching itself is often calculated from concentrations in the soil solution collected below the rooting depth. Results from such experiments should not be extrapolated to the relatively impermeable Gleysols of our site. There, dye-tracer experiments showed a fast, preferential flow of infiltrating precipitation towards drainage trenches (Feyen et al., 1999), and water flow paths with increased NO_3^- concentrations could indeed be detected by microsuction cups (Hagedorn et al., 1999). We therefore proposed that NO_3^- in the runoff from this forest largely corresponds to NO_3^- from rain or snowmelt bypassing the soil matrix. This hypothesis is supported by the high proportion of ^{15}N-labelled NO_3^- leached from the treated catchment, indicating that it comes mainly from the applied N and not from an increased net mineralization of (unlabelled) soil N (Schleppi et al., 1999a).

An end-member mixing analysis (EMMA) also showed that, at peak discharge, more than half of the runoff water in the experimental catchments came directly from precipitation (Hagedorn et al., 2001). For a whole rainfall event, however, this direct contribution appeared to be only approximately 20%. The EMMA model was further able to predict NO_3^- leaching with a high coefficient of determination, but these predictions were systematically too low, reaching approximately 60% of the measured values. Even when the model clearly indicates some direct NO_3^- leaching due to a lack of interaction with the soil ('new' N), the analysis suggests a considerable contribution of pre-event ('old') N. This contradicts the minimal proportion of unlabelled N in the additional leaching induced by the treatment.

In this paper, we present both short- and long-term patterns of NO_3^- leaching with the aim of reconciling these previous findings on the relative contributions of 'new' and 'old' N.

2. Material and Methods

2.1. SITE DESCRIPTION

The Alptal valley is located in central Switzerland. The research site is at an altitude of 1200 m and has a cool, wet climate (6 °C mean temperature and 2300 mm precipitation per year, of which 800 mm is as snow). Atmospheric NO_3^- and NH_4^+ deposition is moderate: 12 and 17 kg N ha^{-1} a^{-1} in bulk and throughfall respectively (Schleppi et al., 1998). Umbric Gleysols occur atop a Flysch substratum (calcareous sandstones with clay-rich schists, a typical formation along the northern edge of the Alps). The slope is about 20% with a west aspect. Depending on the microtopography, the soil bears different humus types: mor (raw humus) on the mounds and anmoor (muck humus) in the depressions, where the water table is high and reducing conditions common.

The trees are predominantly Norway spruce (*Picea abies*), with 15% silver fir (*Abies alba*), and mainly grow on the mounds. The stand is naturally regenerated, with trees up to 250 yr old. With a leaf area index of 3.8, the density of the canopy is low; the basal area is 41 m^2 ha^{-1} for 430 stems ha^{-1} (>10 cm in diameter). The ground vegetation is well developed; different botanical associations form patches according to humus types and light conditions (Schleppi et al., 1999b).

2.2. EXPERIMENTAL CATCHMENTS AND N ADDITION

Two forested catchments, each approximately 1500 m^2 in size, were delimited by trenches (Schleppi et al., 1998). The yearly water budgets for each catchment were found to be approximately balanced, and discharge and water chemistry are comparable with the surrounding, natural catchment of the Erlenbach stream (Schleppi et al., 1998). Element budgets of the artificially delimited catchments can thus be calculated from precipitation and runoff samples. The depth of the water table is automatically measured in two piezometers, one on a mound, one in a slight depression on the slope.

Nitrogen (as NH_4NO_3) was added to rainwater during precipitation events and applied by sprinklers to one of the catchments. This simulated an increased deposition of 25 kg N ha^{-1} a^{-1} to the ground vegetation and the soil. The water used as a vector for the N addition corresponded to a supplementary precipitation of approximately 110 mm per year. During the winter, the automatic irrigation was replaced by the occasional application of a concentrated NH_4NO_3 solution on the snow with a backpack-sprayer. The seasonality of the ambient deposition rates was thereby approximately mimicked. The effects of the treatment were compared with a control catchment receiving only unaltered rainwater and with one year of pre-treatment measurements made from both catchments.

2.3. SAMPLING AND ANALYSES

Bulk deposition and throughfall samples were collected weekly. In both catchments, water discharge was continuously measured with V-notch weirs. Runoff samples were collected proportionally to the discharge and bulked weekly (Schleppi *et al.*, 1998). Additionally, runoff from several rainfall and snowmelt events was monitored with a high temporal resolution (Hagedorn *et al.*, 2000). Soil solution samples were collected from plots (20 m^2 each, 5 replications) located near the catchments and subjected to the same N treatments. Suction plates were used to sample water at 5 and 10 cm depth, suction cups for 30 cm (reduced Bg horizon) (Hagedorn *et al.*, 2000). Anions in water samples were analysed by ion chromatography (Schleppi *et al.*, 1998). Soil cores were taken after 2 and 7 yr on 20 grid points of the treated catchment and analysed for carbon (C) and N contents with a C + N analyser.

2.4. ^{15}N LABELLING

The added N was labelled with ^{15}NH$_4^{15}$NO$_3$ during the first treatment year (Schleppi *et al.*, 1999a). Water samples, pooled quarterly, were concentrated over exchange resins. Anions and cations were eluted separately and NO$_3^-$ was reduced to NH$_4^+$ with Devarda's alloy. Ammonium was converted to NH$_3$ and captured in fibreglass filters enclosed in teflon membranes (adapted from Downs *et al.*, 1999). The filters were analysed by mass spectrometry.

3. Results

3.1. EVENT-BASED ANALYSES OF NO$_3^-$ LEACHING

An initial event-based sampling of runoff water was done in August 1994, during the pre-treatment year and at the end of a wet week (60 mm precipitation in the 6 previous days). During the sampled 12 h rainfall event, NO$_3^-$ concentrations were highest during the first discharge peak (Figure 1). By the end of the runoff peak, these concentrations had progressively declined to values close to those measured before the rain. During this event, NO$_3^-$ concentrations in the runoff remained within the range measured in the topsoil solution (NO$_3^-$-N: 0.05–0.7 mg L^{-1}, Hagedorn *et al.*, 2001). They were also similar to concentrations in the rain itself (0.22 mg L^{-1} NO$_3^-$-N). In the first samples from each catchment, the NO$_3^-$-N/Cl$^-$ ratios were 0.6 and 1.4, respectively. From the second to the last samples, the ratios declined from 0.5 down to below 0.4 in both catchments.

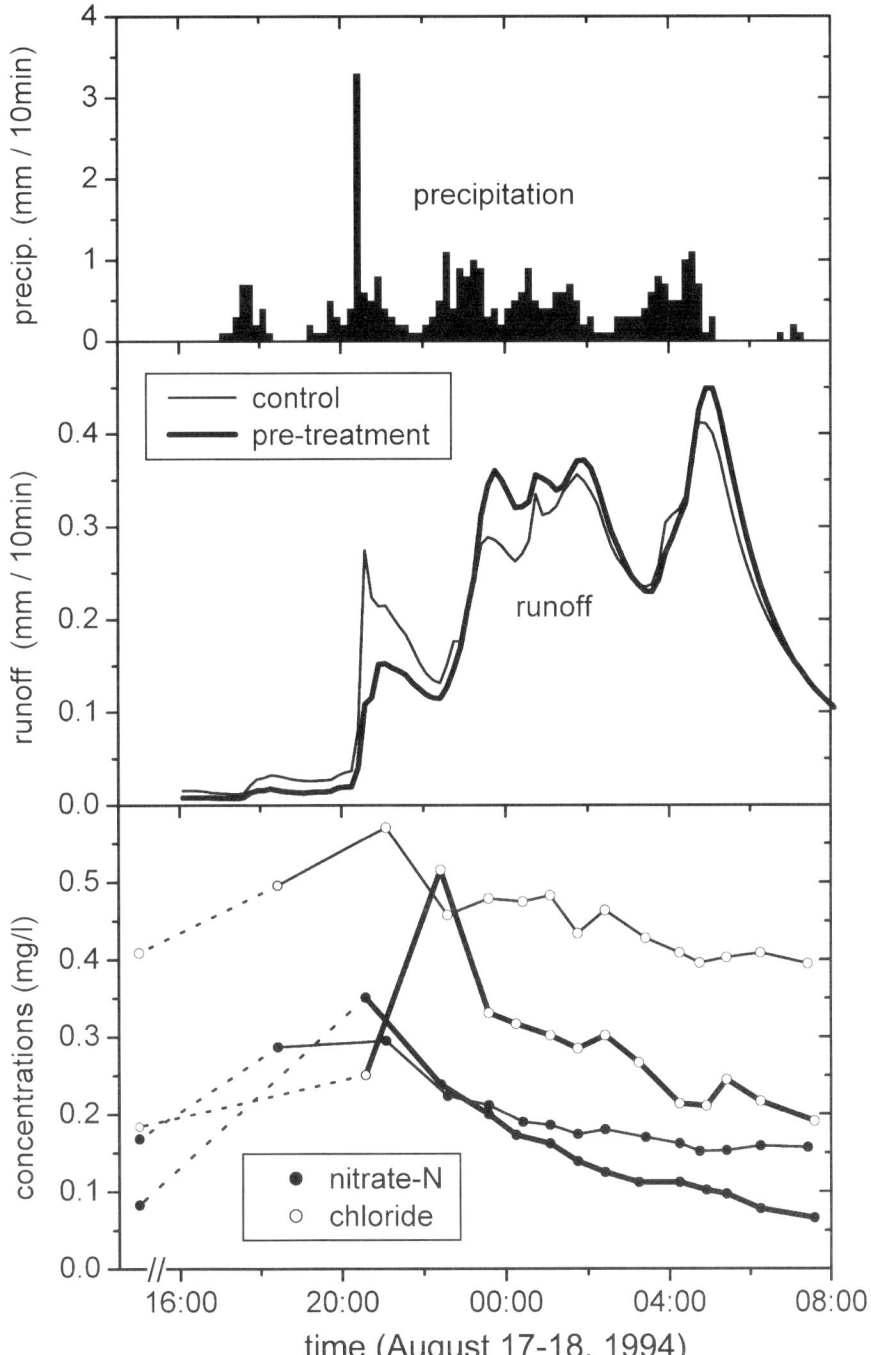

Figure 1. Runoff from the experimental catchments and its concentrations in NO_3^- and chloride during a summer rainfall event, before the N addition started.

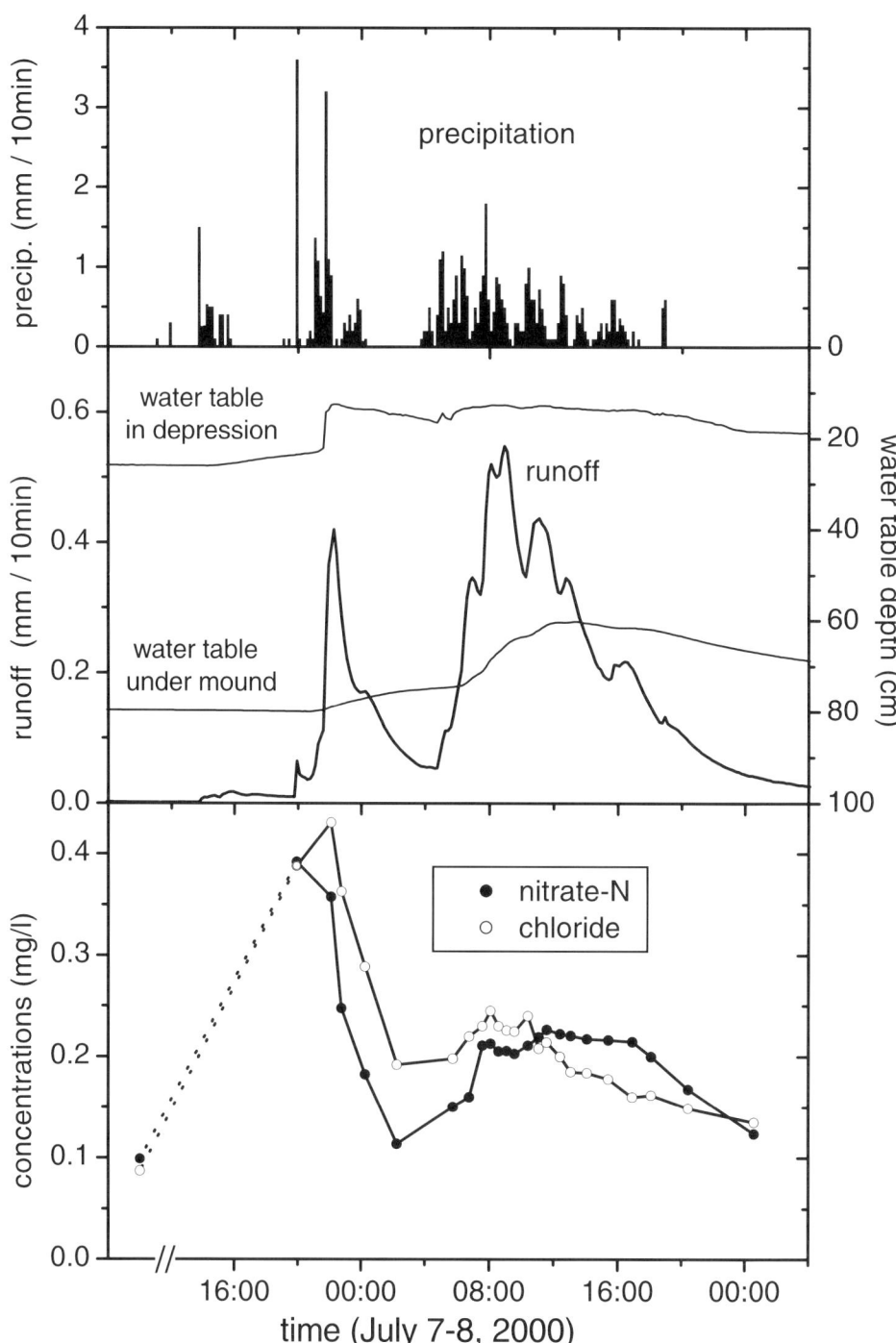

Figure 2. Runoff from the control catchment, its concentrations in NO_3^- and Cl^-, and depth of the water table during a summer rainfall event.

In samples collected in July 2000 from the first runoff peak of another long rainfall event (Figure 2), NO_3^- concentrations increased above values measured in the soil solution, or in the precipitation of this event (NO_3^--N: 0.31 mg L^{-1}). The water table in the depression piezometer also rose sharply during this first runoff peak. During the second part of the event, the water table remained high in the depression, and slowly increased in the mound piezometer. At the same time, a broad NO_3^- peak was measured in the runoff. During the first discharge peak, the ratio NO_3^--N/Cl$^-$ dropped from 1 to 0.6. Then, during the second peak, it increased up to 1.2, before decreasing again in the last three samples down to 0.9.

The first two days of the main snowmelt period were also sampled in April 1995, before the N treatment began (Figure 3). These samples showed a general decrease in NO_3^- concentrations, except for a short peak on the second day. In each catchment, this NO_3^- peak corresponded to the maximum runoff, and also approximately to the maximum water table levels.

The analysis of one rainfall event in July 1998, during the N addition, has previously been presented in Hagedorn *et al.* (2001). There was a NO_3^- peak at the beginning of the event, and NO_3^- concentrations were elevated in the runoff of the treated catchment throughout the event.

Runoff water was further sampled in April 1999 during the 6th and 7th days of a long snowmelt period (Figure 4). Compared to the event sampled in April 1995, the NO_3^- and Cl$^-$ concentrations were low and stable in the runoff from the control catchment. In the N-treated catchment, Cl$^-$ was also low, but NO_3^- concentrations were elevated during the whole event.

3.2. WEEKLY ANALYSES OF NO_3^- LEACHING

Over the seven years of weekly runoff analyses, low NO_3^- concentrations were usually measured in discharge-proportional samples collected during the summer. Higher concentrations were observed when the soil was relatively dry, like in autumn 1995, or during snowmelt, as in winter 1996 (Figure 5). During snowmelt, high concentrations are sometimes coupled to high discharges, leading to considerable NO_3^- fluxes. Over 7 yr, 45% of the losses from the control catchment occurred during three months of the year, between March and May. Based on these observations, a multiple regression analysis was calculated to explain NO_3^- concentrations in weekly runoff samples from the control catchment. The concentration data were first transformed to their logarithms in order to obtain a normal distribution of the residues. Independent variables were: the discharge (as logarithm), the season (as sine and cosine functions with a one year period), the variations of the water table (as the sum of hourly increases), and the bulk N deposition during the week (after subtracting the proportion of snow). The regression model had a coefficient of determination $r^2 = 0.43$ for the control catchment and each factor was highly significant ($p < 0.001$), also in a stepwise model definition. The effect of the discharge was evident but relatively weak: NO_3^- concentrations changed by a factor 10

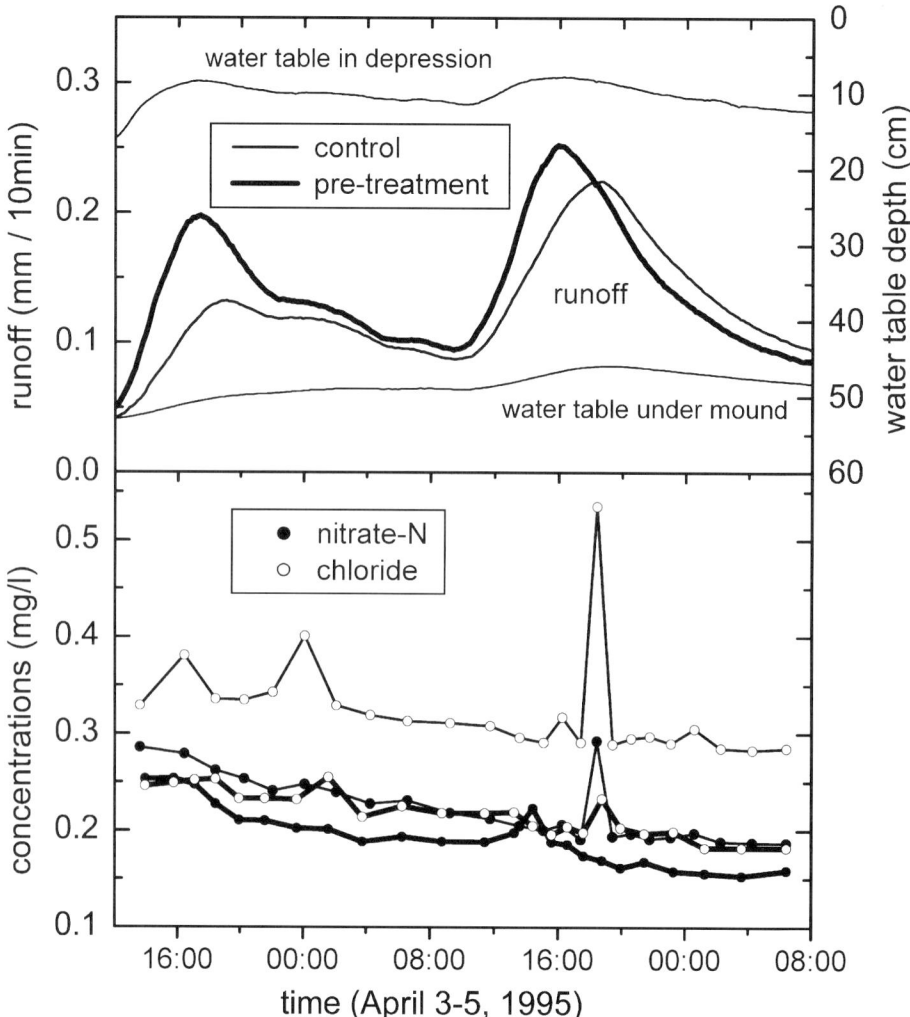

Figure 3. Runoff from the experimental catchments, its concentrations in NO_3^- and Cl^-, and depth of the water table during two days at the beginning of a snowmelt period.

while the discharge changed by more than 4 orders of magnitude. The maximum of the seasonality curve was in April, 2.5 times as high as the minimum in October. Nitrate concentrations further increased by 2% for each cm that the water table rose during the week. In the treated catchment, the effect of the discharge remained, but the seasonality was damped, with a ratio of 1.6 between spring and autumn concentrations. With the addition of N, the effect of the water table fluctuations was no longer significant.

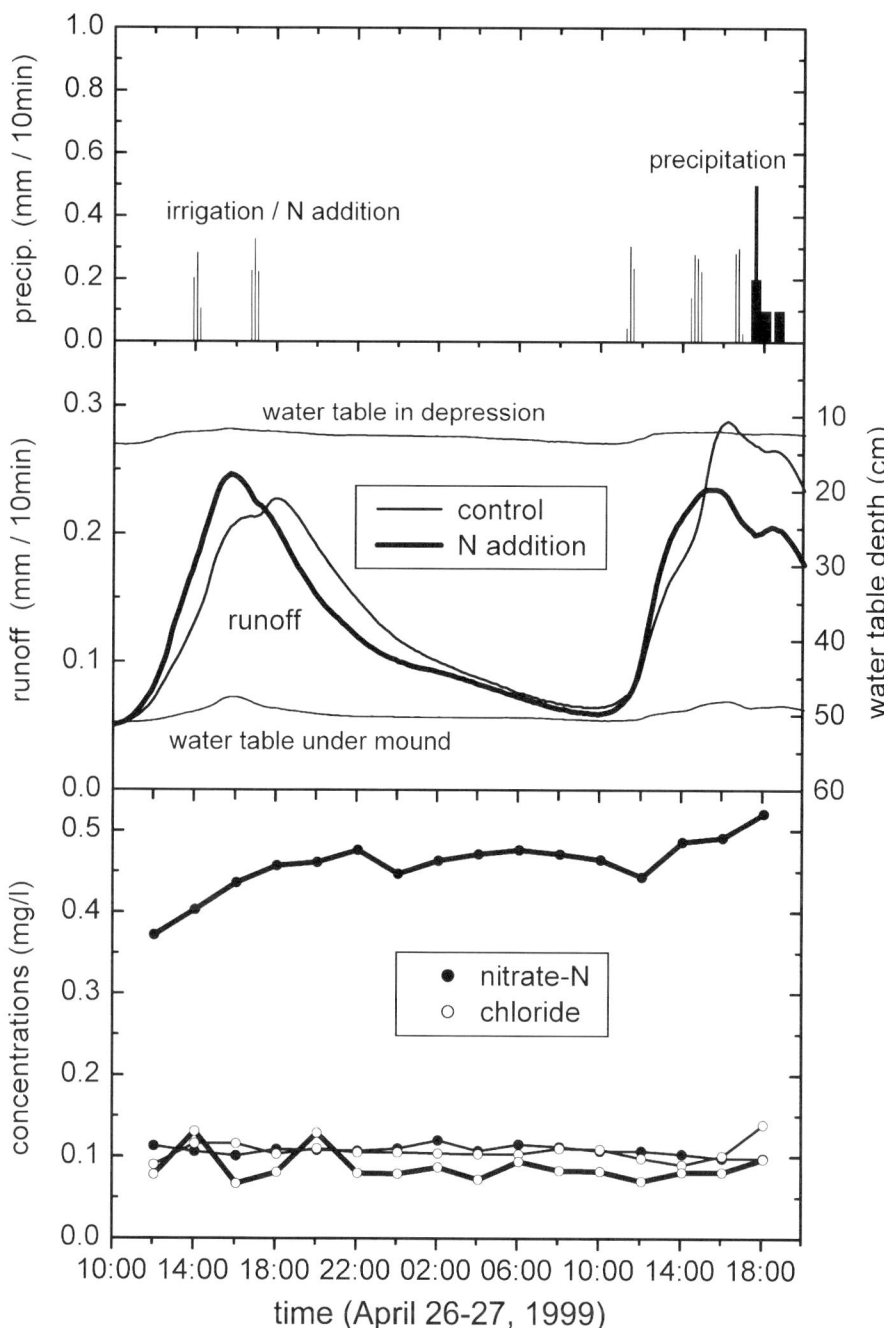

Figure 4. Runoff from the experimental catchments, its concentrations in NO_3^- and Cl^-, and depth of the water table during a longer snowmelt period. In the treated catchment, the water delivered by the irrigation system contained 17 ml L^{-1} of N.

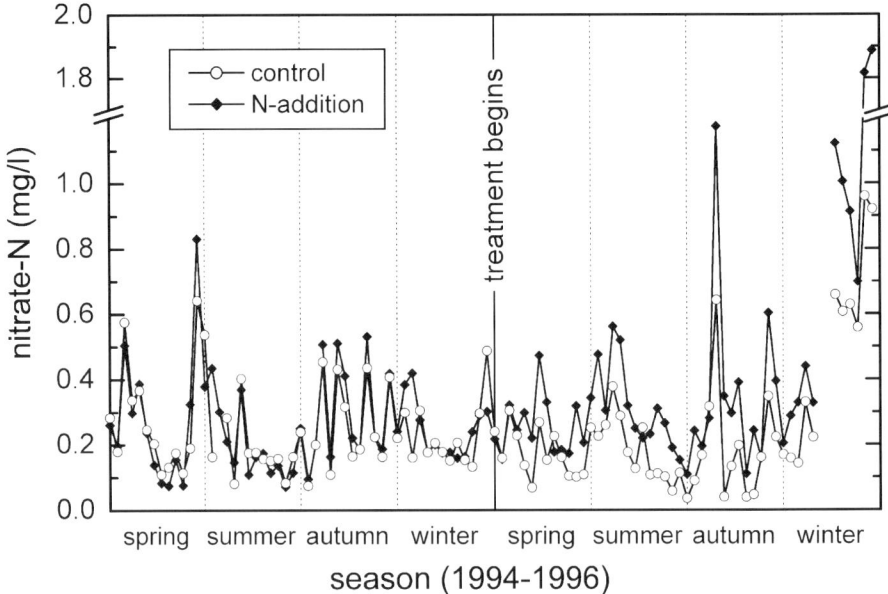

Figure 5. Nitrate concentrations in discharge-proportional runoff samples from the experimental catchments, with the addition of N beginning in the treated catchment after one year of pre-treatment measurements.

The weekly discharge-proportional samples from both catchments were very similar during the pre-treatment year (Figure 5). Within weeks of beginning the N application, the NO_3^- concentrations in runoff doubled in the treated catchment compared to the control catchment. Thereby, the shape of the concentration curve remained almost the same (correlation: $r = 0.73$, $p < 0.001$), only its magnitude was increased.

3.3. Long-term trends

Nitrate-N leaching from the control catchment varied from year to year between 2.5 and 4.8 kg ha^{-1} (Figure 6). It was significantly correlated with the yearly bulk N deposition ($r^2 = 0.63$, $p = 0.02$).

In the first treatment year, NO_3^- leaching was 3.1 kg N higher in the treated catchment than in the control, corresponding to 11% of the N addition. Out of these 3.1 kg, 2.8 ± 0.3 were labelled N. In the following years, the difference between the catchments increased up to 7 kg N, or 30% of the addition. The ^{15}N signal in runoff NO_3^-, however, disappeared within three months from the end of the labelling. The difference between total ^{15}N in lyophilised samples and $^{15}NO_3^-$ did not indicate any labelled dissolved organic N, nor was any ^{15}N signal detected in throughfall after labelling.

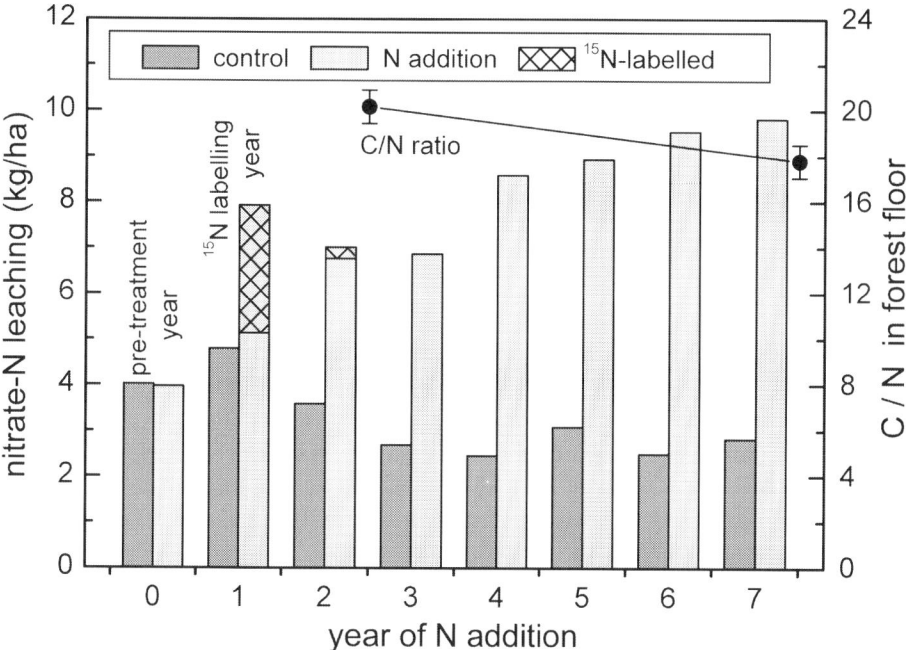

Figure 6. Yearly NO_3^- leaching from the forested experimental catchments during a pre-treatment year (0 davidson = 1994–1995) followed by seven years of N addition (at 25 kg ha^{-1} a^{-1}, with ^{15}N labelling in year 1). C/N mass ratios (± std. error) in the forest floor of the treated catchment after 2 and 7 yr of N addition.

Between 1997 and 2002, the C/N ratio of the forest floor decreased significantly ($p = 0.02$) from 20.2 to 17.8%. If the C content is assumed constant, this corresponds to an accumulation of 23 ± 10 kg N ha^{-1} a^{-1}.

4. Discussion

In winter, the occasional application of NH_4NO_3 to the snow appears to have a sustained effect on NO_3^- concentrations in runoff water. During runoff events, the single N additions by the sprinklers had no distinguishable effect on the concentration curves of NO_3^- in runoff. Finally, the treatment clearly increased weekly NO_3^- concentrations, but did not change the shape of this time-series compared to the control. These observations show that the applications were frequent enough to simulate wet N deposition, even on our site with rapid runoff. This is important since the same amount of nutrient can induce different effects depending on the rate and frequency of application (Persson, 1981), which can then alter the partitioning between plant species, soil, soil microorganisms, and leaching (Johnson, 1992). Our treatment by sprinklers does not allow for a foliar uptake by the trees. This process, however, was shown to be small in the case of wet deposition (Wilson and Tiley, 1998).

Using natural abundances of ^{15}N and ^{18}O from riverine nitrate, Mayer et al. (2002) established that, in large forested catchments, NO_3^- leaching was mainly from nitrification, with only a small possibility of a direct contribution from atmospheric deposition. Our ^{15}N labelling gives opposite results, with 90% of the additional leaching being labelled. If there are interactions between added N and pre-existing ecosystem pools, they must be limited to small N pools and to a relatively short time span (since the ^{15}N signal disappeared from the runoff within three months).

Event-based analyses, including the already published end-member mixing analysis (Hagedorn et al., 2001), show a different pattern. Nitrate peaks usually correspond to a rising water table. This indicates flushing from those soil layers, which become saturated with water (Creed and Band, 1998). Because the water table always rises first in the depressions and only later under the mounds, NO_3^- flushing can also occur as successive NO_3^- peaks during a hydrological event, like in July 2000. This effect is similar to the 'variable source area dynamics' described by Creed and Band (1998). Nitrate peaks observed during the early snowmelt in April 1995 also corresponded to the water table reaching a previously unsaturated soil layer. The highest NO_3^- peak, however, occurred simultaneously with a Cl^- peak. These high concentrations indicate hydrological causes (which affect both anions), without a contribution of net nitrification (which would increase the NO_3^-/Cl^- ratio). During the snowmelt sampled in April 1999, concentrations were consistently low in the control catchment. Concentrations declining during the early snowmelt and staying low through the later snowmelt phase are best explained by the ionic fractionation within the melting snowpack (Waldner et al., 2000). The higher concentrations observed in the control catchment in April 1995 can be explained by the same mechanisms since its snowmelt is always slightly later than in the treated catchment. The absence of a clear NO_3^- peak in April 1999 is probably due, as in the rainfall event of August 1994, to previous flushings, which already depleted NO_3^- from the soil zones reached by the water table.

During some events, like in July 2000, NO_3^- peaks correspond to increased NO_3^-/Cl^- ratios. In this case, some NO_3^- appears to be produced by nitrification. Even if no net nitrification was measurable in the bulk soil of this forest site, Hagedorn et al. (1999) showed that some nitrification takes place along the macropores as they drain out between hydrological events. In summary, NO_3^- is released by the Gleysols at the test catchments by three different mechanisms: (1) NO_3^- from precipitation bypassing the soil by fast preferential and sub-surface-flow, (2) flushing of NO_3^- from previous precipitation events temporarily stored in the soil pores, and (3) flushing of NO_3^- produced by nitrification (in the macropores of the unsaturated soil). In any case, NO_3^- flushed as the water table rises is mainly from recently deposited N, as shown by its strong ^{15}N signal. From the data obtained so far, it is not possible to distinguish between the nitrification of deposited NH_4^+ and nitrification occurring after the mineralisation of assimilated NO_3^-. Because the ^{15}N signal disappears within three months, these release mechanisms are effective on a limited time-scale only. This explains why event-based and long-term contributions of 'new' vs. 'old' N appeared to be in contradiction. Likewise, the discrepancy observed in correlations between NO_3^- and discharge is also a matter of the time-scale: mostly positive within a single event, but negative between weekly samples. The effect of flushing, however, can still be observed in the weekly analyses as NO_3^- concentrations increase with the cumulative movements of the water table during the sampling time. To describe this quantitatively and in more details, it would be necessary to draw a model taking into account the whole dynamics of the water table across the variable zones, which contribute to runoff (mounds vs. depressions).

The observed NO_3^- leaching mechanisms occur preferentially along water-flow paths in the soil (macropores). Because soil heterogeneity is not considered in the N saturation models by Ågren and Bosatta, (1988) or by Aber et al. (1989), our site cannot be described within these concepts. A further limitation of these models was the fact that N inputs to the stable soil organic matter are only possible via litterfall. In their newer hypotheses, Aber et al. (1998) also consider direct immobilisation in the soil, which is clearly what happened with the Alptal forest. Even if it was detectable by ^{15}N analyses, the uptake of deposited N by the trees was insignificant for their nutritional status (Schleppi et al., 1999b) and thus for the return of N as litterfall. Foliar leaching did not play a role either as there was no ^{15}N signal in throughfall after labelling. About 2/3 of the ^{15}N were recovered in the soil (Schleppi et al., 1999a), and the immobilisation in the forest floor was very important (about 15 kg ha^{-1} of labelled N in litter and forest floor). This immobilisation may be abiotic, especially since reducing conditions are frequent in this soil and could enable the redox cycles recently hypothesised by Davidson et al. (2003). The decline in the C/N ratio measured within five years is strong but still in agreement with the ^{15}N partitioning. Comparing different sites, Gundersen et al. (1998) and Emmett et al. (1998) showed the importance of the C/N ratio in the forest floor for its ability to retain N deposition. In our experiment, we observed a progressive increase in NO_3^- leaching within seven years of N addition. According to its disappearing ^{15}N signal, this increased leaching is due to a reduced ability to retain further N inputs rather than to a delayed release of N accumulated in the previous years. We therefore hypothesise that the declining C/N ratio is more and more limiting the ability of the soil to immobilise N from atmospheric deposition.

Acknowledgments

This study received financial support from the Swiss Federal Office of Science and Education and from the Swiss National Fund for Scientific Research. The chemical analyses were done at the central laboratory of the Swiss Federal Institute for

Forest, Snow and Landscape Research (WSL), Birmensdorf (D. Pezzotta). Analyses of ^{15}N were conducted at the Paul Scherrer Institute (PSI), Villigen (Dr. R. Siegwolf, Dr. M. Saurer). Thanks to Melissa Swartz for the English corrections.

References

Aber, J. D., Nadelhoffer, K. J., Steudler, P. and Melillo, J.: 1989, 'Nitrogen saturation in northern forest ecosystems', *BioScience* **39**, 378–386.

Aber, J., McDowell, W., Nadelhoffer, K., Magill, A., Berntson, G., Kamakea, M., McNulty, S., Currie, W., Rustad, L. and Fernandez, I.: 1998, 'Nitrogen saturation in temperate forest ecosystems - Hypotheses revisited', *BioScience* **348**, 921–934.

Ågren, G.I. and Bosatta, E.: 1988, 'Nitrogen saturation of terrestrial ecosystems', *Environ. Pollut.* **54**, 185–197.

Creed, I. F. and Band, L. E.: 1998, 'Export of nitrogen from catchments within a temperate forest: evidence for a unifying mechanism regulated by variable source area dynamics', *Water Resour. Res.* **34**, 3105–3120.

Davidson, E. A., Chorover, J. and Dail, D. B.: 2003, 'A mechanism of abiotic immobilization of nitrate in forest ecosystems: the ferrous wheel hypothesis', *Global Change Biol.* **9**, 228–236.

Downs, M. R., Michener, R. H., Fry, B. and Nadelhoffer, K. J.: 1999, 'Routine measurement of dissolved inorganic ^{15}N in streamwater', *Environ. Monit. Assess.* **55**, 211–220.

Emmett, B. A., Boxman, A. W., Bredemeier, M., Moldan, F., Gundersen, P., Kjønaas, O. J., Schleppi, P., Tietema, A. and Wright, R. F.: 1998, 'Predicting the effects of atmospheric nitrogen deposition in conifer stands: evidence from the NITREX ecosystem-scale experiments', *Ecosystems* **1**, 352–360.

Feyen, H., Wunderli, H., Wydler, H. and Papritz, A.: 1999, 'A tracer experiment to study flow paths of water in a forest soil', *J. Hydrol.* **225**, 155–167.

Gundersen, P., Emmett, B. A., Kjønaas, O. J., Koopmans, C. J. and Tietema, A.: 1998, 'Impact of nitrogen deposition on nitrogen cycling in forests: a synthesis of NITREX data', *For. Ecol. Manage.* **101**, 37–55.

Hagedorn, F., Mohn, J., Schleppi, P. and Flühler, H.: 1999, 'The role of rapid flow paths for nitrogen transformation in a forest soil: a field study with micro suction cups', *Soil Sci. Soc. Am. J.* **63**, 1915–1923.

Hagedorn, F., Schleppi, P., Waldner, P. and Flühler, H.: 2000, 'Export of dissolved organic carbon and nitrogen from Gleysol dominated catchments – the significance of water flow paths', *Biogeochemistry* **50**, 137–161.

Hagedorn, F., Schleppi, P., Bucher, J. B. and Flühler, H.: 2001, 'Retention and leaching of elevated N deposition in a forested ecosystem with Gleysols', *Water, Air, Soil Pollut.* **129**, 119–142.

Johnson, D. W.: 1992, 'Nitrogen retention in forest soils', *J. Environ. Qual.* **21**, 1–12.

Mayer, B., Boyer, E. W., Goodale, C., Jaworski, N. A., van Breemen, N., Howarth, R. W., Seitzinger, S., Billen, G., Lajtha, K., Nadelhoffer, K., van Dam, D., Hetling, L. J., Nosal, M. and Paustian, K.: 2002, 'Sources of nitrate in rivers draining sixteen watersheds in the northeastern U.S.: isotopic constraints', *Biogeochemistry* **57–58**, 171–197.

Persson, H.: 1981, 'The effect of fertilization and irrigation on the vegetation dynamics of a pine-heath ecosystem', *Vegetatio* **46**, 181–192.

Schleppi, P., Muller, N., Feyen, H., Papritz, A., Bucher, J. B. and Flühler, H.: 1998, 'Nitrogen budgets of two small experimental forested catchments at Alptal, Switzerland', *For. Ecol. Manage.* **101**, 177–185.

Schleppi, P., Bucher-Wallin, I., Siegwolf, R., Saurer, M., Muller, N. and Bucher, J. B.: 1999a, 'Simulation of increased nitrogen deposition to a montane forest ecosystem: partitioning of the added ^{15}N', *Water, Air, Soil Pollut.* **116**, 129–134.

Schleppi, P., Muller, N., Edwards, P. J. and Bucher, J. B.: 1999b, 'Three years of increased nitrogen deposition do not affect the vegetation of a montane forest ecosystem', *Phyton* **39**, 199–204.

Stoddard, J. L.: 1994, 'Long-term changes in watershed retention of nitrogen: its causes and aquatic consequences', in L. A. Baker (ed.), *Environmental Chemistry of Lakes and Reservoirs*, Adv. Chem. Ser. **237**, Am. Chem. Soc., Washington, DC, U.S.A., pp. 223–284.

Waldner, P., Schneebeli, M. and Wunderli, H.: 2000, 'Nährstoffaustrag aus einer schmelzenden Schneedecke im Alptal (Kanton Schwyz) am Beispiel von Nitrat', *Schweiz. Z. Forstwes.* **151**, 198–204.

Wilson, E. J. and Tiley, C.: 1998, 'Foliar uptake of wet-deposited nitrogen by Norway spruce: an experiment using ^{15}N', *Atmos. Environ.* **32**, 513–518.

Wright, R. F. and Rasmussen, L.: 1998, 'Introduction to the NITREX and EXMAN projects', *For. Ecol. Manage.* **101**, 1–7.

RADIOGENIC LEAD ISOTOPES AND TIME STRATIGRAPHY IN THE HUDSON RIVER, NEW YORK

STEVEN N. CHILLRUD[1]*, RICHARD F. BOPP[2], JAMES M. ROSS[1],
DAMON A. CHAKY[1], SIDNEY HEMMING[1], EDWARD L. SHUSTER[2],
H. JAMES SIMPSON[1] and FRANK ESTABROOKS[3]

[1] *Lamont-Doherty Earth Observatory of Columbia University, Palisades, New York 10964, U.S.A.;*
[2] *Earth and Environmental Sciences, Rensselaer Polytechnic Institute, Troy, New York 12180, U.S.A.;* [3] *New York State Department of Environmental Conservation, Albany, New York 12233, U.S.A.*
(* author for correspondence, e-mail: chilli@ldeo.columbia.edu; phone: 845-365-8893; fax: 845-365-8155)

(Received 20 August 2002; accepted 18 April 2003)

Abstract. Radionuclide, radiogenic lead isotope and trace metal analyses on fine-grained sediment cores collected along 160 km of the upper and tidal Hudson River were used to examine temporal trends of contaminant loadings and to develop radiogenic lead isotopes both as a stratigraphic tool and as tracers for resolving decadal particle transport fluxes. Very large inputs of Cd, Sb, Pb, and Cr are evident in the sediment record, potentially from a single manufacturing facility. The total range in radiogenic lead isotope ratios observed in well-dated cores collected about 24 km downstream of the plant is large (e.g., maximum difference in $^{206}Pb/^{207}Pb$ is 10%), characterized by four major shifts occurring in the 1950s, 1960s, 1970s and 1980s. The upper Hudson signals in Cd and radiogenic lead isotopes were still evident in sediments collected 160 km downstream in the tidal Hudson. The large magnitude and abrupt shifts in radiogenic lead isotope ratios as a function of depth provide sensitive temporal constraints that complement information derived from radionuclide analyses to significantly improve the precision of dating assignments. Application of a simple dilution model to data from paired cores suggests much larger sediment inputs in one section of the river than previously reported, suggesting particle influxes to the Hudson have been underestimated.

Keywords: contaminant transport, Hudson River sediments, metals, radionuclides, stable lead isotopes, stratigraphy

1. Introduction

The use of fine-grained sediment cores as archives of contaminant loadings and transport processes is largely dependent upon the ability to constrain the age of accumulation of depth sections within a core. For example, well-established sediment layer dates are critical for studies based on paired sediment cores such as determining in situ PCB dechlorination rates (Chillrud, 1996; McNulty, 1997; Bopp *et al.*, in press). For sediment cores collected in lakes and bogs, the radionuclides ^{137}Cs and ^{210}Pb have been widely used to estimate sedimentation rates over the last 50 to 100 yr (Ritchie and McHenry, 1990; Appleby and Oldfield, 1992; and many others). ^{210}Pb, however, has not been used very frequently in large river systems due to relatively low excess ^{210}Pb activities in most riverine sediments (Scott *et al.*, 1985; Beasley *et al.*, 1986; Bush *et al.*, 1987; W. Schell, personal communication, 1994).

It has been known for some time that radiogenic lead isotopes have the potential to provide sensitive time stratigraphic information due to geographical shifts in lead ore production sites over time. In the United States, major shifts in lead ore production occurred in the 1960s and 1970s. By the 1970s less ore was extracted from mines in northwestern states (low ^{206}Pb/^{207}Pb) and more ore from midwestern mines (high ^{206}Pb/^{207}Pb). Observed temporal trends in environmental samples reflect an upward shift in ^{206}Pb/^{207}Pb during the 1960s and 1970s (Graney *et al.*, 1995; Sturges and Barrie, 1987), though not as large as would result from the extreme values of Pb isotopes in these different mining areas. Hurst *et al.* (1996) exploited this temporal trend in Pb isotopes to establish the timing of leaks of leaded gasoline from underground tanks and to estimate the age of depth sections of sediment cores. Marcantonio *et al.* (2002) observed excellent agreement in temporal records of ^{206}Pb/^{207}Pb between Chesapeake Bay sediment cores and data reported for Bahamian corals (Shen and Boyle, 1987).

Upper Hudson sediments are highly contaminated with PCBs, highly chlorinated dioxins and trace metals (Bopp *et al.*, 1998; Chillrud *et al.*, 2003) and display a 10% range in ^{206}Pb/^{207}Pb, characterized by four major shifts occurring in the 1950s, 1960s, 1970s and 1980s that were apparent in cores collected along a 38 km reach of this section of the river (Chillrud *et al.*, 2003). Here, our main objective is to expand the geographical extent of lead isotope stratigraphy to a 160 km reach of the river including both upper and freshwater tidal Hudson sediments. In addition, Cd concentrations are used to examine fine-grained sediment and contaminant transport issues.

1.1. RADIOGENIC LEAD ISOTOPE SIGNAL TO NOISE RATIO

North American ores have lead isotope ratios that differ by as much as 6% to 25%, depending on the isotope pair (Doe and Delevaux, 1972; Doe, 1975; Deloule *et al.*, 1986). Because these ratios can be measured with a precision (2σ) of $\leq 0.05\%$ per atomic mass unit (amu) by mass spectrometry, stable lead isotopes can be especially sensitive tracers. For example, the ratio of maximum potential signal to measurement precision ranges from ca. 500 for ^{206}Pb/^{207}Pb and 60 for ^{207}Pb/^{204}Pb.

1.2. RADIONUCLIDE DATING OF RIVERINE SEDIMENTS

Depth distributions of ^{137}Cs and ^{7}Be were used to approximate dates of particle accumulation for individual depth sections of sediment cores. Cesium-137 from global fallout produced measurable activities in sediments about 1954 and reached a maximum in 1963 (Ritchie and McHenry, 1990). Beryllium-7 is a particle-reactive cosmogenic radionuclide with a half-life of 53 days; consequently, the presence of

^{7}Be at measurable levels in a core section indicates that a significant portion of that sediment accumulated within six months to a year prior to collection of the core, either by sediment deposition or physical mixing. Uncertainties sometimes associated with radionuclide dating of sediment layers include: coring artifacts, diffusion of dissolved ^{137}Cs in pore waters, or bioturbation (see Crusius and Anderson, 1991 and references therein). These uncertainties appear to be minor in areas of the Hudson River with relatively rapid sedimentation rates on the order of 0.5–1 cm a^{-1} or greater (Olsen *et al.*, 1981a; Chillrud, 1996). Cesium-137 has been used to estimate sedimentation rates for many cores from the Hudson River and estuary (Simpson *et al.*, 1976; Bopp *et al.*, 1981, 1982, 1993; Olsen *et al.*, 1981b; Bush *et al.*, 1987; Bopp and Simpson 1989).

2. Experimental Section

Sediment samples discussed here were collected from freshwater reaches of the Hudson River watershed. Locations along the Hudson River are specified as milepoints (mp), defined as the number of statute miles upstream of the southern tip of Manhattan. The Green Island Dam (mp 154) is the upstream limit of tidal stage variations (Figure 1). The main stem of the river upstream of Green Island Dam (the 'upper Hudson') has a series of seven other dams and associated navigational locks. Details of core collection at mp 188.5 (Control Number 1852, collected 19 July 1983) and mp 163.6 (CN R1129, collected 31 July 1997) were described in Chillrud *et al.* (2003). The sediment core collected furthest downstream at mp 88.6 (CN1984) was collected on 21 July 1986 in ca. 9 m of water using a small gravity corer with the same type of polybutyrate liners (diameter ca. 5.6 cm) as the other cores.

Details of sample preparation and analysis are described in Chillrud *et al.*, 2003). Briefly, sediments were dried at 35 °C, ground in a mortar and pestle, then sealed in containers for radionuclide analysis using either an intrinsic germanium or a lithium-drifted germanium gamma detector. Metal concentrations were determined by ICP-MS and flame atomic absorption spectrometry after sediment digestion with strong acids (HNO_3, HF, $HClO_4$, HCl) following procedures of Fleisher and Anderson (1991). For Cd and Pb, replicate digests of sediment reference standards and Hudson River sediments indicate excellent recoveries and precisions between ca. 5 and 10%. Both TIMS and single-collector ICP-MS were used for Pb isotope measurements, and excellent agreement between methods was found (Chillrud *et al.*, 2003).

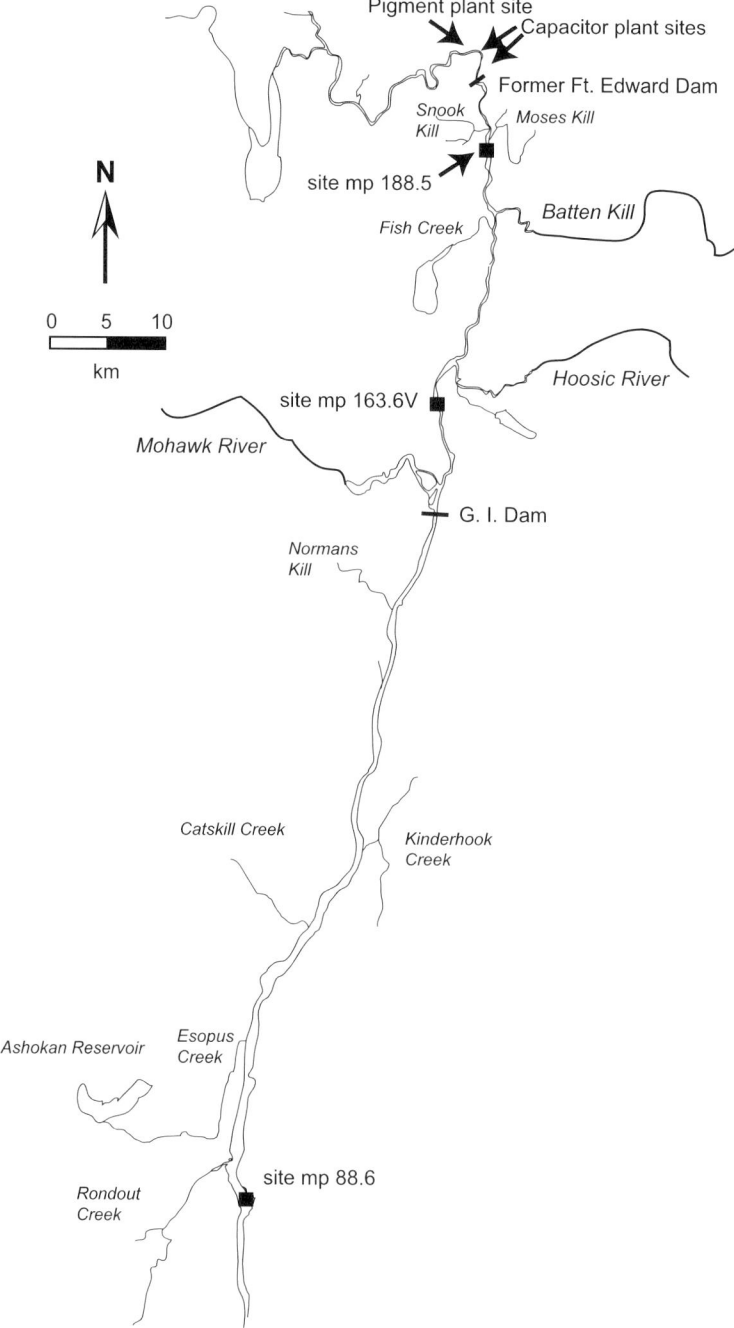

Figure 1. Upper Hudson River location map. Major point sources of metals (Pigment plant) and PCBs (2 capacitor plants) were both upstream of a large dam removed in 1973 (Former Ft. Edward Dam). Locations on the Hudson River have historically been reported as statute miles north of the southern tip of Manhattan (New York City). Coring sites at mile point (mp) 188.5, mp 163.6 and mp 88.6 are also indicated.

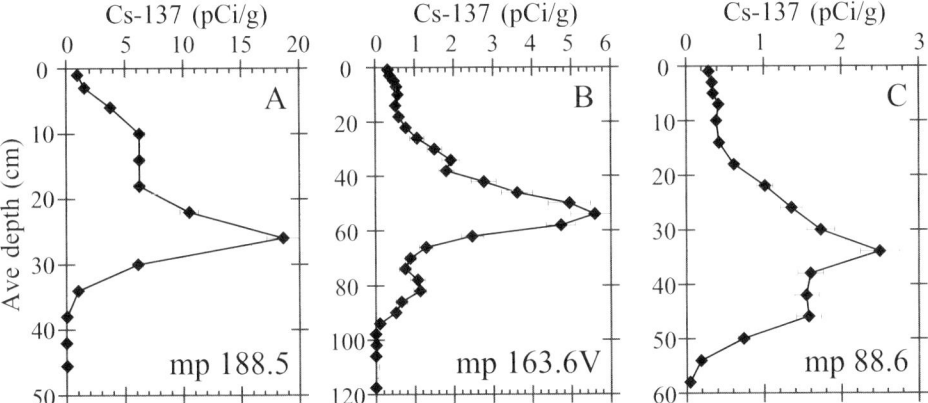

Figure 2. Depth profiles of ^{137}Cs in a sediment cores from site mp 188.5 (Figure 2A), site mp 163.6V (Figure 2B), and site mp 88.6 (Figure 2C). Note scales on both axes differ in each of three plots. Error bars for ^{137}Cs are based on 2 sigma counting statistics.

3. Results and Discussion

3.1. SEDIMENT DATING

Depth profiles of all three cores display a prominent subsurface peak in ^{137}Cs activities and near zero ^{137}Cs activities at depth (Figures 2A–2C), suggesting that each site has received semi-continuous fine particle accumulation over a time period of several decades. Berylium-7 was detected in top sections of each core.

For core mp 188.5, a single sedimentation rate of 1.2 cm a^{-1} is consistent with the timing of all three of the radionuclide marker horizons. However, the plateau in ^{137}Cs activities (see Figure 2A) was previously interpreted as reflecting large sediment transport event(s) initiated by removal of the Ft. Edward Dam in 1973 and probably continued by high spring discharges in 1974 and 1976 (Bopp *et al.*, 1985). Such an interpretation suggests the possibility of at least three mean sedimentation rates: 1.0 cm a^{-1} for the section from the first presence of ^{137}Cs upwards to the plateau (which places the depth section with maximum ^{137}Cs activity in the early-mid 1960s), 2.7 cm a^{-1} for the plateau 'event', and 1.3 cm a^{-1} for the core section between the plateau and the core top.

For core mp 163.6V (Figure 2B), two distinct sedimentation rates were derived between the horizons identified by ^{7}Be and ^{137}Cs – core top to ^{137}Cs maximum (1.59 cm a^{-1}) and ^{137}Cs maximum to first presence of ^{137}Cs (4.44 cm a^{-1}). Such a large temporal change in mean sedimentation rates is surprising; however, these rates are consistent with two additional features of the ^{137}Cs depth profile – a secondary fallout peak in the late 1950's (76 to 84 cm) and a shoulder or plateau in ^{137}Cs activity (32 to 40 cm) consistent with the 'dam removal event' discussed above (Chillrud *et al.*, 2003).

Sedimentation rates for core mp 88.6 (Figure 2D) were estimated to be between 1.5 and 1.8 cm a^{-1} based on radionuclide time horizons (see Chillrud, 1996 for details).

3.2. RADIOGENIC LEAD ISOTOPE STRATIGRAPHY AND SOURCES

Upper Hudson River sediments display large changes in ^{206}Pb/^{207}Pb as a function of depth (time), with four significant shifts in isotope ratios over four decades (Figure 3A). The total increase in ^{206}Pb/^{207}Pb during the 1960s and 1970s was about 10%, which is very large for environmental samples compared to the 1 to 2% change typically observed in other systems that have integrated multiple sources of Pb inputs (e.g., Shen and Boyle, 1987; Graney *et al.*, 1995; Marcantonio *et al.*, 2002). Several lines of evidence indicate that a single pigment manufacturing facility probably dominated contaminant Pb inputs into this reach of the upper Hudson River. These include a basin-wide compilation of discharge permits (Rohmann *et al.*, 1985) and analyses of Pb and other metals in sediments upstream and downstream of the suspected source (Shuster *et al.*, 2002; Bopp *et al.*, *in press*; Chillrud *et al.*, 2003). The large magnitude of the shifts in stable lead isotope ratios is also consistent with a single dominant source. Multiple large contaminant Pb sources to the sediments would tend to diminish the isotopic signal derived from a distinctive single source.

The composition and timing of the shifts during the 1960s and 1970s of lead isotope ratios measured in core mp 188.5 are generally consistent with the shift in major extraction of lead ore from mines in the northwestern United States (low ^{206}Pb/^{207}Pb) to mines in the midwest (high ^{206}Pb/^{207}Pb). Pb isotope ratios measured throughout core mp 188.5 were consistent with mixtures of Pb ores from both these mining areas, although a small local mine in Balmat, New York was another possible end-member source of low ^{206}Pb/^{207}Pb ratios consistent with the sediment data (Chillrud *et al.*, 2003).

Very similar temporal trends of stable Pb isotopes were observed in cores mp 188.5 and mp 163.6V, collected 38 km apart (Figure 3A). The similarity of ^{206}Pb/^{207}Pb variations between the two cores confirms the age assignments based on radionuclides for core mp 163.6V, including the large sedimentation rate change, and indicates that depth profiles of stable Pb isotopes can be used as an additional time stratigraphic tool in upper Hudson River sediment cores. Compared to background Pb-isotope values, the late-1950s to mid-1960s and the mid-1970s to mid-1980s are the time periods when sediment data are most different from background values described below. The timing of the major shifts in lead isotope values, with the exception of the mid-1960s shift, represent time periods when the ^{137}Cs dating method provides only interpolated dates. Clearly, stable Pb isotope depth profiles provide powerful new stratigraphic markers in upper Hudson River sediments.

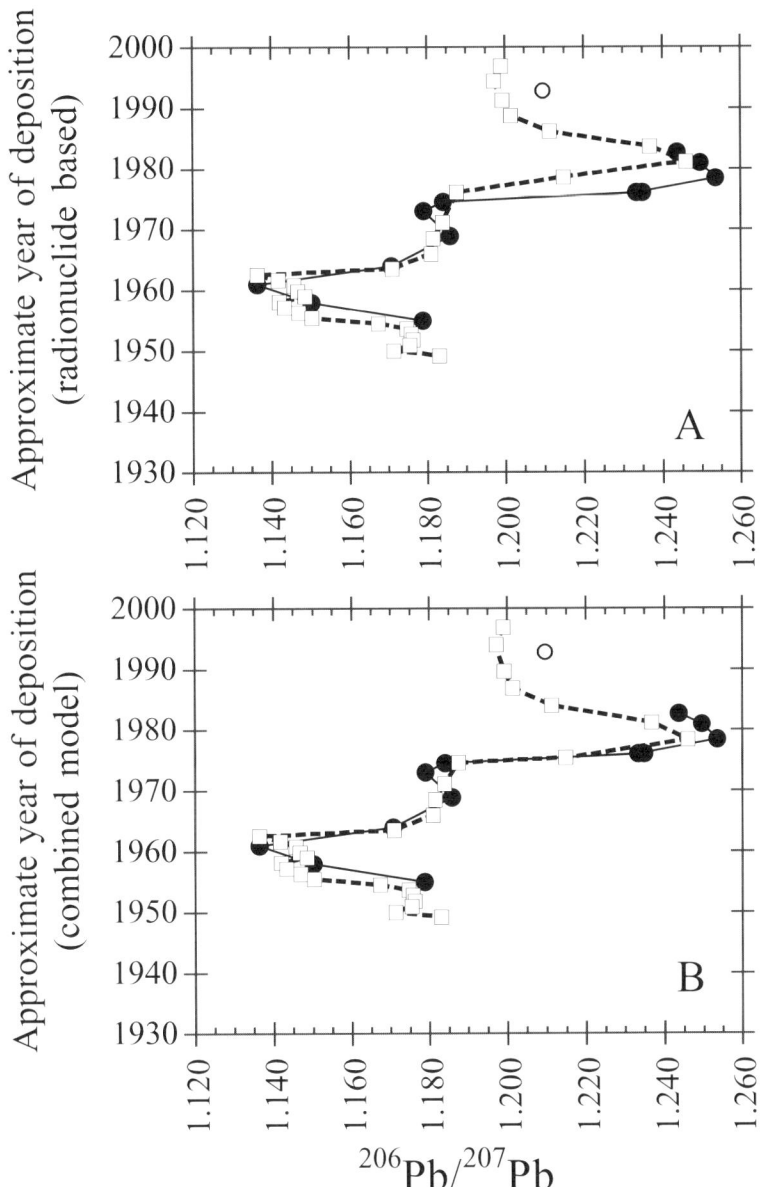

Figure 3. Temporal trends of ^{206}Pb/^{207}Pb in cores collected from sites mp 188.5 and mp 163.6V. In Figure 3A, depth sections are assigned approximate dates of deposition based only on radionuclide data. Samples from coring site mp 188.5 are shown as circles, with open circles being the top 2 cm of a core collected in 1992 and the filled circles being all the sections of a core collected in 1983. Squares display data for core 163.6V collected in 1997. Figure 3B: Temporal trends of ^{206}Pb/^{207}Pb in the sediment cores where approximate dates of deposition were based on both radionuclide and stable lead isotope data. Compared to the dates assigned in Figure 3A, only minor changes were made to assigned dates of certain sections of core mp 163.6V (see text), while no changes were made to core mp 188.5.

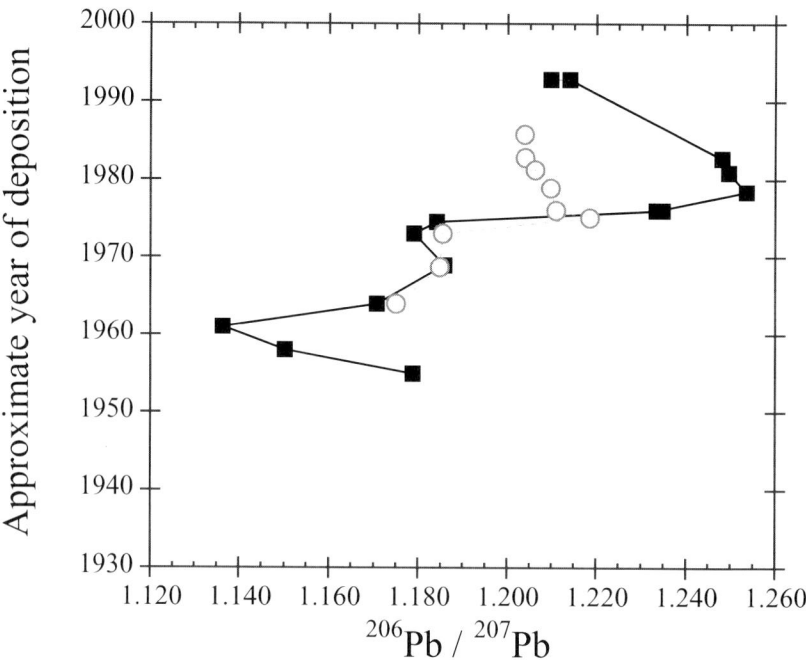

Figure 4. Temporal trend for ^{206}Pb/^{207}Pb ratios for cores mp 188.5 and mp 88.6. Measured ratios are shown as filled squares (mp 188.5) and open circles (mp 88.6). The shift in isotope ratios in the mid 1970s observed in the upstream core was still evident 160 km downstream.

The large down-core variations in lead isotope ratios permit more accurate matching of depth sections from different cores. For example, in Figure 3A, beginning in the mid-1970s, a shift of a few years in the radionuclide-derived dates significantly improves the match in ^{206}Pb/^{207}Pb ratios between cores mp 163.6V and mp 188.5.

Figure 4 compares the ^{206}Pb/^{207}Pb temporal trends of cores mp 188.5 and mp 88.6. Both records show similar ^{206}Pb/^{207}Pb values in the 1960s and early 1970s and display a significant shift toward more radiogenic values in the mid-1970s. Although the shift is less in mp 88.6, similarities in the two records strongly suggest that the upper Hudson River lead isotope signature is still observed at the downstream site and that radiogenic Pb isotope measurements can provide sensitive temporal constraints on sediment deposition over more than a 160-km reach of the Hudson River.

During the time period of the late 1970s and 1980s there was appreciable divergence between Pb isotopic ratios of core mp 188.5 and mp 88.6. Because the measurements were made using total sample digests, the isotopic composition of the background component of Pb, which can include Pb derived from mineral lattices, can significantly influence measured Pb isotope values of environmental samples (Graney *et al.*, 1995). The largest influence of background component of

TABLE I

Maximum trace metal concentrations in core mp 188.5 and estimates of metal levels in uncontaminated fine-grained sediments and tributary sediments

	Ag	Cd	Cu	Pb	Sb	Sn	Zn	Cr
				(μg g^{-1})				
Maximum level in core mp 188.5	3.5	171	173	2400	48	38	1160	2200
Background estimates								
Varved clays[a]	0.06, 0.15	0.13, 0.14	24, 24	34, 34	0.08, 0.09	1.8, 1.8	73, 83	
NY harbor[b]		0.5	25	20			80	60
Average Shale[c]	0.1	0.3	45	20	0.2	6	95	60
Tributary inputs based on sediment core measurements (range of measurements shown below)								
Batten Kill River[d]	0.3–1.8	0.6–1.0	32–48	31–54	1.0–16.7	2.7–7.5	170–1300	
Hoosic River[d]	0.6–4.6	1.0–1.8	32–48	45–74	0.6–3.0	4.9–9.8	110–170	

[a] Measurements made on 2 varved clay samples collected below Thompson Island Pool sediments.
[b] Chillrud (1996).
[c] Turekian and Wedepohl (1961); Marowsky and Wedepohl (1971).
[d] Chillrud Unpublished data.

Pb will occur when contaminant lead levels are relatively low as in the top few depth sections of core mp 88.6. The isotopic composition of the top few samples is quite similar to the estimates for the composition of background Pb based on Ordovician shales (Bock *et al.*, 1989) and measurements of varved clays collected from the upper Hudson River (^{206}Pb/^{207}Pb = 1.209). This suggests that the background component may obscure the upper Hudson River contaminant Pb isotopic signal in core mp 88.6 during the late 1970s and 1980s. Additional sample analysis and particle and contaminant transport modeling can be used to address this issue.

3.3. TRACE METAL CONCENTRATIONS

Core mp 188.5 has very high concentrations of Cd, Sb, Pb, Cr and Zn (Chillrud *et al.*, 2003), summarized here as the maximum concentrations observed in the core (Table I). Maximum contaminant enrichment factors for trace metals (maximum concentration in core divided by concentration of uncontaminated fine-grained sediments (Table I) in this upstream core range from < 10 for Cu to > 100 for Sb and Cd. Detailed information on metal concentrations in these cores can be found elsewhere (Chillrud, 1996; Chillrud *et al.*, 2003).

One potential likely source of a number of these metals, located several kilometers upstream of site mp 188.5, is a pigment factory (Figure 1) that has operated since the early 1900s. This plant was identified as being the largest single point source (outside of the New York City area) of Pb, Cd and hexavalent Cr in the basin, with estimated discharges in 1982 of more than 4.5 kg d^{-1} for Pb and between 0.45 and 4.5 kg d^{-1} for both Cd and hexavalent Cr (Rohmann et al., 1985). Although Sb discharges were not examined by Rohmann et al. (1985), one of the primary uses of Sb is in the manufacturing process of paints, ceramic enamels and glasses (McGraw Hill, 1997).

Cadmium was identified as one of the most sensitive tracers of sediment transport within the upper Hudson River based on the large enrichment over background levels of Cd and the relative lack of downstream Cd sources (Chillrud et al., 2003). The temporal trends in Cd and PCBs in core mp 88.6 (Figure 5) are very similar. Because the dominant source of PCBs to the upper Hudson River was discharge from two capacitor plants located downstream of the pigment plant site (Figure 1), the similarity in temporal trends from the 1950s to the 1980s probably reflects similar transport and accumulation mechanisms. Both contaminants reached maximum concentrations in the early to mid 1970s, consistent with large-scale downstream mobilization of contaminated sediments after the removal of the Ft. Edward dam (Figure 1) in 1973 and with subsequent high-flow events in 1974 and 1976 (Chillrud, 1996). Maximum levels of Cd in core mp 88.6, more than 160 km downstream of the suspected point source, are still more than an order of magnitude above uncontaminated levels of Cd in fine-grained sediment.

The relationship between Cd and PCB concentrations throughout this 160 km section of the river can be explored further by plotting excess Cd versus PCB concentrations. PCB data as Aroclor 1242 component abundance concentrations (see Figure 5 caption) are available from Chillrud (1996) for both mp 188.5 and mp 88.6, but not for mp 163.6V. Excess Cd levels are defined as the measured concentration minus the background concentration in uncontaminated fine-grained Hudson sediments. Data for core mp 88.6 fall on or near a line approximating the trend through the data for mp 188.5 (Figure 6). The similarity of the relationship for the two different cores is consistent with a lack of major sources of PCBs and Cd between mp 188.5 and mp 88.6.

3.4. FINE-GRAINED SEDIMENT FLUXES

The decrease in trace metal levels with distance downstream of the suspected point source allows estimation of inter-site particle inputs through application of a simple dilution model (Chillrud et al., 2003). Particle dilution factors of about four between mp 188.5 and 163.6 agree with estimates based on the monitoring of total suspended solids at stations along the main stem and in the two major tributaries to this reach (Chillrud et al., 2003). Between mp 163.6 and 88.6, the decrease in trace metal levels requires a particle dilution factor of about eight.

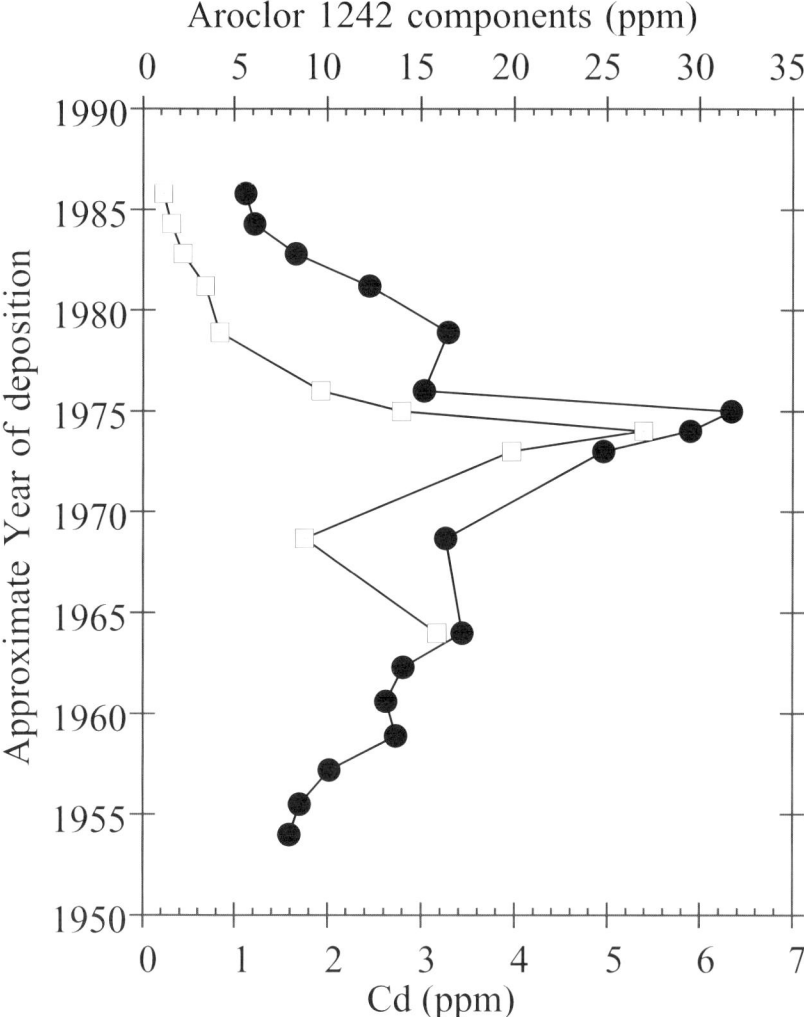

Figure 5. Temporal trends in concentrations of total Cd (filled circles) and Aroclor 1242 components (open squares) in core mp 88.6. Aroclor 1242 component abundances were determined by packed-column chromatography which resolved approximately two dozen distinct peaks. Aroclor 1242 components were calculated as the unweighted mean of the amount of Aroclor 1242 required to produce each of three sample peaks (#4, 6, 8) (Data from Chillrud (1996)).

In this reach, the Mohawk River is the largest tributary source of sediment. Based on USGS suspended sediment and river flow data, Olsen (1979) estimated that the Mohawk River delivers about twice as much sediment as the upper Hudson River accounting for about a factor of three dilution. However, the dilution factor derived from trace metal mass balance calculations suggests a large additional particle source between mp 163.6 and mp 88.6. Potential candidates include minor tributary inputs, bank erosion, resuspension of previously deposited sediments, and

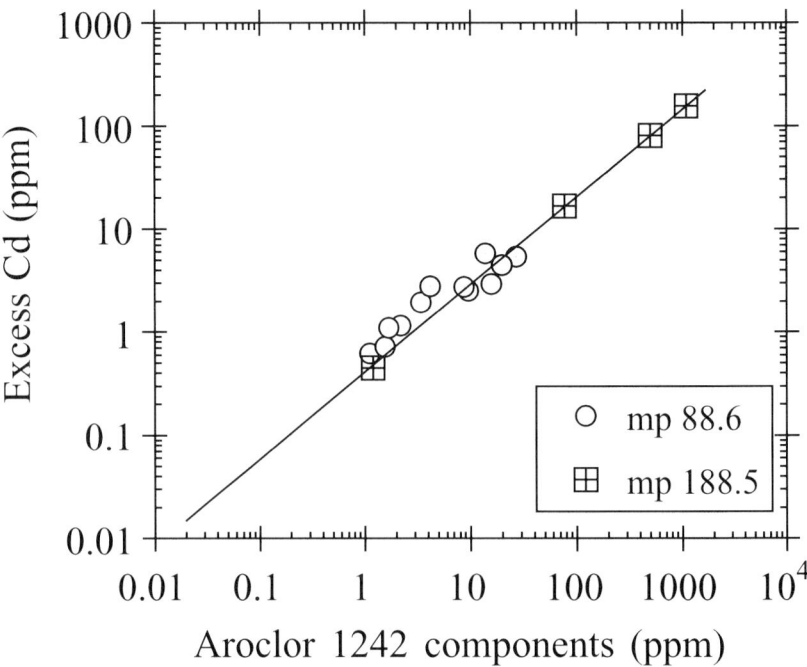

Figure 6. Variation of excess Cd with Aroclor 1242 components in cores mp 188.5 and mp 88.6. Data from Chillrud (1996).

significant underestimation of Mohawk River particle fluxes, potentially associated with short-duration high flow events. USGS sampling of a January 1996 high-flow event indicated that an amount of sediment comparable to the average yearly flux was transported in a few days, while in February 2000, on the order of half the average yearly particle flux was transported during the last two days of the month (Gary Wall, USGS, personal communication).

4. Conclusions

Radiogenic lead isotopes and trace metals are sensitive tracers of upper Hudson River sediments, which can provide critical information relevant to sources of metal contamination and transport processes. The upper Hudson River radiogenic lead isotope ratio signal is still detectable 160 km downstream from the contaminant input area in sediment cores from the tidal Hudson River and was consistent with radionuclide-based dating assignments. The large magnitude and abrupt temporal shifts in radiogenic lead isotope ratios preserved in Hudson River sediment cores provide a way to significantly improve dating models based on radionuclide analyses alone. Application of a simple dilution model to data from paired cores (mp 163.6V and mp 88.6) reveals much larger sediment inputs in this section of the river than previously estimated, suggesting large undocumented sediment fluxes entering the Hudson in that reach of the river.

Acknowledgements

We thank Hudson River Foundation grant 007/94P and NIEHS grants P42 ES07384 and P30 ES09089 for support. This is LDEO contribution 6440.

References

Appleby, P. G. and Oldfield, F.: 1992, 'Application of 210Pb to sedimentation studies', in M. Ivanovich and R. S. Harmon (eds), *Uranium Series Disequilibrium*, Oxford University Press. Oxford, pp. 731–778.
Beasley, T. M., Jennings, C. D. and McCullough, D. A.: 1986, 'Sediment accumulation rates in the lower Columbia River', *J. Environ. Radioact.* **3**, 103–123.
Bock, B., Mclennan, S. M. and Hansonm G. N.: 1998, 'Geochemistry and provenance of the Middle Ordovician Austin Glen member (Normanskill Formation) and the Taconian Orogeny in New England', *Sedimentology* **45**, 635–655.
Bopp, R. F., Simpson, H. J., Olsen, C. R. and Kostyk, N.: 1981, 'Polychlorinated biphenyls in sediments of the tidal Hudson River, New York', *Environ. Sci. Technol.* **15**, 210–216.
Bopp, R. F., Simpson, H. J., Trier, R. M. and Kostyk, N.: 1982, 'Chlorinated hydrocarbons and radionuclide chronologies in sediments of the Hudson River and estuary, New York', *Environ. Sci. Technol.* **16**, 666–676.
Bopp, R. F., Simpson, H. J. and Deck, B. L.: 1985, 'Release of polychlornated biphenyls from contaminated Hudson River sediments', Final Report prepared for the New York State Department of Environmental Conservation, contract NYS C00708.
Bopp, R. F. and Simpson, H. J: 1989, 'Contamination of the Hudson River: The sediment record', in *Contaminated Marine Sediments – Assessment and Remediation*, National Academy Press, Washington, DC., pp. 401–416.
Bopp, R. F., Simpson, H. J., Chillrud, S. N. and Robinson, D. W.: 1993, 'Sediment-derived chronologies of persistent contaminants in Jamaica Bay, NY', *Estuaries* **16**, 608–616.
Bopp, R. F., Chillrud, S. N., Shuster, E. L., Simpson, H. J. and Estabrooks, F. D.: 1998, 'Trends in chlorinated hydrocarbon levels in Hudson River basin sediments: Integrated approaches for studying hazardous substances', *Environ. Health Perspect.* **106**, *Suppl.* **4**, 1075–1079.
Bopp, R. F., Chillrud, S. N., Shuster, E. L. and Simpson, H. J.: 'Contaminant chronologies', in *The Hudson River Estuary*, J. Levinton and J. Waldman (eds), Oxford University Press, *in press*.
Bush, B., Shane, L. A., Whalen, M. and Brown, M. P.: 1987, 'Sedimentation of 74 PCB congeners in the Upper Hudson River', *Chemosphere* **16**, 733–744.
Chillrud, S. N.: 1996, 'Transport and fate of particle-associated contaminants in the Hudson River Basin', *Ph. D. Thesis*, Columbia University, New York. 277 pp.
Chillrud, S. N., Hemming, S., Shuster, E. L., Simpson, H. J., Bopp, R. F. Ross, J. M., Pederson, D. C., Chaky, D., Tolley, L-R. and Estabrooks, F.: 2003, 'Stable lead isotopes, contaminant metals and radionuclides in upper Hudson River sediment cores: Implications for improved stratigraphy and transport processes', *Chemical Geology* **199**, 53–70.
Crusius, J. and Anderson, R. F.: 1991, 'Core compression and surficial sediment loss of lake sediments of high porosity caused by gravity coring', *Limnol. Oceanogr.* **36**, 1021–1030.

Deloule, E., Allegre, C. and Doe, B. R.: 1986, 'Lead and sulfur isotope microstratigraphy in galena crystals from Mississippi Valley-type deposit', *Econ. Geol.* **81**, 1307–1321.

Doe, B. R.: 1975, 'Lead Isotope Data Bank: 2624 Samples and Analyses Cited', USGS Open File Report 76-201, 104 pp.

Doe, B. R. and Delavaux, M. H.: 1972, 'Source of lead in southeast Missouri galena ores', *Econ. Geol.* **67**, 409–425.

Fleisher, M. Q. and Anderson, R.: 1991, 'Particulate matter digestion (from mg to 10's of g) and radionuclide blanks', in D. C. Hurd and D. W. Spencer (eds), *Marine Particlse: Analysis and Characterization*, Geophysical Monograph 63, American Geophysical Union, Washington, DC. pp 221–222.

Graney, J. R., Halliday, A. N., Keeler, G. J., Nriagu, J. O., Robbins, J. A. and Norton, S. A.: 1995, 'Isotopic record of lead pollution in lake sediments from the northeastern United States', *Geochim. Cosmochim. Acta* **59**, 1715–1728.

Hurst, R. W., Davis, T. E. and Chinn, B. D.: 1996, 'The lead fingerprints of gasoline contamination', *Environ. Sci. Technol.* **30**, 304–307.

Marcantonio, F., Zimmerman, A. Xu, Y. and Canuel, E.: 2002, 'A Pb isotope record of mid-Atlantic US atmospheric Pb emissions in Chesapeake Bay sediments', *Mar. Chem.* **77**, 123–132.

Marowsky, G. and Wedepohl, K. H.: 1971, 'General trends in the behavior of Cd, Hg, Tl, and Bi in some major rock forming processes', *Geochim. Cosmochim. Acta* **35**, 1255–1267.

McGraw Hill Encyclopedia of Science and Technology: 1997, 8th Edition, New York. 1: 788.

McNulty, A. K.: 1997, 'In-situ anaerobic dechlorination of polycholorinated biphenyls in Hudson River sediments', *M. S. Thesis*, Rensselaer Polytechnic Institute, Troy, New York, 334 pp.

Olsen, C. R.: 1979, 'Radionuclides, sedimentation and the accumulation of pollutants in the Hudson Estuary', *Ph. D thesis*, Columbia University, New York, New York, 343 pp.

Olsen, C. R., Simpson, H. J., Peng, T.-H., Bopp, R. F. and Trier, R. M.: 1981a, 'Sediment mixing and accumulation rate effects on radionuclide depth profiles in Hudson Estuary sediments', *J. Geophys. Res.* **86(C11)**, 11020–11028.

Olsen, C. R., Simpson, H. J. and Trier, R. M.: 1981b, 'Plutonium, radiocesium and radiocobalt in sediments of the Hudson River estuary', *Earth Planet. Sci. Lett.* **55**, 371–392.

Ritchie, J. C. and McHenry, J. R.: 1990, 'Application of radioactive fallout Cesium-137 for measuring soil erosion and sediment accumulation rates and patterns: A review', *J. Environ. Qual.* **19**, 215–233.

Rohmann, S. O., Miller, R. L., Scott, E. A. and Muir, W. R.: 1985, *Tracing a River's Toxic Pollution, A Case Study of the Hudson*, A.S. McCook (ed.), INFORM Inc., New York.

Scott, M. R., Rotter, R. J. and Salter, P. R.: 1985, 'Transport of fallout plutonium to the ocean by the Mississippi River', *Earth Planet. Sci. Lett.* **75**, 321–326.

Shen, G. and Boyle, E.: 1987, 'Lead in corals: reconstruction of historic industrial fluxes to the surface ocean', *Earth Planet. Sci. Lett.* **326**, 278–280.

Shuster, E., Bopp, R. F., and Zamek E.: 2002, 'Trace metal and dioxin analyses of dated sediment samples from the upper Hudson River', Report to NYS Department of Environmental Conservation, Contract Number C003844.

Simpson, H. J., Olsen, C. R., Williams, S. C. and Trier, R. M.: 1976, 'Man-made radionuclides and sedimentation in the Hudson River estuary', *Science* **194**, 179–183.

Sturges, W. T. and Barrie, L. A.: 1987, 'Lead 206/207 isotope ratios in the atmosphere of North America as tracers of US and Canadian emissions', *Nature* **329**: 144–146.

Turekian, K. K. and Wedepohl, K. H.: 1961, 'Distribution of elements in the earth's crust', *Geol. Soc. Am. Bull.* **72**, 174–192.

MEASURING AEROSOL AND HEAVY METAL DEPOSITION ON URBAN WOODLAND AND GRASS USING INVENTORIES OF ^{210}PB AND METAL CONCENTRATIONS IN SOIL

D. FOWLER[1]*, U. SKIBA[1], E. NEMITZ[1], F. CHOUBEDAR[2], D. BRANFORD[2], R. DONOVAN[4] and P. ROWLAND[3]

[1] *Centre for Ecology and Hydrology, Bush Estate, Penicuik, Midlothian EH26 0QB, U.K.;*
[2] *Department of Physics, University of Edinburgh, James Clerk Maxwell Building, Kings Buildings, Mayfield Road, Edinburgh EH9 3JZ, U.K.;* [3] *Centre for Ecology and Hydrology, Windermere Road, Grange-over-Sands, Cumbria LA11 6JU, U.K.;* [4] *Department of Environmental Sciences, Lancaster University Lancaster LA1 4YQ, U.K.*
(* *author for correspondence, e-mail: dfo@ceh.ac.uk; fax: (44) 131 445 3943*)

(Received 20 August 2002; accepted 14 April 2003)

Abstract. The deposition of aerosols to trees has proved very difficult to quantify, especially in complex landscapes. However, trees are widely quoted to be efficient scavengers of particles from the atmosphere, and a growing proportion of the pollutant burden in the atmosphere is present in the aerosol phase. In this study, the deposition of aerosols onto woodland and grass was quantified at a range of locations throughout the West Midlands of England. The sites included mature deciduous woodland in Edgbaston, and Moseley, and mixed woodland at sites within Sutton Park, a large area of semi-natural vegetation. Aerosol deposition to areas of grassland close to the woodland at each site was also measured. Detailed inventories of ^{210}Pb in soils within the woodland and in grassland soils, together with concentrations in the atmosphere and precipitation, provided the necessary data to calculate the long-term (about 40 years) annual deposition of sub-micron aerosols onto grassland and woodland. The soil inventories of ^{210}Pb under woodland exceeded those under grass, by between 22% and 60%, with dry deposition contributing 24% of the total input flux for grass and 47% for woodland. The aerosol dry deposition velocity to grassland averaged 3.3 mm s^{-1} and 9 mm s^{-1} for woodland. The large deposition rates of aerosols onto woodland relative to grass or other short vegetation ($\times$ 3), and accumulation of heavy metals within the surface horizons of organic soils, leads to large concentrations in soils of urban woodland. Concentrations in the top 10 cm of these woodland soils averaged 252 mg kg^{-1} for Pb with peaks to 400 mg kg^{-1}. Concentrations of Cd averaged 1.4 mg kg^{-1}, Cu, 126 mg kg^{-1}, Ni 23 mg kg^{-1} and Zn 173 mg kg^{-1}. The accumulated Pb in urban woodland soils is shown to be large relative to UK emissions.

Keywords: aerosol, deposition, deposition velocity, heavy metals, ^{210}Pb, woodland

1. Introduction

The focus of current interest in urban air pollutants and their effects on human health is particulate matter, and especially particles smaller than 10 μm diameter (PM$_{10}$) (Holdgate, 1998). Concentrations of PM$_{10}$ in urban air have been shown to be associated with human morbidity and mortality rates in epidemiological studies in North America (Pope *et al.*, 1995) and Europe (Holdgate, 2001). Airborne

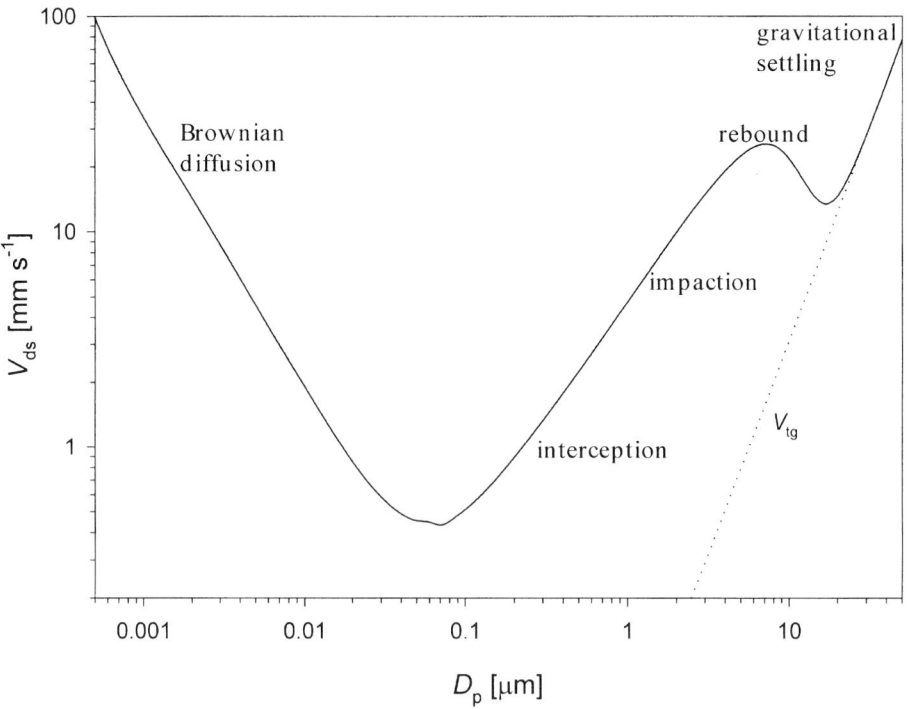

Figure 1. The variation of aerosol deposition to short (height 0.1 m) vegetation with aerosol size, according to the Slinn (1982) model.

particulate matter has a complex source distribution as it is derived from direct emissions from combustion processes and industry and is formed in the atmosphere by condensation and chemical reaction. Particles may also be entrained from the surface into the atmosphere by wind, by vehicles and by mechanical action.

Particles are removed from the atmosphere by precipitation, through direct washout by falling rain or snow. Alternatively, particles may form the condensation nuclei on which cloud droplets develop, or are incorporated into growing cloud droplets by diffusion or phoretic processes (Fowler *et al.*, 1984). Particles also are deposited directly onto terrestrial surfaces following turbulent transfer vertically to the boundary layer and capture by vegetation. The latter process is referred to as dry deposition regardless of the surface, to distinguish it from removal by rain.

The dry deposition of particles onto terrestrial surfaces has been studied for almost 50 yr and the underlying processes that regulate transfer to and capture by terrestrial surfaces has been described (Nicholson, 1988) and reviewed more recently by Gallagher *et al.* (2002).

AEROSOL IMPACTION

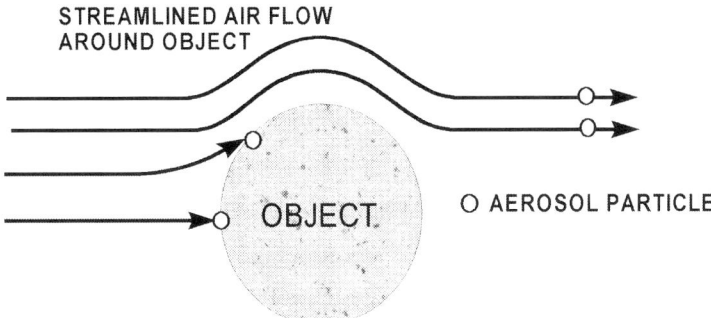

Figure 2. Diagrammatic representation of the inertial impaction of aerosols on terrestrial surfaces.

The rate of aerosol dry deposition to vegetation varies with particle diameter. Dry deposition rates generally are characterized using a deposition velocity ($vd_{(z)}$) according to

$$vd_{(z)} = \frac{f}{\chi_{(z)}} \tag{1}$$

in which f is flux to the surface and χ is the concentration at height (z). For short vegetation (height $\leq$ 1.0 m), such as in grasslands or moorlands, the relationship between deposition velocity and particle size is given in Figure 1, which shows a minimum deposition velocity for particle diameters of typically 0.1 μm.

Deposition velocity increases as particle size decreases below 0.1 μm (diam.), as particle transport through the leaf boundary layer becomes generally more efficient through diffusive processes. For particle diameters greater than 0.1 μm (diam.), deposition velocity increases due to the increasing efficiency of impaction and interception processes with increases in particle size. This effect represents the collision of particles with terrestrial structures, as the increasing inertia of the larger particles prevents them following the streamlines of airflow around surface structures (shown in Figure 2).

For typical urban aerosols, the mass median diameter of particles ranges generally from 0.6 μm to 0.8 μm. This size range is associated with small deposition velocities, in the range 0.5 mm s^{-1} to 3 mm s^{-1} for short vegetation, and if this was the only removal pathway, the particles in the boundary layer would have lifetimes in the atmosphere of a week or more. Most of the aerosol in this size range is removed from the atmosphere by rain or snow. However, it has been suggested that aerodynamically rough vegetation, such as woodland or isolated trees, are efficient surfaces for the capture of particulate air pollutants (Pagel, 1983; Bradshaw *et al.*, 1995; Beckett *et al.*, 1998). A problem in assessing the value of trees as a means of removing particles from urban air is that the results are very variable and few measurements have been made in urban areas. The data available show that at deposition velocities of atmosphere particles onto trees of 0.1 μm to 1.0 μm exceeds the Slinn (1982) model predictions by about an order of magnitude, with deposition velocities of 5 mm s^{-1} to 10 mm s^{-1}. Such large deposition velocities suggest that trees may indeed be efficient scavengers for particles from the atmosphere. The underlying processes that give rise to these results are considered later in the discussion of the measurements presented here. This paper reports new measurements of particle deposition onto urban woodland and grassland at a range of sites in the West Midlands of England. The method averages particle deposition over several decades, and therefore provides a long-term average over a wide range of meteorological conditions. The method is based on the measurement of ^{210}Pb in soil beneath trees and short vegetation, combined with atmospheric measurement of ^{210}Pb concentration in air and precipitation. The objective of the study was to provide quantitative estimates of the deposition rate of urban aerosols onto woodland and short vegetation.

2. Methods

The method used to quantify particle deposition is based on the measurements of soil inventories of the naturally occurring radioisotope ^{210}Pb in soils that have been undisturbed for three or more half-lives of ^{210}Pb ($t^{0.5}$ = 22.3 yr). The ^{210}Pb in the atmosphere arises from the emission of the inert gas Radon (^{222}Rn) from soil, which in turn is a decay product in the radioactive decay series from ^{238}U. The gaseous ^{222}Rn, which has a radioactive half-life of 3.7 days diffuses out of soils into the atmosphere. In the atmosphere the ^{222}Rn decays through a series of short-lived daughter radionuclides to ^{210}Pb, and then through further decay to ^{206}Pb. In the transformation process, the gaseous ^{222}Rn is transformed into a particle that rapidly becomes attached to the aerosol present in the atmosphere. The time taken for the ^{210}Pb to become attached to aerosol varies with temperature and aerosol concentrations, but is typically 50 s (Chamberlain, 1991). Relative to the lifetime of particles in the atmosphere, this is therefore a very rapid process and the atmosphere aerosol is effectively tagged by ^{210}Pb. The ^{210}Pb within atmospheric particles then follows the atmospheric transformations (e.g., particle to cloud droplet to precipitation) on its pathway to the surface. If the particle is deposited onto vegetation and subsequently to soils via leaf fall or removal from foliage by rain, the Pb is very efficiently captured by organic matter in the upper horizons of the soil profile, as shown by Lewis (1977) and by Graustein and Turekian (1989). In clay soils, ^{210}Pb may also be adsorbed, however, beneath an organic rich surface horizon, the sequestration of the Pb from water percolating through the soil effectively captures both the ^{210}Pb and the stable Pb, provided that there is sufficient organic matter. The ^{210}Pb isotope then decays *in situ* and in time, an equilibration is established between the atmospheric input and radioactive decay, as long as the soil is not disturbed. The soil inventory, which must be corrected for any internally generated ^{210}Pb (the supported fraction) may then be used to calculate the annual ^{210}Pb flux since

$$^{210}\text{Pb flux} = \lambda \times I \tag{2}$$

where λ is the decay constant (a^{-1}) and I is the soil inventory of ^{210}Pb (Bq m^{-2}), corrected for the supported fraction. The method has been used to quantify aerosol deposition by Graunstein and Turekian (1989) and to study the spatial variability in wet deposition and the effect of trees in intercepting hill fog (Fowler *et al.*, 1998).

In this study, the West Midlands was chosen as the study area. It is a large urban conurbation comprising 900 km^2, in which trees occupy 4% of the area. There are several locations throughout the area in which old mixed deciduous and conifer woodland are present. Well documented site histories show that some sites have not been disturbed for at least 50 yr, and a substantial accumulation of organic matter in the surface horizon provide the necessary zone in which ^{210}Pb, and stable, elemental lead (^{206}Pb), and other heavy metals, may accumulate.

3. Sites

The locations within the West Midlands range from Sutton Park, on the NE fringe of the conurbation, to Moseley and Edgbaston, much closer to the urban centre.

3.1. SUTTON PARK

Sutton Park, a 650 ha area of semi-natural vegetation approx 10 km NE of Birmingham city centre, (UK national grid reference SP098958) and on the boundary of the conurbation, was established as a Royal park in the 14th century and retains several areas of semi-natural mixed woodland, in which there has been little disturbance during the last 50 yr, and in which the mature woodland canopy has remained fairly constant. The tree species present in the West Woodland include oak (*Quercus robur*), Scots pine (*Pinus sylvestris*), chestnut (*Castanea sativa*), and holly (*Ilex aquifolium*), with an understory of bramble (*Robus fluticosus*). The canopy height was approximately 20 m and the woodland area occupied a 200 m wide block running alongside the A452 at the west edge of Sutton Park. Thus, in addition to the general particulate loading of the air in the conurbation, the emissions from the busy A452 provided a substantial local source of particles upwind in the predominantly westerly airflow over the woodland.

The sampling of soils took place along two transects running parallel to the road, the first 5 m into the woodland from the westerly edge and the second, 100 m from the westerly edge and roughly in the middle of the woodland block. Along each of these transects, five sites were selected, at random, to sample the soil profile.

At each site, four samples 18 × 18 × 20 cm deep were collected and divided into three horizons, 0–5 cm, 5–10 cm, and 10–20 cm. These layers corresponded approximately to the litter and O horizon (0–5 cm), the A1 horizon (5–10 cm), and the A2 horizon (10–20 cm). All samples were analysed separately for ^{210}Pb, to provide the vertical profile in concentration, and the variability at each sampling location, and thus allowing inter-site differences to be calculated. The grassland site within the park was located 1 km E of the woodland in open ground that had not been subject to any management other than mowing once each year, with the cut grass being left on the site.

3.2. EDGBASTON GOLF COURSE AND SSSI

The site is 2 km south of Birmingham city centre (UK national grid ref. SP055 838) adjacent to the campus of Birmingham University. Grassland samples were taken from rough areas of Edgbaston golf course, in areas that have been undisturbed since 1934 except for occasional grass cutting during which the cut grass was left *in situ*. The woodland samples were taken within mature woodland that formed a part of an SSSI in Winterbourne Gardens. The mature trees, 25 m in height included oak, sweet chestnut, holly, and yew with a substantial ground cover of brambles, ivy (*Hedera* spp.), and bluebell (*Scilla nonscripta*). All woodland samples were taken within the central area of the woodland 30 m or more from the edge. The five sampling sites provided four replicate cores, again 18 × 18 cm at the surface and 20 cm deep. Sampling took place during the early spring, when ground flora was active, but before the tree canopy leaves had flushed.

3.3. MOSELEY GOLF COURSE AND BROOK LANE WOODLAND

The site lies 4 km SE of Birmingham city centre (UK grid ref SP 086 819) in a predominantly residential area and within the boundaries of Moseley golf course. All locations within the golf course have remained undisturbed, except for grass cutting for the last 50 yr, the rough grass was longer (25 cm) than at Edgbaston (10 cm). The grassland samples were taken from rough areas of the golf course at locations least disturbed by golfers or course maintenance staff. Samples at five locations in grassland with four replicates, equal in area and depth to the other two main sites were taken. The woodland contained mixed, mainly deciduous species including oak, beech, holly, and chestnut, with a bramble understory and bluebells. Soil samples were taken at five locations within the woodland, all 20 m or more from the edge with four replicates per location.

4. Soil Processing

The soil bulk density was determined by inserting a plastic cylinder of known volume into each sample and recording its fresh weight, taking a subsample for

TABLE I

^{210}Pb inventories of grassland and woodland soils in Sutton Park

Soil horizon	Grassland ^{210}Pb inventory	Westwood coppice Soil horizon	Woodland (middle) Inventory
cm	Bq m^{-2}	cm	Bq m^{-2}
0–7	2341	0–5	2633
7–15	380	5–10	650
15–20	176	10–20	214
Total	**2898**		**3498**

Woodland enhancement +44%. Values are means of the 4 replicates for each horizon.

dry weight after 15 hr at 105 °C in a drying oven. Soil samples were air dried on plastic trays at 30 °C and sieved through a 2 mm mesh. Litter and O horizon soil and vegetation samples were milled to pass through the 2 mm sieve. Samples for ^{210}Pb analysis were ground and placed in standard holders for the detection of gamma-rays from the ^{210}Pb.

The detectors used were high purity geranium (HPGe) with a thin (0.5 mm) Be window. The detector has a relative efficiency of 12.9% at 1.3 MeV and background radioactivity is suppressed using a substantial castle of low activity lead and an anticoincidence shield (Knoll, 1989). The approach not only reduces the combined background, but also significantly reduces a substantial fraction of the Compton interference from radionuclides within the sample. Self-attention of gamma rays within the soil samples was corrected for using the method of Cutshall *et al.* (1983). Counting periods for the soil samples of approximately 24 hr were used to reduce counting errors to acceptable levels. The main gamma energies on which the analysis was focussed were the 46 keV for ^{210}Pb and the 662 keV for ^{137}Cs and 352 keV for ^{214}Pb, the latter being used to correct samples for the supported fraction.

Chemical analyses of the soil samples for the heavy metals Pb, Cd, Cu, V and Zn were made using ICP-MS techniques, to determine the contamination of these soils by traffic and industrial activity, and to allow a comparison between the relative accumulation rates of heavy metals in woodland and grassland soils.

5. Results

The ^{210}Pb inventories for Sutton Park showed the majority of the activity to be located in the surface horizons, with 81% of activity in the grassland with the top 7 cm, and only 6% below 10 cm (Table I). The woodland sites, generally had a much deeper litter and organic horizon, but the fractions of the activity in the surface and

TABLE II

^{210}Pb inventories of grassland and woodland soils at Edgbaston

Soil horizon cm	Grassland ^{210}Pb Inventory Bq m^{-2}	Woodland ^{210}Pb Inventory Bq m^{-2}
0–5	1866	1939
5–10	609	1062
10–20	298	999
Total	**2772**	**4001**

Woodland enhancement +44%.

TABLE III

^{210}Pb inventories of grassland and woodland soils at Moseley

Soil horizon cm	Grassland ^{210}Pb Inventory Bq m^{-2}	Woodland ^{210}Pb Inventory Bq m^{-2}
0–5	1490	2767
5–10	816	1101
10–20	344	384
Total	**2650**	**4253**

Woodland enhancement +60%.

deeper soil horizons remained very similar to those of the grassland. However, the total inventory was substantially larger beneath trees, by 20%.

The Edgbaston comparison, between mature woodland and the rough grass areas of the golf course again show the majority (67%) of ^{210}Pb activity in the surface (0–5 cm) horizon, and a substantial enhancement in the inventory in the woodland soil, relative to grassland (+44%) (Table II).

At Moseley, the absolute value of the ^{210}Pb inventory in grassland soil (2650 Bq m^{-2}) was very similar to that at Edgbaston and Sutton Park, with 56% in the top 5 cm of the soil profile. The inventory beneath the mixed deciduous woodland was larger than in the grassland by 60%, the largest of the three site comparisons (Table III).

5.1. HEAVY METAL CONCENTRATIONS IN WOODLAND AND GRASSLAND SOILS

Subsamples from the soil profiles at all collection sites were analysed for Pb, Cd, Cu, Ni and V. These data are shown in Table IV. The concentration of Pb generally decreases with depth in the soil profile, with concentrations in the litter and O layer

TABLE IV

Heavy metal concentrations in soils of the West Midlands conurbation

Location	Cd mg kg^{-1}	Cu mg kg^{-1}	Ni mg kg^{-1}	Pb mg kg^{-1}	V mg kg^{-1}	Zn mg kg^{-1}
Recommended upper limit	3	135	75	250		300
Sutton Park Pageant area, 0–5 cm	0.53	53	14	100	11	110
Sutton Park Pageant area, 5–15 cm	0.12	33	5	80	16	22
Sutton Park Pageant area, 10–20 cm	0.064	17	1.4	28	13	7.6
Westwood Coppice edge 4, 0–5 cm	2.1	150	28	400	25	410
Westwood Coppice edge 4, 5–10 cm	0.92	140	19	330	30	140
Westwood Coppice edge 4, 10–20 cm	0.26	33	4.3	91	14	38
Westwood Coppice middle 5, 0–5 cm	0.59	150	21	29	29	110
Westwood Coppice middle 5, 5–10 cm	1.2	88	22	14	14	180
Westwood Coppice middle 5, 10–20 cm	0.084	9.2	1.5	13	13	10
Moseley Golf Course, 0–5 cm	0.61	26	14	62	14	130
Moseley Golf Course, 5–15 cm	0.34	44	21	100	36	100
Moseley Golf Course, 10–20 cm	0.39	39	22	84	35	110
Moseley Wood, 0–5 cm	0.58	63	11	120	13	120
Moseley Wood, 5–10 cm	0.43	60	13	160	21	79
Moseley Wood, 10–20 cm	0.33	86	14	200	32	68
Edgbaston Golf Course, 0–5 cm	0.6	130	26	220	33	150
Edgbaston Golf Course, 5–10 cm	0.2	89	13	140	37	48
Edgbaston Golf Course, 10–20 cm	0.11	38	8.7	99	24	37
Edgbaston SSSI, 0–5 cm	2.6	160	31	210	34	480
Edgbaston SSSI, 5–10 cm	2.5	200	35	240	43	480
Edgbaston SSSI, 10–20 cm	1.4	120	23	160	36	270

(0–5 cm) between 62 mg kg^{-1} and 400 mg kg^{-1}. The largest concentrations were found in the woodland soils, declining with depth to 28 mg kg^{-1} to 200 mg kg^{-1} in the 10–20 cm layer. The peak Pb concentrations in woodland soils in Sutton Park exceed recommended upper limit values for soil (250 mg kg^{-1}), while in the Moseley and Edgbaston woodlands the peak concentrations of 200 mg kg^{-1} and 240 mg kg^{-1} approach the recommended upper limit.

It is also interesting that larger concentrations of Pb occur closer to the woodland edge, and hence the road and traffic emission sources in the Westwood coppice. By contrast, the ^{210}Pb inventory values at the woodland edge are smaller (2867 Bq m^{-2} total), by 18% than within the centre of the woodland.

TABLE V
The concentrations of ^{210}Pb in UK rain

Location	Lat.	Long.	Prec. (mm)	Sampling period (a)	Conc. (mBq L^{-1})	Flux (Bq m^{-2} a^{-1})
Brotherswater	54.5	−3	2108		70	147
Milford Haven	51.7	−5	916	3	103	85 ± 13
Plymouth	50.4	−4.2	1240	2	84	68
Edinburgh	55.6	−3.9	930[a]	>1	63 ± 9	68 ± 10

[a] Rain in 1998 measured at the sampling site.

Concentrations of Cd, Cu, Zn and Ni are also large in the woodland soils, and in the case of Cu, exceeding recommended upper limit concentrations for soil (135 mg kg^{-1}, Paterson et al., 2000). For Cd, Cu, Zn and Ni, the concentrations generally decline with depth, with typically an order of magnitude smaller concentrations in the 10–20 cm layer than surface soil. Concentrations of V show weak trends with depth and the woodland values differ little from those of the grassland.

5.2. WET DEPOSITION OF ^{210}PB

The inventory of ^{210}Pb in soil may be used to estimate the annual flux, from Equation 1. The annual flux for each sampling location is then the sum of wet deposition and dry deposition. In principle, the interception of fog or low cloud also contributes to the total input. However, for the West Midlands, the fog frequency is too small for fog deposition to make a significant contribution to the total deposition of the region (CLAG-Fluxes 1997).

To analyse the ^{210}Pb data and interpret the woodland-grassland differences, it is necessary to separate the wet and dry deposited contributions to the inventories in soil presented above. The grasslands and woodlands are in the close proximity (<500 m) at each of the three main sampling sites and at the same altitude and with no complex topography between the sites. The long-term average rainfall can therefore be assumed to be the same in the woodland and grassland at each sampling location.

The concentrations of ^{210}Pb in air and rain vary with the continentality of the site, values being largest in the centres of continents and smallest over the oceans (Graunstein and Turekian, 1983), reflecting the terrestrial sources of ^{210}Pb. Recent measurements in the UK reported by Priess et al. (1996) and Choubedar (2000) show concentrations in rain ranging from 63 mBq L^{-1} to 103 mBq L^{-1} (Table V). The wet deposition flux varies from 68 Bq m^{-2} a^{-1} to 147 Bq m^{-2} a^{-1}, with the largest value for Brotherswater in the Lake District, a very high precipitation (>2000 mm a^{-1}) region of the Lake District in Northern England.

TABLE VI

Total wet and dry deposition of ^{210}Pb containing aerosols to woodland and grass in the West Midlands conurbation

	Sutton Park	Edgbaston	Moseley	Average
Grass				
Total dep. (Bq m^{-2} a^{-1})	89	86.3	82.5	
Dry dep.	24	21	17.5	
V_d (mm s^{-1})	3.8	3.3	2.8	3.3
Woodland				
Total dep. (Bq m^{-2} a^{-1})	108.9	124.6	132.4	
Dry dep.	44.9	59.6	67.4	
V_d (m s^{-1})	7	9.4	10.7	9

TABLE VII

Measured atmospheric concentrations of ^{210}Pb aerosol in surface air (Preiss *et al.*, 1998)

Site		^{210}Pb concentration mBq m^{-3}
Chiltern	Southern England	0.23
Dublin	Eastern Ireland	0.20
Edinburgh	Southern Scotland	0.19
Jutland	W Denmark	0.21
Stavanger	SW Norway	0.17

The wet deposition of ^{210}Pb in Edinburgh and Plymouth is identical at 68 Bq m^{-2}. Rainfall in Birmingham is in the same range as Plymouth and Edinburgh (800–1200 mm). The wet deposition of ^{210}Pb in Birmingham may therefore be assumed to be similar, in the range 60 Bq m^{-2} a^{-1} to 70 Bq m^{-2} a^{-1}, given the small measured range on wet ^{210}Pb deposition across the UK Taking wet deposition of ^{210}Pb for the West Midlands of 65 Bq m^{-2} a^{-1} as the long-term average allows the dry deposition of aerosols to the woodland and grassland to be quantified from the total provided in Tables I–III (Table VI).

The dry deposition flux represents the minor fraction of the annual deposition for all grasslands, contributing typically 20 to 25% of the total. The dry deposition flux may be used to quantify the deposition velocity $Vd_{(z)}$ from a knowledge of the long-term average ^{210}Pb aerosol concentration in air. The data collated by Preiss et al. (1996) for Europe and by Choubedar (2000) for Northern Britain show very similar concentrations of aerosol ^{210}Pb across the UK and surrounding the North Sea of 0.2 mBq m^{-3} with a range of ± 0.03 mBq m^{-3}. Published UK data of air concentrations of ^{210}Pb are summarised in Table VII.

6. Discussion

Quantifying the fate of atmospheric particles has become an important issue as the links between air pollutants and human health have increasingly focussed on aerosols as a major contributor to the human health effects in epidemiological studies (Holdgate et al., 2001). The inter-country exchange of major pollutants, including S, N, and metals, occurs largely in the aerosol phase as particles, the bulk of which are contained in sizes from 0.1 to 1.0 μm (diam.). Thus, the parameterization of deposition rates for particles characteristic of this range is important for source-receptor models of long-range transport of air pollutants.

The measurements reported here show long-term average deposition rates for the aerosols that characterize surface air in the West Midlands. The inventories of ^{210}Pb in woodland soils are larger than those of the grassland, by between 20 and 60% (Tables I to III). The differences between woodland and grass are solely due to dry deposition; the wet deposition may safely be assumed to be identical, being within 1 km, and at the same altitude for each of the sampling locations. The aerosols deposited are those that characterize the urban air in the West Midlands, and these have been shown to have a mass median size of 0.6 to 0.8 μm. The deposition velocities for these aerosols onto grassland and woodland at the three experimental sites range from 2.8 mm s^{-1} for the grassland to 10.7 mm s^{-1} for the woodland. It is now possible to compare the derived deposition velocity for the woodland with literature values.

The values of particle deposition velocity for both short and tall vegetation presented here are large relative to those of Chamberlain (1966), from wind tunnel studies. The measurements reported here have the advantage that they average an extensive range of meteorological conditions, and a range of woodland canopy architecture.

The consistency of the fluxes and deposition velocities, averaging 3.3 mm s^{-1} and ranging from 2.8 mm s^{-1} to 3.8 mm s^{-1} for grassland is encouraging, and this, in part reflects the similarity in the characteristics of the three sites in species composition and canopy height and density. The woodland sites show more variability. This is semi-natural woodland, with considerable variability in tree height and composition, and it was not feasible to quantify the canopy leaf area index at each sampling site, for a range of practical reasons. The replication of sampling locations within the woodland has allowed the difference between woodland and grass to be measured confidently, but the cause of differences in aerosol deposition between the different woodlands is much more difficult to identify. The larger de-

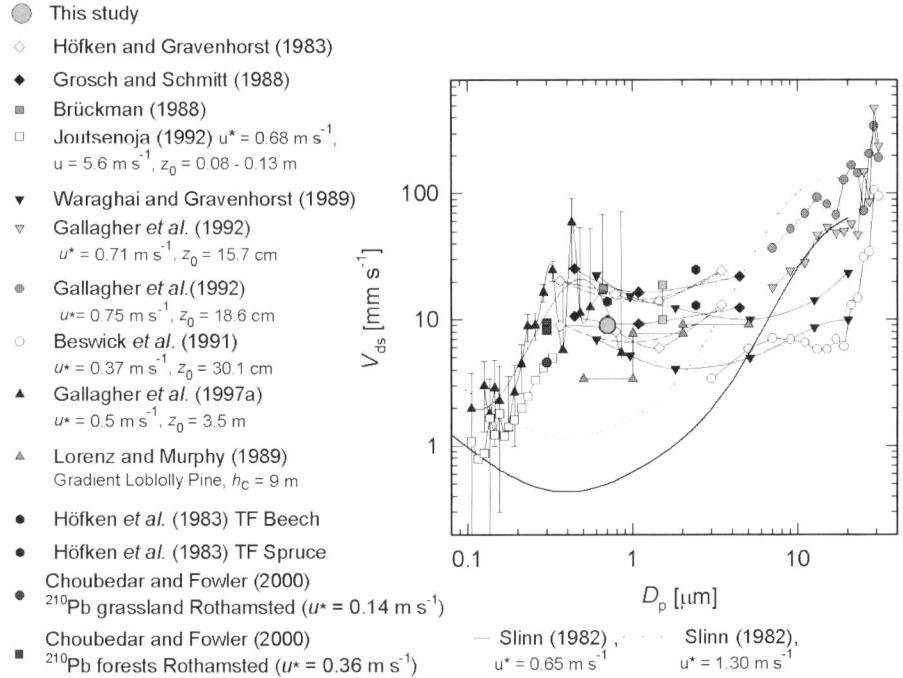

Figure 3. A comparison of the Slinn (1982) model for the variation of deposition velocity of aerosols with aerosol size for woodland with field measurements by a wide range of methods (after Gallagher et al., 1997).

position velocities for Edgbaston and Moseley woodland are most likely the result of taller, larger trees, carrying substantially larger leaf area, but without a detailed measurement exercise at each site, this is somewhat speculative. The leaf area of the canopies at each collecting site would have been a very useful quantity to help interpret differences in the inventories, but this was considered to be impractical at the site, because the available methods to provide canopy leaf areas relative to the soil sampling location have poor resolution. In retrospect, a very careful litter collection at each sampling site for a year would provide a powerful tool to identify the cause of the variations in the deposition velocities to the woodland.

The woodland deposition velocities of 7 mm s^{-1} to 10.7 mm s^{-1} averaging 9 mm s^{-1} show that woodland 'collects' the ambient aerosol at approximately three times the rate of the grassland. The deposition velocity for woodland is very large relative to those suggested by the Slinn (1982) model (<0.5 mm s^{-1}) or by the wind tunnel work as discussed by Garland (1983, 2001). In the absence of other data in support of either larger or small deposition velocities for particles on aerodynamically rough surfaces, it would be difficult to reconcile these differences. However, Gallagher et al. (1997) in reviewing the particle size dependence of aerosol onto natural surfaces, showed that 12 very different methods of measurement showed broadly the same, very large deposition velocities of small particles onto rough surfaces, mainly forest. The Figure summarizing the different studies, and the discrepancy with the widely accepted Slinn (1982) model, has been reproduced here (Figure 3). The Figure includes data for the woodland sites from this study assuming a mass median particle diameter of the ^{210}Pb of 0.7 μm, although with the scatter in this log/log figure, any mass median particle size between 0.4 μm 1.0 μm would fit equally well.

The data for very small particles (0.1 μm) are mainly from eddy co-variance flux measurements, and these are in reasonable agreement with the Slinn model. Similarly for particle sizes greater than 2 μm, the model is in agreement with a significant fraction of the available data. The larger particles (2–10 μm) follow the predicted deposition rates from sedimentation and momentum transfer (Beswick *et al.*, 1991; Wyers *et al.*, 1994). The main discrepancy lies in the size range 0.3 μm to 1.0 μm. This has been the subject of detailed discussion by Gallagher *et al.* (1997) and more recently by Garland (2001). The conclusions by Garland that measurement problems alone are unlikely to result in such consistent model-measurement discrepancy for the submicron aerosols is an important observation. It is probable therefore, that additional mechanisms operate within woodland to affect the transfer of small particles through the quasi-laminar sub-layer to leaf surfaces. Such processes include phoretic mechanisms along electrical potential, or thermal gradients (electrophoresis or thermophoresis). Further, more detailed field measurements are required to test this hypothesis.

There are practical consequences of the large deposition velocities for these aerosols onto forests and woodland, the mitigation of pollutant concentrations in urban areas by planning trees for example. In the case of the West Midlands, the area of trees extends to just 4% of the 900 km^2, and the potential for greatly expanding the area of woodland exists within the conurbation. Such an expansion of woodland would capture more aerosol than the existing land cover and feedback to smaller ambient aerosol concentrations. The detailed modelling of these effects lies beyond the scope of this paper, and is considered in a related modelling paper based on these measurements (Fowler *et al.*, in preparation). The consistent effect of mixed, deciduous woodland on particle deposition provides the basis for quantitative assessment of the benefits of urban tree planting or ambient PM$_{2.5}$ concentrations.

One of the important consequences of increased particle capture by woodland and accumulation of heavy metals within the organic horizons of woodland soil, especially for Pb, Cd, and Zn, is that with time, these soils become contaminated by significant metal concentrations. The data presented in Table IV show Pb concentrations up to 400 mg kg^{-1} and the mean concentration of Pb in the surface 10 cm of woodland soils at these sites is 253 mg kg^{-1}, slightly larger than the recommended upper limit for soil Pb concentrations. The relatively small proportion of the land area of the West Midlands also contains a substantial quantity of Pb in the surface horizon. Taking the measured mean Pb concentration and bulk density of the top 10 cm, and assuming that the sites used are representative of West Midlands woodlands Pb concentrations, the total Pb content of the top 10 cm of woodland soils amounts to 6750 Mg Pb. This value exceeds current UK Pb emissions by more than an order of magnitude and is close to the peak in UK Pb emissions in 1980 of 8212 Mg. Clearly, the large emission and ambient Pb concentrations of earlier decades have been accumulated by these woodland soils since the industrial revolution. This simple extrapolation may be extended to UK urban woodland assuming the woodlands fraction measured in the West Midlands is representative of the UK urban area that covers approximately 10% of the country. The quantity of Pb in urban woodland in the UK may therefore by of the order 100 Gg Pb. While speculative in the wider extrapolation, the exercise shows that Pb accumulation in the surface horizons of urban soils throughout the UK, and especially in the woodland, is a substantial quantity. To refine the calculation, a detailed survey of the region, taking transects away from the major source areas would be necessary. It would also be necessary to subtract the natural Pb present in these samples, which from the data in the current study was a small fraction of the inventory in the urban areas (<5%), but would be substantially larger in rural areas away from roads. The practical implications are that with urban development placing demands on these woodlands for other uses, the potential mobilization of Pb and other heavy metals may represent a threat to human health, among other targets. The arguments have been applied to Pb, but Table IV shows that for Cd, Cu, Ni and Zn the same logic applies. Uncertainties in the simple extrapolation from measurements presented here to the West Midlands or UK urban areas are very large, and difficult to quantify. However, it is clear that urban woodlands remove a large quantity of metals from polluted urban atmosphere, and thus a substantial fraction of the intercepted heavy metals accumulate within urban woodland in organic surface horizons of the soil and may persist for a very long time. The application of urban woodland to remove aerosols and contribute to a reduced population exposure to $PM_{2.5}$ or PM_{10} must therefore be considered together with the long-term fate of the deposited heavy metals, and other persistent pollutants that accumulate in these soils.

Acknowledgment

The authors gratefully acknowledge funding for the work from the NERC URGENT-2 research programme under grant no GST/02/2236.

References

Beckett, K. P., Free-Smith, P. H. and Taylor, G.: 1998, 'Urban woodlands: Their role in reducing the effects of particulate pollution', *Environ. Pollut.* **99**, 347–360.

Beswick, K. M., Hargreaves, K., Gallagher, M. W., Choularton, T. W. and Fowler, D.: 1991, 'Size resolved measurements of cloud droplet deposition velocity to a forest using an eddy correlation technique', *Q. J. R. Meteorol. Soc.* **117**, 623–645.

Bradshaw, A., Hunt, B. and Walmsley, T.: 1995, *Trees in the Urban Landscape: Principles and Practice*, E. and F. N. Spon, London.

Chamberlain, A. C.: 1966, 'Transport of gases to and from grass and grass-like surfaces', *Proc. R. Soc. Lond.* A, 236–265.

Chamberlain, A. C.: 1991, *Radioactive Aerosols*, Cambridge University Press, Cambridge.

Choubedar, F.: 2000, 'The use of radio-nuclides (unsupported ^{210}Pb, ^{7}Be and ^{137}Cs) in air, rain and undisturbed soil as an environmental tool', *Ph.D. Thesis*, University of Edinburgh.

CLAG Fluxes: 1997, 'Deposition fluxes in the United Kingdom: A compilation of the current deposition maps and mapping methods (1992–1994) used for critical loads exceedance assessment in the United Kingdom', Critical Loads Advisory Group, sub-group report on Deposition Fluxes, Institute of Terrestrial Ecology, Penicuik, 45 pp.

Cutshall, N. H., Larsen, I. L. and Olsen, C. R.: 1983, 'Direct analysis of ^{210}Pb in sediments samples: Aelf absorption corrections', *J. Nucl. Instr. Meth.* **206**, 309–312.

Fowler, D.: 1984, 'Transfer to terrestrial surfaces', *Philos. Trans. R. Soc.B* **305**, 281–297.

Fowler, D., Smith, R. I., Crossley, A., Leith, I. D., Mourne, R. W., Brandford, D. D. and Moghaddam, M.: 1998, 'Quantifying fine scale variability in deposition of pollutants in complex terrain using ^{210}Pb inventories in soil', *Water, Air, Soil Pollut.* **105**, 459–470.

Gallagher, M. W., Nemitz, E., Dorsey, J. R., Fowler, D., Sutton, M. A., Flynn, M. and Duyzer, J. H.: 2002, 'Measurements and parameterizations of small aerosol deposition velocities to grassland, arable crops, and forest: influence of surface roughness length on deposition', *J. Geophys. Res.* **107**, D1210.1029/2001 JD000817.

Gallagher, M., Fontan, J., Wyers, P., Ruijgrok, W., Duyzer, J., Hummelsh, P. and Fowler, D.: 1997, 'Atmospheric particles and their interactions with natural surfaces', in S. Slanina (ed.), *Biosphere-Atmosphere Exchange of Pollutants and Trace Substances*, Springer-Verlag, pp. 45–92.

Garland, J. A.: 2001, 'On the size dependence of particle deposition,' *Water, Air, Soil Pollut.: Focus*, **1**, 39–48.

Garland, J. A.: 1983, 'Dry deposition of small particles to grass in field conditions', in H. R. Pruppacher, R. G. Semonin and W. G. N. Slinn (eds), *Precipitation Scavenging, Dry Deposition and Resuspension*, Elsevier, New York, pp. 849–857.

Graunstein, W. C. and Turekian, K. K.: 1983, '^{210}Pb as a tracer of the deposition of submicrometer aerosols', in H. R. Pruppacher, R. G. Semonin and W. G. N. Slinn (eds), *Precipitation Scavenging, Dry Deposition and Resuspension*, Elsevier, New York, pp. 1315–1324.

Graunstein, W. C. and Turekian, K. K.: 1989, 'The effects of forests and topography on the deposition of sub-micrometer aerosols measured by ^{210}Pb and ^{137}Cs in soils', *Agric. For. Meteor.* **47**, 199–220.

Holdgate, S. T.: 1998, 'Quantification of the effects of air pollution on health in the United Kingdom', Department of Health Committee on the Medical Effects of Air Pollutants, The Stationary Office, London.

Holdgate, S. T.: 2001, 'Statement and report on long-term effects of particles on mortality', Department of Health Committee on the Medical Effects of Air Pollutants, The Stationary Office, London.

Knoll, G. F.: 1989, *Radiation Detection and Measurements*, Second Edition, John Wiley, New York.

Lewis, D. M.: 1977, 'The use of ^{210}Pb as a heavy metal tracer in the Susquehanna river system', *Geochim. Cosmochim. Acta* **41**, 1557.

Nicholson, K. W.: 1988, 'The dry deposition of small particles: A review of experimental measurements', *Atmos. Environ.* **22**, 2653–2666.

Pagel, A. A.: 1983, 'Urban relief: A million trees for Los Angeles', *Am. Forests* **89**, 22–58.

Paterson, E., Towers, W. and Langan, S. J.: 2000, 'The use of soil data to predict environmental sensitivity to pollution', in M. J. Wilson and B. Maliszewski-Kordysbach (eds), *Soil Quality, Sustainable Agriculture and Environmental Security in Central and Eastern Europe*, Kluwer Academic Press, Netherlands, pp. 189–206.

Pope, C. A., III, Bates, D. V. and Raizenne, M. E.: 1995, 'Health effects of particulate air pollution: time for resassessment?', *Environ. Health Perspect.* **103**, 472–480.

Preiss, N., Mélières, M-A. and Pourchet, M.: 1996, 'A compilation of data on lead 210 concentration in surface air and fluxes at the air-surface and water-sediment interfaces', *J. Geophys. Res.* **101**, 28,847–28,862.

Slinn, W. G. N.: 1982, 'Prediction for particle deposition to vegetative canopies', *Atmos. Environ.* **16**, 1785–1794.

Wyers, G. P., Vermeulen, A. T., Geysebroek, M., Wayers and A., Mols, J. J.: 1994, 'Deposition of aerosol to deciduous forest', ECN Report December 1994, C-94-051, Netherlands Energy Research Foundation.

THE TRANSIT OF $^{35}SO_4^{2-}$ AND ^3H_2O ADDED *IN SITU* TO SOIL IN A BOREAL CONIFEROUS FOREST

DANIEL HOULE[1,2*], RICHARD CARIGNAN[3] and JEAN ROBERGE[1]

[1] *Ministère des Ressources Naturelles du Québec, Direction de la recherche forestière, 2700 Einstein St., Sainte-Foy, Québec, Canada G1P 3W8;* [2] *St. Lawrence Centre, Environment Canada, 105 McGill Street, Montreal, Québec, Canada H2Y 2E7;* [3] *Université de Montréal, Dept. de sciences biologiques, 90 Vincent-d'Indy Ave., P.O. Box 6128, Stn. A, Montreal, Québec, Canada H3C 3J7*
(* *author for correspondence, address: St. Lawrence Centre, Environment Canada, 105 McGill Street, Montreal, Québec, Canada H2Y 2E7, e-mail: daniel.houle@mrn.gouv.qc.ca; phone: 418-643-7994, ext: 323; fax: 418-643-2165*)

(Received 20 August 2002; accepted 11 April 2003)

Abstract. A solution containing $^{35}SO_4^{2-}$ and ^3H_2O was applied to four plots (5 × 5 m) in a boreal coniferous forest in the Laflamme Lake watershed, Québec, under two contrasting conditions: in summer (plots 1 and 2), and on the snowpack before snowmelt (plots 3 and 4). The transit of both these tracers in the soil solution was then followed through a network of soil lysimeters located at different depths. Four months after the summer application, ^3H_2O had infiltrated the whole soil profile at plot 1, while $^{35}SO_4^{2-}$ was only observed in the LFH and Bhf horizons. A $^{35}SO_4^{2-}$ budget calculated from mid-August to November indicated that 89 and 10.6% of the added $^{35}SO_4^{2-}$ was retained within the LFH and the Bhf layers, respectively. Fifteen months later, the added $^{35}SO_4^{2-}$ was distributed in the following proportions within the soil horizons: LFH (73.7%), Bhf (11.8%) and Bf (12.8%), for a total retention rate of 98.3%. The superficial penetration of ^3H_2O at plot 2 was indicative of a major lateral water movement that prevented the calculation of a $^{35}SO_4^{2-}$ budget. This situation also was observed at plot 4 during snowmelt. At plot 3, ^3H_2O moved freely through the soil profile and a significant fraction of the added $^{35}SO_4^{2-}$ reached the B horizons, where it was presumably adsorbed on aluminum (Al) and ferric (Fe) oxides. The $^{35}SO_4^{2-}$ budget for plot 3 from March to November indicated that 87% of the added $^{35}SO_4^{2-}$ was retained within the soil profile, with most being retained in the B horizons (LFH = 33.1%, Bhf = 33.1%, Bf = 20.8%). The contrasting retention patterns of $^{35}SO_4^{2-}$ within the soil profile following the summer addition and snowmelt likely was caused by the contrasting soil temperatures and soil solution residence times within the different soil layers. The persistence of $^{35}SO_4^{2-}$ in the soil solution of the entire profile long after the initial tracer infiltration, and the relative temporal stability of specific activity of SO_4^{2-}, point to the establishment of an isotopic equilibrium between the added $^{35}SO_4$ and the active S-containing reservoirs within a given soil horizon. Overall, the results clearly illustrate the very strong potential for $^{35}SO_4^{2-}$ retention and recycling in forest soils.

Keywords: coniferous boreal forest, microbial immobilization, organic S, sulfur cycling, sulphur retention, tritiated water

1. Introduction

The acid rain phenomenon caused by elevated atmospheric sulfur (S) deposition has triggered many studies on S cycling in forest ecosystems in the last few decades. These studies are important because SO_4^{2-} is the major anion in the soil solution of temperate forest ecosystems and because it contributes heavily to the soil leaching of aluminum and nutrient cations.

Sulfate adsorption and desorption on aluminum and ferric oxides (Chao et al., 1962; Fuller et al., 1985; Houle and Carignan, 1995) and SO_4^{2-} incorporation into (or SO_4^{2-} mineralization from) the organic matrix (Fitzgerald et al., 1983; Swank et al., 1984; Schindler et al., 1986; Watwood and Fitzgerald, 1988; Houle et al., 2001) have been identified as the main reactions affecting S fluxes in forest soils.

Although there is general recognition of the factors governing S fluxes in forest soils, most of the conclusions have been drawn from laboratory incubations in closed systems. Quantifying the real potential for S retention in natural conditions is a difficult task that involves a necessary quantification of the water's movement through the soil.

In order to determine SO_4^{2-} and water fluxes through the soil in unperturbed conditions, $^{35}SO_4^{2-}$ and tritiated water were applied *in situ* to forested plots (5 × 5 m) in summer ($n = 2$) and to the snowpack ($n = 2$) before snowmelt, at the Laflamme Lake watershed. The transit of both tracers in the soil solution was followed through a network of lysimeters located at different depths to calculate $^{35}SO_4^{2-}$ budgets at the horizon level.

Previous studies on S cycling at this site have shown that SO_4^{2-} fluxes, on the short-term, were controlled by adsorption and desorption reactions within the Bf horizon, but that over the long term, mineralisation of the soil organic S reservoir (1230 kg ha^{-1}) appeared to be responsible for the net release of SO_4^{2-} from the watershed (Houle and Carignan, 1995). The organic soil S reservoir is comprised largely (up to 99%, Houle et al., 2001) of extracellular organic matter that can be decomposed under the action of heterotrophic organisms. Because of the strong potential for SO_4^{2-} adsorption-desorption (Houle and Carignan, 1995) and SO_4^{2-} recycling through organic matter (Houle et al., 2001), it was hypothesized that the major part of the SO_4^{2-} leaving the soil profiles following the *in situ* $^{35}SO_4^{2-}$ addition would be cold (non-radioactive) SO_4^{2-} and that the newly added $^{35}SO_4^{2-}$ would be strongly retained within the soil profile.

2. Methods

2.1. SITE DESCRIPTION

The study was conducted at the Laflamme Lake watershed (68.4 ha, 47°17′ N, 71°14′ W) located 80 km north of Québec City, Québec, Canada (Houle and Carignan, 1992, 1995). The soil is an orthic humo-ferric podzol according to the Canadian Soil Classification System. Based on 29 pits, the average thicknesses for LFH, Ae, Bhf and Bf horizons are 8.8, 4.5, 7.6 and 28.7 cm, respectively (Houle and Carignan, 1992). The forest is composed of 90% balsam fir (*Abies balsamea*) and 10% white birch (*Betula papyrifera*). This region receives mean annual precipitation of 1300 mm (33% as snow). The mean annual temperature is –0.6 °C. The trees were between 50 and 60 yr old and the intermediate vegetation stratum was nearly absent. The forest floor was carpeted by a dense cover of *Oxalis montana* Raf. The four selected plots were representative of the average watershed conditions in terms of slope (average = 9%), soils, and vegetation and they were located not further than 400 m from each other.

2.2. THE FIELD TRACING EXPERIMENT

Tritiated water and $^{35}SO_4^{2-}$ were applied in summer (August 15 and 17, 1988, respectively, on plots 1 and 2) and before snowmelt, directly on the snowcover, on February 14, 1989, on plots 3 and 4. Each plot (5 × 5-m) was divided into 1600-grid elements of 12.5 × 12.5 cm. Two mL of the tracer solution (0.13 mm rain equivalent) were sprayed on each element with a precision glass syringe attached to an atomizer nozzle. Where a tree stem overlapped an element, the solution was sprayed on the edge of the tree, near the soil. Methanol was added to the solution (at a final concentration of 32%) to prevent the syringe from freezing up during the winter application. At this time, the snowpack had accumulated 170 mm of water equivalent and, before springmelt, a further 87 mm (water equivalent) of snow was added.

The ^3H_2O and $^{35}SO_4^{2-}$ loads of the plots corresponded to 4,630 dpm cm^{-2} and 52,991 dpm cm^{-2} and to 54,131 dpm cm^{-2} and 58,838 dpm cm^{-2} for the summer and winter applications, respectively. The increased load of ^3H_2O in winter was due to its expected dilution within the snowpack.

2.3. SOIL SOLUTION SAMPLING AND ANALYSIS

Soil-water was collected at intervals of two to seven days at each plot with three series of tension lysimeters located in the LFH, Bhf, upper Bf (Bf1), and lower Bf (Bf2) horizons, at average depths (for the four plots) of 8, 22, 33 and 48 cm, respectively. Within a given plot (5 × 5 m), the three lysimeter series were located approximatively 2.5 to 3 m from each other. The lysimeters' design and installation have been described elsewhere (Houle and Carignan, 1995). The concentrations of

^3H_2O and $^{35}SO_4^{2-}$ in the samples were determined on 2.5-mL volumes by liquid scintillation using the external standard method to correct for the quenching of emitted photons. All the $^{35}SO_4^{2-}$ (half life = 87.4 d and ^3H_2O (half life = 12.4 yr) concentrations were corrected for radioactive decay taking the dates of tracer application as the zero time. The punctual soil solution concentration values were averaged on a monthly basis. Rodent damage and the presence of frost in some of the tubing were responsible for a few incomplete sample sets, particularly during the 1989 snowmelt. It should be noted that a part of the ^{35}S activity in the soil solution may originate from dissolved organic ^{35}S that would have been formed during the course of the experiment. The proportion of dissolved organic S to total soluble S was 20.1, 14.1 and 8.6% in the soil solution of the LFH, Bhf and Bf horizons, respectively (Houle et al., 2001), suggesting a minor contribution of dissolved organic ^{35}S to the total ^{35}S activity in solution. This is particularly true if we consider that DOS is produced by litter decomposition, and therefore its turnover rate is much lower than SO_4^{2-} and would be expected to have a lower specific activity in the months following the addition of tracers.

2.4. Water and $^{35}SO_4^{2-}$ budget calculation

The water budgets of plots 1 and 2 were monitored intensively during the study. Daily precipitation data were provided by the Forêt Montmorency meteorological station, located less than 500 m from the plots. Throughfall was measured on each plot with 20 123 cm^2 collectors distributed at 1 m intervals along the plot edges at 1.5 m above the ground. The soil volumetric liquid water content was measured with the time-domain reflectometry (TDR) technique (Topp et al., 1980). Water infiltration was calculated from precipitation data and the snowpack water equivalent changes. Standard snow survey methods were used to estimate weekly snowpack water content with 20 cores taken with a Standard Federal snow sampler (Gray and Male, 1981). One-meter deep observation wells at each corner of the plots allowed the monitoring of groundwater levels for plots 1 and 2.

The water budget was calculated with the following basic equation:

$$Q_{in} - \delta S - ET - Q_{out} = 0$$

where Q_{in} and Q_{out} are, respectively, the water inflow and the water outflow for a given soil horizon, δS is the difference in water storage, and ET is the loss due to evapotranspiration. Finally, ET was calculated with the Penman method (1948) adapted to eastern Canada conditions by Mateer (1955) using meteorological data from the Forêt Montmorency meteorological station.

The $^{35}SO_4^{2-}$ budget of each soil layer was calculated monthly by multiplying the amount of water exiting a given layer by the mean monthly $^{35}SO_4^{2-}$ concentration. However, because the Bhf samplers were located within the horizon, their $^{35}SO_4^{2-}$ concentrations would not necessarily be representative of the soil solution exiting the horizon. For this reason, the concentrations of the upper Bf sampler (located

TABLE I

$^{35}SO_4$ budgets for the two plots (1 and 3) where ^3H_2O percolated down the soil profile with no significant lateral transport

	Retention within a given horizon (%)	Cumulative retention (%)
Summer application		
Plot 1 (August to November 1988)		
LFH	89.0	89.0
Bhf	10.6	99.6
Bf	0.3	99.9
Plot 1 (August 1988 to October 1989)		
LFH	73.7	73.7
Bhf	11.8	85.5
Bf	12.8	98.3
Snowpack application		
Plot 3 (March to October 1989)		
LFH	33.1	33.1
Bhf	33.1	66.2
Bf	20.8	87.0

just below the Bhf horizon) were combined with the water fluxes leaving the Bhf horizon and taken as the exportation below the Bhf horizon.

3. Results

3.1. SUMMER EXPERIMENT

At plot 1, from the moment of tracer addition to the beginning of the next winter, ^3H_2O (Figure 1b) had infiltrated the whole profile, reaching the lower Bf (Bf2) lysimeter level. Due to the dispersive nature of fluxes in porous media (Nielsen and Biggar, 1962), the ^3H_2O added in a discrete event spread longitudinally while percolating downward, arriving at a given soil depth in a wave form, which tends to flatten out as depth increases (Figures 1b and 2b). High $^{35}SO_4^{2-}$ concentrations were observed in the LFH and the Bhf layers, while only small amounts had reached the Bf horizon (Figures 1a and 2a). The $^{35}SO_4^{2-}$ budget calculated for plot 1 (Table I), from mid-August to November, indicated that 99.9% of the $^{35}SO_4^{2-}$ was sequestered in the soil profile, with the major proportion being retained in the LFH layer (89%) and the remainder in the Bhf (10.6%). At the following year's

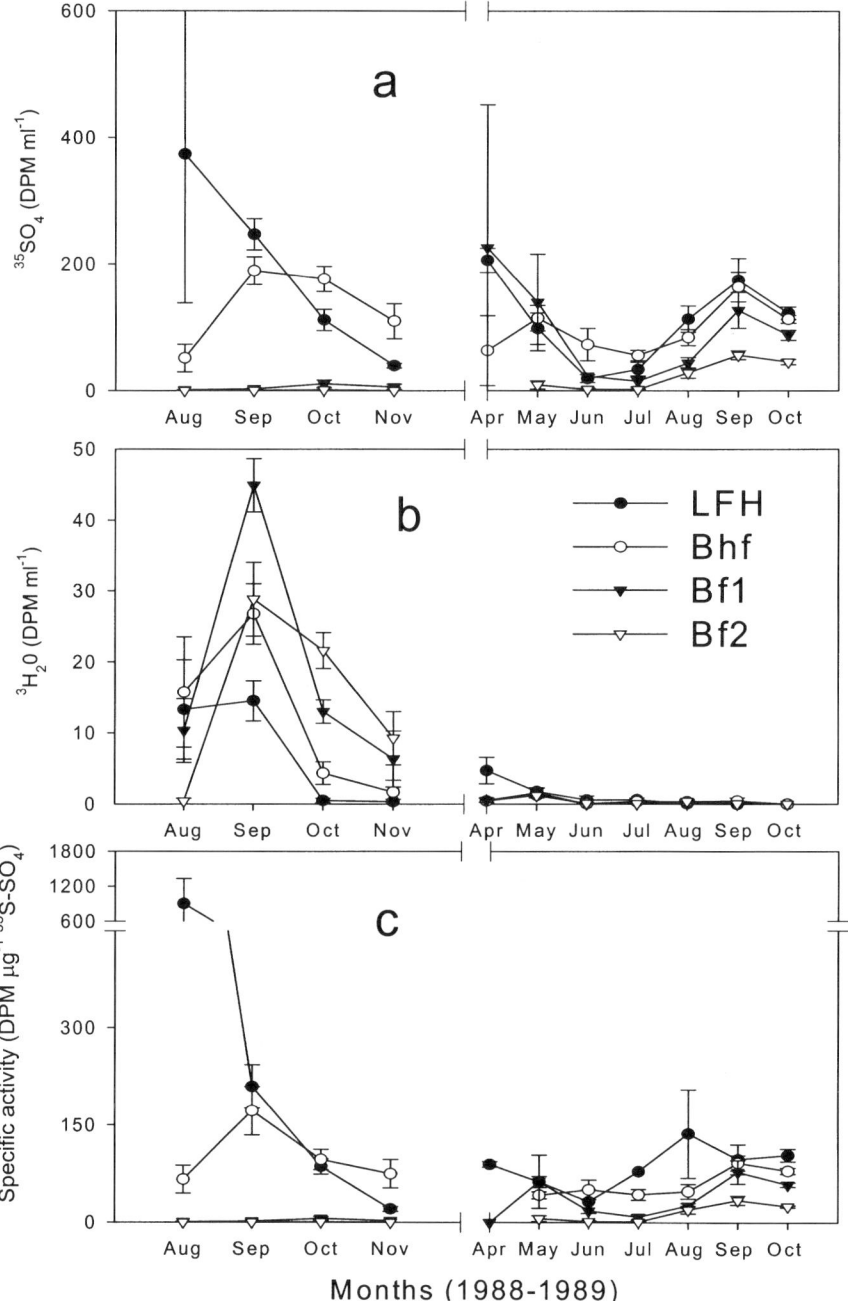

Figure 1. Monthly mean (± standard error, SE) concentrations of $^{35}SO_4^{2-}$ (a), $^{3}H_2O$ (b), and SO_4^{2-} specific activity in the soil solution of LFH, Bhf, upper Bf (Bf1) and lower Bf (Bf2) horizons collected at plot 1 following summer tracer application. All the data are corrected for decay relative to the date of tracer addition.

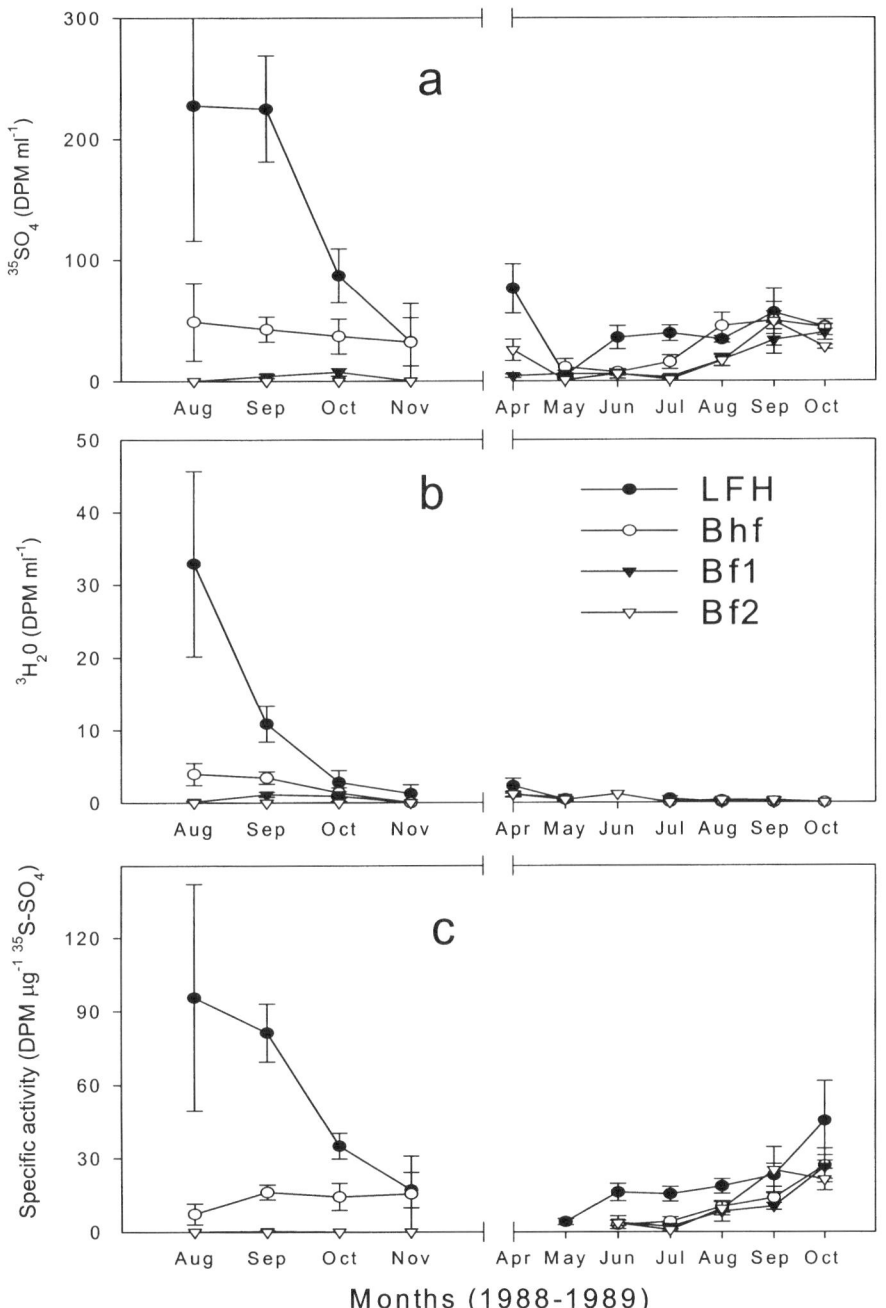

Figure 2. Monthly mean (± SE) concentrations of $^{35}SO_4^{2-}$ (a), ^3H_2O (b) and SO_4^{2-} specific activity in the soil solution of LFH, Bhf, upper Bf (Bf1) and lower Bf (Bf2) horizons collected at plot 2 following summer tracer application.

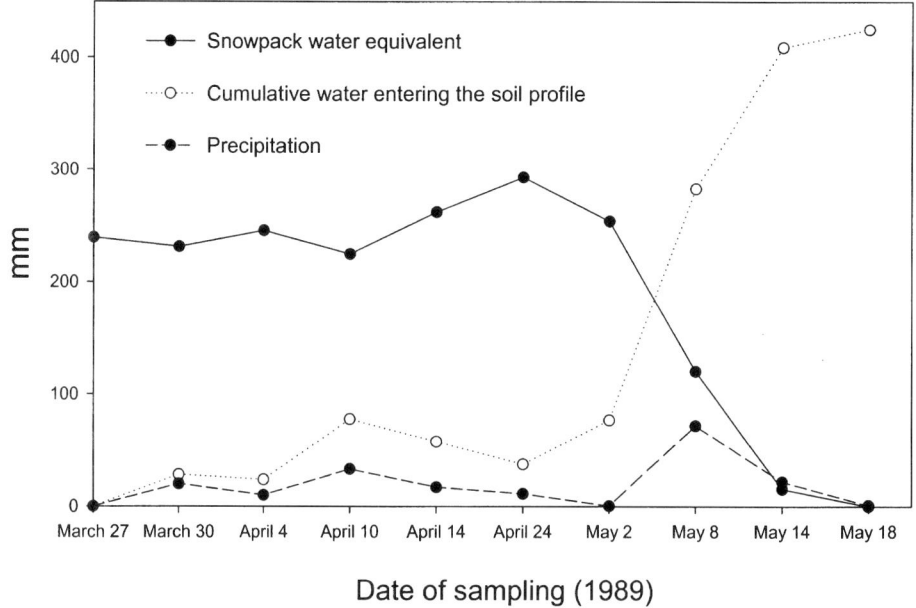

Figure 3. Snowpack water equivalent, precipitation (as rain or snow), and cumulative water amounts that entered the soil profile between March 28 and May 18, 1989.

snowmelt period, $^{35}SO_4^{2-}$ was still present in the LFH and Bhf layers and significant $^{35}SO_4^{2-}$ concentrations were seen in the upper and lower Bf horizon, persisting until October of 1989. Another $^{35}SO_4^{2-}$ budget calculated for the entire 15-month period indicated that the added $^{35}SO_4^{2-}$ was distributed in the following proportions within the soil profile: LFH (73.7%), Bhf (11.8%) and Bf (12.8%), for a total retention rate of 98.3%.

At plot 2, the ^3H_2O occurred at low concentrations in LFH and Bhf samples, emerged weakly in Bf1, and remained absent from all Bf2 samples. From the end of August to mid-October, a major rise in groundwater levels, sometimes reaching the Ae horizon, affected lateral saturated flow in this profile. The ^3H_2O reaching the saturated zone had flowed laterally and most of it would have vanished from the plot in 7 to 14 days. An unknown proportion of $^{35}SO_4^{2-}$ likely was re-routed laterally with ^3H_2O and, consequently, the calculation of a $^{35}SO_4^{2-}$ budget is impossible for plot 2. Without the use of ^3H_2O, it would have been impossible to determine the water's flow pathway and we may have erroneously concluded that the added $^{35}SO_4^{2-}$ was retained within the plot boundaries. Despite the considerable lateral water movement, relatively high $^{35}SO_4^{2-}$ concentrations showed up in the soil profile during the subsequent snowmelt period and eventually reached the lower Bf sampler (Bf2) in August 1989.

3.2. SNOWMELT EXPERIMENT

Contrarily to summer, the input of tracers to the soil were not seen as discrete events during snowmelt. Tracer delivery through the snowcover made the process a continuous, though irregular, one. The melt started slowly and intermittently at the end of March, with an initial water infiltration of 28 mm (Figure 3), and small ^3H_2O concentrations were then observed in the LFH layer at plots 3 and 4 (Figures 4b and 5b). After this event, the melt was weak and intermittent and became rapid and constant at the end of April, which provoked an increase in ^3H_2O concentrations in the soil solution of the two plots. Most of the snow, however, melted in the first two weeks of May, as illustrated by the disappearance of the snowpack (Figure 3). This translated into a massive output of ^3H_2O to the soil profile, resulting in peak concentrations in all the horizons at plot 3 (Figure 4b). The added ^3H_2O at plot 3 likely transited vertically down the soil profile, which was not the case for plot 4, where high ^3H_2O concentrations were observed only in the LFH layer, showing that, similarly to plot 2 in summer, water was rerouted laterally out of the plot boundaries, preventing the calculation of a $^{35}SO_4^{2-}$ budget for plot 4. Between March 27 and May 18, 420 mm of water had infiltrated the soil profile (Figure 3).

During snowmelt, as opposed to the summer experiment, significant $^{35}SO_4^{2-}$ concentrations were observed in the soil solution within the Bf horizon at plot 3, indicating that the $^{35}SO_4^{2-}$ had penetrated deeply into the soil profile along with the tritiated water (Figure 4a). This effect was less apparent at plot 4 (Figure 5a) owing to the lateral water movement that likely contributed to lateral exportations of $^{35}SO_4^{2-}$. After the snowmelt period, between June and October, relatively high $^{35}SO_4^{2-}$ concentrations persisted within the soil solution of all horizons. In addition to the infiltration peak due to snowmelt, a secondary $^{35}SO_4^{2-}$ peak generally was observed between August and October in nearly all the lower Bf samplers. The $^{35}SO_4^{2-}$ budget (Table I) for the whole period indicated that 87% of the added $^{35}SO_4^{2-}$ was retained within the soil profile, with the major part being retained in the B horizons (LFH = 33.1%, Bhf = 33.1%, Bf = 20.8%).

3.3. SPECIFIC ACTIVITY

Following the summer application, the specific activity of SO_4^{2-} in the LFH layer showed a clear decreasing trend from August to November 1988, but was relatively constant in the Bhf horizon (Figures 1c and 2c). Because almost no $^{35}SO_4^{2-}$ reached the Bf horizon, the deeper samplers had a specific activity of close to zero. From April to October 1989, specific activity tended to increase, particularly in the soil solution of the B horizons. During snowmelt, there were major fluctuations in specific activity, particularly in the LFH and Bhf layers at plots 3 and 4 (Figures 4c and 5c). From June to October, specific activity remained fairly stable in nearly all the soil horizons of both plots, except for the higher values observed in June in the Bf2 sampler of plot 4 and the significant fluctuations in the Bf1 and Bf2 samplers of plot 3 in September.

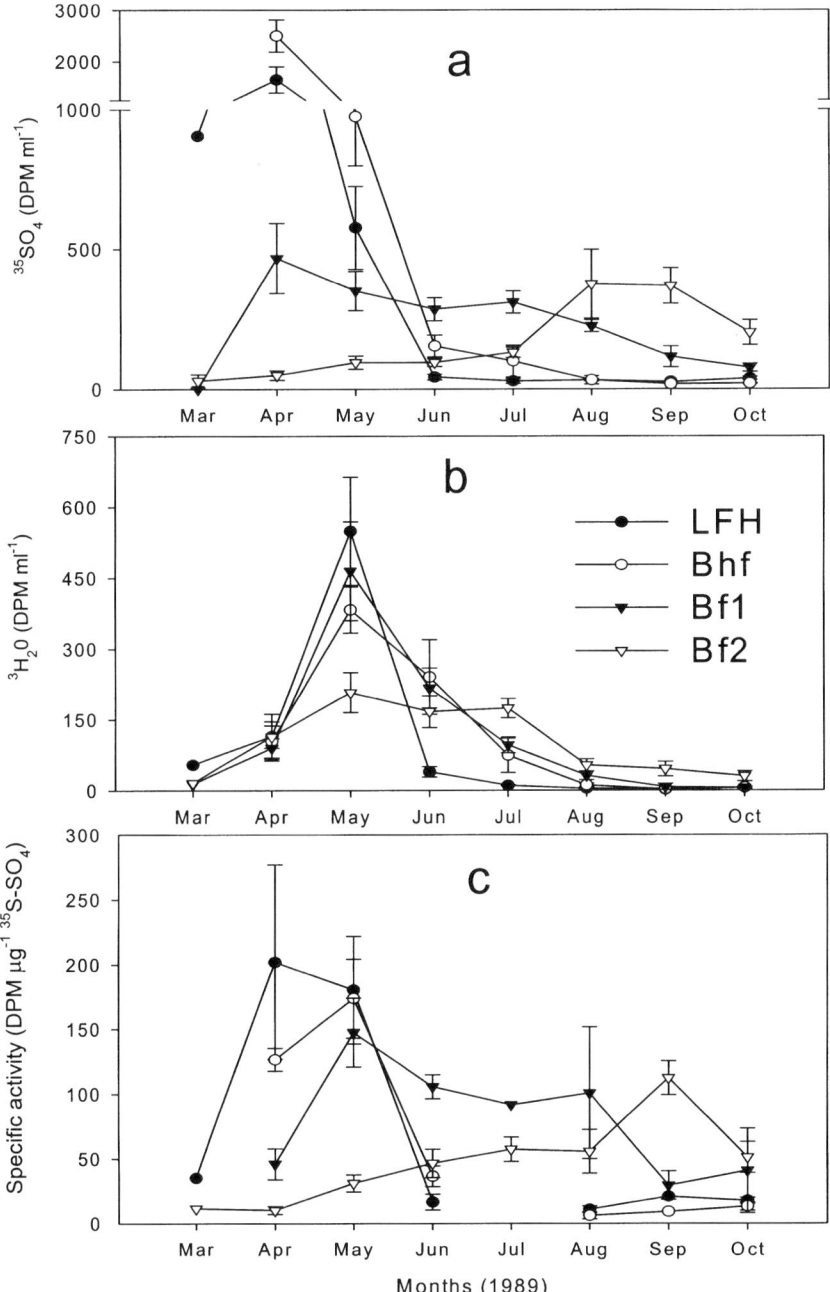

Figure 4. Monthly mean (± SE) concentrations of $^{35}SO_4^{2-}$ (a), ^3H_2O (b), and SO_4^{2-} specific activity in the soil solution of LFH, Bhf, upper Bf (Bf1) and lower Bf (Bf2) horizons collected at plot 3 following winter tracer application.

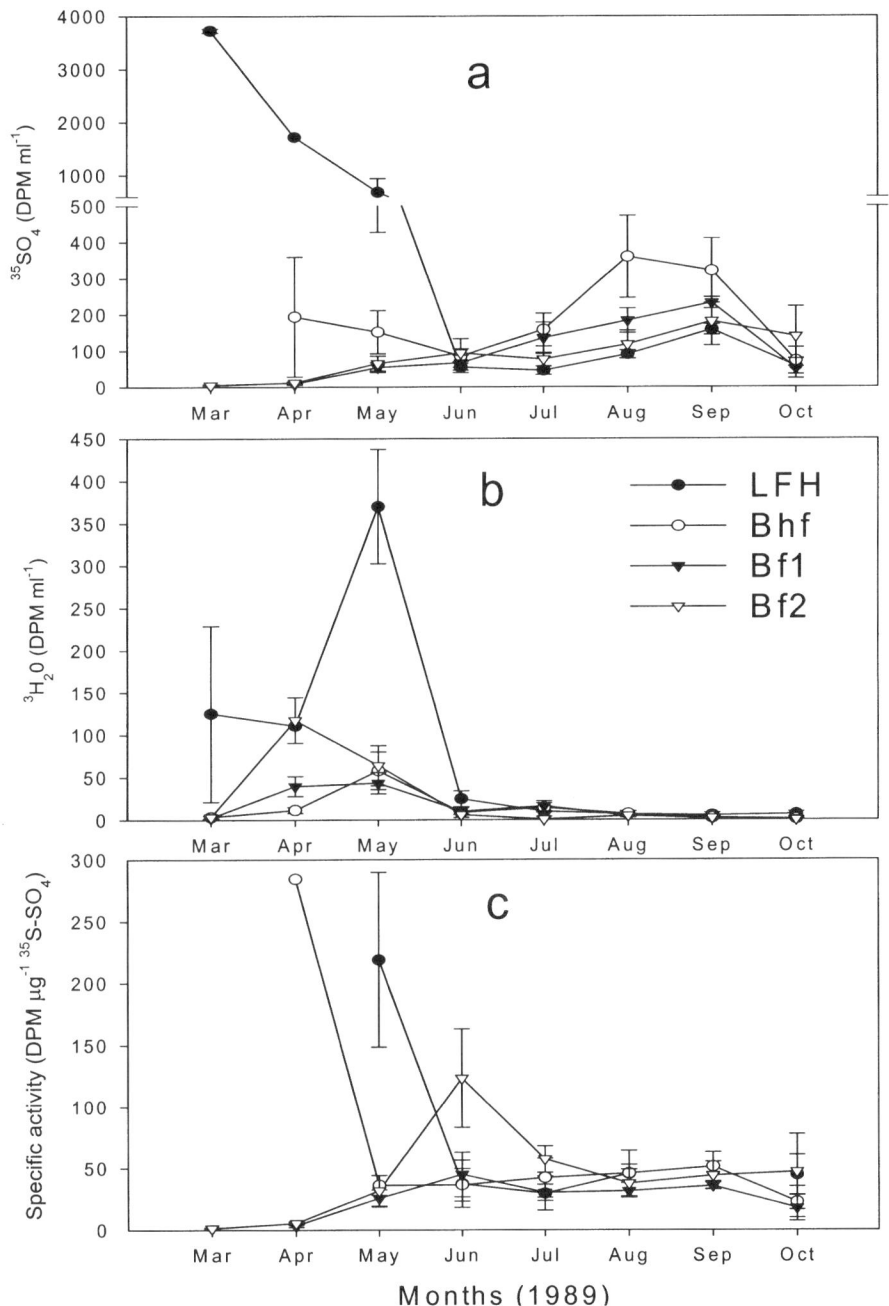

Figure 5. Monthly mean (± SE) concentrations of $^{35}SO_4^{2-}$ (a), ^3H_2O (b), and SO_4^{2-} specific activity in the soil solution of LFH, Bhf, upper Bf (Bf1) and lower Bf (Bf2) horizons collected at plot 4 following winter tracer application.

4. Discussion

During the summer experiment, the comparison of ^3H_2O and $^{35}SO_4^{2-}$ behaviour within the soil solution at plot 1 reveals two distinct patterns: while ^3H_2O moved freely within the soil profile and reached the deeper lysimeters, high $^{35}SO_4^{2-}$ concentrations were observed only in the LFH and Bhf layers. Although small amounts of $^{35}SO_4^{2-}$ had reached the Bf horizon, the observed concentrations were not significant compared to the amounts initially added. These observations clearly demonstrate a strong $^{35}SO_4^{2-}$ accumulation in the top soil horizons. Compared with the summer experiment, the springmelt experiment strongly differs in that relatively high $^{35}SO_4^{2-}$ amounts reached the upper and lower Bf horizon at plot 3 and, to a lesser extent, at plot 4. These differences translated into variant accumulation patterns, the greatest proportion of the added $^{35}SO_4^{2-}$ being found in the LFH layer following the summer application, while the B horizons (Bhf and Bf) were the major locus of accumulation after snowmelt (Table I).

The differing patterns of $^{35}SO_4^{2-}$ migration and retention within the soil profiles at plots 1 and 3 likely were determined by the strongly contrasting hydrologic and thermodynamic conditions that prevailed after application of the tracers. During the summer, the tracers were added on August 15 and 17, days on which rain was expected. However, only 1.6 mm of rain fell on those days and the next heavy rainfall occurred on August 25. Consequently, the added $^{35}SO_4$ stayed in contact with the top of the profile for a 10-day period, during which the LFH temperature averaged 13.2 °C. Previous laboratory experiments on $^{35}SO_4^{2-}$ incorporation have shown that the LFH layer at this site can incorporate over 30 and 50% of the added $^{35}SO_4^{2-}$ within the organic matrix in 48 hr at respective temperatures of 10 and 20 °C. During a 12 day incubation period, the incorporated amounts tend to reach an asymptote around 70 and 80% at the same respective temperatures (Houle et al., 2001). At 13 °C, the amount incorporated would be expected to lie between these values, which is consistent with the 89% observed retention rate for the LFH layer. An unknown part of the added $^{35}SO_4^{2-}$ may also have been incorporated within the aboveground biomass (herbaceous vegetation and balsam fir) via the root system, although incorporation within balsam fir would be limited given that its growth season is mostly finished by mid-August. The $^{35}SO_4^{2-}$ that exited the LFH layer was retained within the Ae and Bhf horizons: both horizons may immobilize 20–30% of added $^{35}SO_4^{2-}$ in the organic fraction within 48 hr at 10 °C, while the Bhf horizon may also adsorb 60–70% of the $^{35}SO_4^{2-}$ under the same conditions (Houle et al., 2001). Other studies also have demonstrated the capacity of the humus layer and of mineral horizons to incorporate significant quantities of $^{35}SO_4^{2-}$ in various soil types (Fitzgerald et al., 1983; Swank et al., 1984; Schindler et al., 1986; Watwood and Fitzgerald, 1988).

In contrast to summer tracing, the input of tracers to the soil during snowmelt was a continuous, though, irregular process. Shortly after the winter application, the tracers formed a ^3H_2O and $^{35}SO_4^{2-}$ layer at the snowpack surface that was later covered over by further snowfall events. The snow's metamorphism, accelerated by the progressive warming of the snowcover in March, would have redistributed the ^3H_2O within the snowpack (Colbeck, 1978). Temperature measurements at the snow-soil interface indicated that by March 25, isothermal conditions had reached the base of the snowpack (Roberge, unpublished data). The first ^3H_2O wave then entered the soil and generated the low concentrations observed in March. Strongly contrasting with the low ^3H_2O delivery from the snowpack, very high $^{35}SO_4^{2-}$ concentrations were observed in March in the LFH layers of plots 3 and 4. As a result, the ratio of $^{35}SO_4^{2-}$ to ^3H_2O was between 15 and 30 in the solution collected in March as compared to a ratio of 1:1 in the added solution. This illustrates the differing behaviour of ^3H_2O and SO_4^{2-} in the snowpack during snow metamorphism, the latter being highly concentrated in the first meltwater because of the ion exclusion mechanism (Davies et al., 1987). After this initial water infiltration, the slow melt became rapid and constant between the end of April and mid-May, which induced a massive delivery of tracers to the soil profile.

Under these conditions, the short water residence time combined with temperatures around zero throughout nearly the entire soil profile certainly limited $^{35}SO_4^{2-}$ incorporation within the LFH layer. In fact, although the LFH layer may immobilize $^{35}SO_4^{2-}$ even at 1 °C in the laboratory (Houle et al., 2001), the amount accumulated after 12 days reached an asymptote to 20% of the added $^{35}SO_4^{2-}$ (Houle et al., 2001). In addition, tree dormancy and the absence of the dense *Oxalis montana Raf.* soil cover prohibited $^{35}SO_4^{2-}$ incorporation within the aboveground biomass.

As a result, a major fraction of the $^{35}SO_4^{2-}$ reached and was retained in the Bhf and Bf horizons. The most probable retention mechanism for $^{35}SO_4^{2-}$ in the B horizons, and particularly in the Bf horizon, given the prevailing conditions, appears to be its adsorption on aluminum and iron oxides (Houle and Carignan, 1992). The water's residence time in the soil profile would have little impact on $^{35}SO_4^{2-}$ adsorption because this reaction is almost instantaneous (Houle and Carignan, 1995). Also, adsorption reactions are less sensitive to low temperatures than biotic reactions, as evidenced by the capacity of the Bf horizon to adsorb 90% of the added $^{35}SO_4^{2-}$ in lab conditions at 1 °C (Houle et al., 2001), the remainder being incorporated within the organic fraction.

Overall, in spite of the conditions that prevailed during snowmelt and the significant $^{35}SO_4^{2-}$ movements in the soil solution, 87% of the added $^{35}SO_4^{2-}$ was retained within the soil profile, compared to 98% after 15 months at plot 1. These results agree well with laboratory studies wherein the $^{35}SO_4^{2-}$ added to soil columns was retained at rates of 99.5% (Dhamala et al., 1990) and 92% (Novák and Přechova, 1995), 20 weeks following application. These results clearly illustrate the very strong potential for soil retention of SO_4^{2-} and show, without a doubt, that SO_4^{2-} may not be qualified as a 'conservative element' in forested ecosystems. The soil profile, however, did not accumulate S-SO_4^{2-} on a net basis during the study period. Rather, it was the source of the excess S (presumably produced from the slow decomposition of the large refractory organic S reservoir) that led to an annual average negative S budget for the Laflamme Lake catchment area (Houle and Carignan, 1995).

Regardless of the contrasting hydrologic conditions that prevailed following the infiltration of the tracers to the soil profiles in summer and the in snowmelt experiment, and despite the major lateral water movements observed in one of the two plots in each season, a general pattern emerged about the $^{35}SO_4^{2-}$ concentration over time: significant amounts of $^{35}SO_4^{2-}$ persisted (and sometimes increased) in the soil solution of nearly all the soil horizons at the four plots long after the initial tracer infiltrations. The persistence of $^{35}SO_4^{2-}$ in the soil solution of the entire profile of plots 1 and 2, and the relative stability of specific activity more than one year after the addition of $^{35}SO_4$, are particularly striking.

Taken together, these observations point to the establishment of an isotopic equilibrium between the added $^{35}SO_4^{2-}$ and the active S-containing reservoirs within a given soil horizon. In this case, the added $^{35}SO_4^{2-}$ would be recycled within the soil S reservoirs at the same rate as the 'cold' SO_4^{2-} with respect to all the reactions affecting S cycling. Such observations agree well with incorporation studies done in the laboratory on soils from the lake catchment area. These studies showed the rapid incorporation of $^{35}SO_4^{2-}$ in the soil's organic and adsorbed SO_4^{2-} reservoirs (Houle *et al.*, 2001) and the attainment of equilibrium between $^{35}SO_4$ and the active S reservoirs within a few days in all the soil horizons. Such a rapid equilibration (*in vitro* and *in situ*) may appear surprising, particularly when the size of the organic S reservoirs (1230 kg ha^{-1}, 200 times the average annual S-SO_4^{2-} deposition) is taken into consideration. However, Houle *et al.* (2001) have shown that only a small proportion (1%) of the organic S reservoir within the soil profile (presumably soil microorganisms) is readily involved in short-term SO_4^{2-} immobilization. The remaining organic S would be composed of refractory extracellular S-containing organic matter.

Of course, the existence of an *in situ* isotopic equilibrium is relative. Compared to what has been observed in the laboratory, field conditions are representative of an open system subjected to strong seasonal changes that are likely inducing pronounced seasonal patterns in S cycling. Many factors can contribute to the punctual variations in specific activity in soil solutions. First, in the LFH layer, atmospheric depositions, canopy inputs (litterfall and foliar exudation), and decomposing herbaceous plants and fine roots constitute potential SO_4^{2-} sources that may affect specific activity at a given time. In the other horizons, the soil solution getting out of the above soil horizons and the decomposition of fine roots (particularly in the Bhf horizon, given the surficial root distribution) are potential factors contributing to variations in specific activity. Finally, the decomposition of the large refractory organic S pool (1230 kg ha^{-1}, Houle and Carignan, 1992) distributed throughout the soil profile and being two or three orders of magnitude larger than the active organic pool (Houle *et al.*, 2001), may easily dilute the specific activity of the soil solution in all the horizons. When all these possibilities are considered, the relative stability of specific activity is remarkable.

Many of the observed trends in specific activity can be explained by the above-described mechanisms. For instance, the strong decrease in specific activity observed in the soil solution of the LFH layer of plots 1 and 2 from August to November 1988 was probably caused by an exceptionally high atmospheric S deposition, along with heavy precipitation, that amounted to 4.4 kg ha^{-1} SO_4^{2-}-S, or 73% of the long-term (1981–2000) average annual deposition (5.9 kg ha^{-1} SO_4^{2-}-S). At plot 3, the gradual increase in the specific activity of the soil solution from June to September 1989 (Figure 4c) in the Bf2 sampler was likely the result of the soil solution infiltrating from the level of the Bf1 sampler and having a higher specific activity. Finally, the rise in specific activity in the soil solution of the LFH layer, which had begun around August 1989 and persisted till October at plots 1 and 2, indicates that internal sources previously had accumulated $^{35}SO_4^{2-}$ at a higher specific activity, and were now returning the stored $^{35}SO_4^{2-}$ to the soil solution. A possible source is fine root decomposition that would be associated with both the beginning of balsam fir dormancy and the fall senescence of herbaceous plants.

5. Conclusion

Overall, then, the simultaneous in situ application of ^3H_2O and $^{35}SO_4^{2-}$ to plots of soil in the Laflamme Lake watershed shows that S cycling is dynamic in nature. For the summer application, nearly 100% of the $^{35}SO_4^{2-}$ was retained within the soil profile, with the largest part being immobilized in the organic LFH fraction (74.7%) at the end of a 15-month period. Following snowmelt, most of the $^{35}SO_4^{2-}$ previously added directly on the snowpack was retained in the B horizons (LFH = 33.1%, Bhf = 33.1%, Bf = 20.8%), leading to an 87% retention rate for the whole soil profile. The contrasting conditions (summer and snowmelt) under which the tracers were tracked illustrated the importance of soil temperature and soil water residence time to determining the short-term conditions of S cycling and, therefore, to determining its accumulation patterns within the soil profile. The present study's results on temporal $^{35}SO_4^{2-}$ concentrations and specific activity variations reinforce previous findings that recent deposition of atmospheric SO_4^{2-} are actively recycled within the soil S reservoirs and rapidly reach steady-state conditions with no net accumulation.

References

Chao, T. T., Harward, M. E. and Fang, S. C.: 1962, 'Adsorption and desorption phenomena of sulfate ions in soils', *Soil Sci. Soc. Am. Proc.* **26**, 234–237.

Colbeck, S. C.: 1978, 'The physical aspects of water flow through snow', in V. T. Chow (ed.), *Advances in Hydrosciences*, Vol. II, Academic Press, New York, New York, U.S.A., pp. 165–206.

Davies, T. D., Brimblecombe, P., Tranter, M., Tsiouris, S. Vincent, C. E., Abrahams, P. and Blackwood, I. L.: 1987, 'The removing of soluble ions from melting snowpack', in H. G. Jones and W. J. Orville-Thomas, (eds), *Seasonal Snowcovers: Physics, Chemistry, Hydrology*, NATO Advanced Science Institutes Series, Series C, Mathematical and Physical, Sciences, vol. 211, D. Reidel, Amsterdam, The Netherlands, pp. 337–392.

Dhamala, B. R., Mitchell, M. J. and Stam, A. C.: 1990, 'Sulfur dynamics in mineral horizons of two northern harwood soils. A column study with ^{35}S', *Biogeochemistry* **10**, 143–160.

Fitzgerald, J. W., Ash, J. T., Strickland, T. C. and Swank, W. T.: 1983, 'Formation of organic S in forest soils: A biologically mediated process', *Can. J. For. Res.* **13**, 1077–1082.

Fuller, R. D., David, M. B. and Driscoll, C. T.: 1985, 'Sulfate adsorption relationships in some forested spodosols of the northeastern U.S.', *Soil Sci. Soc. Am. J.* **49**, 1034–1040,

Gray, D. M. and Male, D. H.: 1981, *Handbook of Snow: Principles, Processes, Management and Use*, Pergamon Press, Toronto, Ontario, Canada, 776 pp.

Houle, D. and Carignan, R.: 1992, 'S speciation and distribution in soils and aboveground biomass of a boreal coniferous forest', *Biogeochemistry* **16**, 63–82.

Houle, D. and Carignan, R.: 1995, 'Role of SO_4^{2-} adsorption and desorption in the long-term S budget of a coniferous catchment on the Canadian Shield', *Biogeochemistry* **28**, 161–182

Houle, D., Carignan, R. and Ouimet, R.: 2001, 'Organic sulfur dynamics in a forested soil', *Biogeochemistry* **53**, 105–124.

Mateer, C. L.: 1955, 'Average insolation in Canada during cloudless day', *Can. J. Technol.* **33**, 12–32.

Nielsen, D. R. and Biggar, J. W.: 1962, 'Miscible displacement: III. Theoretical considerations', *Soil Sci. Soc. Am. Proc.* **26**, 216–221.

Novák, M., and Přechova, E.: 1995, 'Movement and transformation of ^{35}S-labelled sulphate in the soil of a heavily polluted site in northern Czech Republic', *Environ. Geochem. Health* **17**, 83–94.

Penman, H. L.: 1948, 'Natural evaporation from open water, bare soil and grass', *Proc. Royal Soc. London* **193**, 120–145.

Schindler, S. C., Mitchell, M. J., Scott, T. J., Fuller, R. D. and Driscoll, C. T.: 1986, 'Incorporation of ^{35}S-sulfate into inorganic and organic constituents of two forest soils', *Soil Sci. Soc. Am. J.* **50**, 457–462.

Swank, W. T., Fitzgerald, J. W. and Ash, J. T.: 1984, 'Microbial transformation of sulfate in forest soils (North Carolina)', *Science* **223**, 182–184.

Topp, G. C., Davis, G. L. and Annan, A. P.: 1980, 'Electromagnetic determination of soil water content: measurements in coaxial transmission lines', *Water Resour. Res.* **16**, 574–582.

Watwood, M. E. and Fitzgerald, J. W.: 1988, 'S transformations in forest litter and soil: results of laboratory and field incubations', *Soil Sci. Soc. Am. J.* **52**, 1478–1483.

THE MISSING FLUX IN A ^{35}S BUDGET FOR THE SOILS OF A SMALL POLLUTED CATCHMENT

MARTIN NOVÁK[1]*, ROBERT L. MICHEL[2], EVA PŘECHOVÁ[1] and MARKÉTA ŠTĚPÁNOVÁ[1]

[1] *Czech Geological Survey, Geologická 6, 152 00 Prague 5, Czech Republic;* [2] *U.S. Geological Survey, 345 Middlefield Road, MS 434, Menlo Park, CA 94025, U.S.A.*
(* author for correspondence, e-mail: novak@cgu.cz; phone: +4202 51085333; fax: +4202 51818748)

(Received 20 August 2002; accepted 30 May 2003)

Abstract. A combination of cosmogenic and artificial ^{35}S was used to assess the movement of sulfur in a steep Central European catchment affected by spruce die-back. The Jezeří catchment, Krušné Hory Mts. (Czech Republic) is characterized by a large disproportion between atmospheric S input and S output via stream discharge, with S output currently exceeding S input three times. A relatively high natural concentration of cosmogenic ^{35}S (42 mBq L^{-1}) was found in atmospheric deposition into the catchment in winter and spring of 2000. In contrast, stream discharge contained only 2 mBq L^{-1}. Consequently, more than 95% of the deposited S is cycled or retained within the catchment for more than several months, while older S is exported via surface water. In spring, when the soil temperature is above 0 °C, practically no S from instantaneous rainfall is exported, despite the steepness of the slopes and the relatively short mean residence time of water in the catchment (6.5 months). Sulfur cycling in the soil includes not just adsorption of inorganic sulfate and biological uptake, but also volatilization of S compounds back into the atmosphere. Laboratory incubations of an Orthic Podzol from Jezeří spiked with 720 kBq of artificial ^{35}S showed a 20% loss of the spike within 18 weeks under summer conditions. Under winter conditions, the ^{35}S loss was insignificant (<5%). This missing S flux was interpreted as volatilized hydrogen sulfide resulting from intermittent dissimilatory bacterial sulfate reduction. The missing S flux is comparable to the estimated uncertainty in many catchment S mass balances (±10%), or even larger, and should be considered in constructing these mass balances. In severely polluted forest catchments, such as Jezeří, sulfur loss to volatilization may exceed 13 kg ha^{-1} a^{-1}, which is more than the current total atmospheric S input in large parts of North America and Europe.

Keywords: catchment, cosmogenic ^{35}S, gaseous S loss, sulfur

1. Introduction

Sulfur-35, a radioisotope of sulfur (half-life of 87 days), is produced naturally by cosmic-ray spallation of atmospheric argon (Lal and Peters, 1966). It quickly converts to sulfate and is deposited on the earth's surface as either wetfall or dryfall. Once deposited, it follows the pathway of atmospheric sulfate through the catchment. It has proven to be an excellent tracer for determining timescales for migration of atmospheric sulfate in terrestrial ecosystems (Sueker *et al.*, 1999; Michel *et al.*, 2002). Due to its short half-life, it is useful for measuring processes occuring on time-scales of less than one year. Previous studies have shown that sulfur-35 behaves nearly conservatively in alpine-type catchments, with a significant proportion of the atmospheric ^{35}S input being rapidly exported from the system (Michel *et al.*, 2000). It has been unclear whether lower-elevation forested catchments behave similarly, or whether most of the newly deposited sulfur is involved in biogeochemical and physicochemical reactions before leaving the system. In such a case, there will be a negligible amount of ^{35}S in stream discharge.

The cycling of atmospheric sulfur in forest soils can be studied also by means of radiolabelling with artificial sulfur-35. Sulfur-35 can be produced in the laboratory from neutron irradiated potassium chloride. Numerous incubation experiments have used soil spiked with $^{35}SO_4^{2-}$ to infer the rates of biological uptake and mineralization under controlled conditions (Strickland and Fitzgerald, 1984; Houle *et al.*, 2001). Less common are field studies using artificial ^{35}S due to a ban on *in situ* radiolabelling in many countries (Watwood and Fitzgerald, 1988). Sulfur-35 speciation is evaluated relative to the total amount of ^{35}S recovered at the end of the experiment. Often, the decay-corrected recovery of ^{35}S is less than 100%. Little attention has been paid to the relationship between time elapsed since the start of the incubation and the recovery of decay-corrected ^{35}S. Decreasing recovery of ^{35}S over time might suggest the loss of S through an unmeasured flux, such as volatilization of cycled sulfur from moist soils.

We used a combination of cosmogenic and artificial ^{35}S to trace sulfur dynamics in a small polluted catchment in the Czech Republic, Central Europe, which had suffered from severe spruce die-back. Our first objective was to compare the activity of natural cosmogenic ^{35}S in catchment inputs and outputs and thus assess the importance of within-site cycling of the incoming S. Secondly, we conducted a series of laboratory incubations of soil columns spiked with man-made $^{35}SO_4^{2-}$ at summer and winter temperatures. We hypothesized that some of the applied ^{35}S may return to the atmosphere. If so, this sulfur flux should be considered when constructing mass budgets for small catchments.

2. Methods

2.1. STUDY SITE

The 261-ha Jezeří (JEZ) catchment is situated at an elevation between 500 and 900 m in the Krušné hory Mts., northern Czech Republic (Novák *et al.*, 1995, 1996, 2000; Groscheová *et al.*, 1998). Dystric Cambisols to Orthic Podzols are underlain by base-poor, S-difficient orthogneiss. Due to industrial pollution, Norway spruce died back on 60% of the catchment's area 30 years ago (Figure 1). Patches of Norway spruce cover 10% of the studied area, mainly above an elevation of 700 m. European beech covers the remaining 30% at lower elevations. As much as 130 kg S ha^{-1} a^{-1} were deposited on JEZ in the 1980s. Present-day atmospheric deposition is below 20 kg S ha^{-1} a^{-1}. Sulfur export via stream discharge is currently three

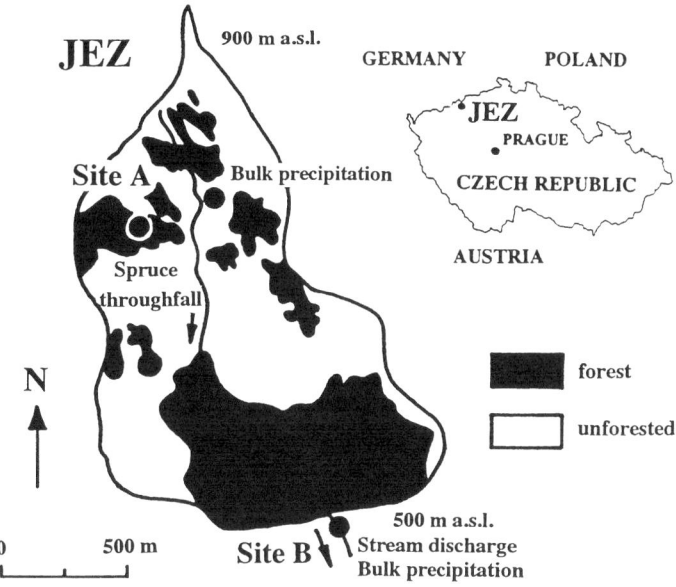

Figure 1. Study site. In unforested areas spruce stands died back due to industrial pollution and were harvested.

times higher than atmospheric input (Novák et al. 2000, 2003). Soil underneath the surviving spruce stands stores 1300 kg S ha^{-1}, of which nearly 80% is in the organic form (Novák and Přechová, 1995). Hydrology of JEZ was described by Peters et al. (1999). An oxygen isotope study of rainwater and streamwater was conducted on the southern slope of the Krušné hory Mts. (3 km west of JEZ; Buzek et al., 1991), and directly at JEZ (Buzek, unpublished data), using the method of Siegenthaler (1971).

2.2. CATCHMENT ^{35}S SAMPLING

Bulk (open area) precipitation, spruce canopy throughfall and stream discharge were sampled for cosmogenic ^{35}S activity measurements in the winter and spring of 2000. Bulk precipitation was sampled at site A (880 m a.s.l.) and site B (490 m a.s.l.; Figure 1). Spruce canopy throughfall was sampled only at the higher elevation (site A). Stream water was sampled at the closing profile of the catchment (site B). At least 20 L of bulk precipitation and spruce throughfall were collected cumulatively over two-to-four week periods. Stream water was collected at the beginning and the end of these periods.

Orthic Podzol collected in the vicinity of spruce throughfall samplers at site A (Figure 1) was used for soil column experiments. Approximately 80 kg from each of the O+A, AE and Bvs horizons were transported to the laboratory. The horizon thickness was 12, 10 and 13 cm, respectively. Stoniness was negligible in O+A, but caused considerable heterogeneity in the AE and Bvs horizons.

2.3. ACTIVITY OF NATURAL COSMOGENIC ^{35}S

Samples of cosmogenic ^{35}S were extracted by passing 20 L of water through an Amberlite IRA-400 ion exchange column. The sulfate was eluted from the column with about 300 mL of 3 M sodium chloride solution. Barium chloride was added to the solution to precipitate barium sulfate. The precipitates were filtered, washed with deionized water and dried. The radioactivity was analyzed in a liquid scintillation counter. Due to the existence of radium in the barium precipitate plus possible contaminate radium being carried through the sampling process, the samples were counted twice about 6 months apart, and the concentration calculated by the difference of the two counts.

2.4. SOIL COLUMN INCUBATIONS

In all, 15 soil colums, 6.5 cm in diameter, 35 cm long, were used for the laboratory experiments. Intact soil horizons O+A were used for the incubations (cf., Novák and Přechová, 1995), while the deeper soil horizons AE and Bvs were pre-treated according to Novák *et al.* (2001) and repacked before the start of the experiments. Oven-drying (60 °C) and sieving (<2 mm) of mineral soil could not be avoided because of mass balance concerns (total mass must be known for each soil horizon in the column). The proportion of stones removed from the soil by sieving was 48.4 and 46.6 vol.% for AE and Bvs, respectively. The density of the used soil fraction was 0.06, 0.7 and 0.8 g cm^{-3} for O+A, AE and Bvs, respectively. Batches of dry soil needed for the AE and Bvs horizons in each column were weighed, sealed in permeable fabric bags (60 mesh) and reburied at the original site and depth. This procedure served to reestablish microbial populations close to their natural state (Sumner, 2000). After 3 weeks, the bags were excavated, and soil columns consisting of the intact O+A horizon and repacked AE and Bvs horizons were set up in a growth chamber in three replicates for each of two treatment durations (13 and 18 weeks) and two temperature regimes (26 °C day, 17 °C night; 3 °C all day and night). Additionally, 3 replicate soil columns were used as a control at time $t = 0$. An activity of 720 kBq of $^{35}SO_4^{2-}$ was applied on the surface of each soil column along with 25 mL of natural spruce throughfall collected at JEZ (40 mg SO_4^{2-} L^{-1}, 10.2 DOC L^{-1}). Within 3 hr, gravitational soil water was collected underneath each column. Then the control soil columns were divided into the three soil horizons, each horizon was homogenized in a planetary mill and frozen. The remaining 12 soil columns were incubated and wetted three times per week with the spruce throughfall (3.3 mm per day, corresponding to 1200 mm a^{-1}; the throughfall precipitation was aerated by a bleeding tube for 1 min before each application). The gravitational soil water was collected following each wetting and pooled within replicate. Air moisture in the growth chamber was kept relatively low thoughout the incubation by means of a ventillator. At the end of each incubation the soil column was divided by horizon (O+A, AE and Bvs), homogenized and frozen. The wet-to-dry conversion ratio was ascertained on a 5 g subsample. Total

TABLE I

Activities of cosmogenic ^{35}S in catchment input and output in the spring of the year 2000. For the location of sites A and B see Figure 1

Sample type	Date	mBq L^{-1}	mg SO$_4^{2-}$	mBq mg^{-1} SO$_4^{2-}$
Bulk precipitation A	29 February–03 April	65 ±3	11.1	5.9
Spruce throughfall A	29 February–03 April	78 ±3	38.2	2.1
Bulk precipitation A	03 April–19 April	25.7±0.7	5.90	4.4
Spruce throughfall A	03 April–19 April	7.7±0.9	21.7	0.4
Bulk precipitation B	03 April–19 April	33.3±0.9	9.00	3.7
Stream discharge B	29 February	5.3±1.9	54.9	0.1
Stream discharge B	19 March	1.6±1.1	44.2	0.04
Stream discharge B	03 April	0.7±1.1	47.3	0.01
Stream discharge B	19 April	−0.3±0.7	43.0	−0.01

sulfur was extracted from a 3 g wet soil subsample by the Eschka's procedure (Chakrabarti 1978; Novák et al., 1994). A Beckman 8098-LS beta spectrometer was used for ^{35}S activity measurements in soil extracts and gravitational soil water. All measured ^{35}S activities in the soil column experiments were corrected for decay. Statistical analyses were performed using the JMP IN package by SAS (Sall and Lehman, 1996). Once the difference in mean ^{35}S activities was found to be significant (ANOVA, F-test, $p < 0.05$), comparisons for each pair of means were performed using the Student's t with the least significant difference.

The laboratory incubations of JEZ soil using artificial ^{35}S as a tracer followed a similar treatment protocol as laboratory incubations of the same soil using ^{34}S as a tracer. The ^{34}S soil incubation study has been reported in a companion paper (Novák et al. 2001), and the results of both studies will be compared here. One of the few differences in both sets of experiments was the concentration of total SO$_4^{2-}$ in the wetting solution. The wetting solution contained 40 and 106 mg SO$_4^{2-}$ L^{-1} in the ^{35}S and ^{34}S experiments, respectively. The total amount of the wetting solution added to soil columns over the 18-week period was the same, regardless of the tracer used. Thus soil in the ^{35}S incubation received about 3 times less total sulfate S than that in the ^{34}S incubation.

3. Results and Discussion

3.1. Cosmogenic ^{35}S in water

The results of sulfur-35 measurements in catchment inputs and outputs are given in Table I. The samples of bulk precipitation and spruce canopy throughfall had relatively high ^{35}S concentrations. The mean ^{35}S concentration in the atmospheric input was 42 mBq L^{-1}. The concentrations of ^{35}S in the atmospheric input at JEZ were similar to the highest concentrations found in mid-continental locations in the United States (Michel *et al.*, 2002). As in the United States, ^{35}S concentrations at JEZ varied greatly from month to month and storm to storm. In general, the highest concentrations of ^{35}S in precipitation are found in late winter and spring due to the mixing of stratospheric air masses with the troposphere, so the high concentrations of ^{35}S at JEZ are reasonable. These results indicate that there is a significant and easily measurable amount of sulfur-35 entering the catchment through precipitation and throughfall. The concentrations of ^{35}S relative to total sulfate (mean of 3.3 mBq mg^{-1} SO$_4^{2-}$) were lower than found in the Rocky Mountains (mean of 56 mBq mg^{-1} SO$_4^{2-}$). This was due to the much higher anthropogenic sulfate concentrations found in the precipitation at JEZ.

The mean ^{35}S concentration in the catchment output was 2 mBq L^{-1}. This was twenty times less than the atmospheric input (Table I). In fact, only the February sample showed any significant concentration of ^{35}S in the discharge (5 mBq L^{-1}). The surface of the soil was frozen throughout winter, at least until the end of February. During the spring snowmelt in March and April 2000 (T > 0°), the amount of natural sulfur-35 in stream discharge was negligible. The mean residence time of water in JEZ calculated from δ^{18}O values of rainwater and streamwater is relatively short (6.5 months), the contribution of instantaneous rainfall to stream discharge is 13% (Buzek, unpublished data). Therefore, it appears that the absence of ^{35}S in stream discharge is not related to a long residence time of atmospheric water in the catchment. We conclude that more than 95% of the atmospheric S deposited onto JEZ in winter and spring is cycled within the catchment and not exported via surface water in the first months after deposition. The large unmeasured sink of atmogenic ^{35}S within JEZ may be represented by adsorption of inorganic sulfate on soil particles and biological uptake (Novák *et al.*, 2000; Likens *et al.*, 2002; Eimers and Dillon, 2002). Additionally, the positive ^{35}S budget (input > output) for JEZ is consistent with significant loss of S in the form of volatile sulfur compounds. We followed this possibility in the soil incubation experiments.

3.2. Artificial ^{35}S in soil

Export of gaseous S forms from forested catchments back to the atmosphere is viewed as negligible in pristine (<10 kg S ha^{-1} a^{-1}) and moderately polluted (<30 kg S ha^{-1} a^{-1}) parts of the world (Mitchell and Fuller, 1988; Krouse and Grinenko, 1991; Mayer *et al.*, 1995; Giesemann *et al.*, 1995; Jenkins *et al.* 2001). At the same time, data on volatile S fluxes from the severely polluted soils of Central Europe (130 kg S ha^{-1} a^{-1}) are scanty (Novák *et al.*, 2001). Most studies assume that volatile S fluxes from forest catchments with well-aerated soils are lower compared to catchments with water-logged peaty soils (Chapman *et al.*, 1996; Alewell and Novák, 2001). Yet, a mechanism of potential S volatilization

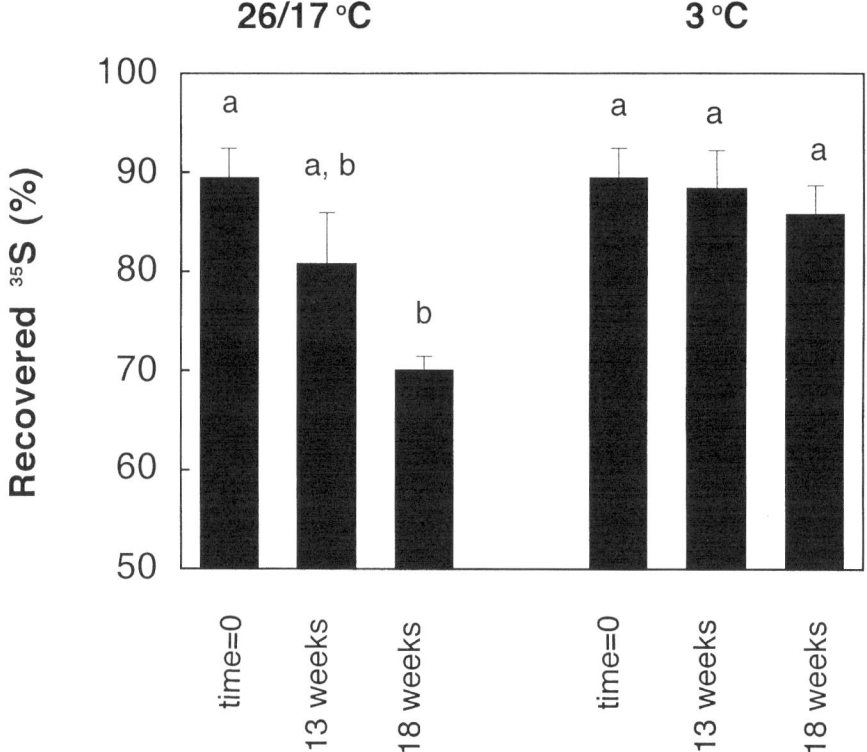

Figure 2. Results of soil column experiments. 720 kBq of $^{35}SO_4^{2-}$ were applied on the surface of the soil at the start of the incubation. The total recovered ^{35}S was calculated as the sum of ^{35}S found in soil horizons O+A, AE, Bvs and in gravitational soil water collected underneath the solum. Three replicates per treatment. Different letters denote means found to be statistically different. All pairs of treatments marked with the same letter exhibit insignificant differences in ^{35}S.

even from soils such as Dystric Cambisols or Podzols is known: anaerobic domains are formed in soil following each precipitation event, and these domains permit dissimilatory bacterial sulfate reduction to take place. The resulting hydrogen sulfide can escape into the atmosphere while the domains persevere for 1 to 2 days following the precipitation event. Figure 2 depicts ^{35}S recovery in our pulse experiments, in which ^{35}S was applied on the soil surface once at the beginning, and the soil surface was then wetted three times per week to model a realistic annual precipitation total. Time $t = 0$ for the control experiment was defined at 3 hr after the radiotracer application. This was necessitated by the need to allow for penetration of the wetting solution through the soil to collect gravitational soil water underneath the columns. The decay-corrected ^{35}S recovery at time $t = 0$ was 90%. We do not know if the remaining 10% were unaccounted for due to analytical uncertainty (0 to 10%, depending on the soil horizon and the extraction method used, cf. Wieder and Lang 1988; Houle *et al.* 2001), or early volatilization of the

TABLE II

Sulfur-35 captured in the individual compartments of the soil column in the experiment. The surface of repacked soil columns was spiked with artificial $^{35}SO_4^{2-}$. Number of replicate soil columns per treatment $n = 3$

Parameters of the incubation experiment	Percentage of found ^{35}S			Gravitational soil water at 26 cm depth
	Soil horizon			
	O+A	AE	Bvs	
time = 0	85	15	0	0
13 weeks, 26/17 °C	74	13	12	1
18 weeks, 26/17 °C	52	8	37	3
13 weeks, 3 °C	39	21	37	3
18 weeks, 3 °C	31	5	42	22

applied ^{35}S. Over the following 18 weeks of incubation, further losses of ^{35}S were insignificant at the lower temperature (3 °C; Figure 2). In contrast, a significant decrease in the ^{35}S recovery was observed at the summer temperature (26/17 °C). The decay-corrected ^{35}S recovery was 81 and 70% after 13 and 18 weeks, respectively. The unmeasured flux of ^{35}S from the soil amounted to 20% (relative to the control experiment) after 18 weeks of incubation under the conditions of a wet Central European summer. We ascribe this missing flux to volatilization of S compounds, such as hydrogen sulfide, from the soil. The higher temperature led to higher rates of biologically mediated S cycling, which is in agreement with the results of previous batch soil experiments (Chapman *et al.*, 1996; Groscheová *et al.*, 2000; Houle *et al.*, 2001).

3.3. DISTRIBUTION OF ^{35}S AMONG SOIL HORIZONS

At time zero, 85% and 15% of the total recovered ^{35}S activity were found in the O+A and AE horizons, respectively, with no activity detected in Bvs and the gravitational soil water (Table II). At the end of the incubations, the distribution of ^{35}S among individual soil horizon differed between the two temperature regimes (Figure 3). At the higher temperature (26/17 °C), more ^{35}S tended to remain in the topmost organic O+A horizons. In contrast, at the lower temperature (3 °C), the entire soil column became more permeable for unreacted ^{35}S and the ^{35}S activites in the gravitational soil water underneath the solum increased (22% vs. 3% after 18 weeks at the winter and summer temperatures, respectively). The smallest amount of ^{35}S was found in the middle horizon AE, relatively poor in organic matter and adsorbed sulfate content (Figure 3). All these data are in agreement with similar

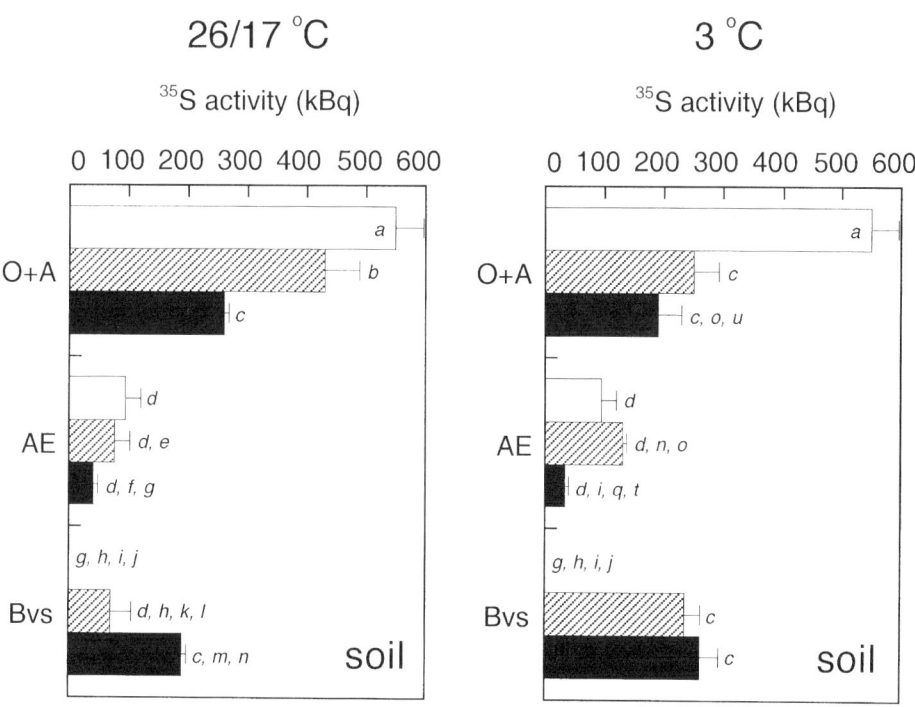
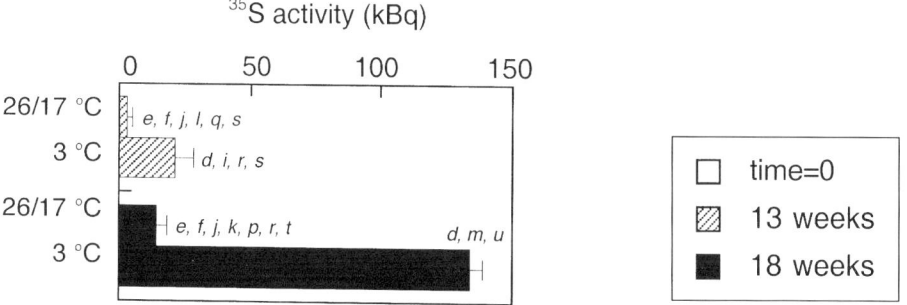

Figure 3. Movement of artificial ^{35}S in the vertical profile through an Orthic Podzol. Number of ^{35}S activity measurements $n = 57$. Different letters denote means found to be statistically different. All pairs of horizons marked with the same letter exhibit insignificant differences in ^{35}S.

experiments using the same soil and the stable isotope ^{34}S as a tracer (Novák et al., 2001).

Sulfur cycling between inorganic and organic species at JEZ had been quantified in another previous study (Novák and Přechová, 1995). Similar to the present study, soil from site A (880 m a.s.l.) was spiked with $^{35}SO_4^{2-}$ and incubated for 13 and 20 weeks at summer temperatures. In contrast to the present study, stones were removed from the profile only at the end of the incubation. Novák and Přechová (1995) found that inorganic free and adsorbed sulfate was the most abundant ^{35}S species (91%). Organic C-bonded, ester-bonded and inorganic reduced sulfur constituted 5, 2 and 2%, respectively, of the activity recovered throughout the incubation. With increasing soil depth, the proportion of organically cycled ^{35}S increased, which is in good agreement with new data from Canadian forest soils (Houle et al., 2001). Whereas in the top horizon organic ^{35}S existed throughout the experiment, in the deeper soil horizons ^{35}S in the C-bonded organic form became undetectable after 8 weeks. Importantly, the amount of ^{35}S in the reduced inorganic form (FeS_x) first increased and then decreased. The presence of reduced inorganic S indicated that dissimilatory bacterial sulfate reduction occured in the studied Podzol.

3.4. COMPARISON OF RECOVERIES IN THE ^{35}S AND ^{34}S TRACER EXPERIMENTS

The soil column experiments reported here (^{35}S as a tracer) and those reported by Novák et al. (2001; ^{34}S as a tracer) employed identical soil, temperature regime, wetting protocol and duration. They differed in the amount of added tracer relative to the size of the S pools in the soil. Whereas the addition of $^{35}S-SO_4^{2-}$ was negligible relative to the soil sulfur pool sizes, the addition of ^{34}S-rich precipitation augmented the existing S pools by as much as 16%. Secondly, the two sets of experiments differed in the timing of tracer application. While ^{35}S was added only once at the start, ^{34}S was being added via wetting solution repeatedly for 13 and 18 weeks. It follows that S recoveries in these two sets of experiments can be more easily compared qualitatively than quantitatively. At summer temperatures, the ^{35}S experiments resulted in a loss of 20% of the tracer. The ^{34}S soil experiments (Novák et al., 2001) also reported a missing S flux. This missing S amounted to 53–74% of the sulfur added by wetting solution and was statistically significant relative to control experiments. This amount is quite large, and again, we suggest volatilization of S from soil as an explanation.

The loss of ^{35}S during our incubation experiments appeared to be linear over the somewhat arbitrarily selected period of 18 weeks (Figure 2). Yet, we do not know if more S is volatilized from the soil after week 18. While further emission of S from the soil may exist even after 18 weeks since deposition at 26/17 °C, such information may be irrelevant with respect to an estimate of the annual S loss to volatilization in a natural setting. The reason is that typical summer temperatures do not last longer than for 18 weeks. If zero S emissions existed for six months, and

20% S emissions existed for the remaining 6 months of the year, then the annual S loss to volatilization would be 10% of the input. In high-deposition areas (130 kg S ha^{-1} a^{-1}) such loss may represent a sizeable S flux. At JEZ, the 10% annual loss of the incoming S back to the atmosphere woud be equal to 13 kg S ha^{-1} a^{-1}. That is more than the total atmospheric S input in large parts of North America and Europe.

A comparison of the results of the ^{35}S and ^{34}S soil experiments at JEZ leads to another implication: All other incubation parameters being similar, the experiment which applied rougly 3 times more *total* S to the soil surface (the ^{34}S experiment) led to a three times higher loss of S, relative to the experiment which added 3 times less *total* sulfur (the ^{35}S experiment). It is well known that S flux through bacterial sulfate reduction increases with increasing availability of sulfate, assuming a sufficent supply of labile organic C. These preliminary data might thus suggest that in high-deposition areas the percentage of volatilized S may increase above the 10% of annual atmospheric inputs estimated in this paper. Further study is needed to test this hypothesis. Part of the study must be careful evaluation of the limitations related to each tracer.

4. Conclusions

We used a combination of natural and man-made radioisotope ^{35}S to obtain insight into S cycling in a heavily stressed Central European forest ecosystem. Both at the catchment level and in a laboratory microcosm, we applied the isotope mass balance approach to investigate pathways of S cycling. Atmospheric input of cosmogenic ^{35}S into the JEZ catchment averaged 42 mBq L^{-1}, a value similar to the largest ^{35}S concentrations reported in North America. In contrast, the catchment output via stream discharge contained very little ^{35}S (2 mBq L^{-1}). These data show that more than 95% of the incoming sulfur are retained in the catchment for at least several months, or removed through an unmeasured flux. The catchment under study systematically exhibits larger export of sulfur than atmospheric input. Stable isotope studies have ruled out bedrock S as the source of the excess sulfate in the stream discharge (Novák et al., 1995, 2000). Stable sulfur isotopes also permit attribution of S in the discharge to various sources (organic and mineral soil as stores of older pollutant S, instantaneous rainfall). Based on stable S isotope constraints, stream discharge contains a large proportion of sulfate from instantaneous rainfall, up to 50%. In this study we have shown that such may not be the case. Young atmospheric sulfate is adsorbed or organically cycled in the soil, while the water flushes out other S, which, being older, contains no ^{35}S. Even at sites with three times higher export of atmogenic S compared to present-day input, and even at such steep catchments as JEZ, a negligible amount of S from the instantaneous rainfall is exported.

Laboratory incubations using intact and repacked soil columns systematically showed low recoveries of artificial ^{35}S, which had been applied to the soil surface. At 1200 mm of rainfall per year and typical summer time temperatures, 20% of the tracer was lost after 18 weeks. We propose, that this missing S flux is represented by volatilization of S compounds from polluted soils following intermittent dissimilatory bacterial sulfate reduction. Since most catchment input/output S mass balances claim an error of ±10% or less, this missing S flux should not be omitted. Moreover, volatilization of S compounds from polluted soils may generate acid rain and contribute to slower recovery from the past acidification of the studied ecosystems (cf., Nriagu et al., 1987).

Acknowledgements

We thank the Czech Grant Agency for financial support (Grants No. 205/93/2404 and 526/02/1061).

References

Alewell, C. and Novák, M.: 2001, 'Spotting zones of dissimilatory sulfate reduction in a forested catchment: The ^{34}S-^{35}S approach', *Environ. Pollut.* **112**, 369–377.

Buzek, F., Hanzlík, J., Hrubý, M. and Tryzna, P.: 1991, 'Evaluation of the runoff components on the slope of an open-cast mine by means of environmental isotopes ^{18}O and T', *J. Hydrol.* **127**, 23–36.

Chakrabarti, J. N.: 1978, 'Analytical procedures for sulfur in coal desulfurization products', In C. Karr Jr. (ed.), *Analytical Methods for Coal and Coal Products*, Academic Press, New York, pp. 279–323.

Chapman, S. J., Kanda, K., Tsuruta, H. and Minami, K.: 1996, 'Influence of temperature and oxygen availability on the flux of methane and carbon dioxide from wetlands: A comparison of peat and paddy soils', *Soil Sci. Plant Nutr.* **42**, 269–277.

Eimers, M. C. and Dillon, P. J.: 2002, 'Climate effects on sulphate flux from forested catchments in south-central Ontario', *Biogeochemistry* **61**, 337–355.

Giesemann, A., Jäger, H. J. and Feger, K. H.: 1995, 'Evaluation of sulphur cycling in managed forest stands by means of stable S-isotope analysis', *Plant Soil* **168–169**, 399–404.

Groscheová, H., Novák, M. and Alewell, C.: 2000, 'Changes in the δ^{34}S ratio of pore-water sulfate in incubated *Sphagnum* peat', *Wetlands* **20**, 62–69.

Groscheová, H., Novák, M., Havel, M. and Černý, J.: 1998, 'Effect of altitude and tree species on δ^{34}S of deposited sulfur (Jezeří Catchment, Czech Republic)', *Water, Air, Soil Pollut.* **105**, 295–303.

Houle, D., Carignan, R. and Ouimet, R.: 2001,'Soil organic sulfur dynamics in a coniferous forest', *Biogeochemistry* **53**, 105–124.

Jenkins, A., Ferrier, R. C. and Wright, R. F.: 2001, 'Assessment of recovery of European surface waters from acidification 1970–2000', *Hydrol. Earth Syst. Sci.* **5**, 273–542.

Krouse, H. R. and Grinenko, V. A.: 1991, *Stable Isotopes. Natural and Anthropogenic Sulphur in the Environment*, SCOPE 43, John Wiley & Sons, New York, pp. 466.

Lal, D. and Peters, B.: 1966, *Cosmic Ray Produced Radioactivity on the Earth, Handbuch der Physik*, Springer-Verlag, New York, U.S.A., pp. 550–612.

Likens, G. E., Driscoll, C. T., Buso, D. C., Mitchell, M. J., Lovett, G. M., Bailey, S. W., Siccama, T. G., Reiners, W. A. and Alewell, C.: 2002, 'The biogeochemistry of sulfur at Hubbard Brook', *Biogeochemistry* **60**, 235–316.

Mayer, B., Feger, K. H., Giesemann, A. and Jäger, H. J.: 1995, 'Interpretation of sulfur cycling in two catchments in the Black Forest (Germany) using stable sulfur and oxygen isotope data', *Biogeochemistry* **30**, 31–58.

Michel, R. L., Campbell, D., Clow, D. and Turk, J. T.: 2000, 'Timescales for migration of atmospherically derived sulphate through an alpine/subalpine watershed, Loch Vale, Colorado', *Water Resour. Res.* **36**, 27–36.

Michel, R. L., Turk, J. T., Campbell, D. H. and Mast, M. A.: 2002, 'Use of natural ^{35}S to trace sulphate cycling in small lakes, Flattops Wilderness Area, Colorado, U.S.A.', *Water, Air, Soil Pollut.: Focus* **2**, 5–18.

Mitchell, M. J. and Fuller, R. D.: 1988, 'Models of sulfur dynamics in forest and grassland ecosystems with an emphasis on soil processes', *Biogeochemistry* **5**, 133–164.

Novák, M., Wieder, R. K. and Schell, W. R.: 1994, 'Sulfur during early diagenesis in *Sphagnum* peat: Insights from δ^{34}S ratio profiles in ^{210}Pb-dated peat cores', *Limnol. Oceanogr.* **39**, 1172–1185.

Novák, M. and Přechová, E.: 1995, 'Movement and transformation of ^{35}S-labelled sulphate in the soil of a heavily polluted site in the Northern Czech Republic', *Environ. Geochem. Health* **17**, 83–94.

Novák, M., Bottrell, S. H., Groscheová, H., Buzek, F. and Černý, J.: 1995, 'Sulphur isotope characteristics of two North Bohemian forest catchments', *Water, Air, Soil Pollut.* **85**, 1641–1646.

Novák, M., Bottrell, S. H., Fottová, D., Buzek, F., Groscheová, H. and Žák, K.: 1996, 'Sulfur isotope signals in forest soils of Central Europe along an air pollution gradient', *Environ. Sci. Technol.* **30**, 3473–3476.

Novák, M., Kirchner, J. W., Groscheová, H., Havel, M., Černý, J., Krejčí, R. and Buzek, F.: 2000, 'Sulfur isotope dynamics in two Central European watersheds affected by high atmospheric deposition of SO$_x$', *Geochim. Cosmochim. Acta* **64**, 367–383.

Novák, M., Jačková, I. and Přechová, E.: 2001, 'Temporal trends in the isotope signature of air-borne sulfur in Central Europe', *Environ. Sci. Technol.* **35**, 255–260.

Novák, M., Buzek, F., Harrison, A. F., Přechová, E., Jačková, I. and Fottová, D.: 2003, 'Similarity between C, N and S stable isotope profiles in European spruce forest soils: Implications for the use of δ^{34}S as a tracer', *Appl. Geochem.* **18**, 765–779.

Nriagu, J. O., Holdway, D. A. and Coker, R. D.: 1987, 'Biogenic sulfur and the acidity of rainfall in remote areas of Canada', *Science* **237**, 1189–1192.

Peters, N. E., Černý, J., Havel, M. and Krejčí, R.: 1999, 'Temporal trends of bulk precipitation and water chemistry (1977–1997) in a small forested area, Krusné hory, northern Bohemia, Czech Republic', *Hydrol. Process.* **13**, 2721–2741.

Sall, J. and Lehman, A.: 1996, *JMP Start Statistics*, Duxbury Press, New York, pp. 656.

Siegenthaler, U.: 1971, 'Sauerstoff-18, Deuterium und Tritium im Wasserkreislauf', unpublished *Ph.D. Dissertation*, University of Bern, Switzerland.

Strickland, T. C. and Fitzgerald, J. W.: 1984, 'Formation and mineralization of organic sulfur in forest soils', *Biogeochem.* **1**, 79–95.

Sueker, J. K., Turk, J. T. and Michel, R. L.: 1999, 'Use of cosmogenic S-35 for comparing ages of water from three alpine-subalpine basins in the Colorado Front Range', *Geomorphology* **27**, 61.

Sumner, E.: 2000, *Handbook of Soil Science*, CRC Press, Boca Raton, pp. 2148.

Watwood, M. E. and Fitzgerald, J. W.: 1988, 'Sulfur transformations in forest litter and soil-results of laboratory and field incubations', *Soil Sci. Soc. Am. J.* **52**, 1478–1483.

Wieder, R. K. and Lang, G. E.: 1988, 'Cycling of inorganic and organic sulfur in peat from Big Run Bog, West Virginia', *Biogeochemistry* **5**, 221.

δ^{13}C OF TREE-RING LIGNIN AS AN INDIRECT MEASURE OF CLIMATE CHANGE

I. ROBERTSON[1,2]*, N. J. LOADER[2], D. McCARROLL[2], A. H. C. CARTER[3], L. CHENG[4] and S. W. LEAVITT[4]

[1] *Quaternary Dating Research Unit, CSIR Environmentek, P.O. Box 395, Pretoria 0001, South Africa;* [2] *Department of Geography, University of Wales Swansea, Swansea SA2 8PP, U.K.;* [3] *Godwin Institute for Quaternary Research, University of Cambridge, Cambridge CB2 3SA, U.K.;* [4] *Laboratory of Tree-Ring Research, University of Arizona, Tucson, AZ 85721, U.S.A.*
(* *author for correspondence, e-mail: i.robertson@swansea.ac.uk; phone: +44 1792 295184; fax: +44 1792 295955*)

(Received 20 August 2002; accepted 10 April 2003)

Abstract. High-resolution paleoclimatic data are an essential requirement for testing numerical models of climate change and the global carbon cycle. If the long tree-ring chronologies, originally established for the purpose of dendrochronology, are to be fully exploited as an indirect measure of past climatic variability, additional techniques are required to obtain this information. The determination of the δ^{13}C value of tree-ring cellulose has been used successfully to reconstruct past climates. However, under both aerobic and anaerobic conditions, the polysaccharide components of vascular plants (mainly cellulose and hemicelluloses) are more prone to rapid degradation than lignin. This has serious implications for the use of carbon isotope values of tree-ring cellulose as an indirect measure of past climates. An absolutely dated ring-width chronology was established for oaks (*Quercus robur* L.) growing at Sandringham Park in eastern England. Carbon isotope values were determined on α-cellulose and 'Klason' lignin isolated from annual latewood samples over the period AD 1895–1999. The carbon isotope values of earlywood lignin are correlated with the latewood carbon isotope values of the previous year, supporting the theory that some of the carbon utilised in earlywood synthesis is assimilated in the previous year. The high-frequency variance in the carbon isotope indices of latewood lignin and cellulose is highly correlated with combined July and August environmental variables, indicating that they were formed at similar times. There was no evidence of secondary lignification. These results demonstrate that the determination of carbon isotope values of latewood lignin offers the potential to obtain unambiguous proxy climatic data covering several millennia.

Keywords: α-cellulose, dendrochronology, diagenesis, lignin, oak, phenology, stable isotopes

1. Introduction

Placing anthropogenic influences upon global climate into a longer-term perspective requires indirect measures of past climatic variability. High-resolution measures of past climate may be obtained from any biological or geological system that exhibits seasonality in growth rates. Tree-ring data have been used extensively to reconstruct past climates from an annual resolution to much lower frequencies depending upon signal strength and site characteristics (e.g., Briffa, 2000; Esper *et al*., 2002). In regions where tree growth is less limited by a dominant climatic factor, stable isotopes have been shown to exhibit a stronger high-frequency climate signal than ring-widths alone (Robertson *et al.*, 1997) and consequently there has been a renewed interest in this area of research.

It has been proposed recently that the analysis of stable isotopes in wholewood or resin-extracted wholewood isolated from tree-rings may be adequate for climatic reconstruction (Borella *et al.*, 1998; Leuenberger *et al.*, 1998; Barbour *et al.*, 2001). This is supported by several studies that have demonstrated that there is a close association between the carbon isotope values of cellulose and wholewood (Bender and Berge, 1982; Leavitt and Long, 1991; Borella *et al.*, 1998). However, if the isotopic composition of the long tree-ring chronologies (Pilcher *et al.*, 1984) is to be used as a proxy measure of past climates, the influence of wood degradation must be investigated. Through time, wood may suffer from partial decay (Becker *et al.*, 1991), making it difficult to measure rings and interpret stable isotope values. Although it has been reported that the preferential decay of cellulose in old wood does not affect hydrogen isotope values (Yapp and Epstein, 1977), the influence of diagenesis requires further investigation.

2. Tree-ring Lignin

Wood is a composite material constructed from a variety of organic polymers. The basic structural material is the complex polymer cellulose with lower molecular weight hemicelluloses acting as a matrix. The third major constituent of wood is lignin. The celluloses and lignin are subject to different biochemical pathways during the photosynthetic assimilation of carbon. Even though the absolute $\delta^{13}C$ values of wood celluloses and lignin may differ, there is no reason to suspect that one will record past climates better than the other will. However, if one constituent contains a higher relative proportion of carbon remobilised from the previous year, the correlation with climate may be diminished. Another class of wood constituents are the solvent-extractable substances, which are essentially resins. Tans *et al.* (1978) used $\Delta^{14}C$ measurements to demonstrate that 4–5% of organic material in an oak moved across the 1962–1963 boundary. In oaks, solvent-extractable substances usually constitute a very small proportion of wholewood (Barbour *et al.*, 2001) while in gymnosperms, the proportion of resin will be considerably higher and hence, resins are usually extracted before isotopic determination.

To minimise potential chemical contamination, the pioneering studies determined the $\delta^{13}C$ values of wholewood alone (Craig, 1954; Farmer and Baxter, 1974a, b; Libby *et al.*, 1976). However, the analysis of a single chemical component was recommended by Epstein *et al.* (1976). As cellulose is the most abundant component of wood, it has become the preferred component for isotope studies. In the following years, numerous studies demonstrated successfully the use of carbon stable isotopes in tree-ring cellulose for climatic reconstruction (Lipp *et al.*, 1991; Anderson *et al.*, 1998; Hemming *et al.*, 1998; McCarroll and Pawellek, 2001).

Although several studies have calculated the isotopic composition of tree-ring lignin (Gray and Thompson, 1977; Leuenberger *et al.*, 1998; Borella *et al.*, 1998, 1999; Barbour *et al.*, 2001), direct measurements are limited (Wilson and Grinsted, 1977, 1978; Grinsted, 1977; Mazany *et al.*, 1980). In one of the few studies on tree-ring lignin, Gray and Thompson (1977) found that the calculated $\delta^{18}O$ values did not record past temperatures. Turney *et al.* (1999) reported results that are more favourable when they observed a link between the $\delta^{13}C$ values of cladodes lignin and vapour pressure deficit for *Phyllocladus alpinus* growing throughout New Zealand. Similarly, vegetation dynamics have been reconstructed from the isotope values of compound-specific lignin biomarkers preserved in sediments (Huang *et al.*, 1999). Despite these limited studies, the physicochemical properties of lignin have not been used extensively in paleoclimatology, but lignin has been the subject of active research in the paper industry, where its removal forms the basis of chemical pulping.

The main function of lignin is that of mechanical support as it renders wood a hard, rigid material able to withstand considerable stress. Lignin also has a significant role in assisting the conduction of water, nutrients and metabolites. Lignin is a highly branched polyphenolic molecule derived from the polymerisation of *p*-coumaryl, coniferyl and sinapyl alcohols. The lignin content of angiosperms and gymnosperms differs. While angiosperm lignin contains a mixture of coniferyl and sinapyl alcohols, gymnosperm lignin is predominantly composed of coniferyl alcohol with a lesser contribution of *p*-coumaryl alcohol. Even within a plant, the relative proportions of these monomers differ (Boudet, 2000).

Although it is generally reported that lignin is resistant to biological degradation (Benner *et al.*, 1987; Spiker and Hatcher, 1987; Schleser *et al.*, 1999b); local environmental conditions dictate the nature and extent of wood decomposition. In most terrestrial environments, the majority of wood decomposition is caused by brown, white, and soft-rot fungi (Blanchette, 2000). Brown rot fungi predominantly degrade cellulose, whereas white rot fungi usually degrade cell wall constituents, including lignin. Soft rot fungi can tolerate a wider range of environmental conditions and are particularly prevalent in moist environments where they preferentially degrade carbohydrates. Within the latter environment, bacteria also are present, degrading carbohydrates with minimal lignin deterioration. Therefore, in most cases encountered with samples dating from the Holocene, wood degradation causes the decomposition of cellulose and hemicelluloses, leading to an increase in the relative proportion of lignin in the residue. Over much longer timescales, it was reported that cellulose decomposition was complete in Tertiary and Mesozoic sub-fossil samples (van Bergen and Poole, 2002).

The degradation of cellulose has serious implications for the use of $\delta^{13}C$ values of tree-ring cellulose as an indirect measure of past climates. Lipp *et al.* (1991) extended the $\delta^{13}C$ values of tree-ring cellulose over the last 1000 years, but due to the differential degradation of cellulose this may not be possible over the entire sub-fossil Holocene record. Therefore, the determination of the $\delta^{13}C$ value of tree-ring lignin may offer the potential to obtain unambiguous proxy climatic data covering several millennia.

3. Methods

A 12 mm diameter core was obtained from an oak (*Quercus robur* L.) growing at Sandringham Park in eastern England (52°50′N, 0°30′E). The core was cross-dated against the site chronology and other independent chronologies (Robertson *et al.*, 1997; Robertson, 1998). Under magnification, the annual earlywood and latewood samples were removed using a razor blade to produce slivers. The slivers were ground in a Wiley Mill until they passed though a 0.85 mm (20 mesh) stainless steel sieve. 'Klason' lignin (with the minimum degree of modification) was isolated by refluxing with 72% sulphuric acid to dissolve carbohydrates (TAPPI, 1988).

The lignin ($\approx$2 mg) was loaded into 125 mm long$\times$7 mm bore borosilicate tubes together with >500 mg of pre-heated copper(II) oxide. The loaded tubes were evacuated to <10^{-3} mbar for about 1 hr, sealed with an oxy-gas torch, and heated in a muffle furnace for >8 hr at 500 °C (Sofer, 1980). The tube was broken in a tube cracker and the resulting gases separated cryogenically. The CO_2 was collected and the isotope ratios were measured on a Finnigan Delta-S mass spectrometer in the Department of Geosciences at the University of Arizona. $\delta^{13}C$ values are expressed relative to the VPDB Standard (Coplen, 1995) with an overall precision of better than $\pm$0.1‰.

Averaged monthly temperature values were calculated using the mean 24 hr maxima and minima daily values read at 09h GMT from the central England temperature (CET) record for AD 1895–1994 (Manley, 1974). Relative humidity values are given for 13h GMT from the combined Cranwell-Waddington record for the period AD 1920–1994. Vapour pressure deficit values were calculated using relative humidity values at 13h GMT from the combined Cranwell-Waddington record and CET values for AD 1920–1994. Rainfall was measured at Sandringham for the period AD 1904–1994 (Robertson, 1998). Phenological data were available from the combined Marsham-Ashtead series (Sparks and Carey, 1995; Sparks *et al.*, 1997; Aykroyd *et al.*, 2001).

4. Results and Discussion

4.1. Intra-ring isotopic variability

With the exception of the work of Bender and Berge (1982), most studies have reported a difference between earlywood and latewood $\delta^{13}C$ values from oak cellulose (Ogle and McCormac, 1994; Robertson *et al.*, 1997; Schleser *et al.*, 1999a). Hill *et al.* (1995) found that carbohydrates formed during the preceding summer

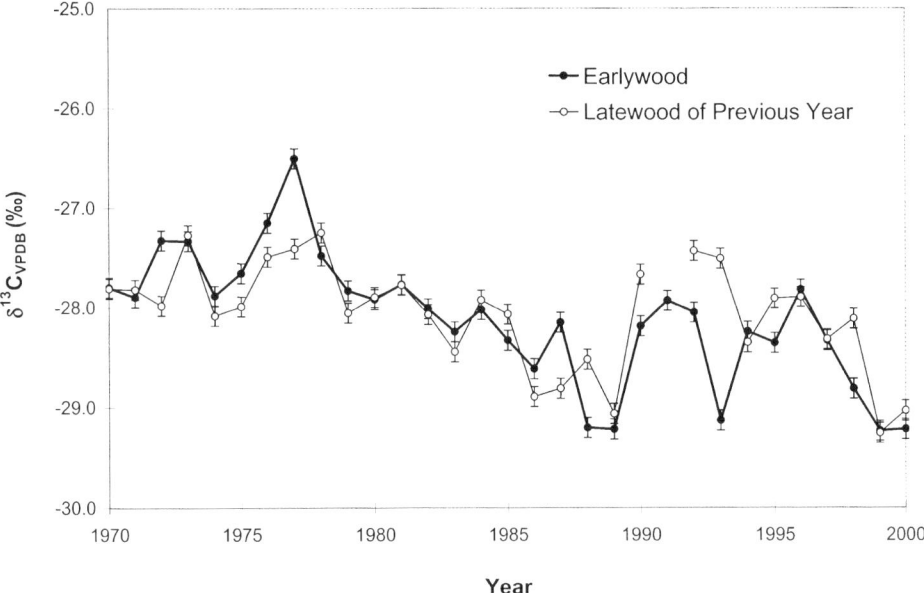

Figure 1. Relationship between $\delta^{13}C$ of earlywood lignin and $\delta^{13}C$ of latewood lignin of the previous year for tree SP19 (AD 1970–2000). Error bars represent overall precision ($\pm 0.1‰$).

were utilised during the synthesis of earlywood cellulose. Figure 1 illustrates that a similar relationship holds for oak lignin. The correlation between the $\delta^{13}C$ values of lignin from earlywood and latewood of the previous year ($r = 0.70$; $p < 0.01$) was similar to the correlation between the $\delta^{13}C$ values of earlywood and the latewood from the same year ($r = 0.66$; $p < 0.01$). However, as part of this relationship may be attributed to low-frequency trends in the time-series, the lignin $\delta^{13}C$ values were standardised by dividing the measured value with a 'fitted' value obtained from a 60 yr Gaussian filter. A 10 yr high-pass Gaussian filter was then applied to the standardised indices to define the high-frequency variance (Robertson *et al.*, 1997). The correlation between the high-frequency carbon isotope indices of lignin from earlywood and latewood of the previous year ($r = 0.36$; $p < 0.05$) remained similar to the correlation between the $\delta^{13}C$ values of earlywood and the latewood from the same year ($r = 0.47$; $p < 0.01$). With the correlation caused by the underlying trend in the time-series removed, there is still supporting evidence that, at least some of the carbohydrates formed during the preceding summer were utilised during earlywood formation (Pilcher, 1995).

Over the period AD 1970–2000, there was no significant difference ($p > 0.01$) between the mean $\delta^{13}C$ value of earlywood lignin ($\bar{x} = -28.08$; $\sigma_{n-1} = 0.65$; $n = 30$) and the mean $\delta^{13}C$ value of latewood lignin ($\bar{x} = -28.07$; $\sigma_{n-1} = 0.54$; $n = 30$) from the same year. However, part of this relationship may be linked to the low-frequency trends in the data. Only one other study has compared the early

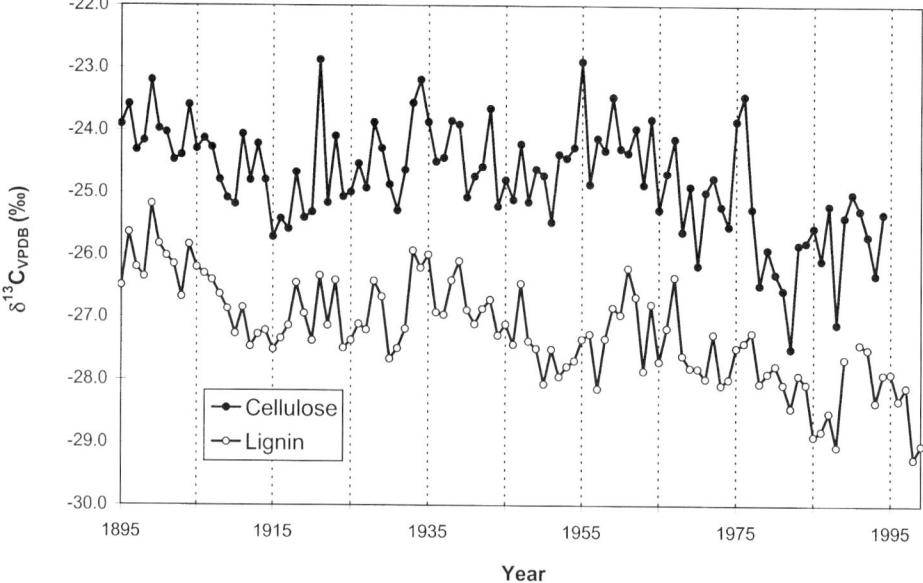

Figure 2. Annual Sandringham δ^{13}C values (AD 1895–1999) for tree SP19. Over the common period (AD 1895–1994), the mean α-cellulose δ^{13}C value = -24.76‰ (n = 100; σ_{n-1} = 0.87) and the mean lignin δ^{13}C value = -27.17‰ (n = 99; σ_{n-1} = 0.76). $\Delta_{\text{cellulose} - \text{lignin}}$ = 2.41‰.

and latewood δ^{13}C values of tree-ring lignin (Grinsted, 1977; Wilson and Grinsted, 1977, 1978). The authors observed that earlywood of Monterey pine (*Pinus radiata* D. Don) was enriched in ^{13}C. For the specimen of Monterey pine growing in the southern hemisphere, earlywood formation corresponds with summer growth, while latewood formation corresponds with winter growth.

Although high-resolution isotopic analysis (Ogle and McCormac, 1994; Loader *et al.*, 1995; Schleser *et al.*, 1999a) has revealed that the physical and isotopic boundaries may not correspond exactly, we adopted a pragmatic approach and analysed the δ^{13}C values of unambiguously assigned latewood (Robertson *et al.*, 1996).

4.2. Relationship between high-frequency carbon isotope indices and environmental variables

Figure 2 shows the annual latewood δ^{13}C values (AD 1895–1999) for lignin isolated from tree SP19 at Sandringham. The annual latewood δ^{13}C values (AD 1895–1994) for α-cellulose extracted from a different core from the same tree also are illustrated (Loader *et al.*, 1997; Robertson *et al.*, 1997). Lignin δ^{13}C values were 2.41‰ more depleted in ^{13}C than α-cellulose. This is similar to previously reported values. Park and Epstein (1961) grew tomato plants in 1.5% CO_2 and observed that the δ^{13}C values of lignin were 0.6‰ lighter than cellulose. Direct measurements

TABLE I

Relationship between averaged monthly 13h GMT relative humidity and high-frequency cellulose and lignin carbon isotope indices (1920–1994) for tree SP19

Averaged monthly relative humidity	Correlation (r) with high-frequency $\delta^{13}C$ index	
	Cellulose	Lignin
January	−0.14	−0.01
February	−0.06	−0.02
March	0.07	−0.02
April	−0.05	−0.13
May	0.01	−0.14
June	−0.27*	−0.26*
July	−0.57**	−0.42**
August	−0.37**	−0.40**
September	−0.31**	−0.30*
October	−0.14	−0.09
November	0.00	−0.09
December	0.02	−0.07

Here, correlations are represented by the Pearson product moment correlation coefficient. Significance level: * $p < 0.05$; ** $p < 0.01$.

on tree-rings have found that lignin is approximately 3‰ isotopically lighter than cellulose (Freyer and Wiesberg, 1974; Grinsted, 1977; Wilson and Grinsted, 1977, 1978; Loader et al., 2003). Mazany et al. (1980) found that the $\delta^{13}C$ values of tree-ring lignin isolated from 10 yr bulk ponderosa pine (*Pinus ponderosa* Laws.) samples were 4.1‰ more depleted than cellulose. Borella et al. (1998) calculated $\delta^{13}C$ values of tree-ring lignin from the measurements of resin-extracted wood and holocellulose. For birch, beech, oak, and spruce, the mean difference between the $\delta^{13}C$ values of cellulose and lignin was 3.25‰.

The anthropogenic decrease in $\delta^{13}C$ values attributed to the burning of fossil fuels and the release of carbon through deforestation (Keeling et al., 1979; Leavitt and Long, 1989) is evident in both the α-cellulose and lignin time-series (Figure 2). These anthropogenic influences and other low frequency trends were removed from the time-series during standardisation (Robertson et al., 1997). Table I illustrates the relationship between monthly averaged relative humidity and high-frequency $\delta^{13}C$ indices over the period AD 1920–1994. For both lignin and α-cellulose, the relative humidity values of July and August were more influential than any other months in determining $\delta^{13}C$ indices. Similar results have been reported for the α-cellulose isolated from oaks in eastern England (Robertson et al., 1997). Therefore,

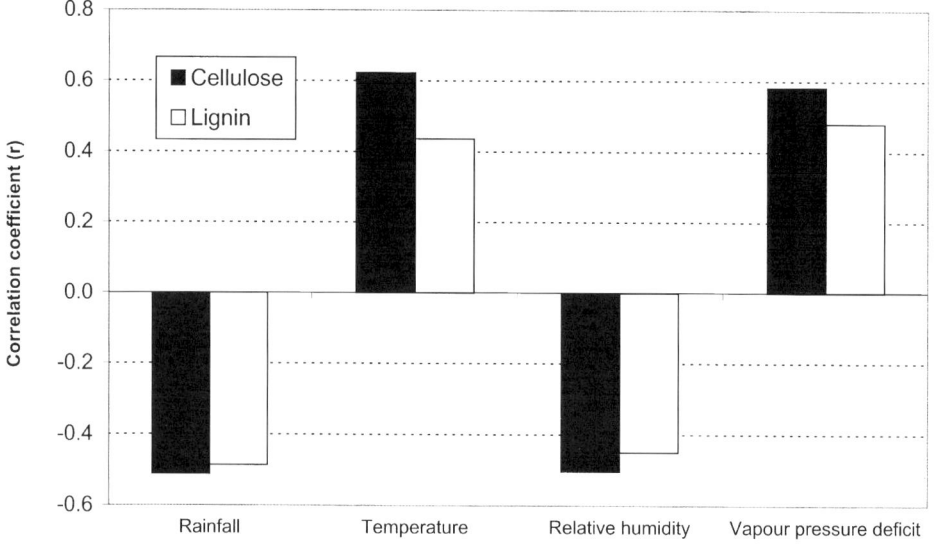

Figure 3. Relationship between high-frequency Sandringham cellulose and lignin carbon isotope indices for tree SP19 and high-frequency environmental variables. All correlations (r) are significant at $p < 0.01$.

the relationship between isotope indices and combined July/August high-frequency environmental variables was investigated.

Figure 3 shows the relationship between high-frequency α-cellulose and lignin carbon isotope indices and environmental variables standardised in a similar manner. The magnitude of the climatic response is reduced as the analyses, in this pilot project, were performed on single cores. It is usually recommended that several cores should be analysed to represent a site with an adequate degree of replication. The number of cores required is frequency-dependent with a greater number of cores necessary to reconstruct lower temporal frequencies (Robertson *et al.*, 1997).

Environmental factors, which affect the rate of photosynthesis or stomatal conductance, will influence tree-ring $\delta^{13}C$ values (Vogel, 1980; Francey and Farquhar, 1982). The inverse association with rainfall may be attributed to water stress causing a decrease in photosynthesis, transpiration, and leaf conductance (Farquhar *et al.*, 1989). Although there is no theoretical basis for the temperature response, many studies have reported an empirically derived relationship with temperature (Heaton, 1999). If temperature is chosen as the limiting growth factor, the response could oscillate around the temperature optimum, giving rise to a positive (Figure 3) or negative response (Schleser *et al.*, 1999a). As the stomatal aperture has been shown to increase in response to an increase in relative humidity, the relationship between relative humidity or the derived vapour pressure deficit with carbon isotope indices has a firm theoretical basis (Hemming *et al.*, 1998), even if the precise mechanism is still unknown (Grantz, 1990). Despite these uncertainties, the relationships between carbon isotope indices and environmental variables are similar to those reported elsewhere (Lipp *et al.*, 1991; Hemming *et al.*, 1998; McCarroll and Pawellek, 2001).

The α-cellulose and lignin δ^{13}C time-series are significantly correlated ($r = 0.73$; $p < 0.01$) indicating that they both record similar changes. However, as part of this relationship may be attributed to low-frequency trends in the time-series, the relationship was investigated at different frequencies (Robertson *et al.*, 1997). The low-frequency trends, evident in the correlation between the time-series ($r = 0.91$; $p < 0.01$) can be attributed partly to the well-documented anthropogenic changes (Figure 2), while the high-frequency variability, reflected in the correlation between the time-series ($r = 0.64$; $p < 0.01$) contains the direct response to climate (Figure 3). Wilson and Grinsted (1977) reported that intra-ring δ^{13}C values from cellulose and lignin followed a similar pattern. The δ^{13}C values of α-cellulose were more sensitive to climate than the δ^{13}C values of lignin (Figure 3). Similar results were reported by Mazany *et al.*, (1980) for 10 yr bulk samples of ponderosa pine. However, the difference between the mean correlation of carbon isotope indices of these wood constituents and environmental variables was not statistically significant ($p > 0.01$), supporting the view that either the δ^{13}C value of tree-ring cellulose or lignin could be used as an indirect measure of past climates. In theory, the δ^{13}C values of wood could also be used as an indirect measure of past climates. However, in addition to the reported problems of diagenesis, the lignin to cellulose ratio varies (Freyer and Wiesberg, 1974). The exact magnitude of this variability both between samples and through time requires more in-depth investigation, but if found to be significant, it would make analysis of the δ^{13}C values of wholewood for paleoenvironmental research of limited value (Grinsted, 1977).

In a similar manner, there is still debate about the preferred wood constituent for oxygen isotopes analysis. Borella *et al.* (1999) found that δ^{18}O values of α-cellulose had a stronger association with the δ^{18}O values of precipitation than that of wood, suggesting that some climatic information may be lost if bulk wood samples are analysed. More recently, Barbour *et al.* (2001) reported that for oaks, the estimated δ^{18}O values of tree-ring lignin had a lower correlation with site parameters than α-cellulose, which was lower than that of wood. Therefore, from this spatial study, they concluded that under specific circumstances the extraction of α-cellulose for paleoclimatic studies may be unnecessary.

4.3. SECONDARY LIGNIFICATION

The deposition of lignin occurs as one of the final stages of cell differentiation after the cessation of cell expansion. Lignin is deposited within the secondary cell wall within the carbohydrate matrix (Donaldson, 2001) and consequently, there is a lag between the formation of cellulose and lignin (Fritts, 1976; Gindl *et al.*, 2000). To investigate the proposed lag between α-cellulose and lignin formation, a moving window was applied to the daily central England temperature data to define the period of optimal tree response to climate (Aykroyd et al., 2001).

For the purposes of comparison, the correlation with combined monthly climatic variables was calculated from the daily temperature data. For the period with both phenological observations and carbon isotope indices, the correlation between the mean temperature for July/August (Julian days 182–243 in non-leap years) and carbon isotope indices was $r = 0.60$ ($p < 0.01$; $n = 90$) for SP19 cellulose and $r = 0.47$ ($p < 0.01$, $n = 90$) for SP19 lignin. This strong association is supported by growth measurements on oaks (Hemming, pers. comm.). As the onset of growth has advanced considerably in recent years (Fitter and Fitter, 2002), the association with temperature for a fixed period after bud burst was investigated using a moving window. This approach accounts for changes in growth patterns that occur independent of the calendar date. Bud burst data were obtained from the combined phenological series for eastern England (Sparks and Carey, 1995; Sparks et al., 1997; Aykroyd et al., 2001).

The optimal association between carbon isotope indices from tree SP19 and temperature was found for a 20 day window, which was applied to daily temperature data for a progressively increasing period after initial bud burst (Figure 4). For the carbon isotope indices from SP19 cellulose, this period was found to begin approximately 89 days after initial bud burst ($r = 0.55$; $p < 0.01$; $n = 90$), whereas for the carbon isotope indices from SP19 lignin, this period was found to begin approximately 88 days after initial bud burst ($r = 0.43$; $p < 0.01$; $n = 90$). For both lignin and α-cellulose, the highest association with temperature coincides with the Lammas growth of foliage, during conditions favourable for photosynthesis (Aykroyd et al., 2001). No evidence was found for the previously reported lag between the start of cellulose and lignin formation (Wilson and Grinsted, 1977, 1978). Similar results are reported by Loader et al. (2003), where a moving window was applied to isotopic data from wood constituents over a 55 yr period, to explore the association with temperature from fixed calendar dates.

5. Conclusions

Lignin $\delta^{13}C$ values were 2.41‰ more depleted in ^{13}C than α-cellulose reflecting the different biochemical pathways followed during the photosynthetic assimilation of carbon. Similar to the results previously reported for α-cellulose, the $\delta^{13}C$ values of earlywood lignin display a positive correlation with the latewood $\delta^{13}C$ values of the previous year, supporting the theory that some of the carbon utilised in earlywood synthesis was assimilated in the previous year. The high-frequency variance in the carbon isotope indices of latewood lignin and cellulose was correlated with combined high-frequency July and August environmental variables indicating that they or their precursors were formed from material fixed at similar

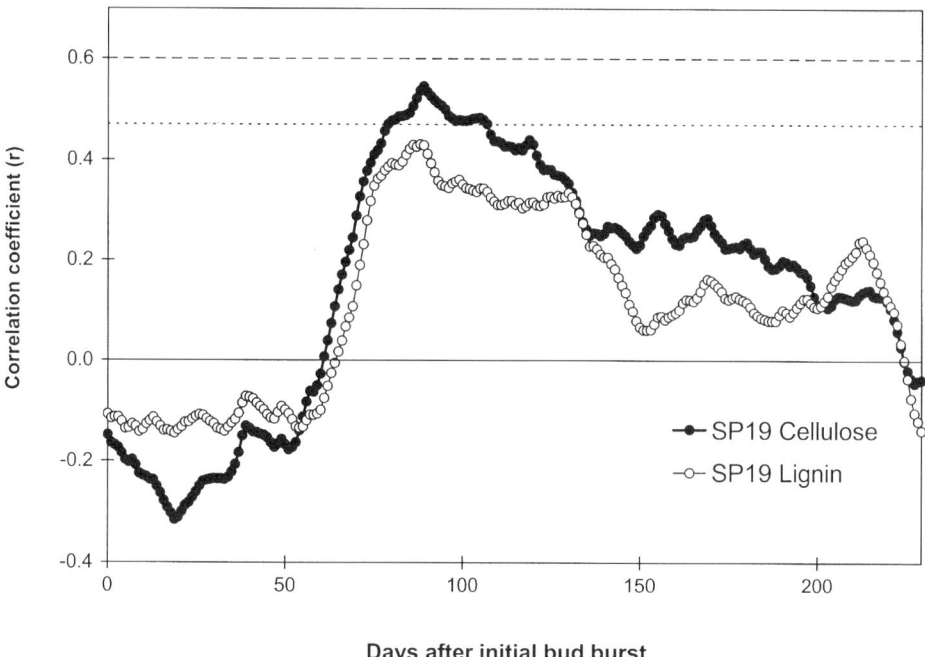

Figure 4. Moving mean 20-day correlation (start-point) between carbon isotope indices (closed circles represent SP19 cellulose; open circles represent SP19 lignin) and daily central England temperature data after initial bud burst each year for the period AD 1895–1994. The dashed line represents the overall correlation between the SP19 cellulose carbon isotope index and the mean temperature for Julian days 182–243 (equivalent to July/August for non-leap years) for the period with both phenological observations and isotope values over the period 1895–1994. Similarly, the dotted line represents the overall correlation between the SP19 lignin carbon isotope index and the mean temperature for Julian days 182–243 (equivalent to July/August for non-leap years) for the period with both phenological observations and isotope values over the period AD 1895–1994.

times. Therefore, $\delta^{13}C$ values of α-cellulose or lignin from latewood could be used as an indirect measure of past climates. Lignin has the advantage that it is more resistant to decay than cellulose under a wide range of environmental conditions.

Acknowledgements

The authors thank Mike Saville and the Royal Estate of Sandringham for permission to sample trees; Chris Turney for technical advice; Chris Eastoe and David Dettman for assistance with mass spectrometry; Phil Jones for providing the updated central England temperature series; Tim Sparks for supplying unpublished phenological data and Debbie Hemming, Cynthia Froyd, Roy Switsur, John Waterhouse and two anonymous referees for constructive advice. This research was supported by the Leverhulme Trust (Grant 19990411), the European Union (EVK2-

CT-2002-00136 and EVK2-CT-2002-00147), the Quaternary Research Association and the award of an Agnese Haury Fellowship at the Laboratory of Tree Ring Research, University of Arizona (IR).

References

Anderson, W. T., Bernasconi, S. M., McKenzie, J. A. and Saurer, M.: 1998, 'Oxygen and carbon isotopic record of climatic variability in tree ring cellulose (*Picea abies*): An example from central Switzerland (1913–1995)', *J. Geophys. Res.* **103**, 31,625–31,636.

Aykroyd, R. G., Lucy, D., Pollard, A. M., Carter, A. H. C. and Robertson, I.: 2001, 'Temporal variability in the strength of proxy-climate correlations', *Geophys. Res. Lett.* **28**, 1559–1562.

Barbour, M. M., Andrews, T. J. and Farquhar, G. D.: 2001, 'Correlations between oxygen isotope ratios of wood constituents of *Quercus* and *Pinus* samples from around the world', *Aust. J. Plant Physiol.* **28**, 335–348.

Becker, B., Kromer, B. and Trimborn, P.: 1991, 'A stable-isotope tree-ring timescale of the Late Glacial/Holocene boundary', *Nature* **353**, 647–649.

Bender, M. M. and Berge, A. J.: 1982, 'Carbon isotope records in Wisconsin trees', *Tellus* **34**, 500–504.

Benner, R., Fogel, M. L., Sprague, E. K. and Hodson, R. E.: 1987, 'Depletion of ^{13}C in lignin and its implications for stable carbon isotope studies', *Nature* **329**, 708–710.

Blanchette, R. A.: 2000, 'A review of microbial deterioration found in archaeological wood from different environments', *Int. Biodeter. Biodegr.* **46**, 189–204.

Borella, S., Leuenberger, M., Saurer, M. and Siegwolf, R.: 1998, 'Reducing uncertainties in δ^{13}C analysis of tree rings: Pooling, milling and cellulose extraction', *J. Geophys. Res.* **103**, 19519–19526.

Borella, S., Leuenberger, M. and Saurer, M.: 1999, 'Analysis of δ^{18}O in tree-rings: Wood-cellulose comparison and method dependent sensitivity', *J. Geophys. Res.* **104**, 19,267–19,273.

Boudet, A.: 2000, 'Lignins and lignification: Selected issues', *Plant Physiol. Biochem.* **38**, 81–96.

Briffa, K. R.: 2000, 'Annual climate variability in the Holocene: Interpreting the message of ancient trees', *Quat. Sci. Rev.* **19**, 87–105.

Coplen, T.: 1995, 'Discontinuance of SMOW and PDB', *Nature* **375**, 285.

Craig, H.: 1954, 'Carbon-13 variations in sequoia rings and the atmosphere', *Science* **119**, 141–144.

Donaldson, L. A.: 2001, 'Lignification and lignin topochemistry – An ultrastructural view', *Phytochemistry* **57**, 859–873.

Epstein, S., Yapp, C. J. and Hall, J. H.: 1976, 'The determination of the D/H ratio of non-exchangeable hydrogen in cellulose extracted from aquatic and land plants', *Earth Planet. Sci. Lett.* **30**, 241–251.

Esper, J., Cook, E. R. and Schweingruber, F. H.: 2002, 'Low-frequency signals in long tree-ring chronologies for reconstructing past temperature variability', *Science* **295**, 2250–2253.

Farmer, J. G. and Baxter, M. S.: 1974a, 'Atmospheric carbon dioxide levels as indicated by the stable isotope record in wood', *Nature* **247**, 273–275.

Farmer, J. G. and Baxter, M. S.: 1974b, 'Drs Farmer and Baxter reply', *Nature* **252**, 757.

Farquhar, G. D., Ehleringer, J. R. and Hubick, K. T.: 1989, 'Carbon isotope discrimination and photosynthesis', *Ann. Rev. Plant Physiol. Plant Mol. Biol.* **40**, 503–537.

Fitter, A. H. and Fitter, R. S. R.: 2002, 'Rapid changes in the flowering time in British plants', *Science* **296**, 1689–1691.

Francey, R. J. and Farquhar, G. D.: 1982, 'An explanation for the ^{12}C/^{13}C variations in tree-rings', *Nature* **297** 28–31.

Freyer, H. D. and Wiesberg, L.: 1974, 'Dendrochronology and ^{13}C content in atmospheric CO_2', *Nature* **252**, 757.

Fritts, H. C.: 1976, *Tree Rings and Climate*, Academic Press, New York.

Gindl, W., Grabner, M. and Wimmer, R.: 2000, 'The influence of temperature on latewood lignin content in treeline Norway spruce compared with maximum density and ring width', *Trees* **14**, 409–414.

Grantz, D. A.: 1990, 'Plant response to atmospheric humidity', *Plant Cell Environ.* **13**, 667–679.

Gray, J. and Thompson, P.: 1977, 'Climatic information from $^{18}O/^{16}O$ analysis of cellulose, lignin and whole wood from tree rings', *Nature* **270**, 708–709.

Grinsted, M. J.: 1977, 'A study of the relationship between climate and stable isotope ratios in tree rings', unpublished *Ph.D. Dissertation*, University of Waikato.

Heaton, T. H. E.: 1999, 'Spatial, species, and temporal variations in the $^{13}C/^{12}C$ ratios of C_3 plants: Implications for palaeodiet studies', *J. Arch. Sci.* **26**, 637–649.

Hemming, D. L., Switsur, V. R., Waterhouse, J. S., Heaton, T. H. E. and Carter, A. H. C.: 1998, 'Climate variation and the stable carbon isotope composition of tree-ring cellulose: An intercomparison of *Quercus robur*, *Fagus sylvatica* and *Pinus sylvestris*', *Tellus* **50B**, 25–33.

Hill, S. A., Waterhouse, J. S., Field, E. M., Switsur, V. R. and Rees, T.: 1995, 'Rapid recycling of triose phosphates in oak stem tissue', *Plant Cell Environ.* **18**, 931–936.

Huang, Y., Freeman, K. H., Eglinton, T. I. and Street-Perrott, F. A.: 1999, 'δ^{13}C analyses of individual lignin phenols in Quaternary lake sediments: A novel proxy for deciphering past terrestrial vegetation changes', *Geology* **27**, 471–474.

Keeling, C. D., Mook, W. G. and Tans, P. P.: 1979, 'Recent trends in the $^{13}C/^{12}C$ ratio of atmospheric carbon dioxide', *Nature* **277**, 121–123.

Leavitt, S. W. and Long, A.: 1989, 'The atmospheric δ^{13}C record as derived from 56 Pinyon trees at 14 sites in the southwestern United States', *Radiocarbon* **31**, 469–474.

Leavitt, S. W. and Long, A.: 1991, 'Seasonal stable-carbon isotope variability in tree-rings: Possible palaeoenvironmental signals', *Chem. Geol.* **87**, 59–70.

Leuenberger, M., Borella, S., Stocker, T., Saurer, M., Siegwolf, R., Schweingruber, F. and Matyssek, R.: 1998, *Stable Isotopes in Tree-rings as Climate and Stress Indicators*, VDF, Zurich.

Libby, L. M., Pandolfi, L. J., Payton, P. H., Marshall III, J., Becker, B. and Giertz-Sienbenlist, V.: 1976, 'Isotopic tree thermometers', *Nature* **261**, 284–288.

Lipp, J., Trimborn, P., Fritz, P., Moser, H., Becker, B. and Frenzel, B.: 1991, 'Stable isotopes in tree ring cellulose and climatic change', *Tellus* **43B**, 322–330.

Loader, N. J., Switsur, V. R. and Field, E. M.: 1995, 'High resolution stable isotope analysis of tree rings: Implications of 'microdendroclimatology' for palaeoenvironmental research', *Holocene* **5**, 457–460.

Loader, N. J., Robertson, I., Barker, A. C., Switsur, V. R. and Waterhouse, J. S.: 1997, 'A modified method for the batch processing of small whole wood samples to α-cellulose', *Chem. Geol.* **136**, 313–317.

Loader, N. J., Robertson, I. and McCarroll, D.: 2003, 'Comparison of stable carbon isotope ratios in the whole wood, cellulose and lignin of oak tree-rings', *Palaeogeogr., Palaeoclim., Palaeoecol.* **196**, 395–407.

Manley, G.: 1974, 'Central England temperatures: Monthly means 1659 to 1973', *Q. J. Royal Met. Soc.* **100**, 389–405.

Mazany, T., Lerman, J. C. and Long, A.: 1980, 'Carbon-13 in tree-ring cellulose as an indicator of past climates', *Nature* **287**, 432–435.

McCarroll, D., and Pawellek, F.: 2001, 'Stable carbon isotope ratios of *Pinus sylvestris* from northern Finland and the potential for extracting a climate signal from long Fennoscandian chronologies', *Holocene* **11**, 517–526.

Ogle, N. and McCormac, F. G.: 1994, 'High-resolution δ^{13}C measurements of oak show a previously unobserved spring depletion', *Geophys. Res. Lett.* **21**, 2373–2375.

Park, R. and Epstein, S.: 1961, 'Metabolic fractionation of C^{13} and C^{12} in plants', *Plant Physiol.* **36**, 133–138.

Pilcher, J. R.: 1995, 'Biological considerations in the interpretation of stable isotope ratios in oak tree-rings', in B. Frenzel, B. Stauffer and M. M. Weiss (eds), *Paläoklimaforschung* 15, European Science Foundation, Strasbourg, France, pp. 157–161.

Pilcher, J. R., Baillie, M. G. L., Schmidt, B. and Becker, B.: 1984, 'A 7272-year tree-ring chronology for western Europe', *Nature* **312**, 150–152.

Robertson, I.: 1998, 'Tree response to environmental change', unpublished *Ph.D. Dissertation*, University of Cambridge.

Robertson, I., Pollard, A. M., Heaton, T. H. E. and Pilcher, J. R.: 1996, 'Seasonal changes in the isotopic composition of oak cellulose', in J. S. Dean, D. M. Meko and T. W. Swetnam (eds), *Tree Rings, Environment and Humanity: Proceedings of the International Conference, Tucson, Arizona, 17–21 May 1994*, Radiocarbon, Tucson, Arizona, pp. 617–628.

Robertson, I., Switsur, V. R., Carter, A. H. C., Barker, A. C., Waterhouse, J. S., Briffa, K. R. and Jones, P. D.: 1997, 'Signal strength and climate relationships in the $^{13}C/^{12}C$ ratios of tree-ring cellulose from oak in east England', *J. Geophys. Res.* **102**, 19,507–19,516.

Schleser, G. H., Helle, G., Lücke, A. and Vos, H.: 1999a, 'Isotope signals as climate proxies: The role of transfer functions in the study of terrestrial archives', *Quat. Sci. Rev.* **18**, 927–943.

Schleser G. H., Frielingsdorf, J. and Blair, A.: 1999b, 'Carbon isotope behaviour in wood and cellulose during artificial aging', *Chem. Geol.* **158**, 121–130.

Sofer, Z.: 1980, 'Preparation of carbon dioxide for stable carbon isotope analysis of petroleum fractions', *Anal. Chem.* **52**, 1389–1391.

Sparks, T. H. and Carey, P. D.: 1995, 'The response of species to climate over two centuries: An analysis of the Marsham phenological record, 1736–1947', *J. Ecol.* **83**, 321–329.

Sparks, T. H., Carey, P. D. and Combes, J.: 1997, 'First leafing dates of trees in Surrey between 1947 and 1996', *London Nat.* **76,** 15–20.

Spiker, E. C. and Hatcher, P. G.: 1987, 'The effects of early diagenesis on the chemical and stable carbon isotopic composition of wood', *Geochim. Cosmochim. Acta* **51**, 1385–1391.

Tans, P. P., De Jong, A. F. M. and Mook, W. G.: 1978, 'Chemical pretreatment and radial flow of ^{14}C in tree rings', *Nature* **271**, 234–235.

Technical Association of the Pulp and Paper Industry (TAPPI): 1988, 'Test Method T222 om-83', Atlanta, U.S.A..

Turney, C. M., Berringer, J., Hunt, J. E. and McGlone, M. S.: 1999, 'Estimating leaf to air vapour pressure deficit from terrestrial plant $\delta^{13}C$', *J. Quat. Sci.* **14**, 437–442.

van Bergen, P. F. and Poole, I.: 2002, 'Stable carbon isotopes of wood: A clue to palaeoclimate?', *Palaeogeogr., Palaeoclim., Palaeoecol.* **182**, 31–45.

Vogel, J. C.: 1980, 'Fractionation of the carbon isotopes during photosynthesis', in *Sitzungsberichte der Heidelberger Akademie der Wissenschaften,* Springer-Verlag, Berlin, pp. 111–135.

Wilson, A. T. and Grinsted, M. J.: 1977, '$^{12}C/^{13}C$ in cellulose and lignin as palaeothermometers', *Nature* **265**, 133–135.

Wilson, A. T. and Grinsted, M. J.: 1978, 'The possibilities of deriving past climate information from stable isotope studies on tree rings', in B. W. Robinson (ed.), *Stable Isotopes in the Earth Sciences*, Department of Scientific and Industrial Research Bulletin, Science Information Division, pp. 61–66.

Yapp, C. J. and Epstein, S.: 1977, 'Climatic implications of D/H ratios of meteoric water over North America (9500–22,000 B.P.) as inferred from ancient wood cellulose C-H hydrogen', *Earth Planet. Sci. Lett.* **34**, 333–350.

ISOTOPIC ASSESSMENT OF SOURCES OF SURFACE WATER NITRATE WITHIN THE OLDMAN RIVER BASIN, SOUTHERN ALBERTA, CANADA

LUC ROCK[1]* and BERNHARD MAYER[1,2]

[1] *Department of Geology and Geophysics, University of Calgary, 2500 University Drive NW, Calgary, Alberta, Canada T2N 1N4;* [2] *Department of Physics and Astronomy, University of Calgary, Canada*
(* author for correspondence, e-mail: lrock@ucalgary.ca; phone: (403) 220 7201; fax: (403) 284 0074)

(Received 20 August 2002; accepted 19 April 2003)

Abstract. Concentrations and isotopic compositions of NO_3^- from the Oldman River (OMR) and some of its tributaries (Alberta, Canada) have been determined on a monthly basis since December 2000 to assess temporal and spatial variations of riverine NO_3^- sources within the OMR basin. For the OMR sites, NO_3^--N concentrations reached up to 0.34 mg L^{-1}, δ^{15}N-NO_3^- values varied between −0.3 and +13.8‰, and δ^{18}O-NO_3^- values ranged from −10.0 to +5.7‰. For the tributary sites, NO_3^--N concentrations were as high as 8.81 mg L^{-1}, δ^{15}N-NO_3^- values varied between −2.5 and +23.4‰, and δ^{18}O-NO_3^- values ranged from −15.2 to +3.4‰. Tributaries in the western, relatively pristine forested part of the watershed add predominantly NO_3^- to the OMR with δ^{15}N-NO_3^- values near +2‰ indicative of soil nitrification. In contrast, tributaries in the eastern agriculturally-urban-industrially-used part of the basin contribute NO_3^- with δ^{15}N-NO_3^- values of about +16‰ indicative of manure and/or sewage derived NO_3^-. This difference in δ^{15}N-NO_3^- values of tributaries was found to be independent of the season, but rather indicates a spatial change in the NO_3^- source, which correlates with land use changes within the OMR basin. As a consequence of tributary influx, δ^{15}N-NO_3^- values in the Oldman River increased from <+3‰ to >+6‰ in the downstream direction (W to E), although [NO_3^--N] increased only moderately (generally <0.5 mg L^{-1}). This study demonstrates the usefulness of δ^{15}N-NO_3^- and δ^{18}O-NO_3^- values in identifying the addition of anthropogenic NO_3^- to riverine systems.

Keywords: δ^{15}N-NO_3^-, δ^{18}O-NO_3^-, nitrate sources, Oldman River Basin, riverine nitrate, stable isotopes

1. Introduction

Water resources in southern Alberta are currently under high strain due to both natural and human influence. The region receives average annual precipitation of about 400 mm a^{-1} compared to a potential evaporation of approximately 800 mm a^{-1} and is hence semi-arid (Morton, 1983). Most of the water in the Oldman River basin (southern Alberta, Canada) is derived from snowmelt in the Canadian Rocky Mountains and is used by agriculture (Alberta Government, 1976). Due to a highly

Water, Air, and Soil Pollution: Focus **4:** 545–562, 2004.
© 2004 *Kluwer Academic Publishers. Printed in the Netherlands.*

variable flow pattern and time distribution of flow for this mountain-fed stream, reservoirs along the Oldman River and some of its tributaries were built to store water for irrigation to ensure agricultural production in a semi-arid climate. In 1977, 793 000 acre-feet of water were diverted from the Oldman River system for irrigation, 14% of which returned to the river system in form of return flow from irrigated lands (OMRSMC, 1978). Hence, agricultural practices including fertilization of crops and feed lot operations may have potentially negative impacts on the water quality of the Oldman River by increasing nutrient and pathogen loads returned to the river system via irrigation canals, tributaries, or non-point source discharge. One of the nutrients that may pose a problem to water quality is N in the form of NO_3^-. Increased NO_3^- loads within an aquatic ecosystem may result in eutrophication, and elevated NO_3^--N concentrations in drinking water pose health problems to both humans and life stock (Bruning-Fann and Kaneene, 1993; Vitousek et al., 1997; Goolsby, 2000). To reduce NO_3^- loading of aquatic systems, it is important to determine the sources of NO_3^- in a watershed.

Repeated concentration analyses alone can only provide information about temporal trends of solutes within aquatic systems. Isotope ratio measurements, however, can provide information about sources and processes affecting a particular solute such as NO_3^- (e.g., Kendall, 1998). The combined analysis of concentration and isotopic composition of NO_3^- thus can provide more detailed information about the sources and potential transformations of NO_3^- in surface water.

The isotopic composition of NO_3^- has proven to be a valuable tool for identifying NO_3^- sources, if both the $\delta^{15}N$ values and the $\delta^{18}O$ values are determined (e.g., Kendall, 1998). Four major sources of NO_3^- can be identified: atmospheric deposition, soil nitrification, synthetic fertilizers, and sewage- or manure-derived NO_3^-. Nitrate from each of these sources is typically characterized by a distinct isotopic signature. Typical $\delta^{15}N$-NO_3^- values for atmospheric NO_3^- range from –5 to +10‰, for NO_3^--containing synthetic fertilizers from –2 to +2‰, from +7 to more than +30‰ for human and animal waste, and from less than –10 to +5‰ for soil NO_3^-. Typical $\delta^{18}O$-NO_3^- values for atmospheric deposition are higher than +25‰, range from +18 to +23‰ in nitrate-containing synthetic fertilizers, and vary from –10 to +15‰ for human and animal waste and soil NO_3^- (e.g., Aravena et al., 1993; Kendall, 1998; Mayer et al., 2001). Most previous studies using the isotopic composition of NO_3^- for identifying sources and/or transformations have focused on groundwater NO_3^- (Gormly and Spalding, 1979; Kreitler, 1979; Flipse et al., 1984; Wells and Krothe, 1989; Böttcher, 1990, Aravena et al., 1993; Exner and Spalding, 1994; Gellenbeck, 1994; Wassenaar, 1995; Whitehead et al., 1999). Only few studies have investigated surface water NO_3^- in rivers (e.g., Lindau et al., 1989; Cravotta, 1995; Kellmann and Hillaire-Marcel, 1998; Campbell et al., 2002; Mayer et al., 2002) and hardly any studies were conducted on a watershed level.

The objective of this ongoing study is to assess the usefulness of the isotopic composition of surface water NO_3^- ($\delta^{15}N$ and $\delta^{18}O$) for identifying its sources in the

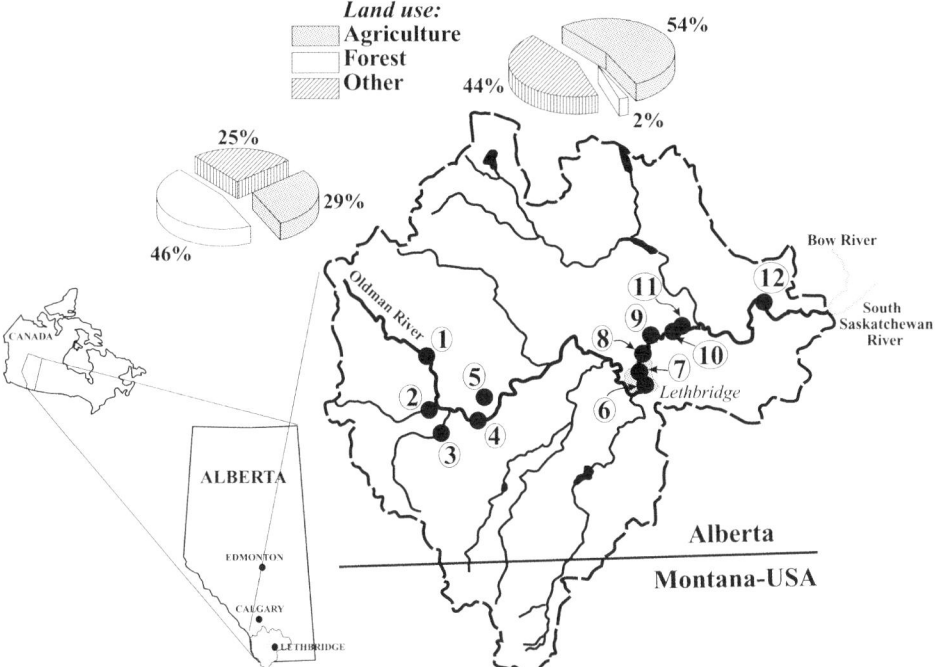

Figure 1. Map of the Oldman River basin, southern Alberta (Canada); also shown location of sampling sites and land use distribution (after OMRWQI, 2002). Sampling sites: 1. Oldman River near Olin Creek (km 0), 2. Crowsnest River (km 19), 3. Castle River (km 24), 4. Oldman River near Brocket (km 39), 5. Beaver Creek (km 62), 6. Six Mile Coulee (km 198), 7. Oldman River at Hwy #3 (km 204), 8. Oldman River SW of Diamond City (km 215), 9. Piyami Drain (km 230), 10. Haney Drain (km 246), 11. Battersea Drain (km 249), 12. Oldman River at Hwy #36 (km 314).

Oldman River basin (Alberta, Canada). A specific goal was to identify the influence of changing land use on riverine NO_3^-.

2. Study Area

The Oldman River basin is located in southern Alberta, Canada (Figure 1), covers an area of about 28 200 km² and provides water resources for over 200 000 people. This basin was chosen as it has almost pristine headwaters in its western part and increased urban/industrial/agricultural activities in its eastern part. Hence, the influence of land use change within the basin on the concentration and isotopic composition of riverine NO_3^- was testable.

The Oldman River (OMR) originates from a small alpine lake on the continental divide in the Canadian Rocky Mountains (Alberta Government, 1976). It then flows eastwards through Paleozoic carbonate rocks with associated evaporites. Subsequently, it cuts through Mesozoic clastic sedimentary rocks of the Foothills before reaching the plains where it cuts mainly through till deposits, containing reduced sulfur species (e.g., pyrite), which overlie Mesozoic and Cenozoic clastic sedimentary rocks (WRD, 1970; Nielsen, 1971; Mossop and Shetsen, 1994). At its mouth in southeastern Alberta, the OMR joins the Bow River to form the South Saskatchewan River, which drains eventually into the Hudson Bay. Water levels rise from March to June due to snowmelt, and recede in July throughout fall and winter (Alberta Government, 1976). Discharge within the OMR ranged from approximately 1.2 m^3 s^{-1} during the winter to more than 40 m^3 s^{-1} in the spring. Discharge in the tributaries rarely exceeded 1 m^3 s^{-1} with the exception of the Castle and Crowsnest Rivers, which can have discharges similar to the OMR. The irrigation season spans the months of May to October.

3. Sampling

Monthly sampling of the main stream of the Oldman River and some of its tributaries commenced in December of 2000. Nine sites were sampled from December 2000 to March 2001. Three were located on the Oldman River [site #4: OMR near Brocket (km 39), site #7: OMR at HWY #3 (km 204), site #12: OMR at HWY #36 (km 314)] and the tributary sites included site #3: the Castle River (km 24), site #6: Six Mile Coulee (km 198), site #10: Haney Drain (km 246), and site #11: Battersea Drain (km 249) (Figure 1). From April 2001 to March 2002, 14 sites were sampled. Two additional sites were located on the Oldman River [site #1: OMR near Olin Creek (km 0), site #8: OMR SW of Diamond City (km 215)] and 3 tributaries sites were added [site #2: Crowsnest River (km 19), site #5: Beaver Creek (km 62), site #9: Piyami Drain (km 230)]. The volume of surface water collected varied from 0.5 to 4 L for the tributary sites, and from 4 to 10 L for the OMR sites. Sampling containers were thoroughly rinsed prior to sampling. Depending on the season, some of the tributary sites were dry (e.g., Beaver Creek from December 2000 to March 2001), and hence no water samples were obtained. Samples were immediately returned to the laboratory in cooled containers and either processed within 72 hr or frozen until further processing.

4. Methods

Nitrate-N concentrations reported in this paper were provided by Alberta Environment, whereas, isotopic measurements were conducted at the Isotope Science Laboratory (ISL) at the University of Calgary (Alberta, Canada). Nitrate-N concentrations were determined by ion chromatography with a detection limit of 0.003 mg L^{-1}.

Nitrogen and oxygen isotope ratios of nitrate were determined by isotope ratio mass spectrometry. Results are expressed in per mil (‰) using the usual delta notation:

$$\delta_{sample}(‰) = \{(R_{sample}/R_{standard}) - 1\} \times 1000,$$

where R is the $^{15}N/^{14}N$ or $^{18}O/^{16}O$ ratio of the sample or an internationally accepted standard. $\delta^{15}N$ values are reported relative to AIR, and $\delta^{18}O$ values with respect to Vienna-Standard Mean Ocean Water (V-SMOW).

$\delta^{15}N\text{-}NO_3^-$ and $\delta^{18}O\text{-}NO_3^-$ were determined using a modified version of the method described by Silva *et al.* (2000). Water samples were filtered through 0.45 μm Micropore filters. After filtration, $BaCl_2$ was added to precipitate $BaSO_4$, $BaCO_3$ and $Ba_3(PO_4)_2$. Subsequently, the samples were acidified with HCl to a pH of less than 4 to remove $BaCO_3$. The remaining precipitates were removed using 0.45 μm Micropore filters. The remaining solutions were passed through pre-filled PolyPrep^{TM} columns with 2 mL AG 50W-X8 cation exchange resin (BIO-RAD), and subsequently through pre-filled PolyPrep^{TM} columns with 2 mL AG 1-X8 anion exchange resin (BIO-RAD) to quantitatively retain NO_3^-. By adding 3×5 mL of 3 M HCl and 2 mL of deionized water to the anion exchange column, NO_3^- was quantitatively eluted and HNO_3 and HCl were collected in a beaker, to which about 8 g of pure and pre-washed Ag_2O (Merck, Darmstadt, Germany) were added.

$$HNO_3 + HCl + Ag_2O \rightarrow AgCl\ (ppc) + H_2O + Ag^+ + NO_3^-.$$

This reaction was complete when the pH of the solution approached 7. After the AgCl precipitate was removed by filtration, only Ag^+ and NO_3^- remained in the solution, which was frozen and subsequently freeze-dried to produce a solid $AgNO_3$ precipitate.

For nitrogen isotope analyses of NO_3^-, 1.5 to 2 mg of $AgNO_3$ were weighed into high purity tin cups. These cups were thermally decomposed in an elemental analyser (Carlo Erba NA 1500) and the resultant N_2 was analyzed by isotope ratio mass spectrometry in continuous-flow mode using a Finnigan MAT delta plus XL. $\delta^{15}N$ values for all samples were normalized against internationally accepted reference materials (IAEA N1 and N2). $\delta^{15}N\text{-}NO_3^-$ values are reported with an overall analytical precision of $\pm 0.3‰$.

Oxygen isotope ratios were determined on CO generated by pyrolysis of high purity sliver cups containing 100 to 300 μg of $AgNO_3$ using a Finnigan MAT TC/EA reactor coupled to a delta plus XL isotope ratio mass spectrometer in continuous flow mode. Accuracy and precision of the measurements was assured by repeated analyses of international reference materials (IAEA-NO-3) and by calibrating all measured oxygen isotope ratios to $\delta^{18}O\text{-}IAEA\text{-}NO\text{-}3 = +23.0‰$. $\delta^{18}O\text{-}NO_3^-$ determinations have an overall analytical precision of $\pm 0.8‰$.

TABLE I

[NO_3^--N], $\delta^{15}N_{nitrate}$ and $\delta^{18}O_{nitrate}$ values for the Oldman River sampling sites

	Western Oldman River sites, predominantly forested							Eastern Oldman River sites, predominantly agriculture								
	Near Olin Creek d/s 0 km Site #1			Near Brocket d/s 39 km Site #4				At HWY #3 d/s 204 km Site #7			SW of Diamond City d/s 215 km Site #8			At HWY #36 d/s 314 km Site #12		
	NO_3^--N (mg L^{-1})	$\delta^{15}N$ (‰)	$\delta^{18}O$ (‰)	NO_3^--N (mg L^{-1})	$\delta^{15}N$ (‰)	$\delta^{18}O$ (‰)		NO_3^--N (mg L^{-1})	$\delta^{15}N$ (‰)	$\delta^{18}O$ (‰)	NO_3^--N (mg L^{-1})	$\delta^{15}N$ (‰)	$\delta^{18}O$ (‰)	NO_3^--N (mg L^{-1})	$\delta^{15}N$ (‰)	$\delta^{18}O$ (‰)
Dec-00	x	x	x	0.061	2.5	−3.4		0.183	8.3	−8.7	x	x	x	0.300	10.3	−7.7
Jan-01	x	x	x	0.128	5.4	−1.6		0.263	13.8	−10.0	x	x	x	0.339	7.7	−9.3
Feb-01	x	x	x	0.101	2.8	−2.7		0.206	5.8	−6.2	x	x	x	0.246	6.6	−7.5
Mar-01	x	x	x	0.122	2.7	−3.7		0.207	6.3	−7.1	x	x	x	0.236	6.1	−6.8
Winter 2001				0.117	3.6	−2.7		0.225	8.6	−7.8				0.274	6.8	−7.9
Apr-01	0.033	2.6	2.5	0.124	3.1	−0.6		0.044	n.d.	n.d.	0.081	n.d.	n.d.	<0.003	n.d.	n.d.
May-01	<0.003	n.a.	n.a.	0.098	n.a.	n.a.		0.009	n.a.	n.a.	0.005	n.a.	n.a.	0.006	n.a.	n.a.
Jun-01	0.038	n.d.	n.d.	0.102	n.d.	n.d.		0.024	n.d.	n.d.	0.073	11.2	−2.4	0.003	n.a.	n.a.
Spring 2001	0.024			0.108				0.026			0.053			0.003		
Jul-01	<0.003	n.d.	n.d.	0.085	1.6	2.2		0.083	n.d.	n.d.	0.003	n.d.	n.d.	<0.003	n.d.	n.d.
Aug-01	0.007	n.d.	n.d.	0.062	−1.0	−0.6		<0.003	n.d.	n.d.	<0.003	n.d.	n.d.	0.006	n.d.	n.d.
Sep-01	0.007	n.d.	n.d.	0.074	5.5	0.0		<0.003	n.d.	n.d.	0.008	n.d.	n.d.	<0.003	n.d.	n.d.
Summer 2001	0.005			0.074	2.0	0.5		0.028			0.004			0.002		
Oct-01	0.012	n.a.	n.a.	0.084	n.a.	n.a.		0.011	n.a.	n.a.	0.014	n.a.	n.a.	0.015	n.a.	n.a.
Nov-01	0.007	n.d.	n.d.	0.112	1.1	−1.9		0.028	n.d.	n.d.	0.019	n.d.	n.d.	0.005	n.d.	n.d.
Dec-01	0.032	n.d.	n.d.	0.132	2.1	−1.1		0.238	5.3	−5.7	0.291	10.4	−5.5	0.329	n.d.	n.d.
Fall 2001	0.017			0.109				0.092			0.108			0.116		
Jan-02	0.051	n.d.	n.d.	0.190	2.6	−0.5		0.227	6.1	−6.4	0.237	7.4	−6.4	0.228	7.4	−6.8
Feb-02	0.083	−0.3	5.7	0.161	2.9	−0.1		0.167	6.2	−6.6	0.220	8.9	−6.0	0.178	6.4	−5.4
Mar-02	0.078	−0.2	5.5	0.175	3.6	0.4		0.221	7.2	−5.6	0.240	6.6	−6.1	0.296	6.4	−6.7
Winter 2002	0.071			0.175	3.0	−0.1		0.205	6.5	−6.2	0.232	7.6	−6.2	0.234	6.7	−6.3

x: Not sampled; n.a.: not analyzed; n.d.: not determined; d/s: km downstream of Oldman River site at Olin Creek. For sampling site location see Figure 1. Note: Samples from May and October 2001 have not yet been analyzed for their isotopic composition.

NITRATE SOURCES IN THE OLDMAN RIVER, CANADA 551

TABLE II

$[NO_3^--N]$, $\delta^{15}N_{nitrate}$ and $\delta^{18}O_{nitrate}$ values for the tributary sampling sites

	Western tributary sites, predominantly forested								Eastern tributary sites, predominantly agriculture									
	Crowsnest River d/s 19 km Site #2			Castel River d/s 24 km Site #3			Beaver Creek d/s 62 km Site #5			Six Mile Coulee d/s 198 km Site #6			Piyami Drain d/s 230 km Site #9			Haney Drain d/s 246 km Site #10			Battersea Drain d/s 249 km Site #11		
	NO_3^--N (mg L^{-1})	$\delta^{15}N$ (‰)	$\delta^{18}O$ (‰)	NO_3^--N (mg L^{-1})	$\delta^{15}N$ (‰)	$\delta^{18}O$ (‰)	NO_3^--N (mg L^{-1})	$\delta^{15}N$ (‰)	$\delta^{18}O$ (‰)	NO_3^--N (mg L^{-1})	$\delta^{15}N$ (‰)	$\delta^{18}O$ (‰)	NO_3^--N (mg L^{-1})	$\delta^{15}N$ (‰)	$\delta^{18}O$ (‰)	NO_3^--N (mg L^{-1})	$\delta^{15}N$ (‰)	$\delta^{18}O$ (‰)	NO_3^--N (mg L^{-1})	$\delta^{15}N$ (‰)	$\delta^{18}O$ (‰)
Dec-00	x	x	x	0.054	2.0	−3.3	x	x	x	7.250	15.4	−14.7	x	x	x	frozen	frozen	frozen	7.290	14.7	−9.3
Jan-01	x	x	x	0.092	1.1	−4.6	x	x	x	2.790	13.1	−14.3	x	x	x	5.820	14.3	−8.0	1.100	14.5	−10.6
Feb-01	x	x	x	0.064	1.2	−3.5	x	x	x	2.740	17.6	−15.2	x	x	x	8.770	13.2	−8.2	1.070	15.6	−9.5
Mar-01	x	x	x	0.088	0.9	n.e.	x	x	x	3.510	17.4	−14.4	x	x	x	6.260	13.4	−8.4	1.220	14.2	−10.7
Winter 2001				0.081	1.1					3.013	16.0	−14.6				6.950	13.6	−8.2	1.130	14.8	−10.3
Apr-01	0.062	n.d.	n.d.	0.127	2.0	−8.5	0.098	n.d.	n.d.	1.220	n.d.	n.a.	0.084	n.d.	n.d.	6.690	15.1	−8.5	3.360	15.4	−11.7
May-01	0.024	n.a.	n.a.	0.088	n.a.	n.a.	<0.003	n.a.	n.a.	2.610	n.a.	n.a.	0.353	n.a.	n.a.	0.506	n.a.	n.a.	0.009	n.a.	n.a.
Jun-01	0.114	n.d.	n.d.	0.090	n.d.	n.d.	0.040	4.9	−3.2	0.129	11.7	−8.1	0.034	n.d.	n.d.	0.998	n.d.	n.d.	0.068	11.4	−5.0
Spring 2001	0.067			0.102			0.046			1.320			0.157			2.731			1.146		
Jul-01	0.038	n.d.	n.d.	0.022	n.d.	n.d.	0.012	n.d.	n.d.	0.038	n.d.	n.d.	0.026	n.d.	n.d.	0.243	n.d.	n.d.	0.005	n.d.	n.d.
Aug-01	0.011	n.d.	n.d.	0.004	n.d.	n.d.	dry	dry	dry	0.063	13.0	−7.9	0.005	−2.5	−4.9	0.615	8.2	−8.6	0.105	9.0	−6.0
Sep-01	0.007	n.d.	n.d.	0.004	n.d.	n.d.	dry	dry	dry	0.098	11.0	−9.6	0.331	23.4	−6.4	0.276	10.4	−6.8	0.006	n.d.	n.d.
Summer 2001	0.019			0.010						0.066			0.121			0.378			0.039		
Oct-01	0.014	n.a.	n.a.	0.011	n.a.	n.a.	dry	dry	dry	0.247	n.a.	n.a.	0.022	n.a.	n.a.	6.630	n.a.	n.a.	0.119	n.a.	n.a.
Nov-01	0.195	3.6	0.7	0.102	n.d.	n.d.	dry	dry	dry	1.570	12.6	−12.8	dry	dry	dry	8.040	13.5	−7.8	2.920	14.2	−9.5
Dec-01	0.472	3.5	2.4	0.073	0.1	1.3	dry	dry	dry	frozen	frozen	frozen	dry	dry	dry	8.300	13.8	−4.5	5.830	14.7	−9.3
Fall 2001	0.227			0.062												7.657			2.956		
Jan-02	0.414	4.4	3.4	0.109	2.0	−1.9	dry	dry	dry	frozen	frozen	frozen	dry	dry	dry	8.810	14.4	−7.0	3.060	15.5	−11.6
Feb-02	0.492	4.6	2.4	0.124	0.9	−3.2	dry	dry	dry	frozen	frozen	frozen	dry	dry	dry	6.470	13.7	−8.9	0.555	14.3	−9.9
Mar-02	0.444	3.9	2.0	0.113	1.1	−3.8	dry	dry	dry	frozen	frozen	frozen	dry	dry	dry	7.620	14.3	−8.8	0.586	14.2	−10.9
Winter 2002	0.450	4.3	2.6	0.115	1.3	−3.0										7.633	14.1	−8.2	1.400	14.7	−10.8

x: Not sampled; n.a.: not analyzed; n.d.: not determined; d/s: km downstream of Oldman River site at Olin Creek. For sampling site location see Figure 1. Note: Samples from May and October 2001 have not yet been analyzed for their isotopic composition.

5. Results

Concentrations and isotopic compositions of NO_3^- from the Oldman River and its tributaries collected between December 2000 and March 2002 are shown in Tables I and II. Ammonium and NO_2^- concentrations represented generally less than 10% of the total dissolved N species, and will not be discussed in this paper. For samples collected in spring or summer, NO_3^--N concentrations were often very low (<0.250 mg L^{-1}) and hence we were unable to determine the isotopic composition of NO_3^-.

5.1. Concentration and isotopic composition of NO_3^- in the Oldman River

In December of 2000, NO_3^--N concentrations ranged from 0.061 to 0.300 mg L^{-1}. Average concentrations at the various sampling locations along the OMR for the 2001 and 2002 seasons were: winter 2001: 0.117 to 0.274 mg L^{-1}; spring 2001: 0.003 to 0.108 mg L^{-1}; summer 2001: 0.002 to 0.074 mg L^{-1}; fall 2001: 0.017 to 0.116 mg L^{-1}; and winter 2002: 0.071 to 0.234 mg L^{-1} (Table I). Nitrate-N concentrations were lowest during the summer months (July, August, September) and highest during winter (January, February, March) (Figure 2a). Between May and October, NO_3^--N concentrations were <0.1 mg L^{-1} independent of stream km. Between November and April, NO_3^--N concentrations were higher (up to 0.4 mg L^{-1}) and dependent upon stream km. They were low in the upstream portion (0–100 km) and increased continuously with increasing flow distance (Figure 2a).

In December of 2000, $\delta^{15}N$-NO_3^- values ranged from +2.5 to +10.3‰, and $\delta^{18}O$-NO_3^- values varied between –3.4 and –8.7‰. Average $\delta^{15}N$-NO_3^- values at the various sampling locations along the OMR for the winter of 2001 were +3.6 to +8.6‰ and –2.7 to –7.9‰ for $\delta^{18}O$-NO_3^- values, and for the winter of 2002 they were –0.3 to +7.6‰ and –6.3 to +5.6‰, respectively (Table I). Even though no average values can be reported for the spring, summer, and fall 2001, the few observations obtained appear to indicate that there is no significant difference in the isotopic composition of riverine $\delta^{15}N$-NO_3^- with season (Table I, Figure 2b). At the OMR site near Brocket, 39 km downstream of the site near Olin Creek, $\delta^{15}N$-NO_3^- values varied typically between +1 and +4‰ and $\delta^{18}O$-NO_3^- values were >–3‰. At the eastern site SW of Diamond City, 215 km downstream of Olin Creek, $\delta^{15}N$-NO_3^- values were generally higher than +7‰ and $\delta^{18}O$-NO_3^- values were below –3‰. Hence, $\delta^{15}N$-NO_3^- values were lowest in the western upstream portion of the basin (<+5‰ in the uppermost 100 km) and increased with increasing flow distance (Figure 2b). $\delta^{18}O$-NO_3^- values were consistently higher in the upstream portion and lower in the downstream portion of the basin (Table I).

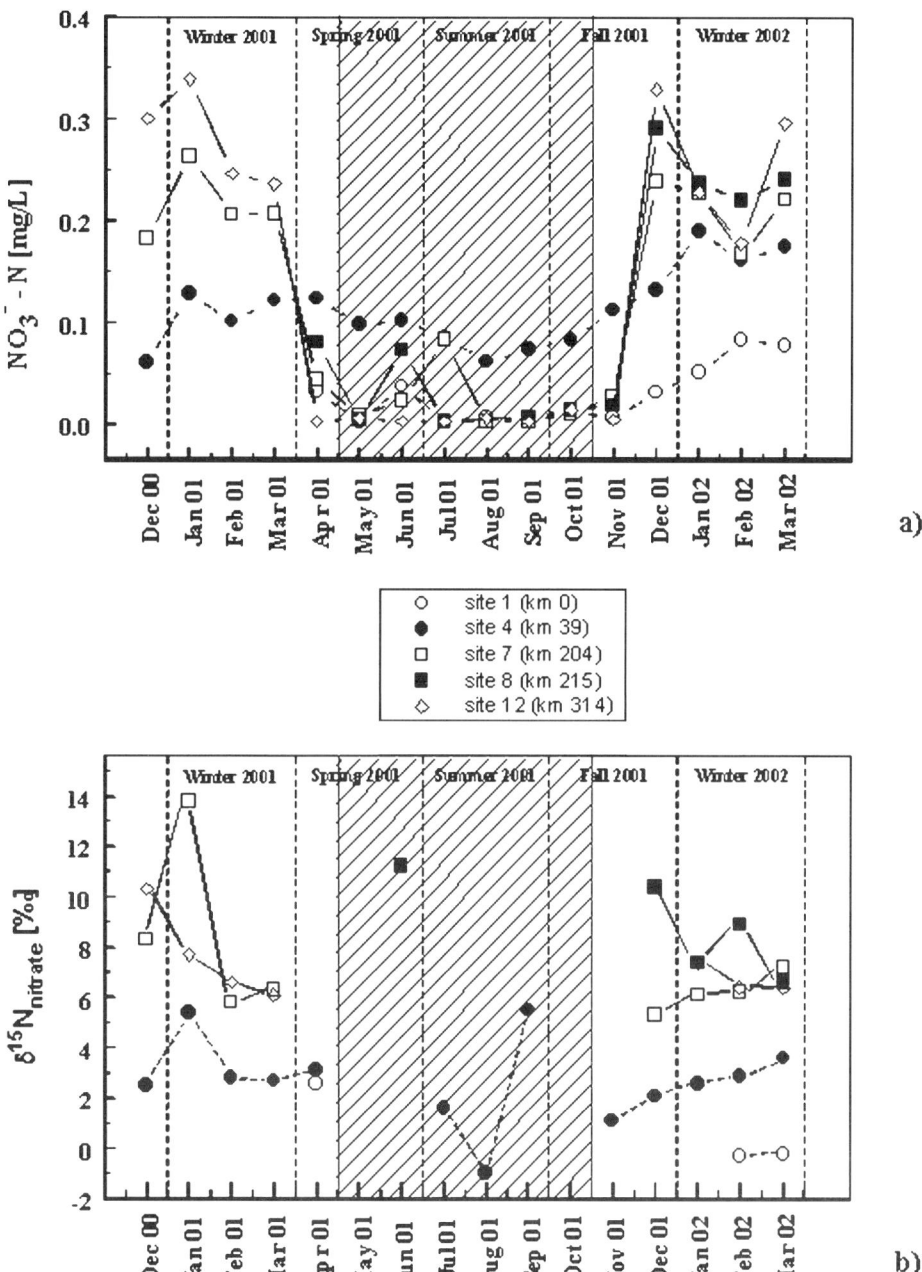

Figure 2. (a) [NO$_3^-$-N] versus sampling month for the OMR sites; (b) δ^{15}N-NO$_3^-$ values versus sampling month for the OMR sites; hatched area represents irrigation season. Data are listed in Table I. For site location see Figure 1.

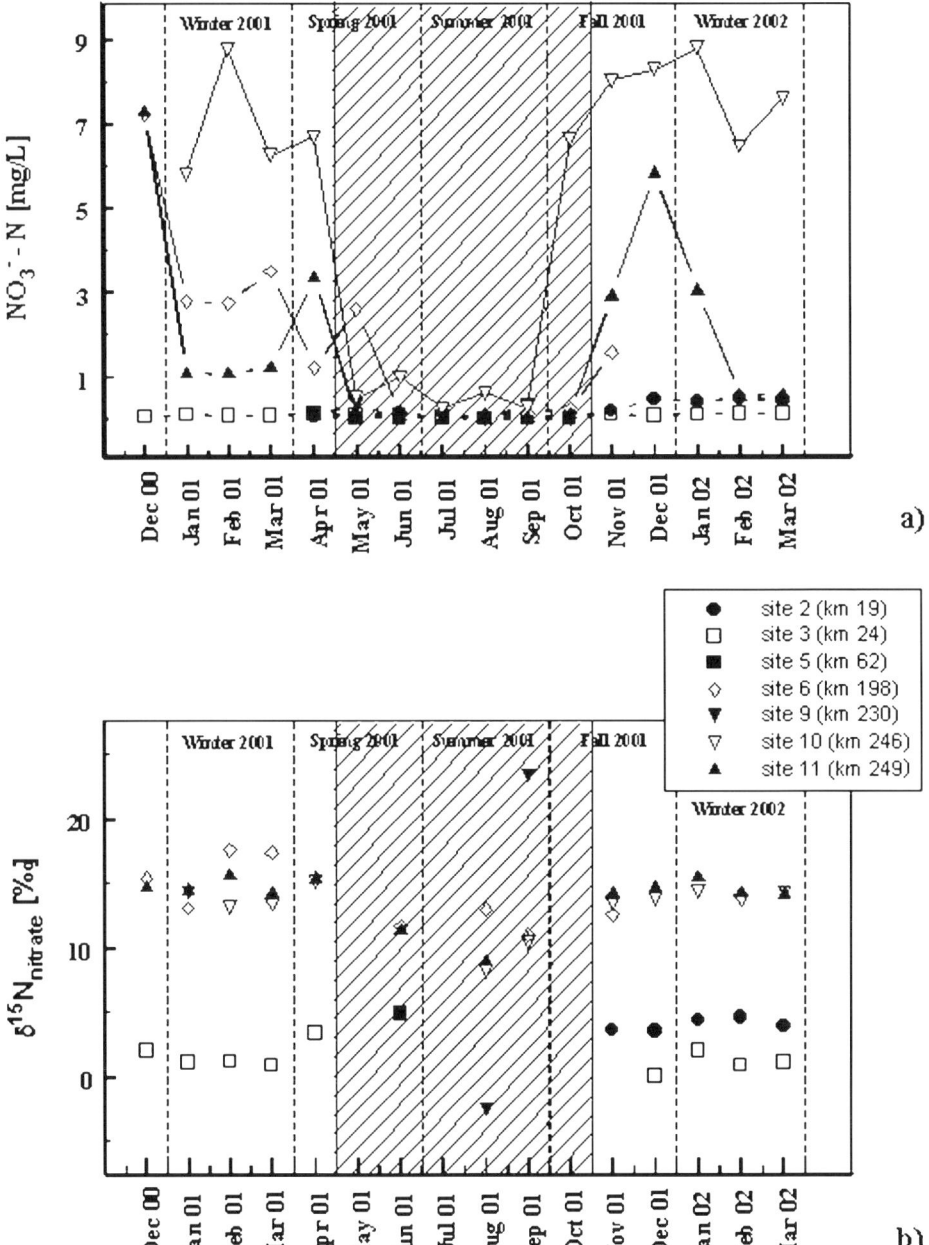

Figure 3. (a) [NO_3^--N] versus sampling month for the tributary sites; (b) δ^{15}N-NO_3^- values versus sampling month for the tributary sites; hatched area represents irrigation season. Data are listed in Table II. For site location see Figure 1.

5.2. CONCENTRATION AND ISOTOPIC COMPOSITION OF NO_3^- IN TRIBUTARIES

In December of 2000, NO_3^--N concentrations ranged from 0.054 to 7.290 mg L^{-1}. Average NO_3^--N concentrations at the various tributary sampling locations for the 2001 and 2002 seasons were: winter 2001: 0.081 to 6.950 mg L^{-1}; spring 2001: 0.046 to 2.731 mg L^{-1}; summer 2001: 0.010 to 0.378 mg L^{-1}; fall 2001: 0.062 to 7.657 mg L^{-1}; and winter 2002: 0.115 to 7.633 mg L^{-1} (Table II). Nitrate-N concentrations were generally lowest during the summer months and highest during winter (Figure 3a). Between June and September, NO_3^--N concentrations were <1 mg L^{-1} independent of location. Between October and May, NO_3^--N concentrations varied considerably (up to 9 mg L^{-1}) dependent upon location. The lowest NO_3^--N concentrations of less than 0.5 mg L^{-1} were consistently observed in the upstream portion (0–100 km). The highest NO_3^--N concentrations often exceeding 5 mg L^{-1} were found in the downstream portion of the watershed (Figure 3a).

In December of 2000, $\delta^{15}N$-NO_3^- values ranged from +2.0 to +15.4‰, and $\delta^{18}O$-NO_3^- values varied between −14.7 and −3.3‰. Average $\delta^{15}N$-NO_3^- values for the various tributary sampling locations varied between +1.1 and +16.0‰ in winter of 2001, and between −4.0 and −14.6‰ for $\delta^{18}O$-NO_3^- values (Table II). Similar ranges of isotopic compositions of NO_3^- were observed for the spring, summer and fall of 2001. Hence, the available data indicate that $\delta^{15}N$-NO_3^- and $\delta^{18}O$-NO_3^- values do not vary with season. It is, however, important to note that the western-upstream and eastern-downstream tributary sites were characterized by distinct isotopic compositions of NO_3^- (Table II, Figure 3b). Tributaries in the uppermost 100 km of the OMR basin were characterized by $\delta^{15}N$-NO_3^- values <+5‰ and $\delta^{18}O$-NO_3^- values of >−2‰. Tributaries in the downstream portion of the OMR basin (>190 km) had $\delta^{15}N$-NO_3^- values typically higher than +9‰ and $\delta^{18}O$-NO_3^- values of less than −5.0‰.

6. Discussion

Plotting NO_3^--N concentrations versus $\delta^{15}N$-NO_3^- values (Figure 4a) and $\delta^{18}O$-NO_3^- values (Figure 4b) for the Oldman River and tributary samples revealed that at NO_3^--N concentrations <1 mg L^{-1}, $\delta^{15}N$-NO_3^- values varied between −2 and +14‰ and $\delta^{18}O$-NO_3^- values ranged between −10 and +5‰. At NO_3^--N concentrations >1 mg L^{-1}, $\delta^{15}N$-NO_3^- values were consistently around +15±1‰ and $\delta^{18}O$-NO_3^- values near −10‰. The combination of high $\delta^{15}N$-NO_3^- and low $\delta^{18}O$-NO_3^- values observed for all samples with NO_3^--N concentrations >1 mg L^{-1} is indicative of NO_3^- from either sewage or manure (Figure 5). Samples with NO_3^--N concentrations <1 mg L^{-1} originated either from soil nitrification, sewage, manure, or a combination of these three sources (Figure 5). Trends of increasing $\delta^{15}N$-NO_3^- and $\delta^{18}O$-NO_3^- values with decreasing NO_3^--N concentrations were not observed

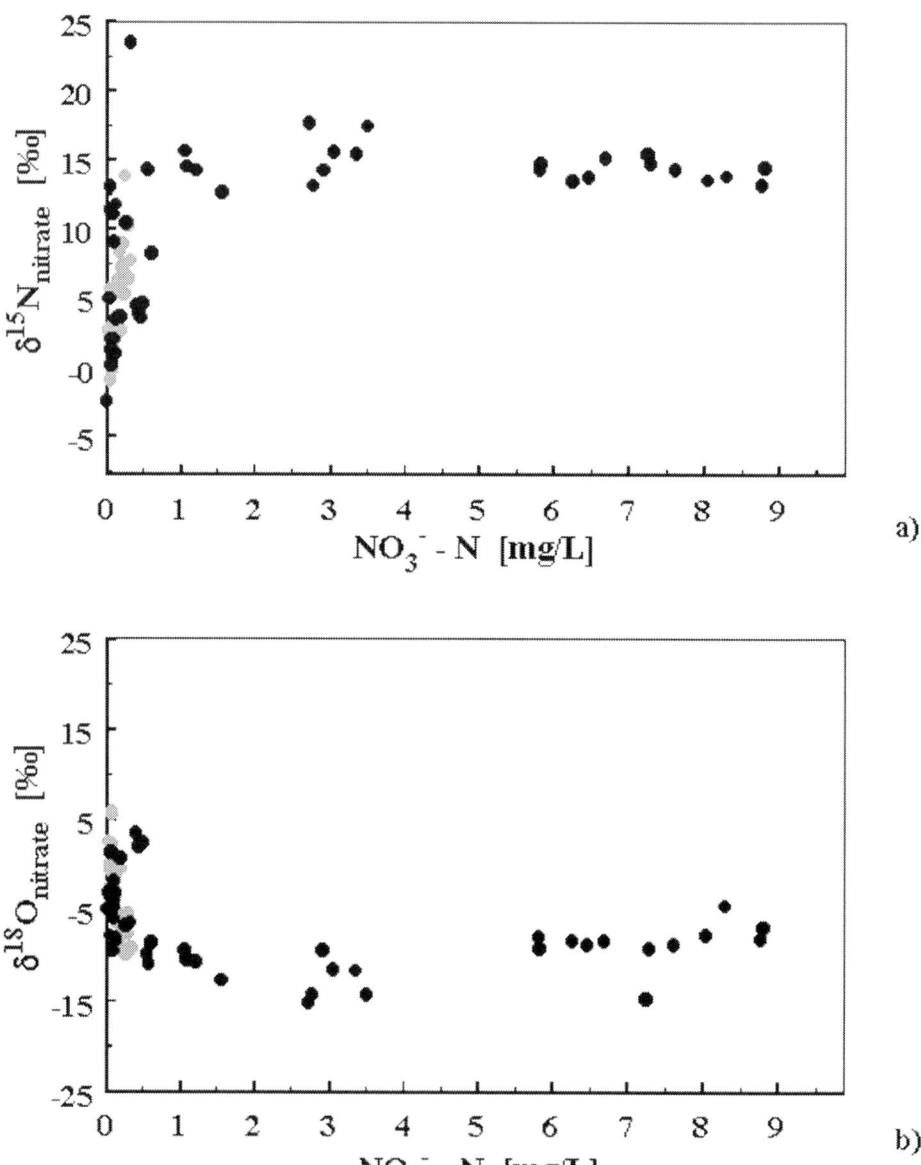

Figure 4. (a) δ^{15}N-NO_3^- values versus [NO_3^--N] for the Oldman River (gray circles) and tributary (black circles) sites; (b) δ^{18}O-NO_3^- values versus [NO_3^--N] for the Oldman River (gray circles) and tributary (black circles) sites.

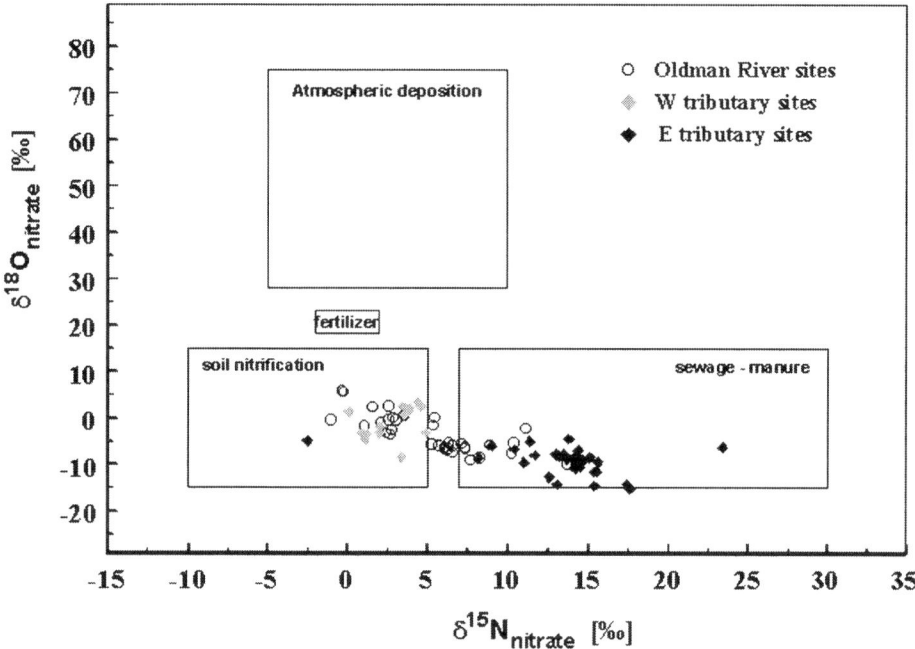

Figure 5. δ^{18}O-NO$_3^-$ versus δ^{15}N-NO$_3^-$ values from December 2000 to March 2002 for the Oldman River (open circles), western-upstream tributaries (gray diamond) and eastern-downstream tributaries (black diamonds) sites; also shown are typical isotopic ranges of nitrate sources (modified after Kendall, 1998).

(Figure 4). Hence, the isotopic composition of NO$_3^-$ seems to predominantly reflect NO$_3^-$ sources rather than NO$_3^-$ transformation processes, such as denitrification.

6.1. SPATIAL CHANGES IN CONCENTRATION AND ISOTOPIC COMPOSITION OF NO$_3^-$

Based on concentrations and isotopic compositions of NO$_3^-$, the tributaries can be subdivided into two distinct groups: (a) the NO$_3^-$ source being mainly soil nitrification with low NO$_3^-$-N concentrations (<1 mg L^{-1}) and δ^{15}N-NO$_3^-$ values (<+5‰) and (b) the NO$_3^-$ source being mainly sewage or manure with higher NO$_3^-$-N concentrations and δ^{15}N-NO$_3^-$ values ($\geq$+9‰) (Figure 5). These two groups are clearly spatially separated within the OMR basin. Tributaries with soil nitrification as the main NO$_3^-$ source are all located in the western part of the OMRB, whereas tributaries with NO$_3^-$ derived from sewage or manure are all located within the eastern part (Table II, Figure 5).

Nitrate within the OMR is derived either from soil nitrification, from sewage or manure, or from a mixture of the two former sources (Figure 5). The western most upstream (<100 km) sites have typically δ^{15}N-NO$_3^-$ values <+4‰ and NO$_3^-$-N concentrations <0.15 mg L^{-1}, which is indicative of NO$_3^-$ derived from soil nitrification. Further downstream, δ^{15}N-NO$_3^-$ values increased by 3 to 5‰ (Table I). Hence, some of the eastern downstream sites (>100 km) had δ^{15}N-NO$_3^-$ values between +5 and +7‰ indicating a mixture of NO$_3^-$ derived from soil nitrification and sewage or manure. Other eastern sites had δ^{15}N-NO$_3^-$ values >7‰, which indicates that NO$_3^-$ was mainly derived from sewage or manure. Despite this input of anthropogenic NO$_3^-$, NO$_3^-$-N concentrations increased only moderately (generally <0.5 mg L^{-1}) with increasing flow distance.

This distinct west to east separation in terms of the NO$_3^-$ source for both the tributaries and the OMR sites appears to be a reflection of the specific land use within the OMR basin, as illustrated by data for samples collected in March 2002 (Figure 6). In the predominantly forested headwater portion of the watershed, the OMR site at km 0 (site #1) had a NO$_3^-$-N concentration of 0.08 mg L^{-1} and a δ^{15}N-NO$_3^-$ value of –0.2‰. These parameters increased to approximately 0.30 mg L^{-1} and +6.4‰ at OMR km 314 (site #12), in the mainly agriculturally used part of the watershed. The increasing NO$_3^-$-N concentrations and δ^{15}N-NO$_3^-$ values in the Oldman River seem to be mainly caused by the influx of water from tributaries draining irrigated and manured fields in the eastern part of the basin (Figure 6). Tributaries at km 246 and 249 had NO$_3^-$-N concentrations of 7.62 mg L^{-1} and a δ^{15}N-NO$_3^-$ value of +14.3‰, and NO$_3^-$-N concentrations of 0.59 mg L^{-1} and a δ^{15}N-NO$_3^-$ value of +14.2‰, respectively. This seems to suggest that manure-derived NO$_3^-$ is predominantly responsible for the increasing NO$_3^-$-N concentrations and δ^{15}N-NO$_3^-$ values in the downstream portion of the Oldman River. To which extent sewage-derived NO$_3^-$, which has a similar nitrogen isotope composition, contributes to increasing NO$_3^-$-N concentrations and δ^{15}N-NO$_3^-$ values in the OMR remains to be determined. The latter is also a potential contributor to the riverine NO$_3^-$ load, as several sewage treatment plants are located downstream of OMR km 205.

In general, western tributaries add NO$_3^-$ to the OMR having δ^{15}N-NO$_3^-$ values around +3‰, and eastern tributaries add NO$_3^-$ with δ^{15}N-NO$_3^-$ values near +15‰ resulting in δ^{15}N-NO$_3^-$ values of riverine nitrate in the Oldman River between +2‰ (upstream) and +7‰ (downstream). These values agree with those reported by Mayer *et al.* (2002) for the outlets of 16 watersheds in the northeastern U.S. They determined that typical δ^{15}N-NO$_3^-$ values at the watershed outlets with predominantly forested areas are <+5‰, and found δ^{15}N-NO$_3^-$ values between +5 to +8‰ for areas with significant agricultural and urban land use. Similarly, our isotopic results also appear to correlate with land use changes within the OMR basin, which has almost pristine waters in its forested western part and increased urban-industrial-agricultural activities in its eastern portion.

It remains to be seen whether land use changes are solely responsible for the change in concentration and isotopic composition of NO$_3^-$ in the Oldman River. Isotope and mass balances are currently performed to determine to which extent processes such as NO$_3^-$ assimilation influence riverine NO$_3^-$ in the OMR basin. Furthermore, the high δ^{15}N-NO$_3^-$ values might not only be indicative of manure-

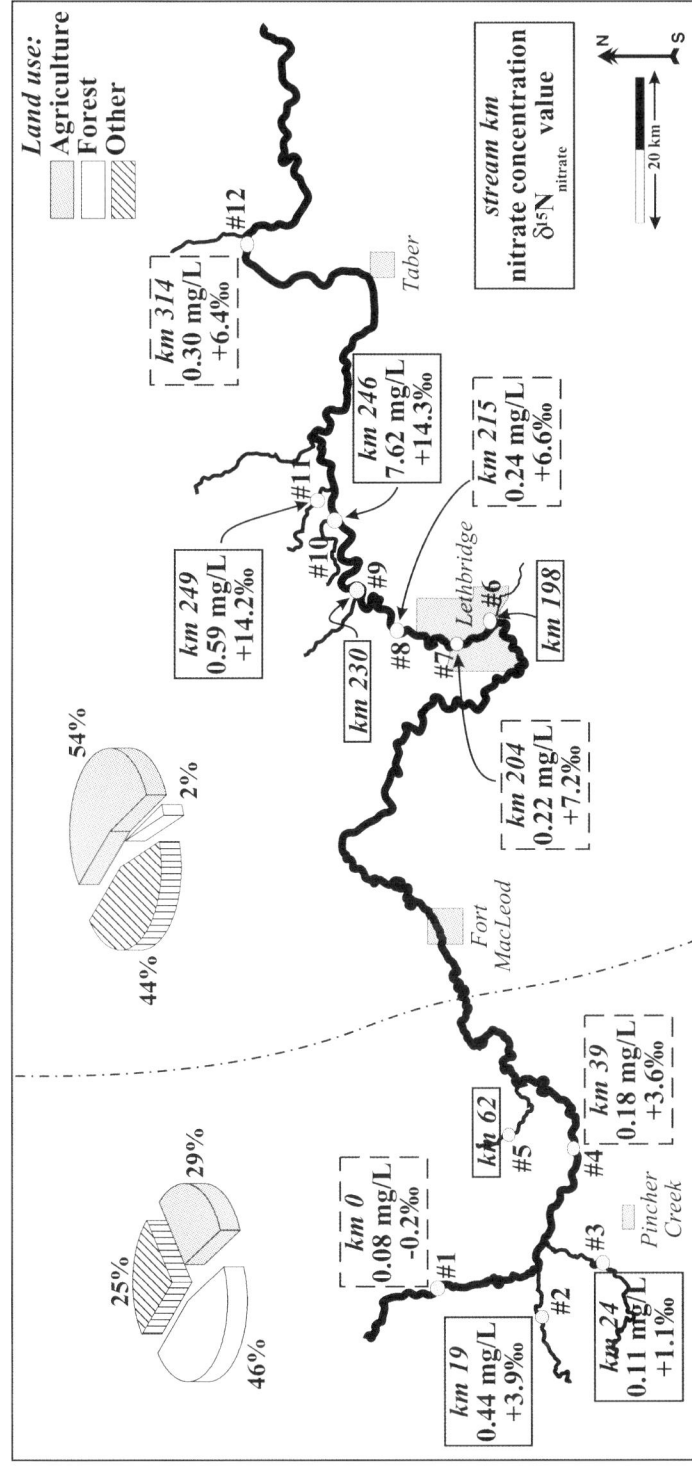

Figure 6. Map view of the Oldman River and some of its tributaries with NO_3^- concentrations and δ^{15}N-NO_3^- for OMR sites (broken line box) and some of its tributaries (solid line box) for samples obtained in March 2002; also shown land use within the Oldman River Basin.

or sewage-derived NO_3^-. Another potential source is 'natural' NO_3^- formed by oxidation of till derived NH_4^+. Hendry *et al.* (1984) documented that groundwater about 100 km SE of Lethbridge, Alberta, contains 'natural' NO_3^-, which had δ^{15}N-NO_3^- values ranging from +8 to +26‰. Future work will have to assess whether the high δ^{15}N-NO_3^- values are solely indicative of NO_3^- from manure, or whether NO_3^- from sewage or oxidation of till-derived NH_4^+ constitute additional sources.

6.2. Temporal changes in concentration and isotopic composition of NO_3^-

Isotope data appear to indicate that NO_3^- sources do not change seasonally, however, NO_3^--N concentrations vary significantly with season (Figures 2 and 3). Whereas western tributaries had low NO_3^--N concentrations (≤ 0.5 mg L^{-1}) throughout the year, eastern tributaries were characterized by high NO_3^--N concentrations (up to 9 mg L^{-1}) in the non-irrigation season and low concentrations throughout the irrigation period (Table II, Figure 3a). We suspect that full irrigation canals during the irrigation season act as 'losing streams' preventing agricultural return flow to reach the drains. At the end of the irrigation season when water levels in the irrigation canals are low, the hydraulic gradient is reversed and the drains become 'gaining streams' from November to March. Hence, agricultural return flow infiltrates into the tributaries resulting in an increase in NO_3^--N concentration. Another explanation for the decrease in NO_3^--N concentrations during the summer might be dilution due to the increased flow of water from irrigation.

The eastern Oldman River sites (>200 km) display a similar trend than the eastern tributaries with elevated NO_3^- concentrations in the winter and low concentrations in the summer (Figure 2). This suggests that influx of water from the tributaries influences the NO_3^- concentration in the Oldman River. Nevertheless, additional processes such as assimilation or dilution from snowmelt water might be partly responsible for the decrease in NO_3^--N concentrations observed in the Oldman River during the summer, which will be further investigated.

7. Conclusions

The presented data indicate that δ^{15}N-NO_3^- and δ^{18}O-NO_3^- values represent primarily NO_3^- sources and not NO_3^- transformation processes within the Oldman River basin. In general, tributaries in the western part of the watershed add NO_3^- to the Oldman River having δ^{15}N-NO_3^- values around +2‰ indicative of soil nitrification, and eastern tributaries add NO_3^- with δ^{15}N-NO_3^- values near +15‰ indicative of manure-derived NO_3^-. The isotopic composition of NO_3^- in the Oldman River is mainly determined by that of tributary NO_3^-, and hence increased from less than +3‰ to more than +6‰ with increasing flow distance. Nitrate sources at any given site do not vary seasonally. Sources of NO_3^- change spatially dependent upon land use within the Oldman River basin. Nitrate concentrations are influenced by irrigation practices, resulting in low NO_3^- concentrations during the irrigation season and high NO_3^- concentrations in the non-irrigation season, when agricultural return flow reaches the tributaries. This study shows the usefulness of $\delta^{15}N$-NO_3^- and $\delta^{18}O$-NO_3^- values in identifying the sources of riverine NO_3^- and their dependence on land use.

Acknowledgements

We thank D. Duncan, R. Eade, and R. Walker from Alberta Environment for help with collecting the samples; D. Allison, O. Mahler, K. Saffran and R. Walker from Alberta Environment, and L. Schinkel from Alberta Agriculture Food and Rural Development for providing hydrometric and chemical data. We gratefully acknowledge technical assistance by N. Kuzmak, N. Lozano, J. Overend-Pontoy, E. Strusievicz, and S. Taylor from the Isotope Science Laboratory at the University of Calgary. The National Science and Engineering Research Council of Canada (NSERC), the Alberta Agricultural Research Institute (AARI), and the Canadian Water Network (a member of the Networks of Centers of Excellence of Canada) provided financial support for this study.

References

Aravena, R., Evans, M. L. and Cherry, J. A.: 1993, 'Stable istopes of oxygen and nitrogen in source identification of nitrate from septic systems', *Ground Water* **31**, 180–186.

Alberta Government: 1976, 'Oldman River flow regulation preliminary planning studies', Vol. 1, *Main Report*, Alberta Environmental Engineering Support Services, Planning Division.

Böttcher, J., Strebel, O., Voerkelius, S. and Schmidt, H.-L.: 1990, 'Using isotope fractionation of nitrate-nitrogen and nitrate-oxygen for evaluation of microbial denitrification in a sandy aquifer', *J. Hydrol.* **114**, 413–424.

Bruning-Fann, C. S. and Kaneene, J. B.: 1993, 'The effects of nitrate, nitrite and N-nitroso compounds on human health: A review', *Vet Hum. Toxicol.* **35**, 521–538.

Campbell, D. H., Tonnesen, K. A. Kendall, C., Chang, C. C. Y. and Silva, S. R.: 2002, 'Pathways for nitrate release from an alpine watershed: Determination using $\delta^{15}N$ and $\delta^{18}O$', *Water Resour. Res.* **38**, 101–109.

Cravotta III, C. A.: 1995, 'Use of stable isotopes of carbon, nitrogen, and sulfur to identify sources of nitrogen in surface waters in the lower Susquehanna River basin, Pennsylvania', *Open-File Report*, U.S. Geological Survey.

Exner, M. E. and Spalding, R. F.: 1994, 'N-15 identification of nonpoint sources of nitrate contamination beneath cropland in the Nebraska Panhandle: Two case studies', *Appl. Geochem.* **9**, 73–81.

Flipse Jr., W. J., Katz, B. G., Lindner, J. B. and Markel, R.: 1984, 'Sources of nitrate in groundwater in a sewered housing development, central Long Island, New York', *Ground Water* **22**, 418–426.

Gellenbeck, D. J.: 1994, 'Isotopic compositions and sources of nitrate in ground water from western Salt River Valley, Arizona', Water-Resources Investigations, U.S. Geological Survey.

Goolsby, D. A.: 2000, 'Mississippi Basin nitrogen flux believed to cause Gulf hypoxia', *EOS, Trans. Am. Geophys. Union* **81**, 321–327.

Gormly, J. R. and Spalding, R.: 1979, 'Sources and concentrations of nitrate-nitrogen in ground water of the central Platte Region, Nebraska (Buffalo, County, Hall County, Merrick County)', *Ground Water* **17**, 291–301.

Hendry, M. J., McCready, R. G. L. and Gould, W. D.: 1984, 'Distribution, source and evolution of nitrate in a glacial till of southern Alberta, Canada', *J. Hydrol.* **70**, 177–198.

Kellmann, L. and Hillaire-Marcel, C.: 1998, 'Nitrate cycling in streams: Using natural abundances of NO_3^--$\delta^{15}N$ to measure *in situ* denitrification', *Biogeochemistry* **43**, 273.

Kendall, C.: 1998, 'Tracing nitrogen sources and cycling in catchments', in C. Kendall and J. J. McDonnell (eds), *Isotope Tracers in Catchment Hydrology*, Elsevier, pp. 521.

Kreitler, C. W.: 1979, 'Nitrogen-isotope ratio studies of soils and groundwater nitrate from alluvial fan aquifers in Texas', *J. Hydrol.* **42**, 147–170.

Lindau, C. W., Delaune, R. D., Patrick Jr., W. H. and Lambremont, E. N.: 1989, 'Assessment of stable nitrogen isotopes in fingerprinting surface water inorganic nitrogen sources', *Water, Air, Soil Pollut.* **48**, 489–496.

Mayer, B., Hütter, B., Veizer, J., Bollwerk, S. M. and Mansfeldt, T.: 2001, 'The oxygen isotope composition of nitrate generated by nitrification in acid forest floors', *Geochim. Cosmochim. Acta* **65**, 2743.

Mayer, B., van Breemen, N., Howarth, R. W., Seitzinger, S., Billen, G., Lajtha, K., Nadelhoffer, K., van Dam, D., Hetling, L. J., Nosil, M., Paustian, K., Boyer, E. W., Goodale, C. and Jaworski, N. A.: 2002, 'Sources of nitrate in rivers draining sixteen watersheds in the northeastern U.S.: Isotopic constraints', *Biogeochemistry* **57**, 171–197.

Morton, F. I.: 1983, 'Operational estimates of lake evaporation', *J. Hydrol.* **66**, 77–100.

Mossop, G. and Shetsen, I.: 1994, 'Geological atlas of the Western Canada sedimentary basin', Canadian Society of Petroleum Geologists, Calgary, AB, Canada.

Nielsen, G. L. R.: 1971, 'Hydrogeology of the irrigation study basin, Oldman river drainage, Alberta, Canada', Brigham Young University Research Studies, Geology Series 18, Part 1, pp. 3–98.

OMRSMC: 1978, 'Oldman River Final Report', Oldman River Study Management Committee.

OMRWQI: 2002, 'Water quality and land use interactive map and data CD', Oldman River Basin Water Quality Initiative.

Silva, S. R., Kendall, C., Wilkinson, D. H., Ziegler, A. C., Chang, C. C. Y. and Avanzino, R. J.: 2000, 'A new method for collection of nitrate from freshwater and the analysis of nitrogen and oxygen isotope ratios', *J. Hydrol.* **228**, 22–36.

Vitousek, P. M., Matson, P. A., Schindler, D. W., Schlesinger, W. H. and Tilman, D. G., Aber, J. D., Howarth, R. W. and Likens, G. E.: 1997, 'Human alteration of the global nitrogen cycle: sources and consequences', *Ecol. Appl.* **7**, 737–750.

Wassenaar, L. I.: 1995, Evaluation of the origin and fate of nitrate in the Abbotsford aquifer using the isotopes of ^{15}N and ^{18}O in NO_3^- ', *Appl. Geochem.* **10**, 391–405.

Wells, E. R. and Krothe, N. C.: 1989, 'Seasonal fluctuation in $\delta^{15}N$ of groundwater nitrate in a mantled karst aquifer due to macropore transport of fertilizer-derived nitrate', *J. Hydrol.* **112**, 191–201.

Whitehead, E., Hiscock, K. and Dennis, P.: 1999, 'Evidence for sewage contamination of the Sherwood Sandstone aquifer beneath Liverpool, U.K.', in B. Ellis (ed), *Impacts of Urban Growth on Surface Water and Groundwater Quality*, IAHS Publication No. 259, pp. 179–185.

WRD: 1970, 'South Saskatchewan River basin study, bedrock geology', Water Resources Division, Department of Agriculture, Edmonton, Alberta, Canada.

ASSESSMENT OF HEAVY METAL AND PAH CONTAMINATION OF URBAN STREAMBED SEDIMENTS ON MACROINVERTEBRATES

GARY BEASLEY and PAULINE E. KNEALE*

School of Geography, University of Leeds, Leeds LS2 9JT, U.K.
(* author for correspondence, e-mail: p.e.kneale@leeds.ac.uk; phone: 0113 343 3340;
fax: 0113 3433308)

(Received 20 August 2002; accepted 12 April 2003)

Abstract. The results from measuring PAH and metal contamination together with macroinvertebrate communities at 62 headwater stream sites gives a significant insight into the range and scale of contamination. Monitoring streambed sediments at 62 sites from rural to inner city and in industrial locations presented a unique opportunity to distinguish the conditions that enhance pollution runoff at sites that are less obviously 'at risk' and to compare these results with sites of expected high contamination, for example in industrial areas and at motorway junctions. We used pCCA (partial Canonical Correspondence Analysis) to tease out the relationships between individual macroinvertebrate families and specific metal and PAH contaminants, and showed that it is not always the metals and PAHs with the greatest total concentrations that are doing the damage to the ecology. Ni and Zn are the critical metals, while benzo(b)fluoranthene, anthracene and fluoranthene are the most contaminating PAHs. The results identify previously unrecognized 'high risk' pollution sources, lay byes used for commercial parking, on-street residential parking areas, and the junctions at the bottom of hills with traffic lights, where surface runoff feeds rapidly to the streams. While this study looks at sites across Yorkshire, UK, it clearly has a broader significance for understanding contamination risks from diffuse runoff as a prerequisite for effective sustainable urban drainage system (SUDS) agendas and the protection of urban stream ecology.

Keywords: diffuse urban runoff, macroinvertebrates, metal contamination, PAHs, stream bed sediments

1. Introduction

Despite efforts to control water quality through targeting point sources of major contaminants, many rivers and streams experience biological quality below that suggested by their environmental characteristics. Streambed sediments can be expected to accumulate have metal- and oil-based contamination (polycyclic aromatic hydrocarbons, PAHs) in urban areas where runoff is associated with vehicle trafficking, pavements, roofs, guttering and industry (Marsalek *et al.*, 1999; van Metre *et al.*, 2000). Metals and oils preferentially attach to fine particles (Estèbe *et al.*, 1997; Lee *et al.*, 1997), which are entrained in surface runoff and deposited as streambed sediments. The sediment contamination is often a magnitude greater than in the overlying water column (Power and Chapman, 1992). These contaminants exert a persistent and wide-reaching stress on the freshwater ecosystem leading to the impairment of tolerant species and the disappearance of the sensitive macroinvertebrate species and, through their accumulation, damage species in the higher trophic levels (Field and Pitt, 1990; Beyer *et al.*, 2000). The lack of a detailed knowledge of these interrelationships, and therefore the ability to identify diffuse contamination risks, has critical implications for the intelligent implementation of sustainable urban drainage system (SUDS) agendas (Environment Agency, 2000; D'Arcy and Frost, 2001; Harremoes, 2002).

This study reports the empirical results of monitoring the ecology and sediment chemistry in a significant number of streams that receive only diffuse urban runoff from a variety of surfaces. The analysis compares the field results with RIVPACS forecasts for pristine streams (Wright, 2000) and the risks of each heavy metal and PAH to individual macroinvertebrate families are isolated using pCCA (partial Canonical Correspondence Analysis).

2. Methodology

A detailed examination of surface sewer network maps and verification from field investigations identified 62 sites on 27 first order streams in west and south Yorkshire, United Kingdom, for sampling in May and September 1999 (Figure 1; Beasley, 2001; Beasley and Kneale, 2002). In this paper, the September results are discussed. Sediment and macroinvertebrate samples were collected 25 m above and below surface storm water inflows in rural, residential, industrial, and motorway subcatchments. In the analysis, sites were subjectively arranged in order of hypothesised increasing contaminant risk, rural through to motorway land use. In order to ensure comparability with data for the Environment Agency (EA) RIVPACS system model, site data waere collected on each visit using EA procedures (Environment Agency, 1997). At each site, data were obtained for: altitude, distance from source, slope, stream width, depth, discharge class, percentage particle cover from silt to boulders, dissolved oxygen, electrical conductivity, and pH. Similarly, sediment and macroinvertebrate sampling followed EA protocols.

2.1. STREAM SEDIMENT COLLECTION

Representative triplicate random samples were taken from the upper 0–5 cm layer and composited to minimise heterogeneity in heavy metal concentrations and sampling variance (Argyraki *et al.*, 1995). Trials showed that sediment collected from the near surface (0–5 cm) with a plastic trowel caused minimum disturbance, low risk of cross contamination, and retention of the finest particles. The sediments for metal determinations were placed in airtight, zip-sealed polythene bags, and double bagged. The same field procedures were implemented for the collection of PAH sediment samples, which were packed tightly into 500 mL glass bottles with glass stoppers and covered with aluminium foil to deter light penetration (Greenberg *et*

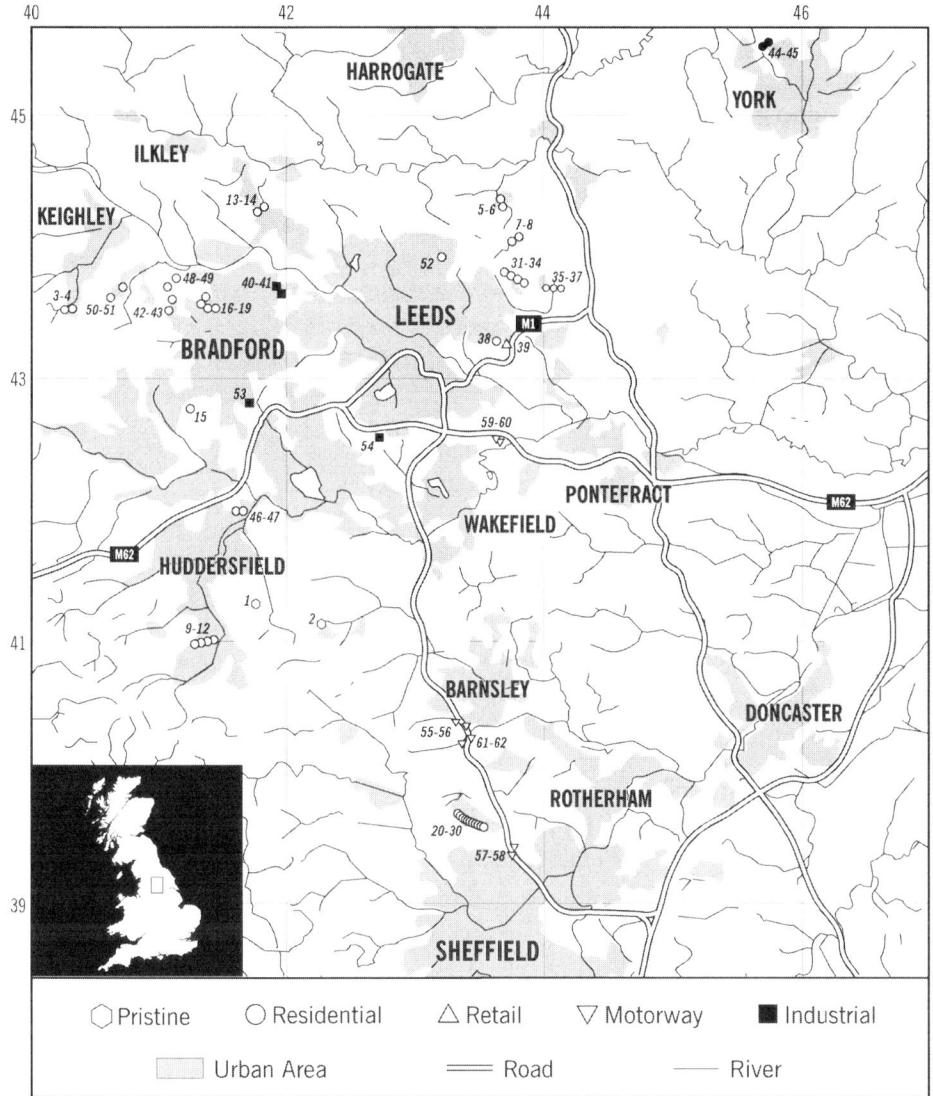

Figure 1. Site locations, Yorkshire, UK.

al., 1992). Samples were refrigerated at 4 °C, not frozen, as PAH determinations with frozen sediments have been linked to reduction in concentrations of up to 99% (Fox *et al.*, 1991).

2.2. MACROINVERTEBRATE COLLECTION

Standard macroinvertebrate collection involved three 1-min kick samples, each from 1 m^2 of streambed. Samples were decanted into 1100 mL polypropylene bottles with just enough water to maintain dampness, which reduced damage and retarded carnivore activity during transportation. They were preserved using 95% ethanol, sorted for identification to family level, and recorded using EA audit sheets. Examples of each taxon were placed in vials containing ethanol for quality assurance checking with EA scientists. The abundance of each family was recorded using EA audit sheets.

2.3. LABORATORY ANALYSIS

2.3.1. *PAHs*

Air-dried PAH samples were ground, passed through a riffle box to ensure homogenisation and reground to pass a 212 :m aperture sieve. 3 g of sieved sediment was extracted with 15 mL of dichloromethane in glass vials spiked with d14-Terphenyl for 16 hr on a mechanical shaker. The extract was analysed by gas chromatography mass spectrometry (GCMS) using a Varian 3800 series gas chromatograph, connected to a Saturn 2000 ion trap mass spectrometer. The GC oven was programmed from 50 °C (held for 1 min) to 290 °C at a rate of 10 °C min^{-1}. Temperature was held constant at 290 °C for 6 min (total run time 31 min). The injector was held at a constant temperature of 300 °C with a split ratio of 50:1. The carrier gas was helium supplied at a constant flow rate of 1.5 mL min^{-1} for 6 min. Duplicate samples were taken for quality control in conjunction with a CRM (Coal carbonised site soil LGC 6138).

2.3.2. *Heavy metals*

Following standard drying processes (Mudroch and MacKnight, 1994) subsamples were taken by coning and quartering, and then ground lightly using an agate mortar and pestle. Each subsample was passed through a 2000 microns synthetic polymer woven screen to minimise contamination (Mudroch and Azcue, 1995), while retaining only the sand and silt-sized particles. Analytical-grade (AnalaR) acids were used for all extraction solutions and cleaning procedures. The metals (Cd, Cr, Cu, Fe, Pb, Ni and Zn) were extracted from 0.5 g of each size fraction in 50 mL PTFE vials using a three-step sequential extraction technique (Quevellier, 1997). This method identifies the metals from the three geochemical phases; exchangeable, reducible and oxidisable (Beasley, 2001). The supernatant produced after extraction was acidified to pH 2 to prevent any adsorption or precipitation of metals. Metal concentrations were determined using a Jarrell-Ash Inductively Coupled Plasma Optical Emission Spectrometer (IAP-AES). Duplicate samples, blanks and CRM 601 were incorporated in each run for quality assurance.

2.4. ANALYSIS

2.4.1. *RIVPACS*

Comparison of observed stream biological quality with that expected assuming no contamination assists in identifying the severity of contamination and ecological risks between and within land uses. Therefore, the predicted biological quality at each site was forecast with the EA RIVPACS model (Wright, 2000). The RIVPACS model, (Wright, 2000) uses river and catchment environmental characteristic data to forecast the expected Biological Monitoring Working Party (BMWP) biotic index under non-polluted conditions, providing a baseline comparison with measured results. The model uses data collected at each site for altitude, distance from source, slope, stream width, depth, discharge class, percentage particle cover from silt to boulders, dissolved oxygen, electrical conductivity, and pH.

2.4.2. *pCCA*

Partial canonical correspondence analysis (pCCA) used here followed the method of ter Braak (1987, 1994) to determine the relative importance of contaminants (bioavailable metals and PAHs) in explaining the variability in the macroinvertebrate community composition. In order to do so, the explanatory variables were subdivided into a set of covariables and a set of variables-of-interest. The covariables represent habitat variables are not the prime focus of the research and as such did not enter the synthetic gradients (ter Braak and Verdonschot, 1995). The variables-of-interest used to construct the synthetic gradients are the streambed sediment heavy metal and PAH concentrations. With the covariables representing gradients that are already extracted having been partialled out (ter Braak, 1996), the ordination diagrams display the unimodal relationships (optimum abundance) between macroinvertebrates, heavy metals and PAHs.

The analysis was carried out using the programme CANOCO 4.0 following the data set conversion into CANOCO 4.0 format using the utility program CanoImp (Beasley, 2001). In Run 1, the first extraction, bioavailable metals were incorporated against macroinvertebrates recorded for September. In Run 2, the influence of PAHs on the macroinvertebrates was assessed. Essentially, the data are displayed as rankings (from the 'forward selection' option in CANOCO; Table I) and explored further using ordination diagrams. Table I shows the top 10 ranked variables that explain community composition by metals and PAHs. Interpretation of the ordination diagrams facilitates the ranking of families in relation to each element identifying those that are tolerant and sensitive (Tables II and III).

On ordination diagrams, the variables with long arrows are more strongly related to the pattern of variation in species composition than those with short arrows. Families whose points are projected close to the arrow tips are largely restricted to streams with those characteristics and vice versa for families whose endpoints project onto the lower end of the arrow.

3. Results and Discussion

As has been discussed elsewhere (Beasley, 2001; Beasley and Kneale, 2002) the results support the general hypothesis that as suburbanisation increases, the concentrations of metals and PAHs can be expected to rise, agreeing with Andoh,

TABLE I

Ranked pCCA results for two runs using unrestricted Monte Carlo significance test ($p < 0.05$)

pCCA RUN 1			pCCA RUN 2		
Top 10 rankings for weighted bioavailable metal concentrations and water chemistry variables			Top 10 rankings for PAHs and water chemistry variables		
Variable	F	p	Variable	F	p
Ni	2.01	0.020*	Benzo(b)fluoranthene	2.93	0.005*
Zn	1.50	0.045*	Electrical conductivity	1.85	0.030*
PH	1.31	n.s	Anthracene	1.61	0.035*
Pb	1.05	n.s	pH	1.61	0.030*
Cu	0.91	n.s	Fluoranthene	1.57	n.s
Electrical conductivity	0.92	n.s	Napthalene	1.45	n.s
Dissolved oxygen	1.00	n.s	Dibenz(a,h)anthracene	1.17	n.s
Cd	1.20	n.s	Indeno(1,2,3-cd)pyrene	1.02	n.s
Cr	0.66	n.s	Acenaphthene	0.89	n.s
Fe	0.53	n.s	Fluorene	0.93	n.s

(1994), Ellis *et al.* (1997), Marsalek *et al.* (1999), and Lee and Bang, (2000). Careful analysis of the metals data indicates that motorway runoff is not always the worst culprit (Figure 2; n.b. the sites were subjectively arranged in order of hypothesised increasing contaminant risk, rural through to motorway land use). Zn contamination is found at several of the residential sites, most noticeably site 16 (and corresponding downstream sites 18 and 19) and sites 20–30. These sites were all found to have serious traffic-related sources. Zinc is the major metal in vehicle tires. It is postulated that the high sediment Zn levels are the result of runoff from heavily trafficked, steep 'A' roads where deceleration causes tires to wear with concurrent increased release of Zn, agreeing with the findings of Kim *et al.* (1998) and Draper *et al.* (2000). Sites 59 and 60 have the highest concentrations overall because runoff derives from a major junction roundabout and motorway exit lanes that are hot spots for vehicle braking. Sites 40 and 41 receive drainage from an industrial outlet where heavy trafficking by lorries and where tight turning restrictions induces tire wear, which is thought to account for the enhanced Zn levels, but some building-related contamination may occur here as well.

Site 54 showed very high Cu concentrations in September 1999, although not in the previous May, and further investigation suggests that this was due to spillage or illegal dumping. The persistence of such short-term variations on catchment sediment chemistry requires further investigation. The patterns between May and

September vary, but not very substantially, the overall patterns being consistent and of more significance than the fluctuations.

The PAH results also showed an increasing concentration gradient from rural through residential to industrial land uses supporting the link with vehicles found by Maltby *et al.* (1995) and van Metre *et al.* (2000). However, results from this study show clearly that this is not always true and depends on the individual element (Beasley, 2001; Beasley and Kneale, 2002). Napthalene does appear to adhere to this general pattern; the highest concentrations were recorded in stream sediments below industrial and motorway inflows (Figure 2). However, certain residential subcatchments, for example sites 46 and 47, posed a greater contamination risk than either the industrial or motorway source areas. The reason lies in specific catchment characteristics. These residential areas with roadside parking accumulate higher concentrations of total PAHs largely because of high concentrations of fluoranthene, phenanthrene and pyrene (Figure 3), which originate from crankcase oil drippings leaked onto the road surface (Latimer *et al.*, 1990). Supporting evidence for leaked oil rather than traffic volume being responsible for high concentrations of PAHs derives from sites 45 and 59 that drain heavily utilised lay-byes.

Understanding contamination risk is further complicated by the stream or river characteristics into which the runoff drains. If the stream has a silt-laden streambed due to low flow allowing fine particles to settle, it would be expected that contaminant concentrations would be greater than a gravel bed stream receiving identical contaminant input. Differences in stream characteristics have undoubtedly influenced the levels of contamination in this study and one would presume the ecological quality. It is hypothesised that a silt laden, shallow, slow flowing stream would support a relatively depauperate macroinvertebrate community structure. However, the characteristics of the streams in this investigation are all shown to be capable of supporting a 'good' biological quality, as evidenced by in the RIVPACSs model forecasts. The RIVPACs forecast incorporates detailed information about altitude, geology, stream bed sediment size, and channel characteristics (Wright, 2000). Actual biological quality is inversely related to the observed contamination levels, with a general reduction in macroinvertebrate diversity as contamination increases. The predicted RIVPACS, scores range from 102–155 and always exceed the actual scores. Actual BMWP values ranged from 25–95 except for one score of 123 in the September samples.

3.1. TOTAL METAL CONTAMINATION

Evaluating the toxicity of the concentrations of the sediment associated metals is limited by the absence of UK or European standards, but we can compare the results with the Ontario Ministry of Environment (OME) total metal sediment toxicity guidelines (Persaud *et al.*, 1989) (Figure 2). For each metal there are some sites that exceed the critical thresholds, but the largest number of excedences were for

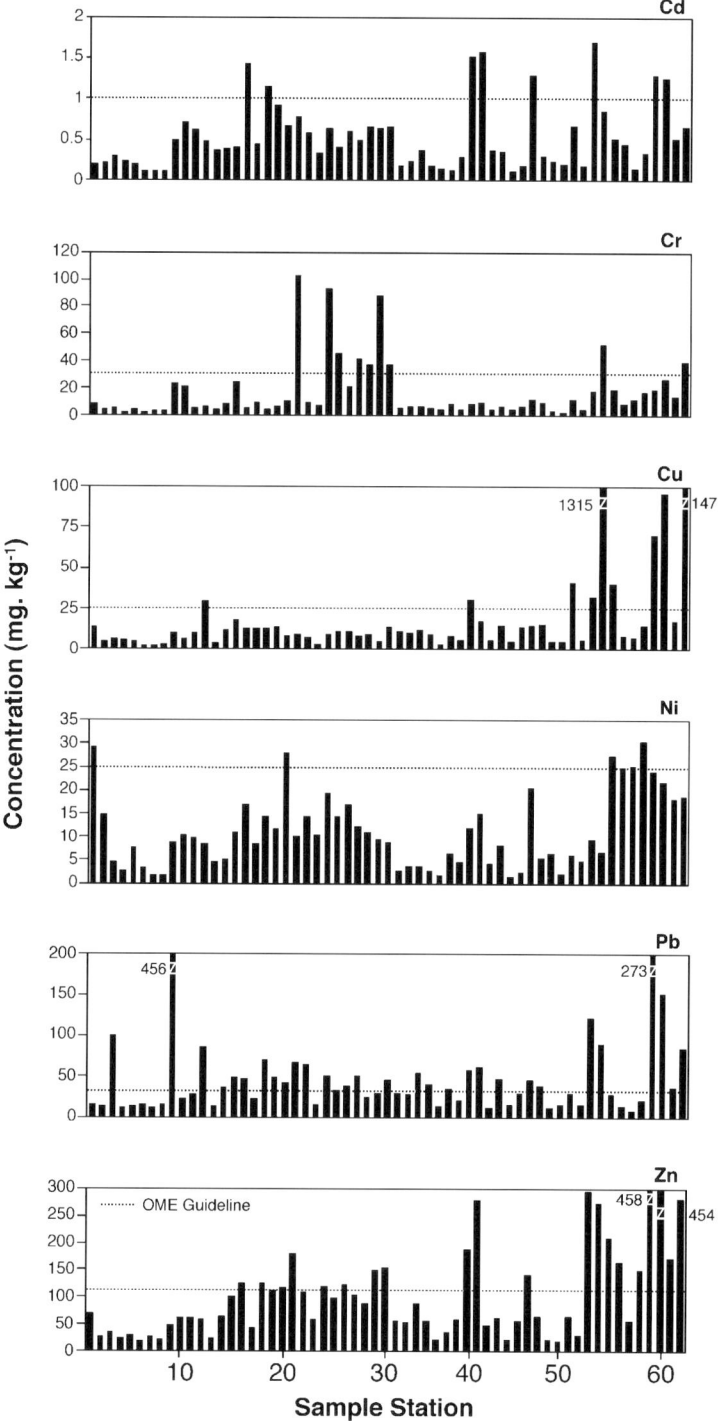

Figure 2. Total metal sediment toxicity compared with the OME toxicity guidelines.

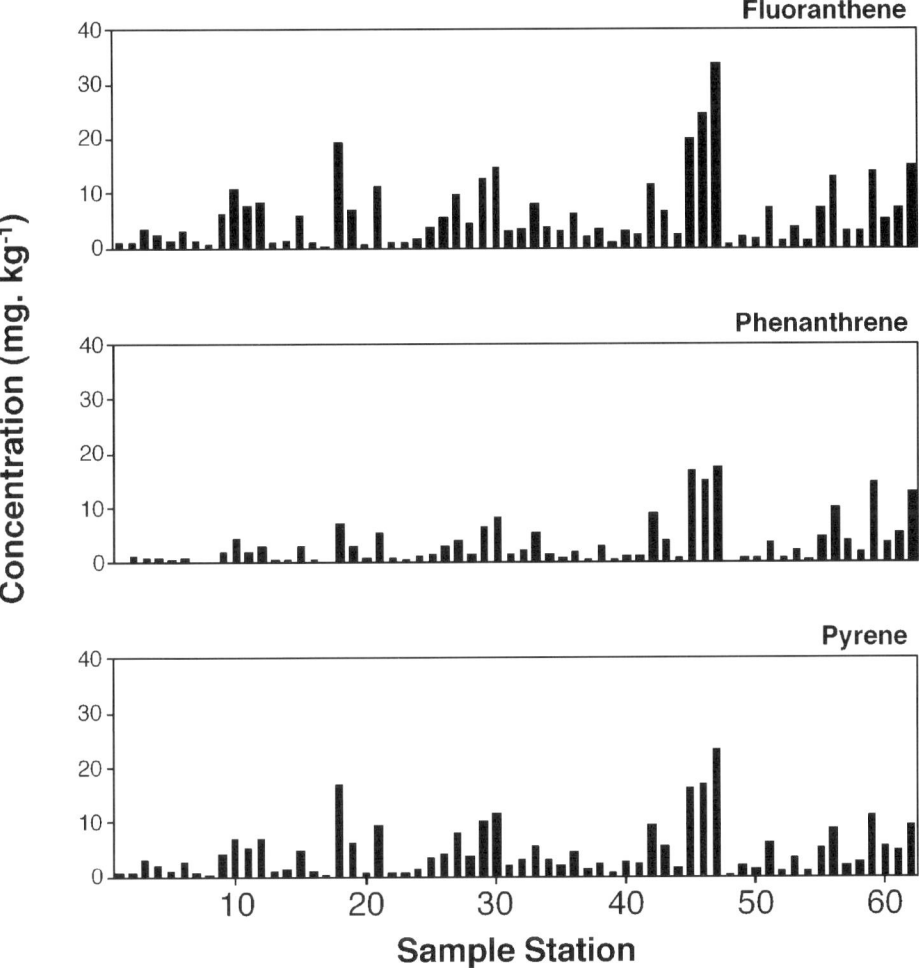

Figure 3. Streambed sediment concentrations of three PAHs.

Pb and Zn. This classifies the sediments as potentially toxic to macroinvertebrates and fish. The association of Pb and Zn from vehicle emissions and Zn from vehicle tires can be linked to the more contaminated and heavily trafficked sites, such as at site 16 and the motorway sites 59, 60 and 62, all of which exceed the toxicity guidelines. Extrapolating Canadian sediment guidelines to Yorkshire streams must be done cautiously, but is nonetheless indicative of problems in headwaters that have been expected to be much cleaner. It would, of course, be helpful to have UK sediment chemistry toxicity guidelines that are based on bioavailable rather than total metals. Such guidelines would be more sensitive than those based on total metal extractions, although their application would still require sensitivity to the local hydrological and catchment circumstances.

TABLE II
Families determined as tolerant or sensitive to metals and water chemistry variables, Run 1

Variable	Five most tolerant families (tolerance ranked left to right)	Five most sensitive families (sensitivity ranked left to right)
Ni	Phil, Perl, Chlo, Rhya, Hept	Plan, Ephemeri, Valv, Hydrob, Dyti
Zn	Chlo, Phys, Hali, Hydroph, Rhya	Hydrom, Plan, Nemo, Ephemeri, Ephemere
pH	Plan, Ephemeri, Valv, Phys, Hali	Perl, Phil, Hydrom, Leuc, Hept
Pb	Chlo, Phys, Hali, Hydroph, Rhya	Hydrom, Nemo, Plan, Ephemeri, Ephemere
Cu	Chlo, Phys, Hali, Hydroph, Rhya	Hydrom, Plan, Nemo, Ephemeri, Ephemere
Electrical conductivity	Chlo, Phys, Hali, Plan, Lymna	Perl, Phil, Hydrom, Nemo, Leuc
Dissolved oxygen	Plan, Ephemeri, Valv, Hydrob, Dyti	Phil, Perl, Chlo, Rhya Hept
Cadmium	Chlo, Phys, Hali, Rhya, Hydroph	Hydrom, Plan, Ephemeri, Nemo, Ephemere
Cr	Chlo, Phys, Rhya, Hydroph, Hali	Hydrom, Plan, Ephemeri, Nemo, Ephemere
Fe	Phil, Perl, Chlo, Rhya, Hydroph	Plan, Ephemeri, Hydrom, Valv, Limne

Asel, *Asellidae*; Chlo, *Chloroperlidae*; Dyti, *Dytiscidae*, Ephemere, *Ephemerellidae*; Ephemeri, *Ephemeridae*; Erpo, *Erpobdellidae*; Hali, *Haliplidae*; Hept, *Heptageniidae*; Hydrob, *Hydrobiidae*; Hydrom, *Hydrometridae*; Hydroph, *Hydrophilidae*; Leptoc, *Leptoceridae*; Leuc, *Leuctridae*; Lymna, *Lymnaeidae*; Nemo, *Nemouridae*; Odon, *Odontoceridae*; Perl *Perlodidae*; Phil, *Philopotamidae*; Phys, *Physidae*; Plan, *Planorbidae*; Rhya, *Rhyacophilidae*; Seri, Sericostomatidae; Simu, *Simulidae*; Spha, *Sphaeriidae*; Valv, *Valvatidae*

3.2. Bioavailable metal contamination

In the pCCA analyses, the bioavailable rather than total metal data were used, showing that Zn and Ni were consistently the most significant metals in determining macroinvertebrate community structures (Table II). Although total concentrations of Ni were much lower than for example Cu or Pb, Ni is more readily bioavailable and it is this understanding of metal speciation that will assist in the longer term in assigning appropriate catchment mitigation measures. Although Cu and Pb are found in higher total concentrations, their bioavailable concentrations are insufficient to pose a toxic threat in these streams. By targeting the source areas of those metals with known high bioavailabilities and establishing bioavailability toxicity thresholds, concerted inroads could be made into reducing contaminant-induced ecological stress.

TABLE III

Families determined as tolerant or sensitive to metals and water chemistry variables, Run 2

Variable	Five most tolerant families (tolerance ranked left to right)	Five most sensitive families (sensitivity ranked left to right)
Benzo(b)fluoranthene	Asel, Valv, Ephemere, Spha, Hydrob	Hydrom, Leptoc, Phil, Odon, Hept
Electrical conductivity	Hydroph, Ephemere, Asel, Spha, Phys	Leptoc, Hydrom, Phil, Ephemeri, Hali
Anthracene	Valv, Asel, Spha, Phys, Hydrob	Phil, Hydrom, Odon, Hept, Leptoc
pH	Valv, Hydrob, Asel, Spha, Phys	Hydrom, Phil, Odon, Hept, Leuc
Fluoranthene	Asel, Valv, Ephemere, Spha, Hydrob	Hydrom, Leptoc, Phil, Odon, Hept
Naphthalene	Hydroph, Ephemere, Asel, Perl, Spha	Leptoc, Ephemeri, Hydrom, Phil, Leptop
Dibenz(a,h,)anthracene	Hydroph, Ephemere, Asel, Spha, Phys	Leptoc, Leptop, Hali, Ephemeri, Seri
Indeno(1,2,3-cd)pyrene	Valv, Asel, Spha, Phys, Hydrob	Hydrom, Phil, Odon, Hept, Leptoc
Acenaphthene	Valv, Asel, Spha, Phys, Ephemere	Hydrom, Phil, Leptoc, Odon, Hept
Fluorene	Valv, Asel, Spha, Phys, Ephemere	Hydrom, Phil, Leptoc, Odon, Hept

Run 2 (Table II) shows that it is not only those PAHs that are found in the highest concentrations that have the greatest impact on stream ecology. The group represented by fluoranthene was ranked fifth despite being the most prevalent in terms of concentration. The most significant PAHs are those represented by benzo(b)fluoranthene, which although present in lower concentrations, have carcinogenic impacts on macroinvertebrates. The pCCA indicate that it is not total contamination that matters in terms of either the metals or the PAHs, the stresses on stream ecology are more subtle.

3.3. INDICATOR FAMILIES

The pCCA allows us to tease out the more detailed relationships between the contaminants and macroinvertebrate families (Figures 4 and 5). The literature suggests metal contamination leads to reduction in specific Ephemeroptera (mayfly) families (Kiffney and Clements, 1994; Gower *et al.*, 1994; Clements *et al.*, 2000), and their replacement by chironomids and oligacheates. This effect is confirmed in these Yorkshire streams, particularly by the absence of *Leptoceridae* and *Ephemerellidae*. The widespread distribution of *Baetidae* and its central position in

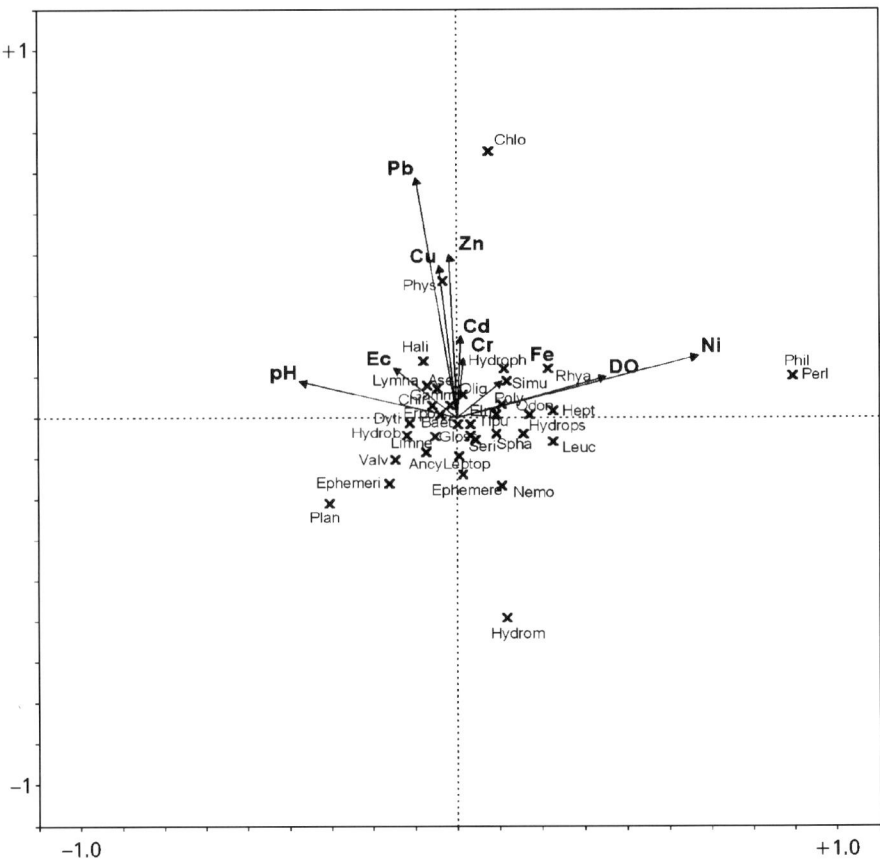

Figure 4. Species – Environment biplot based on pCCA Run 1. Families are represented by crosses.

the species-environment ordination diagrams supports the findings of Gower *et al.*, (1994) in that is has moderate metal tolerance. Similarly, there is variation in the tolerance of Plecoptera (stonefly) families, *Leuctridae* being slightly more tolerant than *Nemouridae*, whereas *Chloroperlidae* appear here to be particularly tolerant to metals. Tricoptera (caddis fly) families display slightly greater tolerances, especially *Polycentropodidae*, *Rhyacophilidae*, and *Hydropsychidae*, with the latter two families preferring the sites with high dissolved oxygen concentrations. The cased caddis flies, *Limnephilidae* and *Sericostomatidae* are more sensitive. *Asellidae* are seen here to be moderately tolerant to a number of heavy metals.

The rankings (Table II) generated from the ordination plot (Figure 3) uncover several indicator families that were at risk from bioavailable metal pollution, notably *Ephemerellidae*, *Ephemeridae*, *Nemouridae*, and *Planorbidae*. Looking at the risk from PAH, *Hydrometridae* and *Ephemeridae* were most at risk. *Leptoceridae*, *Philopotamida*, *Ondontoceridae*, and *Heptageniidae* were not as sensitive

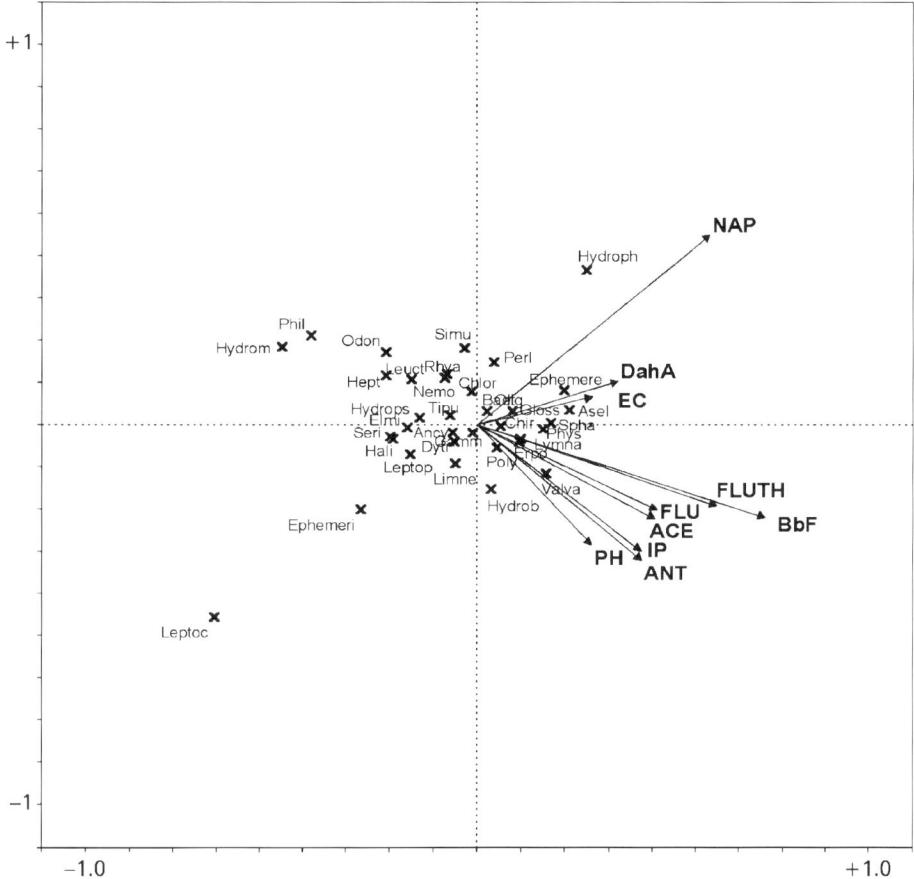

Figure 5. Species – Environmental biplot based on pCCA Run 2. Families are represented by crosses.

to metals, but were particularly sensitive to enhanced PAH levels (Table IV). In the Plecoptera (stonefly) families, the *Leptoceridae* were particularly sensitive, much more so than the *Nemouridae*, *Chloroperlidae* and *Perlodidae*. Of the Tricoptera larva (caddis fly), only the caseless *Polycentropodidae* exhibited moderate tolerance towards PAHs (Figure 5). The *Ephemerellidae*, although sensitive to metals, were shown to tolerate PAH pollution.

4. Conclusions

The SUDS (sustainable urban drainage system) agenda indicates that runoff management should be targeted to mitigate contamination of streams so that the ecological quality is retained or restored. Identifying streams, or sections of streams, at risk is complicated by local circumstances. Whereas previous studies have concentrated on *a priori* 'worst case' sites, this study has looked at a suite of sites. The results from 62 sites are inevitably noisy, but indicative. They provide a broader context for management consideration. Surface runoff from lay byes used for commercial goods parking and on-street parking in residential areas were shown to generate more PAH contamination than a motorway slip road. Runoff from a junction at the foot of a hill with traffic lights, where braking is hard and regular, was a bigger problem than runoff from roads elsewhere. Overnight on-street parking led to highly contaminated runoff from the street surfaces.

We have shown that looking at 'total' metals or 'total' PAH results can be very misleading. Bioavailable metals are much more useful and the contaminants that have the greatest abundance do not necessarily cause the most biological impairment. The pCCA results pull out the greater importance of bioavalable Ni and Zn, although their total concentrations are less than for Pb, and the importance of the benzo(b)fluoranthene and anthracene groups.

Catchments are heterogeneous, so catchment-specific characteristics are crucial to our understanding. These results suggest that risk management would benefit from: 1) controlling runoff from high hydrocarbon collection areas, such as parking areas, regardless of their size or traffic flow data; 2) identifying catchment source areas for bioavailable metals, such as Ni and Zn, from road junctions, roundabouts, exit lanes, and roads with steep gradients, and 3) developing bioavailable metal and PAH sediment toxicity guidelines and thresholds.

References

Andoh, R. Y. G.: 1994, 'Urban runoff: Nature, characteristics and control', *J. Inst. Water Eng. Managers* **8**, 371–378.

Argyraki A., Ramsey M. H. and Thompson M.: 1995, 'Proficiency testing in sampling: Pilot study on contaminated land', *Analyst* **120**, 2799–2803.

Beasley, G. E.: 2001, 'Investigating the relations between streambed sediment contaminants and macroinvertebrate assemblages in urban headwater streams in Yorkshire', Unpublished PhD thesis, University of Leeds.

Beasley, G. and Kneale, P. E.: 2002, 'Reviewing the impact of metals and PAHs on macroinvertebrates in urban watercourses', *Prog. Phys. Geogr.* **26**, 236–270.

Beyer, W. N., Day, D., Melancon, M. J. and Sileo, L.: 2000, 'Toxicity of Anacostia River, Washington DC, USA, sediment fed to mute swans (*Cygnus olor*)', *Environ. Toxicol. Chem.* **19**, 731–735.

Clements, W. H., Carlisle, D. M., Lazorchak, J. M. and Johnson, P. C.: 2000, 'Heavy metals structure benthic communities in Colorado mountain streams', *Ecol. Appl.* **10**, 626–638.

D'Arcy, B. and Frost, A.: 2001, 'The role of best management practices in alleviating water quality problems associated with diffuse pollution', *Sci. Total Environ.* **265**, 359–367.

Draper, D., Tomlinson, R. and Williams, P.: 2000, 'Pollutant concentrations in road runoff: Southeast Queensland case study', *J. Environ. Eng.* **126**, 313–320.

Ellis, J. B., Revitt, D. M. and Llewellyn, N.: 1997, 'Transport and the environment: Effects of organic pollutants on water quality', *J. Chartered Inst. Water Environ. Manage.* **11**, 170–177.

Environment Agency: 1997, *Assessing Water Quality – General Quality Assessment (GQA) Scheme for Biology*, Environment Agency Fact Sheet, Environment Agency, Bristol.

Environment Agency: 2000, *Sustainable Urban Drainage Systems*, Environment Agency, Bristol.

Estèbe, A., Boudries, H., Mouchel, J.M. and Thevenot, D.R.: 1997, 'Urban runoff impacts on particulate metal and hydrocarbon concentrations in the river Seine, suspended solid and sediment transport', *Water Sci. Technol.* **36**, 185–195.

Field, R. and Pitt, R. E.: 1990, 'Urban storm-induced discharge impacts: US Environmental Protection Agency research program review', *Water Sci. Technol.* **22**, 1–7.

Fox, M. E., Murphy, T. P., Thiessen, L. A. and Khan, R. M.: 1991, 'Potentially significant underestimation of PAHs in contaminated sediments', *Proceedings 7th Eastern Region Conference*, Quebec, Canadian Association on Water Pollution Research and Control, pp. 80–86.

Gower, A. M., Myers, G., Kent, M. and Foulkes, M. E.: 1994, 'Relationships between macroinvertebrate communities and environmental variables in metal-contaminated streams in south-west England', in D. M. Harper and A. J. D. Ferguson (eds), *The Ecological Basis for River Management*, John Wiley & Sons, New York, New York, USA, pp. 181–191.

Greenberg, A. E., Clesceri, L. S. and Eaton, A. (eds): 1992, *Standard Methods for the Examination of Water and Wastewater*, 18th edition, American Public Health Association, Washington, DC, USA.

Harremoes, P.: 2002, 'Integrated urban drainage, status and perspectives', *Water Sci. Technol.* **45**, 1–10.

Kiffney, P. M. and Clements, W. H.: 1994, 'Effects of heavy metals on a macroinvertebrate assemblage from a Rocky Mountainstream in experimental microcosms', *J. N. Am. Benthol. Soc.* **13**, 511–523.

Kim, K. W., Myung, J. H., Ahn, J. S. and Chon, H. T.: 1998, 'Heavy metal contamination in dusts and stream sediments in the Taejon area, Korea', *J. Geochem. Explor.* **64**, 409–419.

Latimer, J. S., Hoffman, E. J., Hoffman G., Fasching, J. L. and Quinn, J. G.: 1990, 'Sources of petroleum hydrocarbons in urban runoff', *Water, Air, Soil Pollut.* **52**, 1–21.

Lee, D. K., Touray, J. C., Bailif, P. and Ildeonse, J. P.: 1997, 'Heavy metal contamination of settling particles in a retention pond along the A71 motorway in Sologne, France', *Sci. Total Environ.* **201**, 1–15.

Lee, J. H. and Bang, K. W.: 2000, 'Characterisation of urban stormwater runoff', *Water Res.* **34**, 1773–1780.

Maltby, L., Boxall, A. B. A., Forrow, D. M., Calow, P. and Betton, C. I.: 1995, 'The effects of motorway runoff on freshwater ecosystems: 2. Identifying major toxicants', *Environ. Toxicol. Chem.* **14**, 1093–1101.

Marsalek, J., Rochfort, Q., Brownlee, B., Mayer, T. and Servos, M.: 1999, 'An explanatory study of urban runoff toxicity', *Water Sci. Technol.* **39**, 33–39.

Mudroch, A. and Azcue, J. M.: 1995, *Manual of Aquatic Sediment Sampling*, Lewis Publishers, Ann Arbor, Michigan, USA.

Mudroch, A. and MacKnight, S. D.: 1994, *Handbook of Techniques for Aquatic Sediments Sampling*, 2nd edition, Lewis Pubishers, Ann Arbor, Michigan, USA.

Persaud, D., Jaagumagi, R. and Hayton, A.: 1989, *Development of Provincial Sediment Quality Guidelines*, Ontario Ministry of the Environment and Energy, Toronto, Ontario, Canada.

Power, E. A. and Chapman, P. M.: 1992, 'Assessing sediment quality', in G. A. Burton, Jr. (ed.), *Sediment Toxicity Assessment*, Lewis Publishers, Ann Arbor, Michigan, USA, pp. 1–18.

Quevellier, P. H., Rauret, G., Lopez-Sanchez, J. F., Rubio, R., Ure, A. and Muntau, H.: 1997, 'The certification of the EDTA-extractable contents (mass fractions) of Cd, Cr, Ni, Pd and Zn in sediment following a three-step sequential extraction procedure – CRM 601', Office for Official Publications of the European Communities, Luxembourg.

ter Braak, C. F. J.: 1987, *CANOCO – A FORTRAN Program for Canonical Community Ordination by [partial][detrended][canonical] Correspondence Analysis, Principal Components Analysis and Redundancy Analysis (Version 2.1)*, DLO-Agricultural Mathematics Group, Wageningen, The Netherlands.

ter Braak, C. F. J.: 1994, 'Canonical community ordination. Part I: Basic theory and linear methods', *Ecoscience* **1**, 127–140.

ter Braak, C. F. J.: 1996, *Unimodal Models to Relate Species to Environment*, Agricultural Mathematics Group, Wageningen, The Netherlands.

ter braak, C. F. J. and Verdonschot, P. F. M.: 1995, 'Canonical correspondence analysis and related multivariate methods in aquatic ecology,' *Aquat. Sci.* **57**, 255–289.

van Metre, P. C., Mahler, B. J. and Furlong, E. T.: 2000, 'Urban sprawl leaves its PAH signature', *Environ. Sci. Technol.* **34**, 4064–4070.

Wright, J. F.: 2000, 'An introduction to RIVPACS', in J. F. Wright, D. W. Sutcliffe and M. T. Furse (eds), *Assessing the Biological Quality of Fresh Waters: RIVPACS and other Techniques*', The Freshwater Biological Association, Ambleside, pp. 1–24.

POST-GLACIAL LEAD DYNAMICS IN A FOREST SOIL

CHRIS E. JOHNSON[1], ROBERT J. PETRAS[1], RICHARD H. APRIL[2] and THOMAS G. SICCAMA[3]

[1] Department of Civil & Environmental Engineering, 220 Hinds Hall, Syracuse University, Syracuse, New York 13244, U.S.A.; [2] Department of Geology, Colgate University, Hamilton, New York 13346, U.S.A.; [3] School of Forestry & Environmental Studies, Yale University, New Haven, Connecticut 06511, U.S.A.
(* author for correspondence, e-mail: cejohns@mailbox.syr.edu; phone: +315-443-4425; fax: +315-443-1243)

(Received 20 August 2002; accepted 10 April 2003)

Abstract. The use of alkyl-Pb additives in gasoline during the 20th century resulted in widespread Pb pollution. The objective of this study was to determine the relative importance of atmospherically deposited Pb and Pb released through weathering to soil Pb pools at the Hubbard Brook Experimental Forest, New Hampshire. We employed a selective extraction method to estimate the amount of Pb that was: water-soluble + exchangeable (EX); inorganically bound (IB); organically bound (ORG); bound to amorphous oxides (AMOX); and bound in crystalline minerals (RES). After normalizing crystalline-Pb concentrations to the immobile element Ti, we estimated that 14.1 kg ha^{-1} of Pb has been weathered from Hubbard Brook soils in the 12,000–14,000 yr since deglaciation – a long-term average release of 1.0–1.2 g ha^{-1} a^{-1}. Analysis of Ti-normalized total Pb concentrations indicated a net post-glacial decrease of 7.2 kg ha^{-1} in the total Pb pool – consisting of a net accumulation of 4.9 kg ha^{-1} in the O horizon, and a net loss of 12.1 kg ha^{-1} from mineral soil. Atmospheric deposition of Pb between 1926 and 1989 (estimated as 8.7 kg ha^{-1}) was a major source of Pb in the post-glacial period. Together, long-term weathering release and 20th century atmospheric deposition could account for all of the Pb in the EX, IB, ORG, and AMOX fractions. Lead from gasoline appears to constitute a major fraction of the total Pb burden in Hubbard Brook soils. Periodic analysis of soil Pb fractions may be useful in monitoring the fate of Pb in forest soils.

Keywords: atmospheric deposition, forest ecosystems, fractionation, lead pollution, northern hardwoods, organic matter, soil chemistry, spodosol, trace metals, weathering

1. Introduction

For much of the 20th century, gasoline in the United States and Canada contained alkyl-Pb anti-knock additives. Increasing consumption of gasoline resulted in widespread Pb pollution, stretching even to the polar regions (Sturges and Barrie, 1989). Studies of forest soils in the United States and Europe in the 1970s and 1980s documented high concentrations of Pb, especially in organic surface horizons (e.g., Heinrichs and Mayer, 1977; Van Hook *et al.*, 1977; Andresen *et al.*, 1980; Tyler, 1981). Many of these researchers concluded that Pb was accumulating in O horizons, and that Pb was likely to remain in O horizons for centuries, or longer.

Concerns about Pb pollution and possible environmental and health effects led the U.S. Congress to pass legislation in the mid-1970s mandating that new automobiles use unleaded gasoline. Subsequently, atmospheric deposition of Pb has declined dramatically. For example, Pb deposition at the Hubbard Brook Experimental Forest in New Hampshire declined by 97% between 1976 and 1989 (Johnson *et al.*, 1995). The concentration and amount of Pb in O horizons also has declined since the mid-1970s, at Hubbard Brook (Johnson *et al.*, 1995) and elsewhere in the northeastern U.S. (Friedland *et al.*, 1992). The decline in forest floor Pb on decadal time scales suggests that Pb may be more mobile within forest soils than previously thought.

As Pb moves through soils, it may bind to soil constituents in a variety of ways, including exchange processes, complexation with soil organic matter, and adsorption or co-precipitation with oxide minerals (Adriano, 1986). Many investigators have attempted to quantify the distribution of Pb among various soil fractions through selective chemical extraction (e.g., Miller and McFee, 1983; Johnson and Petras, 1998). Application of these procedures to forest soils may aid in determining the fate of anthropogenic Pb as it moves into the mineral soil. Selective extraction studies in soils near point sources of metal pollution have documented high concentrations of Pb in near-surface soils, and in relatively labile fractions (Miller and McFee, 1983; Hogan and Wotton, 1984). In contrast, studies of trace metal fractions in areas remote from point sources have been rare. For example, Johnson and Petras (1998) found that only 14–25% of the total Pb in mineral soils at Hubbard Brook resided in 'labile' pools (exchangeable, adsorbed, organically bound, and bound in amorphous oxides). Atmospheric inputs of Pb during the period of heaviest use of leaded gasoline could account for 35% of the Pb found in the exchangeable and organic fractions of the soil, 14% if the amorphous-oxide fraction was included (Johnson and Petras, 1998).

Soluble Pb in forest soils is also produced through mineral weathering. Thus, the importance of atmospherically deposited Pb to the total Pb burden of the soil depends on the rate of weathering release of Pb. In this research, we estimated the post-glacial weathering release of Pb at Hubbard Brook by examining depletion patterns in the soil profile. Our objective was to estimate the relative importance of gasoline-derived Pb to soil Pb pools by comparing post-glacial weathering release to the estimated atmospheric inputs of Pb during the 20th century.

2. Materials and Methods

2.1. STUDY SITE

The Hubbard Brook Experimental Forest (HBEF) is located in the southern White Mountain region of New Hampshire. The climate at the HBEF is cool and humid, with mean January and July temperatures of $-9\ °C$ and $19\ °C$, respectively. An average of 139 cm of precipitation falls annually, with 25–33% as snow (Federer *et al.*, 1990).

Vegetation at the study site is predominantly northern hardwood, with approximately equal amounts of American Beech (*Fagus grandifolia* Ehrh.), sugar maple (*Acer saccharum* Marsh.), and yellow birch (*Betula alleghaniensis* Britt.). Upper elevations and exposed ridge-tops contain stands of red spruce (*Picea rubens* Sarg.), balsam fir (*Abies balsamea* (L.) Mill), and paper birch (*Betula papyrifera* var. *cordifolia* (Marsh.) Regel). Much of the forest was logged between 1909–1917, and there is no evidence of recent fire (Bormann and Likens, 1979).

Soils at the HBEF are primarily well-drained Haplorthods, Fragiorthods, and Dystrochrepts of the Becket, Berkshire, Lyman, Skerry, and Tunbridge series. The soils have developed in till of varying depth (0–8 m), deposited during the Wisconsinan glacial period, which ended in New England 12,000–14,000 years ago. Glacial till is largely derived from local bedrock – pelitic schists and gneisses of the Rangeley Formation. An O horizon averaging 6.0 cm in depth overlies mineral soil averaging 59 cm to C horizon or bedrock (Johnson *et al.*, 1991a). Hubbard Brook soils are highly acidic, with mean pH in water ranging from 3.4 in the O horizon to 4.7 in the C horizon (Johnson *et al.*, 1991b). Cation exchange capacity, measured by neutral-salt extraction, ranges from 0.3 $cmol_c$ kg^{-1} (C horizon) to 18 $cmol_c$ kg^{-1} (Oa horizon), and is almost entirely contributed by organic matter (Johnson, 2002).

2.2. SAMPLING AND ANALYSIS

Details of soil sampling and the analysis of trace metal fractions may be found in Johnson and Petras (1998), and are described here only briefly. The soil samples were collected from watershed 5 (W5; area: 23 ha, elevation: 510–750 m) at the HBEF in the summer of 1983, prior to a clear-cutting experiment. Nine randomly selected samples of each of the major pedogenic horizons (Oa, E, Bh, Bs1, Bs2, C) were analyzed for trace metal fractionation.

We used a modified version of the selective extraction procedure developed by McLaren and Crawford (1973). Our procedure resulted in five fractions (Johnson and Petras, 1998):

EX – water-soluble + exchangeable – 0.05 M $CaCl_2$ extraction;

IB – inorganically bound – 2.5% (v/v) CH_3COOH;

ORG – organically bound – 0.1 M $K_4P_2O_7$ (potassium pyrophosphate);

AMOX – amorphous-oxide bound – acid oxalate (pH 3.0);

RES – 'residual', or crystalline-bound – concentrated HCl-HF digestion.

Details of reagent preparation, soil-to-solution ratios, and extraction times may be found in Johnson and Petras (1998). Selective dissolution procedures such as the one we used have been criticized for being non-specific, extracting more, or less, metal than expected (Nirel and Morel, 1990). Nevertheless, any study of soil metals requires some chemical extraction methods. Despite the pitfalls, these methods yield valuable information about soil pools of varying binding strength.

The concentrations of immobile elements increase during weathering because of the loss of more mobile elements. Estimates of element depletion, therefore, must be made using a relatively immobile element as an index. Zirconium and titanium are often used for this purpose (e.g., Taylor and Blum, 1995; Jersak *et al.*, 1997). In this study, we used Ti as an immobile index element. However, we were unable to measure Ti concentrations on the same samples that were used for Pb fractionation. X-ray fluorescence analysis of soils from nine pits excavated approximately 200 m west of W5 resulted in average Ti concentrations (mg g^{-1}) of 4.14 (E horizon), 3.86 (Bh), 3.48 (Bs1), 3.49 (Bs2), and 2.65 (C). Digestions of samples from 48 pits in watershed 1, approximately 250 m east of W5, yielded values (in mg g^{-1}) of 2.24 (Oa horizon), 4.15 (E), 3.06 (Bh), 3.60 (Bs1), 3.79 (Bs2), and 2.16 (C) (C. Nezat, Univ. of Michigan, personal communication, 2002). We used average values from these two studies for our analysis.

2.3. INTERPRETATION AND COMPUTATIONS

We interpreted the EX, IB, ORG, and AMOX fractions as being pedogenic and 'labile' (Johnson and Petras, 1998). In contrast, we assumed that Pb in the RES, or crystalline, fraction was tightly bound, and released to solution through mineral weathering. This fraction includes Pb bound in silicate mineral lattices as well as Pb bound in crystalline oxide minerals such as magnetite.

Based on these assumptions, we estimated the amount of Pb released in each horizon through mineral weathering by estimating the depletion of Pb from the crystalline (RES) fraction. For the estimation of weathering release of Pb, both crystalline-Pb and Ti concentrations were expressed per unit mineral mass, under the assumption that Ti and crystalline-Pb were found exclusively in the mineral fraction:

$$C_{min} = \frac{C_{soil}}{(1 - f_{LOI})} \tag{1}$$

where C_{min} is the concentration of Pb or Ti per unit mineral mass, C_{soil} is the concentration of Pb or Ti per unit soil mass, and f_{LOI} is the loss-on-ignition (overnight, 500 °C), expressed as a fraction of soil mass.

The loss or gain of crystalline Pb was calculated for each soil horizon as:

$$\Delta (Pb_R)_i = \left[\left(Pb_R / Ti \right)_i - \left(Pb_R / Ti \right)_C \right] (Ti)_i (M_s)_i (1 - f_{LOI})_i \tag{2}$$

where Pb_R refers to Pb in the crystalline (RES) fraction; $(\Delta Pb_R)_i$ is the loss (if negative) or gain (if positive), in kg ha^{-1}, of crystalline-bound Pb in horizon i; $(Pb_R/Ti)_i$ and $(Pb_R/Ti)_C$ are the dimensionless Pb_R:Ti ratios in the ith and C horizons, respectively; $(Ti)_i$ is the Ti concentration in the ith horizon, in mg kg^{-1};

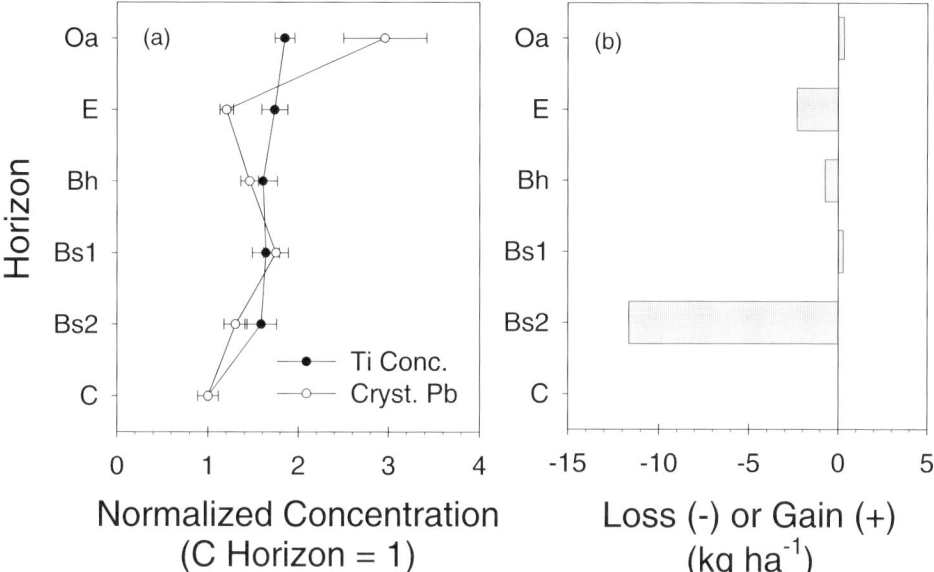

Figure 1. Gain or loss of crystalline-Pb from Hubbard Brook soils. Concentrations of Ti and crystalline-bound Pb, expressed per unit mineral mass and normalized to the C horizon, are shown in (a). Net accumulation or loss from the crystalline-Pb pool is shown in (b).

and $(M_s)_i$ is the soil mass (<2 mm) of the ith horizon, in Mg ha^{-1}. The values of $(M_s)_i$ were taken from Johnson and Petras (1998). Total weathering loss of Pb was estimated by summing the $(\Delta Pb_R)_i$ terms for all horizons. If the assumption that Ti is immobile is incorrect, then Equation (2) will lead to an underestimation of the weathering release of Pb. Therefore, our estimates represent minimum values.

The calculation in (2) was repeated for total Pb (Pb_T) to estimate the net gain or loss of total Pb in each horizon, and by summation, the soil profile. In considering the accumulation or loss of total Pb, Equation (1) was not used. Total Pb equaled the sum of the various Pb fractions. All of the terms in Equation (2) contain uncertainty. The largest uncertainty is in the fraction concentrations – typically with standard errors of 10–20% of the mean.

3. Results and Discussion

3.1. WEATHERING RELEASE OF Pb

When normalized to the C horizon, concentrations of crystalline Pb were lower than concentrations of Ti in all mineral horizons except the Bs1 (Figure 1a), indicating that weathering has resulted in the mobilization of crystalline-bound Pb from soil minerals. Weathering release of Pb was greatest in the Bs2 horizon, followed by the E and Bh horizons (Figure 1b; Table I).

TABLE I

Determination of the loss or accumulation of lead from the crystalline fraction of Hubbard Brook soils

Horizon	Dry mass[a]	Loss-on-ignition[b]	Ti	Crystalline-Pb (Pb_R)	Pb_R/Ti	Pb Loss (−) or Gain (+)	Crystalline Pb
	mg ha^{-1}	g kg^{-1}	— mg (kg min)$^{-1}$ —			kg ha^{-1}	kg ha^{-1}
Oi + Oe	22	871		0[c]		0	
Oa	65	516	4630	42.6	0.00920	+0.33	1.34
E	261	46	4340	20.9	0.00480	−2.32	5.19
Bh	225	138	4010	24.1	0.00601	−0.73	4.68
Bs1	276	136	4100	29.6	0.00723	+0.28	7.07
Bs2	2492	81	3960	22.4	0.00566	−11.7	51.3
C		38	2500	17.4	0.00694		

[a] Johnson and Petras (1998)
[b] Johnson et al. (1991b)
[c] The Oi and Oe horizons were assumed to contain no mineral matter.

Summing for all horizons, we estimated a total post-glacial weathering release of 14.1 kg Pb ha^{-1} (Table I). Averaged over 12,000–14,000 years, this results in a mean annual weathering release of Pb of 1.0–1.2 g ha^{-1} a^{-1}. This long-term average is much lower than the short-term average of 7 g ha^{-1} a^{-1} estimated for Hubbard Brook by Johnson et al. (1995). Many studies have reported similarly large differences between recent base-cation weathering fluxes, estimated using streamwater export, and long-term fluxes estimated by soil depletion patterns (e.g., April et al., 1986; Bain et al., 2001). The weathering release reported in Johnson et al. (1995) is almost certainly an overestimate, as it was based on the Pb/Na ratio in the C horizon, multiplied by the net Na export in stream water. Sodium is readily lost during weathering processes, so it is unlikely that the Pb/Na ratio in C horizon soil reflects the present-day stoichiometry of weathering release.

The large loss of crystalline-Pb from the Bs2 horizon was largely due to the greater mass of this horizon; the Bs2 horizon accounts for approximately 75% of the soil mass above the C horizon at Hubbard Brook (Table I). When expressed as a percentage of the crystalline-Pb pool, post-glacial weathering loss from the Bs2 horizon was only 18% of the pool. The fractional loss of Pb through weathering was greatest in the E horizon (Figure 2).

The crystalline-Pb to Ti ratio was greater in the Oa and Bs1 horizons than in the C horizon (Table I), resulting in an estimated increase in crystalline Pb in these horizon (Figure 1b). This suggests that Pb has accumulated in the crystalline fraction of the Oa and Bs1 horizons (Figure 1b). In the case of the Bs1 horizon, the difference is small, and may be due to incomplete extraction of Pb by the oxalate reagent used to estimate the AMOX fraction. This reagent is presumed to dissolve

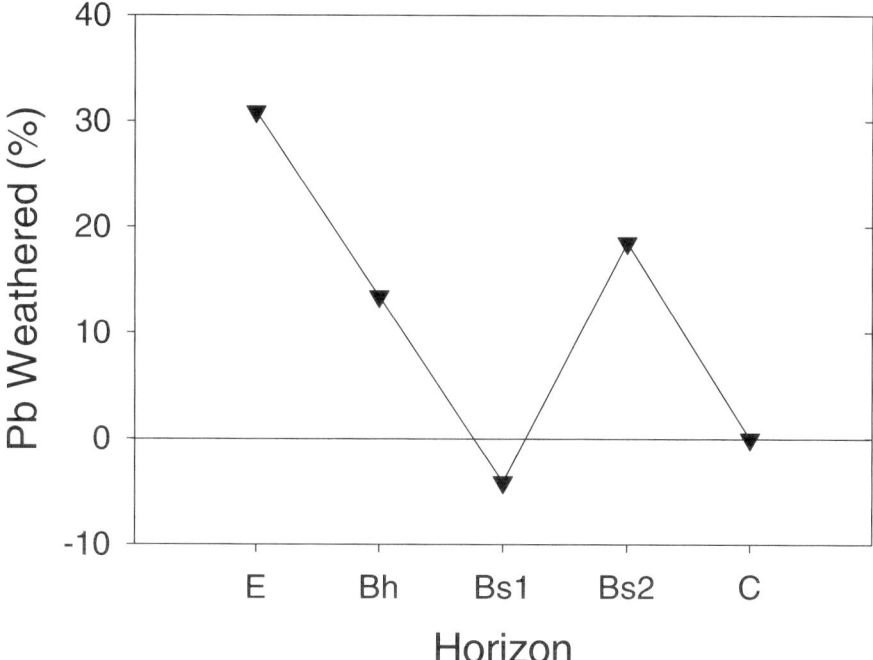

Figure 2. Weathering release of Pb, expressed as a percentage of crystalline-bound Pb.

amorphous Fe and Al oxides, thereby releasing adsorbed or co-precipitated Pb. Incomplete dissolution of amorphous oxide minerals would have the greatest effect in the Bs1 horizon, where podzolization results in the greatest accumulation of Fe and Al in Spodosols (Buol *et al.*, 1997).

Alternatively, it is possible that Pb has migrated into the crystalline fraction from other soil pools. Teutsch *et al.* (2001) studied the Pb-isotopic ratios of Pb fractions in Israeli soils, and concluded that 10–20% of the anthropogenic Pb in the soil had migrated into their 'aluminosilicate' fraction. The mechanism for this migration is not clear, but much of it is likely associated with exchange involving the interlayers of layered aluminosilicates. Soils at Hubbard Brook have very low clay contents, though analyses using x-ray diffraction have indicated the presence of smectite and hydroxy-interlayered vermiculite in the clay-sized fraction of upper soil horizons (R. H. April, unpublished data). Our data suggest that Pb migration into the crystalline fraction may be occurring in soil minerals of the Oa horizon. The magnitude of this process could be substantial; the apparent gain represents 25% of the crystalline Pb pool in the Oa horizon (Table I).

3.2. POST-GLACIAL Pb ACCUMULATION AND LOSS

The normalized concentration of total Pb was considerably greater than Ti in the

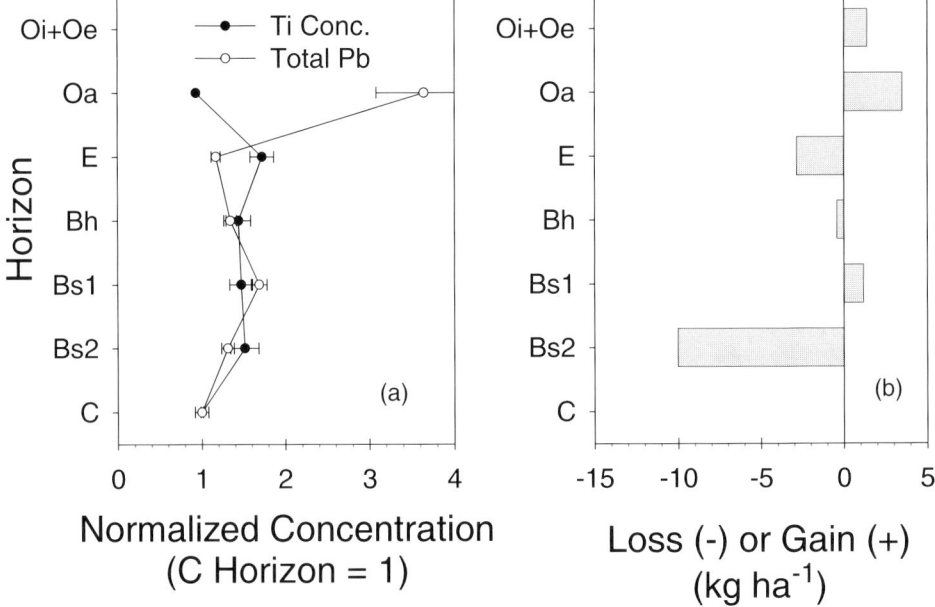

Figure 3. Gain or loss of total Pb from Hubbard Brook soils. Concentrations of Ti and total Pb, expressed per unit soil mass and normalized to the C horizon, are shown in (a). Net accumulation or loss from individual horizons is shown in (b).

Oa horizon, indicating a significant accumulation of Pb (Figure 3a). For the profile as a whole, we estimated a net decrease of 7.2 kg Pb ha^{-1} in the total Pb pool (Table II, Figure 3b). This can be divided into a net accumulation of 4.9 kg Pb ha^{-1} in the O horizon, and a net loss of 12.1 kg Pb ha^{-1} in the mineral soil.

Jersak *et al.* (1997) studied the geochemistry of trace metals in three Wisconsinan-aged Spodosols of northern New York and Vermont. Using a similar computational approach, they estimated net losses of 20 and 24 kg Pb ha^{-1} from mineral soils of the Adams and Becket series, respectively – soils similar to those at Hubbard Brook. The difference between our estimated net loss and those reported by Jersak *et al.* (1997) can be explained by differences in the distribution of Pb among various fractions. Jersak *et al.* (1997) reported organically bound Pb concentrations ranging from <0.1 to 0.44 mg kg^{-1}, and Fe-oxide-bound Pb concentrations ranging from <0.1 to 2.08 mg kg^{-1}, in E and B horizon soils of the Adams and Becket series. In contrast, E and B horizons of Hubbard Brook soils contain 1.6 to 5.1 mg kg^{-1} of organically bound Pb, and 1.27 to 3.37 mg kg^{-1} of amorphous-oxide-bound Pb (Johnson and Petras, 1998). Therefore, it appears that the Spodosols at Hubbard Brook have retained Pb in pedogenic fractions to a significantly greater degree than the Spodosols studied by Jersak *et al.* (1997).

TABLE II

Determination of the loss or accumulation of total lead in Hubbard Brook soils

Horizon	Dry mass[a]	Ti	Total Pb (Pb$_T$)	Pb$_T$/Ti	Pb Loss (−) or Gain (+)	Total Pb[†]
	mg ha^{-1}	mg (kg soil)$^{-1}$	mg (kg soil)$^{-1}$		kg ha^{-1}	kg ha^{-1}
Oi + Oe	22		64.1		+1.41	1.41
Oa	65	2240	72.5	0.0324	+3.51	4.71
E	261	4150	23.3	0.00562	−2.87	6.09
Bh	225	3460	26.7	0.00772	−0.43	6.01
Bs1	276	3540	33.6	0.00949	+1.19	9.26
Bs2	2492	3640	26.1	0.00717	−10.02	65.1
C		2410	19.9	0.00827		

[a] Johnson and Petras (1998).

3.3. Sources and fate of Pb in Hubbard Brook soils

Our data suggest that gasoline-derived Pb is a major source of Pb to the Hubbard Brook ecosystem. We estimated that 14.1 kg Pb ha^{-1} has been released through weathering in the 12000–14000 yr since deglaciation. In an earlier paper, we estimated that 8.66 kg Pb ha^{-1} was deposited in precipitation during the period 1926–1989 (Johnson *et al.*, 1995). Therefore, the amount of Pb deposited in 20th century precipitation represents more than 60% of the Pb released through weathering in the entire post-glacial period. The fact that precipitation Pb inputs at Hubbard Brook declined by 97% between 1976 and 1989 (after the passage of legislation restricting the use of leaded gasoline) suggests that virtually all of the Pb in precipitation was anthropogenic.

We estimated that 4.9 kg Pb ha^{-1} has accumulated in the O horizons at Hubbard Brook (Table II). The apparent increase in the crystalline Pb pool of the Oa horizon (Table I) suggests that there was little net release of Pb from mineral weathering in the forest floor. Thus, the post-glacial Pb accumulation in the forest floor is predominantly of atmospheric origin. Nevertheless, the 4.9 kg ha^{-1} of atmospheric Pb accumulated in the O horizons represents only 57% of the 20th century Pb deposition. This suggests that much of the anthropogenic Pb deposited during the 20th century has migrated into the mineral soil. This is consistent with the findings of Miller and Friedland (1994), who estimated that the storage time of Pb in organic forest soil horizons was on the order of decades rather than centuries, as previously suggested (e.g., Van Hook *et al.*, 1977; Smith and Siccama, 1981).

Together, the post-glacial Pb release from the crystalline fraction and the Pb deposited in precipitation during the leaded-gasoline period of the 20th century provided 22.8 kg Pb ha^{-1} to the soil solution. This amount is nearly identical to the total labile Pb in the soil – the sum of the EX, IB, ORG, and AMOX fractions

(22.1 kg Pb ha^{-1}; Johnson and Petras, 1998). Therefore, post-glacial weathering and atmospheric deposition can account for all of the pedogenic Pb in Hubbard Brook soils.

This analysis does not account for the input of Pb from atmospheric deposition prior to the use of leaded fuels. Nor does it account for the losses of Pb in drainage waters. The input of Pb from pre-industrial precipitation is unclear. A 'pristine' value for Pb in precipitation is elusive because even precipitation in polar regions contains pollutant Pb (Sturges and Barrie, 1989). In the late 19th century some small-scale Pb smelters operated in Vermont, upwind of Hubbard Brook. However, sediments deposited throughout the 19th century in Black Pond, 20 km from Hubbard Brook, contain only background levels of Pb (Johnson *et al.*, 1995).

As for streamwater, Wang *et al.* (1995) showed that drainage losses of Pb at Hubbard Brook in the 1990s were very low, at least an order of magnitude lower than precipitation inputs. Stream export during the leaded-gasoline period was somewhat higher (Smith and Siccama, 1981; Johnson *et al.*, 1995), but still two orders of magnitude lower than precipitation inputs. Our most recent data indicate that streamwater Pb concentrations at Hubbard Brook in 1999–2001 are in the range of 100–300 ng L^{-1}, while precipitation Pb has steadily declined to 250–750 ng L^{-1} (C.E. Johnson, unpublished data). Considering that this precipitation likely contains anthropogenic Pb of non-gasoline origin, as well as gasoline-Pb from non-US sources, it appears likely that pre-industrial Pb inputs and outputs at Hubbard Brook were in approximate steady-state. Shotyk *et al.* (1998) estimated pre-historic Pb deposition in the Jura Mountains of Switzerland to be on the order of 0.1–1.0 g ha^{-1} a^{-1}, somewhat lower than present-day streamwater fluxes at Hubbard Brook, which likely includes some Pb of anthropogenic origin. This too suggests that pre-historic inputs and outputs may have been at or near steady-state.

A fuller accounting of the sources and fate of Pb in these soils may be possible through isotopic analyses. Numerous investigators have used the Pb-isotopic composition of soil digests to estimate the amount of anthropogenic Pb in the soil (e.g., Bacon *et al.*, 1995, Erel *et al.*, 1997; Hansmann and Köppel, 2000). However, the isotopic signature of the total digest of a soil sample does not represent the signature of the Pb most likely to leach into drainage waters or groundwater. The Pb in different soil fractions (i.e., exchangeable or organically bound) is a mixture of anthropogenic and rock-derived Pb. The analysis of the isotopic composition of various soil Pb pools may yield valuable insight into the fate of Pb inputs. However, there have been very few studies that have combined soil chemical fractionation with isotopic analyses (Steinmann and Stille, 1997; Teutsch *et al.*, 2001; Emmanuel and Erel, 2002). Teutsch *et al.* (2001) demonstrated that the isotopic composition of Pb fractions could be used to infer the fate of anthropogenic Pb in semi-arid soils in Israel. Similar analyses would complement the budgetary approach presented here.

It is important to note that Hubbard Brook is distant from point sources of Pb pollution, and from major urban areas (Boston is 160 km to the southeast). Despite the remote location, anthropogenically-derived Pb constitutes a substantial fraction of the labile Pb in Hubbard Brook soils. Therefore, Pb derived from gasoline and industrial sources is likely to be even more important in other soils of the northeastern United States. Monitoring changes in the Pb fractions over time could provide useful information regarding the mobility and fate of Pb in forest soils.

Acknowledgements

This work was sponsored by the National Science Foundation, Long-Term Ecological Research Program. We thank Pat Koepple and Dianne Keller for assistance with XRF analyses. Joel Blum provided helpful comments on the manuscript. This is a contribution of the Hubbard Brook Ecosystem Study. The Hubbard Brook Experimental Forest is owned and operated by the USDA Forest Service, Northeast Research Station, Newtown Square, Pennsylvania, U.S.A.

References

Adriano, D. C.: 1986, *Trace Elements in the Terrestrial Environment*, Springer-Verlag, New York, New York, U.S.A., 533 pp.

Andresen, A. M., Johnson, A. H. and Siccama, T. G.: 1980, 'Levels of lead, copper and zinc in the forest floor in the northeastern United States', *J. Environ. Qual.* **9**, 293–296.

April, R., Newton, R. and Coles, L. T.: 1986, 'Chemical weathering in two Adirondack watersheds: Past and present-day rates', *Geol. Soc. Am. Bull.* **97**, 1232–1238.

Bacon, J. R., Berrow, M. L. and Shand, C. A.: 1995, 'The use of isotopic composition in field studies of lead in upland Scottish soils (U.K.)', *Chem. Geol.* **124**, 125–134.

Bain, D. C., Roe, M. J., Duthie, D. M. L. and Thomson, C. M.: 2001, 'The influence of mineralogy on weathering rates and processes in an acid-sensitive granitic catchment', *Appl. Geochem.* **16**, 931–937.

Bormann, F. H. and Likens, G. E.: 1979, *Pattern and Process in a Forest Ecosystem*, Springer-Verlag, New York, New York, U.S.A.

Buol, S. W., Hole, F. D., McCracken, R. J. and Southard, R. J.: 1997, *Soil Genesis and Classification*, 3rd ed., Iowa State University Press, Ames, Iowa, U.S.A.

Emmanuel, S. and Erel, Y.: 2002, 'Implications from concentrations and isotopic data for Pb partitioning processes in soils', *Geochim. Cosmochim. Acta* **66**, 2517–2527.

Erel, Y., Veron, A. and Halicz, L.: 1997, 'Tracing the transport of anthropogenic lead in the atmosphere and in soils using isotopic ratios', *Geochim. Cosmochim. Acta* **61**, 4495–4505.

Federer, C. A., Flynn, L. D., Martin, C. W., Hornbeck, J. W. and Pierce, R. S.: 1990, 'Thirty Years of Hydrometeorological Data at the Hubbard Brook Experimental Forest, New Hampshire', USDA Forest Service Gen. Tech. Rep. NE-141, Washington, DC, U.S.A.

Friedland, A. J., Craig, B. W., Miller, E. K., Herrick, G. T., Siccama, T. G., and Johnson, A. H.: 1992, 'Decreasing lead levels in the forest floor of the northeastern U.S.A.', *Ambio* **21**, 400–403.

Hansmann, W. and Köppel, V.: 2000, 'Lead-isotopes as tracers of pollutants in soils', *Chem. Geol.* **171**, 123–144.

Heinrichs, H. and Mayer, R.: 1977, 'Distribution and cycling of major and trace elements in Two central European forest ecosystems', *J. Environ. Qual.* **6**, 402–407.

Hogan, G. D. and Wotton, D. L.: 1984, 'Pollutant distribution and effects in forests adjacent to smelters', *J. Environ. Qual.* **13**, 377–382.

Jersak, J., Amundson, R. and Brimhall Jr., G.: 1997, 'Trace metal geochemistry in spodosols of the northeastern United States', *J. Environ. Qual.* **26**, 511–521.

Johnson, C. E.: 2002, 'Cation exchange properties of acid forest soils of the northeastern U.S.A.', *Eur. J. Soil Sci.* **53**, 271–282.

Johnson, C. E., Johnson, A. H., Huntington, T. G. and Siccama, T. G.: 1991a, 'Whole-tree clear-cutting effects on soil horizons and organic-matter pools', *Soil Sci. Soc. Am. J.* **55**, 497–502.

Johnson, C. E., Johnson, A. H. and Siccama, T. G.: 1991b, 'Whole-tree clear-cutting effects on exchangeable cations and soil acidity', *Soil Sci. Soc. Am. J.* **55**, 502–507.

Johnson, C. E. and Petras, R. J.: 1998, 'Distribution of zinc and lead fractions within a forest spodosol', *Soil Sci. Soc. Am. J.* **62**, 782–789.

Johnson, C. E., Siccama, T. G., Driscoll, C. T., Likens, G. E. and Moeller, R. E.: 1995, 'Changes in lead biogeochemistry in response to decreasing atmospheric inputs', *Ecol. Applic.* **5**, 813–822.

McLaren, R. G. and Crawford, D. V.: 1973, 'Studies on soil copper I: The fractionation of copper in soils', *J. Soil Sci.* **24**, 172–181.

Miller, E. K. and Friedland, A. J.: 1994, 'Lead migration in forest soils: Response to changing atmospheric inputs', *Environ. Sci. Technol.* **28**, 662–669.

Miller, W. P. and McFee, W. W.: 1983, 'Distribution of cadmium, zinc, copper, and lead in soils of industrial northwestern Indiana', *J. Environ. Qual.* **12**, 29–33.

Nirel, P. M. V. and Morel, F. M. M.: 1990, 'Pitfalls of sequential extractions', *Water Res.* **24**, 1055–1056.

Shotyk, W., Weiss, D., Appelby, P. G., Cheburkin, A. K., Frei, R., Gloor, M., Kramers, J. D., Reese, S. and van der Knaap, W. O.: 1998, 'History of atmospheric lead deposition since 12,730 ^{14}C yr BP from a peat bog, Jura Mountains, Switzerland', *Science* **281**, 1635–1640.

Smith, W. H. and Siccama, T. G.: 1981, 'The Hubbard Brook Ecosystem Study: Biogeochemistry of lead in the northern Hardwood Forest', *J. Environ. Qual.* **10**, 323–333.

Steinmann, M. and Stille, P.: 1997, 'Rare earth element behavior and Pb, Sr, Nd isotope systematics in a heavy metal contaminated soil', *Appl. Geochem.* **12**, 607–623.

Sturges, W. T. and Barrie, L. A.: 1989, 'Stable lead isotope ratios in arctic aerosols: Evidence for the origin of arctic air pollution', *Atmos. Environ.* **23**, 2513–2519.

Taylor, A. and Blum, J. D.: 1995, 'Relation between soil age and silicate weathering rates determined from the chemical evolution of a glacial chronosequence', *Geology* **23**, 979–982.

Teutsch, N., Erel, Y., Halicz, L. and Banin, A.: 2001, 'Distribution of natural and anthropogenic lead in Mediterranean Soils', *Geochim. Cosmochim. Acta* **65**, 2853–2864.

Tyler, G.: 1981, 'Leaching of metals from the A-horizon of a spruce forest soil', *Water, Air, Soil Pollut.* **15**, 353–369.

van Hook, R. I., Harris, W. F. and Henderson, G. S.: 1977, 'Cadmium, lead, and zinc distributions and cycling in a mixed deciduous forest', *Ambio* **6**, 281–286.

Wang, E. X., Bormann, R. H., and Benoit, G.: 1995, 'Evidence of complete retention of atmospheric lead in the soils of northern hardwood forested ecosystems', *Environ. Sci. Technol.* **29**, 735–739.

RESIDUAL CADMIUM AND LEAD POLLUTION AT A FORMER SOVIET MILITARY AIRFIELD IN TARTU, ESTONIA

ÜLO MANDER*, AIN KULL and JANE FREY

Institute of Geography, University of Tartu, 46 Vanemuise St, 51014 Tartu, Estonia
(* author for correspondence, e-mail: mander@ut.ee; phone: +372-7-375819;
fax: +372-7-375825)

(Received 20 August 2002; accepted 6 April 2003)

Abstract. This paper presents data on the levels and dynamics of cadmium (Cd) and lead (Pb) concentration in the plants, soil, and groundwater of the landing corridor and airfield of a former Soviet military air base in Estonia, immediately at the end of its 40-year service in 1992 and over the following 8 yr. In 1991–92 we found high Cd concentrations in the meadow plants Trifolium pratense and Dactylis glomerata (up to 56 mg kg^{-1}). In 1993, the Cd concentration had dropped to 0.12–0.19 mg kg^{-1}, and stabilized in 1997–2000 at 0.04 mg kg^{-1}. Cd concentration in plants decreased significantly with increasing distance from the landing strip. Elevated Cd concentration (0.012 mg L^{-1}) was found in the fuel of the TU-22M (Backfire) strategic bombers. In 1991 and 1993, leaded fuel influenced the mean Pb concentration in plants (1.8–4.2 mg kg^{-1}). Average Pb concentration in both topsoil and the 30–40 cm soil horizon decreased between 1991 and 2000 from 28 to 6.5 and from 13.5 to 4.3 mg kg^{-1}, respectively. Cd concentration in the topsoil of the landing corridor showed a significant increase between 1991 and 1993 (0.07–0.3 and 0.3–1.2 mg kg^{-1}, respectively), but stabilized later on the level of 0.04 mg kg^{-1}. The concentrations of both Pb and Cd in the soil were higher closer to the fuel bunkers. Current assessment of the movement of these metals from the vegetation to the soil and to groundwater is linked to potential leaching to the surrounding environment.

Keywords: cadmium, *Dactylis glomerata*, groundwater, lead, river sediments, soil, *Trifolium pratense*

1. Introduction

The impact of aircraft activity on heavy metal pollution on the ground has rarely been studied and there is little published material on heavy metal contamination in military aerodromes. The former Soviet military aerodrome in Tartu (about 800 ha, also known as the Tartu-Raadi airfield), one of the largest transportation airfields of the former Warsaw Pact countries, was established in 1953 and was used intensively for 40 yr (Figure 1). In addition to the airfield, many smaller military bases supporting the aerodrome were scattered in the surroundings of the town of Tartu (population: 100,000 in September 2002). After the withdrawal of the Soviet/Russian troops in 1992–93, it was found that parts of the airfield (about 120 ha) were heavily polluted with petrol, kerosene, nitric acid, different rocket fuel components and mixtures, phenols, nitrates, chlorine, heavy metals, and other chemicals (Auer and Raukas, 2002). Oil and heavy metal pollution also was pre-

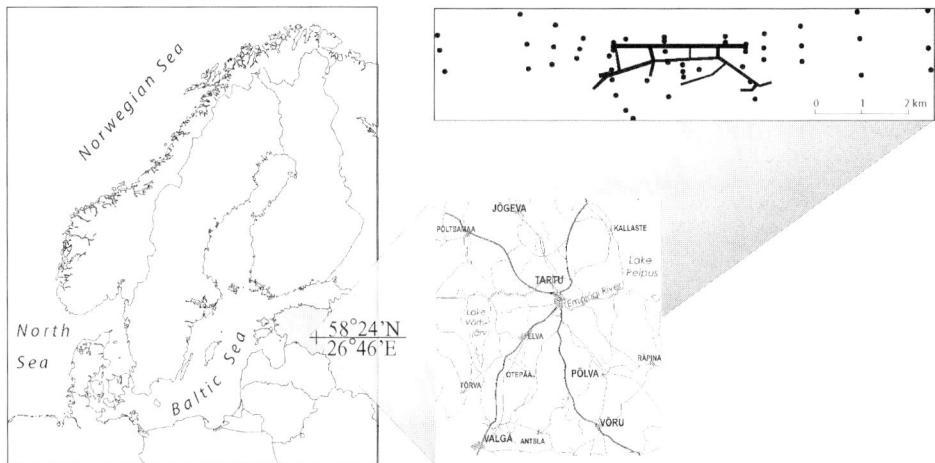

Figure 1. Location of the former military airfield in Tartu, Estonia.

sumed in the landing corridors. In the surroundings of Tartu, higher heavy metal concentration in mosses has been detected (Rühling et al., 1992; Mäkinen, 1993). Although some studies have considered the potential heavy metal pollution load of airfields (e.g., high Pb concentration in mosses and lichens in the Helsinki-Vantaa Airport area, Finland (Mäkinen, et al., 1980), elevated Pb concentration in the air at Heathrow Airport, London, U.K. (Nichols et al., 1981), high Cd content in feral pigeons living in the Heathrow Airport area (Hutton and Goodman, 1980) 2–5-fold higher Cd concentration in mushrooms of the Prague Airport area in Cerny Val, Czech Republic, than in non-polluted forests (Sova et al., 1991), and significantly higher Cd content in feathers, heart, liver and muscles of laughing gulls living in the John F. Kennedy Airport area, New York (Gochfeld et al., 1996)), there are very few published reports on heavy metal contamination of military areas (Svoma, 1993). This paper examines the effect of long-term exploitation of a former military aerodrome in Estonia by the Soviet air force and the decline over time in residual Cd and Pb levels in the landing corridor and airfield since its abandonment.

2. Study Area and Methods

The former Soviet military airfield of Tartu-Raadi lies within the southeast Estonian moraine plain (40 to 55 m above mean sea level). It is slightly undulating (slopes achieve 3–4%), and dissected by primeval valleys filled with glacio-fluvial gravels and sands. The bedrock (Devonian sandstone and clays) is covered by a 2–8 m till of the Weichselian glaciation and by glacio-lacustrine sands and gravels. Groundwater table depth (0.5–5 m) varies depending on relief and geomorphologic conditions. The upland soils are mostly podzoluvisols and luvisols on loamy sands and sandy loams, with the surface soil organic matter content and pH values of

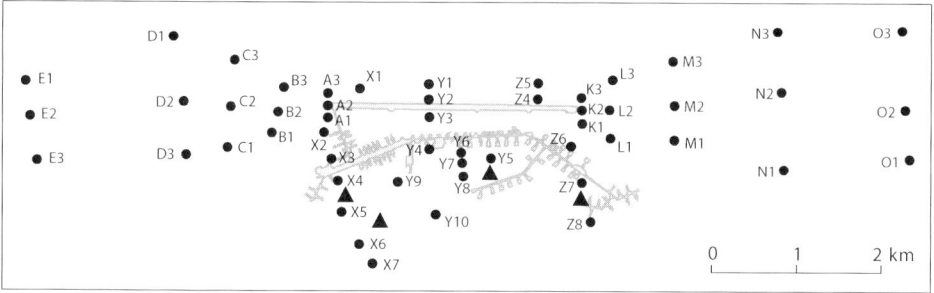

Figure 2. Location of plant, soil, and groundwater sampling sites in the airfield area. Black triangles indicate location of main fuel storages.

1.6–1.9% and 5.6–6.5, respectively. In lower-lying areas, gleysols and peatlands dominate.

In 1991, 1992, 1993, 1997 and 2000, soil samples from the 0–10 and 30–40 cm horizons, and samples of two meadow plants (lower 1/3 parts of *Dactylis glomerata* stems, and inflorescences of *Trifolium pratense*) were collected in late autumn (October-November) in funnel-shaped transects along the landing corridors at increasing distances (0–5, 300–400, 900–1200, 1600–2500, and 3300–4000 m to the west, and 20–75, 700–750, 1300–1600, 2400–3000 and 3800–4300 m to the east) from the paved landing track (Figure 2). *D. glomerata* and *T. pratense* were the most common herbaceous species spread all over the airfield area. In the literature, there is comparative data available on Cd and Pb content in *D. glomerata* (Olsen and Chong, 1991; Kabata-Pendias and Dudka, 1991; Esser, 1996) and in *T. pratense* (Dudka *et al.*, 1996; Krauss and Diez, 1997). We took soil (using a 30 mm corer) and plant samples in three sampling sites at each distance, in three replicates. In each sampling site, three 50 mm PVC pipes were installed for shallow (0.4–2.4 m) groundwater sampling. Dominating soil types, soil-texture classes, and organic matter contents in the soil are presented in Table I.

In 1992, access to the airfield territory was permitted. This enabled soil and plant sampling at 20 sites on the airfield to be carried out. Our results were compared with the heavy metal concentrations of the Emajõgi River sediments reported by Laanemets *et al.* (1997). In 1996, these bottom sediment samples were taken as 5 replicates from 8 sampling sites upstream of drainage channels inflows from the airfield and 5 replicates from 8 downstream sampling sites, and analysed for heavy metal contents (Laanemets *et al.*, 1997). Pb and Cd contents in soils, river sediments, and plants, were measured in the laboratory of the Estonian Environmental Research Center, Ministry of the Environment of Estonia, using atomic-absorption spectrophotometry (AAS; graphite-furnace technique). Cd and Pb from soil samples were extracted using HNO_3; plant tissues were digested by HNO_3 and 30% H_2O_2 (Jones, 1991). Dried soil samples (150 g) were sieved and the fraction <1 mm was used for the analysis. Soil organic material concentration was measured as ignition loss; soil types were described according to both FAO/ISRIC and

TABLE I

Soil types (FAO/ISRIC, in brackets USDA classification), soil-texture classes, and organic material content (average ± standard deviation) of the sampling sites. See location in Figure 2

Sampling site series	Soil type(s)	Soil-texture classes	Organic material (%)
E	Eutric Gleysol (Aquept)/ Peatland soil (Histosol)	Loamy sand/peat	5.4 ± 4.6
D	Stagnic Luvisol (Ochriglossudalf)/ Podzoluvic Gleysol (Aqualf)	Loamy sand on sandy loam	1.8 ± 0.4
C	Stagnic Luvisol (Ochriglossudalf)/ Calcari-Luvic Gleysol (Argiaqualf)	Loamy sand on sandy loam	2.1 ± 0.7
B	Stagnic Luvisol (Ochriglossudalf)/ Podzoluvic Gleysol (Aqualf)	Loamy sand on sandy loam	2.2 ± 1.2
A	Gleysol (strong anthropogenic disturbance)	Sandy loam on loam	2.7 ± 2.5
X	Gleysol (anthropogenic disturbance)/ Stagnic Luvisol (Ochriglossudalf) Eutric Gleysol (Aquept)	Loamy sand, sand, gravel, peat	2.2 ± 7.6
Y	Stagnic Luvisol (Ochriglossudalf) Podzoluvic Gleysol (Aqualf)/ Eutric Gleysol (Aquept)	Loamy sand on sandy loam, loam, gravel	2.9 ± 6.8
Z	Podzoluvic Gleysol (Aqualf) Technogenic substrate with thin humic layer	Loamy sand (on) sandy loam	0.7 ± 1.2
K	Technogenic substrate	Course sand/loamy sand	0.4 ± 0.5
L	Podzoluvic Gleysol (Aqualf) Calcari-Luvic Gleysol (Argiaqualf)	Loamy sand on sandy loam	1.2 ± 0.7
M	Stagnic Luvisol (Ochriglossudalf)/ Podzoluvic Gleysol (Aqualf)	Loamy sand on sandy loam	1.8 ± 0.6
N	Stagnic Luvisol (Ochriglossudalf)/ Eutric Gleysol (Aquept)	Loamy sand on sandy loam, peat	2.0 ± 0.9
O	Peatland soil (Histosol)/ Gleyic luvisol	Loamy sand, sandy loam, peat	6.2 ± 12.8

USDA classifications. Plant and soil concentrations are expressed on a g^{-1} or kg^{-1} dry mass basis. Before laboratory analysis, groundwater samples were filtered, conserved with 0.1 M HNO$_3$ and frozen. Water and fuel samples were analysed for Pb and Cd concentration in the laboratory of the Department of Water and Environmental Research, Linköping University, Sweden using the AAS technique.

For data interpolation, the Kriging method was used. This is one of the more flexible methods and is well suited for irregular datasets if there is also a need to extrapolate grid values beyond the collected data range. Kriging with the linear variogram model was found to adequately represent the measured values. To eliminate the effect of sampling along the transects, the following data search parameters were used: number of search sectors, 4; search ellipse radius, a = 4000 m and b = 1000 m; and search ellipse angle, $\alpha = 5°$. Normality of variables was checked using the Lilliefors test. In most cases, the distribution differed from the normal, and, therefore, non-parametric tests were performed. We used the nonparametric Duncan test to check the significance of differences between the Cd and Pb concentrations in plants and soil at different sampling sites and times. The level of significance $\alpha = 0.05$ was accepted in all cases. Spearman rank order correlation analysis of relationships between different variables was performed.

3. Results and Discussion

We found very high Cd concentrations in *T. pratense* and *D. glomerata*: 8.0 ± 6.8 and 16.2 ± 14.6 mg kg^{-1}, respectively in 1991, and 5.0 ± 6.5 and 6.9 ± 13.0 mg kg^{-1} in 1992, respectively (Table II). Maximal values of Cd concentration reached 56 mg kg^{-1}, which is comparable with the Cd concentration in metallophytic plants growing on highly contaminated soils (Dahmani-Muller *et al.*, 2000). At the same time, the topsoil (0–10 cm) showed only slight contamination with cadmium: 0.16 ± 0.16 and 0.31 ± 0.16 mg kg^{-1} in 1991 and 1992, respectively. This is, however, 38–63 times lower than the average Cd concentration in urban lawn soils in Warsaw (Czarnowska, 1999), and 4–9 times lower than the average level of Cd in street dust (tyre wear, asphalt erosion, fine soil) in relatively less polluted Nordic cities like Oslo: 1.4 ± 0.2 mg kg^{-1} (de Miguel *et al.*, 1997).

In 1993, the Cd content of plants dropped significantly, to 0.19 ± 0.36 and 0.12 ± 0.20 mg kg^{-1} in *T. pratense* and *D. glomerata*, respectively. In the topsoil, however, the Cd concentration increased 3-fold to 1.0 ± 1.6 mg kg^{-1}. In the 30–40 cm layer, the Cd content increased from 0.14 ± 0.11 in 1991 to 0.55 ± 0.23 mg kg^{-1} in 1993 (Table II).

Cadmium concentration in plants decreased slightly with increasing distance from the landing strip. However, significantly differing values were found only just at the end of the landing strip in an eastward direction. Likewise, *T. pratense* and topsoil (0–10 cm) samples from the airfield showed significantly higher Cd content

TABLE II

Average ± standard deviation values of Cd and Pb in soil and in plant tissues

	Cd in soil (mg kg^{-1})		Cd in plants (mg kg^{-1})		Pb in soil (mg kg^{-1})		Pb in plants (mg kg^{-1})	
	0–10 cm	30–40 cm	Trifolium	Dactylis	0–10 cm	30–40 cm	Trifolium	Dactylis
1991	0.16 ± 0.16^b	0.14 ± 0.11^b	8.0 ± 6.8^a	16.2 ± 14.6^a	28.1 ± 30.8^a	13.5 ± 11.1^a	1.8 ± 0.7^a	2.7 ± 1.8^a
1992	0.31 ± 0.16^b		5.0 ± 6.5^a	6.9 ± 13.0^b				
1993	1.0 ± 1.6^a	0.55 ± 0.23^a	0.19 ± 0.36^b	0.12 ± 0.20c	11.4 ± 20.3^b	4.4 ± 1.3^b	3.5 ± 6.5^a	4.2 ± 6.4^a
1997	0.18 ± 0.15^b		0.04 ± 0.02^c	0.05 ± 0.04^c	8.7 ± 4.0^c		0.6 ± 0.5^b	0.3 ± 0.2^b
2000	0.12 ± 0.09^b	0.11 ± 0.12^b	0.04 ± 0.02^c	0.04 ± 0.02^c	6.5 ± 2.8^c	4.3 ± 1.4^b	0.4 ± 0.2^b	0.4 ± 0.6^b

[a,b,c] Significantly differing values ($p < 0.05$) according to the Duncan Test.

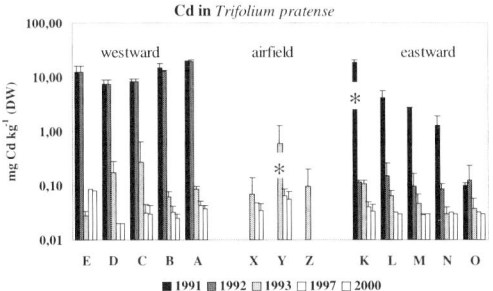
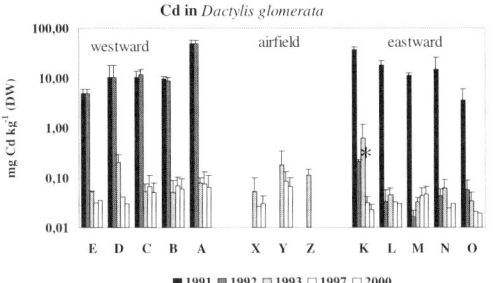
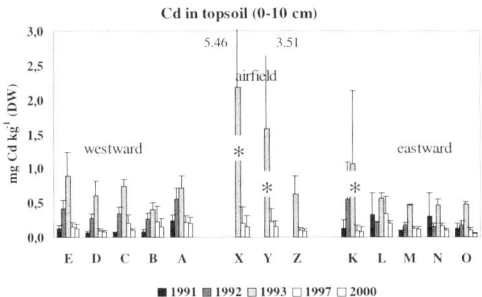
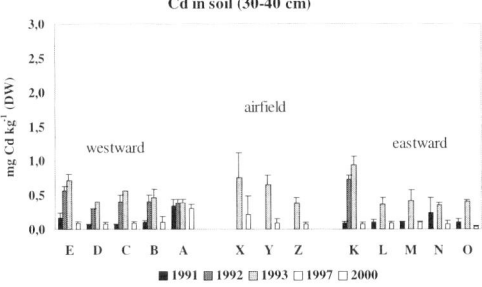

Figure 3. Dynamics of cadmium concentration (average ± standard deviation) in inflorescences of *Trifolium pratense*, stems of orchard grass *Dactylis glomerata*, and in soil (0–10 and 30–40 cm). See sampling site (E-O) location in Figure 2. * Significantly differing values ($p < 0.05$) within the data series of the same year according to the Duncan Test.

in 1993 (Figure 3). In the 30–40 cm soil layer, no significant differences between sampling sites were found.

Both temporal and spatial variations in Cd concentration in plants and soils were interpolated by the Kriging method and are presented in Figures 4 and 5. The very high Cd concentration in plants in 1991 can be explained by several factors. The most probable of them is the influence of aircraft exhausted gases. The following facts support this version: first, we found no significant correlation between the Cd content in the soil and in both plant species (Table III); second, the concentration decreases with increasing distance from the landing strip (Figures 3–5), and third, a significantly higher Cd concentration (0.022 mg L^{-1}) was found in the fuel used by the TU-22M (Backfire) strategic bombers. Cd concentration in the normal plane fuel and rocket fuel was 0.005 and 0.006 mg L^{-1}, respectively. Pb concentration in the fuels analysed varied from 0.014 to 0.025 mg L^{-1}, however the differences were non-significant. The Backfires left Tartu military airfield in 1992. In the following years, the pollution load of plants reduced significantly.

We found significant Spearman rank order correlation between the Cd and Pb content in both plant species and also between both metals in the soil (Table III). However, we did not find significant rank correlation between the soil humus content (see Table I) and any other variables.

A decrease in average Pb concentration in *Trifolium* (1.8 and 3.5 mg Pb kg^{-1} in 1991 and 1993, respectively) and *Dactylis* (2.7 and 4.2 mg Pb kg^{-1} in 1991 and 1993, respectively) was found only directly at the edge of the landing strip (Figure 6). In 1993, a significant decrease in plant and soil Pb concentration with increasing distance from the landing strip (Figure 6) was found. Significant differences were found also in plant samples taken in the vicinity of fuel bunkers at the airfield. Nevertheless, the Pb content in plant tissues of both species decreased significantly in time: in 2000 it was 0,4 mg Pb kg^{-1} (Table II, Figure 6). In soil samples, the Pb content also decreased in time. Average Pb content in both topsoil and at a depth of 30–40 cm decreased between 1991 and 2000 from 28 to 6.5, respectively, and from 13.5 to 4.3 mg kg^{-1}, respectively (Table II).

In contrast to the high heavy metal concentrations in meadow plants and some soil samples, the measured heavy metal concentration in shallow groundwater was relatively low. In 1992 we measured 0.28 and 4.9 μg L^{-1} of Cd and Pb respectively, which is 5–35 times lower than that in the soil water at sludge application sites (Richards *et al.*, 1998). In 2000, the concentration decreased to 2.6 and 0.06 μg L^{-1} respectively. However, for Pb the decrease was non-significant.

We found no significant rank correlation between the Cd and Pb concentration in shallow groundwater and the concentration of these metals in the soil and in plant tissues. Low concentration of Cd and Pb in groundwater supports the idea that the heavy contamination of meadow plants by Cd, found in 1991 and 1992 was a short-term effect caused either by an accidental event or by exhaust gases from the aircraft (Backfire-type strategic bombers). The results suggest that the exhaust

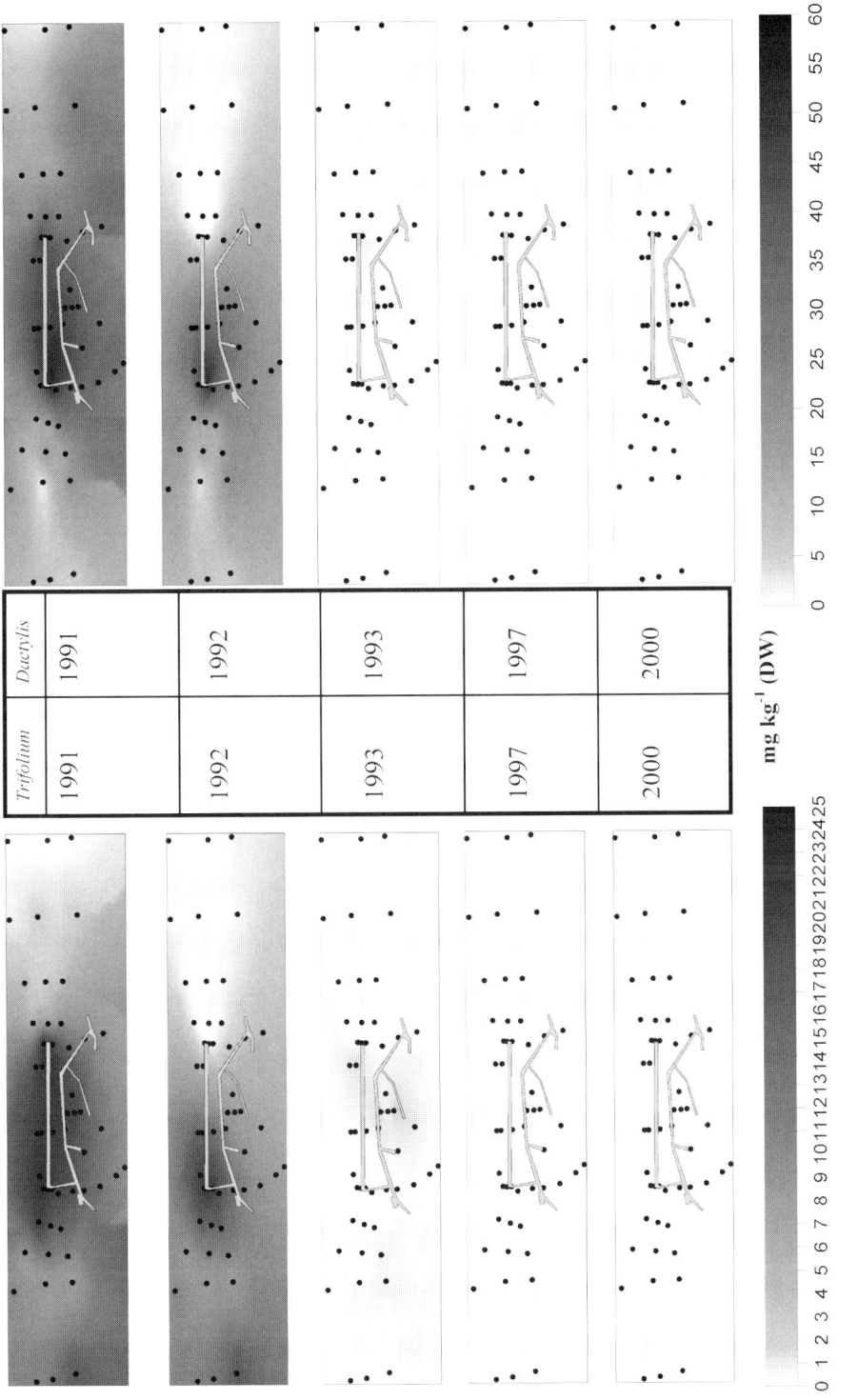

Figure 4. Dynamics of the spatial distribution of the Cd content in *Trifolium pratense* and *Dactylis glomerata*.

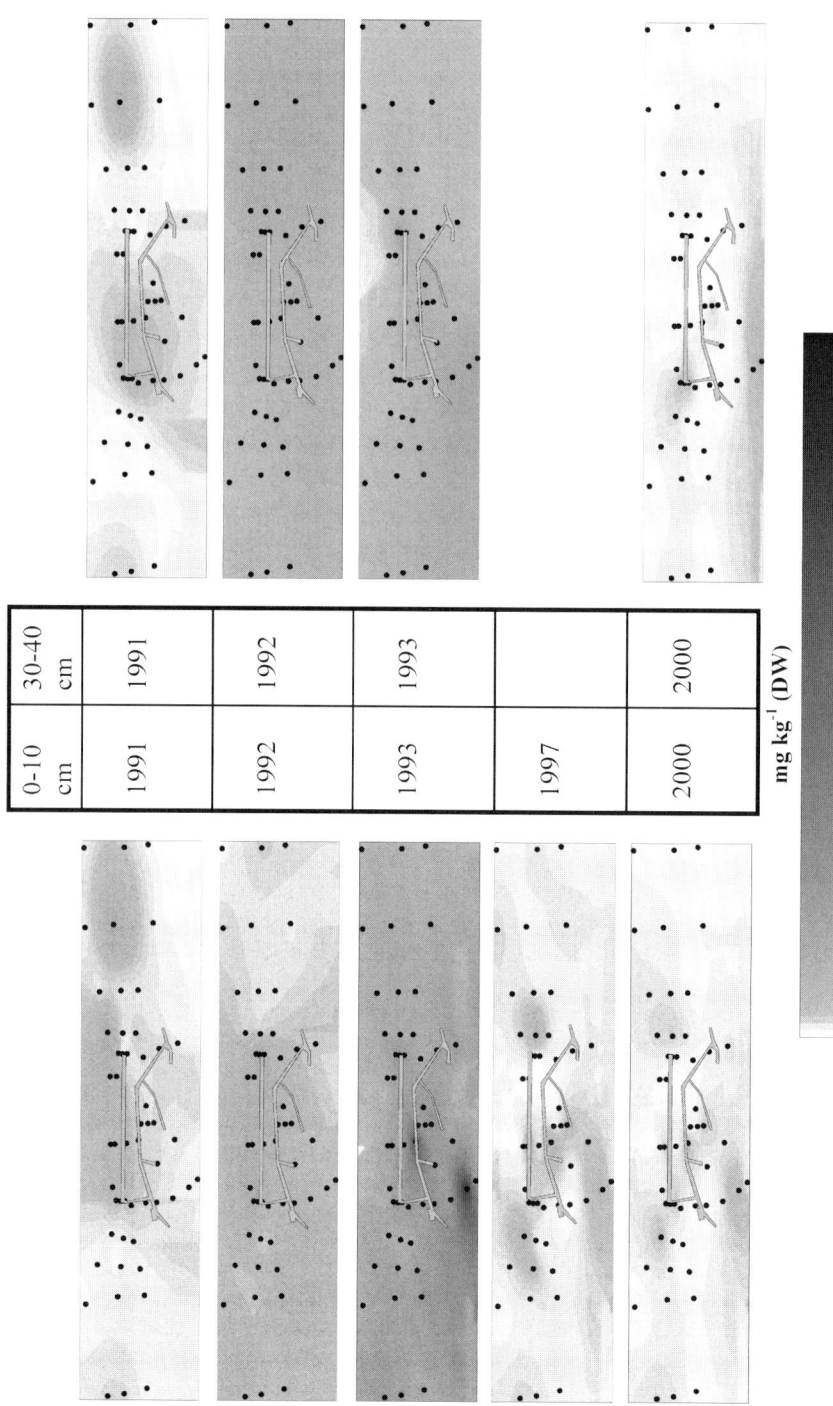

Figure 5. Dynamics of the spatial distribution of the Cd content in soil (0–10 cm and 30–40 cm).

TABLE III

Spearman rank order correlation values (r) between the variables studied

	Cd in soil (0–10 cm)	Cd in soil (30–40 cm)	Cd in Trifolium	Cd in Dactylis	Pb in soil (0–10 cm)	Pb in soil (30–40 cm)	Pb in Trifolium	Pb in Dactylis
Cd in soil (0–10 cm)		0.88^b	0.09	0.05	0.28^a	−0.02	0.47^b	0.44^b
Cd in soil (30–40 cm)			0.11	−0.06	0.19^a	0.03	0.50^b	0.42^b
Cd in Trifolium				0.64^b	0.52^b	0.60^b	0.64^b	0.63^b
Cd in Dactylis					0.55^b	0.63^b	0.50^b	0.60^b
Pb in soil (0–10 cm)						0.62^b	0.25^a	0.25^a
Pb in soil (30–40 cm)							0.27^a	0.36^b
Pb in Trifolium								0.81^b
Pb in Dactylis								

^{a} $p < 0.05$.
^{b} $p < 0.001$.

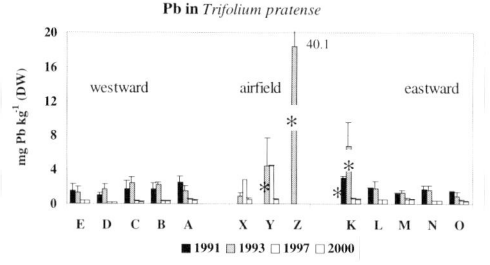
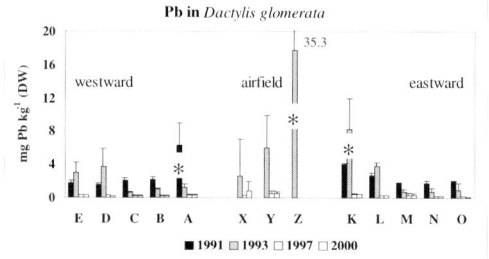
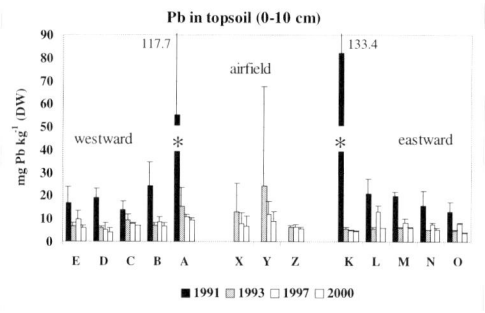
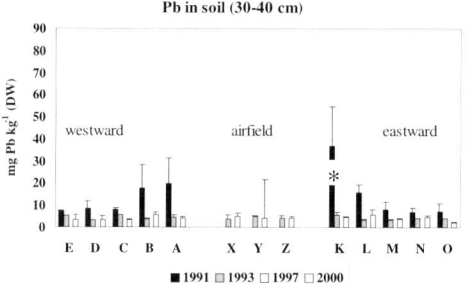

Figure 6. Dynamics of lead concentration (average ± standard deviation) in inflorescences of *Trifolium pratense*, stems of *Dactylis glomerata*, and in soil (0–10 and 30–40 cm). See sampling site (E-O) location in Figure 2. * Significantly differing values ($p < 0.05$) within the data series of the same year according to the Duncan Test.

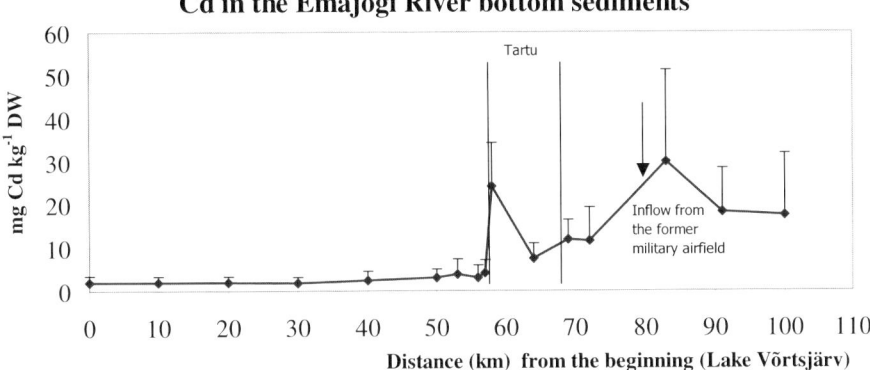

Figure 7. Cadmium concentration (average ± standard deviation) in the Emajõgi River sediments (according to Laanemets *et al.*, 1997).

gases affected the plant Cd concentration directly as surface contamination, and the root Cd uptake was probably of minor importance.

Elevated Cd and Pb concentrations in plant tissues and soil can also be caused by tyre wear of motor vehicles (de Miguel *et al.*, 1997; Czarnowska, 1999). Most probably, tyre wear of landing aircraft added some contamination to the samples taken from landing strip edges (A and K series) and from the vicinity of fuel bunkers and connecting roads/runways at the airfield (Y series; Figures 3–6). In addition, the cessation of the use of leaded gasoline in Estonia (50–350 mg Pb L^{-1} and 0.005–0.1 mg Cd L^{-1} in 1980's; Mander, 1985) since 1992–93 may play an important role in decreasing Pb contamination (Environment Information Centre, 2001).

On the other hand, high concentrations of Cd in the bottom sediments of the Emajõgi River, downstream of the airfield drainage channels (up to 30 mg kg^{-1}) may originate from the airfield (Figure 7). This value is comparable with the heavy metal content in contaminated alluvial sediments of large rivers in old industrial regions such as the Elbe, Oker and Vistula (Gröngröft *et al.*, 1998; Kabata-Pendias, 2000), but exceeds by many times the Cd concentration in less contaminated rivers (Blanc *et al.*, 1999) and channels (Stephens *et al.*, 2001). Seemingly, the airborne Cd has been washed out from plant residues and soil profiles relatively quickly without significant pollution of the groundwater. This explanation is supported by investigations on the vertical mobility of Cd and Pb in sandy and sandy loam soils near smelters in France (Sterckeman *et al.*, 2000). According to this study, Cd leaching was several times faster than that of Pb. Our data coincide with this finding, because the speed of Pb decontamination of plants and soils in the Tartu-Raadi airfield area has been remarkably lower than that of Cd (Table II, Figures 3–6). Especially in the disturbed soils of the military airfield (see Table I), disturbances may enhance the leaching of heavy metals and may significantly alter the ability of humus to adsorb metals (*cf.* Ukonmaanaho *et al.*, 2001). Another explanation, besides rapid leaching, is that lateral water movement during snowmelt and soil erosion (direct passage of soil particles to the drainage system), resulted in higher Cd concentrations in river sediments. Some Cd is most probably bound in soil particles, and our soil extraction method was not powerful enough to resolve all the Cd potentially originating from the airfield. However, further investigations of different fractions of drainage channels and river sediments should be undertaken to find the pathways of potential leaching or lateral transport of Cd to the surrounding environment.

4. Conclusions

Very high Cd contamination found in 1991–92 in *D. glomerata* stems and *T. pratense* inflorescences sampled from the former Tartu military airfield decreased significantly in the following years. In 1993, it resulted in a slight but significant increase in Cd concentration in the topsoil and in the 30–40 cm soil horizon. However, in 1997 and 2000 the concentration stabilized at the background levels of natural areas. Cd concentration in plants decreased significantly with increasing distance from the landing strip. At the same time, in comparison with the area to the east of the landing strip, there was significantly higher Cd contamination to the west of the landing strip.

Both Cd and Pb are airborne, but have different sources. Pb concentration in plants and soils also showed a significant decrease from 1991 to 2000. However, in plants, Pb concentration was influenced less by aircraft than by local pollution hot spots, such as fuel bunkers and gas stations. Based on the high Cd concentration in the fuel of Backfire-type strategic bombers we have assumed that the Cd originated from exhaust gases of this aircraft. Non-significant correlation between the Cd content in soil and plants also supports this assumption. Thus, the exhaust gases affected the plant Cd concentration directly, as surface contamination, and the Cd uptake by roots was probably of minor importance. High concentrations of Cd determined in the bottom sediments of the Emajõgi River downstream the airfield drainage channels allow us to speculate that this Cd originated from the airfield. Analogous situations may be detected and may require protective measures to be undertaken in other former Soviet military airfields in Central and Eastern Europe.

Acknowledgments

This study was supported by the Stockholm Environment Institute (SEI)-Tallinn, Estonian Environmental Research Centre, Estonian Science Foundation grants No 5247 and 5261, and the Estonian Ministry of Education targeted financing scheme TBGGG0549. The authors thank Dr. Peter J. Petridis from Intratech Ltd (Athens, Greece) for financial assistance, and students of the University of Tartu Ms. Eva Kauksi and Ms. Kairit Tipp for their help in field studies and cartographic work.

References

Auer, M. R. and Raukas, A.: 2002, 'Determinants of environmental clean-up in Estonia', *Environ. Plan. C – Gov. Policy* **20**, 679–698.
Blanc, G., Lapaquellerie, Y., Maillet, N. and Anschutz, P.: 1999, 'A cadmium budget for the Lot-Garonne fluvial system (France)', *Hydrobiologia* **410**, 331–341.
Czarnowska, K.: 1999, 'Heavy metals in lawn soils of Warsaw', *Soil Sci. Ann. Poland* **50**, 31–39.
Dahmani-Muller, H., van Oort, F., Gelie, B. and Balabane, M.: 2000, 'Strategies of heavy metal uptake by three plant species growing near a metal smelter', *Environ. Pollut.* **109**, 231–238.
de Miguel, E., Llamas, J. F., Chacon, E., Berg, T., Larssen, S., Røyset, O. and Vadset, M.: 1997, 'Origin and patterns of distribution of trace elements in street dust: unleaded petrol and urban lead', *Atmos. Environ.* **31**, 2733–2740.
Dudka, S., Piotrowska, M. and Terelak, H.: 1996, 'Transfer of cadmium, lead, and zinc from industrially contaminated soil to crop plants: a field study', *Environ. Pollut.* **94**, 181–188.
Esser, K. B.: 1996, 'Reference concentrations for heavy metals in mineral soils, oat, and orchard grass (*Dactylis glomerata*) from three agricultural regions in Norway', *Water, Air, Soil Pollut.* **89**, 375–397.
Gochfeld, M., Belant, J. L., Shukla, T., Benson, T. and Burger, J.: 1996, 'Heavy metals in laughing gulls: Gender, age and tissue differences', *Environ. Toxicol. Chem.* **15**, 2275–2283.
Gröngröft, A., Jähnig, U., Miehlich, G., Lüschow, R., Maass, V. and Stachel, B.: 1998, 'Distribution of metals in sediments of the Elbe estuary in 1994', *Water Sci. Technol.* **37**, 109–116.
Hutton, M. and Goodman, G. T.: 1980, 'Metal contamination of feral pigeons *Columba livia* from the London, England, U.K. area. 1. Tissue accumulation of lead, cadmium and zinc', *Environ. Pollut., Ser. A* **22**, 207–218.
Jones, Jr., J. B.: 1991, 'Plant tissue analysis in micronutrients', in J. J. Mortvedt, F. R. Cox, L. M. Shuman, and R. M. Welch (eds), *Micronutrients in Agriculture*, 2nd edition (SSSA Book Series, No 4), Soil Society of America, Madison, Wisconsin, U.S.A., pp. 477–521.
Kabata-Pendias, A.: 2000, *Trace Elements in Soils and Plants*, 3rd edition, Boca Raton, Florida, U.S.A., CRC Press, 456 p.
Kabata-Pendias, A. and Dudka, S.: 1991, 'Base-line data for cadmium and lead in soils and some cereals of Poland', *Water, Air, Soil Pollut.* **57**, 723–731.
Krauss, M. and Diez, T.: 1997, 'Uptake of heavy metals by plants from highly contaminated soils', *Agrobiological Research – Z. für Agrarbiologie, Agrikulturchemie, Ökologie* **50**, 343–349.
Laanemets, A., Kruus, U., Kivistik, M, Uri, M. and Sults, Ü.: 1997, 'Emajõe seisund enne Tartu puhastusseadmete I etapi valmimist', South-Estonian Environmental Laboratory, 80 p. [State of the Emajõgi River before the exploitation of the 1st block of sewage treatment plant in Tartu; in Estonian].
Mäkinen, A., Heikkinen, S. and Toikka, A.: 1980, 'Liikenteen ja teollisuuden vaikutus ilman puhtauteen Helsinki-Vantaan lentoaseman alueella', Kasviekologian laboratorio, Helsinki, 54 p.+13 app. [Air quality impact of traffic and industry in the Helsinki-Vantaa airport area; in Finnish].
Mäkinen, A.: 1993, 'Biomonitoring of atmospheric heavy metal deposition in Estonia; A chemical survey of mosses in 1989', Ministry of Environment, Environmental Policy Dept. Helsinki, Report 1, 56 pp.
Mander, Ü.: 1985, 'Roads as pollution sources', *Eesti Loodus* **28**, 307–316 (in Estonian, summary in English).
Nichols, T. P., Leinster, P., McIntyre, A. E., Lester, J. N. and Perry, R.: 1981, 'Survey of air pollution in the vicinity of Heathrow Airport, London, England, U.K.', *Sci. Total Environ.* **19**, 285–292.
Olsen, F. J. and Chong, S. K.: 1991, 'Reclamation of acid coal refuse', *Landscape Urban Plann.* **20**, 309–313.

Richards, B. K., Steenhuis, T. S., Peverly, J. H. and McBride, M. B.: 1998, 'Metal mobility at an old, heavily loaded sludge application site', *Environ. Pollut.* **99**, 365–377.

Rühling, Å., Brumelis, G., Goltsova, N. Kvietkus, K, Kubin, E., Liiv, S., Magnusson, S., Mäkinen, A., Pilegaard, K., Rasmussen, L., Sander, E. and Steinnes, E.: 1992, 'Atmospheric heavy metal deposition in Northern Europe 1990', *Nord* **12**, 1–41.

Stephens, S. R., Alloway, B. J., Parker, A., Carter, J. E and Hodson, M. E.: 2001, 'Changes in the leachability of metals from dredged canal sediments during drying and oxidation', *Environ. Pollut.* **114**, 407–413.

Sterckeman, T., Douay, F., Proix, N. and Fourrier, H.: 2000, 'Vertical distribution of Cd, Pb and Zn in soils near smelters in the North of France', *Environ. Pollut.* **107**, 377–389.

Sova, Z., Cibulka, J., Szakova, J., Miholova, D., Mader, P. and Reisnerova, H.: 1991, 'Contents of cadmium and lead in mushrooms from two areas in Bohemia', *Sb. Agron. Fak. Ceskych-Budejovicich Zootech. Rada* **8**, 13–29.

Svoma, J.: 1993, 'Investigation and decontamination of soil and groundwater at former Soviet army bases in Czechoslovakia', in F. Arendt, G. J. Annokkee, R. Bosman and W. J. van den Brink (eds), *Contaminated Soil '93. Fourth International KfK/TNO Conference on Contaminated Soil, 3–7 May 1993, Berlin, Germany, Vol. I*, Kluwer Academic Publishers, Dordrecht/Boston/London, pp. 747–754.

Ukonmaanaho, L., Starr, M., Mannio, J. and Ruoho-Airola, T.: 2001, 'Heavy metal budgets for two headwater forested catchments in background areas of Finland', *Environ. Pollut.* **114**, 63–75.

MERCURY AND METHYLMERCURY IN RUNOFF FROM A FORESTED CATCHMENT – CONCENTRATIONS, FLUXES, AND THEIR RESPONSE TO MANIPULATIONS

JOHN MUNTHE* and HANS HULTBERG

IVL Swedish Environmental Research Institute, P.O. Box 47086, SE-402 58 Göteborg, Sweden
(* author for correspondence, e-mail: john.munthe@ivl.se; phone: +46 31 725 62 00;
fax: +46 31 725 62)

(Received 20 August 2002; accepted 26 April 2003)

Abstract. Measurements of TotHg (total mercury) and MeHg (methylmercury) in runoff from the covered catchment G1 and the reference catchment F1 at Lake Gårdsjön, Sweden, have been performed and evaluated. The roof over the covered catchment limits atmospheric deposition input of TotHg and MeHg and a response in runoff concentrations and transport was expected. Based on data from 10 yr of monitoring, no statistically significant change in runoff flux of TotHg or MeHg can be observed. A slight decrease in MeHg output in the covered catchment was observed after 2 yr of the experiment. This can be explained as a temporary effect caused by the roof construction. The main conclusion is that release of TotHg and MeHg from the forest soil is controlled by factors other that wet deposition input, for example the mineralisation of organic matter. Furthermore, there is no indication of a depletion of the Hg pool in the soil. In spring 1999, the reference catchment F1 was affected by forestry machinery significantly disturbing the forest soil layer in a limited area. This caused MeHg concentrations to increase dramatically in runoff and led to an increase of the annual transport by at least a factor of 3. This indicates that forestry and other activities that disturb forest soils may be important for controlling MeHg fluxes to aquatic ecosystems.

Keywords: Catchment, Lake Gårdsjön, mercury, methylmercury, runoff

1. Introduction

Contamination of terrestrial and aquatic ecosystems by mercury (Hg) has been a prioritised environmental issue in Sweden and many other countries for over 20 yr. Large research efforts have been made to explain the links between Hg emissions to the atmosphere and Hg accumulation in freshwater fish. The link between emissions, long-range transport, deposition, and resulting accumulation of MeHg in fish was established in coordinated research efforts in the 1980's (Lindqvist *et al.*, 1991). Emissions and resulting atmospheric deposition of Hg have been reduced in Europe over the period of 1980 to 2000 (Iverfeldt *et al.*, 1995; Munthe *et al.*, 2001a). During the period of 1990 to 1995, a decreasing trend in Hg concentrations in freshwater fish in Sweden was observed and was attributed to the decrease in deposition (Johansson *et al.*, 2001).

A number of research projects on TotHg and MeHg dynamics in forested catchments and wetlands have been performed in Sweden (Iverfeldt, 1991; Hultberg *et*

TABLE I

Summary of TotHg and MeHg fluxes in the Gårdsjön catchments. Data from Munthe et al. (1995a, b) and Hultberg et al. (1995)

	TotHg (μg km^{-2})	MeHg (μg km^{-2})
Open field wet deposition	11.2	0.34
Throughfall wet deposition	16	0.20
Litterfall deposition	23	0.55
Run-off transport	2.2 (F1) 1.0 (G1)	0.1 (F1) 0.05 (G1)

al., 1994; Munthe et al., 1995a, b; Lee et al., 1994a, 2000), in North America (St. Louis et al., 1996; Driscoll et al., 1998), and in Germany (Schwesig et al., 1999; Schwesig and Matzner, 2000). Most of these studies have focused on input/output budgets and on relationships between TotHg and MeHg behaviour and hydrology, DOC etc. In Table I, a summary of deposition and runoff fluxes in the Gårdsjön catchments during the period of 1990 to 1995 is presented. Based on decreases in open field deposition of TotHg observed within the National Swedish Monitoring Network (1994 to present), the current wet deposition of TotHg in Gårdsjön area can be expected to be somewhat lower than the numbers presented in Table I.

2. Experimental

2.1. SITE DESCRIPTION

Lake Gårdsjön is located about 15 km from the west coast of Sweden. The forest cover is mainly mature Norway Spruce with some Scots Pine. The region is characterised by a low weathering rate bedrock and thin soils and is thus sensitive to acid deposition. Annual deposition is about 1150 mm and runoff is about 550 mm. Acidification research started in the late 1960s with a focus on effects of acidification and liming. Since then, a large number of research projects related to acidification, effects of nitrogen deposition, cycling of heavy metals and POPs, have been carried out in this area. A detailed site description can be found in (Andersson et al., 1998).

The reference area, F1, is a 37,000 m^2 subcatchment at Lake Gårdsjön. A long-term record of runoff chemistry has been maintained since 1979 and F1 is currently monitored within the UN ECE CLRTAP (The United Nation Economic Commission for Europe, Convention on Long-Range Transboundary Air Pollution) programme Integrated Monitoring. The covered catchment, G1, is 6,300 m^2 and is located about 1 km from F1. The covered catchment project was initiated in the late 1980's with the main aim to investigate the time dependence of recovery of severely acidified forest soil. A plastic roof was erected over catchment G1 in April 1991, completely removing the wet deposition input. The roof cover was removed, after 10 complete years of experiment, in June 2001. During the experiment, the forest floor was irrigated with artificial throughfall with a chemical composition similar to pre-industrial times. A detailed description of the experimental setup and initial results can be found in Hultberg and Skeffington (1998). In parallel to the acidification and recovery focus of the covered catchment project, the cycling of Hg and MeHg in the forest ecosystem has been a major issue.

2.2. SAMPLING AND ANALYSES

Runoff samples for analyses of Hg and MeHg were collected manually using 125 mL acid leached Teflon bottles. Samples were immediately brought to the IVL laboratory and preserved with 0.5 mL of 30% HCl (Merck Suprapur). For analysis of total mercury, the samples were oxidised using BrCl for at least 8 hr. Excessive BrCl was removed by addition of 0.5 mL of NH_2OH. A sample aliquot was transferred to a bubbler flask where $SnCl_2$ was added. The produced HgO was purged using purified N_2 and was collected on a gold trap. After 20 min. the trap was transferred to the analytical system where it was connected to an argon gas stream leading to the CVAFS detector. Collected Hg was desorbed by heating the gold trap.

Samples for analysis of MeHg were distilled to 80% and transferred to a reaction bottle where an ethylating reagent was added (sodium tetraethylborate). The formed ethylated mercury species were purged from the reaction bottle by a N_2 gas stream and collected on Carbotrap adsorbents. The collected mercury species were then thermally desorbed on to a GC column. After separation, the organomercury species were pyrolysed and detected using CVAFS. A detailed description of the analytical procedures for MeHg can be found in Lee *et al.* (1994b).

Mercury fluxes in runoff were calculated by interpolating monthly or bi-weekly concentration data to daily values and multiplying with daily water transport. Water fluxes were measured using calibrated weirs.

2.3. QUALITY ASSURANCE/QUALITY CONTROL

During storage and transport to the sampling site, the sampling bottles were filled with a diluted HCl solution (0.5 mL 30% HCl; Merck Suprapur), which was discarded before sampling. Plastic gloves were used during sampling and whenever handling the bottles with one or both of the plastic bags open.

The detection limit (defined as 3 σ of the method blank value) was determined for each sample batch and was normally 0.06 ng L^{-1} for both TotHg and MeHg. Blank values were usually below 0.03 ng L^{-1} for TotHg and below 0.1 ng L^{-1} for MeHg. The instrument was calibrated for TotHg using aqueous mercury solutions (0.6 μg L^{-1}) made from a NIST standard reference material diluted with MilliQ water (>18 M Ω resistivity) and 5 mL L^{-1} 30% Suprapur HCl (Merck). For MeHg

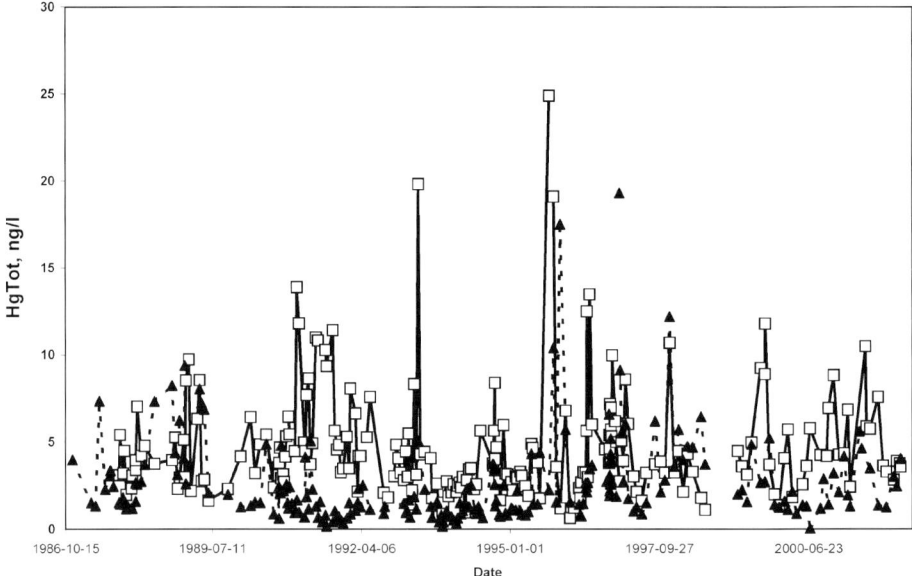

Figure 1. Concentration of total mercury in run-off from reference F1 (open squares, solid line) and covered catchment G1 (filled triangles, dashed line), at Lake Gårdsjön.

calibration, a stock solution of CH_3HgCl (1000 mg L^{-1}) was prepared. This stock solution was then stepwise diluted to a working standard with a concentration of 1.0 μg L^{-1}. Certified reference materials for TotHg (CRM-280, Lake sediment) and MeHg (DORM-2, dogfish-muscle) were used on a weekly basis.

2.4. MERCURY AND METHYLMERCURY IN RUNOFF

The runoff output from the covered catchment (G1) and the reference catchment (F1) were monitored from 1988 to 2001. The complete dataset for TotHg is represented in Figure 1 and for MeHg in Figure 2.

The concentration of TotHg was relatively constant in time, but showed considerable variability in individual events, mainly positively correlated with water flow. There were no general trends in concentrations in the reference area (F1) or experimental area (G1).

Concentrations of MeHg showed a much higher variability over time. Higher concentrations generally were found in both the G1 and F1 catchments during the period of 1987 to 1994. Elevated concentrations around 1990 to 1992 may have been caused by the construction of new dams in both catchments and by construction of the roof in G1. Between 1995 and 1999, much lower and more consistent concentrations were found. From early 1999, a number of high-concentration events appeared especially in the reference area (F1). These events were linked to damage caused by forest machinery (tractors) crossing the catchment (see discus-

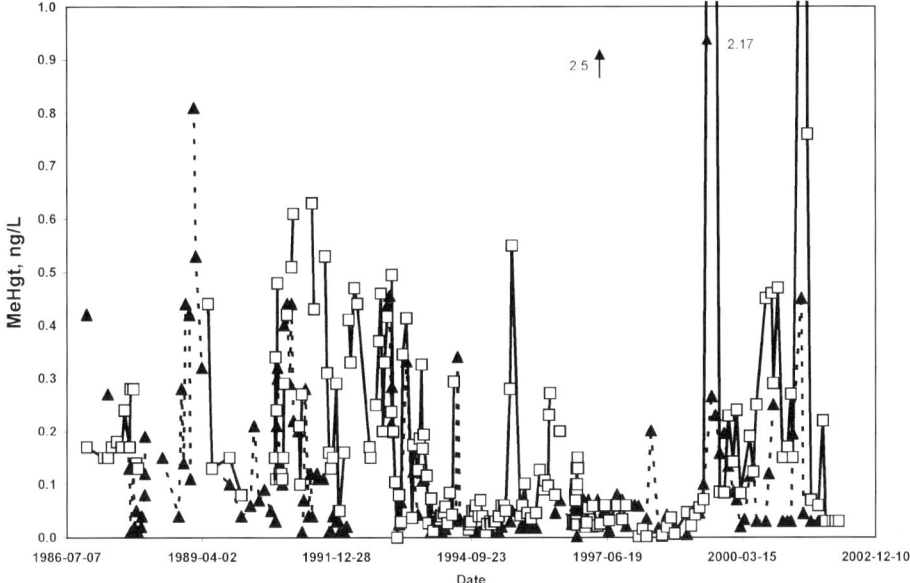

Figure 2. Concentration of methylmercury in run-off from reference catchment F1 (open squares, solid line) and covered catchment G1 (filled triangles, dashed line), at Lake Gårdsjön. Elevated concentrations at the end of period are discussed separately below.

sion below). The average concentrations of TotHg and MeHg in the two catchments are presented in Table II.

2.5. IMPACT OF A DECREASED DEPOSITION (COVERED CATCHMENT)

In Figure 3, the accumulated transport of TotHg in catchments F1 and G1 is presented. The two lines cross at the same time as the roof was erected, indicating a change in mobilisation and transport in the G1 experimental catchment. However, when considering the trend during the year before the roof construction, it is clear that this change started earlier and that it cannot be attributed to the decreased deposition input. From around early 1994, the two lines are more or less parallel indicating that no changes have occurred in TotHg transport.

The accumulated mobilisation and transport of MeHg is shown in Figure 4. The accumulated transport of MeHg behaves rather differently from that of TotHg. In the years before the construction of the roof and continuing about one year after, the output of MeHg was similar in the two catchments. Relatively sharp changes in output flux occured.

The impact of a decreased wet deposition input of Hg and MeHg was evaluated in two earlier studies. Based on data from the first two years of the experiment, a significant decrease of MeHg was noted (Hultberg *et al.*, 1995). With a larger database available, Munthe *et al.* (1998) concluded that no statistically significant

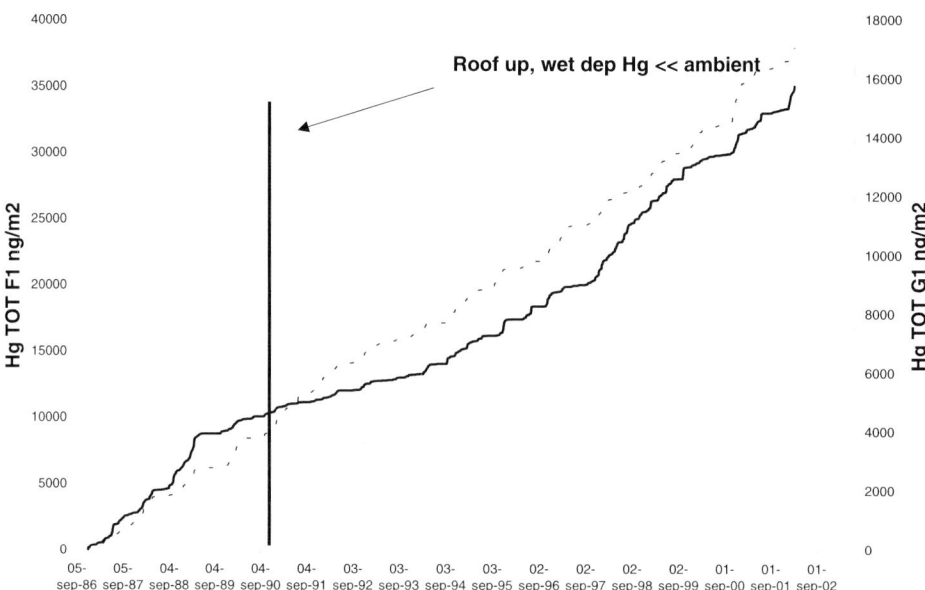

Figure 3. Accumulated mobilisation and transport of total mercury in reference catchment F1 (thin line) and experimental catchment G1 (thick line).

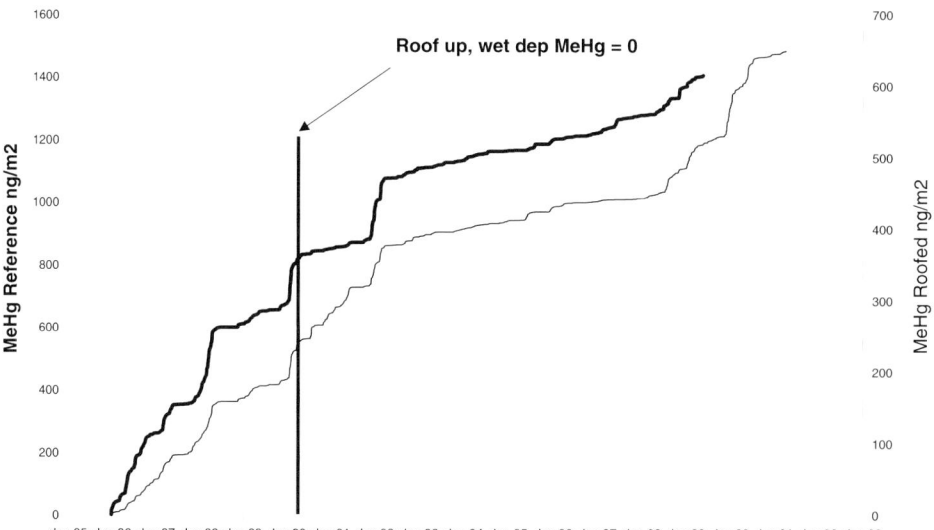

Figure 4. Accumulated mobilisation and transport of methylmercury in reference catchment F1 (thin line) and experimental catchment G1 (thick line). The increased transport from F1 after 1999 is discussed in Section 3.3.

TABLE II

Summary of TotHg and MeHg fluxes in the Gårdsjön catchments. Data from Munthe et al. (1995a, b) and Hultberg et al. (1995)

	F1 HgTot (ng L^{-1})	F1 MeHg (ng L^{-1})	G1 HgTot (ng L^{-1})	G1 MeHg (ng L^{-1})
Average	4.67	0.20	2.57	0.10
Median	3.85	0.10	1.73	0.05
Max	24.90	2.55	19.30	0.81
Min	0.63	0.002	0.03	0.002
SD	3.19	0.34	2.47	0.12

difference between the experimental catchment (G1) and the reference catchment (F1) could be observed. Based on the complete dataset represented in Figures 3 and 4, there continues to be no significant difference in MeHg and TotHg concentrations or flux between the open (F1) and covered (G1) catchments. No clear explanation can be found for the initially observed decrease in MeHg output. It is possible that the construction of the roof, and prior to that the dam, may have disturbed the soil and caused temporary changes in TotHg and MeHg flux. The subsequent decrease in atmospheric input of TotHg and MeHg also may have led to a temporary effect in runoff transport. After an extended period (2–4 yr), the soilwater-soil reached a new 'steady-state', with no further changes in concentrations or transport. However, there is no direct evidence supporting this or other phenomena and it cannot be excluded that the observed changes were not related to the decreased wet deposition. The evaluation is further complicated by the fact that a simultaneous decrease of TotHg deposition has occured due to emission reductions in Europe (Iverfeldt et al., 1995). It is possible that the decreased atmospheric mercury deposition to F1 to some extent has diminished the expected changes between F1 and G1. However, deposition data for MeHg do not indicate any decrease (Munthe et al., 2001b). Any influence of deposition changes outside the covered catchment experiment would thus have to be related to processes occuring within the catchment.

2.6. IMPACT OF LOGGING ROAD

In the spring of 1999, a clear-cutting operation was initiated in a forested area adjacent to reference catchment F1. To reach the clear-cutting area, heavy forestry machinery (tractors, transport vehicles) were driven through the upper area of reference area F1. The vehicles caused significant disturbance of the forest soil

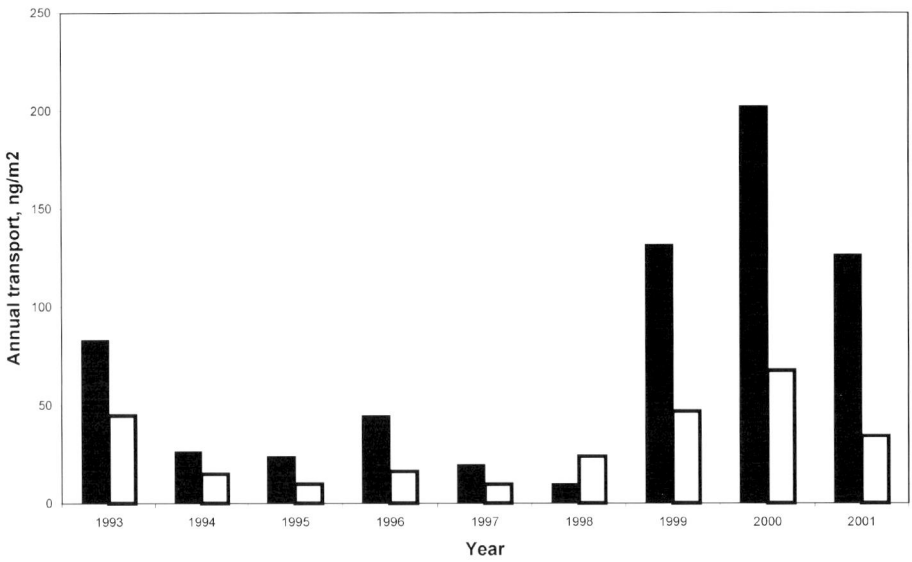

Figure 5. Annual transport of MeHg from reference catchment F1 (filled bars) and experimental catchment G1 (open bars).

and a temporary logging road was constructed (soil, branches, etc. were used to stabilise the road). The temporary road blocked the water flow in the small stream in the center of the catchment and led to the formation of a small dam (< 50 m^2) upstream. The blockage was removed immediately after discovery and the stream flow returned to normal within a few weeks after the incident. However, the natural soil conditions were altered and the resulting change in water flow paths remained.

The temporary logging road led to sharp increases in the concentrations of MeHg in runoff sampled at the catchment outlet some 100 m downstream (Figure 2). In contrast to this, TotHg was fairly stable (Figure 1). In Figure 5 and Figure 6, the annual transport of MeHg and TotHg, respectively, for the period 1993 to 2001 are presented.

The increase in MeHg transport is significant for 1999 (when the incident occured) but also for 2000 and 2001. This indicates that a permanent damage to the forest soil and the water flowpaths has occurred. During the same period, a slight increase in MeHg transport from the experimental catchment G1 also occurred. In 1999 and 2000, TotHg fluxes in F1 also were considerably higher than previous years. The reason for this is most likely that the years 1998, 1999 and 2000 were unusually wet, with considerably higher runoff than the preceeding years. During these years, runoff from reference area F1 was 748, 827 and 749 mm, respectively, whereas the average for the period of 1994 to 2001 was 547 mm. In Table III, the average annual fluxes for the period 1994 to 1998 are compared with the period 1999 to 2001.

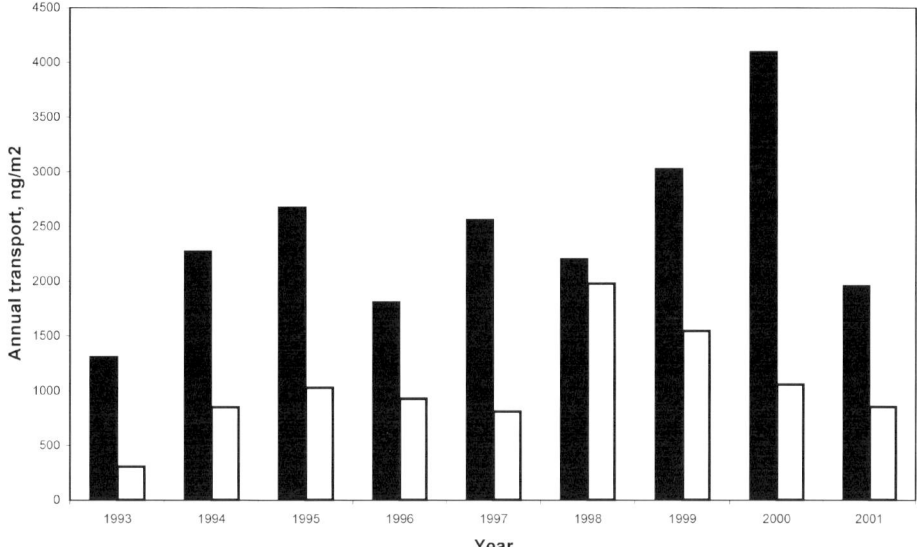

Figure 6. Annual transport of TotHg from reference catchment F1 (filled bars) and experimental catchment G1 (open bars).

TABLE III

Impact of logging road in reference area F1. Average annual fluxes of MeHg and TotHg in both catchments

	Average fluxes				
	F1 MeHg (ng m^{-2})	F1 TotHg (ng m^{-2})	G1 MeHg (ng m^{-2})	G1 TotHg (ng m^{-2})	F1 Runoff (mm)
1994–1998 Before logging	25	2308	15	1119	547
1999–2001 After logging	154	3031	49	1152	662
Ratio	6.2	1.3	3.3	1.0	1.2

3. Discussion

No significant differences in concentrations or transport of MeHg and TotHg can be observed in the two catchments, F1 and G1, between 1991 and 1998. Thus it can be concluded that the decreased input of atmospheric mercury does not have a long-term effect on runoff concentrations. This indicates that MeHg and TotHg in runoff are regulated by processes other than wet deposition input. In a recent study of TotHg and MeHg mobility in forest soils using stable isotopes (Munthe et al., 2001b), TotHg was shown to be tightly bound in the forest soils. The same study also indicated some production of MeHg after additions of inorganic Hg. This effect decreased with time. Large amounts of TotHg (13–18 kg km^{-2}) and MeHg (86–123 g km^{-2}) (Lee et al., 1994a) are accumulated in the soils in the Gårdsjön catchments. A large fraction of this is of anthropogenic origin. The results from the covered catchment experiments suggest that runoff transport is not directly affected by current input. The pools in the forest soil levels are sufficiently large to act as a source of TotHg and MeHg in runoff. Further discussion of the results from the covered catchment experiment can be found in Munthe et al. (1998).

The impact of the temporary logging road suggests that forestry and other activities that lead to hydrological changes in forest soils may cause mobilistion of the MeHg pools in the soil. The increased output may be explained by the changing water flowpaths mobilising the MeHg already present in the soil. The logging road blocked the water flow thereby creating a small 'reservoir'. This may also have led to increased methylation via biotic or abiotic processes. However, the logging road was cleared after a few weeks allowing the water a clear, but altered due to the disturbance, flow-path. Despite this, the MeHg levels continued to be elevated indicating that the initial 'reservoir' phase was of minor importance.

Clear-cutting in catchments previously has been shown to influence MeHg levels in zooplankton (Garcia and Carignan, 1999) and TotHg levels in pike (Garcia and Carignan, 2000). This study suggests that the changing hydrological pathways and damage of forest soils may be an important factor for the MeHg loading on streams and lakes. Because no clear-cutting was performed in the catchment where the effects were observed, it appears that tree removal is not an important factor. Similar results were presented in a study of MeHg and TotHg transport from a clear-cut catchment in Finland (Porvari et al., 2003). In the Finnish study, significant increases in fluxes of TotHg as well as MeHg were observed, in contrast to this study where mainly MeHg was affected. The incresase in TotHg was mainly related to the increased water flows after clear-cutting, because no increase in concentrations were observed.

The presently available information is not sufficient to make overall estimates of the role of forestry activities in MeHg loading on freshwater ecosystems. The results are, however, sufficient to suggest that further study of this phenomenon is warranted. To date, no other factors have been shown to efficiently mobilise TotHg or MeHg from forest soils (except flooding for hydroelectric reservoirs) to an extent comparable to the effects observed in this study.

Acknowledgements

This study was partly funded by the Nordic Council of Ministers. The covered catchment project was funded by Vattenfall AB, National Power, Power Gen, and the Foundation of the Swedish Environmental Research Institute (SIVL). The analytical skills of Ms. Emma Lord and Ms. Pia Spandow are gratefully acknowledged. Many thanks also to Mr. Jan Tobisson who collected most of the samples and Dr. Filip Moldan who assisted in the flux calculations.

References

Andersson, B. I., Bishop, K. H., Borg, G. C., Giesler, R., Hultberg, H., Huse, M., Moldan, F., Nyberg, L., Nygaard, P. H. and Nyström, U.: 1998, 'The covered catchment site: A description of the physiography, climate and veghetation of three small coniferous forest catchments at Gårdsjön, South-west Sweden', in H. Hultberg and R. A. Skeffington (eds), *Experimental Reversal of Acid Rain effects. The Gårdsjön Roof project*. JohnWiley & Sons Ltd., Chichester, England.

Driscoll, C., Holsapple, J., Schofield, C. and Munson, R.: 1998, 'The chemistry and transport of mercury in a small wetland in the Adirondack region of New York, U.S.A.', *Biogeochemistry* **40**, 137–146.

Garcia, E. and Carignan, R.: 2000, 'Mercury concentrations in northern pike (Esox lucius) from boreal lakes with logged, burned, or undisturbed catchments', *Can. J. Fish. Aquat. Sci.* **57** (Suppl. 2), 129–135.

Garcia, E. and Carignan, R.: 1999, 'Impact of wildfire and clear-cutting in the boreal forest on methyl mercury in zooplankton', *Can. J. Fish. Aquat. Sci.* **56**, 339–345.

Hultberg, H. and Skeffington, R. (eds): 1998, *Experimental Reversal of Acid Rain Effects: The Gårdsjön Roof Project*. John Wiley & Sons Ltd., Chichester, England.

Hultberg, H., Iverfeldt, Å. and Lee, Y.-H.: 1994, 'Methylmercury input/output and accumulation in forested catchments and critical loads for lakes in Southwestern Sweden', in C. J. Watras and J. Huckabee (eds), *Mercury Pollution. Integration and Synthesis*. Lewis Publishers, CRC Press, Inc., Boca Raton, Florida, U.S.A., pp. 313–322.

Hultberg, H., Munthe, J. and Iverfeldt, Å.: 1995, 'Cycling of methyl mercury and mercury – responses in the forest roof catchment to three years of decreased atmospheric deposition', *Water, Air, Soil Pollut.* **80**, 415–424.

Iverfeldt, Å.: 1991, 'Mercury in forest canopy throughfall water and its relation to atmospheric deposition', *Water, Air, Soil Pollut.* **56**, 553–542.

Iverfeldt, Å., Munthe, J., Brosset, C. and Pacyna, J.: 1995, 'Long-term changes in concentration and depositon of atmospheric mercury over Scandinavia', *Water, Air, Soil Pollut.* **80**, 227–233.

Johansson, K., Bergbäck, B. and Tyler, G.: 2001, 'Impact of atmospheric long range transport of lead, mercury and cadmium on the Swedish forest environment', *Water, Air, Soil Pollut.: Focus* **1**, 279–297.

Lee, Y. H., Bishop, K. and Munthe, J.: 2000, 'Do concepts about catchment cycling of methylmercury and mercury in boreal catchments stand the test of time? Six years of atmospheric inputs and runoff export at Svartberget, northern Sweden', *STOTEN* **260**, 11–20.

Lee, Y.-H., Borg, G. Ch., Iverfeldt, Å. and Hultberg, H.: 1994a, 'Fluxes and turnover of methylmercury: Mercury pools in forest soils', in C. J. Watras and J. Huckabee (eds), *Mercury Pollution. Integration and Synthesis*, Lewis Publishers, CRC Press, Inc., Boca Raton, Florida, U.S.A., pp. 343–354.

Lee, Y-.H., Munthe, J. and Iverfeldt, Å.: 1994b, 'Experiences on analytical procedures for the determination of methylmercury in environmental samples', *Appl. Organomet. Chem.* **8**, 659–664.

Lindqvist, O., Johansson, K., Aastrup, M., Andersson, A., Bringmark, L, Hovsenius, G., Hakanson, L., Iverfeldt, Å, Meili, M. and Timm, B.: 1991, 'Mercury in the Swedish Environment – Recent research on causes, consequences and corrective methods', *Water, Air, Soil Pollut.* **55**, 23–32.

Munthe, J., Hultberg, H. and Iverfeldt, Å.: 1995a, 'Mechanisms of deposition of mercury and methylmercury to coniferous forests', *Water, Air, Soil Pollut.* **80**, 363–371.

Munthe, J., Hultberg, H., Lee, Y.-H., Parkman, H., Iverfeldt, Å. and Renberg, I.: 1995b, 'Trends of mercury and methylmercury in deposition, run-off water and sediments in relation to experimental manipulations and acidification', *Water, Air, Soil Pollut.* **85**, 743–748.

Munthe, J., Lee, Y. H., Hultberg, H., Iverfeldt, Å., Borg, G. Ch. and Andersson, I.: 1998, 'Cycling of mercury and methyl mercury in the Gårdsjöm catchments', in H. Hultberg and R.A. Skeffington (eds), *Experimental Reversal of Acid Rain Effects: The Gårdsjön Roof Project*. John Wiley & Sons, pp. 261–276.

Munthe, J., Kindbom, K., Kruger, O., Petersen, G., Pacyna, J. and Iverfeldt, Å.: 2001a, 'Examining source-receptor relationships for mercury in Scandinavia – modelled and empirical evidence', *Water, Air, Soil Pollut.: Focus* **1**, 299–310.

Munthe, J., Lyvén, B., Parkman, H., Lee, Y-H., Iverfeldt, Å., Haraldsson, C., Verta, M. and Porvari, P.: 2001b, 'Mobility and methylation of mercury in forest soils. Development of an in-situ stable isotope tracer technique and initial results', *Water, Air, Soil Pollut.: Focus* **1**, 385–393.

Porvari P., Verta M., Munthe J. and Haapanen, M.: 2003, 'Forestry practices increase mercury and methylmercury output from boreal forest catchments', *Environ. Sci. Technol.*, in press.

Schwesig, D., Ilgen, G. and Matzner, E.: 1999, 'Mercury and methylmercury in upland and wetland acid forest soils of a watershed in NE-Bavaria, Germany', *Water, Air, Soil Pollut.* **113**, 141–154.

Schwesig, D. and Matzner, E.: 2000, 'Pools and fluxes of mercury and methylmercury in two forested catchments in Germany', *Sci. Total Environ.* **260**, 213–223.

St. Louis, V. L., Rudd, J. W. M., Kelly, C. A., Beaty, K. G., Flett, R. J. and Roulet, N. T.: 1996, 'Production and loss of methylmercury and loss of total mercury from boreal forest catchments containing different types of wetlands', *Environ. Sci. Technol.* **30**, 2719–2729.

DEPOSITION AND FATE OF LEAD IN A FORESTED CATCHMENT, LESNI POTOK, CENTRAL CZECH REPUBLIC

TOMÁŠ NAVRÁTIL[1,2]*, MAREK VACH[1], PETR SKŘIVAN[1],
MARTIN MIHALJEVIČ[2] and IRENA DOBEŠOVÁ[1]
[1] *Geological Institute, Academy of Science, Czech Republic;* [2] *Faculty of Science, Charles University, Prague*
(* author for correspondence, e-mail: navratilt@gli.cas.cz; phone: +420-2-33087222; fax: +420-2-20922670)

(Received 20 August 2002; accepted 8 April 2003)

Abstract. The deposition of trace elements and their fate in a forest ecosystem has been monitored at the experimental site, Lesni Potok catchment (LP), with granite bedrock. The catchment is located 30 km ESE from Prague. Annual bulk Pb-deposition flux F_{Pb} was 3.41 kg km^{-2} a^{-1} in 1994 and gradually decreased to 0.49 kg km^{-2} a^{-1} in 2001. The decrease is comparable with those observed in Germany and in the U.S.A. in the 1970s and 1980s. The total sales ban of leaded gasoline in the Czech Republic since January 2001 was accompanied by a pronounced decrease of F_{Pb} in a single year. The residual Pb-deposition flux is assigned to both the long-range transport of fine-grained vehicular lead aerosol (with a long residence time in the atmosphere) and to the emissions from power plant boilers burning lignite mined in the Czech northwest coal basin. The F_{Pb} of lead correlates strongly with those of As, Cd, Cu, Zn and Be, the typical metals in coal fly ash, at two monitored sites. Topsoil horizons contain elevated concentrations of Pb (53–67 mg kg^{-1}), which are of anthropogenic origin. Soils in the riparian areas contain increased concentrations of Pb when compared to soils on the hillslope areas. Significant amounts of Pb were found on a stream substrate and Fe-precipitate sampled from the stream. Low concentrations of Pb in bark and bole wood suggest that the uptake of Pb by vegetation is negligible. The very small surface water outputs (average of 0.002 kg km^{-2} a^{-1}) compared to inputs (average of 1.890 kg km^{-2} a^{-1}) from the LP catchment indicate an ongoing accumulation of Pb in a forested landscape.

Keywords: biogeochemical cycle, catchment, central Czech Republic, deposition, lead, monitoring, Pb, precipitation

1. Introduction

The entire solid surface of our planet is contaminated by anthropogenic (mostly vehicular) lead (Pb). Recurring problems with this toxic trace metal in our environment are well known and widely discussed in the literature. Research of Antarctic ice indicates that the current concentration of Pb is 2 pg Pb g^{-1}, whereas during most of Holocene it was approximately 0.4 pg Pb g^{-1} (Boutron and Patterson, 1986). The authors postulate that >99% of the tropospheric Pb in mid-1960s in the northern hemisphere originated from human activities. Worldwide sources of anthropogenic Pb emissions to the atmosphere (in descending order) are vehicular sources (by far the largest), non-ferrous metal production, coal combustion, steel and iron manufacturing, and cement production (Nriaghu and Pacyna, 1988). Coal-derived Pb emissions dominated over Pb from both smelting and traffic in the Czech Republic (Novák et al., 2003).

The input of Pb in precipitation declined between 1976 and 1989 by 97% at Hubbard Brook Experimental Forest (HBEF) in New Hampshire (USA) (Johnson et al., 1995). The consumption of Pb in gasoline declined in the USA by >90% between 1976 and 1989. Consequently the atmospheric concentration of Pb decreased during these years by >50% in urban areas.

The Pb concentration in litter horizons of forest soils was studied by Suchara and Sucharová (2000) throughout the Czech Republic. The reported mean content of Pb in the litter from Central Bohemia was 151 mg kg^{-1} (d.w.). Lead isotopic data from soils of two catchments in the Czech Republic (Nacetin and Salacova Lhota) enabled Emmanuel and Erel (2002) to conclude that anthropogenic Pb was stored primarily in the organic horizons. The anthropogenic Pb was associated with surface-bound and organic matter fractions. The proportions of these fractions to total Pb were 33 to 50% and 23 to 47%, respectively. Isotopic ^{210}Pb dating has been used for the evaluation of historical rates of Pb deposition over past 200 yr in the Czech Republic (Vile et al. 2000).

The uptake by vegetation has been estimated to be a minor flux in the Pb budget in a number of studies (Heinrichs and Mayer 1977, Friedland and Johnson 1985, Johnson et al., 1995). The distribution of Pb in tree tissues of red spruce (*Picea rubens*) decreased in order: twigs > bark > roots > needles > wood (Friedland and Johnson 1985).

Deposition of vehicular Pb at the Lesni Potok (LP) catchment between 1989 and 1991 corresponded to the mean values reported for rural sites of central Europe. The Pb deposition decreased with time and it was positively correlated with precipitation amount (Skřivan and Vach, 1993; Skřivan et al., 2000a). The aim of the present study was to undertake a more thorough assessment of the deposition, cycling, and fate of anthropogenic lead in the experimental landscape through a longer time span.

2. Materials and Methods

The Lesni Potok catchment (LP) is located in a rural countryside approximately 30 km ESE from Prague, Czech Republic (Figure 1). The area of the catchment is 99% afforested. The vegetation of the catchment is 46% coniferous (mainly Norway spruce; *Picea abies*) and 53% deciduous (mainly European beech; *Fagus sylvatica*). The area received an annual average of 626 mm of precipitation in the period 1990–2001, and the annual average runoff from the catchment was 90 mm in the period 1994–2001. The significant difference between the water input and output was due to ~80% evapotranspiration.

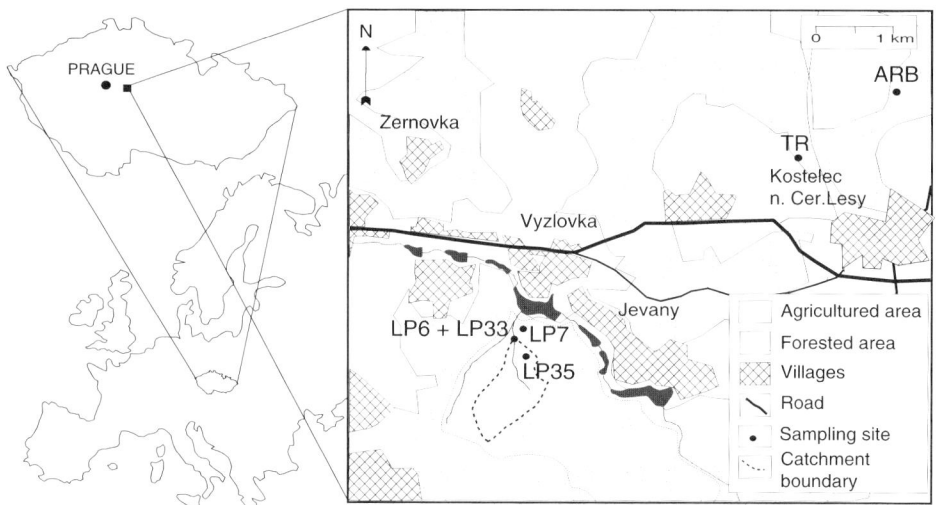

Figure 1. Location of Pb sampling sites and localities at Lesni Potok catchment, Czech Republic.

Due to 99% afforestation of the catchment, the chemical composition of bulk deposition was monitored at two near by sites (TR and ARB), which were 4 km from the catchment and about 1 km apart from each other (Figure 1). The site TR was affected by the presence of a local road about 100 m distant. Identical polyethylene (PE) collectors were installed at the two sites. The outer larger PE funnel was equipped with a smaller glass funnel, which collects the sample and prevents adsorption of Pb onto the PE. Prior to installation, the glass funnel was checked for Pb content by leaching the glass with diluted 1:10 Merck Suprapur HNO_3 for 24 hr. No Pb was detected in this acid blank. An amount of 2.5 mL of diluted HNO_3 was inserted into the precipitation sampler every month to prevent losses of analytes by adsorption onto the walls of PE sample container. Resulting pH of samples was always <2, thus the determined Pb concentration represents the sum of dissolved and acid-soluble aerosol forms of lead in deposition.

Two types of throughfall (beech and spruce) were sampled monthly at LP catchment at the two adjacent sites LP6 and LP7 (Figure 1). Sampling of throughfall fluxes started in 1997. The throughfall samples were acidified 24 hr before the filtration and filtered through 0.45 μm cellulose membranes prior to analysis.

Two soil profiles were studied at the site LP (locations LP33 and LP35, Figure 1). The <2 mm fraction was used for analysis of total and leachable Pb concentrations. The labile soil Pb concentrations were evaluated by two methods – leaching by either 0.1 M HNO_3 or 0.5 M NH_4NO_3. The leaching was performed under room temperature, for 24 hr and with volume to solid ratio of 200.

Soil water was collected from two sets of zero-tension lysimeters installed at LP in the vicinity of the studied soil profiles LP33 and LP35. Site LP33 was situated in the riparian zone of the LP brook and LP35 is situated on a hillslope. The lysimeters have been sampled monthly since 2000. The soil water samples were filtered through a 0.45 μm membrane and stabilized with 1 mL of HNO_3 per 100 mL of sample.

The output of stream water at LP was measured by means of a Thompson weir located at the site LP6. The surface water at the LP has been sampled monthly since 1993, while the Pb analyses started in 1994. Shallow groundwater was sampled from a well equipped with plastic casing and located in close vicinity of the Thompson weir.

The assimilatory organs of selected tree species were removed by plastic scissors and immediately sealed in PE bags. Wood cores were extracted from tree boles up to the tree center with an increment borer (Haglöf, Sweden, PTFE coated) after the rough bark and phloem were removed. All samples of organic matter were dried in the laboratory to constant weight at room temperature in a flow box. The stem wood and bark were digested by hot concentrated nitric acid (Merck, Suprapur) under pressure in PTFE crucibles in a microwave oven. Samples of the assimilatory organs were slowly mineralized and digested by concentrated nitric acid (Merck, Suprapur) at room temperature for 14 d in glass volumetric flasks. The reaction products were diluted using double-distilled water and filtered through a previously leached nitrocellulose membrane filter (Sartorius, pore size 0.45 μm). The filter was then rinsed with double-distilled water and the filtrate was filled up to a known volume.

All the Pb concentrations were determined by ETA atomic adsorption spectrometry (ETA AAS). During the water year 2000, all Pb concentrations were determined in samples preconcentrated by sub-boiling evaporation. Solute precipitation was prevented by means of HNO_3 addition prior to pre-concentration. Each step of the sample collection, processing and analyses was accompanied by blank determinations to check for any possible loss of analyte or sample contamination. Detailed methods of sampling were described in Skřivan *et al.* (2000b, 2002).

3. Results and Discussion

3.1. Annual deposition fluxes and trends

The annual Pb fluxes in and out of LP catchment are in Table I. The time span of our monitoring covers the decade of the Czech Republic economical transition from centrally planed to market economy, which started in November 1989. The changes in the Czech economy and environmental policies have been accompanied by an attenuation of heavy industry that had little pollution control and by desulphurization of the large power plants burning lignite coal. The increased concern for the environment caused broader introduction of unleaded gasoline (Skřivan *et al.*, 2000a) and the final ban of leaded gasoline sales in January 2001. The range of calculated annual fluxes between 3.56 and 0.49 kg Pb km^{-2} a^{-1} at site TR

TABLE I

Annual fluxes of Pb (kg km^{-2} a^{-1}) (FPb) at monitored localities in and around Lesni Potok catchment

Location	TR	TR	ARB	LP7	LP6	LP6	LP6	LP6
Type	Bulk	Bulk	Bulk	THS[a]	THB[a]	Input[a]	Stream	Stream
year	PA [mm]	Pb flux	Pb flux	Pb flux	Pb flux	Pb flux	output [mm]	Pb flux
1990	417	2.43						
1991	518	2.08						
1992	645	2.29						
1993	609	3.24						
1994	667	3.41					113	
1995	804	2.46					122	
1996	689	2.83	1.50				151	0.066
1997	727	2.28	1.09	1.23	0.50	0.83	98	0.046
1998	671	1.60	0.71	1.15	0.83	0.97	35	0.018
1999	532	0.81	0.31	0.53	0.27	0.39	73	0.029
2000	442	1.28	0.50	0.97	0.38	0.65	67	0.019
2001	794	0.49	0.47	0.61	0.26	0.42	63	0.030
Decrease%		*78%*	*43%*	*37%*	*48%*	*41%*		

[a] THB = throughfall beech, THS = throughfall spruce, INPUT = due to afforestation (0.46 * THS + 0.53 * THB).

corresponds with the mean values of <5 kg Pb km^{-2} a^{-1} presented by the Czech Hydrometeorological Institute (Fiala and Ostatnicka, 1997). Lower Pb deposition fluxes of 0.47–1.50 kg Pb km^{-2} a^{-1} were detected at the more pristine site ARB in the period 1996–2001.

These rates are considerably lower than the estimated Pb-deposition rates for 1992 evaluated from peat cores of eight sites in the mountainous border region of the Czech Republic (7 to 32 kg Pb km^{-2} a^{-1}, Vile et al., 2000). However the mountainous parts of the Czech Republic may receive as much as double the precipitation amount than the areas in central Czech Republic. An interesting feature of Table I is the general increase of Pb fluxes at all sites in 2000 (accompanied with the increase of fluxes of other elements, such as As, Cd, Zn at locations TR and ARB), in spite of extremely low precipitation.

3.2. Pb EMISSION SOURCES

Lead deposition is generated mostly from well-mixed higher layers of the atmosphere that contain fine aerosol particles transported from more distant areas. Table II presents results of the correlation analyses of monthly precipitation amount (PA, mm mo^{-1}) and monthly deposition bulk fluxes of selected elements (μg m^{-2}

TABLE II

Correlation coefficients (r) of monthly elemental fluxes at sites TR and ARB. Bold values are significant correlations on level $p > 0.001$

	n	PA/TR	Cu/TR	Zn/TR	Pb/TR	As/TR	Cd/TR	PA/ARB	Cu/ARB	Zn/ARB	Pb/ARB	As/ARB	Cd/ARB
PA/TR	62	1.00											
Cu/TR	31	0.40	1.00										
Zn/TR	44	0.38	**0.58**	1.00									
Pb/TR	59	0.44	**0.73**	0.52	1.00								
As/TR	42	0.06	**0.74**	0.44	**0.60**	1.00							
Cd/TR	39	0.46	**0.68**	**0.60**	**0.71**	0.26	1.00						
PA/ARB	56	**0.95**	**0.74**	0.34	0.47	0.25	0.47	1.00					
Cu/ARB	32	0.45	**0.59**	0.41	**0.73**	**0.69**	**0.66**	0.43	1.00				
Zn/ARB	43	0.36	0.29	**0.55**	0.43	0.43	0.27	0.33	**0.78**	1.00			
Pb/ARB	51	**0.65**	**0.79**	0.52	**0.88**	**0.57**	**0.70**	**0.63**	0.38	**0.57**	1.00		
AS/ARB	29	0.49	0.21	0.27	**0.81**	**0.83**	0.08	0.46	**0.75**	**0.58**	**0.70**	1.00	
Cd/ARB	34	**0.57**	**0.87**	0.52	**0.86**	**0.67**	**0.91**	**0.57**	−0.38	0.40	**0.88**	**0.66**	1.00

mo^{-1}). The fluxes were not calculated when the Pb concentration was below the analytical detection limit. The deposition flux of the Pb strongly correlates ($p < 0.001$) with PA at location ARB. This indicates that most of the Pb was washed out from the atmosphere by rainwater. However at the location TR, the Pb deposition fluxes did not correlate significantly with PA, probably due to vehicular Pb aerosols deposition. The samples of such easily soluble aerosols were collected at location TR. A significant correlation was also found for Pb fluxes with the fluxes of trace elements such as As, Cd, Cu, and Zn (Table II). Elevated concentrations of all these elements, together with Pb, are present in the fly ash of the Czech lignite coal. The correlation analysis suggests that a significant portion of the lead deposited at TR and ARB originates in coal burning emissions.

The correlation of fluxes at TR and ARB indicate that these deposition fluxes vary dependently even under the impact of a local road in case of TR. However, the deposition fluxes of Pb at the site ARB were lower during the whole monitored period by 47–61%, until year 2001 when the difference was only 4%. An explanation for this difference in magnitudes between the Pb fluxes at these two sites is due to the proximity of TR to the local road (Figure 1). The small difference (4%) between the Pb fluxes in 2001 is likely due to the ban of leaded gasoline sales in the Czech Republic.

3.3. Pb IN THE THROUGHFALL

Tree canopy type affects throughfall fluxes. The beech throughfall fluxes were lower during the entire monitored period than the fluxes below the spruce trees. The difference is attributed to different properties of the canopy surfaces and to the absence of the beech canopy in winter. The throughfall fluxes of Pb declined through time under both beech and spruce. Beech throughfall Pb flux declined from 0.50 to 0.26 kg km^{-2} a^{-1} and spruce from 1.23 to 0.61 kg km^{-2} a^{-1} between 1997 and 2001 (Table I). The throughfall fluxes of Pb were lower than the bulk fluxes at the site TR, except for the year 2001. On the other hand, the spruce throughfall was always higher than the bulk fluxes at the site ARB. The decline of the Pb throughfall deposition was comparable to the decline at the pristine location ARB (Table I).

3.4. Pb IN SOIL AND SOILWATER

Soil in upland settings in the catchment was classified as Dystric Cambisol and soil in riparian zones was classified as Gleyic Cambisol (Deckers *et al.*, 1998). The pedological classification of individual layers in both studied profiles is shown in Figure 2.

The riparian profile LP33 contained significantly higher concentrations of total Pb than the hillslope profile LP35 (Figure 2). The concentrations of acid-leachable Pb were comparable at both profiles. The NH_4NO_3-leachable concentrations were higher in the entire thickness of the LP33 profile and it correlated with content of

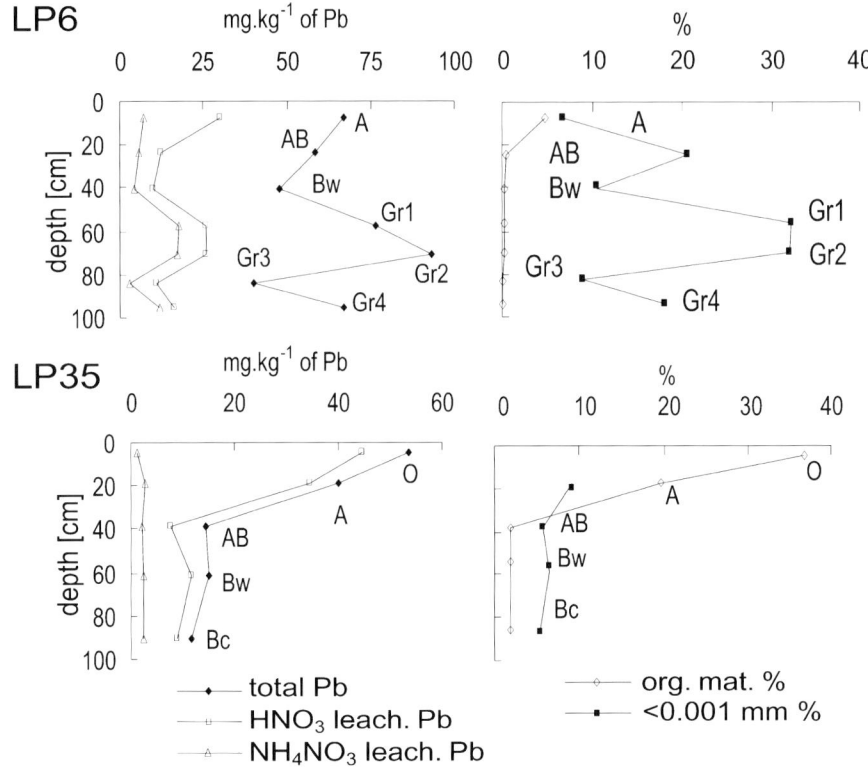

Figure 2. Data from two soil profiles at LP catchment: (left) – total and leachable Pb concentrations; (right) – content of organic material and finest particles (<0.001 mm).

fine particles (<0.001 mm). The NH_4NO_3-leachable concentration was low in the O-horizon (1.3 mg kg^{-1}) and virtually constant (about 2.5 mg kg^{-1}) through the rest of profile LP35.

The increased proportion of total Pb in the riparian soil (Figure 2) was perhaps associated with higher content of the silicate fraction. The increased acid-leachable concentrations in the two uppermost horizons (O, A) of the hillslope profile LP35 were attributed to the increased concentration of organic matter (Figure 2). The contrasting higher concentrations of NH_4NO_3-leachable Pb (interpreted as exchangeable + surface bound) in the riparian profile may be caused by consumption of organic substances complexing Pb in soil solution, changes of the soil environment to more reducing conditions, or increased content of finest fraction particles (Figure 2).

The soil water from the top 5 cm of profile LP33 (∼middle of A horizon) exhibited an average Pb concentration of 8.5 μg L^{-1} in the period 2001–2002. The average Pb concentration in the soilwater decreased to 5.0 μg L^{-1} at 15 cm depth (below the A horizon). Deeper soilwater from the depth of 45 cm (below the first 15 cm of G horizon) exhibited even lower average Pb concentration of 2.1

μg L^{-1}. The DOC concentration also decreased with depth suggesting that Pb is transported as an organic complex species. The average concentrations of Pb in soilwaters at profile LP35 were 6.1 μg L^{-1} at 5 cm depth below the O horizon and 3.0 μg L^{-1} at 15 cm depth in the middle of A horizon. The soilwaters from the top LP35 layers were rich in DOC.

The soil water sampled from below the organic rich horizons contained concentrations of Pb between 1.1 and 14.3 μg L^{-1}. The lower Pb concentrations between 1.1 and 3.4 μg L^{-1} in soilwater from the deeper horizon are due to loss of organic substances from the solution.

It is inferred from the deposition fluxes that the topsoil layers contain the anthropogenic lead. This lead is bound to organic matter and may be mobilized by the acid leaching together with the exchangeable and surface-bound fraction of Pb. In the riparian zone, the lower soil layers contained increased concentrations of surface bound + exchangeable Pb due to increased content of the finest particles (layers G1 and G2). The origin and the movement of Pb in soils of the catchment is possible to assess in detail by means of similar methods as were used by Emmanuel and Erel (2002).

3.5. Pb IN SURFACE WATER

The Pb output in streamwater from the catchment has been monitored since 1994. The concentrations of DOC in the LP surface water are about 5 mg L^{-1}. This low concentration of DOC is consistent with the low concentrations of exported Pb. Most of the monthly Pb measurements in the original unfiltered samples of surface water during the seven years of monitoring were below the detection limit of the AAS (0.5 μg L^{-1}). The mean value obtained from preconcentrated samples was approximately 0.1 μg L^{-1}. The highest concentration found in the LP surface water was 1.8 μg L^{-1}.

An acidification experiment of the LP surface stream was performed during a baseflow period with discharge about 2.7 L s^{-1} in the summer of 2001 (Navrátil et al., 2003). The experiment was designed to evaluate the buffering capacity of the streambed. An amount of 22 moles of HCl were added to the stream during 4 hr, lowering the streamwater pH from 4.9 to 3.7. The Pb concentration increased from an initial 0.7 μg L^{-1} to 1.8 μg L^{-1} during the experiment. This supports the presence of mobile lead in the stream bottom sediment, which can be released by the increased H$^+$ input. The main reasons for the mobilization of Pb may be either the charge changes of the stream substrate surface and the consequent release of the adsorbed ions, and/or the concurrent dissolution of the Al and Fe precipitates.

The HNO$_3$-leachable Pb concentration in the sampled stream bottom sediment was 56.5 mg kg^{-1}. The sediment sample was treated identically to the soil samples. The concentration of Pb in stream bottom sediment was comparable to the concentrations found in the upper soil horizons. The composition of the stream sediment was determined by qualitative X-ray diffraction; more than 50% was organic mat-

TABLE III

Concentrations of Pb in beech and spruce tissue at Lesni Potok catchment

All data of Pb in mg kg^{-1} d.w.				
		Spruce		**Beech**
	Wood	2.498		0.423
	Bark	2.234		2.230
Needles/leaves				
Date of smapling		**Spruce**	Date of smapling	**Beech**
04/25/2000		<0.16	04/25/2000	<0.33
05/10/2000		<0.20	05/10/2000	<0.17
08/02/2000		<0.12	08/02/2000	<0.18
10/04/2000		<0.16	10/04/2000	0.32
11/01/2000		<0.18	11/01/2000	0.53
* 25.4.2000		<0.18		
** 25.4.2000		<0.18		
*** 25.4.2000		<0.13		
**** 25.4.2000		<0.28		

* Number of stars describes needle age class (e.g. * = 1 yr old needle).

ter, followed by quartz and illite. Another significant amount of Pb in the stream was detected on Fe precipitate, which may be found in the LP stream during the low discharge periods. The precipitate occurred as clouds of suspended matter in the stream. It contained 15% dry weight of Fe and Pb concentration of 60 mg kg^{-1}. These materials (stream sediment and Fe precipitate) found in the stream represent a significant source of Pb, which may be released during episodes with low pH values, as suggested by the stream acidification experiment. The flux of exported Pb depends primarily on the amount of water discharge from the catchment (Table I).

3.6. Pb IN SHALLOW GROUNDWATER

Shallow groundwater was collected from the well situated close to the LP6 Thompson weir during year 2000. Very low concentrations of Pb were found in the shallow groundwater of the LP with an average pH of 6.7. The mean concentration of Pb in samples originating approximately 2 m below the groundwater level was 0.037 μg L^{-1}. This suggests that Pb does not move with the groundwater.

3.7. Pb IN VEGETATION

Several differences were detected in Pb content of stem wood, bark and assimilatory organs of the European beech and Norway spruce of the LP catchment (Table III). The bark of both species contained similar Pb concentrations of 2.23

mg kg^{-1} (all values in this paper are dry weight). The stem wood of beech contains almost 6-fold lower concentration of Pb than the spruce wood (Table III). We were not able to detect Pb in the spruce needles by means of the chosen analytical method (Table III). On the other hand, beech leaves were enriched with Pb through time in growing season up to 0.53 mg kg^{-1}.

The enrichment of Pb in the stem wood of spruce may be the result of its shallow rooting in comparison with beech. The shallow root systems of spruce are in soil with increased Pb concentrations relative to the shallow soils of beech habitat. This may result in enhanced Pb uptake. The accumulation of Pb on the leaves of beech trees may be the cause of decreased beech throughfall fluxes. Due to the low Pb concentrations in tree tissues of the LP catchment we conclude that the pool of Pb stored in the vegetation is small.

4. Conclusions

The deposition of Pb has decreased in central Bohemia 30 km ESE from Prague during the past 10 yr at least by 78%. The diminishing differences between sites TR and ARB reveal that deposition of vehicular Pb (which affected site TR) has diminished after the ban of leaded gasoline sales in 2001. Burning of brown coal in Czech power plants probably still represents a significant part of anthropogenic Pb deposition. The deposited Pb remains in the uppermost soil layers, especially in horizons rich in organic matter. Soilwater from surface horizons contained up to 14.3 μg L^{-1} of Pb. The concentrations of dissolved forms of Pb in surface water are low; the mean value is less than 0.1 μg L^{-1}. Average Pb concentration in shallow groundwater is even lower, at approximately 0.04 μg L^{-1}. Increased Pb concentrations were found in the stream bottom sediment and in the ferrous precipitate collected during low discharge. These materials become sources of mobilized Pb upon acidification of streamwater, as suggested by an experimental acidification of the surface stream. The root uptake of Pb by forest trees is low. Relative to beech, higher uptake occurs in the shallow-rooting spruce, this is reflected in increased concentrations in the stem wood and assimilatory organs. Some Pb is trapped by the beech canopy and may be the cause of the low Pb concentrations in beech throughfall.

Acknowledgments

Funding for the study was provided by the grants No. CEZ Z 3-013-912 of the Academy of Sciences of the Czech Republic (AS CR), by the project No. A3013603 of the Grant Agency of the AS CR and by project No. B3013203 of the Grant Agency of the AS CR. We thank two anonymous reviewers for helpful comments. For more information about the Lesni potok catchment please visit http://www.gli.cas.cz/lesnipotok/.

References

Boutron, C. F. and Patterson, C. C.: 1986, 'Lead concentration changes in Antarctic ice during the Wisconsin/Holocene transition', *Nature* **323**, 222–225.

Deckers, J. A., Spaargaren, O. C., Nachtergaele, F. O., Oldeman, L. R. and Brinkman, R.: 1998, 'World reference base for soil resources', World Soil Resources Reports, **84**, Food and Agriculture Organization of the United Nations, Rome.

Emmanuel, S. and Erel. Y.: 2002, 'Implications from concentrations and isotopic data for Pb partitioning processes in soils', *Geochim. Cosmochim. Acta*, **66**, 2517–2527.

Fiala, J. and Ostatnicka, J. (eds): 1997, *Air Pollution in the Czech Republic*, Czech Hydrometeorological Institute, Prague.

Friedland, A. J. and Johnson, A. H.: 1985, 'Lead distribution and fluxes in a high-elevation forest in northern Vermont', *J. Environ. Qual.* **14**, 332–336.

Hamelin, B., Grousset, F. E., Biscaye, P. E., Zindler, A. and Prospero, J. M.: 1989, 'Lead isotopes in trade wind aerosols at Barbados: The influence of European emissions over the North Atlantic', *J. Geophys. Res.* **94**, 243–250.

Heinrichs, H. and Mayer, R.: 1977, 'Distribution and cycling of major and trace elements in two central European forest ecosystems', *J. Environ. Qual.* **6**, 402–407.

Johnson, C. E., Siccama, T. G., Driscoll, C. T., Likens, G. E. and Moeller, R. E.: 1995, 'Changes in lead biogeochemistry in response to decreasing atmospheric inputs', *Ecol. Appl.* **5**, 813–822.

Navrátil, T., Vach, M., Norton, S. A., Skřivan, P., Hruška, J. and Maggini, L.: 2003, 'Chemical response of a small stream in a forested catchment (central Czech Republic) to a short-term in-stream acidification', *Hydrology and Earth System Sciences* **7**(3), 411–423.

Novák, M., Emmanuel, S., Vile, M. A., Erel, Y., Verón, A., Pačes, T., Wieder, R. K., Vaněček, M., Štěpánová, M., Břízová, E. and Hovorka, J.: 2003, 'Origin of lead in eight central European peat bogs determined from isotope ratios, strengths, and operation times of regional pollution sources', *Environ. Sci. Technol.* **37**, 437–445.

Nriagu, J. O. and Pacyna, J. M.: 1988, 'Quantitative assessment of worldwide contamination of air, water and soils by trace metals', *Nature* **333**, 134–149.

Skřivan, P. and Vach, M.: 1993, 'Decreasing emissions of vehicular lead in central Bohemia?', *Acta Univ. Carol. (Geologica)* **37**, 45–55.

Skřivan, P., Navrátil, T. and Burian, M.: 2000a, '10 yr of monitoring the atmospheric inputs at the Cernokostelecko region, central Bohemia', *Sci. Agric. Bohem.* **31**, 139–154.

Skřivan, P., Minařík, L., Burian, M., Martínek, J., Žigová, A., Dobešová, I., Kvídová, O., Navrátil, T. and Fottová, D.: 2000b, 'Biogeochemistry of beryllium in an experimental forested landscape of the Lesní potok catchment in Central Bohemia, Czech Republic', *GeoLines* **12**, 41–62.

Skřivan, P., Navrátil, T., Vach, M., Sequens, J., Burian, M. and Kvídová, O.: 2002, 'Biogeochemical cycling of metals in the environment: Factors controlling their content in the tissues of selcted forest tree species', *Sci. Agric. Bohem.* **33**, 71–78.

Suchara, I. and Sucharová, J.: 2000, 'Distribution of long-term accumulated atmospheric deposition loads of metal and sulphur compounds in the Czech Republic determined through forest floor humus analyses', Acta Pruhoniciana, VUSTKOZP, Pruhonice.

Vile, M. A., Wieder, R. K. and Novák, M.: 2000, '200 yr of Pb deposition throughout the Czech Republic: Patterns and sources', *Environ. Sci. Technol.* **34**, 12–21.

TRACE METALS IN DIFFERENT CROP/CULTIVATION SYSTEMS IN GREECE

E. VAVOULIDOU*, E. J. AVRAMIDES, P. PAPADOPOULOS and A. DIMIRKOU

NAGREF, Soil Science Institute of Athens, S. Venizelou 1, 141 23 Lykovrissi, Greece
(* author for correspondence, e-mail: ssia@otenet.gr; phone: +30-10-2816974;
fax: +30-10-2842 129)

(Received 20 August 2002; accepted 26 April 2003)

Abstract. Typical soils in Greece are neutral or alkaline and frequently are lime-rich, conditions that favour the accumulation of trace elements. The traditional use of metal-based fungicides in orchards and vineyards may have led to the accumulation of trace metals. Concentrations of Fe, Cu, Mn, Zn (aqua regia digestion) and some other soil parameters were measured in organically and conventionally cultivated soils (0–30 cm) from vineyards, olive groves and citrus groves of varying ages, and in uncultivated soils. Many vineyards and olive groves are situated in hilly or mountainous areas with sloping ground or terraces in contrast to citrus, which is cultivated in lower lying areas. Due to the difficulty of access, these crops often are cultivated extensively in both systems. Trace metal concentrations were found to lie within the ranges expected for the predominant soil types. Cu concentrations were relatively high (>100 mg kg^{-1}) in a few samples, but were not correlated with the age of the cultivation. A two-way ANOVA analysis showed larger differences in the mean concentrations of Cu, Mn and Zn between different crops ($p \leq 0.001$ for Cu, $p \leq 0.05$ for Zn, and $p \leq 0.1$ for Mn) than between different cultivation systems (no significant differences). The crop by cultivation interaction was not statistically significant for any metal ($p > 0.8$). Strong correlations ($p \leq 0.001$) were found between Fe, Mn and Zn and both clay concentration and CEC, although these relationships were not uniform throughout the different crop and cultivation systems. Concentrations of Cu were related to clay concentrations only for vineyards and to CEC only for citrus. Correlations were not found with organic matter or pH.

Keywords: citrus, copper fungicides, Cu, Fe, grapes, Mn, olives, organic agriculture, Zn

1. Introduction

The need for environmental protection and consumer demand for high quality agricultural products have led to greatly increased interest in the application of organic farming methods in Europe in recent years. European Union legislation on the organic production of agricultural products, including details of products authorised for soil fertilization and plant protection, was laid down in 1991 with Council Regulation EEC/2092/91. With the introduction of this regulation and subsidies supporting farmers converting to organic agriculture, many farmers in Greece have started to apply organic practices in the past few years (Figure 1), especially in the production of olives and grapes for wine making, the country's main crops. However, organic fertilisers are not widely used, partly because of the

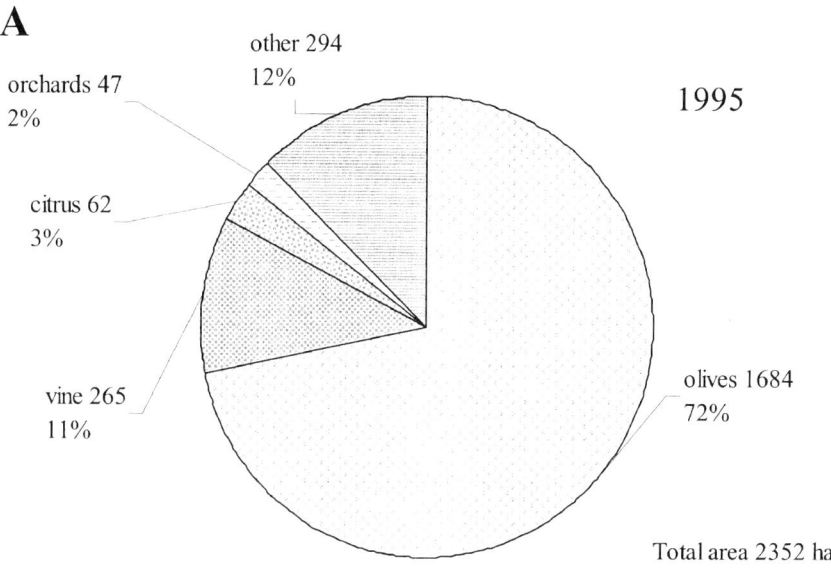

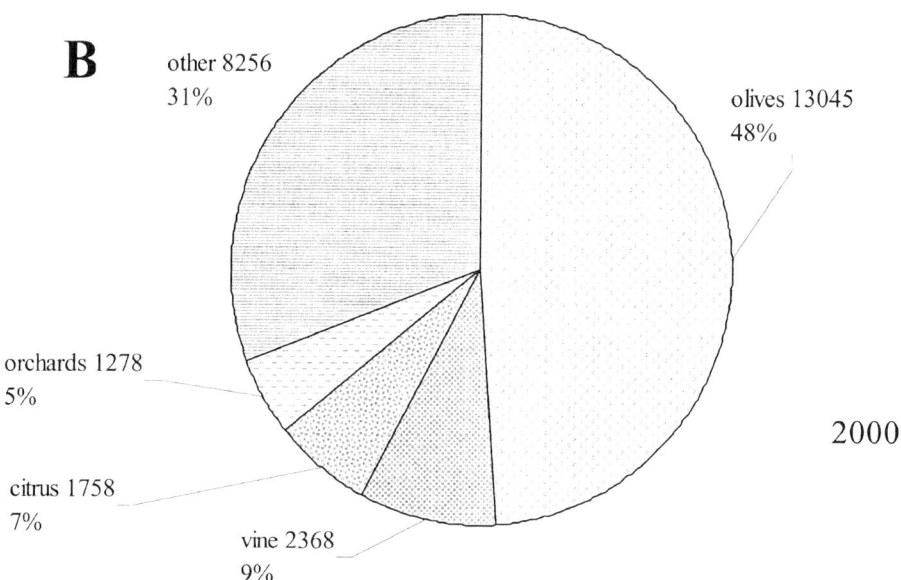

Figure 1. Areas (ha) in Greece cultivated with organic farming systems (Ministry of Agriculture statistics) by crop: (A) in 1995 (total 2352 ha) by crop 1995, and (B) in 2000 (total 26,706 ha).

difficult access in mountainous areas where olives and vines are often cultivated, and partly because of their cost and lack of availability in areas where there is no animal husbandry. Even in conventional agriculture, the cultivation of olives is frequently extensive, with emphasis on the control of insects and weeds. During the conversion period from conventional to organic, there is a need both for control and assessment of the benefits in terms of soil and crop quality. One aspect of soil quality is the potential for the accumulation or depletion of trace elements in soil and their availability to plants. The use of copper fungicides, such as Bordeaux mixture, is currently still allowed in organic agriculture, though this is a highly debated issue. Several studies in other countries (e.g., Kunisch and Hurle, 1986; Aoyama and Nagumo, 1997) have shown that substantial quantities of metals may accumulate in the topsoil layer as a result of the long-term use of such metal-based fungicides.

The trace element composition of soil, although initially inherited from the parent material, is subsequently influenced by the predominating pedogenic and anthropogenic processes. Clay minerals, hydrated metal oxides mainly of Fe, Mn and Al, and organic matter are considered to be the most important soil components contributing to and competing for the sorption of trace elements. In addition, carbonates can be an important trace metal sink in some soils (Kabata-Pendias, 2001). Trace element migration in soil profiles is affected by chemical, physical and biological soil properties, including soil redox potential, soil pH and buffering capacity, cation exchange capacity (CEC), amount and quality of organic matter, soil moisture and temperature, plant species, and microorganisms. Clay soils, both neutral and alkaline, such as those commonly found in Greece, provide good storage for trace elements and will supply them to plants at a slow rate, in contrast to acid soils from which several elements are easily leached (Kabata-Pendias, 2001). However, alkaline clay soils are vulnerable to the accumulation of heavy metals when inputs through atmospheric deposition and agricultural chemicals are high (Kabata-Pendias, 2001). Sources of trace metal pollution other than agriculture are not widespread in Greece, and significant accumulation due to industrial sources is unlikely. However, copper fungicides traditionally have been used in vineyards and orchards in Greece for many decades.

Little data are available for the trace element content of Greek soils. The objective of this work was to determine concentrations of Fe, Cu, Mn and Zn and other physical and chemical soil parameters in organic, conventional and uncultivated soils in vineyards, olive groves, and citrus groves of varying ages. This will provide a database against which future trends following the changeover to organic agriculture may be measured, will determine background levels in the main types of Greek agricultural soils, and will test for contamination occurring as the result of the long-term use of metal-based fungicides under Greek climatic conditions (characterised by dry summers and winter rainfall).

TABLE I

Means and standard deviations (sd) for soil properties at study sites in different Greece regions

Region	No. of samples			Statistical Data	Clay: <0.002 mm (%)	Silt: 0.002–0.05 mm	Sand: 0.05–2 mm	Type	pH water sat.	Total CaCO$_3$	Organic matter (mg g^{-1})	Total N (mg g^{-1})	P (Olsen) (mg kg^{-1})	Exchangeable K (mg g^{-1})	Exchangeable Mg (mg g^{-1})	CEC (meq 100 g^{-1})	Total Fe (mg g^{-1})	Total Cu (mg kg^{-1})	Total Mn (mg g^{-1})	Total Zn (mg kg^{-1})
	Vine	Olive	Citrus																	
Central Greece	15	–	3	Mean	43	20	37	C	7.6	20.5	20	1.47	35	0.37	0.37	24	0.34	39	0.84	67
				sd	13	5	13		0.2	16.9	10	0.55	33	0.27	0.32	12	0.13	14	0.39	20
Peloponesse	18	37	16	Mean	37	26	37	CL	7.0	13.3	27	1.76	29	0.18	0.18	18	0.24	50	0.88	59
				sd	10	8	13		0.9	18.9	14	0.81	33	0.11	0.09	8	0.09	21	0.40	25
Western Greece	–	3	8	Mean	50	27	23	C	7.2	12.2	28	1.81	35	0.17	0.19	25	0.36	55	0.92	71
				sd	23	14	12		0.4	10.7	11	0.70	41	0.09	0.06	12	0.19	26	0.16	28
Aegean Islands	15	–	–	Mean	15	12	73	SL	5.8	1.5	13	0.90	49	0.10	0.13	7	0.19	25	0.37	35
				sd	3	7	9		1.2	0.7	8	0.46	57	0.05	0.09	2	0.10	17	0.27	16
Crete	–	10	7	Mean	34	21	45	CL	7.2	8.9	35	2.31	40	0.25	0.28	10	0.29	36	0.55	67
				sd	7	7	9		0.5	16.3	24	1.11	47	0.15	0.18	2	0.11	13	0.40	41
Overall	48	50	34	Mean	36	23	40		6.9	11.4	26	1.71	32	0.55	0.25	17	0.27	44	0.79	60
				sd	14	9	17		0.7	16.5	15	0.87	39	0.44	0.25	9	0.11	21	0.46	28

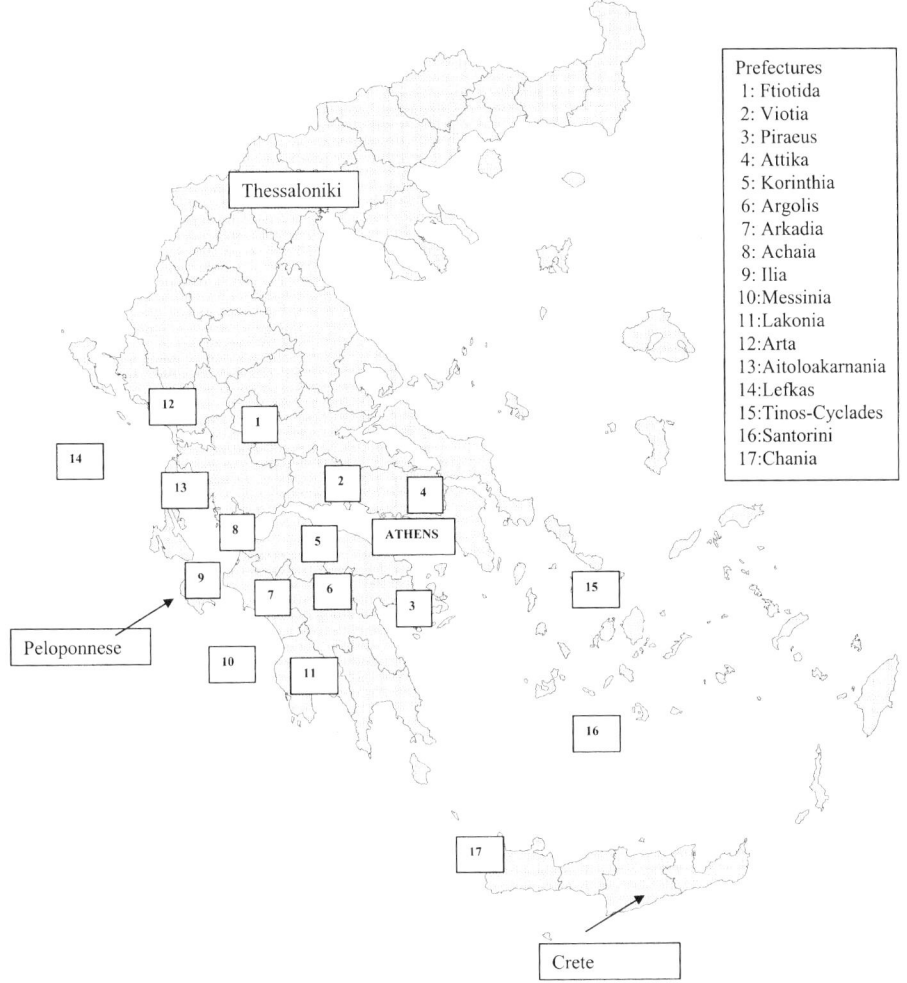

Figure 2. Regional sampling sites.

2. Materials and Methods

The soil samples were taken from vineyards, olive groves and citrus groves in different regions of Greece from October 1999 to April 2000 (Table I, Figure 2). Many of the cultivations were more than 20 yr old and some olive groves were more than 100 yr old. With the exception of citrus, most sites were in hilly or mountainous areas with sloping ground or terraces. Sampling was conducted using a riverside auger to a depth of 0–30 cm for all soils. For each site, 10 core samples were collected from different locations in the field and these were combined into a single bulk sample on site. For the organic practice sites, two different samples were collected: one from beneath the trees or vines and the other, as a control, from

TABLE II

F values form a two-way ANOVA analysis with crop-cultivation system interaction

	Degrees of freedom	Fe	Cu	Mn	Zn
Crop	2	1.400	11.350[c]	2.572[a]	3.247[a]
System	3	0.511	0.212	14.544	0.3
Crop-system	6	0.478	0.191	0.122	0.151

[a] Significant F value for $p \leq 0.1$.
[c] Significant F value for $p \leq 0.001$.

the spaces between the rows where it was unlikely that fertilizers or fungicides had been directly applied. For each organic practice sample (or group of samples where these were from neighbouring fields) samples were also taken from a nearby field in which the same crop was cultivated using conventional practice and from a field that had lain uncultivated for many years. Thus, four samples (organic, control, conventional and fallow) were typically taken from a given location.

On arrival at the laboratory, the samples were air-dried and crushed to pass a 2-mm sieve. Analyses were carried out for total Fe, Cu, Mn, and Zn (aqua regia digestion, method ISO/DIS 11466), and additionally for particle size (pipette method, Kilmer and Alexander, 1949), pH (in a water-saturated soil paste), total $CaCO_3$ equivalent (Bernard calcimeter), active $CaCO_3$ (Alexiadis, 1976), organic matter (Walkley-Black method, Nelson and Sommers, 1982), total nitrogen (Kjeldahl method, Bremmer, 1966), cation exchange capacity by soil saturation with sodium acetate at pH 8 and the uptake of Na^+ released with 1 M ammonium acetate, exchangeable K^+, Na^+, Ca^{2+}, and Mg^{2+} following extraction by 1 M ammonium acetate, and available P (Olsen et al., 1954).

3. Results and Discussion

Table I gives the means and standard deviations for some physical and chemical parameters of soils collected from different regions of the country. The soils were mostly clay soils with low organic matter content (≤ 30 mg g^{-1} for 75% of soils) and were largely neutral or alkaline. On the Greek mainland, 18 of the 27 locations sampled were rich in total $CaCO_3$ (8 with concentrations <2% and 10 with <20%). For soils from the islands (Lefkas, Tinos and Santorini) nearly all contained <2% total $CaCO_3$. Available P was adequate (>5 mg kg^{-1}) in all but 2 organic, 6 control and 6 fallow soils. However, 71 of the 132 samples contained low exchangeable K^+ (<0.2 mg g^{-1} for soils with CEC ≥ 10 meq 100 g^{-1} and <0.06 mg g^{-1} for soils with CEC < 10 meq 100 g^{-1}) and 59 contained low total N (<1.5 mg g^{-1}). Soils from the Aegean islands were particularly poor in organic matter and total N.

Concentrations of the trace metals Fe, Cu, Mn and Zn were in the range expected for the predominant types of soils (Kabata-Pendias, 2001). Cu concentrations were relatively high (>100 mg kg^{-1}) in a few samples, but no correlation was observed between the concentration and the age of the cultivation. Table II shows the results of a 2-way ANOVA analysis with crop and cultivation system as main effects and the crop by cultivation system interaction. This indicates that there are larger differences in mean concentration of Cu, Zn and Mn between different crops than between different cultivation systems. The crop by cultivation interaction was not statistically significant for any metal ($p > 0.8$). The differences between crops are likely, in part, to be due to regional soil differences. For example, citrus are grown mainly in the Peloponnese and Western Greece, where soils are generally richer in Cu (Table I).

Table III shows data for the trace metal concentrations and some other soil parameters for each crop/cultivation system group. Most soil parameters are similar between organic soils and their controls, which is indicative of the extensive agriculture commonly applied in the mountainous areas where vines and olives are often cultivated, and where many of the conversions to organic practices have occurred. Table IV gives statistical data for the correlation of trace metal concentrations with clay and CEC for these crop/cultivation system groups and also for all samples combined. No correlation was found between metal concentrations and pH, although it should be noted that fewer than 20% of the soils were acidic (pH $\leq$ 6.0), or between metal concentrations and organic matter; data for these have not been included in Table IV. As shown in Table IV, strong correlations ($p \leq 0.001$) were found between Fe, Mn and Zn and clay concentration and CEC, although these relationships were not uniform throughout the different crops and cultivation systems. For example, the correlation of Mn with CEC was much stronger in the citrus group, while this is the only group for which Fe was not correlated with clay. Clay minerals have a high ability to fix Fe^{3+} ions; the ability of the soil to exchange cations is one of the most important soil properties governing the cycling of trace elements in the soil (Kabata-Pendias, 2001). The lack of any significant correlation between Cu and the soil components generally considered to make the most important contributions to the sorption of trace metals is noticeable. Concentrations of Cu were related to clay only for vineyards, and to CEC only for citrus. For vineyards only, Cu also was correlated with the concentrations of the other three metals ($r = 0.45$, $p \leq 0.01$ with Fe, $r = 0.51$, $p \leq 0.001$ with Zn and $r = 0.45$, $p \leq 0.01$ with Mn) and with organic matter ($r = 0.32$, $p < 0.05$).

4. Conclusions

Concentrations of the trace elements Fe, Cu, Mn and Zn are in the range expected for the predominant soil types included in this study, indicating a long-term balance between input, mainly from agricultural chemicals, and output by slow leaching

TABLE III

Means ± sd for physical and chemical parameters for samples grouped according to the crop and cultivation system

Crop	Cultivation system	No. of samples	Clay <0.002 mm (%)	Silt 0.002–0.05 mm (%)	Sand 0.05–2 mm (%)	pH water sat. paste	Organic matter (mg g^{-1})	Total N (mg g^{-1})	Exchangeable Ca (mg g^{-1})	Exchangeable Mg (mg g^{-1})	CEC (meq 100g^{-1})	Total Fe (mg g^{-1})	Total Cu (mg kg^{-1})	Total Mn (mg g^{-1})	Total Zn (mg kg^{-1})
Vine	Conventional	10	30 ± 16	19 ± 8	51 ± 23	6.1 ± 1.6	13 ± 7	1.1 ± 0.4	4.3 ± 4.5	0.14 ± 0.07	15 ± 7	21 ± 9	31 ± 18	0.55 ± 0.42	46 ± 22
	Organic	13	37 ± 17	20 ± 8	44 ± 22	6.8 ± 1.0	17 ± 8	1.2 ± 0.5	5.7 ± 4.2	0.25 ± 0.15	21 ± 12	24 ± 11	38 ± 18	0.70 ± 0.41	51 ± 22
	Control	12	39 ± 17	21 ± 8	41 ± 23	6.8 ± 1.1	14 ± 6	1.0 ± 0.5	5.7 ± 4.0	0.20 ± 0.13	21 ± 12	27 ± 13	37 ± 16	0.80 ± 0.42	54 ± 24
	Fallow	13	34 ± 17	18 ± 9	48 ± 23	7.1 ± 0.6	26 ± 14	1.6 ± 0.6	5.2 ± 4.0	0.24 ± 0.14	22 ± 12	26 ± 11	38 ± 14	0.86 ± 0.61	54 ± 28
Olive	Conventional	11	37 ± 17	22 ± 9	40 ± 16	6.4 ± 1.3	31 ± 13	2.2 ± 0.9	4.5 ± 3.6	0.15 ± 0.08	19 ± 11	31 ± 13	40 ± 11	0.64 ± 0.30	65 ± 25
	Organic	13	38 ± 15	23 ± 10	39 ± 11	7.0 ± 0.8	31 ± 9	2.0 ± 0.6	5.5 ± 3.2	0.19 ± 0.09	19 ± 10	29 ± 14	43 ± 22	0.70 ± 0.29	68 ± 32
	Control	12	36 ± 9	23 ± 11	41 ± 11	7.1 ± 0.7	24 ± 10	1.5 ± 0.6	5.4 ± 3.2	0.15 ± 0.06	16 ± 7	26 ± 12	46 ± 22	0.76 ± 0.23	63 ± 39
	Fallow	14	40 ± 16	23 ± 11	37 ± 15	7.2 ± 0.5	42 ± 30	2.3 ± 1.4	6.0 ± 3.4	0.22 ± 0.19	19 ± 10	29 ± 16	42 ± 26	0.81 ± 0.48	63 ± 39
Citrus	Conventional	8	36 ± 10	28 ± 7	36 ± 7	7.3 ± 0.4	26 ± 8	1.8 ± 0.4	7.5 ± 1.8	0.24 ± 0.08	16 ± 4	23 ± 8	59 ± 24	0.84 ± 0.23	56 ± 25
	Organic	10	33 ± 8	31 ± 7	36 ± 12	7.5 ± 0.2	25 ± 8	1.8 ± 0.5	8.0 ± 2.1	0.23 ± 0.07	16 ± 6	26 ± 10	59 ± 15	0.91 ± 0.42	65 ± 21
	Control	9	35 ± 10	27 ± 7	37 ± 15	7.4 ± 0.2	26 ± 7	1.7 ± 0.5	8.1 ± 2.4	0.19 ± 0.06	16 ± 7	25 ± 7	57 ± 24	0.89 ± 0.42	63 ± 20
	Fallow	7	35 ± 14	27 ± 7	38 ± 15	7.5 ± 0.2	32 ± 14	2.5 ± 1.4	8.3 ± 1.7	0.43 ± 0.53	20 ± 13	30 ± 12	55 ± 25	1.02 ± 0.51	68 ± 23

TABLE IV
Correlation coefficients (r) for trace metal concentrations with some soil parameters

Crop/system	Parameter	n	Cu	Mn	Zn	Fe
Overall	Clay	130	0.20[a]	0.45[c]	0.50[c]	0.63[c]
	CEC	81	0.10	0.35[c]	0.70[c]	0.81[a]
Vine	Clay	48	0.53[c]	0.52[c]	0.71[c]	0.61[c]
	CEC	25	0.27	0.18	0.88[c]	0.81[c]
Olive	Clay	50	0.02	0.44[c]	0.35[b]	0.77[c]
	CEC	34	0.12	0.40[a]	0.58[c]	0.87[c]
Citrus	Clay	32	0.26	0.45[b]	0.57[c]	0.27
	CEC	22	0.58[b]	0.66[c]	0.86[c]	0.77[c]
Organic	Clay	34	0.20	0.45[b]	0.53[c]	0.65[c]
	CEC	22	0.03	0.47[a]	0.79[c]	0.83[c]
Control	Clay	33	0.31	0.63[c]	0.36[a]	0.49[b]
	CEC	21	0.22	0.31	0.54[a]	0.78[c]
Conventional	Clay	29	0.21	0.49[b]	0.57[b]	0.69[c]
	CEC	17	0.05	0.27	0.24	0.87[c]
Fallow	Clay	34	0.13	0.35[a]	0.55[c]	0.68[c]
	CEC	21	0.31	0.38	0.90[c]	0.79[c]

[a] Significant for $p \leq 0.05$.
[b] Significant for $p \leq 0.01$.
[c] Significant for $p \leq 0.001$.

from the generally neutral or alkaline and frequently lime-rich soils. No differences were observed in the total concentrations of Fe, Cu, Mn, and Zn in organic practice soils compared to conventional and fallow soils, but significant differences were found for Cu, Zn, and Mn between soils cultivated with different crops. Taking all soil samples as a single group, concentrations of Fe, Zn, and Mn, but not Cu, were strongly correlated with clay concentration and CEC. Correlations generally were not observed with organic matter or pH. In soils from vineyards only, however, Cu was correlated with both clay and organic matter concentrations. The relatively high values found for Cu in a few samples are likely to be due to the long-term use of copper-based fungicides, which are the only products allowed for the control of fungal diseases in organic cultivation. Given the current debate over the desirability of their continued use in organic agriculture and the differences observed in Cu concentrations between crops, further studies are being carried out on the Cu content of soils for a wider variety of land uses and growing conditions.

Acknowledgements

This work was supported by NAGREF and the Ministry of Agriculture within the framework of the EU programme 'Examination and comparative evaluation of the application of the programme for Organic Farming'.

References

Alexiadis, K.: 1976, *Physical and Chemical Analysis of Soil*, Aristotelian University, Thessalonica, pp. 165–167.
Aoyama, M. and Nagumo, T.: 1997, 'Effects of heavy metal accumulation in apple orchard soils on microbial biomass and microbial activities', *Soil Sci. Plant Nutr.* **43**, 601–612.
Bremmer, J. M. and Keeney D. R.: 1966, 'Determination and isotope-ratio analysis of different forms of nitrogen in soils: 3. Exchangeable ammonium, nitrate, and nitrite by extraction-distillation methods', *Soil Sci. Soc. Am. Proc.* **30**, 577–582.
Council Regulation (EEC) No 2092/91, Offic. J. E. C. No. L198, 1991.
Method ISO/DIS 11466: 1994, 'ISO Standards Compendium. Environment - Soil Quality', 1st edition, International Organization for Standardization, Geneva, Switzerland, pp. 309–314.
Kabata-Pendias, A.: 2001, *Trace Elements in Soils and Plants*, 3rd edition, CRC Press, Boca Raton, Florida, U.S.A.
Kilmer, V. J. and Alexander L. T.: 1949, 'Methods of making mechanical analyses of soils', *Soil Sci.* **68**, 15–24.
Kunisch, M. and Hurle, K.: 1986, 'Kupfergehalte in Weinbergsböden: Konsequenzen für das Pfanzenwachstum', Verhandlungen der Gesellschaft für Ökologie, Band XIV, University of Hohenheim, Stuttgart, 98 pp.
Nelson, D. W. and Sommers, L. E.: 1982, 'Total carbon, organic carbon, and organic matter', in A. L. Page *et al.* (eds), *Method of Soil Analysis, Part 2*, 2nd edition, Agronomy Monographs 9, American Society for Agronomy and Soil Science Society of America, Madison, Wisconsin, U.S.A., pp. 539–579.
Olsen, S. R., Cole, C. V., Watanabe, F. S. and Dean, L. A.: 1954, *Estimation of Available Phosphorus in Soils by Extraction with Sodium Bicarbonate*, USDA Circular 939, 19 pp.

ASSESSING CHANGES IN PHOSPHORUS CONCENTRATIONS IN RELATION TO IN-STREAM PLANT ECOLOGY IN LOWLAND PERMEABLE CATCHMENTS: BRINGING ECOSYSTEM FUNCTIONING INTO WATER QUALITY MONITORING

HELEN P. JARVIE*, COLIN NEAL and RICHARD J. WILLIAMS

Centre for Ecology and Hydrology, Wallingford, Oxfordshire, OX10 8BB, U.K.
(* author for correspondence, e-mail: hpj@ceh.ac.uk; phone: +44 (0)1491 838800; fax: +44 (0)1491 692424)

(Received 20 August 2002; accepted 18 April 2003)

Abstract. Changes in concentrations of soluble reactive phosphorus (SRP), excess partial pressure of carbon dioxide ($EpCO_2$), and chlorophyll-*a* were examined for two rivers in the in the upper Thames catchment: the main river Thames at Wallingford and a chalk stream tributary, the River Kennet. Sampling began in the spring of 1997 and has covered extremes in river flow conditions. During the sampling period there was a dramatic reduction in phosphorus (P) inputs from the introduction of effluent P-treatment at sewage treatment works, as a result of the EU Urban Wastewater Treatment Directive. Despite major reductions in baseflow SRP concentrations in the River Kennet, from around 700 μg-P L^{-1} to around 100 μg-P L^{-1}, observations of aquatic plant communities indicate overall degradation in ecological quality since effluent P-treatment was introduced. The degradation was associated with a spring and summer decline in growth of *Ranunculus*, a macrophyte of high conservation value in chalk streams, particularly from 2000 onwards, linked to shading by epiphytic algae. Although the $EpCO_2$ records indicate a reduction in primary productivity since effluent P-treatment, the River Kennet may have become more sensitive to epiphyte blooms. Episodes of epiphyte proliferation appear to be linked temporally to small increases in SRP concentrations (typically above a 100 μg-P L^{-1} threshold) under summer baseflow conditions. The in-stream system is highly complex and individual processes and causality are difficult to resolve, particularly given changes in river flows linked to background climatic variability and limited availability of biological data. This study demonstrates the need for integrated long-term biological and chemical monitoring of river systems subject to major perturbations to assess timescales required to produce new dynamic equilibria in ecosystem response.

Keywords: chalk stream, epiphyte, Kennet, macrophyte, nutrients, phosphorus, *Ranunculus*, sewage effluent, Thames

1. Introduction

The Thames in south-east England is a major UK river basin, providing an important nutrient flux to the North Sea (Jarvie *et al.*, 1997). Permeable aquifer systems (Cretaceous Chalk and Jurassic Limestones) form the dominant geologies in the study catchments as well as large areas of lowland Britain, and are subject to intense pressures for groundwater extraction and effluent disposal. Nutrient inputs from intensive agricultural production and increasing population pressures in the rural catchments of southern England are compounded by trends toward greater climate variability, with particular risk of low-flow extremes as a result of low rainfall inputs and relatively high evapotranspiration rates of around 60%. The trend toward lower summer baseflow conditions reduces the capacity for dilution of sewage effluents, resulting in nutrient enrichment and increasing risk of eutrophication. There are major concerns about the ecological quality of lowland UK rivers, linked to increases in nutrient status over the last 50 yr (Mainstone and Parr, 2002). Concerns range from deterioration of the characteristic Chalk-river plant and animal communities, with reductions in species diversity and proliferation of nutrient tolerant species in the smaller tributaries, to increased rates of phytoplankton productivity in the larger, deeper, slower-flowing river reaches. Excessive plant growth and microbial breakdown of plant biomass may result in depletion of dissolved oxygen, which is detrimental to invertebrate and fish populations (Fisher *et al.*, 1995). The upper River Kennet, a Chalk groundwater-fed tributary of the Thames, is designated a Site of Special Scientific Interest (SSSI) and exemplifies many issues relating to the sustainable management of lowland permeable rivers of high conservation value. *Ranunculus penicillatus* var. calcareous (R. W. Butcher) C. D. K. Cook (water crowfoot) is the dominant macrophyte within the Kennet and other lowland Chalk streams, and is of high ecological importance owing to the habitat it provides for fish, particularly brown trout. Abundant growth of *Ranunculus* is regarded as an indicator of good ecological quality within Chalk streams (Mainstone and Parr, 2002) and *Ranunculus* is designated a priority habitat under the European Union Habitats Directive (192/43/EEC). However, there have been increasing concerns about poor growth of *Ranunculus* and associated proliferation of nutrient-tolerant epiphytic diatoms over the last ten years (Flynn *et al.*, 2002). Epiphytic blooms have been of particular concern during the protracted droughts that occurred in 1991–2 and 1996–7, when water extraction pressures and reduced capacity for dilution of sewage effluent resulted in elevated nutrient concentrations, particularly of phosphorus (P). A core target for managing eutrophication in lowland rivers has been reductions in phosphorus inputs from sewage effluent discharges. The implementation of the European Union (EU) Urban Wastewater Treatment Directive (Council of the European Communities, 1991) means that all sewage treatment works (STWs) serving populations of >10,000 population equivalents (p.e.) and discharging into sensitive waters are now designated as 'Qualifying Discharges' for tertiary effluent treatment to remove phosphorus.

This paper reports results from water quality monitoring studies that began in the spring of 1997 at two sites within the Thames basin: the upper River Kennet at Mildenhall and the main River Thames at Wallingford (Figure 1). The paper brings together new information on aquatic ecology (indicators of primary productivity and observations of aquatic plant communities) to examine changes in aquatic ecology in relation to changes in dissolved inorganic phosphorus (soluble reactive phosphorus, SRP) and river flows, since both factors can have important controls

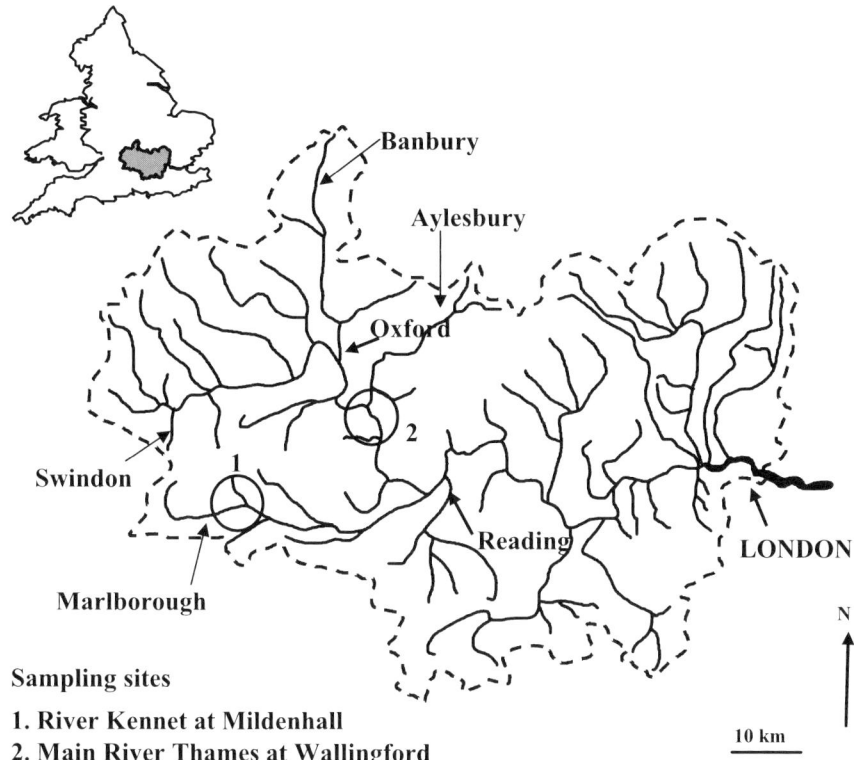

Figure 1. Map of the Thames catchment, showing locations of the sampling sites and major towns.

on freshwater plant growth (Hecky and Kilham, 1988; Sand-Jensen *et al.*, 1989). Earlier studies have shown the more general patterns in water quality for the two sites (Neal *et al.*, 1998, 2000a,b, 2002; Jarvie *et al.*, 2002b). These studies have demonstrated that point-source P reductions on the River Kennet have resulted in SRP concentrations approaching critical target levels for running freshwaters subject to eutrophication (100 μg-P L^{-1} annual average concentration; Department of the Environment, 1993), and the Environment Agency's target for Chalk streams (60 μg-P L^{-1} maximum concentration; Environment Agency, 2000).

2. Study Area and Data Collection

The River Thames at Wallingford drains a largely rural agricultural catchment (3500 km^2), with a few major settlements (Oxford, with a population of 176,000 population equivalents (p.e.); Swindon, 192,000 p.e.; Aylesbury, 89,000 p.e.; and Banbury, 89,000 p.e.) (Figure 1). The geology of the Thames catchment is mixed sedimentary rocks of mainly Jurassic age, which provide important groundwater aquifer sources (Institute of Hydrology, 1998). The upper River Kennet at Mildenhall drains a rural Chalk catchment of 142 km^2. This site is located approximately 2 km downstream of the market town of Marlborough with a population of 12,000 p.e. The choice of these rivers facilitates comparison of a large slow-flowing deep river system subject to large inputs of sewage effluent from major centres of population, with a rural shallow headwater Chalk stream.

The sampling period reported here, March 1997 to July 2002, encompassed some major in-stream perturbations in both river systems. Point-source P reductions were undertaken at the Oxford, Abingdon, Swindon and Aylesbury sewage treatment works (STWs) in the upper Thames catchment in late 1998, and in the upper Kennet catchment at the Marlborough STW in August 1997. By examining the ratios of SRP to boron (B), a conservative tracer of sewage effluent, we calculated that these point-source P reductions resulted in significant declines in in-stream SRP loadings from sewage by about 66 % in the Kennet and 50% in the Thames (Jarvie et al., 2002 a,b). In addition, the last 4 yr has seen unprecedented extremes in climate variability within the southeast of Britain (Marsh and Dale, 2002), resulting in major contrasts in river flow conditions in the Thames and Kennet (Figure 2), from extreme low flows in the summer of 1997 to exceptionally high flows in the winter/spring of 2000/2001. These flow extremes are particularly dramatic on the groundwater-fed upper River Kennet (at Marlborough gauging station), where daily gauged flows in each month from March to November 1997 were the second lowest on record since 1972, reflecting very low groundwater recharge over the preceding winter. By contrast, the 2000/2001 winter recharge generated record flows in the Kennet: runoff at the Marlborough gauging station between May 2000 and April 2001 was 465 mm, whereas the previous maximum for this time span was 291 mm. The main river Thames (gauging station at Days Weir) also experienced very low flows during summer 1997, with June to September flows about 55% of average. The River Thames responds more rapidly to rainfall events than the River Kennet, and the high flows of winter 2000/2001 in the Thames were sustained over a shorter period. However, the October–April runoff in the Thames was the highest on record in a series from 1938. The introduction of effluent P-treatment on the River Kennet therefore also coincided with a major increase in river flows. Summer baseflows from 1998 to 2000 have been at least an order of magnitude higher than the summer of 1997. This is a result of plentiful winter groundwater recharge, especially during the winter of 2000/2001.

Measurements of SRP were undertaken on a weekly basis at both sites (full details of sampling and analytical methods are provided by Neal et al., 2000a). In terms of biological data, three data sources were brought together:

(i) New data are presented for chlorophyll-a and excess partial pressure of carbon dioxide ($EpCO_2$) on the Thames and Kennet. These were measured on a weekly basis to provide simple indicators of changes in-stream biological status (Jarvie and Neal, 1998; Williams et al.., 2000). Suspended chlorophyll-a concentrations were measured using the method of Marker (1994). $EpCO_2$

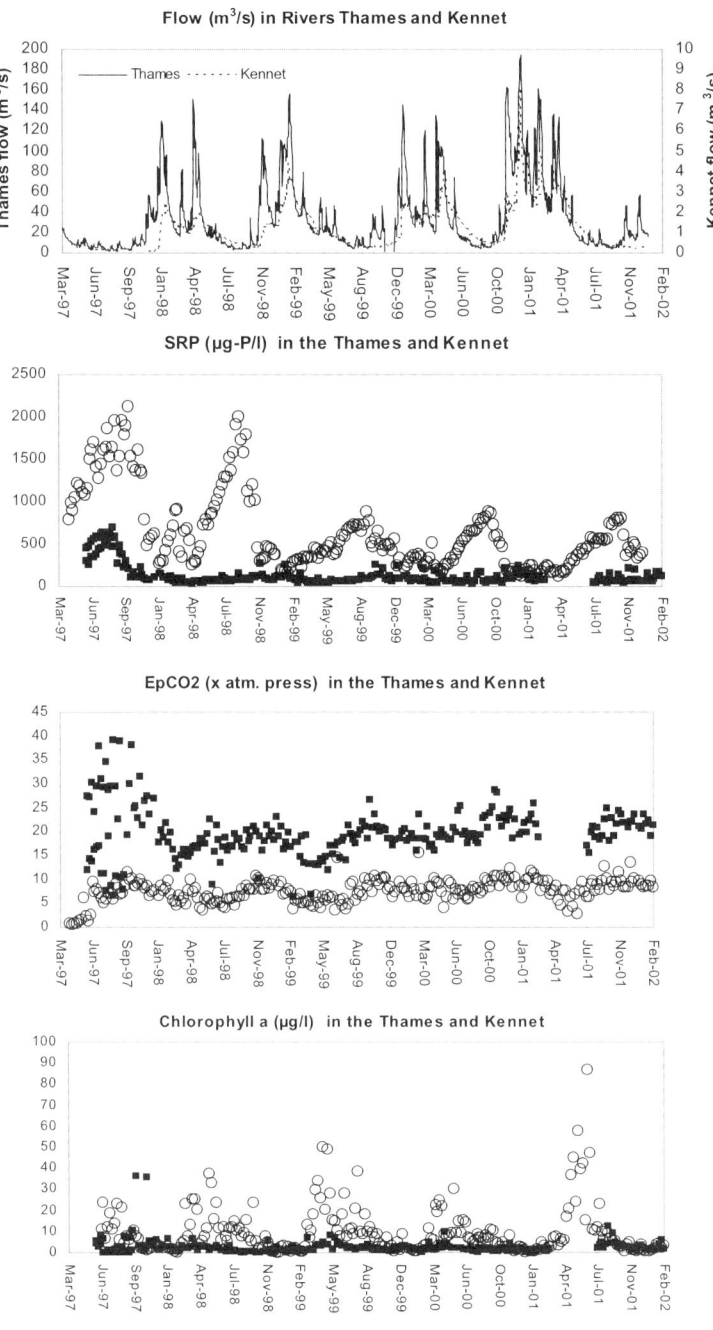

Figure 2. Timeseries showing flow, soluble reactive phosphorus concentration (SRP), excess partial pressure of carbon dioxide ($EpCO_2$) and chlorophyll-*a* on the River Kennet at Mildenhall and the River Thames at Wallingford.

is the ratio of the partial pressure of CO_2 in river water to the equilibrium partial pressure of CO_2 in the atmosphere and can be calculated thermodynamically from pH, temperature and alkalinity data (Neal et al., 1998). An $EpCO_2$ value of 1 corresponds with equilibrium with the atmosphere, whereas 10 and 0.1 correspond to 10 times and one-tenth atmospheric pressure respectively. $EpCO_2$ shows diurnal fluctuations linked to changes in the relationships between: (a) photosynthesis (uptake of CO_2 by aquatic plants during daylight hours), (b) respiration (release of CO_2 by aquatic plants at night and heterotrophic organisms, particularly those breaking down organic matter in the stream), and (c) exchange of CO_2 at the water-air interface. These diurnal patterns can be used to calculate rates of net primary productivity in river water (Williams et al., 2000).

(ii) Direct measurements of in-stream plants, including fortnightly measurements of macrophyte species cover, macrophyte biomass and epiphyte biomass, recorded between October 1998 and September 2000 at Mildenhall (Flynn et al., 2002), and in-stream macrophyte surveys carried out at two other sites downstream of Marlborough STW in 1996/7 and 1998/9 (Wright et al., 2002).

(iii) Observations of changes in plant ecology, recorded by anglers, river keepers and local residents on the upper Kennet, published in the Action for the River Kennet (ARK) Newsletters.

3. Results

3.1. Variations in Nutrient Water Quality

The River Thames showed a pronounced seasonality in SRP concentrations, with peak concentrations corresponding with lowest flows during the summer months (Figure 2). Peak SRP concentrations were considerably higher in the Thames than in the Kennet, with maximum SRP concentrations in 1997 and 1998 in the Thames exceeding 2000 μg-P L^{-1} (before point source P-reductions were introduced), compared with a maximum of 700 μg-P L^{-1} for the Kennet. Following effluent P-reductions on the Thames, maximum summer low flow SRP concentrations reached about 800 μg-P L^{-1}. SRP concentrations in the Kennet were dominated by a period of high concentrations (up to 700 μg-P L^{-1}) in the summer of 1997, a period prior to P-removal at Marlborough STW and a time when river flows, and thus dilution of the point source input, were lowest. Subsequently, SRP concentrations in the River Kennet remained close to or below 100 μg-P L^{-1}, with the exception of intermittent spikes in SRP concentrations. Many of these spikes were linked to high-flow events, during which diffuse sources of SRP were mobilised. No such diffuse sources of SRP could be discerned for the Thames, where the SRP concentrations remained strongly influenced by point-source inputs, even after P-removal at the major STWs.

3.2. VARIATIONS IN WATER QUALITY 'INDICATORS' OF IN-STREAM BIOLOGICAL ACTIVITY

Timeseries of weekly spot-sampled chlorophyll-a concentrations and $EpCO_2$ levels for the Thames and Kennet are shown in Figure 2. During the initial period of weekly sampling on the River Kennet (June to September 1997), sampling was undertaken twice on each sampling day, firstly at dawn, before the onset of daytime plant photosynthesis, and secondly at midday, when rates of photosynthesis and thus CO_2 uptake were likely to be maximal. During the remainder of the sampling period, samples were collected between mid-morning and midday. The effects of twice-daily sampling were clearly observed for the $EpCO_2$ record on the River Kennet, where there was a very high diurnal range in $EpCO_2$ during the summer of 1997, linked to the night-time release of CO_2 by microbial respiration and daytime net uptake of CO_2 by photosynthesis. The greatest difference between dawn and midday samples occurred in mid-August, with $EpCO_2$ levels ranging from 39 times atmospheric pressure at dawn to 6 times atmospheric pressure at midday, reflecting high rates of photosynthesis and respiration. After the summer of 1997, there were a few intermittent daytime $EpCO_2$ minima on the Kennet, which fell below 10 times atmospheric pressure. These minima were linked to high flows, and probably reflect near-surface runoff (which has a lower $EpCO_2$ level than groundwaters or deep soil water), rather than intense photosynthetic activity.

$EpCO_2$ levels under baseflow conditions after point-source P-removal on the Kennet did not show such pronounced daytime minima compared with the summer of 1997 and tended to fall between 15 and 25 times atmospheric pressure. This suggests that there was a reduction in primary productivity in the River Kennet after the summer of 1997 (which might indicate ecological improvement), but that since 1998, there were no major changes in net primary productivity.

$EpCO_2$ levels on the Thames were considerably lower than on the Kennet, with most values lying between 3.5 and 15 times atmospheric pressure because the River Kennet is groundwater-fed and thus has high CO_2 levels from direct input of groundwater from the Chalk aquifer. $EpCO_2$ concentrations decreased as the residence time in the river increased, due to degassing. Therefore, in a larger scale river system like the Thames, the river water has a longer period for degassing on transport from the groundwater source.

The Thames also showed lower variability in $EpCO_2$ values. This was in part due to sampling bias, as only one daytime sample was collected during the summer of 1997, $i.e.$ there were no equivalent dawn high-level values for the Thames, compared with the Kennet. However, during this period, $EpCO_2$ minima in the Thames fell to 0.4 times atmospheric pressure, indicating rapid rates of photosynthetic uptake of CO_2. At this time there was also a crash in silicon concentrations, indicating a diatom bloom (Neal et al., 1998). The lower scatter in the timeseries for the Thames reveals seasonality, which was not so clearly apparent for the Kennet data. On the Thames, there were summer minima in $EpCO_2$ levels. This pattern

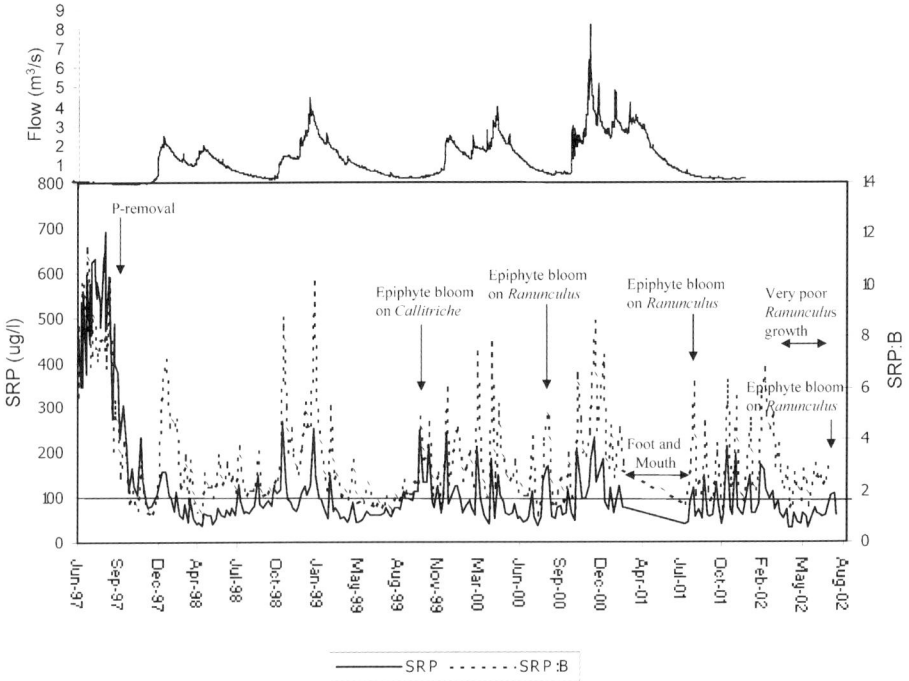

Figure 3. Timeseries showing soluble reactive phosphorus concentration (SRP), the ratio of SRP to boron (SRP:B) in the River Kennet at Mildenhall in relation to river flow (The timing of effluent P-treatment and biological events are also recorded.)

appeared to be strongly linked to maxima in chlorophyll-a concentrations (Figure 2). The Thames showed a stronger link between $EpCO_2$ and chlorophyll-a because primary production in the deep, slow-flowing river system was primarily linked to phytoplankton, whereas in the shallow River Kennet, uptake of CO_2 reflected a combination of photosynthetic activity from proportionally larger macrophyte and benthic populations, as well as phytoplankton.

Spring and summer chlorophyll-a concentrations were also considerably higher on the Thames, with weekly values typically peaking in April–June, with values greater than 30 μg L^{-1} and up to 87 μg L^{-1} in June 2001. For the Kennet, maximum chlorophyll-a concentrations were not restricted to the spring: in 1997, the highest recorded chlorophyll-a concentrations occurred in October and November (around 35 μg L^{-1}), reflecting mobilisation of in-stream algae under high flows following a period of epiphytic and benthic blooms. In 2001, peak chlorophyll-a concentrations occurred during baseflow conditions during late August/September. After 1997, peak chlorophyll-a concentrations on the Kennet did not exceed 13 μg L^{-1}.

3.3. CHANGES IN IN-STREAM FLORA OF THE RIVER KENNET IN RELATION TO CHANGING FLOW AND WATER QUALITY

The major features of the changing flora of the upper Kennet downstream of Marlborough STW were as follows (Figure 3):

(a) Increases in *Ranunculus* growth in the spring/summer of 1998 and 1999 and the spring and early summer of 2000 (Flynn *et al.*, 2002; Wright *et al.*, 2002) corresponding with dramatic reductions in epiphyte growth.

This was attributed to the increase in magnitude and frequency of river high flows, because *Ranunculus* preferentially colonises faster flowing waters and prefers deeper water environments (Dawson, 1976). The ARK Newsletter (Winter 2000/2001) reports that 'The effect (of high flows) on the health of the river has been gratifying. Water Crowfoot (*Ranunculus*) which is so vital to the ecology of Chalk streams is dependent upon good clean water flow and its splendid growth this year has been a wonderful demonstration of this'.

Anecdotal evidence also indicates a dramatic reduction in epiphytic diatoms, which may be attributed to reductions in SRP concentrations. However, the higher flows also reduced the capacity for colonisation of epiphytic colonies on the macrophytes (Sand-Jensen *et al.*, 1989). The coincidence of reduced SRP inputs from Marlborough STW with markedly higher river flows makes it difficult to deconvolute the relative importance of hydrochemical and physical factors (such as hydrodynamic effects of changing flow velocities and changes in solar radiation) on *Ranunculus* and epiphyte growth.

(b) Epiphyte bloom on *Callitriche Obtusangular* (Le Gall) in October 1999.

Flynn *et al.* (2002) identified an epiphyte bloom at a time when natural seasonal *Ranunculus* growth dynamics were such that there was little *Ranunculus* present within the river and the dominant taxon upon which the epiphytes developed was another aquatic macrophyte, *Callitriche* (water starwort).

(c) Epiphyte blooms on *Ranunculus* in August 2000 and May – July 2001.

In the third week of August 2000, a dramatic change in ecological balance within the river was recorded by local residents, with a rapid onset of epiphytic algal growth. While most prevalent within the slower flowing reaches, the epiphyte inundation spread downstream over the following few weeks. This effect was clearly recorded by Flynn *et al.* (2002) at the end of their monitoring period, with an increase in mid-channel *Ranunculus* periphyton (epiphytic algae plus bacteria, fungi, inorganic and organic detritus covering the plant), increasing from <5 g m^{-2} in June and July 2000, to about 22 gm^{-2} in the first 2 weeks in August and about 40 g m^{-2} in the second half of August 2000 (whereas, during the preceding summer of 1999, mid-channel *Ranunculus* periphyton did not exceed 10 g m^{-2}).

In the spring and early summer of 2001, good *Ranunculus* growth was, again, reported on the upper Kennet: '(2001) started full of promise with high flows of clear water and, as a result, abundant growth of that essential to the health of all Chalk streams, *Ranunculus*..... .Indeed there has been more abundant growth than we have seen for many years resulting in the need for serious weed-cutting. In April and May, we had a really healthy looking Kennet....' (ARK Newsletter, Winter 2001/2002). However, this was followed by a marked decline in the ecological status of the river: '...towards the end of May (2001), the water developed an opalescent quality, not only below Marlborough, but also above Clatford. Worse still, below Marlborough sewage treatment works there began to appear algal growth and diatom activity which, by July, covered the river bed and all the weed with thick brown coating excluding the sun and thereby preventing any further growth of *Ranunculus*... The appearance of the River was reminiscent of the years immediately prior to the advent of phosphate removal from the sewage effluent by Thames Water at Marlborough STW' (ARK Newsletter, Winter 2001/2002).

(d) Dramatically reduced *Ranunculus* growth during the spring and summer of 2002, with epiphyte colonisation in July 2002.

The spring and early summer of this year (2002) were characterised by extremely poor growth of *Ranunculus*, in contrast to 1998–2001, where spring and early summer *Ranunculus* growth was abundant. This poor growth continued to the time of writing, but was further degraded by epiphytic growth since mid-July, with the result that the in-stream vegetation on the upper Kennet below Marlborough is now largely absent, a catastrophic change in the ecological status of the river.

The large-scale and progressive degradation of the classic Chalk stream ecology in the River Kennet since 1998 took place against a backdrop of a large reduction in SRP concentrations following effluent P-reduction in the late summer of 1997. All the epiphyte blooms occurred under stable low flow conditions when SRP concentrations were approximately one-seventh of those during the extreme low flows in the Kennet prior to effluent P-treatment. In addition, weekly daytime $EpCO_2$ values from summer 2000 to present have remained consistently high. This feature clearly shows that, even during the epiphyte 'blooms' after effluent P-removal, net in-stream productivity has been considerably lower than during the summer of 1997.

The SRP concentrations during the baseflow periods after effluent P-treatment, when the upper River Kennet was subject to epiphyte invasions (in 2000, 2001 and 2002) are examined in closer detail in Figure 3. In all cases, the proliferation of epiphytes coincided with an elevation in SRP concentrations above a 'threshold' of c. 100 μg-P L^{-1} (which coincidentally is the target level for SRP concentrations in running waters subject to eutrophication). SRP spikes of similar magnitude, which were linked to high flow events, also occurred during the period after effluent P-treatment. These high-flow SRP spikes were related to diffuse-source delivery from agricultural catchment surface and were not linked to any marked change in in-stream flora, presumably because the lower water residence times and higher flow velocities inhibit epiphyte colonisation.

Because of the UK foot and mouth disease outbreak, water quality sampling could not be undertaken during the spring and early summer of 2001. Therefore, there are no records of SRP concentrations on the River Kennet for this early summer period, corresponding with the onset of the major epiphyte bloom described above. Sampling recommenced on 25 July 2001, and during the following five months through to December, flows have remained low, gradually declining from around 0.7 m^3 s^{-1} to 0.3 m^3 s^{-1}. However, the upper Kennet has been subject to short-lived 'spikes' in elevated SRP concentrations exceeding 100 μg-P L^{-1} and up to around 200 μg-P L^{-1} during this period. The origins of the spikes in baseflow SRP, and the other short-lived elevations in SRP concentrations over 100 μg-P L^{-1} under baseflow linked to epiphytic proliferation, are difficult to ascertain. These SRP spikes may be associated with short-lived runoff events, which have very little impact in terms of overall flow, but have a relatively high SRP loading. Alternatively, they may result from increased inputs of SRP from sewage effluent. However, SRP:B ratios increased over these high SRP events (Figure 3), which indicates that the high SRP concentrations do not simply result from an increase in the loading of sewage effluent *per se* or a reduction in effluent dilution under reduced river flows, because in both cases there would be no increase in SRP:B. Instead, this enrichment of SRP relative to the boron sewage signal may arise from: (i) reductions in the efficiency of the P-treatment plant at Marlborough STW, resulting in discharge of effluent with a higher SRP:B ratio, (ii) diffuse inputs of SRP from, for example, catchment surface runoff or in-stream sources, or (iii) some combination of (i) and (ii).

4. Conclusions

The Rivers Kennet and Thames exemplify key issues of ecological sensitivity of rivers draining lowland permeable catchments to perturbations arising from climate variability, the balance between sewage inputs, diffuse-source nutrient supplies, and river flows. Following the dramatic reductions in SRP concentrations and higher river flows on the Rivers Kennet and Thames since 1998, water quality 'indicators' of biological status, such as $EpCO_2$, point to a general improvement in the ecological state of both the Thames and Kennet in terms of a reduction in overall net primary productivity. However, direct field observations of the plant ecology in the River Kennet indicate a far more complex response. Episodes of epiphyte proliferation appear to be temporally linked to small short-lived increases in SRP concentrations (to just above 100 μg L^{-1}) under baseflow conditions. However, spikes in SRP concentration, which are linked to high flows and diffuse-source SRP delivery, do not produce a marked epiphyte response. No causality between SRP concentrations and epiphyte response can be established here, given the limited biological data available, but it would appear that small increases in SRP concentrations under baseflow conditions now may be linked with a disproportionate response from the in-stream plant ecology. One hypothesis is that, after P-stripping, the river ecology may have become sensitized to small increases in SRP. The short-lived nature of the increases in SRP concentrations and the rapid development of epiphyte blooms indicate that this is a highly dynamic and responsive system. Aquatic plant data collected for these lowland rivers are typically seasonal or annual plant ecology surveys, which are insufficient to adequately assess the ecological status of sensitive lowland rivers like the Kennet. Conventional 'indicators' of biological status such as $EpCO_2$, also fail to resolve the degradation associated with changes in ecosystem balance associated with the 'post-impacted' river. This clearly makes the case for long-term intensive monitoring programmes, which combine water quality measurements with direct field observations of aquatic plant ecology to assess the effects of changes in hydrochemical and physical factors on the ecological quality of rivers.

This research has important implications in terms of ecosystem response following point-source P mitigation. Since the introduction of the Urban Wastewater Treatment Directive in 1991, there has been major capital investment in reducing phosphorus inputs from STWs discharging into biologically sensitive rivers. Between 2001 and 2005, it is estimated that £250 million ($150 million) will be spent by the UK water companies on point-source P reductions (Pretty *et al.*, 2003). Although effluent P-treatment programmes have been highly successful in reducing phosphorus concentrations and loads in major lowland river systems, in many cases there has been limited beneficial effect in terms of in-stream ecology (Jarvie *et al.*, 2002a,b). The main problem has been that reductions in point-source P inputs to rivers have not brought baseline phosphorus concentrations down to levels sufficient to improve in-stream ecology. Therefore, increasingly attention is being directed to the role of upstream sources such as smaller sewage treatment works and diffuse agricultural inputs, which are much more difficult and costly to manage and control, with potentially large implications for agricultural practice.

Control of P-inputs from a single dominant point-source input on the upper Kennet has brought SRP concentrations down to levels regarded as critical for aquatic plant response (<100 μg-P L^{-1}). However, rather than observing ecological improvement, we have seen a deterioration in the plant ecology of the River Kennet in the aftermath of point-source P reduction. This is counterintuitive and the reasons are not immediately clear. As phosphorus levels are reduced to critical thresholds, the in-stream biology and the in-stream cycling of nutrients may be also be reaching a critical new 'post-impacted' state, with enhanced ecological sensitivity to change. The within-stream system is complex in terms of physical, chemical and biological interactivity, and individual processes are difficult to resolve, especially given the background climatic instability that the UK is presently encountering. The complexities in chemical, hydrological and biological interactions could well introduce feedback/fractal mechanisms, which are currently poorly understood in relation to the biological interrelationships between 'predators', 'prey' and grazers as well as the competition between plant communities with different nutrient tolerances and nutrient requirements. New dynamic models are needed to link water quality with biological responses within an ecosystem framework, incorporating complexities of inter-species competition and food chain interactions. A model of phosphorus-epiphyte-macrophyte relationships on the River Kennet (Wade et al., 2001, 2002a,b) has made an important first step. However, this must be coupled with: (a) wider studies to assess whether the observations on the River Kennet have implications for other river systems and what the controlling process mechanisms are, and (b) integrated and quantifiable biological monitoring alongside water quality monitoring over both short time scales for more rigorous statistical assessment of event-scale dynamic responses and over longer timescales to examine responses linked to climate variability and the lag times required to produce new dynamic equilibria within ecosystems.

Acknowledgements

The authors wish to extend their thanks to Paul Lidgett, John Sutton and Graham Scholey at the Environment Agency for providing a forum for discussion of issues surrounding the water quality status of the upper Kennet. The local Action for the River Kennet group has provided a wealth of experience, enthusiasm and anecdotal information on long-term trends in the ecological status of the upper River Kennet. Particular thanks are extended to Roger De Vere, Jack Ainslie and John Hounslow for supplying back issues of the Action for the River Kennet Newsletter and providing regular updates on changes in appearance and ecology of the river.

References

Action for the River Kennet (ARK): 2001, *Newsletter, Winter 2000/2001*, ARK, 'Pennings', Mildenhall, Marlborough, Wiltshire, United Kingdom.
Action for the River Kennet (ARK): 2002, *Newsletter, Winter 2001/2002*, ARK, 'Pennings', Mildenhall, Marlborough, Wiltshire, United Kingdom.
Council of the European Communities: 1991, Council Directive Concerning Urban Wastewater Treatment (91/271/EC).
Dawson, F. H.: 1976, 'The annual production of the aquatic macrophyte *Ranunculus penicillatus var. calcareous* (R. W. Butcher) C. D. K. Cooke', *Aquat. Bot.* **2**, 51–73.
Department of the Environment: 1993, Ministry of Agriculture Fisheries and Food, Welsh Office, 'Methodology for identifying sensitive areas (Urban Wastewater Directive) and designating vulnerable zones (Nitrates Directive) in England and Wales', Consultation Document.
Environment Agency: 2000, 'Aquatic Eutrophication in England and Wales: A Management Strategy', Environment Agency, Bristol, 32 pp.
Fisher, T. R., Melack, J. M., Grobbelaar, J. U. and Howarth, R. W.: 1995, 'Nutrient limitation of phytoplankton and eutrophication of inland, estuarine and marine waters', in N. Tiessen (ed.) *Phosphorus in the Global Environment*, SCOPE 54, John Wiley and Sons Ltd., pp. 301–322.

Flynn, N. J., Snook, D. L., Wade, A. J. and Jarvie, H. P.: 2002, 'Macrophyte and periphyton dynamics in a UK Cretaceous Chalk stream: the River Kennet, a tributary of the Thames', *Sci. Tot. Environ.* **282-283**, 143–157.

Hecky, R. E. and Kilham, P.: 1988, 'Nutrient limitation of phytoplankton in freshwater and marine ecosystems: a review of recent evidence on the effects of enrichment', *Limnol. Oceanogr.* **33**, 796–822.

Institute of Hydrology: 1998, 'Hydrological Data UK. Hydrometric Register and Statistics 1991–1995', Institute of Hydrology (now Centre for Ecology and Hydrology), Wallingford, Oxfordshire, United Kingdom.

Jarvie, H. P., Neal, C. and Tappin, A. D.: 1997, 'European land-based pollutant loads to the North Sea: an analysis of the Paris Commission data and review of monitoring strategies', *Sci. Total Environ.* **194-195**, 39–58.

Jarvie, H. P. and Neal, C.: 1998, 'Autotrophs versus heterotrophs: spatial and temporal variations in the inorganic carbon chemistry of east coat UK rivers', in H. Wheater and C. Kirby (eds), *Hydrology in Changing Environment*, Volume I, Wiley, Chichester, pp. 573–584.

Jarvie, H. P., Lycett, E., Neal, C. and Love, A.: 2002a, 'Patterns in nutrient concentrations and biological quality indices across the upper Thames river basin, UK', *Sci. Total Environ.* **251-252**, 263–294.

Jarvie, H. P., Neal, C., Williams, R. J., Neal, M., Wickham, H. D., Hill, L. K., Wage, A. J., Warwick, A. and White, J.: 2002b, 'Phosphorus sources, speciation and dynamics in the lowland eutrophic River Kennet, UK', *Sci. Total Environ.* **251-252**, 175–204.

Mainstone, C. P. and Parr, W.: 2002, 'Phosphorus in rivers - ecology and management', *Sci. Total Environ.* **282-283**, 25–47.

Marsh, T. J. and Dale, M.: 2002, 'The UK Floods of 2000–2001: A Hydrometeorological Appraisal', *J. Chartered Inst. Water Environ. Manage.* **16**, 180–188.

Neal, C., Harrow, M. and Williams, R. J.: 1998, 'Dissolved carbon dioxide and oxygen in the River Thames: Spring-summer 1997', *Sci. Total Environ.* **210-211**, 205–218.

Neal, C., Jarvie, H. P., Howarth, S. M., Whitehead, P. G., Williams, R. J., Neal, M., Harrow, M. and Wickham, H.: 2000a, 'The water quality of the River Kennet: initial observations on a lowland chalk stream impacted by sewage inputs and phosphorus remediation', *Sci. Total Environ.* **251-252**, 477–496.

Neal, C., Williams, R. J., Neal, M., Bhardwaj, L. C., Wickham, H., Harrow, M. and Hill, L. K.: 2000b, 'The water quality of the River Thames at a rural site downstream of Oxford', *Sci. Total Environ.* **251-252**, 441–458.

Neal, C., Watts, C., Williams, R. J., Neal, M., Hill, L. and Wickham, H.: 2002, 'Diurnal and longer term patterns in carbon dioxide and calcite saturation for the River Kennet south-eastern England', *Sci. Total Environ.* **282-283**, 205–231.

Pretty, J. N., Mason, C. F., Nedwell, D. B., Hine, R. E., Leaf, S. and Dils, R.: 2003, 'Environmental costs of freshwater eutrophication in England and Wales', *Environ. Sci. Technol.* **32**, 201–208.

Sand-Jensen, K. A. J., Jeppesen, E., Nielsen, K., van der Bijl, L, Kjermind, L, Wiggers Nielsen, L. and Iversen, T. M.: 1989, 'Growth of macrophytes and ecosystem consequences in a lowland Danish stream', *Freshwat. Biol.* **22**, 15–32.

Wade, A. J., Whitehead, P. G., Hornberger, G. M., Jarvie, H. P. and Snook, D. L.: 2002b, 'On modelling the flow controls on macrophyte and epiphyte dynamics in a lowland permeable catchment: The River Kennet, southern England', *Sci. Total Environ.* **282-283**, 375–394.

Wade, A. J., Whitehead, P. G., Hornberger, G. M., Jarvie, H. P. and Flynn, F.: 2002a, 'On modelling the impacts of phosphorus stripping at sewage works on in-stream phosphorus and macrophyte-epiphyte dynamics: a case study for the River Kennet', *Sci. Total Environ.* **282-283**, 395–416.

Wade, A. J., Hornberger, G. M., Whitehead, P. G., Jarvie, H. P. and Flynn, N.: 2001, 'On modeling the mechanisms that control in-stream phosphorus, macrophyte, and epiphyte dynamics: An assessment of a new model using general sensitivity analysis', *Water Resour. Res.* **37**, 2777–2792.

Williams, R. J., White, C., Harrow, M. L. and Neal, C.: 2000, 'Temporal and small scale spatial variations of dissolved oxygen in the Rivers Thames, Pang and Kennet, UK', *Sci. Tot. Environ.* **251/252**, 477–495.

Wright, J. F., Gunn, R. J. M., Winder, J. M., Wiggers, R., Vowles, K., Clarke, R. T. and Harris, I.: 2002, 'A comparison of the macrophyte cover and invertebrate fauna at three sites on the River Kennet in the mid 1970s and late 1990s', *Sci. Total Environ.* **282–283**, 121–142.

REMOVAL OF PHOSPHORUS IN CONSTRUCTED WETLANDS WITH HORIZONTAL SUB-SURFACE FLOW IN THE CZECH REPUBLIC

JAN VYMAZAL

Říčanova 40, 169 00 Praha 6, Czech Republic
(* author for correspondence, e-mail: vymazal@yahoo.com; phone: +420 2 3335 0180;
fax: +420 2 3335 0180)

(Received 20 August 2002; accepted 2 April 2003)

Abstract. The major processes responsible for phosphorus (P) removal in constructed wetlands with horizontal sub-surface flow (HSF CWs) are adsorption, precipitation and plant uptake if the biomass is harvested. The filtration materials frequently used in HSF CWs, i.e., gravel or crushed rock, provide only limited adsorption and plants are not regularly harvested. As a result, the removal of P in HSF CWs is usually low and typically amounts to only 40 to 60% during the treatment of municipal or domestic sewage. The average inflow and outflow P concentrations for vegetated beds of Czech HSF CWs were 6.6 mg L^{-1} and 3.6 mg L^{-1}, respectively. The average P removal was 45.7%. Despite a wide fluctuation of inflow phosphorus loading rates in the Czech CWs (10.9 – 356 g P m^{-2} a^{-1}) the retention of P is well predictable. The CWs in the Czech Republic are relatively new and therefore, it is not possible to evaluate long-time performance of P removal. However, results from systems that have been in operation for longer periods (maximum of 9 years) indicate that P removal decreases over years probably as a result of limited sorption capacity. The amount of P removed by aboveground *Phragmites* biomass is very low and usually does not exceed 5% of the total removed P in the beginning of operation. As the sorption decreases over years and macrophyte biomass increases at the same time the importance of P bound in biomass becomes higher but it rarely exceeds the level of 20% of the total P removed.

Keywords: constructed wetlands, Czech Republic, horizontal flow, nutrient uptake phosphorus, *Phragmites australis*

1. Introduction

Most wetland studies have shown that soil/litter compartment is the major (>95%) long-term storage pool for phosphorus (P) (Richardson and Marshall, 1986; Verhoeven, 1986; Faulkner and Richardson, 1989) and that wetlands are not particularly effective as P sinks when compared with terrestrial ecosystems (Richardson, 1985). Soil adsorption and peat accumulation, i.e., storage in organic matter, control long-term P sequestration in wetlands (Richardson and Marshall, 1986). The adsorption and retention of P that does occur in wetland soils is controlled by the interaction of redox potential, pH, Fe, Al and Ca minerals, and the amount of native soil P (Faulkner and Richardson, 1989). In acid soils, inorganic P is adsorbed on hydrous oxides of Fe and Al and may precipitate as insoluble Fe- and Al-phosphates. Precipitation as insoluble Ca-P can occur at pHs greater than

7.0 (Qualls and Richardson, 1995). Adsorption of P is greater in mineral than in organic soils.

Redox potentials below +250 mV will cause the reduction of Fe^{3+} to Fe^{2+}, releasing associated P (Faulkner and Richardson, 1989). On the other hand, declining redox potential caused by flooding can cause the transformation of crystalline Al and Fe minerals to the amorphous form; and amorphous Al and Fe hydrous oxides have higher P sorption capacity than crystalline oxides due to their larger number of singly-coordinated surface hydroxyl ions (Patrick and Khalid, 1974). Faulkner and Richardson (1989) reported that the most important retention mechanisms are claimed to be ligand exchange reactions, where phosphate displaces water or hydroxyls from the surface of Fe and Al hydrous oxides to form monodentate and binuclear complexes within the coordination sphere of the hydrous oxide. Plants contain only a small amount of the total P that occurs in wetlands; thus the uptake capacity of macrophytes in wetlands is limited (Verhoeven, 1986; Brix, 1994; Vymazal, 1995, 2001).

Several types of constructed wastewater-wetland systems have been developed, including systems with horizontal sub-surface flow (HSF CWs, Figure 1) that usually do not provide high P removal. HSF CWs systems apparently are not effective at P removal because the filtration medium that often is used (pea gravel, crushed stones) usually does not actively adsorb P, mostly because it does not contain adequate concentrations of Ca, Fe or Al. In addition, P removal is limited because the litter formed by decomposing vegetation stays on the surface of the substrate and is thus not in contact with wastewater that typically is kept 10 to 15 cm below the surface. In HSF CWs the plants are usually not harvested, so P is leached out after plant senescence. Moreover, it has been suggested that plants in HSF CWs can accumulate only less than 15% of the removed P at the peak standing stock; this amount does not constitute a major portion of removed phosphorus (Vymazal, 2001).

The purpose of this paper is to present results on P removal from the HSF CWs in the Czech Republic during the period 1992–2001 with respect to overall P removal treatment efficiency, seasonal efficiency, the dependence of P removal on temperature, and the amount of P that potentially could be removed by harvesting vegetation.

2. Study Sites

The data used in this paper were obtained from 25 operational HSF CWs designed to treat domestic or municipal sewage. Design parameters of the Czech HSF CWs have been described in detail elsewhere (Vymazal, 1998, 2002) and, therefore, only major design features will be mentioned here. All systems include a pretreatment step that consists of screens and septic or Imhoff tanks. Pretreatment in systems treating wastewater from combined sewerage (i.e., combined sewage and storm-

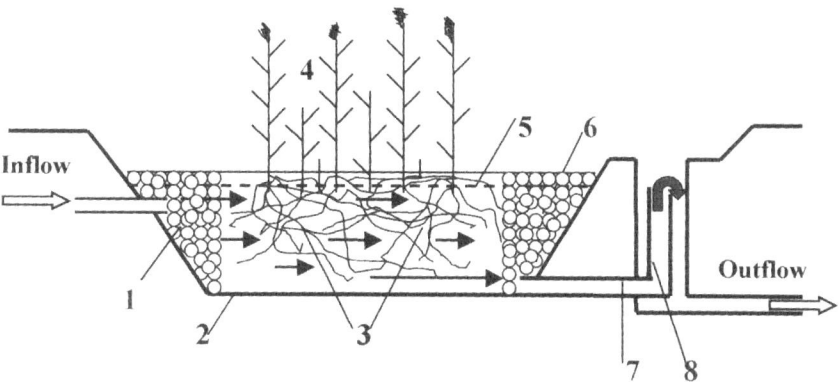

Figure 1. Schematic representation of a constructed wetland with sub-surface horizontal flow. 1 – Distribution zone with large stones, 2 – impermeable liner (usually PVC or HDPE), 3 – filtration substrate (gravel or crushed rock), 4- vegetation, 5 – water level in the bed, 6 – collection zone with large stones, 7 – collection drainage pipe, 8 – outlet structure for maintaining of water level in the bed. The arrows indicate only a general flow pattern (modified from Vymazal, 2001).

water runoff) includes also grit chambers. However, only a limited amount of P is removed in pretreatment. Most systems use coarse substrates (pea gravel or crushed stones) with sizes of 4–8, 8–16 or even 16–32 mm, with few systems using finer media (sand, 0–8 mm). The depth of the filtration bed varies between 0.6 and 0.8 m for most systems. The majority of the systems use common reed (*Phragmites australis*) as a vegetation cover for the treatment beds. In several systems reed canarygrass (*Phalaris arundinacea*) was planted alone or together with *Phragmites* in strips perpendicular to the flow. The size of the constructed wetlands taken into consideration ranges from 75 to 900 PE (population equivalent) with vegetated beds ranging from 300 m² to 4500 m² (specific area ca. 5 m² per 1 PE).

According to Czech law, the major criterion for the evaluation of any wastewater treatment plant is the comparison of effluent parameters with discharge limits. Therefore, it is not mandatory to take samples of the inflowing untreated wastewater. However, many local authorities also sample influent to evaluate the treatment efficiency expressed as removal percentage (referred to throughout this paper as change in total P concentration between influent and effluent water). In our study, only systems where samples were taken from both inflowing untreated wastewater (i.e., raw sewage) and outflowing water (i.e., effluent from vegetated beds) were included. Several systems are monitored more closely including sampling wastewater after pretreatment. This additional sampling point allows the possibility of calculating loadings into the vegetated beds of constructed wetlands. Most systems were sampled on a monthly basis.

3. Results and Discussion

3.1. REMOVAL EFFICIENCY

Figure 2 shows the relationship between total P concentration (TP) in inflow untreated sewage and TP concentrations in water flowing out of vegetated beds of HSF CWs. The average inflow TP concentration of raw wastewater was 6.0 mg L^{-1} (SD = 3.0 mg L^{-1}) while outflow TP concentration was 3.0 mg L^{-1} (SD = 4.0 mg L^{-1}) giving an average treatment efficiency of 49.9% (SD = 29.1%). However, in the Czech Republic, there is no discharge limit for TP for wastewater treatment plants <5,000 PE and, therefore constructed wetlands, which usually treat wastewater from small sources of pollution, are not dimensioned for P removal. In the literature, there is little information on the P removal in constructed wetlands including pretreatment. Most studies focus only on the treatment effect of the vegetated beds which, in fact, are only a part of the whole treatment system. Mæehlum and Jenssen (1998) reported the average removal of TP in HSF CWs as high as 77.0% (SD = 23.0%) from 10 systems in Norway. The discharge limits for P are very strict and, therefore, special media with high P adsorption capacity (e.g., iron-rich sand, manufactured light-weight ceramic particle aggregate) are often used for vegetated beds in Norway.

Figure 3 shows the relationship between mechanically pretreated wastewater entering vegetated beds and the final outflow from the beds. With one exception, TP concentrations in inflowing water were greater than TP in the outflow suggesting TP removal, but to various extents. The average inflow TP concentration of pretreated wastewater entering beds was 6.6 mg L^{-1} (SD = 3.9 mg L^{-1}) while outflow TP concentration was 3.6 mg L^{-1} (SD = 3.6 mg L^{-1}) giving an average treatment efficiency of 45.7% (SD = 26.0%). The regression equation is very similar to that developed by Kadlec and Knight (1996) for 90 HSF CWs in the U.S., Australia, the U.K. and Denmark: $C_o = 0.51 C_i^{1.10}$ ($R^2 = 0.64$). The negligible difference between the treatment efficiency of the whole system and vegetated beds reflects the fact that commonly used pretreatment units usually do not remove P. Contrary to information about the TP removal in the whole system, there are many references concerning vegetated beds only (Table I).

Results presented in Table I indicate that TP removal from municipal sewage in HSF CWs usually does not exceed 50% and the outflow concentrations are well above 1 mg L^{-1}, which is required for larger treatment plants. Knight (1996) reported an average P removal of 56.0% (SD = 27.7%) based on the results from 31 free water surface (FWS) wetlands around the world. Kadlec and Knight (1996) reported average inflow and outflow TP concentrations from 68 FWS CWs in North America of 3.8 mg L^{-1} and 1.6 mg L^{-1}, respectively, giving the treatment efficiency of 57.1%. The higher removal in FWS wetlands may be influenced by contact of wastewater with the litter layer on the surface of the substrate, however,

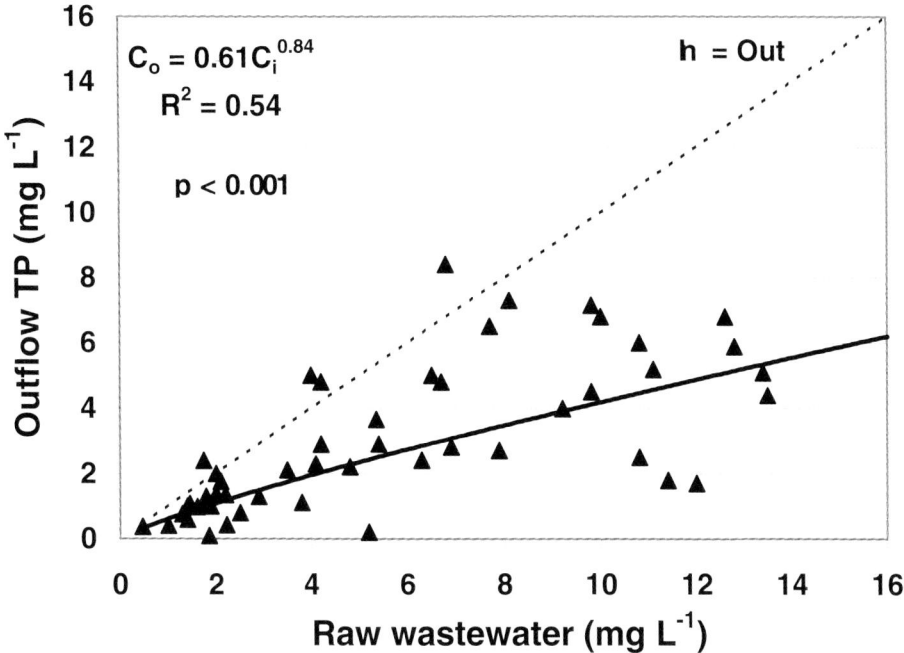

Figure 2. Relationship between concentrations of total phosphorus in inflowing untreated wastewater (C_i) and water outflowing from vegetated beds of constructed wetlands (C_o). Points represent annual mean concentrations for each system.

TABLE I

Removal of phosphorus in HSF systems treating municipal sewage. Concentrations are in mg L^{-1}, loadings in g m^{-2} a^{-1}, EFF = efficiency in %, REM = removed load

	Average concentration				Average loading				
	IN	OUT	EFF	n	IN	OUT	REM	EFF	n
Austria[1]	5.1	2.0	60.8	3	93.7	17	76.7	81.9[a]	3
Czech Republic	6.56	3.57	45.6	44	116	73	43	37.1	42
Denmark + UK[2]	8.6	6.3	26.7	67	120	95	25	20.8	50
Germany-LS[3]	11.4	3.99	65.0	26					
N. America[4]	4.41	2.97	32.7	8	188	146	42	22.3	8
Poland[5]	7.65	4.10	46.4	5	99	58	41	41.4	5
Sweden[6]	5.03	2.10	46.4	3	148	57	91	61.5	3

[1] Haberl and Perfler (1990), [2] Brix (1994b), [3] Börner *et al.* (1998) LS = Lower Saxony, [4] Kadlec and Knight (1996), [5] Kowalik and Obarska-Pempkowiak (1998), [6] Sundblad (1998).
[a] Allows for evapotranspiration.

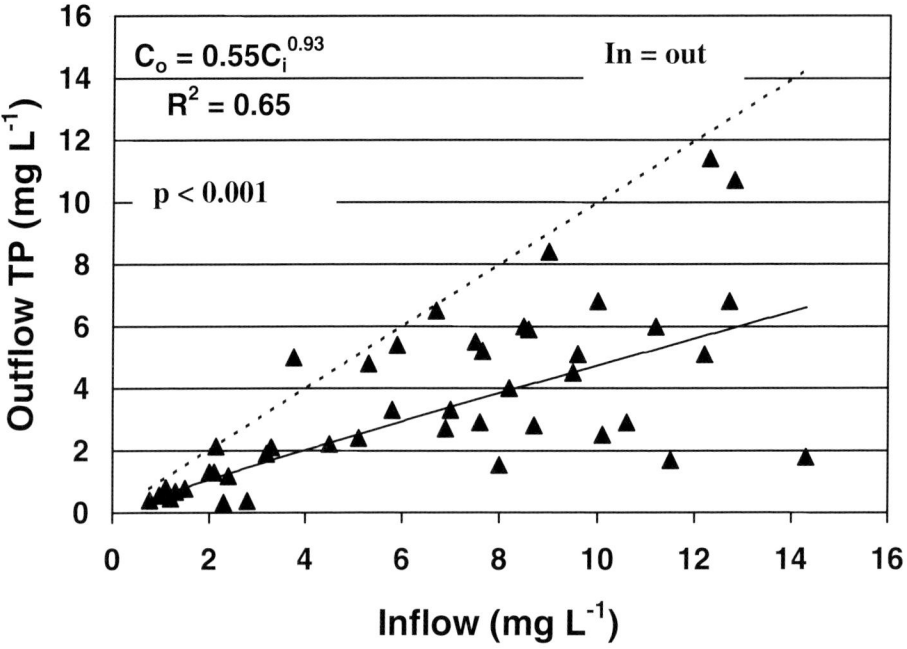

Figure 3. Relationship between concentrations of total phosphorus in mechanically pretreated wastewater, i.e., water entering vegetated beds (C_i) and water outflowing from vegetated beds of constructed wetlands (C_o). Points represent annual mean concentrations for each system.

in FWS systems there is considerably less contact of wastewater with the substrate than in HSF systems.

Removal of P in HSF CWs is primarily influenced by wetland size and filtration materials. All HSF CWs in the Czech Republic and most wetlands around the world are sized for sufficient removal of organics (usually as BOD_5) and suspended solids (Vymazal, 2001). The following equation, first proposed by Kickuth (1977), has been widely used for sizing of HSF systems for domestic or municipal sewage treatment:

$$A_h = Q_d (\ln C_i - \ln C_o) / K_{BOD},$$

where A_h = surface area of bed (m^2), Q_d = average flow (m^3 d^{-1}), C_i = influent BOD_5 (mg L^{-1}), C_o = effluent BOD_5 (mg L^{-1}), and K_{BOD} = rate constant (m d^{-1}). There has been considerable discussion on the K_{BOD} value. The formerly proposed value of 0.19 m d^{-1} (69.4 m a^{-1}) by Kickuth (1977) results in undersizing of the bed, and consequently a low treatment effect. At present, using field measurements from many operational systems, the value of 0.1 m d^{-1} (36.5 m a^{-1}) (Cooper *et al.*, 1996) is considered as sufficient. Based on the field results from HSF systems in North America Kadlec and Knight (1996) developed an areal rate constant K_{TP} adjusted to 20 °C of 12 m a^{-1} (0.033 m d^{-1}). Brix (1998) reported K_{TP} value based

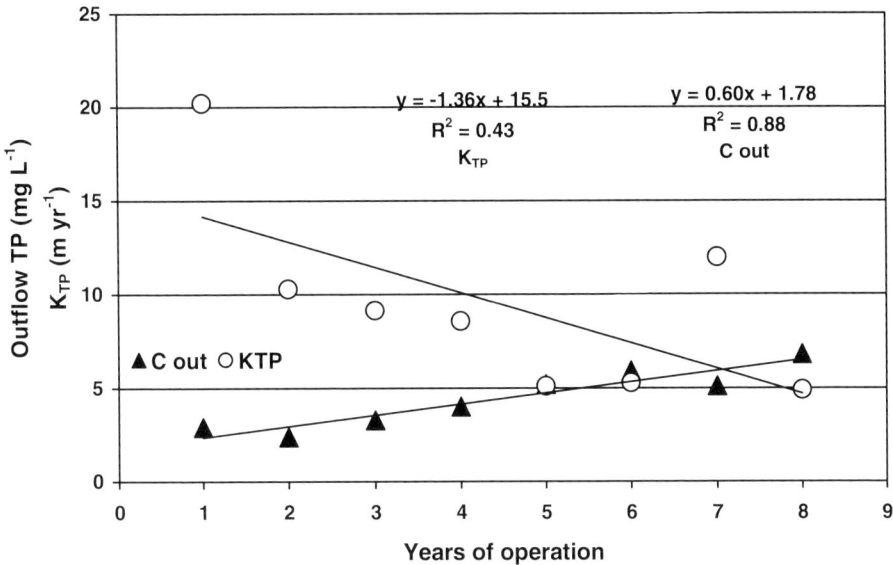

Figure 4. Changes in K_{TP} and outflow TP concentration during the operation of HSF CW at Kolodeje.

on the Danish experiments to be 9 m a^{-1} (0.025 m d^{-1}), but this value is dependent on flow rate, loading rate and time of operation. The average value from the Czech HSF systems is 10.1 m a^{-1} (0.028 m d^{-1}). These results indicate that if the beds were sized for TP removal they should be about 3 to 4 times larger.

Adsorption and precipitation of P to the substrate is a major removal mechanism in HSF CWs and therefore, the capacity of a HSF system to remove P may depend on the contents of Ca, Al and Fe in the substrate. It seems that sandy materials with high content of Ca, Fe, Al or Mg provide the highest P removal (Brix *et al.*, 2001). However, sandy materials have a much lower hydraulic permeability than either gravel or crushed rock, which are commonly used to maintain subsurface flow. Because of that, various artificial media have been tested: LECA (light-weight expanded clay aggregates (Mæehlum and Jenssen, 1998), granulated laterite (Wood and McAtamney, 1996), shale (Drizo *et al.*, 1997) or crushed marble (Gervin and Brix, 2001). Also, it has been shown that P removal due to sorption decreases over time because of a finite P-sorption capacity of the bed (Vymazal, 1999). This has been referred to as the "aging phenomenon" in wetlands that receive wastewater (Kadlec, 1985). This phenomenon is shown in Figure 4 – outflow concentrations gradually increase over years of operation while K_{TP} decreases. However, based on the experience from systems that have been in operation for 5 or more years around the world, even though the sorption capacity becomes gradually exhausted, the removal at lower rates will continue.

The relationship between inflow and outflow TP loadings (Figure 5) is much stronger than the relationship between inflow and outflow concentrations (Figure

3). The average inflow and outflow loadings were 116 ± 87 and 73 ± 84 g m^{-2} a^{-1} with the average removal rate of 43.0 ± 32.9 g m^{-2} a^{-1} ($n = 42$). Corresponding to the increase in outflow TP concentration over time the amount of removed load also decreases over time. This phenomenon is documented in Figure 6 using data from constructed wetlands at Ondrejov and Kolodeje, two systems that have been in operation since 1991 and 1993, respectively. The amount of removed load in the Czech HSF CWs is very similar to that reported from Poland and North America (Table I). Removal rates are very high in comparison to natural, low loaded wetlands. Schlesinger (1978) estimated P accumulation of 0.15 g m^{-2} a^{-1} in Georgia's Okefenokee Swamp and Richardson and Marshall (1986) estimated P accumulation rate in a central Michigan fen to be between 0.2 and 0.5 g m^{-2} a^{-1}. General estimates vary among authors but stay quite low as compared to treatment wetlands. Nichols (1983) estimated the rate at which P is accumulated in peat ranges between 0.005 and 0.22 g m^{-2} a^{-1} in moderate to cold climates and possibly up to 0.5 g m^{-2} a^{-1} in warm, highly productive areas. In nutrient-enriched wetlands long-term P accretion may reach nearly 1 g m^{-2} a^{-1} (Craft and Richardson, 1993). Craft and Richardson (1998) reported P accumulation rates in organic soil freshwater wetlands in the United States in the range of 0.06 to 0.90 g m^{-2} a^{-1}. Research to date would suggest that permanent storage of P in natural wetlands is below 1 g m^{-2} a^{-1} and usually averages around 0.5 g m^{-2} a^{-1} (Nichols, 1983; Richardson, 1985; Richardson and Marshall, 1986; Johnston, 1991; Craft and Richardson, 1993).

It is important to realize that P loadings of natural wetlands that do not receive wastewater are several orders of magnitude lower than loadings to constructed wetlands treating municipal sewage. For example, Richardson (1989b) presented results from 11 natural wetland ecosystems from North America and Europe; the inflow P loading was between 0.01 and 0.47 g m^{-2} a^{-1}. The inflow loadings for HSF wetlands in present study ranged from 10.9 to 356 g m^{-2} a^{-1}; in Poland (Kowalik and Obarska-Pempkowiak, 1998) and U.S.A. (Reed, 1993) the inflow P loadings ranged between 54.8 and 179 g m^{-2} a^{-1} and 52.6 and 274 g m^{-2} a^{-1}, respectively. Constructed wetlands receiving industrial or agricultural wastewaters may be loaded even more heavily. Prystay and Lo (1998) reported the inflow load of 759 g m^{-2} a^{-1} for a wetland receiving greenhouse wastewaters.

3.2. REMOVAL OF P VIA PLANT HARVESTING

At present, HSF CWs in temperate and cold regions are not designed with the primary purpose of removing P from wastewater by plant uptake and subsequent harvesting despite some promising calculations reported in the literature (Reddy and Smith, 1987). However, numbers for maximal nutrient standing stocks, i.e., the amount of a given element in plant dry biomass per given area, are frequently calculated from the maximal values of plant biomass and maximal nutrient concentration in the biomass. These numbers, however, could be misleading because it has been

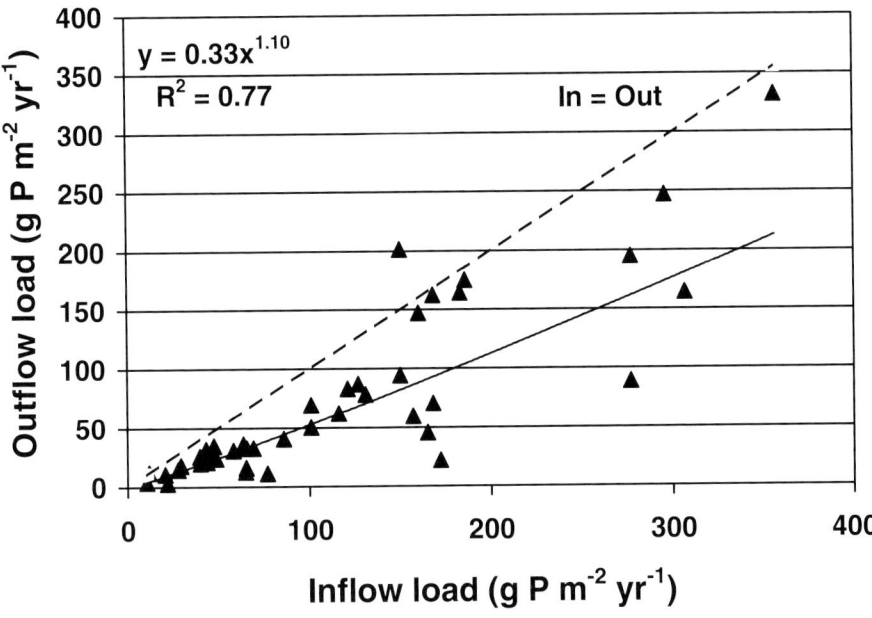

Figure 5. Relationship between inflow and outflow phosphorus loadings in the Czech HSF CWS. Each point represents an annual average. Data from 14 systems.

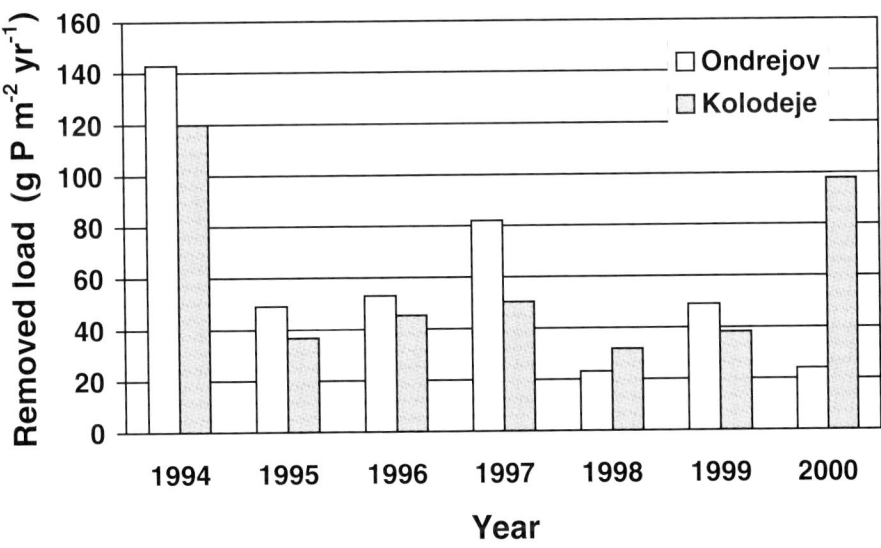

Figure 6. Total removed phosphorus load in CWS at Kolodeje and Ondrejov during the period of 1994–2000.

well established that maximal plant biomass and maximal nutrient concentration do not occur at the same time of the year (e.g., Boyd, 1970; Dykyjová, 1973). In addition, some species used in HSF CWs, and especially *Phragmites australis*, cannot be harvested during the growing season because it would seriously damage the plants. This is because carbohydrate reserves are translocated to the rhizomes only in late fall. Therefore, the potential amounts of P that could be removed by biomass harvests are overestimates of the amount that can actually be removed.

In order to determine how much P could be removed by harvesting plants, biomass was harvested in HSF CWs at Ondrejov and Kolodeje during years 1994 to 2000 in late summer at the time of peak standing stock. P-standing stocks in the aboveground biomass were quite stable (approximately between 2 and 5 g P m^{-2}), but because of decreasing amount of removed load (Figure 6) the percentage of P sequestered in plant biomass increased (Figure 7) from less than 3% to 17% and 14.3% in Ondrejov and Kolodeje, respectively. Even though the P standing stock values were high in comparison to literature data for *Phragmites australis* (Vymazal 1995; Vymazal *et al.*, 1999) they represent only small portion of P removed by the systems. *Phragmites* was harvested at the end of August/beginning of September and at this time P concentration in aboveground biomass varied only little between 2.0 and 2.44 mg g^{-1} dry biomass. However, during the winter, when *Phragmites* could be harvested, the concentration of P in the biomass decreases to about 1.0 mg g^{-1} dry mass in February and therefore, the amount of P removed by biomass harvesting is about 50% lower than values obtained during the peak standing stock in late summer.

Results from Ondrejov and Koloděje are in a good agreement with literature data from HSF CWs. Haberl and Perfler (1990) reported from a HSF system in Mannersdorf in Austria the amount of P in *Phragmites* aboveground biomass between 2.0 and 2.6 g m^{-2} and represented 2.3 to 5.4% of the overall P removal. Obarska-Pempkowiak (1999) reported P standing stock in *Phragmites* growing in HSF beds at Darzlubie in Poland 3.8 g m^{-2} representing only 1.4% of removed P. Vymazal (1999) reported that in HSF at Chmelná in the Czech Republic, the P standing stock in *Phalaris arundinacea* aboveground biomass of 1.5 g P m^{-2} accounted for 7.5% of the total P removed. Limited plant potential for P uptake has also been reported for *Typha*. Herskowitz (1986) reported that in the FWS CW at Listowel in Ontario the P standing stock in *Typha* biomass during 1981-1984 varied between only 0.18–1.67 g m^{-2}. The cattail harvest represented annual P removal of 5.6% of the total annual phosphorus removed. The situation is different under tropic climatic conditions. Okurut (2001) reported that nutrient uptake by *Cyperus papaus* and *Phragmites mauritanus* growing as a floating mat in a constructed wetland in Uganda accounted for 33% and 61% of the total P s removed, respectively. This was achieved by keeping the plants in growth phase by regular harvesting up to four times a year. When plants were in steady state growth phase the contribution of plant uptake to P removal dropped to only 3.2%.

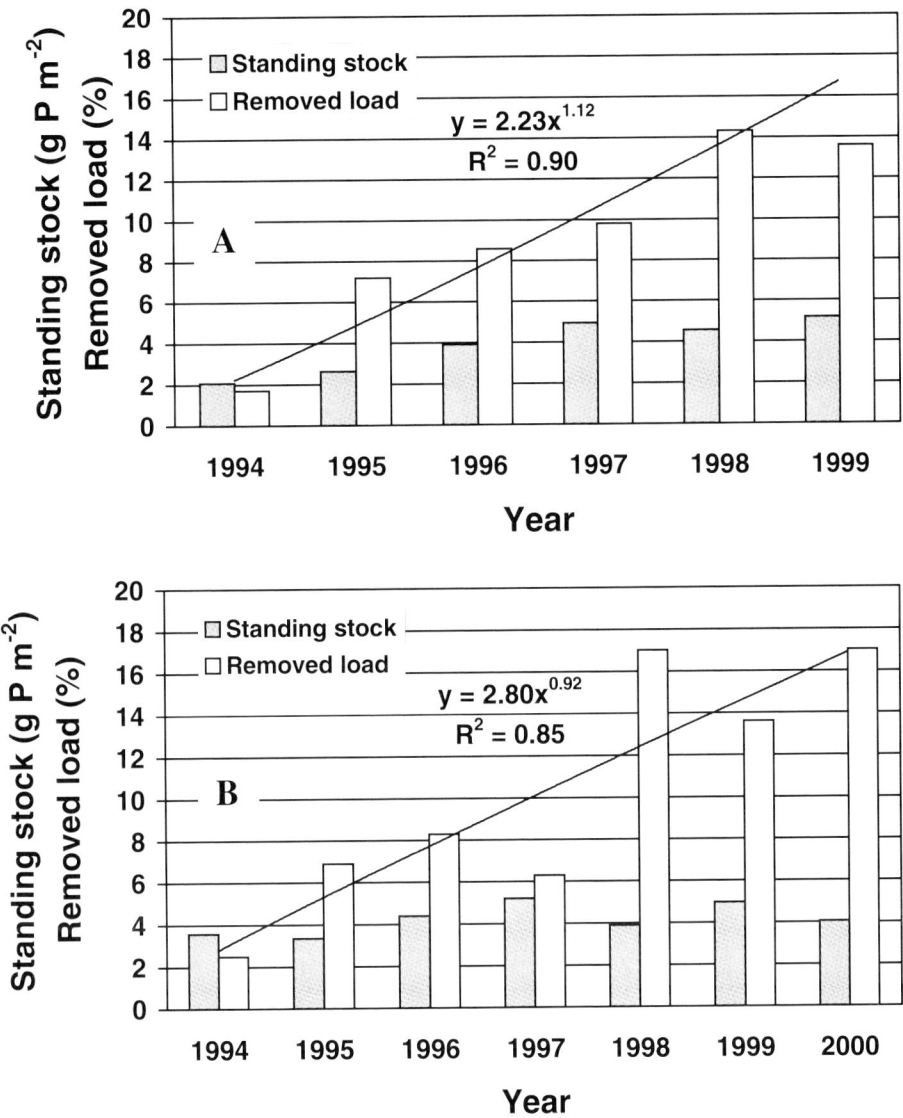

Figure 7. Phosphorus standing stock at the peak biomass of *Phragmites australis* and percentage of P removed through harvesting. A = Kolodeje, B = Ondrejov.

4. Summary

Phosphorus removal in the Czech HSF CWs is generally low – <50% of the input with an average outflow concentrations of 3.6 mg L^{-1} – but comparable with literature data on HSF CWs. Low removal is primarily due to the low P adsorption capacity of the substrates that are used in the vegetated beds (e.g., pea gravel and crushed stones). Inflow loadings to HSF CWs for municipal or domestic wastewater are usually 2 to 3 orders higher as compared to natural unpolluted wetlands. In this study, the loading rates varied between 10.9 and 356 g m^{-2} a^{-1} with the mean value of 116 g m^{-2} a^{-1}. The inflow and outflow loading rates are well correlated ($R^2 = 0.77$).

Amount of P sequestered in aboveground biomass of *Phragmites australis* during the peak standing stock is low and steady from year to year (approximately between 2 and 5 g P m^{-2}) and P removal through plant uptake represents only a small portion of P removal by the treatment system. Due to usual decrease in P removal over years the amount of P in the biomass during the peak standing stock can increase up to 17% of removed P.

Acknowledgements

The study was partially supported by grant No. 204/94/1821 Ecological Role of Wetlands in the Landscape and Their Use for Wastewater Treatment and grant No. 206/02/1036 Processes Determining Mass Balance in Overloaded Wetlands from the Grant Agency of the Czech Republic.

References

Börner, T., von Felde, K., Gschlössl, T., Gschlössl, E., Kunst, S. and Wissing, F. W.: 1998, 'Germany', in J. Vymazal, H. Brix, P. F. Cooper, M. B. Green and R. Haberl (eds), *Constructed Wetlands for Wastewater Treatment in Europe*, Backhuys Publishers, Leiden, The Netherlands, pp. 169–190.

Boyd, C. E.: 1970, 'Production, mineral accumulation and pigment concentrations in *Typha latifolia* and *Scirpus americanus*', *Ecology* **51**, 285–290.

Brix, H.: 1994a, 'Functions of macrophytes in constructed wetlands', *Water. Sci. Technol.* **29**, 71–78.

Brix, H.: 1998, 'Denmark', in J. Vymazal, H. Brix, P. F. Cooper, M. B. Green and R. Haberl (eds), *Constructed Wetlands for Wastewater Treatment in Europe*, Backhuys Publishers, Leiden, The Netherlands, pp. 119–149.

Brix, H., Arias, C. A. and del Bubba, M.: 2001, 'Media selection for sustainable phosphorus removal in subsurface flow constructed wetlands', *Water Sci. Technol.* **44**, 47-54.

Cooper, P. F., Job, G. D., Green, M. B. and Shutes, R. B. E.: 1996, *Reed Beds and Constructed Wetlands for Wastewater Treatment*, WRc Publications, Medmenham, Marlow, U.K., 184 pp.

Craft, B. C. and Richardson, C. J.: 1993, 'Peat accretion and phosphorus accumulation along a eutrophication gradient in the northern Everglades', *Biogeochemistry* **22**, 133–156.

Craft, B. C. and Richardson, C. J.: 1998, 'Recent and long-term organic soil accretion and nutrient accumulation in the Everglades', *Soil Sci. Soc. Am. J.* **62**, 834–843.

Drizo, A., Frost, C. A., Smith, K. A. and Grace, J.: 1997, 'Phosphate and ammonium removal by constructed wetlands with horizontal subsurface flow, using shale as a substrate', *Water Sci. Technol.* **35**, 95–102.

Dykyjová, D.: 1973, 'Content of Mineral Macronutrients in Emergent Macrophytes During Their Seasonal Growth and Decomposition', in S.Hejny, (ed.), *Ecosystem Study on Wetland Biome in Czechoslovakia*, Czechoslovak IBP/PT-PP Report No. 3, Trebon, Czech Republic, pp. 163–172.

Faulkner, S. P. and Richardson, C. J.: 1989, 'Physical and chemical characteristics of freshwater wetland soils', in D. A. Hammer (ed.), *Constructed Wetlands for Wastewater Treatment. Municipal, Industrial and Agricultural*, Lewis Publishers, Chelsea, Michigan, pp. 41–72.

Gervin, L. and Brix, H.: 2001, 'Removal of nutrients from combined sewer overflows and lake water in a vertical-flow constructed wetland system', *Water Sci. Technol.* **44**, 171–176.

Haberl, R. and Perfler, R.: 1990, 'Seven years of research work and experience with wastewater treatment by a reed bed system', in P. F. Cooper and B. C. Findlater (eds), *Constructed Wetlands in Water Pollution Control*, Pergamon Press, Oxford, UK, pp. 205–214.

Herskowitz, J.: 1986, 'Town of Listowel Artificial Marsh Project Final Report', Project No. 128 RR, Ontario Ministry of the Environment, Toronto, Canada, 253 pp.

Johnston, C. A.: 1991, 'Sediments and nutrient retention by freshwater wetlands: Effects on surface water quality', *CRC Crit. Rev. Environ. Control* **21**, 491–565.

Kadlec, R. H.: 1985, 'Aging phenomena in wastewater wetlands', in P. J. Godfrey, E. R. Kaynor, S. Pelczarski and J. Benforado (eds), *Ecological Considerations in Wetlands Treatment of Municipal Wastewaters*, Van Nostrand Reinhold Company, New York, pp. 338–350.

Kadlec, R. H. and Knight, R. L.: 1996, *Treatment Wetlands*, CRC Press/Lewis Publishers, Boca Raton, Florida, U.S.A., 893 pp.

Kickuth, R.: 1977, 'Degradation and incorporation of nutrients from rural wastewaters by plant rhizosphere under limnic conditions', in *Proceedings of the International Conference on Utilization of Manure by Land Spreading*, Commission of the European Community, EUR 5672e, London, U.K., pp. 335–343.

Knight, R. L.: 1996, *Free Water Surface Wetlands for Wastewater Treatment: A Technology Assessment*, CH2M Hill, Gainesville, Florida, 104 pp.

Kowalik, P. and Obarska-Pempkowiak, H.: 1998, 'Poland', in J. Vymazal, H. Brix, P. F. Cooper, M. B. Green and R. Haberl (eds), *Constructed Wetlands for Wastewater Treatment in Europe*, Backhuys Publishers, Leiden, The Netherlands, pp. 217–225.

Mæehlum, T. and Jenssen, P. D.: 1998, 'Norway', in J. Vymazal, H. Brix, P. F. Cooper, M. B. Green and R. Haberl (eds), *Constructed Wetlands for Wastewater Treatment in Europe*, Backhuys Publishers, Leiden, The Netherlands, pp. 207–216.

Nichols, D. S.: 1983, 'Capacity of natural wetlands to remove nutrients from wastewater', *J. Water Pollut. Control Fed.* **55**, 495–505.

Obarska-Pempkowiak, H.: 1999, 'Nutrient cycling and retention in constructed wetland system in Darzlubie near Puck Bay, Southern Baltic Sea', in J. Vymazal (ed.), *Nutrient Cycling and Retention in Natural and Constructed Wetlands*, Backhuys Publishers, Leiden, The Netherlands, pp. 41–48.

Okurut, T. O.: 2001, 'Plant growth and nutrient uptake in a tropical constructed wetland', in J. Vymazal (ed.), *Transformations of Nutrients in Natural and Constructed Wetlands*, Backhuys Publishers, Leiden, The Netherlands, pp. 451–462.

Patrick, W. H., Jr. and Khalid, R. A.: 1974, 'Phosphate release and sorption by soils and sediments: Effect of aerobic and anaerobic conditions', *Science* **186**, 53–55.

Prystay, W. and Lo, K. V.: 1998, 'Assessment of constructed wetlands for the reduction of nitrogen and phosphorus from greenhouse wastewaters', in S. M. Tauk-Tornisielo and E. Salati Filho (eds), *Proceedings of the 6th International Conference on Wetland Systems for Water Pollution Control*, Aguas de São Pedro, Brazil, Brazil, pp. 101–114.

Qualls, B. C. and Richardson, C. J.: 1995, 'Forms of soil phosphorus along a nutrient enrichment gradient in the northern Everglades', *Soil Sci.* **160**, 183–198.

Reddy, K. R. and Smith, W. H. (eds): 1987, *Aquatic Plants for Water Treatment and Resource Recovery*, Magnolia Publishing, Orlando, Florida, 1032 pp.

Reed, S. C.: 1993, *Subsurface Flow Constructed Wetlands for Wastewater Treatment: A Technology Assessment*, U.S. EPA Office of Water, EPA 832-R-93-008, 50 pp. + app.

Richardson, C. J.: 1985, 'Mechanisms controlling phosphorus retention capacity in freshwater wetlands', *Science* **228**, 1424–1427.

Richardson, C. J.: 1989, 'Biochemical cycles: regional', in B. C. Patten (ed.), *Wetlands and Shallow Continental Water Bodies*, SPB Academic Publishing, The Hague, The Netherlands, pp. 259–279.

Richardson, C. J. and Marshall, P. E.: 1986, 'Processes controlling movement, storage and export of phosphorus in a fen peatland', *Ecol. Monogr.* **56**, 279–302.

Schlesinger, W. H.: 1978, 'Community structure, dynamics, and nutrient cycling in Okefenokee cypress swamp', *Ecol. Monogr.* **48**, 43–65.

Sundblad, K.: 1998, 'Sweden', in J. Vymazal, H. Brix, P. F. Cooper, M. B. Green and R. Haberl (eds), *Constructed Wetlands for Wastewater Treatment in Europe*, Backhuys Publishers, Leiden, The Netherlands, pp. 251–259.

Verhoeven, J. T. A.: 1986, 'Nutrient dynamics in minerotrophic peat mires', Aquat. Bot. 25, 117–167.

Vymazal, J.: 1995, *Algae and Element Cycling in Wetlands*, CRC Press/Lewis Publisher, Boca Raton, Florida, 689 pp.

Vymazal, J.: 1998, 'Czech Republic', in J. Vymazal, H. Brix, P. F. Cooper, M. B. Green and R. Haberl (eds), *Constructed Wetlands for Wastewater Treatment in Europe*, Backhuys Publishers, Leiden, The Netherlands, pp. 90–117.

Vymazal, J.: 1999, 'Removal of phosphorus in constructed wetlands with horizontal sub-surface flow in the Czech Republic', in J. Vymazal (ed.), *Nutrient Cycling and Retention in Natural and Constructed Wetlands*, Backhuys Publishers, Leiden, The Netherlands, pp. 73–83.

Vymazal, J.: 2001, 'Types of constructed wetlands for wastewater treatment: Their potential for nutrient removal', in J. Vymazal (ed.), *Transformations of Nutrients in Natural and Constructed Wetlands*, Backhuys Publishers, Leiden, The Netherlands, pp. 1–93.

Vymazal, J.: 2002, 'The use of sub-surface constructed wetlands for wastewater treatment in the Czech Republic: 10 years experience', *Ecol. Eng.* **18**, 633–646.

Vymazal, J., Dušek, J. and Květ, J.: 1999, 'Nutrient uptake and storage by plants in constructed wetlands with horizontal sub-surface flow: A comparative study', in J. Vymazal (ed.), *Nutrient Cycling and Retention in Natural and Constructed Wetlands*, Backhuys Publishers, Leiden, The Netherlands, pp. 85–100.

Wood, R. B. and McAtamney, C. F.: 1996, 'Constructed wetlands for waste water treatment: The use of laterite in the bed medium in phosphorus and heavy metal removal', *Hydrobiologia* **340**, 323–331.

SCALING AND MAPPING REGIONAL CALCULATIONS OF SOIL CHEMICAL WEATHERING RATES IN SWEDEN

CECILIA AKSELSSON[1], JOHAN HOLMQVIST[2], MATTIAS ALVETEG[1], DANIEL KURZ[3] and HARALD SVERDRUP[1]

[1] *Department of Chemical Engineering II, Lund Institute of Technology, P.O. Box 124, SE-221 00 Lund, Sweden;* [2] *SWECO International AB, Geijersgatan 8, SE-216 18 Malmö, Sweden;* [3] *EKG Geo-Science, Ralligweg 10, CH-3012 Bern, Switzerland*
(* author for correspondence, e-mail: cecilia.akselsson@chemeng.lth.se; phone: +46 46-222 48 66; fax: +46 46-14 91 56)

(Received 20 August 2002; accepted 1 April 2003)

Abstract. Weathering rates of base cations are crucial in critical load calculations and assessments of sustainable forestry. The weathering rate on a single site with detailed geological data can be modelled using the PROFILE model. For environmental assessments on a regional scale, the weathering rates for sites are scaled into regional maps. The step from sites to regional level requires focus on the spatial variation of weathering rates. In this paper, a method is presented by which weathering rates are calculated for 25589 Swedish sites with total elemental analysis for the soil. Based on a part of the results, a methodology for creating area covering maps by geostatistical analysis and kriging is described. A normative reconstruction model was used to transform total elemental analysis to mineralogy. Information from the Swedish Forest Inventory database and other databases were used to derive texture and other important information for the sites, e.g. climate, deposition and vegetation data. The calculated weathering rates show a regional pattern that indicates possibilities for interpolation of data in large parts of Sweden. Geostatistical analysis of an area in southern Sweden shows different properties for different base cations. Kriging was performed for potassium to demonstrate the method. It was concluded that different base cations and different regions have to be analysed separately, in order to optimise the kriging method.

Keywords: base cations, geostatistics, kriging, mineralogy, PROFILE, Sweden, weathering

1. Introduction

Estimations of weathering rates are needed in the nutrient budget calculations in calculations of critical loads of acidity (Posch *et al.*, 2001), as well as for assessing sustainable harvesting in forest ecosystems (Holmqvist, 2001). There are a few methods to estimate weathering rates. The historic weathering rate can be estimated by comparing the mineralogy at soil depths affected by weathering with not weathered layers (Olsson *et al.*, 1993). Current weathering rates can be estimated on a catchment scale through mass balance studies in catchments (Pačes, 1986). For sites with detailed information on mineralogy, soil texture, and land use, the current weathering rate also can be estimated using the PROFILE model (Sverdrup and Warfvinge, 1993, 1995; Warfvinge and Sverdrup, 1995; Sverdrup *et al.*, 1996).

The PROFILE model has been used earlier to calculate weathering rates and critical loads for 1883 Swedish forest soils (Warfvinge and Sverdrup, 1995; Bertills and Lövblad, 2002) within the Convention on Long-range Transboundary Air Pollution (CLRTAP). As such calculations were part of the scientific basis in the emission reduction negotiations within the Convention, results typically were aggregated to either the 50 km by 50 km or the 150 km by 150 km resolution used by the European Monitoring and Evaluation Programme (EMEP). While such resolutions may be useful when sharing the burden of reducing emissions between countries, a much higher resolution, and consequently many more sites, is needed to estimate the sustainable harvesting of sawlogs, pulpwood and biofuels. It has been shown that at the EMEP50 and EMEP150 resolutions, a high proportion of the uncertainty in the aggregated results is due to the limited number of sites used (Barkman and Alveteg, 2001). Recently, total elemental analyses for soil from 26754 Swedish sites were made available to us, and because a plausible mineralogy can be calculated from a total elemental analysis, this drastically improves the estimation of weathering rates in Sweden.

The aim of this paper is to describe a method by which weathering rates can be calculated for these 26754 sites and to show how geostatistical analysis can be used to create kriged maps of weathering rates in Sweden.

2. Methods

2.1. THE PROFILE MODEL

PROFILE is a steady-state soil chemistry model, originally developed to calculate the effect of acid rain on soil chemistry (Sverdrup and Warfvinge, 1993, 1995; Warfvinge and Sverdrup, 1995; Sverdrup *et al.*, 1996). PROFILE includes process-oriented descriptions of chemical weathering of minerals, leaching and accumulation of dissolved chemical components and solution equilibrium reactions. In PROFILE, the soil is divided into soil layers with different properties, preferably based on the naturally occurring soil stratification. Weathering is calculated using transition state theory and the geochemical properties of the soil system, such as soil wetness, mineral surface area, hydrogen, cation and organic acid concentrations, temperature and mineral composition.

2.2. REQUIRED INPUT

The basis for this study is 22940 till sites with total elemental analyses of soil at a depth of 1 metre, supplied by the Swedish Geological Survey (SGU), together with total elemental analyses of soil from 1897 sites from the National Forest Inventory coordinated by the Swedish University of Agricultural Sciences (SLU) and 1917 sites from Terra Mining (total of 26754 sites). Total elemental analyses were transformed to mineralogy using a normalisation model (Section 2.2.1).

TABLE I

Input data parameters to PROFILE

Parameter	Unit	Source	95-perc	5-perc
Mineral composition	Weight fraction	SGU[a]	–	–
Mean annual temp.	°C	SMHI[b]	6.8	0.35
Precipitation	m yr^{-1}	SMHI	0.95	0.65
Runoff	m yr^{-1}	SMHI	0.57	0.16
SO_4^{2-} deposition	mmol$_c$ m^{-2} yr^{-1}	SMHI & IVL[c]	56	12
NO_3^- deposition	mmol$_c$ m^{-2} yr^{-1}	SMHI & IVL	47	7.6
NH_4^+ deposition	mmol$_c$ m^{-2} yr^{-1}	SMHI & IVL	53	7.0
Cl^- deposition	mmol$_c$ m^{-2} yr^{-1}	SMHI & IVL	83	5.4
Na^+ deposition	mmol$_c$ m^{-2} yr^{-1}	SMHI & IVL	78	5.8
Ca^{2+} deposition	mmol$_c$ m^{-2} yr^{-1}	SMHI & IVL	17	4.2
Mg^{2+} deposition	mmol$_c$ m^{-2} yr^{-1}	SMHI & IVL	21	1.9
K^+ deposition	mmol$_c$ m^{-2} yr^{-1}	SMHI & IVL	5.8	1.6
BC uptake	mmol$_c$ m^{-2} yr^{-1}	SLU[d]	30	10
N uptake	mmol$_c$ m^{-2} yr^{-1}	SLU	30	10
BC in litterfall	mmol$_c$ m^{-2} yr^{-1}	SLU	140	38
N in litterfall	mmol$_c$ m^{-2} yr^{-1}	SLU	170	49
Texture class	code	SLU	–	–
Soil type	code	e	–	–
Moisture class	code	SLU	–	–
O-layer thickness	m	e	–	–
E-layer thickness	m	e	–	–
Soil depth	m	e	–	–

[a] SGU, Swedish Geological Survey.
[b] SMHI, Swedish Meteorological and Hydrological Institute.
[c] IVL Swedish Environmental Research Institute.
[d] SLU, Swedish University of Agricultural Science.
[e] Default value, see further descriptions in sections 2.2.7 and 2.2.9.

For other required input parameters (Table I), no measurements for the sites were available, but data from other sources could be used. For some parameters, existing area-covering maps were used. For others, the Surfer software was used to analyse the geostatistical properties of different national datasets and, where applicable, make area-covering maps using kriging interpolations. Values were then assigned to the 26754 sites using the ESRI ArcView software. 704 sites in the southernmost and in the mountain region in north-northwest were excluded, because these areas were not covered by the texture class information. Another 461 sites were excluded after the mineral normalisation modelling due to inconsistent mineralogies, leaving a total of 25589 sites to which PROFILE was applied.

Because not all parameters needed by PROFILE are available at the regional level, the same transfer algorithms and default values as used in Swedish critical load calculations (Bertills and Lövblad, 2002) were used in this study. CO_2 partial pressure, aluminium solubility and soil solution DOC are some examples where these data were used.

2.2.1. *Mineral Composition*

The Swedish bedrock is mainly igneous and consists of gneisses and granites and to a less extent intrusive bedrocks such as diabase and porphyr. Common soil minerals are quartz, K-feldspar, plagioclase and muscovite. There are, however, several regions with dark minerals, like hornblende, epidote, biotite and chlorite. Calcite-rich till is found mainly in regions with sedimentary bedrock, such as Scania (the southernmost part of Sweden) and the island Gotland in the east. In the study by Warfvinge and Sverdrup (1995), Sweden was divided into four different provinces with different mineralogical properties. The same provinces were used in this study, although Province 4, which in the former study consisted solely of the calcite-rich island Gotland, was extended to include all sites with a calcium oxide content larger than 4%.

The mineral normalisation modelling involved three different steps (SAEFL, 1998). In the first step the soil chemistry was transformed to base compounds, i.e. a set of normative, stoichiometrically ideal compounds from which real mineral stoichiometries can be formed as linear combinations. The base compounds were then changed into real primary minerals using specified mass balance formulas, i.e. linear combinations, based on prior knowledge of expected mineralogy and mineral stoichiometry in the area. Different mass balance formulas were used for different provinces. The resulting minerals together with the remainder of the base compounds were then used to calculate the amount of secondary minerals formed by weathering. The normalisation modelling showed an excess of base compounds for 461 sites. These sites were excluded in further calculations.

2.2.2. *Mean Temperature*

Mean annual soil temperature is used to adjust soil solution equilibrium constants, weathering rate coefficients and the nitrification rate coefficient. Mean temperature was derived from a grid map with 50 by 50 km grids, used in Warfvinge and Sverdrup (1995).

2.2.3. *Precipitation and Runoff*

Precipitation and runoff data were used to calculate convective water flux through the soil profile. Maps with yearly means for 1961–1990 from the Swedish Meteorological and Hydrological Institute (SMHI), with isolines every 100 mm for precipitation, and every 30 mm for runoff, were used. The isoline maps were converted to grid maps in ArcView.

TABLE II
Base cations uptake ($mmol_c$ m^{-2} yr^{-1}) in different parts of Sweden

Part of Sweden	County codes	BC uptake
North	AC, BD, Z	10
Central-north + southeast	H, I, S, X, Y, W	20
South-central	AB,C, D, E, F, G, K, N, OPR, T, U	30
Scania (the southernmost county)	LM	40

2.2.4. *Deposition*

Deposition of different components affects soil chemistry. The deposition of sulphur, nitrogen, chloride, and base cations was based on work by SMHI and the Swedish Environmental Research Institute (IVL) (Bertills and Lövblad, 2002), where deposition was assigned for 1883 forest inventory sites. Deposition modelled by SMHI was used as a basis, but because the model only calculated a mean deposition for all land use classes and for anions only, a Swedish deposition monitoring network coordinated by IVL was used as a complement. Deposition data are from 1997.

Deposition was kriged using only the sites where the dominant species was Norway spruce (*Picea abies*) (759 sites), because local differences caused by different stand types are not desirable in this kind of mapping, where the regional trends rather than local variations are studied.

2.2.5. *Base Cation and Nitrogen Uptake*

The uptake data are needed for the mass balance calculations. The uptake from the database in Warfvinge and Sverdrup (1995) was used, where uptake was calculated for 1883 forest inventory sites by SLU. The geostatistics of the nitrogen and base cation uptake dataset showed a large local variation such that these data were not suitable for kriging interpolation. Differences between different parts of Sweden could, however, be observed on the maps. Sweden was therefore divided into different regions with different uptake values, based on the dominating uptake level (Tables II and III).

2.2.6. *Base Cations and Nitrogen in Litterfall*

Litterfall mainly affects the soil chemistry and hence the weathering rate in the upper layers. Litterfall was based on the database with 1883 sites used in Warfvinge and Sverdrup (1995). Naturally it differs between tree species, and therefore only the spruce sites were used for kriging.

TABLE III

Nitrogen uptake (mmol$_c$ m^{-2} yr^{-1}) in different parts of Sweden

Part of Sweden	County codes	N uptake
North	AC, BD, Y, Z, W	10
Central-south	SB, C, D, E, F, G, H, I, OPR, S, T, U, X	20
South	K, LM, N	30

2.2.7. *Texture Class and Soil Type*

PROFILE needs the mineral surface area to calculate the weathering rate. Mineral surface area measurements are, however, not available on a regional level in Sweden. In the study by Warfvinge and Sverdrup (1995), mineral surface area was therefore based on soil texture classification using a total of nine soil texture classes.

Most parts of Sweden consist of sandy-loamy till, but there are several regions where finer as well as coarser till dominates. The best available data for this mapping are kriged data (5 by 5 km grids) from the Swedish National Survey of Forest Soils and Vegetation performed by the Department of Forest Soils, SLU, based on data from 23500 sites. The till classes are divided into fine till, normal till and coarse till which represent loamy till (class 6), sandy-loamy till (class 5) and sandy till (class 3) in the classification from Warfvinge and Sverdrup (1995).

Soil type was used to assign a value of the nitrification rate constant. In this study, due to lack of soil type information, the nitrification rate for iron podsol, the most common class (60 %) in the dataset from Warfvinge and Sverdrup (1995), was used for all sites.

2.2.8. *Moisture Class*

Moisture class was used to calculate the volumetric water content. In this study, soil moisture was classified according to Warfvinge and Sverdrup (1995). The best available data are kriged data (5 by 5 km grids) from the Swedish National Survey of Forest Soils and Vegetation performed by the Department of Forest Soils, SLU, based on data from 23500 sites. This dataset contains four moisture classes, which represent classes 2–5 in Warfvinge and Sverdrup (1995).

2.2.9. *Soil Depth and Thickness of Different Layers*

The soil depth was set to 1 m and was divided into four layers. The humus layer thickness was set to 10 cm, the E-layer to 5 cm and the two remaining layers to 20 and 65 cm.

TABLE IV
Statistics of weathering rates (mmol$_c$ m^{-2} yr^{-1})

Base cation	Mean	Median	St. dev.	95-perc.	5-perc.
Calcium (Ca)	38	31	a	55	17
Magnesium (Mg)	17	15	9	33	5
Potassium (K)	15	14	6	25	8
Sodium (Na)	40	37	15	70	22

[a] The standard deviation for Ca weathering is not presented, since the distribution is bi-modal, which makes the standard deviation value less interesting. The soil is calcareous on less than 1% of the sites, and there the calcium weathering rates are up to two orders of magnitude higher than on the rest of the sites. For further statistical and geostatistical analysis these sites have to be treated separately.

3. Results

The mean weathering rates for calcium (Ca), magnesium (Mg), potassium (K) and sodium (Na) for the sites are 38, 17, 15 and 40 mmol$_c$ m^{-2} yr^{-1} (Table IV).

K weathering varies between less than 10 and more than 50 mmol$_c$ m^{-2} yr^{-1} and distinct areas with different levels of weathering can be distinguished (Figure 1). There are two very densely sampled regions, in the south and in the northeast. In general, the weathering rates are lower in the northeastern region than in the southern region. The regional differences in weathering rates reflect to a large extent different soil mineralogies. K is interesting because recent studies in Sweden have shown a connection between tree vitality and K availability (Rosengren-Brinck *et al.*, 1998).

Omnidirectional semivariograms for the area in southern Sweden indicated in Figure 1 illustrate how the spatial dependence varies with distance between different sites for the different base cations (Figure 2). The semivariograms for K and Mg show increased semivariance to about 15 km after which the semivariance stays on a constant level. Thus there is spatial dependence to up to 15 km, which is information that can be used to perform kriging interpolation in the area. In Figure 3, K weathering rate has been kriged using a spherical model illustrated in Figure 2, to demonstrate the method. The semivariance of Ca weathering increases with distance to 20–25 km, after which the semivariance stays at a constant level. For Na weathering, the semivariance increases continuously, which indicates a spatial dependence within the distance of 40 km covered by the semivariogram. However, the gradient of the curve decreases at about 15 km, and then increases at 30 km. This indicates that there is a spatial trend in the dataset, and it should be detrended before kriging, which is exemplified in Figure 2 (filled circles).

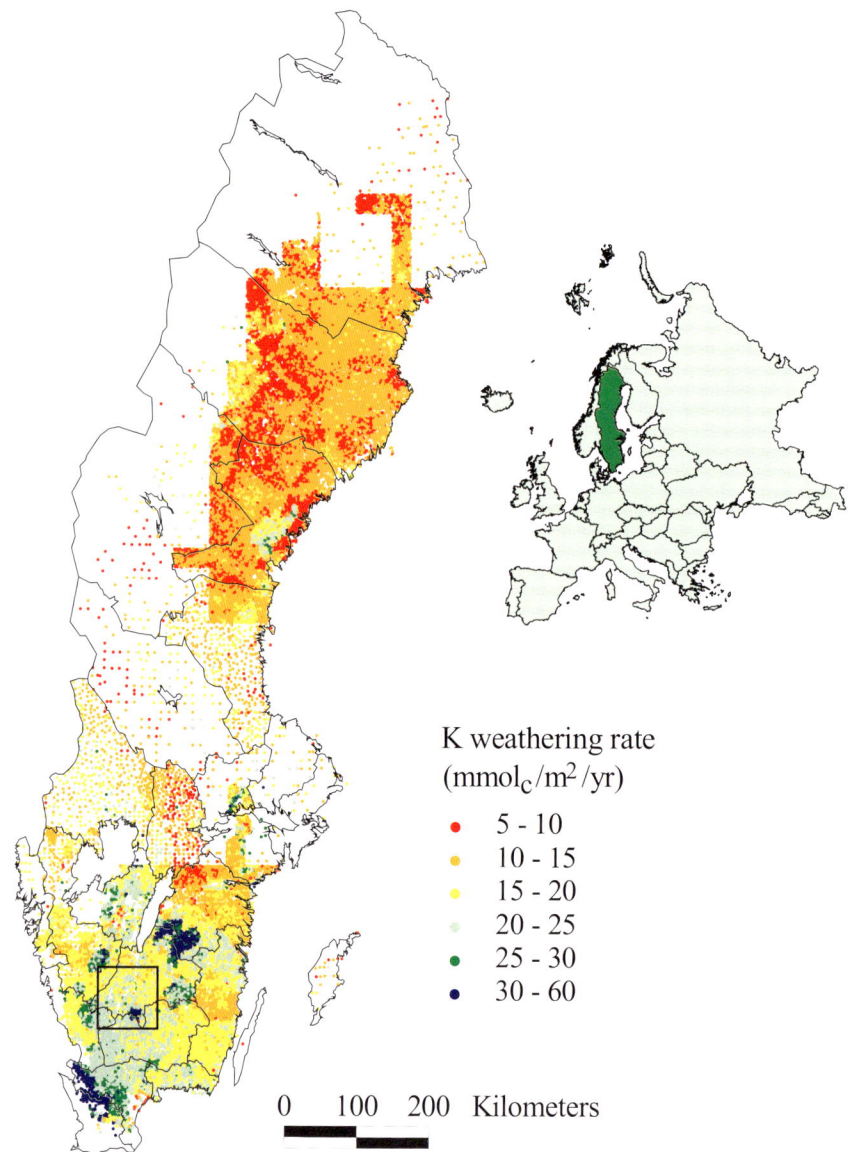

Figure 1. K weathering on 25589 sites, calculated with PROFILE. The marked area in southern Sweden is used for geostatistical analysis.

4. Discussion

In this study, long-term annual average weathering rates were modelled on 25589 sites, and the geostatistical properties of the modelled weathering rates show that valid area covering maps can be constructed using kriging interpolation. The weathering rates for the different base cations have different geostatistical properties, and

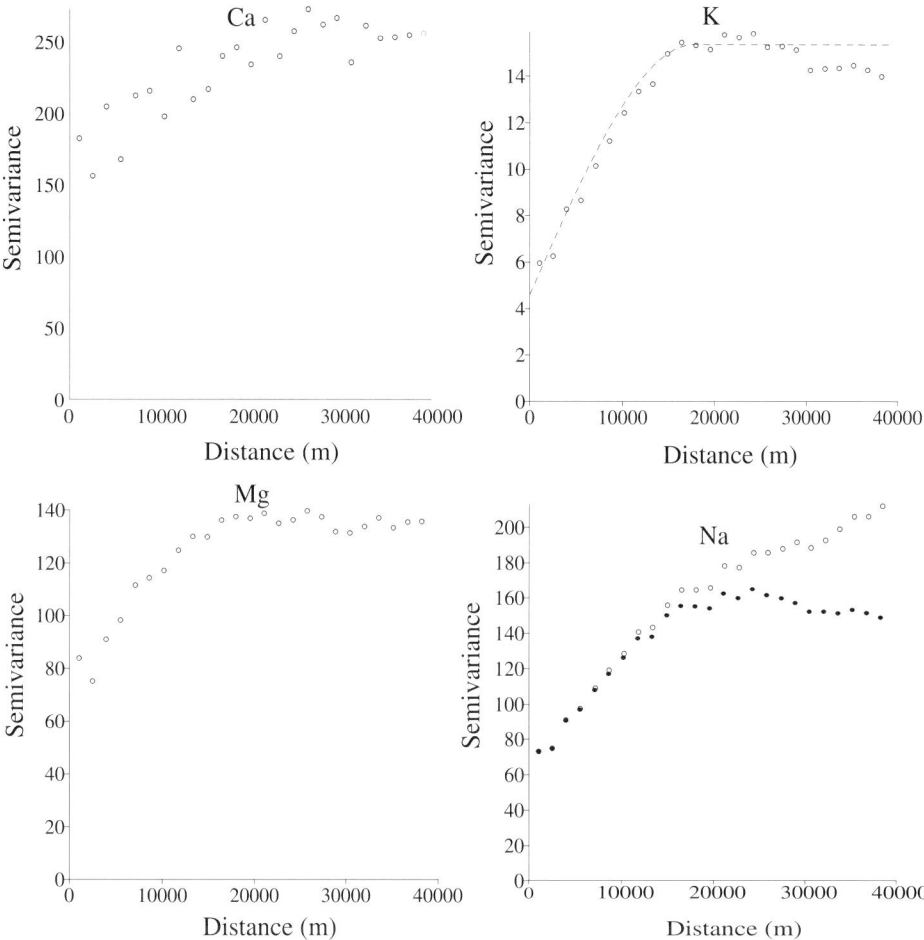

Figure 2. Semivariograms for weathering rates of Ca, K, Mg and Na for an area in southern Sweden (Figure 1). The dashed line in the K semivariogram is a spherical function used for kriging (Figure 3). The filled circles in the Na semivariogram show semivariance after linear detrending.

they consequently have to be analysed separately in order to optimally apply the kriging method.

The weathering data for the 25589 sites show a regional pattern, with distinct areas with different weathering rates. Density of sites as well as geostatistical properties differ between regions, which implies that different regions have to be geostatistically analysed separately. In areas where the mineralogy varies substantially over short distances, the variation in weathering rates may be too large to be captured by the sites, i.e. no spatial dependence will exist. In such areas, more detailed information is needed.

PROFILE is a widely used model and extensive uncertainty analyses have been made (cf. Barkman and Alveteg, 2001). The input data used for this study have

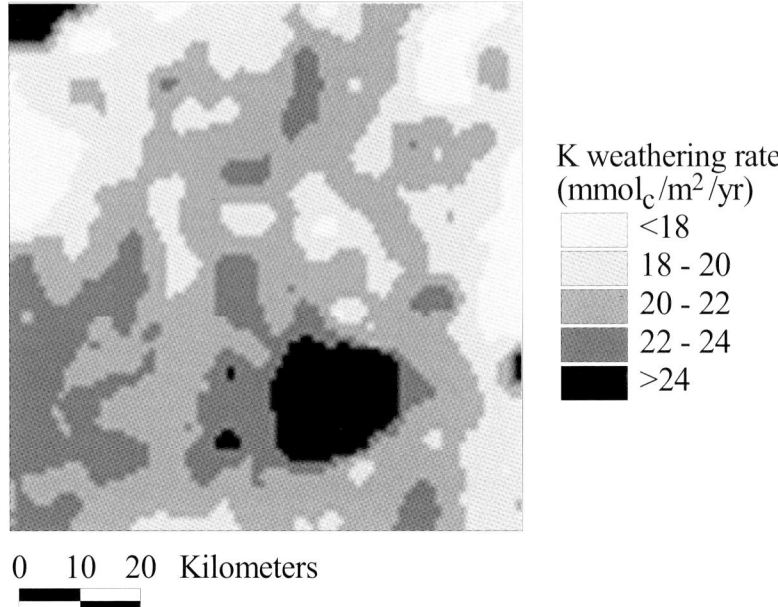

Figure 3. Kriged K weathering in an area in southern Sweden (compare Figure 1).

different spatial resolution, which is important to keep in mind when using the resulting weathering maps. They should be used on a national and regional level, but not locally, because local variations, in texture and deposition for example, are not captured by the input data.

Although this mapping was successful, there are several opportunities for improvements. We used deposition, litterfall and uptake data for spruce forest only. In future mappings, calculations of weathering will be made for pine and deciduous forest, as well. In these calculations, the soil depth will be adapted to the root depth of the different tree species. Other species-dependent variables, like deposition, also will influence the calculated weathering rate.

Mineralogical data are critical for calculating weathering in PROFILE and the mineralogy normalisation model is being developed further in Switzerland. Field measurements in Sweden will improve the knowledge about chemistry-mineralogy connections. Another key parameter is texture, and attempts will be made to optimise the texture data by combining different sources (maps, field measurements, etc.). The results from this study will be further analysed, in order to define other critical parameters that should be prioritised for improvements.

5. Conclusions

The described methodology can be used to create area-covering maps of weathering rates in Sweden. The geostatistical analysis has to be carried out separately for different areas and different base cations in order to optimise the kriging method. In some areas, where the local variation is too large to be captured by the currently available sites, more information will be needed to create area-covering maps.

Acknowledgements

This study was financed by ASTA (International and National Abatement Strategies for Transboundary Air Pollution) and SUFOR (Sustainable Forestry in Southern Sweden), two research programs in Sweden. Part of the dataset used in this publication is based on the Swedish National Survey of Forest Soils and Vegetation, performed by the Department of Forest Soils, SLU. Other data sources were SGU (Swedish Geological Survey), IVL Swedish Environmental Research Institute, SMHI (Swedish Meteorological and Hydrological Institute) and Terra Mining. The authors are solely responsible for the interpretation of these data.

References

Barkman, A. and Alveteg, M.: 2001, 'Identifying potentials for reducing uncertainty in critical load calculation using the PROFILE model', *Water, Air, Soil Pollut.* **125**, 35–54.

Bertills, U. and Lövblad, G.: 2002, 'Kritisk belastning för svavel och kväve', *Technical Report, Rapport 5174*, Naturvårdsverket, Stockholm (in Swedish).

Holmqvist, J.: 2001, 'Modelling chemical weathering in different scales', *Ph.D. Thesis*, Department of Chemical Engineering II, Lund University.

Olsson, M., Rosén, K. and Melkerud, P-A.: 1993, 'Regional modelling of base cation losses from Swedish forest soils due to whole-tree harvesting', *Appl. Geochem. Suppl. Issue* **2**, 189–194.

Pačes, T.: 1986, 'Weathering rates of gneiss and depletion of exchangeable cations in soils under environmental acidification', *J. Geol. Soc.* **143**, 673–677.

Posch, M., de Smet, P. A. M., Hettelingh, J-P. and Downing, R. (eds): 2001, 'Modelling and Mapping of Critical Thresholds in Europe. Status Report 2001', *Technical Report, RIVM Report No. 259101010*, Coordination Center for Effects, Bilthoven, The Netherlands.

Rosengren-Brinck, U., Nihlgård, B., Bengtsson, M. and Thelin, G.: 1998, 'Samband mellan barrförlust, barrkemi och markkemi i Skåne', *Technical Report*, Barrförlust och luftföroreningar, Rapport 4890, Naturvårdsverket, Stockholm.

SAEFL: 1998, 'Critical loads of acidity for forest soils', *Technical Report, Environmental Documentation No. 88, Air/Forest*, Swiss Agency for the Environment, Forest and Landscape (SAEFL), Bern, Switzerland.

Sverdrup, H. and Warfvinge, P.: 1993, 'Calculating field weathering rates using a mechanistic geochemical model (PROFILE)', *J. Appl. Geochem.* **8**, 273–283.

Sverdrup, H. and Warfvinge, P.: 1995, 'Estimating field weathering rates using laboratory kinetics', in A. F. White and S. L. Brantly (eds), *Chemical Weathering of Silicate Minerals*, Mineralogical Society of America, Washington DC, Reviews in Mineralogy, Vol. 31, pp. 485–541.

Sverdrup, H., Warfvinge, P. and Rosén, K.: 1996, 'Critical loads of acidity and nitrogen, based on multiple criteria for different Swedish ecosystems', *Water, Air, Soil Pollut.* **85**, 2375–2380.

Warfvinge, P. and Sverdrup, H.: 1995, 'Critical Loads of Acidity to Swedish Forest Soils', *Technical Report*, Department of Chemical Engineering II, Lund University, Lund.

FINNISH LAKE SURVEY: THE ROLE OF CATCHMENT ATTRIBUTES IN DETERMINING NITROGEN, PHOSPHORUS, AND ORGANIC CARBON CONCENTRATIONS

MIITTA RANTAKARI, PIRKKO KORTELAINEN, JUSSI VUORENMAA,
JAAKKO MANNIO and MARTIN FORSIUS

Finnish Environment Institute, P.O. Box 140, 00251 Helsinki, Finland
(* author for correspondence, e-mail: miitta.rantakari@environment.fi; phone: +358 9 40300385;
fax: +358 9 40300390)

(Received 20 August 2002; accepted 27 April 2003)

Abstract. This study is based on a Finnish lake survey conducted in 1995, a dataset of 874 statistically selected lakes from the national lake register. The dataset was divided into subgroups to evaluate lake water-catchment relationships in different geographical regions and in lakes of different size. In the three southernmost regions, the coefficients of determination in multiple regression equations varied between 0.40 and 0.53 for total nitrogen (TN) and between 0.37 and 0.53 for total phosphorus (TP); the best interpreters were agricultural land and water area in the catchment. In the two northernmost regions, TN concentrations in lake water were best predicted by the proportion of peatlands in the catchment, the catchment slope, and TP concentrations by lake elevation and latitude. Coefficients of determination in multiple regression equations in these northern regions varied between 0.39 and 0.67 for TN and between 0.41 and 0.52 for TP. For all the subsets formed, the best coefficients of determination explaining TN, TP, and total organic carbon (TOC) were obtained for a subset of large lakes (>10 km^2), in which 72–83% of the variation was explained. This was probably due to large heterogeneous catchments of these lakes.

Keywords: catchment, Finland, lake survey, nitrogen, organic carbon, phosphorus

1. Introduction

The concentrations of total nitrogen (TN), total phosphorus (TP), and total organic carbon (TOC) of lake waters are influenced by many in-lake, catchment, and atmospheric processes. The complexity of these processes has led researchers to seek empirical relationships between water quality and catchment characteristics (e.g., Rapp *et al.*, 1985; Rasmussen *et al.*, 1989; Kortelainen, 1993; D'Arcy and Carignan, 1997; Herlihy *et al.*, 1998; Kopáček *et al.*, 2000; Mander *et al.*, 2000). Subsets of a large database usually produce regression equations with significantly smaller standard errors than equations formed for the whole database (Rasmussen *et al.*, 1989). However, if regional models are formed for areas with very homogeneous land use, the explanatory power of the model may be weaker than that of a model formed for a wider and more heterogeneous area (Herlihy *et al.*, 1998).

There are almost 30 000 lakes with a surface area greater than 0.04 km^2 in Finland. Finnish lakes typically are coloured by high concentrations of TOC. Nutrients (N,P) in forested and pristine areas are transported in an organic form (Kortelainen and Saukkonen, 1998). Finland is a large sparsely populated country and habitation is concentrated mainly in southern coastal regions. Municipal waste loading and industry, as well as non-point loading from agriculture, increase nutrient concentrations in lakes in these more densely populated areas.

The large survey dataset of this study, consisting of 874 randomly selected lakes, enabled us to test whether the empirical relationships between land use and water quality differ in different geographical regions and in lakes of different size. The results provide new information about possibilities to extrapolate regression equations for different kinds of lakes and geographical regions. We also compared the predictability of certain water quality parameters (TOC, TN and TP) on the basis of catchment attributes.

2. Materials and Methods

The data set is based on a Finnish lake survey conducted in the autumn of 1995 (Henriksen *et al.*, 1996). Altogether 874 lakes were selected using stratified random sampling with unequal sampling fractions from the national lake register, with the requirement that the minimum of 1% of the population of lakes within any county/region was included. The other selection requirement was that the proportions of lakes in size classes 0.04–0.1, 0.1–1, 1–10, and 10–100 km^2 were 1:1:4:8, respectively. All of the lakes >100 km^2 were included. One water sample was taken from the middle of each lake shortly after the autumn overturn and 24 chemical parameters were measured. Total nitrogen was analysed colourimetrically after oxidation to NO_3^--N, the sum of NO_3^--N and NO_2^--N colourimetrically by autoanalyser after reduction to NO_2^--N, and NH_4^+-N colourimetrically with hypochlorite and phenol. Total organic nitrogen (TON) was calculated as the difference between total and inorganic nitrogen. TP was measured by a colorimetric method after oxidation and PO_4^{3-}-P by spectrophotometric determination. TOC was analysed by oxidation to CO_2 followed by IR-measurement (Henriksen *et al.*, 1996; Mannio *et al.*, 2000).

The catchment areas of the lakes were determined on topographic maps. Catchment boundaries were digitized and combined with land use data based on satellite images using the ArcView georeferencing software. Catchment area, lake area and proportions of peatland, forest on mineral soil, agricultural land, water consisting of the lake itself and the upstream water bodies, and built-up area in the catchments were determined, as well as lake elevation, mean catchment altitude, mean catchment slope, and the ratio of catchment area to lake area. For the largest lakes (>10 km^2), mean catchment altitude and mean slope were not determined due to the very large size of the catchments. The location of each lake in the south-north direction was expressed as latitude (geographical coordinates, minutes and seconds).

The relationships between the catchment attributes and the lake TN, TP and TOC concentrations were examined using Pearson's correlation coefficients and stepwise multiple linear regression analyses. In the regression analyses, all cases with an absolute value of the Studentized residual exceeding 3 were excluded; only the independent variables with p-values less than 0.05 were included in the model (PROC REG, SAS Institute Inc., 1989). Because of the missing mean catchment altitude and mean slope for the largest lakes, two alternate stepwise multiple regression equations were performed for each region and size group: one with all the catchment parameters as possible independent variables and another without mean catchment altitude and mean slope. If there was no significant difference (0.02) in the coefficient of determination between these two regression equations, the equation without mean catchment altitude and mean slope was chosen, because it also included the largest lakes. The water quality and the catchment parameters were $\log_e$ or square root transformed if it was necessary to improve the normality of the distribution.

In order to study the relationships between catchment characteristics and the lake water in different parts of Finland, the country was divided into 5 geographical regions (Figure 1). Region 1 is characterised by coastal lowlands with low lake density and clay soils with intensive agriculture. Especially in northern parts of the region, scattered zones of forested upland with bedrock outcrops occur. This region has the highest population density in Finland. Region 2 is called the lakedistrict due to its high number of lakes, and especially large chain-like watercourses with forested catchments and intensive forestry. Region 3 is topographically very flat and has a low lake density. High proportion of peatlands and cultivated clay soil characterise this area. Region 4 has a combination of the characteristics of Regions 2, 3, and the southern parts of Region 5. Region 5 has the coldest climate and the lowest human impact. Large peatlands in southern areas, and bedrock outcrops and a high density of small lakes in the northern parts, are characteristic for this region (Mannio *et al.*, 2000). The underlying bedrock in all five regions is predominantly base-poor igneous and metamorphic rock.

In order to study whether better relationships are achieved between water quality and catchment characteristics in lakes located in similar geographical regions or in lakes of a certain size, two subgroups according to the lake size were separated: 0.04–0.1 km^2 and >10 km^2.

TABLE I

The median values for the lake and catchment characteristics in the five regions, in lakes in size groups <0.1 km² and >10 km² and in the whole data set

	Whole data set	Lakes		Region				
		<0.1 km²	>10 km²	1	2	3	4	5
	$n = 874$	$n = 305$	$n = 89$	$n = 146$	$n = 357$	$n = 67$	$n = 141$	$n = 163$
Lake area (LA) (km²)	0.22	0.06	82	0.30	0.28	0.23	0.21	0.14
Catchment area (CA) (km²)	4.6	1.2	1500	5.2	5.8	3.9	4.6	3.3
Water (%)	8.8	6.3	15	10	9.3	5.1	8.2	7.4
Agricultural land (%)	2.3	0.0	5.0	4.8	5.3	2.2	1.4	0.0
Forest (%)	65	73	62	68	69	53	56	62
Peatland (%)	14	11	17	5.6	8.8	30	30	19
Built-up area (%)	0.1	0.0	0.4	0.3	0.2	0.2	0.0	0.0
CA/LA	16	20	20	12	16	20	16	20
TOC (mg L^{-1})	7.7	8.8	6.6	8.7	9.7	14	7.4	5.4
TN (μg L^{-1})	430	450	390	490	470	650	335	270
(NO$_3^-$-N + NO$_2^-$-N + NH$_4^+$-N)/TN (%)	7.7	6.9	16	16	9.4	5.8	5.4	3.4
TP (μg L^{-1})	13	14	11	14	15	30	15	7.0
PO$_4$/TP (%)	15	15	17	13	16	11	17	17
pH	6.6	6.5	7.0	6.5	6.5	6.4	6.8	6.8
Alkalinity (μeq L^{-1})	120	100	160	130	110	80	120	120
Conductivity (mS m^{-1})	3.0	3.0	4.6	4.4	3.8	3.2	2.3	2.5

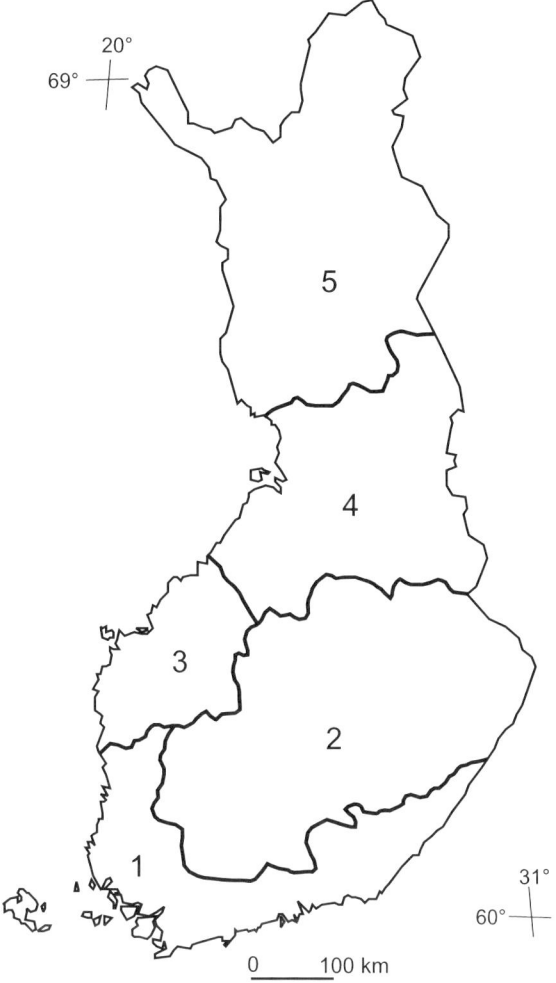

Figure 1. Geographical regions in Finland: Region 1, southern coast; Region 2, lake district; Region 3, western coast; Region 4, Oulu Region; Region 5, northern Finland.

3. Results and Discussion

3.1. Nitrogen

The median concentration of TN in different regions ranged between 270 and 650 μg L^{-1}, being highest on the western coast (Region 3) and lowest in Lapland (Region 5) (Table I). The median concentrations of inorganic nitrogen (NO$_2^-$ + NO$_3^-$ + NH$_4^+$) ranged from 8 to 67 μg L^{-1}, with the highest concentrations in southern Finland (Region 1) and the lowest in Lapland. The water quality of the lakes in different regions is more completely described in Mannio *et al.* (2000).

TABLE II

Regression equations predicting TN in the whole dataset, in Regions 1, 2, 3, 4 and 5, and in the small lakes (<0.1 km²) and the large lakes (>10 km²). TN is $\log_e$ transformed in all equations. Abbreviations and units for the catchment variables: LA = lake area (ha), CA = catchment area (ha), LAT = latitude (min, sec), SLOPE = catchment slope (%), ALT = altitude (m), ELEV = elevation (m), FIELD = agricultural land (%), WATER = water (%), PEAT = peatland (%), BUILT = built up area (%), FOREST = forest (%)

	n	Intercept											R^2
All	697	11.7	-0.0640	LAT	-0.220	√SLOPE	0.0739	ln FIELD	-0.0974	√WATER	-0.167	ln ALT	0.58
1)	125	6.64	0.130	√FIELD	-0.122	√WATER	-0.126	√SLOPE					0.53
2)	283	8.93	0.0864	√FIELD	-0.142	√WATER	-0.470	ln ALT	-0.135	ln SLOPE			0.50
3)	66	24.0	0.124	√FIELD	-0.263	ln ALT	-0.260	LAT					0.40
4)	99	6.66	-0.300	ln SLOPE	-0.0849	√WATER	0.0536	ln BUILT					0.39
5)	153	6.14	0.160	√PEAT	-0.0947	ln CA/LA	-0.141	ln SLOPE	0.123	ln BUILT	-0.0684	ln LA	0.67
Small	304	12.4	-0.093	LAT	-0.227	ln SLOPE	0.0840	ln FIELD					0.58
Large	88	11.6	0.160	√FIELD	-0.240	√WATER	-0.0810	LAT					0.77

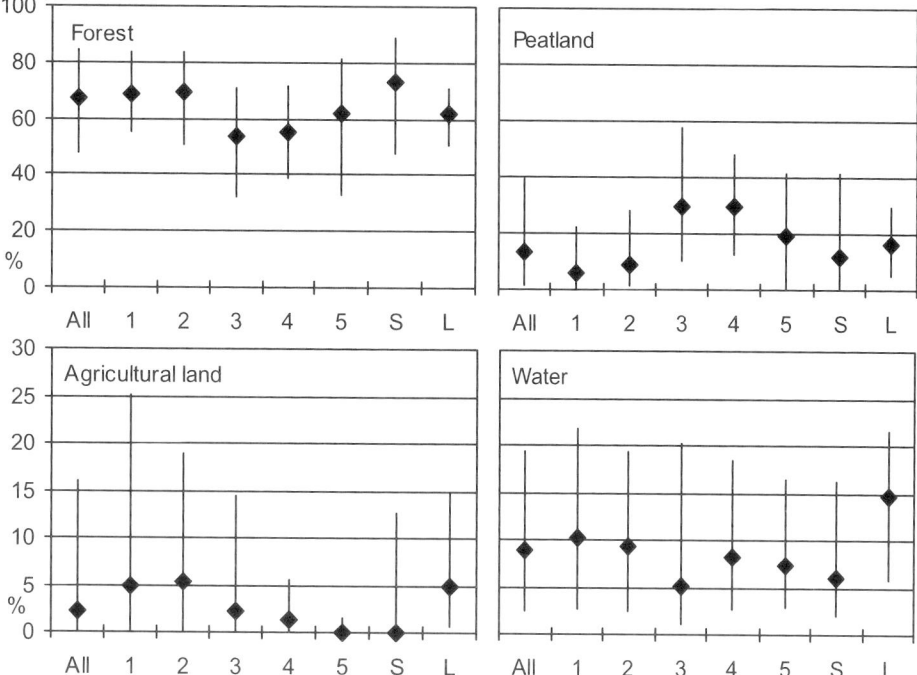

Figure 2. Medians and the difference between the 10th and 90th percentiles for forest, peatland, agricultural land, and water proportions in the catchment in the whole dataset, in Regions 1, 2, 3, 4 and 5, and in the small lakes (S) (<0.1 km^2) and the large lakes (L) (>10 km^2).

Catchment characteristics explained 39–67% of the variation of the lake water TN in different regions. In the whole dataset, the three best interpreters for TN in the multiple regression equation were latitude, mean catchment slope, and the proportion of agricultural land in the catchment (Table II). In the three southernmost regions (Regions 1, 2 and 3), the catchment area covered by agricultural land was the main factor explaining the variation of TN, as well as of organic and inorganic nitrogen. In these regions the coefficients of determination for TN in the regression equations varied from 0.40 to 0.53 (Table II). Farming in Finland is concentrated in the southern part of the country (Figure 2) and it has been shown in previous studies that agriculture is an important source of nitrogen in watercourses (Rekolainen, 1989; Arhaimer and Lidén, 2000; Mander et al., 2000). The proportion of water area in the catchment was another important factor explaining variation of TN in southern Finland. Water proportion was more strongly negatively correlated with the organic fraction of nitrogen than with the inorganic fraction, probably due to more effective sedimentation of organic matter compared to inorganic nutrients.

In southern Finland, organic nitrogen was related to agricultural land, while in northern Finland, the peatland proportion of the catchment correlated best with organic nitrogen. This suggests that in southern Finland, organic nitrogen is predominantly either transported from the fields or is autochthonous nitrogen bound to algae, as lakes surrounded by fields tend to be more eutrophic than lakes surrounded by peatlands. Recently, it has also been reported that notable amounts of organic matter and DOC can be leached from agricultural lands (Correll et al., 2001; McTieran et al., 2001), undoubtedly including organic nitrogen. The TOC/TON ratio of lake water in southern Finland (Regions 1 and 2) was negatively related to the proportion of agricultural land in the catchment ($r = -0.44$ and $r = -0.46$; $p < 0.0001$) and positively related to the peatland proportion ($r = 0.53$ and $r = 0.54$; $p < 0.0001$), probably indicating the different nature of organic matter originating from peatlands and in-lake processes or fields.

In the two northernmost regions (Regions 4 and 5), the mean slope of the catchment and the peatland proportion had the best correlation with TN, respectively. In Region 5, the multiple regression analyses resulted in an equation including 5 variables, which together explained 67% of the variation (Table II). It has been suggested that in Finland, peatlands are a source of organic nitrogen (Kortelainen et al., 1997). Also, TOC was strongly related to the peatlands in Region 5. There is very little agriculture in Region 5 and fields in the catchment had little effect on the TN concentrations of lakes ($r = 0.25$; $p < 0.01$). Where the human impact on nitrogen transport is minor, the natural sources of nitrogen are emphasized. In the southern regions, the nitrogen leached from agricultural land dominates over the natural sources of nitrogen, and therefore no connection between nitrogen concentration of the lake water and the peatland proportion, and only a very weak connection between catchment slope and lake water nitrogen, was detected.

Most of the nitrogen (97% of TN) in lakes in northern Finland was organic (compared with 84% in Region 1). However, in northern Finland, the proportion of water in the catchment did not correlate negatively with lake water nitrogen, as was the case in the three southernmost regions. Most of the organic nitrogen in the north is bound to dissolved organic carbon leached from the peatlands, whereas in southern regions, the lakes are typically more eutrophic and a greater part of organic nitrogen is probably particulate, originating from autochthonous processes, and thus sediments more easily. This is supported by the observation that in southern Finland, in the subgroup of lakes located in agricultural areas (agricultural land, 15–40%; peatlands, 0–6%), TN was strongly related to the water percentage ($r = -0.66$; $p < 0.0001$), but in more peatland and forest dominated areas (peatlands, 11–34%; agricultural land, 0–7%), the relationship with water was weak ($r = -0.23$; $p < 0.29$). This type of exercise cannot be made for northern Finland, because in Region 5, field percentage was >4% in only seven lakes.

The ice cover period is also longer in northern Finland than in southern parts of country and the degradation of organic matter can be expected to be more efficient in the south. The higher deposition of nitrogen in the south may enhance the degradation of organic matter and thus the upper water bodies regulate the concentrations of TOC, TN, and TON. Lepistö et al. (1995) observed that catchments with high proportions of organic soil had high retention of NO_3^- and concluded that one possible explanation might be the high C/N ratio of soil organic matter, due to immobilization of inorganic N and increased decomposition of organic matter as a result of microbial activity. In northern Finland, probably the low deposition of nitrogen and the colder climate both restrict the decomposition of organic matter, decreasing organic nitrogen retention in watercourses.

In Region 4, catchment slope was the main predictor for TN concentrations in lakes. The multiple regression analyses of the data resulted in an equation with three variables included and an R^2 of 0.39 (Table II). Catchment slope correlated negatively with peatland percentage of the catchment, as peatlands typically are formed in topographically flat areas. Part of the negative correlation between TN in lakes and mean catchment slope can undoubtedly be explained by the greater proportion of peatlands in flat areas. Catchment slope, however, also appears to have a direct influence on lake water TN concentrations. In Region 4, the peatland proportion of the catchment did not replace the catchment slope in the regression model if the slope was left out of the model, and the coefficient of determination was significantly lower. On steep slopes, the vegetation is often sparse, especially if the soil cover is thin, and only little organic nitrogen is leached (Kopáček, 2000). On the other hand, when slopes are steep there is no formation of waterlogged soil. This enables the water to percolate deeper into the soil, where at least part the inorganic nitrogen is adsorbed to the mineral soil.

Finland is a long country; the distance between the southernmost and northernmost points is almost 1200 km. Latitude explains reasonably well the TN concentrations for the whole dataset and in different lake size groups. The TN concentrations decrease towards the north in Finland, as agriculture, habitation as well as aerial deposition of nitrogen are concentrated in the southern parts of the country and decrease toward the north.

For the large lakes (>10 km^2), the percentage of agricultural land in the catchment was the best predictor of TN. The multiple regression analyses yielded an equation with three variables, which explained together 77% of the variation (Table II). There were also significant correlations between TN and built-up area, latitude, and proportion of water in the catchment, $R^2 = 0.25$, $R^2 = 0.22$, and $R^2 = 0.17$ ($p < 0.0001$), respectively. For the smallest lakes (<0.1 km^2), the regression equation was somewhat different. Latitude, mean catchment slope, and the proportion of fields in the catchment explained together only 58% of the variation of TN, and the correlation between agricultural land and TN was low ($R^2 = 0.26$; compared to $R^2 = 0.55$ in the large lakes).

3.2. PHOSPHORUS

The median TP concentrations ranged from 7 to 30 μg L^{-1} in the five regions, with a median for the whole data set of 13 μg L^{-1} (Table I). Similar to the regional distribution of TN, TP concentrations were highest in Region 3 and lowest in Lapland (Region 5). Leaching of TP from unmanaged forested catchments has been shown to be significantly lower than from managed forested catchments (Mattsson et al.,

2003) or from agricultural land (Rekolainen, 1989). In Lapland, the catchments are more pristine and the effects of agriculture are minor compared to the rest of the country.

The regression equations explaining TP concentrations in lakes in different geographical regions differed more from each other than those for TN. Whereas the equation of TN for the whole dataset explained 58% of the variation, for TP the equation for the whole dataset explained only 49% of the variation (Table III).

The proportion of agricultural land in the catchment explained best the variation of TP concentrations in Region 1, and the multiple regression analyses resulted in an equation with four variables included and an R^2 of 0.53 (Table III). The weakest explanatory power was in the multiple regression equation for Region 2, with an R^2 only 0.37. In Region 2, the average proportion of the agricultural land is as high as in Region 1 (Table I), but in this area there are also naturally phosphorus-rich soils (Koljonen, 1992) and the impact of agriculture on the phosphorus status of the lakes is probably partly obscured by natural sources of phosphorus. Moreover, the agricultural land in this area is commonly used for pasture and grass crops and less than in Region 1 for growing cash crops (Agricultural Census, 2000). In Region 2, the cultivated fields are less frequently situated on clay soils than in southern Finland and the leaching of phosphorus by erosion is probably minor compared to Region 1. Arheimer and Lidén (2000) obtained supporting results from agricultural catchments in southern Sweden, where TP in streamwater was not correlated to the fraction of arable land in the catchment, but instead was strongly correlated to the soil texture. In Region 2, the water percentage of the catchment was the most important characteristic explaining the variation of TP. In the lake district (Region 2), chain-like watercourses are very typical and this confirms the earlier finding that at least particulate phosphorus is sedimented in the upper watercourses (Arheimer and Lidén, 2000).

The proportion of peatlands in the catchment correlated positively with lake water phosphorus in the two northern regions, especially in Lapland (Region 5). In Region 5, the multiple regression equation explained 52% of the variation (Table III). Peatlands have been reported in some studies to be greater sources of phosphorus than mineral soils (Dillon *et al.*, 1991). Dillon and Molot (1997) suggested that phosphorus is transported from peatlands combined with humic matter. On the other hand, the phosphorus absorption capacity of peatlands, at least in nutrient-poor, acid *Sphagnum* peats, is very weak due to their low Fe and Al content, and a great part of the phosphorus entering the peatland may be leached to the watercourses (Nieminen and Jarva, 1996).

In Regions 3 and 4, the lake elevation was an important predictor for TP concentrations in lakes (Table III). The high proportion of peatlands and cultivated clay soil characterise the coastal plains of these regions. These regions are situated between the southern Finland, where the agricultural land is an important source of phosphorus in lakes, and the northernmost region (5), where the peatlands appear to be the main sources of phosphorus. The lake elevation in these regions

TABLE III

Regression equations predicting TP in the whole dataset, in Regions 1, 2, 3, 4 and 5, and in the small lakes (<0.1 km²) and the large lakes (>10 km²). TP is log$_e$ transformed in all equations. Abbreviations and units for the catchment variables, cf. Table II

	n	Intercept									R^2
All	810	11.2	LAT	-0.105	√WATER	-0.265	FOREST	-0.0172	ln FIELD	0.0700	0.49
1)	123	4.11	√FIELD	0.141	√WATER	-0.204	FOREST	-0.0104	ELEV	-0.00354	0.53
2)	286	8.18	√WATER	-0.371	FOREST	-0.0172	ln ALT	-0.640			0.37
3)	66	38.2	ELEV	-0.00880	LAT	-0.528	FOREST	-0.0117	√FIELD	0.168	0.52
4)	101	15.4	ELEV	-0.00200	ln BUILT	0.198	PEAT	0.0155	LAT	-0.187	0.41
5)	159	15.8	LAT	-0.196	√PEAT	0.138	ln BUILT	0.179	ln FIELD	0.0896	0.52
Small	300	12.2	LAT	-0.127	FOREST	-0.0152	√WATER	-0.259	ln FIELD	0.0728	0.54
Large	88	5.78	√WATER	-0.635	√FIELD	0.220	FOREST	-0.0220			0.72

probably combines the impacts of these two catchment variables and explains TP concentrations in lakes better than the proportion of agricultural land or peatland alone.

For the subset of the large lakes (>10 km^2), water percentage of the catchment strongly controlled the concentration of TP, and the multiple regression equation with three variables explained 72% of the variation (Table III). The large lakes have large catchments, and in many cases numerous basins, upstream. Supposedly most of the phosphorus leached from the catchment is consumed, decomposed or sedimented in upper basins. As with TN, fields in the catchment also appeared to increase TP concentrations in large lakes. For the small lakes (<0.1 km^2), latitude explained best the variation of TP and the coefficient of the determination was 0.54 (Table III). Phosphorus concentrations in lakes decrease in Finland towards the north, as habitation, agriculture and industry all are concentrated in southern Finland. This can probably be seen better in the subset of small lakes than in the subset of large lakes, as the small lakes are more evenly distributed over the whole country, whereas the majority of the large lakes are situated south of Region 5.

3.3. TOTAL ORGANIC CARBON

The median TOC in this data set was 7.7 mg L^{-1}, ranging between 5.4 and 14 mg L^{-1} in the five regions (Table I) (Mannio et al., 2000). The catchment characteristics explained 31–68% of the variation of TOC of the lake water in the different regions. In southern Finland (Regions 1, 2 and 3), the catchment area covered by water was the main feature explaining the variation of TOC, whereas in northern Finland (Region 5), peatland proportion of the catchment alone explained 60% of the variation of TOC (Table IV). In northern Finland, most of the peatlands in the catchments are still in a pristine state. In southern Finland, there are some pristine peatlands as well as catchments with large areas of ditched peatlands. This probably makes it more difficult to determine the relationship between TOC concentrations and peatlands of the catchment compared to northern Finland, as was suggested previously on the basis of a lake survey dataset from 1987 (Kortelainen, 1993). In Region 3, peatlands explained only 10% of the variation of TOC. This region is characterized by a high proportion of peatlands in almost every catchment (Figure 2). As the topography of the area is very flat, the lakes are correspondingly shallow and their volume is low. This results in highly coloured waters, the water colour median being 155 Pt mg L^{-1}. Also the TOC concentrations are much higher in this area than in other parts of Finland (Table I).

For the large lakes (>10 km^2; $n = 88$), the catchment variables explained as much as 83% of the variation of TOC (Table IV). The proportion of water in the catchment was the best predictor of TOC in the subset of large lakes. Most of these lakes are drainage lakes with one or more upstream water bodies. Decomposition of TOC in upstream lakes is most probably the explanation for negative correlation between the TOC concentration and the proportion of water. According to Cole and

TABLE IV

Regression equations predicting TOC in the whole dataset, in Regions 1, 2, 3, 4 and 5, and in the small lakes (<0.1 km²) and the large lakes (>10 km²). TOC is $\log_e$ transformed in all equations except in the equation for Region 3 where TOC is untransformed. Abbreviations and units for the catchment variables, cf. Table II

	n	Intercept							R^2
All	714	10.3	-0.195 √WATER	-0.119 LAT	0.0767 √PEAT	-0.163 √SLOPE			0.62
1)	124	3.20	-0.206 √WATER	-0.212 √SLOPE	0.0559 √PEAT				0.68
2)	287	3.14	-0.238 √WATER	-0.190 ln SLOPE	0.0439 √PEAT				0.57
3)	66	15.3	-2.376 ln WATER	1.71 √FIELD					0.43
4)	141	2.05	0.0144 PEAT	-0.153 √WATER					0.31
5)	163	3.36	0.265 √PEAT	-0.0493 LAT	0.0103 FOREST	-0.0575 ln CA/LA			0.65
Small	305	10.0	-0.123 LAT	0.0979 √PEAT	-0.225 ln WATER				0.54
Large	88	12.5	-0.288 √WATER	-0.164 LAT	0.204 √PEAT				0.83

Caraco (2001), even very old terrestrial carbon can be decomposed in the aquatic ecosystem rather rapidly.

For the small lakes (<0.1 km^2; $n = 305$), the variables in the multiple regression equation were the same, but the order was different: latitude, peatlands, and water explained together 54% of the variation.

3.4. DIFFERENCES IN RELATIONSHIPS OF CATCHMENT ATTRIBUTES AND WATER QUALITY BETWEEN SUBGROUPS OF THE LAKES

The highest coefficients of determination in southern Finland (Regions 1 and 2) were obtained for regression equations explaining variation of TOC. The regression equations explaining TOC also reached the same or higher coefficient of determination by fewer variables than equations explaining TN or TP. This also was true for the subset of the large lakes that are mostly situated in southern Finland. The better explanatory power of regression equations explaining TOC is probably attributable to the more uniform origin of TOC compared to nitrogen or phosphorus. In Finland, TOC is suggested to originate mainly from natural sources in the catchments (Kortelainen, 1993). Inorganic nitrogen can, in southern Finland, originate to a great extent from deposition, and both phosphorus and nitrogen are added to the catchments as fertilizers. Organic forms of N and P also behave differently in the catchments than the soluble inorganic forms of these nutrients. Therefore TOC can probably be more reliably predicted by catchment variables.

The coefficients of determination for TN and TOC were both rather high in Region 5. This is most likely attributable to the uniform origin of total nitrogen and total organic carbon in this region. Most of the nitrogen in lake water in northern Finland is organic and the correlation coefficient between TN and TOC in Region 5 is 0.80. In Regions 3 and 4, the coefficients of determination in the regression equations for TN and TOC were lower than in other subsets. The topography is very flat in the costal areas of these regions and the lakes are correspondingly very shallow and their volume is small. This may create high concentrations in the lake water, even with only moderate loading from the catchment. Therefore, the predictability of water quality by the catchment attributes is weak.

Rapp *et al.* (1985) suggested that the extremes tend to balance each other to produce 'average' concentrations as watershed size increases. The smallest watersheds are most likely to be dominated almost entirely by a single factor. This can also be seen in our data set. The catchments of the large lakes are heterogeneous, containing all the determined catchment factors, whereas the smallest lakes may be situated in, for example, forested catchments with no habitation and agriculture. Different factors determine the water quality of these small lakes with very dissimilar catchments, and it is difficult to develop a single regression equation that could satisfactorily explain the variation of water quality for both types of lakes. The same problem may occur in the case of equations developed for lakes in a certain region, because the coefficients of regression were not very high in some of the regional regression equations.

The sampling of the surveyed lakes included only one water sample per lake. This could be assumed to cause more unexplained variability for the large lakes than for the small lakes. The water quality of the large lakes was, however, better predicted by the catchment variables than the water quality of the small lakes. For large lakes, it is easier to find a single common equation capable of explaining the variation of water quality, as all of the catchments have the same catchment characteristics, but in different proportions. The higher coefficients of determination for the data set of large lakes may partly derive from the nature of the data. The five land cover classes of the data are a combination of 64 classes, and this may cause inaccuracy in some of the classes. For example in the class 'forests in mineral soil', the forest can be pine, spruce or deciduous, and the trees in the area can be clear cut, saplings or mature. With large catchments, the differences within the class probably balance each other, resulting in average conditions, whereas in small catchments the class 'forests in mineral soil', can mean very different conditions in different catchments. The regression equations developed for the whole dataset appeared to be some kind of 'average equation' of all of the equations for the subgroups, because the equations differed considerably between the subgroups. When developing the regression equation for the whole dataset, it also was necessary to exclude more cases on the basis of absolute value of the Studentized residual compared to the separate regional or size group equations, due to greater dispersion in the whole dataset.

4. Conclusions

The regression equations for the entire data set could reasonably well explain the variation of the TN, TP and TOC in lake water in the five regions. However, the equations obtained for a specific region could not as easily be applied to another region due to differences in the relationships between catchment characteristics and the lake water quality. The multiple regression equations for the whole dataset formed a kind of 'average' of all the separate equations. The highest coefficient of determination was, however, obtained for the subgroup containing the largest lakes (>10 km^2) of the dataset, most probably because the large lakes have large heterogeneous catchments, and because of the considerable differences between the catchments. Small lakes in Finland are more homogeneous than large lakes and most of them are situated in mainly forested catchments with little variation in land use.

In southern Finland, agriculture had a strong influence on TN and TP concentrations in lakes. In northern parts of the country, peatlands as sources of N and P were emphasized, as the human impact in the catchments was minor. There are numerous chain-like watercourses in Finland and most lakes have one or more upstream basins. The sedimentation and decomposition of organic carbon, nitrogen and phosphorus in the lake itself, and in the upstream water bodies, appeared to decrease concentrations especially in southern Finland. Upstream water bodies also reduce the land area in the catchment, which act as a source of carbon, nitrogen and phosphorus. In northern Finland, lake percentage of the catchment had very little explanatory power for TOC, TN and TP. As the ice cover period is longer in northern Finland compared to southern parts of the country, the decomposition of organic matter can be expected to be more efficient in southern Finland.

Acknowledgements

We thank the Finnish Global Change Research Programme (FIGARE) of the Academy of Finland for funding this research.

References

Agricultural Census: 2000, 'SVT Agriculture, forestry and fishery 2002:51', Information Centre of the Ministry of Agriculture and Forestry, Helsinki, Finland.
Arheimer, B. and Lidén, R.: 2000, 'Nitrogen and phosphorus concentrations from agricultural catchments – Influence of spatial and temporal variables', *J. Hydrol.* **227**, 140–159.
Cole, J. J. and Caraco, N. F.: 2001, 'Carbon in the catchments: Connecting terrestrial carbon losses with aquatic metabolism', *Mar. Freshwat. Res.* **52**, 101–110.
Correll, D. L., Jordan, T. E. and Weller, D. E.: 2001, 'Effects of precipitation, air temperature, and land use on organic carbon discharges from Rhode river watersheds', *Water, Air, Soil Pollut.* **128**, 139–159.
D'Arcy, P. and Carignan, R.: 1997, 'Influence of catchment topography on water chemistry in southeastern Québec Shield lakes', *Can. J. Fish. Aquat. Sci.* **54**, 2215–2227.
Dillon, P. J. and Molot, L. A.: 1997, 'Effect of landscape form on export of dissolved organic carbon, iron, and phosphorus from forested stream catchment', *Water Resour. Res.* **33**, 2591–2600.
Dillon, P. J., Molot, L. A. and Scheider, W.: 1991, 'Phosphorus and nitrogen export from forested stream catchments in Central Ontario', *J. Environ. Qual.* **20**, 857–864.
Henriksen, A., Skjelvåle, B. L., Lien, L., Traaen, T. S., Mannio, J., Forsius, M., Kämäri, J., Mäkinen, I., Berntell, A., Wiederholm, T., Wilander, A., Moiseenko, T., Lozovik, P., Filatov, N., Niinioja, R., Harriman, R. and Jensen, J. P.: 1996, 'Regional Lake Surveys in Finland-Norway-Sweden-Northern Kola – Russian Karelia – Scotland – Wales 1995. Coordination and design', *Acid Rain Research Report* **40**, NIVA, Oslo.
Herlihy, A., Stoddard, J. and Johnson, C.: 1998, 'The relationship between stream chemistry and watershed land cover data in the Mid-Atlantic region, U.S.', *Water, Air, Soil Pollut.* **105**, 377–386.
Koljonen, T. (ed.): 1992, *Geochemical Atlas of Finland, Part 2: Moraine*, Geological Survey of Finland, Espoo, Finland.
Kopáček, J., Stuchlík, E., Straškrabová, V. and Pšenáková, P.: 2000, 'Factors governing nutrient status of mountain lakes in the Tatra Mountains', *Freshwat. Biol.* **43**, 369–383.
Kortelainen, P.: 1993, 'Content of total organic carbon in Finnish lakes and its relationship to catchment characteristics', *Can. J. Fish. Aquat. Sci.* **50**, 1477–1483.
Kortelainen, P. and Saukkonen, S.: 1998, 'Leaching of nutrients, organic carbon and iron from Finnish forestry land', *Water, Air, Soil Pollut.* **105**, 239–250.

Kortelainen, P., Saukkonen, S. and Mattsson, T.: 1997, 'Leaching of nitrogen from forested catchments in Finland', *Global Biogeochem. Cycles* **11**, 627–638.

Lepistö, A., Andersson, L., Arheimer, B. and Sunblad, K.: 1995, 'Influence of catchment characteristics, forestry activities and deposition on nitrogen export from small forested catchments', *Water, Air, Soil Pollut.* **84**, 81–102.

Mander, U., Kull, A. and Kuusemets, V.: 2000, 'Nutrient flows and land use change in a rural catchment: A modelling approach', *Landscape Ecol.* **15**, 187–199.

Mannio, J., Räike, A. and Vuorenmaa, J.: 2000, 'Finnish lake survey 1995: Regional characteristics of lake chemistry', *Verh. Int. Ver. Limnol.* **27**, 362–367.

Mattsson, T., Finér, L., Kortelainen, P. and Sallantaus, T.: 2003, 'Brook water quality and background leachinxg from unmanaged forested catchments in Finland', *Water, Air, Soil Pollut.* **147**, 275–297.

McTieran, K., Jarvis, S., Scholefield, D. and Hayes, M.: 2001, 'Dissolved carbon losses from grazed grasslands under different management regimes', *Water Res.* **35**, 2565–2569.

Nieminen, M. and Jarva, M.: 1996, 'Phosphorus adsorption by peat from drained mires in Southern Finland', *Scand. J. For. Res.* **11**, 321–326.

Rapp, G., Allert, J., Liukkonen, B., Ilse, J., Loucks, O. and Glass, G.: 1985, 'Acid deposition and watershed characteristics in relation to lake chemistry in northeastern Minnesota', *Environ. Int.* **11**, 425–440.

Rasmussen, J., Godbout, L. and Scallenberg, M.: 1989, 'The humic content of lake water and its relationship to watershed and lake morphometry', *Limnol. Oceanogr.* **34**, 1336–1343.

Rekolainen, S.: 1989, 'Phosphorus and nitrogen load from forest and agricultural areas in Finland', *Aqua Fenn.* **19**, 95–107.

SAS Institute Inc.: 1989, *SAS/STAT User's Guide, Version 6, Edition 4. Vol. 2*, SAS Institute Inc., Cary, North Carolina, U.S.A.

CARBON STORAGE IN TAGUS SALT MARSH SEDIMENTS

ISABEL CAÇADOR[1]*, ANA LUÍSA COSTA[1] and CARLOS VALE[2]

[1] *Institute of Oceanography, Faculty of Sciences, University of Lisbon, Rua Ernesto de Vasconcelos Campo Grande 1749-016 Lisbon, Portugal;* [2] *Institute for Sea and Fisheries Research (IPIMAR) Av. Brasília 1449-006, Lisbon, Portugal*
(* author for correspondence, e-mail: icacador@fc.ul.pt; phone: 351-21 75 00 104; fax: 351-21 75 000 09)

(Received 20 August 2002; accepted 7 April 2003)

Abstract. Seasonal variation of above ground and belowground biomass of *Spartina maritima* and *Halimione portulacoides*, decomposition rates of belowground detritus in litterbags, and carbon partitioning in plant components and sediments were determined in two Tagus estuary marshes with different environmental conditions. Total biomass was higher in the saltier marsh from 7,190 to 6,593 g m^{-2} dw and belowground component contributed to more than 90%. Litterbag experiment showed that 30 to 50% of carbon is decomposed within a month (decomposition rate from 0.024 to 0.060 d^{-1}). Slower decomposition in subsequent periods agrees with accumulation of carbon concentration in sediment. Atmospheric carbon annually transferred to the plant belowground biomass is stored more efficiently in sediments of Corroios than Pancas.

Keywords: carbon, *Halimione portulacoides*, litterbags, salt marsh, *Spartina maritima*, Tagus estuary

1. Introduction

The nature and extent of the routes followed by primary production in marine communities have important implications in carbon consumption and preservation in marine ecosystems and global carbon budget (Cebrian, 2002). Salt marshes typically exhibit high rates of productivity (Mitsch and Gosselink, 2000) and are excellent carbon sinks as they take CO_2 from the atmosphere and store it in living plant tissue (Williams, 1999). The importance of these aspects is recognised with the creation and restoration of wetlands degraded or destroyed (Craft *et al.*, 1999).

Much has been written about whether salt marshes export materials to the adjacent seawaters (*cf.* Weinstein and Kreeger, 2000). The outwelling hypothesis, which states that organic matter exported from coastal marshes fuels food chains in adjacent waters (Teal, 1962; Nixon, 1980, Odum *et al.*, 1995), has stimulated research on net aerial primary production (Bouchard and Lefeuvre, 2000). The export varies geographically, inter-annually, with local differences related to tidal movement, the primary medium of transfer of aerial organic matter (Gross *et al.*, 1990; Morris and Haskin, 1990; Bouchard and Lefeuvre, 2000). Most studies were concentrated on the aboveground standing crop; the contribution of belowground

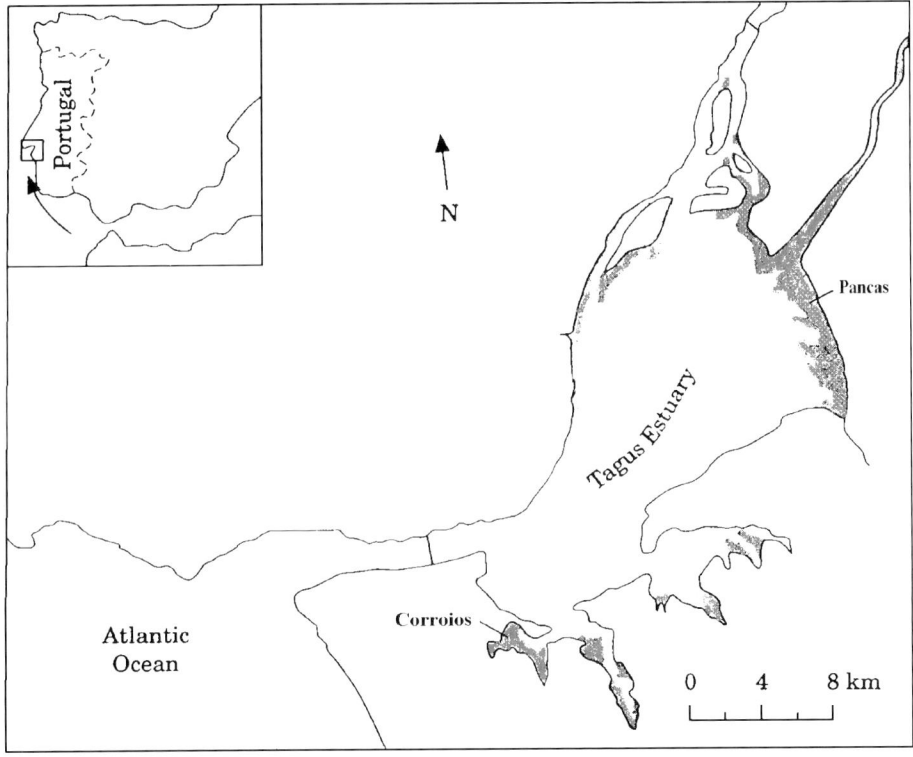

Figure 1. Location of the Tagus estuary and sampling tidal marshes.

components to nutrient and energy budgets of the adjacent coastal ecosystems often has been neglected (Groenendiijk and Vink-Lieavaart, 1987). Belowground production in salt marshes may exceed aboveground production (Kostka *et al.*, 2002). Due to this high belowground allocation, decomposition of roots contributes greatly to carbon turnover in salt marsh sediments (Scheffer and Aerts, 2000). The oxidation of organic matter depends on the chemical composition of halophytes (Hemminga and Buth, 1991), the delivery of atmospheric oxygen at the rhizosphere and the complex root-sediment interactions (Caçador *et al.*, 1996). Tidal inundation affecting the sediment environmental conditions also influences the decomposition rate (Foote and Reynolds, 1997).

The aim of this study was to estimate the aboveground and belowground production of *Spartina maritima* and *Halimione portulacoides* in two marshes of the Tagus estuary, detritus production, and the decomposition rate of belowground biomass using the litterbag method. On the basis of production data and carbon concentrations in plant components, sediment cores and decomposing litter, we examined the carbon accumulation rate in vegetated sediments of the two salt marshes.

2. Site Description

The Tagus estuary is one of the largest estuaries on the Atlantic coast of Europe, covering about 320 km^2. The southern and eastern parts contain extensive intertidal mud flats harbouring salt marsh plant communities dominated by *Spartina maritima* (Poales: Poaceae), *Halimione portulacoides* (Caryophyllales: Chenopodiaceae) and *Arthrocnemum fruticosum* (Caryophyllales: Chenopodiaceae). A typical zonation is visible, with homogeneous stands of the pioneer species *S. maritima* colonising the bare muds of low marshes and natural depressions, *H. portulacoides* occupying the banks of creeks, and *A. fruticosum* being found largely in the upper parts of the salt marshes. The study was carried out in two marshes: Pancas and Corroios (Figure 1). The marshes were chosen due to their contrasting characteristics: Pancas is a young salt marsh with extensive mudflats (800 ha) located in the Tagus Nature Reserve; Corroios is older, smaller (400 ha) and located in the proximity of urbanised and industrial areas. Due to the highly branched system of channels and the high tidal ranges (max. 4 m), the sites where the study took place experienced two tidal flushing twice each day.

3. Materials and Methods

3.1. SAMPLING

Two pure stands of *S. maritima* and *H. portulacoides* were surveyed between June 1998 and April 1999 in Pancas and Corroios. Aboveground biomass was determined every two months by clipping the vegetation at ground level in three squares of 0.3 × 0.3 m. At each period, detritus deposited on the sediment of the same plots was removed by hand and transported to the laboratory in plastic bags. After cutting the aboveground material and removing the detritus, two sediment cores were taken at each study site using a 7 cm diameter, 100 cm long tube. In one core, belowground biomass of each plant was sorted out from the sediment cores to a depth of 25 cm depth where most of roots are present. The second core was sliced at intervals of 0–5, 5–15, 15–25, 25–35, 35–45 and 45–55 cm and total carbon was determined. Three sediment cores were collected from non-vegetated areas of the sites and sliced at the same intervals.

3.2. *IN-SITU* MEASUREMENTS

Sediment temperature was measured at low tide at 15 cm depth using a digital thermometer. Approximately at the same depth, redox potential (Eh) and pH were measured *in situ* using a Crison pH/mV meter with a platinum electrode. Pore water of rooted sediments was removed by suction with a syringe equipped with a tube that was introduced at 15 cm depth in the sediment. Salinity was measured in pore waters using a refractometer.

3.3. DETERMINATION OF BIOMASS AND DETRITUS

The collected aboveground plant material was transported to the laboratory, rinsed with demineralised water, separated into stems and leaves, and dried to constant weight at 80 °C. Belowground material in the upper 25 cm was separated from the sediment using a 212 μm-mesh sieve and demineralised water. The remaining plant material was dried at 80 °C for 48 hr and weighed (Gross *et al.*, 1991). The detritus collected on the sediment surface was treated in similar way.

3.4. LITTERBAG FIELD EXPERIMENT

Belowground biomass of *S. maritima* and *H. portulacoides* was collected from several locations at the Pancas and Corroios marshes in February 1999. The samples were rinsed and then dried. Approximately 5 g of root material were placed in 24, 10 × 10 cm nylon mesh bags with 450-μm diameter holes. The bags were buried at 10 cm depth in their respective environments in order to mimic as closely as possible their natural habitat. A set of three bags was collected monthly for each plant species between February and September. In the laboratory, the plant material was removed from the litterbags, rinsed with distilled water, dried at 80 °C for 48 hr, weighed and analysed for total carbon. Before burying litterbags, subsamples of each species were saved to determine initial carbon concentrations. Decay rates (k) were determined through this first-order decay function, $X_t = X_o e^{-kt}$ (Bouchart and Lefeuvre, 2000).

3.5. DETERMINATIONS OF GRAIN SIZE AND TOTAL CARBON

Sediment samples were air-dried and cleaned of roots with tweezers, passed through a 0.25 mm mesh. After destroying organic matter by loss on ignition at 550 °C and carbonates by HCl dissolution, particle size was determined according to Stoke's law, in a 1000 mL measuring cylinder beaker with distilled water (Gee and Bauder, 1986). For carbon measurements, sediments and biological material were ground and homogenised. Total carbon (C) concentration was determined on all materials using a CHNS/O analyser (Fisons Instruments Model EA 1108).

3.6. STATISTICAL ANALYSIS

The obtained data were statistically analyzed using one-way analysis of variance following Sokal and Rohlf (1981).

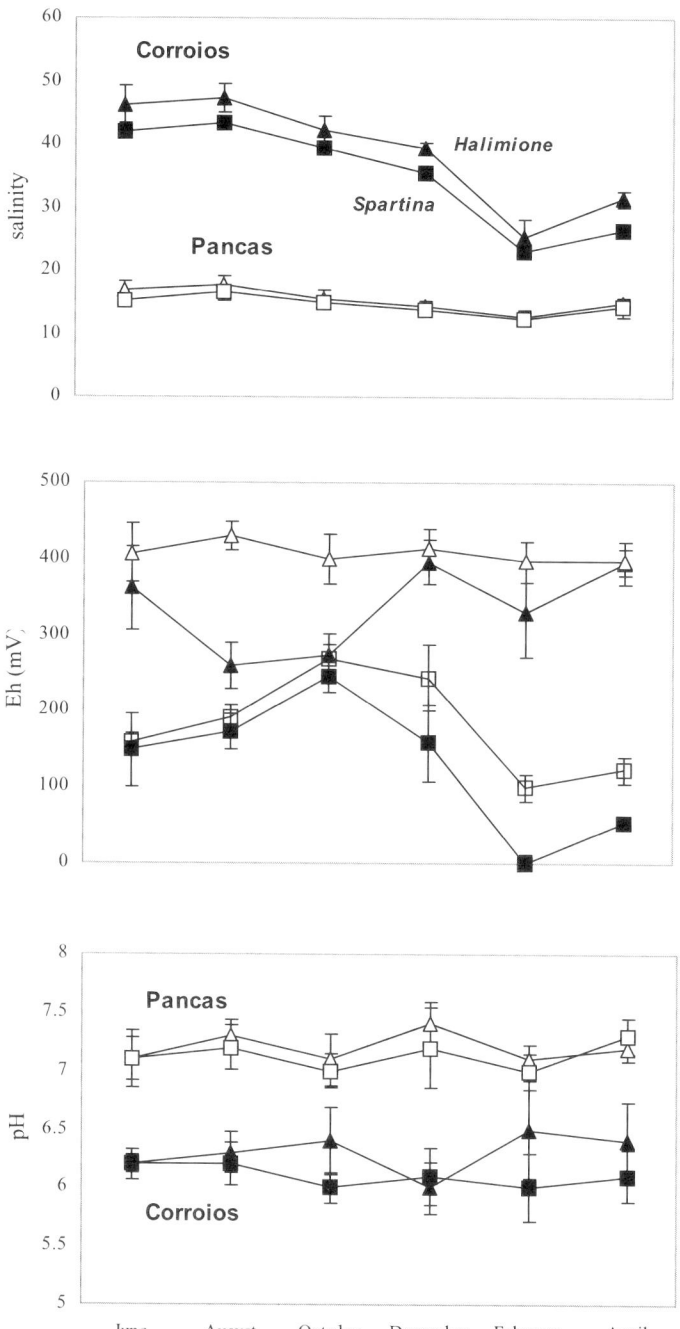

Figure 2. Seasonal variation of salinity, redox potential and pH in rooted sediments of *Spartina maritima* and *Halimione portulacoides* in Pancas and Corroios salt marshes (means ± standard deviations).

4. Results and Discussion

4.1. PHYSICAL AND CHEMICAL CHARACTERISTICS OF ROOTED SEDIMENTS

Sediments of both salt marshes were comprised mainly of fine particles, about 60% silt and 38% clay. Sediment temperature was similar in the two marshes and varied seasonally between 15 and 26 °C. Rooted sediments were saltier in Corroios (salinity between 23 and 47 ppt) than in Pancas (13 to 19 ppt) reflecting the salinity gradient along the estuary (Figure 2). The salinity remained relatively constant in Pancas and decreased substantially in Corroios, as freshwater moved seaward in winter. The redox potential was higher in sediments colonised by *H. portulacoides* (max. 430 mV) than by *S. maritima* (max. 270 mV) in the two marshes, and decreased in winter (0–100 mV) when temperature was lower and the plants became less active (Caçador *et al.*, 2000). pH in rooted sediments was lower at Corroios (6.0–6.5) than at Pancas (7.0–7.4) and did not differ considerably between the two plant communities or seasonally.

4.2. ABOVEGROUND AND BELOWGROUND BIOMASS

Belowground biomass exceeded aboveground biomass at both sites (Figure 3). The highest values were in Corroios with 7,190 g m^{-2} of *S. maritima* roots and 6,593 g m^{-2} of *H. portulacoides* roots contributing 96 and 90% of the total biomass, on average respectively. Belowground biomass in Pancas accounted, on average, for 73% (*S. maritima*) and 54% (*H. portulacoides*) of the total biomass. Among the aboveground components, stems of *H. portulacoides* in Pancas had the highest biomass (510 to 1,920 g m^{-2}) and leaves of *S. maritima* in Corroios lowest biomass (75 to 135 g m^{-2}). Considering all the data, plant biomass of the two marshes differed significantly ($p < 0.05$). Whereas leaf and stem biomass of *S. maritima* and *H. portulacoides* in Corroios were relatively constant throughout the year, seasonal variations were observed in Pancas, with maximum values in August (*S. maritima*) and October (*H. portulacoides*), and a minimum in February. Belowground biomass of the two species varied seasonally: values at Corroios increased from February to June and at Pancas between February and October. Presumably the prolonged growth in Pancas reflects lower salinity conditions (cf. Mendelssohn and Morris, 2000).

4.3. CARBON IN LEAVES, STEMS, AND BELOWGROUND BIOMASS

Leaf C concentrations ranged between 280 and 440 mg g^{-1} throughout the year. These values are in accordance with the ranges of C concentration reported in plants from other tidal marshes (Groenendijk and Vink-Lievaart, 1987, Zawislanski *et al.*, 2001). For each tissue type, C concentrations did not differ between species or with season ($p > 0.05$).

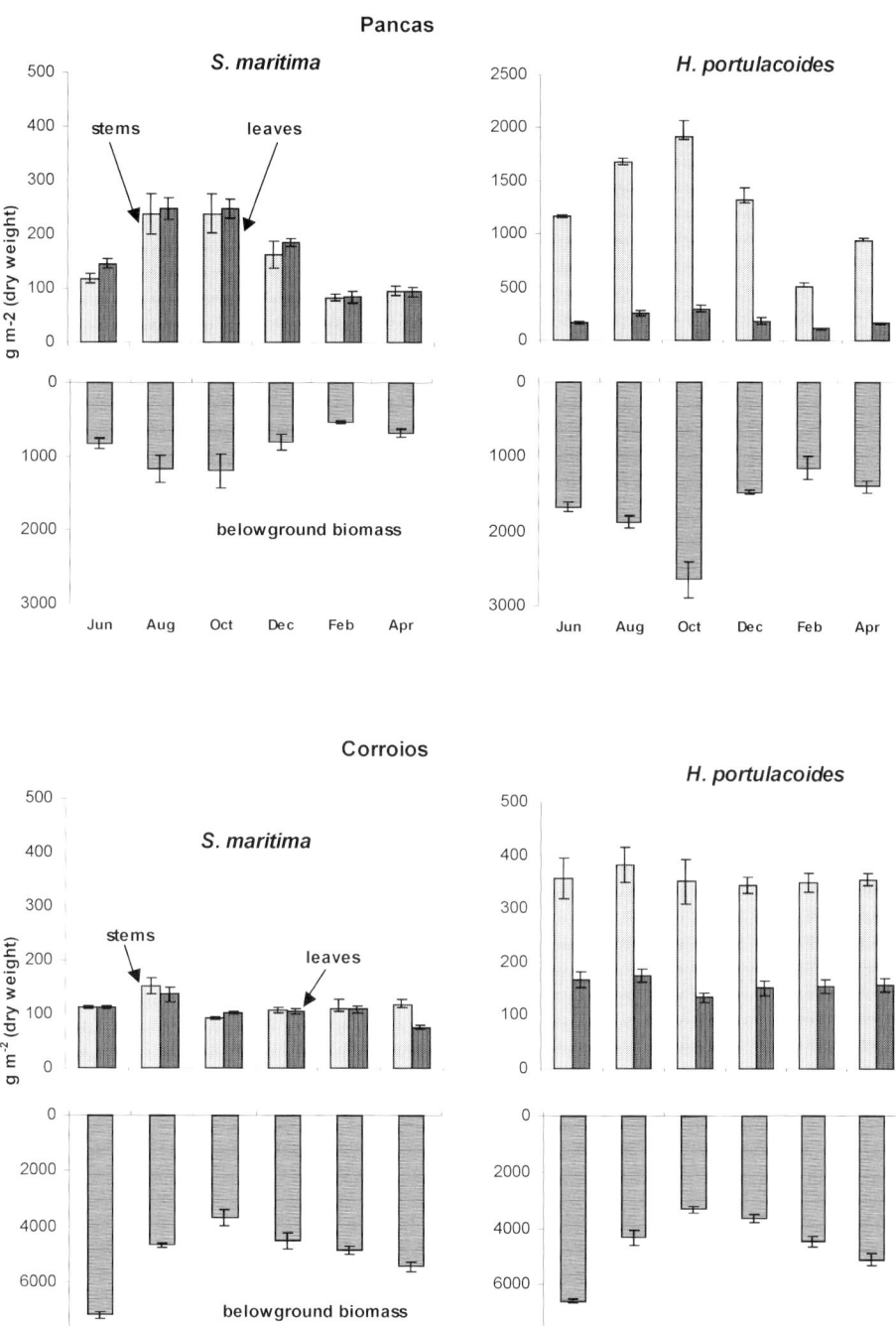

Figure 3. Seasonal variation of aboveground and belowground biomass of *Spartina maritima* and *Halimione portulacoides* in Pancas and Corroios (means ± standard deviations).

4.4. NET PRIMARY PRODUCTION

Aboveground and belowground net primary production (g C m^{-2}) were estimated as the difference between maximum and minimum biomass, expressed in C units, over the annual period of study:

$$NPP_{Above} = \max(\textit{leaf biomass} \times [C]_l + \textit{stem biomass} \times [C]_s) - \min(\textit{leaf biomass} \times [C]_l + \textit{stem biomass} \times [C]_s)$$

and

$$NPP_{Below} = \max(\textit{belowground biomass} \times [C]_b) - \min(\textit{belowground biomass} \times [C]_b)$$

where $[C]_l$, $[C]_s$ and $[C]_b$ are the concentrations of carbon in leaf, stem and belowground biomass, respectively.

At Corroios, NPP_{Below} was 22 and 32 times higher than NPP_{Above} for *S. maritima* and *H. portulacoides*, respectively (Table I). For both species, NPP_{Above} was higher and NPP_{Below} was lower at Pancas than at Corroios, such that the ratios of NPP_{Below} to NPP_{Above} were only 2 and 0.9 for *S. maritima* and *H. portulacoides*, respectively (Table I). NPP_{Above} at Pancas is comparable to values reported for several tidal marshes in Europe (Bouchard and Lefeuvre, 2000) and in North America (Mendelssohn and Morris, 2000). Total NPP ($NPP_{Above} + NPP_{Below}$) values for *S. maritima* and *H. portulacoides* of 358 and 1,281 g C m^{-2} a^{-1} at Pancas and 1,054 and 929 g C m^{-2} a^{-1} at Corroios, respectively, are comparable to values reported for several tidal marshes (Wolff *et al.*, 1979) and exceed the values reported for microtidal Mediterranean coastal marshes (Ibañez *et al.*, 2000).

Carbon allocated to belowground biomass at Pancas was between 48 and 65% of total C uptake, while at Corroios it exceeded 95%. These results are in line with studies of *Spartina alterniflora* in US salt marshes (Kostka *et al.*, 2002), and of other species in fen ecosystems (Scheffer and Aerts, 2000). Changes in the proportion of belowground and aboveground biomass may reflect adaptive responses of the plants in stressed environments (Groenendijk and Vink-Lieavaart, 1987). Because Corroios is an old marsh, the increased allocation of biomass to roots of *S. maritima* and *H. portulacoides* may result from intense competition for nutrients.

The ratio between primary production and maximum biomass provides an estimate of annual C turnover rate in plants (Table I). Values for Pancas were higher than for Corroios, indicating that a larger proportion of the biomass is replaced every year in that marsh. Considering the combined areas of the two studied marshes (400 + 800 = 1200 ha), we estimate that 1–10 Gg of C is taken up annually from the atmosphere.

TABLE I

Maximum and minimum content of C (g C m^{-2}) in aboveground and belowground parts of *S. maritima* and *H. portulacoides* from Pancas and Corroios and *NPP* (g C m^{-2}) estimated as the difference between maximum and minimum values. The turnover rate was calculated as the ratio between primary production (NPP_{Above} + NPP_{Below}) and maximum content of C

	Pancas	Corroios
S. maritima		
Maximum aboveground biomass (g C m^{-2})	187	114
Minimum aboveground biomass (g C m^{-2})	69	67
NPP_{Above} (g C m^{-2})	119	46
Turnover rate (%)	63	40
Maximum belowground biomass (g C m^{-2})	445	2610
Minimum belowground biomass (g C m^{-2})	207	1602
NPP_{Below} (g C m^{-2})	239	1008
Turnover rate (%)	54	39
H. portulacoides		
Maximum aboveground biomass (g C m^{-2})	922	210
Minimum aboveground biomass (g C m^{-2})	257	183
NPP_{Above} (g C m^{-2})	665	28
Turnover rate (%)	72	13
Maximum belowground biomass (g C m^{-2})	1100	2281
Minimum belowground biomass (g C m^{-2})	485	1380
NPP_{Below} (g C m^{-2})	616	901
Turnover rate (%)	56	40

4.5. Litter production

In the two marshes, litter on the sediment surface follows a seasonal pattern with higher values at the end of summer and a decrease in winter, especially evident for *H. portulacoides* litter at Pancas (Figure 4). The quantity derived from aerial parts of *H. portulacoides* greatly exceeded that from *S. maritima*. Litter production may be estimated by direct or indirect methods (Bouchard and Lefeuvre, 2000), although all procedures have limitations (Hopkinson et al., 1978). We choose the simplest method, viz. the difference between maximum and minimum litter pools at different times during a year (Smalley, 1959; Linthurst and Reimold, 1978). Comparing litter production with the corresponding NPP_{Above}, we calculated that only 5% of the NPP_{Above} at Corroios remained at the site versus between 4 and

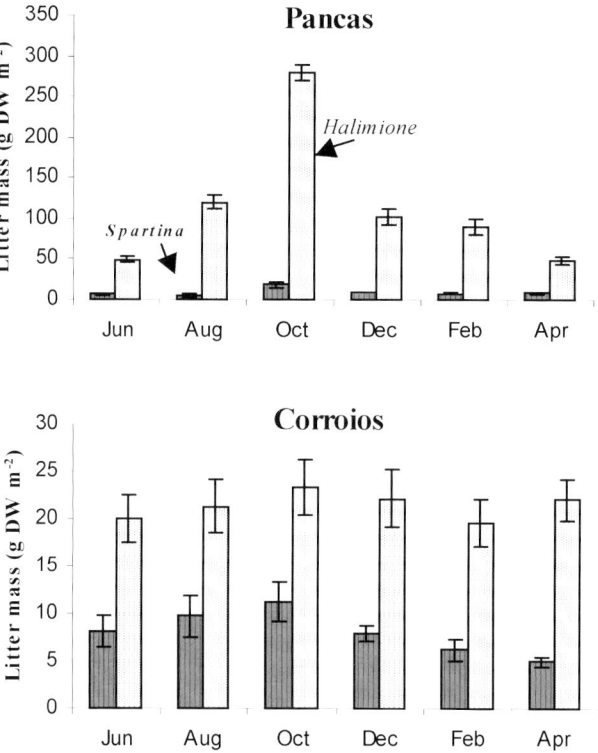

Figure 4. Changes in the litter mass on sediments colonised by *Spartina maritima* and *Halimione portulacoides* in Pancas and Corroios salt marshes (means ± standard deviations).

14% at Pancas. This means that more than 86% of the produced C was exported to the estuary. Detritus being washed over from the marsh by the daily tidal flushing explains the high export in comparison to other studies (Dame, 1982; Wolff et al., 1979; Bouchard and Lefeuvre, 2000). Consequently this exported organic matter is transformed elsewhere in the estuary or coastal zone.

4.6. Litterbag experiments

At Pancas and Corroios, belowground litter decomposed rapidly in sediments (Table II). After one month, approximately 50 and 30% of the litter had decomposed at Pancas and at Corroios, respectively, corresponding to k values ranging from 0.024 to 0.060 d^{-1}). At each site, k values for *S. maritima* roots were lower than for *H. portulacoides* roots. Decomposition was slower in Corroios indicating a tendency for higher accumulation of organic matter than at Pancas. After 4 months, 50 to 70% of the belowground litter still remained in the sediment. Decomposition rates in the Tagus salt marsh sediments are faster than values reported for the salt marshes of Ebre Delta (Curcó et al., 2002) and slower than values reported for

TABLE II

Decomposition rates (k values; $k = \ln(x_0-x_t)/t$, where x_0 = initial dry weight (g) and x_t = final dry weight (g) for *S. maritima* and *H. portulacoides* in Pancas and Corroios saltmarshes

	Month of collection	Incubation period, t (d)	k (d^{-1})
Pancas			
S. maritima	March	31	0.037
	April	59	0.020
	May	87	0.012
	June	118	0.011
H. portulacoides	March	31	0.060
	April	59	0.031
	May	87	0.024
	June	118	0.016
Corroios			
S. maritima	March	31	0.024
	April	63	–0.002
	May	93	0.009
	June	119	–0.001
H. portulacoides	March	31	0.045
	April	63	0.018
	May	93	0.011
	June	119	0.012

Spartina in North America marshes (Foote and Reynolds, 1997 and references therein). Carbon concentration of the decomposing belowground material ranged within the same interval (32–40%) that was observed in plant parts. These results are in agreement with other works (Valiela *et al.*, 1985; Benner *et al.*, 1991) indicating that total C concentrations in decaying litter decrease only slightly.

4.7. CARBON CONCENTRATIONS IN SEDIMENTS

The amount of organic C transferred yearly to sediments and the decomposition rate of organic matter are reflected in the vertical profiles of C concentrations in sediments. Depth variation of C differed considerably between vegetated to non-vegetated sediments (Figure 5). Carbon concentrations in non-vegetated sediments were approximately 20 mg g^{-1} at the surface and decreased gradually with depth. Vegetated sediments exhibited a subsurface C enrichment in layers of higher root biomass, with concentrations reaching 3.3 mg g^{-1} C at Pancas and 7.5 mg g^{-1} at Corroios. This pattern was observed in all surveys and indicates retention of

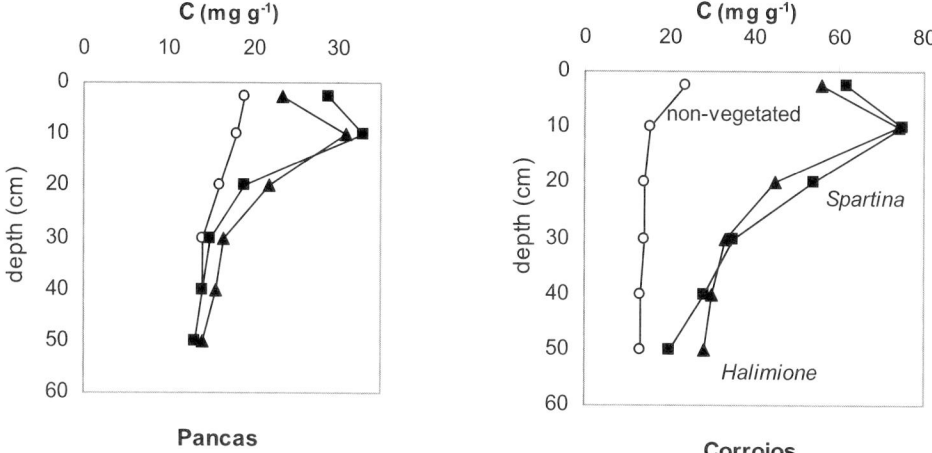

Figure 5. Vertical profiles of total C in non-vegetated sediments and sediments colonised by *Spartina maritima* and *Halimione portulacoides* in Pancas and Corroios (means ± standard deviations).

organic matter as root-derived material. Where belowground biomass was higher and degradation rate slower (Corroios) C accumulation in sediments was greater. Sediment layers of higher root biomass exhibited an increase in C concentration from October to February and a decrease in April. This seasonal variation resulted from the slow decay of belowground components of the plants in winter, along with an increasing allocation of C to rooted sediments. The decrease in C concentration in summer probably results from faster oxidation of particulate organic matter and the escape of dissolved organic constituents.

4.8. ESTIMATION OF CARBON STORED IN SALT MARSH SEDIMENTS

On the basis of the difference in the quantity of carbon in the top 50 cm of the sediment profile of vegetated and non-vegetated sites, one may estimate that 9 to 10 kg C m^{-2} is stored in Pancas and 21–22 kg C m^{-2} in Corroios due to plant activity. Assuming a sedimentation rate for both marshes of 0.8 cm y^{-1} (Vale, 1990; Caçador *et al.*, 1996), the 50 cm depth corresponds to 62.5 yr. The mean annual amounts of carbon incorporated in the sediments are: 0.15 kg C m^{-2} y^{-1} (9.5 kg C m^{-2} divided by 62.5 yr) in Pancas, and 0.34 kg C m^{-2} y^{-1} (21.5 kg C m^{-2}/62.5 yr) in Corroios. These quantities are comparable to the reported values of carbon content retained in the upper sediments of coastal marshes in Florida (Choi *et al*, 2001), and exceeded largely the sequestration rate of peatlands (Roulet, 2000). Overall, C retention in sediments differed between the two marshes, with greater accumulation in the older marsh (Corroios), attributable to higher NPP_{Below} along with slower root decomposition.

Acknowledgements

This study was supported by PRAXIS/CTE/11025/98 and Programa Operacional Ciência, Tecnologia, Inovação do Quadro Comunitário de Apoio III.

References

Benner, R., Fogel, M. L. and Sprague, E. K.: 1991, 'Diagenesis of belowground biomass of *Spartina alterniflora* in salt-marsh sediments', *Limnol. Oceanogr.* **36**, 1358–1374.

Bouchard, V. and Lefeuvre, J. C.: 2000, 'Primary production and macro-detritus dynamics in a European salt marsh: Carbon and nitrogen budgets', *Aquat. Bot.* **67**, 23–42.

Caçador I., Vale C. and Catarino, F.: 1996, 'Accumulation of Zn, Pb, Cu and Ni in sediments between roots of the Tagus estuary salt marshes, Portugal', *Estuar. Coast. Shelf. Sci.* **42**, 393–403.

Caçador, I., Vale, C. and Catarino, F.: 2000, 'Seasonal variation of Zn, Pb, Cu and Cd concentrations in the root-sediment system of *Spartina maritima* and *Halimione portulacoides* from Tagus estuary salt marshes', *Mar. Environ. Res.* **49**, 279–290.

Cebrian, J.: 2002, 'Variability and control of carbon consumption, export, and accumulation in marine communities', *Limnol. Oceanogr.* **47**, 11–22.

Choi, Y., Wang, Y., Hsieh, Y-P. And Robinson, L.: 2001, 'Vegetation succession and carbon sequestration in a coastal wetland in northwest Florida: Evidence from carbon isotopos', *Global Biogeochem. Cycles* **15**, 311–319.

Craft, C., Reader, J., Sacco, J. N. and Broome, S. W.: 1999, 'Twenty-five years of ecosystem development of constructed *Spartina alterniflora* (Loisel) Marshes', *Ecol. Appl.* **9**, 1405–1419.

Curcó, A., Ibàñez, C., Day, J. W. D. and Prat, N.: 2002, 'Net primary production and decomposition of salt marshes of the Ebre Delta (Catalonia, Spain)', *Estuaries* **3**, 309–324.

Dame, R. F.: 1982, 'The flux of floating macrodetritus in the North Inlet estuarine ecosystem', *Mar. Ecol. Prog. Ser.* **16**, 161–171.

Foote, A. L. and Reynolds, K. A.: 1997, 'Decomposition of saltmeadow cordgrass (*Spartina patens*) in Louisiana Coasatal Marshes', *Estuaries* **90**, 579–588.

Gee, G. W. and Bauder, J. W.: 1986, 'Particle size analysis', in Madison, U.S.A.: American Society of Agronomy-Soil Science Society of America (eds), Methods of soil analysis. Part I – Physical and mineralogical method, *Agronomy Monograph* **9**, 383–411

Groenendijk, A. M. and Vink-Lieavaart, M. A.: 1987, 'Primary production and biomass on a Dutch salt marsh: emphasis on the belowground component', *Vegetatio* **70**, 21–27.

Gross, M. F., Hardisky, M. A. and Klemas, V.: 1990, 'Inter-annual variability in the response of *Spartina alterniflora* biomass to amount of precipitation; *J. Coast. Res.* **6**, 949–960.

Gross, M. F., Hardisky, M. A., Wolf, P. L. and Klemas, V.: 1991, 'Relationship between aboveground and belowground biomass of *Spartina alterniflora* (smooth cordgrass)', *Estuaries* **14**, 180–191.

Hemminga, M. A. and Buth, G. J. C.: 1991, 'Decomposition in salt marsh ecosystems of the S.W. Netherlands: the effects of biotic and abiotic factors', *Vegetatio* **92**, 73–83.

Hopkinson, C. S., Gosselink, J. G. and Parrondo, R. T.: 1978, 'Aboveground production of seven marsh plants species in coastal Louisiana', *Ecology* **59**, 760–769.

Ibañez, C., Curcó, A., Day, J. W. D. and Prat, N.: 2002, 'Structure and productivity of microtidal Mediterranean coastal marshes' in M. P. Weinstein and D. A. Kreeger (eds), *International Symposium: Concepts and Controversies in Tidal Marsh Ecology*, Kluwer Academic Publishers, Dordrecht, pp. 107–136.

Kostka, J. E., Gribsholt, B., Petrie, E., Dalton, D., Skelton, H. and Kristensen, E.: 2002, 'The rates and pathways of carbon oxidation in bioturbed saltmarsh sediments', *Limnol. Oceanogr.* **47**, 230–240.

Linthurst, A. and Reimold, R. J.: 1978, 'An evaluation of methods for estimation the net primary production', *Ecology*, **49**, 147–149.

Mendelssohn, I. A. and Morris, J. T.: 2000, 'Eco-physiological controls on the productivity of *Spartina Alterniflora* Loisel', in M. P. Weinstein and D. A. Kreeger (eds), *International Symposium: Concepts and Controversies in Tidal Marsh Ecology*, Kluwer Academic Publishers, Dordrecht, pp. 59–80.

Mitsch, W. J. and Gosselink, J. G.: 2000, *Wetlands*, John Wiley & Sons, New York, 920 pp.

Morris, J. T., Haskin, B.: 1990, 'A 5 yr record of aerial primary production and stand characteristics of *Spartina alterniflora.*', *Ecology* **71**, 2209–2217.

Nixon, S. W.: 1980, 'Between coastal marshes and coastal waters – a review of twenty years of speculation and research on the role of salt marshes in estuarine productivity and water chemistry', in P. Hamilton and K. B. Macdonald (eds), *Estuarine and Wetland Processes with Emphasis on Modelling*, Plenum Press, New York, pp. 437–525.

Odum, W. E., Odum, E. P, and Odum, H. T.: 1995, 'Nature's pulsing paradigm', *Estuaries* **18**: 547–555.

Roulet, N. T.: 2000, 'Peatlands, carbon storage, greenhouse gases, and the Kyoto Protocol: Prospects and significance for Canada', *Wetlands* **20**, 605–615.

Scheffer, R. A. and Aerts, R.: 2000, 'Root decomposition and soil nutrient and carbon cycling in two temperate fen ecosystems', *Oikos* **91**, 541–549

Sokal, R. R. and Rohlf, F. J.: 1981, *Biometry*, 2nd edition, W. H. Freeman and Co., San Francisco.

Smalley, A. E.: 1959, 'The growth cycle of *Spartina* and its relation to the insect populations in the marsh', in R. A. Ragotzie, J. M. Teal, L. R. Pomeroy and D. C. Scott (eds), *Proc. Salt Marsh Conf.*, Athens, University of Georgia, pp. 96–100.

Teal, J. M.: 1962, 'Energy flow in the salt marsh ecosystem of Georgia', *Ecology* **43**, 614–624.

Valiela, I., Teal, J., Allen S. D., Etten, R. Goehringer and Volkmann, S.: 1985, 'Decomposition in salt marsh ecosystems: the phases and major factors affecting disappearance of above-ground organic matter', *J. Exp. Mar. Biol. Ecol.* **89**, 29–54.

Vale, C.: 1990: 'Temporal variations of particulate metals in the Tagus river estuary' *Sci. Total Environ.* **97/98**, 137–154.

Weinstein, M. P. and Kreeger, D. A.: 2000, *Concepts and Controversies in Tidal Marsh Ecology*, Kluwer Academic Publishers, Dordrecht, 875 pp.

Williams, J. R.: 1999, 'Addressing global warming and biodiversity through forest restoration and coastal wetlands creation', *Sci. Total Environ.* **240**, 1–9.

Wolff, W. J., Van Eeden, M. J. and Lammens, E.: 1979, 'Primary production and import of particulate organic matter on a salt marsh in the Netherlands', *Neth. J. Sea Res.* **13**, 242–255.

Zawislanski, P. T., Chau, S., Mountford, H., Wong, H. C. and Sears, T. C.: 2001, 'Accumulation of selenium and trace metals on plant litter in a tidal marsh', *Estuar. Coast. Shelf Sci.* **52**, 589–603.

LEVELS AND CHARACTERISTICS OF TOC IN THROUGHFALL, FOREST FLOOR LEACHATE AND SOIL SOLUTION IN UNDISTURBED BOREAL FOREST ECOSYSTEMS

MICHAEL STARR* and LIISA UKONMAANAHO

Finnish Forest Research Institute, Vantaa Research Centre, P.O. Box 18, FIN-01301 Helsinki, Finland
(* author for correspondence, e-mail: michael.starr@metla.fi; phone: +358 102112450; fax: +358 102112206)

(Received 20 August 2002; accepted 23 May 2003)

Abstract. Total organic carbon (TOC) concentrations and fluxes in throughfall, forest floor leachate, soil solution (15 and 35 cm depths), and groundwater for coniferous forest sites in the boreal zone throughout Finland are described. Eight upland forest stands and one peatland forest stand are included in the study and the samples were collected during 1991–1997. Carbon (C) pools in the living tree biomass and soil compartments are presented, and the hydrophobic/hydrophilic and acidic components of dissolved organic carbon (DOC) in samples collected in autumn 1999 and spring 2000 from two of the sites are compared. Biomass (aboveground and belowground) pools of C averaged 88 Mg ha^{-1} and soil (humus layer + 20 cm soil layer) averaged 55 Mg ha^{-1}. Stand throughfall TOC monthly mean concentrations ranged from 4.0 to 18.6 mg L^{-1} and annual fluxes averaged 4.0 g m^{-2} yr^{-1}. TOC concentrations in the water passing through the forest floor and soil decreased with depth. Plot mean concentrations at 35 cm depth values ranged from 4.1 to 21.2 mg L^{-1} and fluxes averaged 3.7 g m^{-2} yr^{-1}. Throughfall TOC concentrations were lowest during the winter, snowfall period and highest during the growing season. No monotonic trends in throughfall TOC concentrations over the 1991–1997 period were found. Soil solution TOC concentrations varied considerably, both within and between years. DOC in throughfall, forest floor, and soil solutions and in both autumn and spring seasons was dominated by hydrophobic fractions, particularly acids. Spruce canopies and litter appear to be important sources of soluble organic carbon, particularly acidic and hydrophobic compounds. Further studies on the nature and dynamics of organic carbon fluxing through coniferous, boreal forest ecosystems are needed.

Keywords: boreal, dissolved organic carbon, DOC fractionation, soil solution, throughfall

1. Introduction

Boreal coniferous forest ecosystems contain large amounts of carbon in the biomass and soil (Kauppi *et al.*, 1997; Finér *et al.*, 2003) and leach considerable amounts of organic carbon to surface waters (Kortelainen and Saukkonen, 1998). Total organic carbon (TOC) concentrations in Norwegian and Swedish lake and stream waters reportedly have increased during the 1990s. These increases have been attributed to climate change, although this trend has not been clearly found in Finland (Skjelkvåle *et al.*, 2001). However, knowledge about the fluxes of organic carbon (C) associated with throughfall and soil solution in boreal forest ecosystems is limited compared to that known about forests in the temperate zone (Michalzik *et al.*, 2001; Piirainen *et al.*, 2002).

In this paper we report on the levels and trends in TOC concentrations in throughfall and soil solution collected at nine boreal, forest stands over a seven-year period. The hydrophobic and acidic character of the dissolved organic carbon (DOC) in throughfall and soil solution samples collected in autumn and spring at two of the stands also is described. But first we describe the pools of C in the tree biomass and soil at the stands. One of the stands is on peatland (histosols) and TOC concentrations at the top of the water table are reported rather than those in water from the unsaturated zone. The nine stands are unmanaged, old stands have differing species composition and are located throughout Finland. The study is unique in that it deals with C in undisturbed forests in the boreal zone and because it is a multiple site, long-term and integrated study in which the same sampling and analytical procedures have been used (Bergström *et al.*, 1995).

2. Materials and Methods

2.1. STUDY SITES

The data presented in this paper have been collected from four catchments (Figure 1). These headwater catchments belong to the network of UN-ECE International Co-operative Programme on Integrated Monitoring (Bergström *et al.*, 1995) and, nowadays, also to the network of UN-ECE ICP-Forests/EU Intensive Monitoring sites set up under the 1979 Convention on Long-Range Transboundary Air Pollution. They are located in extensive areas of protected forest (national parks) having semi-natural, undisturbed old-growth forests. The terrain in each catchment was formed by the last glaciation and the landscape is characterised by areas of upland coniferous or mixed forest, peatland, seepage lakes and ponds, and a discharge lake with stream. No agriculture is carried out in the vicinity and there are no sources of air pollution emissions nearby. The annual mean temperature ($°C$) is 3.1 at Valkea-Kotinen, 2.0 at Hietajärvi, –0.5 at Pesosjärvi, and –1.9 at Vuoskojärvi. In the same order, the annual precipitation (mm) is 618, 592, 571 and 395. The number of days per year with snow cover (in the open) increases from about 130 at Valkea-Kotinen to about 210 at Vuoskojärvi.

In each catchment, a number (7–9) of permanent monitoring plots, varying in size from 25×25 to 40×40 m, have been established (1988–1989) in the main habitat types (stands). In this paper only those plots at which throughfall and soil solution have been monitored are included. The stands are composed of varying proportions of Scots pine, Norway spruce, and deciduous (mainly birch) trees (Table I). For further a description and details about the catchments and plots, see Bergström *et al.* (1995).

TABLE I

Soil type and stand (1991–1992) characteristics of the nine study plots

Catchment	Plot	Elevation	Soil type	Stem volume	Scots pine	Norway spruce	Deciduous[a]	Standing dead	Age
		m a.s.l.		m³ ha⁻¹	%	%	%	%	dominant trees, years
Valkea-Kotinen	VK2	156	Terric Histosol	433	8	85	7	2	190
	VK3	161	Dystric Cambisol	579	4	54	41	2	155
Hietajärvi	HJ1	168	Haplic Podzol	201	86	0	14	1	100
	HJ4	167	Haplic Podzol	232	91	0	9	0	230
Pesosjärvi	PJ1	263	Carbic Podzol	154	2	69	29	1	320
	PJ2	270	Haplic Podzol	184	48	36	16	6	240
	PJ3	293	Haplic Podzol	144	24	51	25	1	240
Vuoskojärvi	VJ2	146	Haplic Podzol	16	3	0	97	–	150
	VJ3	158	Haplic Podzol	50	98	0	2	1	180

[a] Mainly Birch, *Betula* spp., but also Aspen, *Populus tremula* and, at Vuoskojärvi, Mountain birch (*B. pubescens* spp. *czerepanovii*).

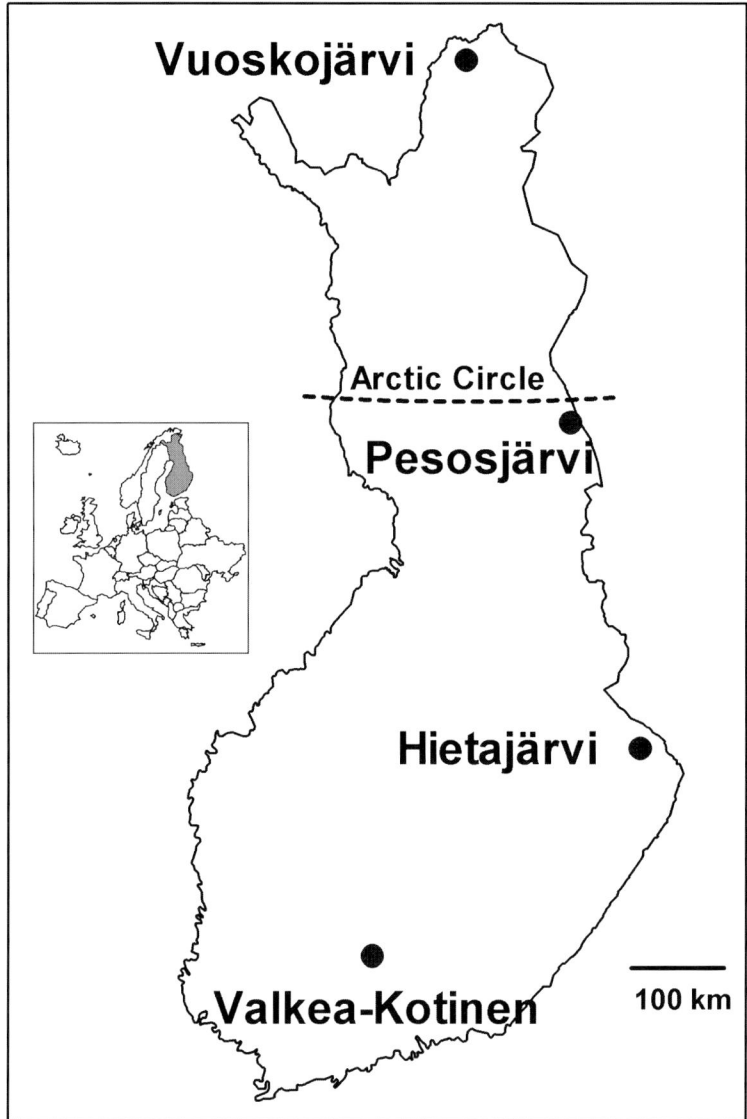

Figure 1. Map showing location of the four study catchments.

2.2. BIOMASS AND SOIL C POOLS

All trees (living and dead, both snags and fallen) on the plots have been mapped measured for breast height diameter, top height, canopy height, and crown projection. Forest stand C pools were calculated from estimates of biomass fractions of all the living trees on the plot (kg ha^{-1}) and assuming a C content of 52% for all fractions. The biomass fractions for Scots pine, Norway spruce, and birch (*Betula pendula* and *B. pubescens*) were calculated using the allometric functions developed by Marklund (1988) and for mountain birch (*B. pubescens* ssp. *czerepanovii*) at Vuoskojärvi using those developed by Starr *et al.* (1998). Marklund's functions provide estimates of stem (over bark), needles (but not birch leaves), living and dead branches, and stump and coarse roots, while the functions for mountain birch provide estimates of stem (over bark), living and dead branches, and foliage (at full development) but not stump and roots.

With the exception of plot 2 at Valkea-Kotinen (VK2), which is on peat, four parallel composite soil samples of the organic (humus) layer (Of + Oh), 0–5 and 5–20 cm layers were taken systematically at each plot. At plot VK2, the 0–5, 5–10 and 10–20 cm layers were sampled. Carbon concentrations in the air-dried humus layer, peat (milled) and mineral soil (<2 mm) samples were determined using a LECO-CHN analyser. The soil C pools for each sampled layer were calculated from the mean soil sample C concentration, layer thickness, bulk density, and stone content values. For further details, see Starr and Ukonmaanaho (2001).

2.3. Throughfall and soil solution C concentrations and fluxes

Throughfall was collected using permanently open (bulk) collectors (12–16 during snow-free period, 6–8 during winter) located systematically round the plots (Ukonmaanaho, 2001). Soil solution at depths of 15 and 35 cm depth in the mineral soil was sampled using suction cup lysimeters (initially 3 at each depth and then 6) installed along one edge of the plots. A suction of 60 kPa was used to draw the samples. At the Hietajärvi plots, in addition to the suction lysimeters, three zero-tension lysimeters were installed at 15 and 35 cm depths and also directly under the humus layer. Of the zero-tension lysimeter data, only those from the humus layer, which collect forest floor leachate, has been used in this paper. Lysimeters were not installed in the peatland plot at Valkea-Kotinen, VK2, but groundwater wells (5) were.

A weekly sampling interval during the snow-free period (usually May–November) was used to collect and analyse all the water samples and monthly volume-weighted mean TOC concentration values were calculated. From winter 1994/1995, monthly snowfall sampling during the winters was initiated. Throughfall TOC concentration values for the winter months during 1991–1994 were given the mean values calculated from the 1995–1997 data and corresponding hydrologic fluxes (mm) estimated using a regression model based on the strong correlation between bulk deposition collected in the open (Finnish Meteorological Institute) and throughfall ($r = 0.93$, $n = 593$). Monthly throughfall concentration values were multiplied by the hydrologic flux to give monthly TOC flux values, which were then summed to give annual values. To calculate forest floor leachate and soil solution annual TOC fluxes, the available monthly volume-weighted TOC concentration values were averaged and multiplied by a modelled annual hydrologic flux value (Starr, 1999). For the peatland plot, VK2, we calculated the mean TOC concentration of the groundwater from weekly samples, which are only available for May–October 1995.

Organic carbon concentrations in the water samples were measured using a Shimadzu TOC-5000 Analyzer. Prior to analysis, the samples were filtered (Schleicher & Schuell 589/2). Because dissolved organic carbon (DOC) is defined as organic carbon in water passing through a membrane filter of pore size 0.45 μm, we refer to our organic carbon concentrations as total organic carbon (TOC) concentrations.

2.4. Throughfall, forest floor and soil solution DOC fractionation

As part of a Nordic project (Vogt et al., 2001), the DOC in throughfall, forest floor, and soil solution samples taken from Valkea-Kotinen (VK3) and Hietajärvi (HJ4) in September–October (Autumn) 1999 and in April–May (Spring) 2000 following snowmelt was characterized by fractionation. The fractionation was made using adsorption/exchange resins (XAD-8 nonionic, MSC-1 cation exchange and Duolite A-7 anion exchange resins) based on the procedure described by Qualls and Haines (1991) that separates DOC into hydrophilic and hydrophobic acids, bases and neutrals. The samples were collected weekly, but bulked by season and kept refrigerated until analysed in September 2000, when a subsample of 200–250 mL was filtered through a membrane filter of pore size 0.45 μm for analysis. Carbon concentrations in the filtrates were measured using the same Shimadzu TOC-5000 Analyzer as for the other water samples.

3. Results and Discussion

3.1. Soil and forest stand C pools

The size of the soil C pool (humus layer + 0–20 cm layer, corrected for stone content) at the plots ranged from 29 to 112 Mg ha^{-1}, and averaged 55 Mg ha^{-1} (Figure 2). As could be expected, the peat plot (VK2) had the largest soil pool. The smallest pools were associated with the Vuoskojärvi plots, the northernmost catchment, where the primary production and therefore input of organic matter to the soil is low. Our values are comparable to those reported by Tamminen (2000), Liski and Westman (1995, 1997), Kauppi et al. (1997), and Finér et al. (2003) for Finnish forest soils. Earlier studies showed that our soil C contents were correlated to soil pH, exchangeable Ca^{2+} + Mg^{2+} contents, exchangeable acidity, and heavy metal contents (Starr and Ukonmaanaho, 2001).

The tree stand C pools ranged from 11 to 200 Mg ha^{-1} and averaged 88 Mg ha^{-1} (Figure 2). The stem compartment accounted for 50% or more of the total biomass pool. The coarse root fraction at the mountain birch plot, VJ2, could not be estimated but the belowground C pool (i.e. soil plus coarse roots) at the other catchments accounted for half or more of the combined soil and stand C pool. The

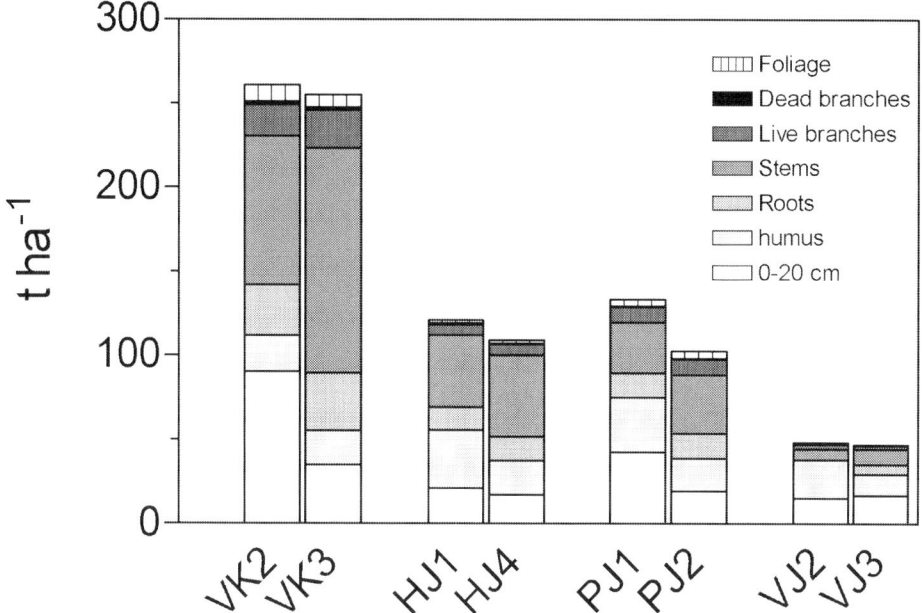

Figure 2. Carbon pools in the stand and soil at each study plot (values for the root compartment at VJ2 are not available).

biomass of fine roots, although having a high turnover, is small (Finér *et al.*, 2003). It should be noted that we have not included the C associated with dead trees (snags and fallen stems). This pool has not yet been calculated for our sites, but the volume of coarse woody debris (CWD) in Norway spruce dominated old-growth forests in Fennoscandia ranges from 50 to 120 m^3 ha^{-1}, and accounts for 28% of total stand volume (living + dead) on average (Siitonen, 2001). Corresponding CWD volumes for Scots pine dominated old-growth stands are 20 to 120 m^3 ha^{-1}, which accounts for 25% of total stand volume. Standing and fallen dead stems and fallen branches accounted for 13% of the total (living + dead) above-ground biomass compartment in a Norway spruce dominated old-growth stand in eastern Finland (Finér *et al.*, 2003).

3.2. THROUGHFALL, FOREST FLOOR AND SOIL SOLUTION TOC CONCENTRATIONS AND FLUXES

Plot average throughfall TOC concentrations collected during the 1991–1997 period ranged from 4.2 to 18.6 mg L^{-1} (Table II). This range falls within that presented by Michalzik *et al.* (2001) for temperate forests in their review of 42 case studies. Our throughfall TOC concentrations generally decreased northwards. This trend might be related to a similar trend in stand volume (Table I), which would result in a reduced degree of interaction between precipitation and canopy. Rainfall passing

TABLE II

Descriptive statistics (mean, standard deviation and number of samples) of TOC concentrations (mg L^{-1}) in monthly bulk deposition, throughfall, forest floor and soil solution samples collected from each study plot during 1991–1997

	Plot								
	VK2	VK3	HJ1	HJ4	PJ1	PJ2	PJ3	VJ2	VJ3
Bulk deposition[a]									
Mean	3.7	3.7	3.1	3.1	3.8	3.8	3.8	3.6	3.6
St. dev.	1.2	1.2	0.8	0.8	1.8	1.8	1.8	1.9	1.9
n	12	12	12	12	12	12	12	12	12
Throughfall									
Mean	18.6	14.0	7.4	8.8	7.9	4.9	n.d.	4.0	8.4
St. dev.	9.5	7.5	5.0	5.5	4.9	4.0		3.7	7.1
n	58	59	53	57	54	54		53	53
Forest floor									
Mean	29.1[b]		45.7	53.1	n.d.	n.d.	n.d.	n.d.	n.d.
St. dev.	10.0		17.3	28.4					
n	103		29	31					
Soil solution 15 cm									
Mean	n.d.	20.1	11.9	13.6	n.d.	36.0	7.2	17.2	23.3
St. dev.		17.2	12.7	7.4		56.2	4.8	17.3	33.5
n		41	47	38		38	47	15	56
Soil solution 35 cm									
Mean	n.d.	9.5	8.3	6.4	n.d.	14.0	4.1	8.5	21.2
St. dev.		9.4	22.2	7.9		27.5	3.3	4.4	34.1
n		51	46	37		44	47	17	55

[a] Monthly samples collected outside the forest during 1997.
[b] Mean weekly groundwater May–September 1995.
n.d. = Not determined.

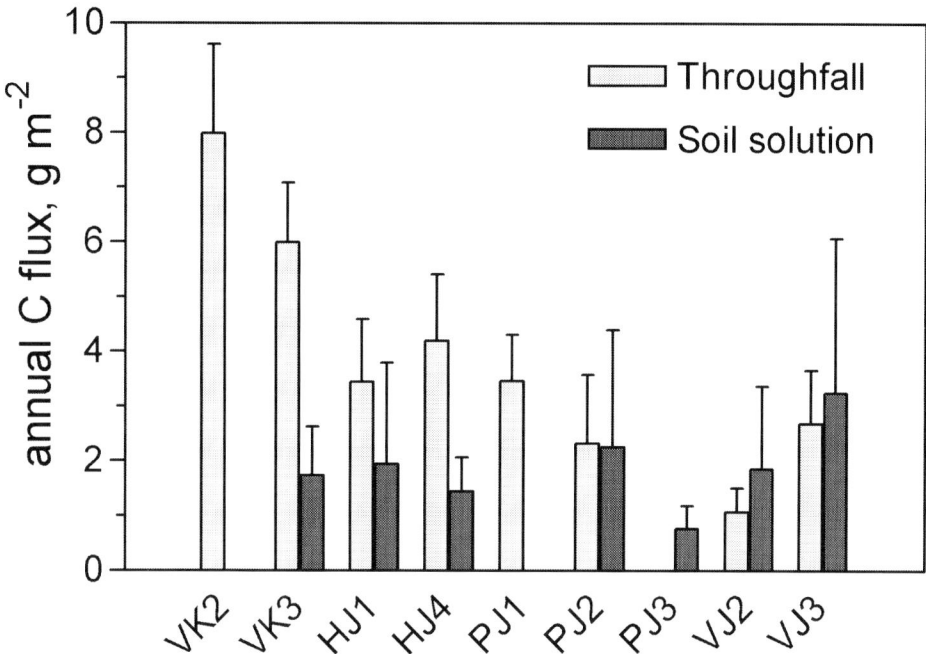

Figure 3. Annual mean (1991–1997) carbon fluxes in throughfall and soil solution (40 cm depth) for each study plot (See Table I for plot abbreviations). Error bars are standard deviations.

through the canopy is known to dissolve slightly soluble and soluble organic acids both from the foliage itself and from dry deposition accumulated on foliar surfaces (Thurman, 1985). Bulk precipitation collected in 1997 at a nearby open area in each catchment has been analysed for TOC (Table II). The enrichment of the throughfall in organic C is indicated by the ratio of throughfall to bulk precipitation TOC concentrations. Plot values of this ratio calculated from the mean concentration values for 1997 ranged from 2 to 7, being the highest for the Valkea-Kotinen plots. But besides having the highest stand volumes, the Valkea Kotinen stands also have the highest proportion of Norway spruce present (Table I). The higher TOC concentrations at Valkea-Kotinen may therefore reflect a species effect on the leaching of organic C from forest canopies, with spruce being more susceptible than other species.

The mean annual throughfall TOC fluxes ranged from 1.2 to 8.0 g m^{-2} yr^{-1} and averaged 4.0 g m^{-2} yr^{-1} (Figure 3). These values are similar to those reported by Piirainen *et al.* (2002) for a Norway spruce dominated stand in eastern Finland, but are clearly less than those reported for temperate forests (Michalzik *et al.*, 2001). Because our throughfall mean TOC concentrations were similar to those reported by Michalzik *et al.* (2001), the smaller fluxes at our boreal stands compared to those in the temperate zone must be due to a difference in the amounts of precipitation. The throughfall TOC flux at our sites decreases northwards, as do TOC concentra-

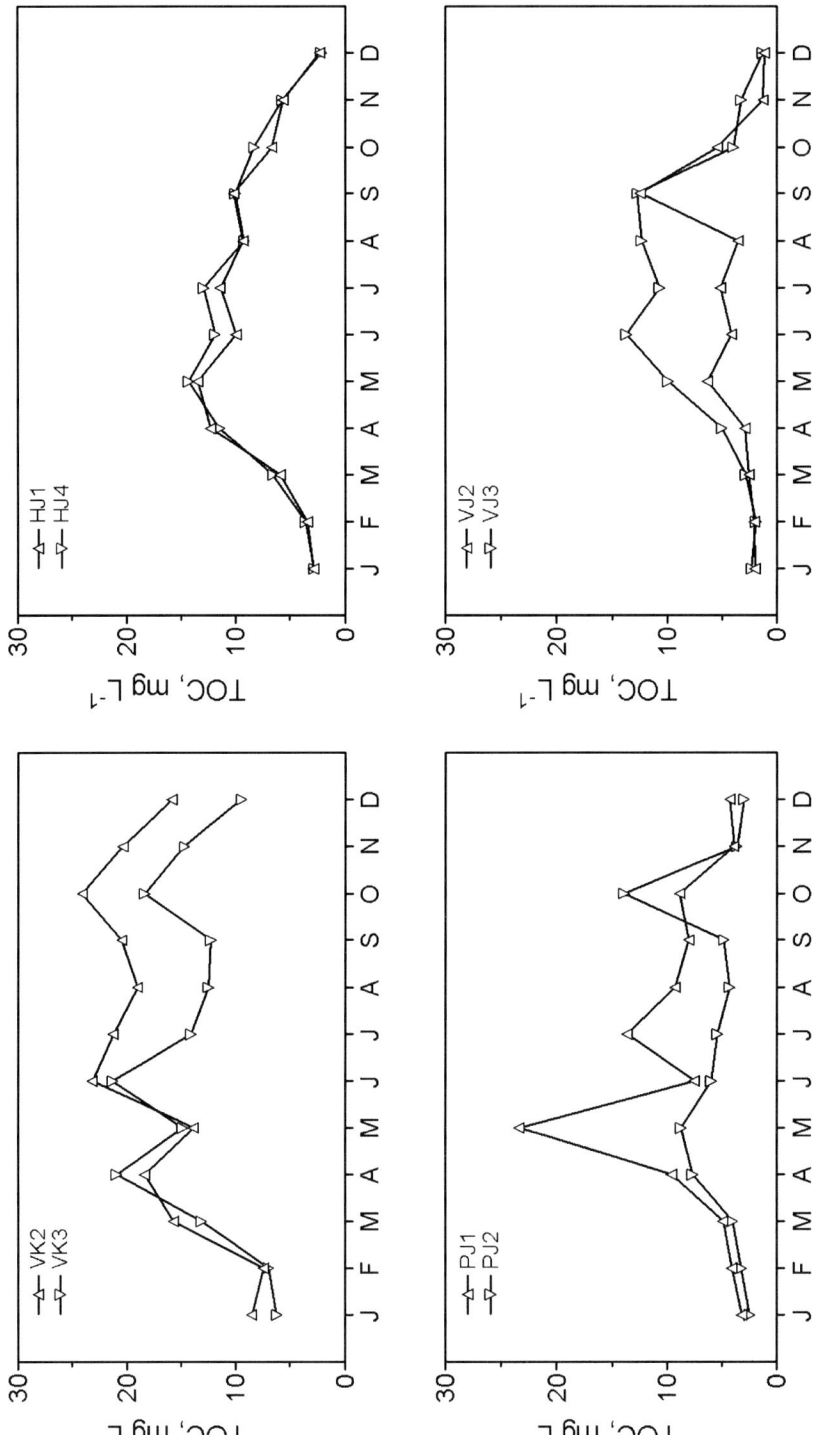

Figure 4. Monthly mean throughfall TOC concentrations by stand during 1991–1997.

tions and amount of precipitation (Table I). Furthermore, a greater proportion of the annual precipitation falls as snow when going northwards, and snow has less leaching efficiency than rainfall. Thus throughfall TOC concentrations tended to peak during the growing season and to be the lowest during the winter months when the stands were dormant and precipitation fell as snow (Figure 4). Seasonal-Kendall slope estimates of the monotonic trend indicated that there was no significant ($p >$ 0.123) trend in throughfall TOC concentrations at any of the plots over the period 1991–1997.

Monthly soil solution TOC concentrations varied considerably, both within and between plots (Table II). The mean of the monthly concentration values available ranged from 7.2 to 36.0 mg L^{-1} for the soil solution samples collected at 15 cm depth and from 4.1 to 21.2 mg L^{-1} for those collected at 35 cm depth. These values are similar to reported concentrations in soil solutions taken with zero-tension lysimeters for a Podzol soil in eastern Finland (Piirainen et al., 2002), but somewhat lower than DOC concentrations in soil solution samples extracted by centrifugation from Nordic Podzol soils (Van Hees et al., 2000a; Riise et al., 2000). At Hietajärvi, forest floor leachates were yellow-coloured, indicating the presence of humic substances. The TOC concentrations were 6 times higher, on average, than TOC concentrations in throughfall and 4 times those at 15 cm depth and 7 times those at 35 cm depth. Soil solution TOC concentrations at 35 cm depth at all plots were generally lower than those at 15 cm depth, indicating a removal of C from soil solution as it moves downwards. Such a removal can take place through microbial decay, adsorption to Fe and Al sesquioxides, which accumulate in the B-horizon of Podzols, and coagulation and precipitation (Thurman, 1985; Riise et al., 2000). Unlike throughfall, there was no relationship between soil solution TOC concentrations and latitude, indicating that local soil factors were more important in controlling soil solution TOC concentrations than climate factors.

Plot soil solution fluxes of TOC at 40 cm depth ranged from 2.4 to 4.9 g m^{-2} yr^{-1} and averaged 3.7 g m^{-2} yr^{-1} (Figure 3). Our highest values were somewhat lower than B-horizon seepage fluxes reported for temperate sites by Michalzik et al. (2001). Our values were also smaller than B-horizon fluxes from a Podzol in eastern Finland reported by Piirainen et al. (2002). But the hydrologic fluxes in their study by were based on zero-tension lysimeter catches, which were probably underestimates of the true hydrologic flux. The relatively high soil solution fluxes at Vuoskojärvi may be the result of modelled hydrologic fluxes that were too high. There were too many missing monthly values to allow reliable Seasonal-Kendall trend slope estimates to be made for soil solution TOC concentrations. Simple visual inspection of the data indicates that forest floor leachate TOC concentrations increased with the hydrologic flux, which is higher during spring snowmelt and in autumn when lowered evapotranspiration losses result increased inputs to the soil. Soil solution TOC concentrations appeared to be less affected by the amount of percolation. However, these observations need further testing.

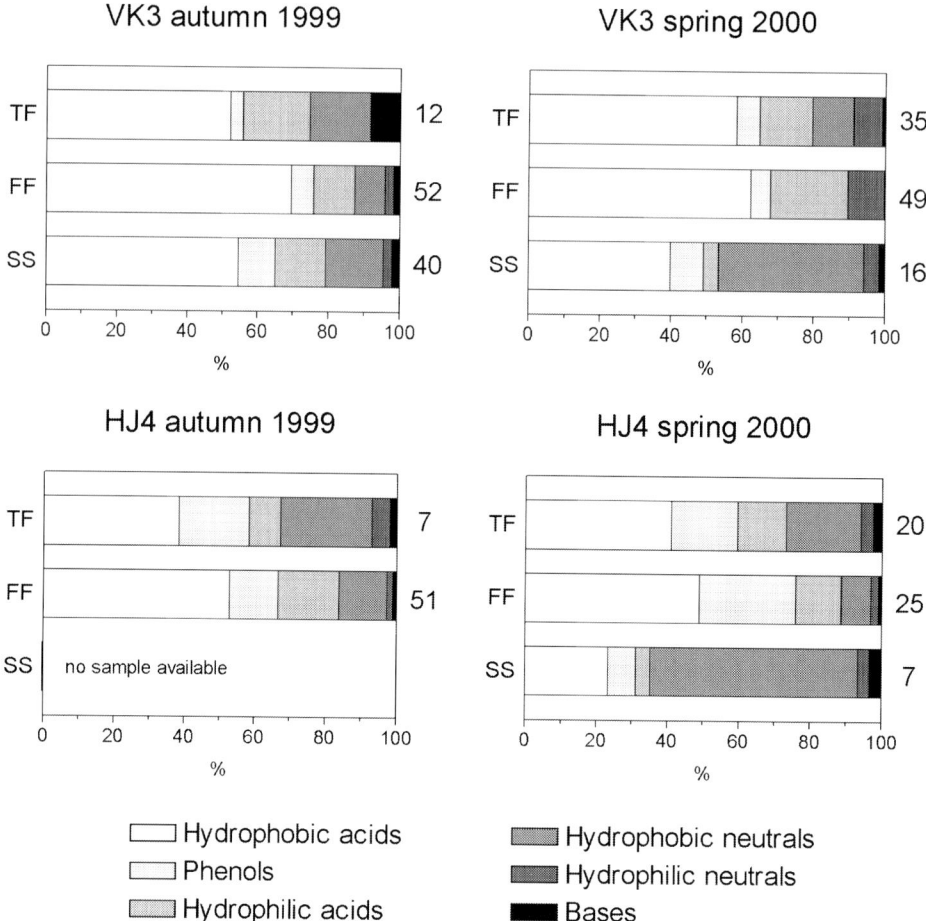

Figure 5. Distribution of various DOC fractions in Valkea-Kotinen (VK3) and Hietajärvi (HJ4) throughfall and soil solution samples taken in autumn 1999 and spring 2000. Numbers at end of bars are the total DOC concentrations (mg L^{-1}).

3.3. CHARACTERISATION OF THROUGHFALL AND SOIL SOLUTION DOC

The results of the DOC fractionation of Valkea-Kotinen and Hietajärvi throughfall and soil solution samples are presented in Figure 5. Because the hydrophobic base fraction was so small it has been combined with the hydrophilic base fraction, which consists of proteins and amino acids, and so easily biodegradable. The phenol fraction consists of weak hydrophobic acids, weak because of the presence of phenolic rather than carboxylic acid groups, and inhibits microbial activity. The hydrophobic acid fraction (humic and fluvic acids, i.e. aquatic humic substances) is relatively resistant to decay while easily degradable carbohydrates dominate the neutral fractions (Qualls and Haines, 1991; Thurman, 1985). The hydrophilic acid fraction contains low molecular weight organic acids (Thurman, 1985; Qualls and

Haines, 1991), which are important to the mobilization and transport of Al and Fe in Podzol soils (Van Hees *et al.*, 2002b).

Although DOC concentrations in throughfall were less than those in the forest floor solutions at both sites and in both seasons, the composition of both types of water was similar, being dominated by hydrophobic acids (>38%) and with bases forming only a minor part (<8%). The combined acids fraction (hydrophobic, phenols and hydrophilic) accounted for 67 to 90% of the DOC in the throughfall and forest floor leachates. The dominance of acid fractions, particularly the hydrophobic acid fraction, was also found in throughfall and Oa horizon solutions from an oak-hickory stand by Qualls and Haines (1991). The combined hydrophobic fraction dominated (68 to 90%) the DOC of all samples. This contrasts with the results presented by Kaiser *et al.* (2001), who found that the DOC in forest floor leachate from both Scots pine and beech stands to be dominated by the hydrophilic fraction. The hydrophilic fraction has been shown to leach in larger amounts from deciduous litter than from coniferous litter (Hongve, 1999). The phenol fraction in our study, especially in the samples from Hietajärvi, was clearly more important than in the study by Qualls and Haines (1991). These differences are probably due to the dominance of coniferous tree species in our stands. Thus, while DOC in temperate deciduous forests is dominated by hydrophilic and neutral compounds the DOC in boreal, coniferous forests is dominated by less easily decomposable acidic and hydrophobic compounds. The dominance of hydrophobic and acidic fractions in the samples from Valkea-Kotinen in particular indicates that spruce needles are an important source of these fractions.

Throughfall DOC concentrations were greater in spring than in autumn, as was also seen in our main throughfall TOC data (Figure 4), while the forest floor leachate and soil solution DOC concentrations showed the opposite. There was no clear seasonal difference in the composition of DOC, except that the hydrophobic neutrals fraction, for an unknown reason, was absent from the forest floor Valkea-Kotinen spring sample and appeared to dominate the composition of springtime DOC in the 15 cm soil solution.

4. Conclusions

Below-ground (soil+roots) C pools accounted for half or more of the ecosystem C pool (aboveground and belowground forest stand biomass, humus layer, and 0–20 cm soil layer) in these undisturbed, boreal, old-growth forest ecosystems. Precipitation passing through the canopy was enriched in organic carbon, with annual TOC fluxes being 2–7 times greater than corresponding bulk deposition collected in the open. Although throughfall TOC concentrations were similar to those reported for temperate zone forests, annual fluxes were smaller, as were also soil solution TOC fluxes, reflecting lower precipitation amounts in the boreal zone. TOC concentrations are substantially increased as throughfall passed through the forest floor, but then decline as the water continues to percolate downwards through the soil. At 35 cm depth, soil solution TOC concentrations approached those in throughfall. Monthly throughfall TOC concentrations were higher during the growing season but no long-term (1991–1997) trends could be discerned. Hydrophobic compounds, particularly hydrophobic acids, dominated the DOC in throughfall, forest floor leachate and soil solution. The fraction of phenols appears to be more important in boreal, coniferous forest ecosystems than in temperate, deciduous forests. Spruce canopies and litter appear to be important sources of soluble organic carbon, particularly acidic and hydrophobic compounds. Further studies on the nature and dynamics of organic carbon fluxing through coniferous, boreal forest ecosystems are needed.

Acknowledgements

We wish to thank Dr. Veikko Kitunen, Central Laboratory, Vantaa Research Centre, for performing the DOC fractionation analysis and Mr. Markus Hartman, Vantaa Research Centre, for applying Marklund's biomass functions.

References

Bergström, I., Mäkelä, K. and Starr, M. (eds): 1995, 'Integrated Monitoring Programme in Finland. First National Report', Ministry of the Environment, Environmental Policy Department; *Report 1*, Ministry of the Environment, 138 pp.

Finér, L., Piirainen, S., Mannerkoski, H. and Starr, M.: 2003, 'Carbon and nitrogen pools in an old-growth, Norway spruce-mixed forest in eastern Finland and changes associated with clear-cutting', *For. Ecol. Manage.* **174**, 51–63.

Hongve, D.: 1999, 'Production of dissolved organic carbon in forested catchments', *J. Hydrol.* **224**, 91–99.

Kauppi, P. E., Posch, M., Hänninen, P., Henttonen, H. M., Ihalainen, A., Lappalainen, E., Starr, M. and Tamminen, P.: 1997, 'Carbon reservoirs in peatlands and forests in the boreal regions of Finland', *Silva Fenn.* **31**, 13–25.

Kaiser, K., Guggenberger, G., Haumaier, L. and Zech, W.: 2001, 'Seasonal variations in the chemical composition of dissolved organic matter in organic forest floor layer leachates of old-growth Scots pine (*Pinus sylvestris* L.) and European beech (*Fagus sylvatica* L.) stands in northeastern Bavaria, Germany', *Biogeochemistry* **55**, 103–143.

Kortelainen, P. and Saukkonen, S.: 1998, 'Leaching of nutrients, organic carbon and iron from Finnish forestry land', *Water, Air, Soil Pollut.* **105**, 239–250.

Liski, J. and Westman, C. J.: 1995, 'Density of organic carbon in soil at coniferous forest sites in southern Finland', *Biogeochemistry* **29**, 183–197.

Liski, J. and Westman, C. J.: 1997, 'Carbon storage in forest soil of Finland. 2. Size and regional patterns', *Biogeochemistry* **36**, 261–274.

Marklund, L.: 1988, 'Biomassafunktioner för tall, gran och björk i Sverige. Summary: Biomass functions for pine, spruce and birch in Sweden', Department of Forest Survey, Swedish University of Agricultural Sciences, *Report 45*, 73 pp.

Michalzik, B., Kalbitz, K., Park J.-H., Solinger S. and Matzner, E.: 2001, 'Fluxes and concentrations of dissolved organic carbon and nitrogen – A synthesis for temperate forests', *Biogeochemistry* **52**, 173–205.

Piirainen, S., Finér, L., Mannerkoski, H. and Starr, M.: 2002, 'Effects of forest clear-cutting on the carbon and nitrogen fluxes through podzolic soil horizons', *Plant Soil* **239**, 301–311.

Qualls, R. G. and Haines, B. L.: 1991, 'Geochemistry of dissolved organic nutrients in water percolating through a forest ecosystem', *Soil Sci. Soc. Am. J.* **55**, 1112–1123.

Riise, G., Van Hees, P. A. W., Lundström, U. S. and Strand, L. T.: 2000, 'Mobility of different size fractions of organic carbon, Al, Fe, Mn and Si in podzols', *Geoderma* **94**, 237–247.

Siitonen, J.: 2001, 'Forest management, coarse woody debris and saproxylic organisms: Fennoscandian boreal forests as an example', *Ecol. Bull.* **49**, 11–41.

Skjelkvåle, B. L., Mannio, J., Wiklander, A. and Andersen, T.: 2001, 'Recovery from acidification of lakes in Finland, Norway and Sweden 1990–1999', *Hydrol. Earth Syst. Sci.* **5**, 327–337.

Starr, M.: 1999, 'WATBAL: A model for estimating monthly water balance components, including soil water fluxes', in S. Kleemola and M. Forsius (eds), *8th Annual Report 1999 UN ECE ICP Integrated Monitoring*, Finnish Environment Institute, Helsinki, Finland, The Finnish Environment 325, pp. 31–35.

Starr, M., Hartman, M. and Kinnunen, T.: 1998, 'Biomass functions for mountain birch in the Vuoskojärvi Integrated Monitoring area', *Boreal Environ. Res.* **3**, 297–303.

Starr, M. and Ukonmaanaho, L.: 2001, 'Results from the first round of the Integrated Monitoring soil chemistry subprogramme', in L. Ukonmaanaho and H. Raitio (eds), *Forest Condition in Finland, National Report 2000, Finnish Forest Research Institute, Research Papers* 824, pp. 140–157.

Tamminen, P.: 2000, 'Soil factors', in E. Mälkönen (ed.), *Forest Condition in a Changing Environment – The Finnish Case*, Kluwer Academic Publishers, Dordrecht, The Netherlands, pp. 72–86.

Thurman, E. M.: 1985, *Organic Geochemistry of Natural Waters*, Nijhoff/Dr. W. Junk Publishers, Dordrecht, The Netherlands.

Ukonmaanaho, L.: 2001, 'Canopy and soil interaction with deposition in remote boreal forest ecosystems: A long-term integrated monitoring approach', *Finnish Forest Research Institute, Research Papers* 818.

Van Hees, P. A. W., Lundström, U. S. and Giesler, R.: 2000a, 'Low molecular weight organic acids and their Al-complexes in soil solutionc- compostion, distribution and seasonal variation in three podzolized soils', *Geoderma* **94**, 173–200.

Van Hees, P. A. W., Lundström, U. S., Starr, M. and Giesler, R.: 2000b, 'Factors influencing aluminium concentrations in soil solution from podzols', *Geoderma* **94**, 289–310.

Vogt, R. D., Gjessing, E., Andersen, D. O., Clarke, N., Gadmar, T., Bishop, K., Lundstrøm, U. and Starr, M.: 2001, 'Natural Dissolved Organic Material (NOM) in the Nordic countries. 1. TOC Intercalibration 2. Physico-chemical characteristics of DOM', *NORDTEST Report TR 479*, 150 pp.

BIOMASS AND NUTRIENT DYNAMICS OF RESTORED NEOTROPICAL FORESTS

ARIEL E. LUGO[1]*, WHENDEE L. SILVER[2] and SANDRA MOLINA COLÓN[3]

[1] *International Institute of Tropical Forestry, USDA Forest Service, Jardín Botánico Sur, 1201 Calle Ceiba, Río Piedras, PR 00926-1119, Puerto Rico;* [2] *Ecosystem Sciences Division, Department of Environmental Science, Policy, and Management, 151 Hilgard Hall 3110, University of California, Berkeley, California 94720, U.S.A.;* [3] *Department of Biology, Pontificia Universidad Católica de Puerto Rico, 2250 Las Américas Suite 570, Ponce, Puerto Rico 00731-6382*
(* *author for correspondence, e-mail: alugo@fs.fed.us; phone: 787 766 5335; fax: 787 766 6263*)

(Received 20 August 2002; accepted 28 April 2003)

Abstract. Restoring species-rich tropical forests is an important activity because it helps mitigate land deforestation and degradation. However, scientific understanding of the ecological processes responsible for forest restoration is poor. We review the literature to synthesize the current state of understanding of tropical forest restoration from a biogeochemical point of view. Aboveground biomass and soil carbon accumulation of restored tropical forests are a function of age, climate, and past land use. Restored forests in wet life zones accumulate more biomass than those in moist or dry life zones. Forests restored on degraded sites accumulate less aboveground biomass than forests restored on pastures or agricultural land. Rates of aboveground biomass accumulation in restored forests are lower than during natural succession, particularly during the first decades of forest establishment. Rates of litterfall, biomass production, soil carbon accumulation, and nutrient accumulation peak during the first few decades of restored forest establishment and decline in mature stages. Changes in species composition and canopy closure influence the rate of primary productivity of older restored stands. Species composition also influences the rate and concentration of nutrient return to the forest floor. The ratio of primary productivity to biomass is high in young restored forests and low in mature stands irrespective of climate. The ratio is low when past land use has little effect on biomass accumulation, and high when past land uses depresses biomass accumulation. This effect is due to a high rate of litterfall in restored forests, which helps restore soil by circulating more nutrients and biomass per unit biomass accumulated in the stand. The degree of site degradation and propagule availability dictates the establishment and growth of tree species. Reestablishment of forest conditions and the enrichment of sites by plant and animal species invasions lead to faster rates of succession, aboveground primary productivity, and biomass accumulation in restored forests. Our review demonstrates that nutrient cycling pathways and nutrient use efficiency are critical for interpreting the suitability of tree species to different conditions in forest stands undergoing restoration.

Keywords: carbon cycling, Luquillo Experimental Forest, neotropical forest restoration, nutrient cycling, organic matter dynamics, Puerto Rico, tree plantations

1. Introduction

The restoration of tropical forests is a matter of concern due to the importance of maintaining land productivity and sustaining the economies of tropical countries (Brown and Lugo, 1994). We use the term restoration to mean the restoration of forests, irrespective of species composition, on deforested lands by rehabilitating forest conditions and site productivity (Brown and Lugo, 1994). There is a clear tendency in the tropics for addressing the problem of degraded lands through reforestation and afforestation projects. For example, the area of tree plantation in the 90 countries monitored by the Food and Agriculture Organization (FAO, 1993) increased by 150% at an annual rate of 2.6 million ha between 1980 and 1990 and 2.0 million ha a^{-1} between 1990 and 2000 (FAO, 2001b). The purpose of many of these plantations is for land protection or rehabilitation. Numerous literature reports of successes of both small and large-scale restoration projects in the tropics further demonstrate the increasing importance of tropical forest restoration (e.g., Field, 1996; Parrotta and Turnbull, 1997; FAO, 2001a). Nevertheless, the area subjected to restoration and rehabilitation is still small (13%) in comparison to the area deforested annually in the tropics (FAO, 2001b).

There are many techniques for restoring tropical forests, and several authors have reviewed them (e.g., Brown and Lugo, 1994; Parrotta and Turnbull, 1997; Ashton et al., 2001). The many techniques for restoring species-rich tropical forests on degraded lands involve two basic approaches. The first approach involves planting one or more tree species, and allowing them to develop into species-rich forests. A second approach is allowing natural regeneration and succession, either accelerated or not by humans, to proceed to mature states. Usually, management objectives, site conditions, and availability of resources determine which approach to use. However, where soil degradation is extreme, as occurs after mining activities, tree planting is required to restore forest conditions because arrested succession prevents natural forest regeneration (Parrotta and Turnbull, 1997).

The scientific understanding of tropical forest restoration lags behind the practice of restoring such forests. Much progress is possible without scientific understanding when restoring tropical forests in deforested lands because tropical succession can rapidly reestablish forest cover on most abandoned sites. However, overcoming arrested succession requires human intervention and technical knowledge to restore forests. The complexity of tropical forests is a major reason for the knowledge gap about their ecology and silviculture. Most of the literature addresses the species aspects of the activity (what species to plant or favor, how to promote or accelerate species richness, how to deal with alien species, etc.) and there is very little work on biogeochemistry. Moreover, it takes about 100 years for planted or fallowed forest stands to reach maturity in the tropics (Brown and Lugo, 1990), and we lack long-term studies of the successional process and biogeochemistry in these systems. As a result, a definitive synthesis of organic matter and nutrient dynamics in restored forests is not possible at this time.

We present case-study information on rates of organic matter and nutrient accumulation of tropical forests that were either restored through planting on degraded lands and allowed to mature as mixed species forests, or restored naturally and matured after abandonment of various types of land use. The case studies include a species-rich forest restored in Puerto Rico through planting on pastures on a moist site (La Condesa in the Luquillo Experimental Forest, LEF), and species-rich forests restored through both planting and natural succession after abandonment of various land uses on dry (Guánica Forest) and wet (LEF and the Amazon) forest sites. We compare the results from these case studies with forest regeneration (natural and planted) elsewhere in the tropics. Our objective is to identify the range of rate processes associated with restored tropical forests and some of the primary mechanisms that might affect those rates. We ask if rates of organic matter and nutrient dynamics in restored forests are different from rates observed in natural forests of similar age and growing under similar climates and soils.

2. Description of the Study Sites

We use data from three locations: the Colombian and Venezuelan Amazon (Saldarriaga, 1994), the LEF (Lugo et al., 1990b; Lugo, 1992; Cuevas and Lugo, 1998), and the Guánica Forest Puerto Rico (Molina Colón, 1998; Lugo and Fu, 2002). Each location had several forest stands. In Colombia and Venezuela, Saldarriaga (1994) described a chronosequence with six age classes of Tierra Firme Forest in the tropical wet lowlands. Stands were naturally recovering from slash and burn agricultural land use. At the LEF, Silver et al. (in press) studied the organic matter dynamics of La Condesa, a subtropical moist forest (sensu Holdridge, 1967) planted on degraded pasturelands some 60 years earlier. At the LEF, Lugo (1992) also studied a chronosequence of paired tree plantations and secondary forests established on abandoned agricultural lands in the subtropical wet forest life zone (8 stands). Two plantations were coniferous (*Pinus caribaea*) and two were broad leaved (*Swietenia macrophylla*). In addition, Lugo et al. (1990b), and Cuevas and Lugo (1998) studied litter dynamics in 10 tree species planted on abandoned subsistence agricultural lands in the LEF arboretum (Francis, 1989). At Guánica Forest, Molina Colón (1998) studied naturally restored dry forests following abandonment of various land uses (4 stands). She compared these with an adjacent mature dry forest stand. We also present nutrient concentration data for *Swietenia mahagoni* leaves from plantations in Guánica Forest established on sites previously used for agriculture (Lugo and Fu, 2002). We compare these data with unpublished information for leaves of *Dacryodes excelsa*, a primary forest species in the LEF.

In summary, we present data for 22 stands in the three locations (the Amazon, LEF, and Guánica Forest). Seven of the stands were restored through planting and the other 15 were restored through natural regeneration. Stands represent five types of past land uses (farming, pastures, housing, baseball park, and charcoal production), three life zones (dry, moist, and wet), and ranged in age from 4 to about 200 yr.

Because we are reporting data for forest stands that are not in steady state, we cannot estimate the rate of turnover with available data. Instead, we use the ratio of an organic matter production flux (stemwood biomass accumulation, litterfall, or their sum) to the corresponding biomass at a particular time in the development of the forest. This ratio, expressed in percent per year, is the percentage of standing biomass fixed annually through primary productivity at a given age. The absolute value of this ratio varies with the flux used (stemwood biomass production, litterfall or their sum) and the biomass compartments included in the denominator. For the purpose of this article, we standardized net aboveground primary productivity (NAPP) to mean aboveground stem biomass accumulation plus litterfall. This flux was divided by total live biomass (aboveground plus roots). We also used aboveground stem biomass accumulation rate when litterfall data were not available. In those instances, we divided by the aboveground biomass, which excluded roots. Net primary productivity (NPP) was the sum of aboveground biomass accumulation, litterfall, and root productivity. When NPP data were available, we divided by total live biomass.

3. Organic Matter Dynamics

The pattern of organic matter production and accumulation in secondary tropical forests is well known (Brown and Lugo, 1990). Silver *et al.* (2000) found higher aboveground biomass in reforested tropical systems in wet than in moist or dry life zones. Net primary productivity is higher early in the development of the forest, and declines as it reaches maturity (Figure 1a). Forest biomass accumulates steadily through the life of the forest (Figure 1b). In the absence of disturbances, the forest continues to accumulate biomass as it approaches steady state asymptotically. Old forest stands accumulate most of their biomass in soil and large trees (Brown and Lugo, 1992), tree that are usually absent in young forests, particularly those planted when restoring sites after conversion to non-forest. In some instances, large trees remain on site after land abandonment, and these trees have important ecological functions during forest restoration (Toh *et al.*, 1999; Silver *et al.*, in press).

An aspect that differentiates restored forests from stands undergoing natural succession is the effect that past land use has on the rate of biomass accumulation (Brown and Lugo, 1990). Silver *et al.* (2000) showed that aboveground biomass production in rehabilitated forests on abandoned land was a function of the type of land degradation. Biomass production was fastest in abandoned agriculture, slower in abandoned pastures, and slowest in cleared land with arrested succession. All of these rates are slower, however, than those measured in natural successions (Aide *et al.*, 1995; Silver *et al.*, 2000).

We found that ratios of organic matter production to biomass based on NAPP and total biomass were higher than ratios based on aboveground biomass accumulation rate and aboveground biomass (Figure 2). These differences are artifacts of the computation, as both estimates resulted in the same patterns. The pattern of the organic matter production to biomass ratio in restored forests has three characteristics (Figure 2). The ratio is high early after planting or natural regeneration,

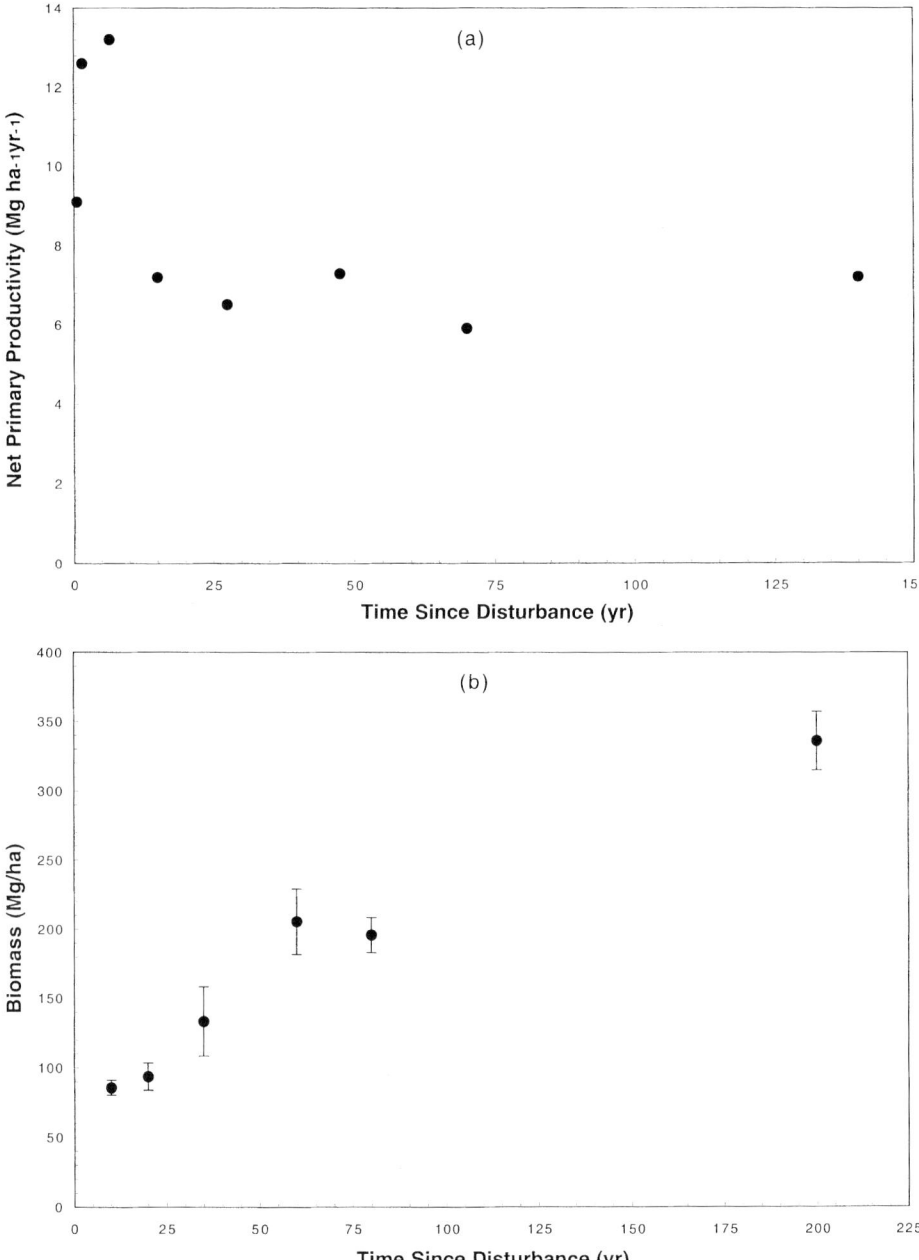

Figure 1. Net primary productivity (a), and total biomass (b) in a chronosequence of sites recovering from agricultural use in Tierra Firme forests of the Colombian and Venezuelan Amazon. Total biomass includes aboveground live and dead aboveground plus roots (100-cm depth), while net productivity includes root productivity. Standard error bars are based on $n = 3$ (age class 60) or 4. All data are from Saldarriaga (1994).

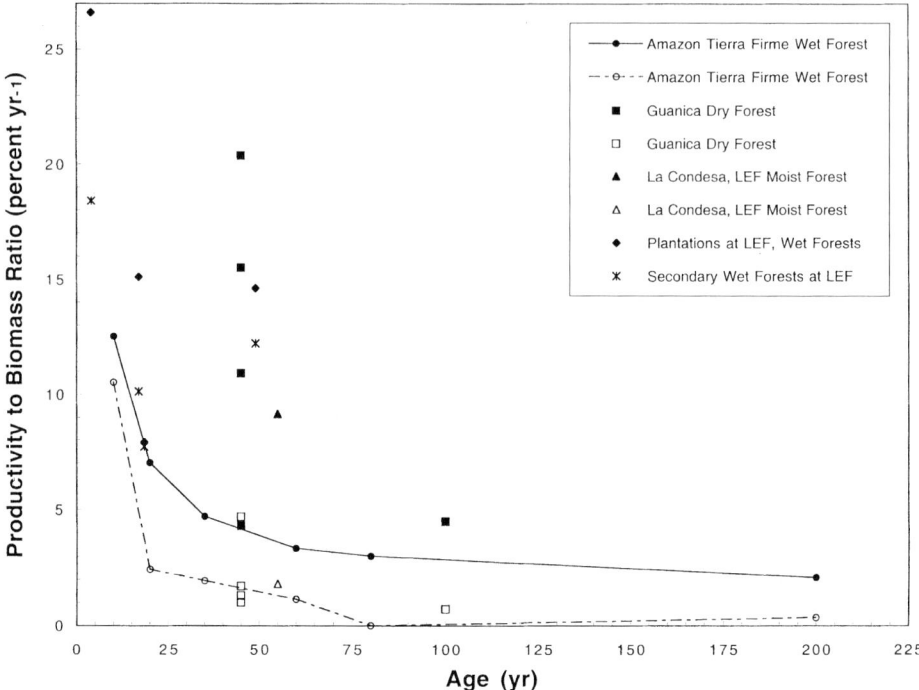

Figure 2. Ratio of net aboveground primary productivity (NAPP) to total biomass (solid line), and ratio of aboveground biomass accumulation rate to aboveground biomass (dotted line). Net aboveground primary productivity is the sum of litter fall rate and aboveground biomass accumulation rate. Total biomass includes aboveground and live root biomass. Aboveground biomass includes stems and leaves. Closed symbols represent ratios of NAPP/total biomass and open symbols represent ratios of aboveground biomass accumulation rate to aboveground biomass. The connected points highlight the chronosequence of Saldarriaga (1994). Other data are from Lugo (1992), Lugo and Murphy (1986), Molina Colón (1998), and Silver *et al.* (in press). Luquillo Experimental Forest is LEF.

and decreases as the forest matures. However, even mature restored forests (La Condesa) had ratios between 2 and 9% a^{-1}. The organic matter production to biomass ratio does not appear to be life-zone-dependent as are the rates of production and biomass accumulation (Brown and Lugo, 1982). Notice for example, that dry forests in Guánica exhibit ratios similar in magnitude to those of the wet and moist forests at LEF or the Amazon.

The dry forest stands in Guánica, each representing a different past land use at age 45 yr, exhibited a wide range in the ratio of organic matter production to biomass, suggesting that the intensity of past land use affects the relation between production and biomass matter. The mechanism is that past land use affected the accumulation of biomass to a higher degree than it did the primary productivity of the site. Those sites with low biomass (former baseball park, houses, and farms) had higher organic matter production to biomass ratios than high biomass charcoal pits and mature forest. The level of litterfall, a major component of primary

TABLE I

Nutrient accumulation (kg ha^{-1}) in vegetation (above and belowground) of various rehabilitated forest stands in Puerto Rico (Lugo, 1992), and mature undisturbed stands of montane rain forests in New Guinea (Edwards and Grubb, 1982). The average aboveground mineral content of four undisturbed mature tropical sites is included for comparison (Edwards, 1982). The age of secondary forests is approximate

Site and age (yr)	N	P	K
Planted stands			
Pinus caribaea (4)	360	13	53
Pinus caribaea (17)	1450	55	514
Swietenia macrophylla (18.5)	565	29	364
Swietenia macrophylla (49)	1001	37	415
Natural regeneration			
Secondary Forest (4 yr)	227	8	165
Secondary Forest (18 yr)	619	35	484
Secondary Forest (17 yr)	498	36	468
Secondary Forest (49 yr)	479	29	369
Undisturbed mature forest			
New Guinea	820	43	854
Four mature tropical forests	1599	92	1218

productivity, was not significantly different among sites (Molina Colón, 1998). Thus, even though a site might not be accumulating much biomass, it could exhibit high productivity and could be transferring organic matter at relatively fast rates, a situation that influences site conditions through soil and litter (Brown and Lugo, 1990).

4. Nutrient Dynamics

Young tree plantations and secondary forests on degraded lands had a lower aboveground nutrient capital than mature undisturbed forests (Table I). Planted forest usually had a larger nutrient pool than secondary forests of similar age and, as they aged, accumulated as much N as undisturbed mature stands (Table I). Planted forests are of known age, so it is possible to estimate a rate of nutrient accumulation in biomass. For those in Table I, the rates of nutrient accumulation decreased

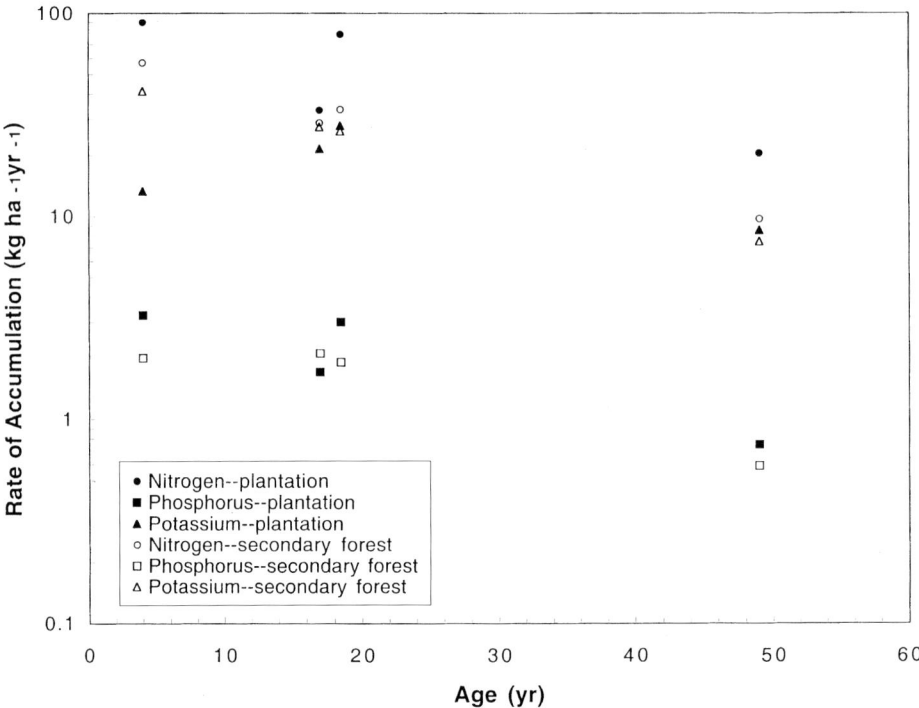

Figure 3. Rate of nutrient accumulation with age in four plantations (solid circles), and four paired secondary forests (open circles) of similar age in the Luquillo Experimental Forest, Puerto Rico Data are derived from Table I.

with age (Figure 3). The reduction in rate was particularly notable in the oldest plantation. For example, the rates of P accumulation (kg ha^{-1} a^{-1}) were 3.3, 3.0, 1.7, and 0.8 for plantations at ages 4, 19, 17, and 49 yr. These dynamics reflect the tendency of trees for high uptake of nutrients from the soil early in their life cycle and to depend on retranslocation during later stages of the life cycle (Bowen and Nambiar, 1984; Attiwill and Leeper, 1987).

Trees 'dilute' their overall initial nutrient capital by increasing organic matter storage at a faster rate than the storage of nutrients. This process allows trees to survive in infertile soils, provided they successfully become established. Increased nutrient retranslocation and recycling with age allows older forest stands to sustain a larger biomass pool with similar or lower soil nutrient uptake as younger stands. This behavior contributes to the function of forests as nutrient and carbon sinks. Sequestration of carbon and nutrients occurs in woody tissue with low nutrient concentration, while production of new tissue uses nutrients acquired by retranslocation and soil uptake.

The key to the success of restored forest function is the capacity of the first tree crop to acquire and concentrate sufficient nutrient capital from the soil to establish forest conditions, i.e., a canopy cover, understory microclimate, and modification

TABLE II

Ratio of annual nutrient return to nutrient accumulation in litter of ten tree species growing as plantations under the same conditions on abandoned agricultural land in the Arboretum of the Luquillo Experimental Forest. Units are in a^{-1} and data for the estimates are from Lugo et al. (1990), and Cuevas and Lugo (1998). Species are arranged in rank order of annual litter production rate. Annual litter production rates for species with the same letters are not statistically different at $p < 0.05$

Species	N	P	K	Ca	Mg
Pinus caribaea[a]	0.40	0.45	0.75	0.53	0.66
Hibiscus elatus[a]	1.46	1.75	2.91	1.16	2.13
Eucalyptus saligna[ab]	0.73	0.76	1.43	0.59	1.15
Pinus elliottii[b]	0.30	0.36	0.51	0.32	0.52
Eucalyptus patentinervis[b]	0.82	1.17	1.12	0.46	0.84
Khaya nyasica[b]	0.70	0.84	1.69	0.65	0.93
Swietenia macrophylla[bc]	0.53	0.66	1.27	0.55	0.80
Terminalia ivorensis[bc]	1.70	1.95	2.97	1.29	2.25
Hernandia sonora[bc]	1.59	1.36	2.81	1.90	2.18
Anthocephalus chinensis[c]	0.57	0.83	1.09	0.50	0.72

of soil structure and chemistry. Once forest conditions are established, other tree and non-tree plant and animal species invade the site and succession processes change in response to the biota.

Our research with tree plantations and natural forest stands in Puerto Rico shows that the capacity to acquire soil nutrients from soil, return nutrients via litterfall, accumulate nutrients in soil, and rate of decomposition are species-specific (Frangi and Lugo, 1985; Lugo et al., 1990a, b; Wang et al., 1991; Cuevas and Lugo, 1998). Others report similar observations on species-specific nutrient dynamics in tropical tree plantations established on degraded lands (Bernhard-Reversat and Loumeto, 2002; Montagnini, 2001, 2002). Table II illustrates species-specific differences in terms of the ratio of nutrients in litterfall to nutrients stored in litter of 23 to 26 yr-old plantations. Three species, *Hibiscus elatus, Terminalia ivorensis,* and *Hernadia sonora* had high ratios for all nutrients, suggesting that the litter of these species recycles nutrients very rapidly. In contrast, *Pinus caribaea* and *Pinus elliottii* consistently had low ratios, suggesting nutrient accumulation in their litter compartment. Other species exhibited variation in the ratios according to the nutrient.

Figure 4 shows a contrasting pattern of nutrient concentration during leaf development and decomposition in *Dacryodes excelsa*, a primary forest species, and *Swietenia mahagoni*, a species used for site restoration in dry climates. Leaves

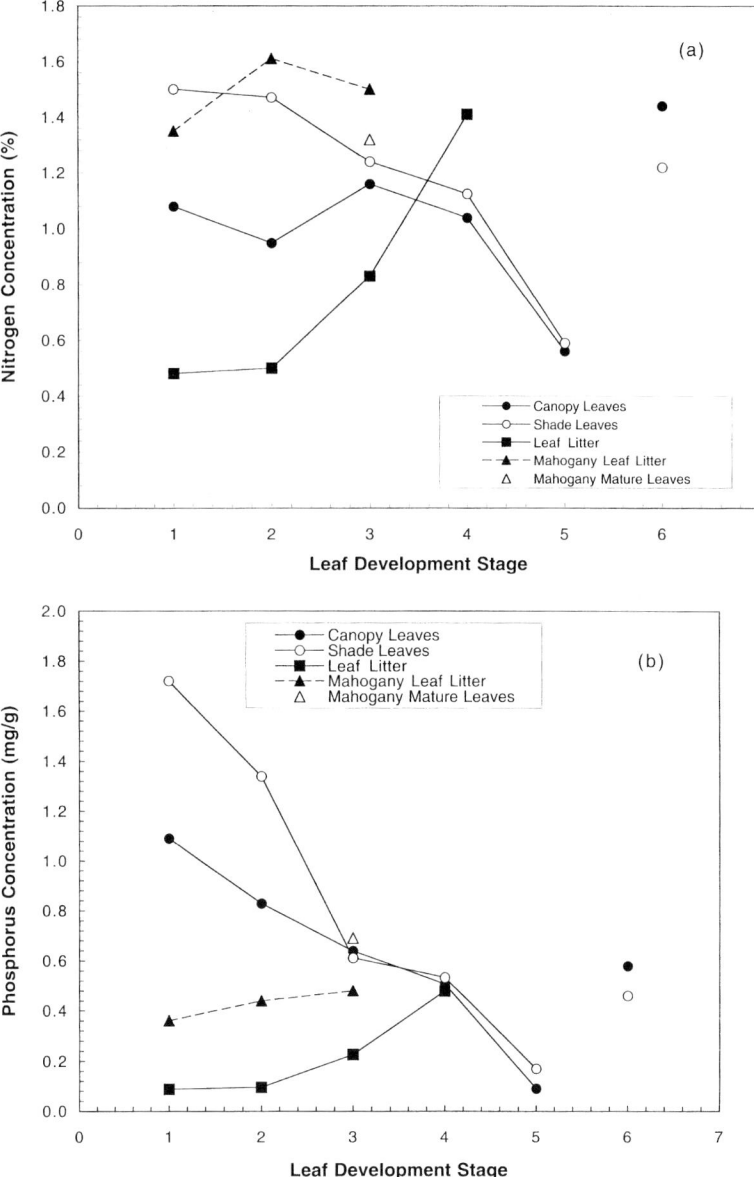

Figure 4. Change in nutrient concentration (N, P, K, and Ca) of leaves and leaf litter of *Dacryodes excelsa* (solid lines) and leaf litter of *Swietenia mahagoni* (dotted lines). Each point represents a different stage of leaf development or decomposition. For canopy and shade leaves, 1 = expanding leaves, 2 = young leaves, 3 = mature leaves, 4 = old leaves, 5 = senescent leaves, and 6 = green leaves blown or knocked down to the forest floor. For *Dacryodes* leaf litter, 1 = yellow leaves, 2 = whole brown leaves, 3 = fragmented brown leaves, and 4 = highly decomposed leaves. For *Swietenia* leaf litter, 1 = recently fallen leaves, 2 = leaves buried half way into the litter layer, and 3 = leaves in the bottom of the litter layer (Lugo and Fu, 2002). Open triangles correspond to the nutrient concentration of mature *Swietenia* leaves (Sánchez *et al.*, 1997).

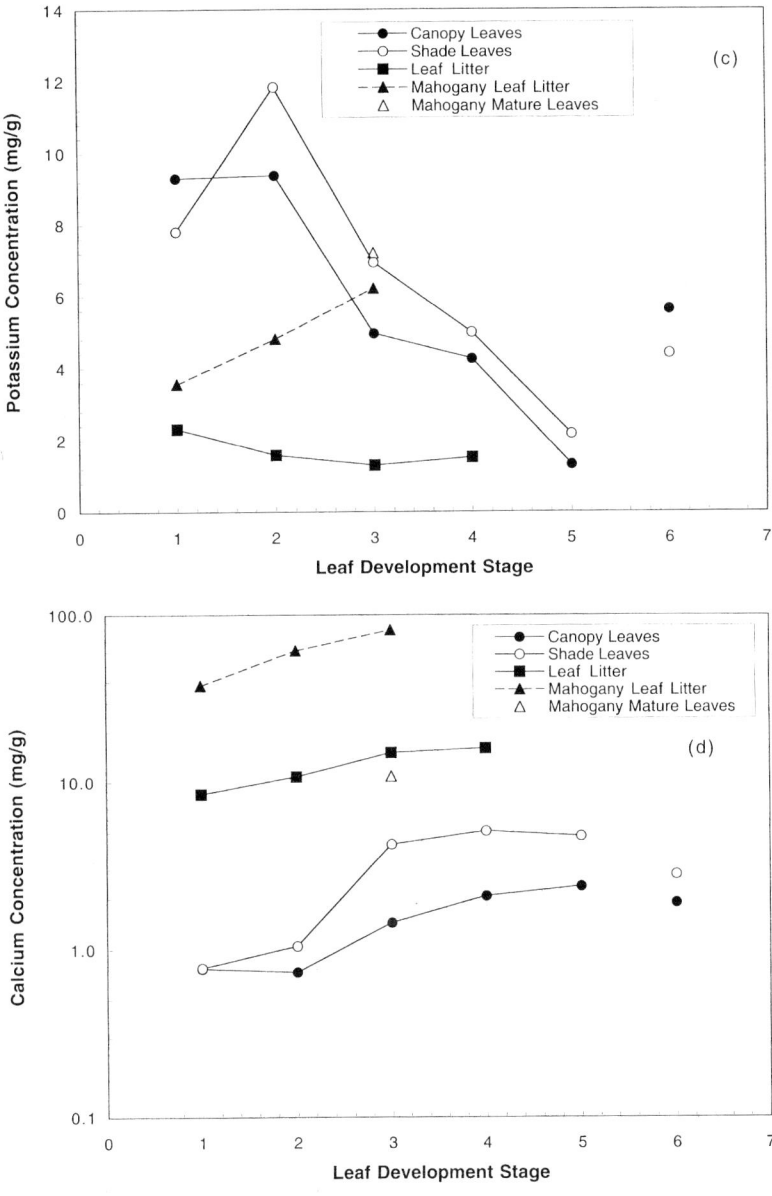

Figure 4. (Continued).

of *Dacryodes* constantly change their nutrient concentration through their development and decomposition stages. Canopy leaves have lower concentrations than shade leaves, but follow the same pattern of reduction in nutrient concentration with age. At the time of leaf fall, the leaf is at its lowest concentration due to retranslocation. Lugo (1992) and Medina and Cuevas (2002) found similar patterns in *Swietenia macrophylla* leaves. Cuevas and Lugo (1998) found that the nutrient

TABLE III

Examples of how tree species influence nutrient cycling attributes of stands. The text contains literature citations with examples

Nutrient cycling attribute	Implications for restoration
Uptake rate	Capacity to grow in the site.
Retranslocation rate	Regulates the quality of litterfall, reduces the uptake requirement.
Return to the forest floor	Opportunity for recycling and improvement of site fertility.
Accumulation in biomass	Sink function and retention of nutrients on site.
Distribution between above and belowground compartments	Determines opportunity for building soil fertility (belowground) vs. circulating nutrients aboveground.
Quality of tissue	Influence on decomposition and consumption rates by fungi, bacteria, and soil organisms.
Efficiency of recycling	High efficiency favors living plants (re use), low efficiency makes more nutrients available for the rest of the system.
Efficiency of storage	High efficiency favors the sink function.
Episodic return	Introduces pulses of nutrient availability.
Episodic retranslocation	Causes periodic changes in the quality of litterfall.
Episodic mast production	Can dominate the nutrient return pathway and favor particular nutrient cycling pathways.
Episodic change in use efficiency	Causes periodic changes in the quality of plant tissue.

concentration in leaf litterfall of mahogany and other species changed throughout the year, thus causing temporal change in the quality of fresh litter.

During decomposition, nutrient concentrations increased to levels similar to those of mature leaves (Figure 4). Leaves that fall green have higher nutrient concentration than senescent leaves. In comparison with *Dacryodes*, *Swietenia* leaves exhibit higher nutrient concentrations throughout their decomposition stages, and a lower difference between the concentration of recently fallen leaves and mature leaves. This suggests a lower rate of retranslocation in *Swietenia* than *Dacryodes*, a higher quality litter for decomposers and soil organisms, and a greater quantity of nutrient return to the forest floor per gram of leaf litterfall.

The dynamics of nutrient cycling in restored stands will change depending on the species that occupy the site, their abundance, and dominance. A particular species can affect nutrient cycling in different ways depending on the process, its magnitude, timing, and efficiency (Table III). In addition, soil biota plays a significant role in nutrient dynamics and its species composition changes depending on the quality of available litter and soil substrates (Zou and González, 2002).

5. Discussion

From a biogeochemical point of view, restored tropical forests contend with degraded sites where nutrient supply, soil structure, and soil water are not optimal for normal tree regeneration and growth. The level of land degradation limits the capacity of forest stands to acquire nutrients and accumulate biomass. When land degradation is extreme, tree planting is necessary and selection of species, and even fertilization and watering, are needed because tree species differ in their water- and nutrient-use efficiency, and in their capacity for fostering other species (Lugo *et al.*, 1990a, b; Wang *et al.*, 1991; Parrotta, 1995; Cuevas and Lugo, 1998; Montagnini, 2001, 2002). In addition, the success of tree establishment depends on the ability to concentrate nutrients and water from the soil. Nepstad *et al.* (2001) showed that roots of secondary forest species in the Amazon had the capacity of reaching up to 8 m into the soil profile within 10 to 15 yr of establishment on high bulk density soils. They also had a higher rate of mycorhizal infestations than species of mature forests. The outcome was that within this short time, these secondary forest species were able to restore normal functioning in terms of nutrient and water uptake and recycling.

Once a forest is established, it appears that restored forests attain high ratios of organic matter production to biomass (Figure 2) and their nutrient accumulation and cycling rates (Figure 3) reach levels similar to those of native forests. However, restored forests are young in relation to primary forests and lag mature forests in structural complexity, including coarse woody debris, habitat diversity, biomass (Figure 1), nutrient pools (Table I), and species richness.

Rates of organic matter production and nutrient accumulation are high early in the life of a forest stand and decrease with age or time since disturbance (Figures 1a and 3). The first decade of forest establishment is a time of rapid exchange of organic matter and nutrients between vegetation and soil. Such high rates of exchange influence soil fertility and soil organic matter (Reddy, 2002), as was demonstrated by Silver *et al.* (2000) in a literature review and by Silver *et al.* (in press) at La Condesa. In that forest (Table IV), it took several decades for trees to reverse the soil carbon balance from a source to a sink, and allow a significant acceleration of aboveground primary productivity.

The measured changes at La Condesa were associated with a change in tree species composition from the 13 initially planted on 1937 to 75 species at age 55. It is apparent that species composition is an important aspect of the establishment and functioning of restored forests. Part of the reason is that under degraded conditions the initial pool of species available for colonization is small and usually alien. Restorationists have opportunities to match species to sites, particularly when site conditions are extreme. Because species differ in the acquisition, retranslocation, and nutrient return capacity, they can either accelerate nutrient cycling or slow it down. As species richness increases, natural forces and self-organization come into play and it is not practical to control the species composition of restored sites.

TABLE IV

Characterization of a 55 yr-old subtropical moist forest restored through planting of 13 tree species on a degraded pasture at La Condesa, Luquillo Experimental Forest, Puerto Rico. Data are from Silver *et al.* (in press). Data are based on trees >9.1 cm in diameter at breast height and a forest area of 4.64 ha

Parameter	Value
State variable (Mg ha^{-1})	
Soil organic matter	204
Aboveground biomass	160
Root biomass	2.5
Annual rates (Mg ha^{-1})	
Accumulation of tree species	1 species a^{-1}
Accumulation of aboveground biomass	2.8
Litterfall	10.6 to 12.9
Aboveground net primary productivity	14.9
Root productivity	0.3
Net primary productivity	15.4
Soil organic matter accumulation	1.8
Net soil organic matter sink	1.1

As a result, nutrient and biomass dynamics approach those of natural forests. The time required to attain high rates of primary productivity and nutrient cycling takes decades and depends on the severity of the original land degradation.

Acknowledgements

This study is in collaboration with the University of Puerto Rico (UPR). It is part of the USDA Forest Service contribution to the National Science Foundation Long-Term Ecological Research Program at the Luquillo Experimental Forest (Grant BSR-8811902 to the Institute for Tropical Ecosystem Studies of the UPR and the International Institute of Tropical Forestry, USDA Forest Service). We thank G. Reyes and M. Alayón for their help in the production of the manuscript and J. Morales, T. Hueth, and F. Scatena for their review of the manuscript.

References

Aide, T. M., Zimmerman, J. K., Herrera, L., Rosario, M. and Serrano, M.: 1995, 'Forest recovery in abandoned tropical pastures in Puerto Rico', *For. Ecol. Manage.* **77**, 77–86.

Ashton, M. S., Gunatilleke, C. V. S., Singhakumara, B. M. P. and Gunatilleke, I. A. U. N.: 2001, 'Restoration pathways for rain forest in southwest Sri Lanka: A review of concepts and models', *For. Ecol. Manage.* **154**, 409–430.

Attiwill, P. M. and Leeper, G. W.: 1987, *Forest Soils and Nutrient Cycles*, Melbourne University Press, Australia, pp. 202.

Bernhard-Reversat, F. and Loumeto, J. J.: 2002, 'The litter system in African forest-tree plantations', in V. M. Reddy (ed.), *Management of Tropical Plantation-forests and Their Soil-litter System*, Science Publishers Inc., Enfield, New Hampshire, U.S.A., pp. 11–39.

Bowen, G. D. and Nambiar, E. K. S.: 1984, *Nutrition in Plantation Forests*, Academic Press, New York, New York, U.S.A., pp. 516.

Brown, S. and Lugo, A. E.: 1982, 'The storage and production of organic matter in tropical forests and their role in the global carbon cycle', *Biotropica* **14**, 161–187.

Brown, S. and Lugo, A. E.: 1990, 'Tropical secondary forests', *J. Trop. Ecol.* **6**, 1–32.

Brown, S. and Lugo, A. E.: 1992, 'Aboveground biomass estimates for tropical moist forests of the Brazilian Amazon', *Interciencia* **17**, 8–18.

Brown, S. and Lugo, A. E.: 1994, 'Rehabilitation of tropical lands: A key to sustaining development', *Restor. Ecol.* **2**, 97–111.

Cuevas, E. and Lugo, A. E.: 1998, 'Dynamics of organic matter and nutrient return from litterfall in stands of ten tropical tree plantation species', *For. Ecol. Manage.* **112**, 263–279.

Edwards, P. J.: 1982, 'Studies of mineral cycling in a montane rain forest in New Guinea, V, Rates of cycling in throughfall and litter fall', *J. Ecol.* **70**, 807–827.

Edwards, P. J. and Grubb, P. J.: 1982, 'Studies of mineral cycling in a montane rain forest in New Guinea, IV, Soil characteristics and the division of mineral elements between the vegetation and soil', *J. Ecol.* **70**, 649–666.

Field, C. D. (ed.): 1996, *La Restauración de Ecosistemas de Manglar*, Sociedad Internacional Para Ecosistemas de Manglar, College of Agriculture, University of Ryukyus, Okinawa, Japan, pp. 278.

Food and Agriculture Organization: 1993, 'Forest resources assessment 1990, Tropical countries', FAO Forest Paper 112, Rome, Italy, 61 pp. + appendices.

Food and Agriculture Organization: 2001a, 'Rehabilitation of degraded sites', *Unasylva* **52**, 1–60.

Food and Agriculture Organization: 2001b, 'State of the World's Forests 2001', Rome, Italy, pp. 181.

Francis, J. K.: 1989, 'The Luquillo Experimental Forest Arboretum', Res. Note SO-358, Southern Forest Experiment Station, USDA Forest Service, New Orleans, Louisiana, pp. 8.

Frangi, J. L. and Lugo, A. E.: 1985, 'Ecosystem dynamics of a subtropical floodplain forest', *Ecol. Monogr.* **55**, 351–369.

Holdridge, L. R.: 1967, *Life Zone Ecology*, Tropical Science Center, San José, Costa Rica.

Lugo, A. E.: 1992, 'Comparison of tropical tree plantations with secondary forests of similar age', *Ecol. Monogr.* **62**, 1–41.

Lugo, A. E., Wang, D. and Bormann, F. H.: 1990a, 'A comparative analysis of biomass production in five tropical tree species', *For. Ecol. Manage.* **31**, 153–166.

Lugo, A. E., Cuevas, E. and Sánchez, M. J.: 1990b, 'Nutrients and mass in litter and top soil of ten tropical tree plantations', *Plant Soil* **125**, 263–280.

Lugo, A. E. and Murphy, P. G.: 1986, 'Nutrient dynamics of a Puerto Rican subtropical dry forest', *J. Trop. Ecol.* **2**, 55–76.

Lugo, A. E. and Fu, S.: 2002, 'Structure and dynamics of mahogany plantations in Puerto Rico', in A. E. Lugo, J. Figueroa Colón and M. Alayón (eds), *Big-leaf Mahogany: Genetics, Ecology, and Management*, Springer-Verlag, New York, New York, U.S.A., pp. 288–328.

Medina, E. and Cuevas, E.: 2002, 'Comparative analysis of the nutritional status of mahogany plantations in Puerto Rico', in A. E. Lugo, J. Figueroa Colón and M. Alayón (eds), *Big-leaf Mahogany: Genetics, Ecology, and Management*, Springer-Verlag, New York, New York, U.S.A., pp. 129–145.

Molina Colón, S.: 1998, 'Long-term recovery of a Caribbean dry forest after abandonment of different land uses in Guánica, Puerto Rico', *Ph.D. Thesis*, University of Puerto Rico at Río Piedras, Puerto Rico, pp. 271.

Montagnini, F.: 2001, 'Nutrient considerations in the use of silviculture for land development and rehabilitation in the Amazon', in M. E. McClain, R. L. Victoria and J. E. Richey (eds), *The Biogeochemistry of the Amazon Basin*, Oxford University Press, England, pp. 106–121.

Montagnini, F.: 2002, 'Tropical plantations with native trees: Their function in ecosystem restoration', in V. M. Reddy (ed.), *Management of Tropical Plantation-Forests and their Soil-litter System*, Science Publishers Inc, Enfield, New Hampshire, U.S.A., pp. 73–94.

Nepstad, D., Moutinho, P. R. S. and Markewitz, D.: 2001, 'The recovery of biomass, nutrient stocks, and deep soil functions in secondary forests', in M. E. McClain, R. L. Victoria and J. E. Richey (eds), *The Biogeochemistry of the Amazon Basin*, Oxford University Press, England, pp. 139–155.

Parrotta, J. A.: 1995, 'Influence of overstory composition on understory colonization by native species in plantations on a degraded tropical site', *J. Veg. Sci.* **6**, 627–636.

Parrotta, J. A. and Turnbull, J. W. (eds): 1997, 'Catalyzing native forest regeneration on degraded tropical lands', *For. Ecol. Manage.* **99**, 1–290.

Reddy, V. M. (ed.): 2002, *Management of Tropical Plantation-Forests and their Soil-litter System*, Science Publishers Inc., Enfield, New Hampshire, U.S.A., pp. 422.

Saldarriaga, J. G.: 1994, 'Recuperación de la selva de "Tierra Firme" en el alto río Negro Amazonia colombiana-venezolana', *Estudios en la Amazonia Colombiana* **5**, 1–201, Editorial Presencia, Colombia.

Sánchez, M. J., López, E. and Lugo, A. E.: 1997, 'Chemical and physical analyses of selected plants and soils from Puerto Rico (1981–1990)', USDA Forest Service, International Institute of Tropical Forestry, *Research Note IITF-RN-1*, pp. 112.

Silver, W. L., Ostertag, R. and Lugo, A. E.: 2000, 'The potential for carbon sequestration through reforestation of abandoned tropical agricultural and pasture lands', *Restor. Ecol.* **8**, 394–407.

Silver, W. L., Kueppers, L. M., Lugo, A. E., Ostertag, R. and Matzek, V.: 'Carbon sequestration and plant community dynamics following reforestation of tropical pasture', *Ecological Applications* (in press).

Toh, I., Gillespie, M. and Lamb, D.: 1999 'The role of isolated trees in facilitating tree seedling recruitment at a degraded sub-tropical rainforest site', *Restor. Ecol.* **7**, 288–297.

Wang, D., Bormann, F. H., Lugo, A. E. and Bowden, R. D.: 1991, 'Comparison of nutrient-use efficiency and biomass production in five tropical tree taxa', *For. Ecol. Manage.* **46**, 1–21.

Zou, X. and González, G.: 2002, 'Earthworms in tropical tree-plantations: Effects of management and relations with soil carbon and nutrient use efficiency', in V. M. Reddy (ed.), *Management of Tropical Plantation-forests and their Soil-litter System*, Science Publishers Inc., Enfield, New Hampshire, U.S.A., pp. 289–301.

LIST OF REVIEWERS

A.T. Abdallah
Rein Aerts
Julian Aherne
Brian Amiro
M.R. Ashmore
Mark Ashton
Björn Berg*
Peter Blaser
Simon Bottrell
Virginie Bouchard
Scott Bridgham
Steven J. Burian
John Neil Cape
Patrick Crill
J. Cristophe-Clement
Etienne Dambrine
Tanguy Daufresne
Peter J. Dillon*
Nancy Dise
Jose Dorea
M. Elizalde-Gonzales
Simon Emmanuel
Bridget Emmett*
Keith Eshleman
Chris D. Evans
John Farmer
Ivan Fernandez
Jan Fiala
Martin Forsius
Christina Forti
Vincent Gauci
Gerhard Gebauer
Anette Giesemann
Pierre Girard
Ian D. Green
Maria Greger
Per Gundersen*
John Gunn

Terry Haines
John Hamilton-Taylor
Mark Harmon
John Harrison
Lorin Hatch
F. Haubrich
T.H.E. Heaton
Jan Helešič
Rachel Helliwell
Arne Henriksen
Alan Herlihy
Atle Hindar
Jana Hladíková
Martin Hodnett
Jenýk Hofmeister
Richard Hooper
Diana Hope
Michael Hornung
Ben Houlton
Daniel Houle
Christian Huber
Michael K. Hughes
Iva Hůnová
Lars Hylander
Dean Jeffries
L.E.T. Jenkin
Alan Jenkins
Chris E. Johnson
Karsten Kalbitz
A.D. Karathanasis
Chev Kellogg
Heike Kempter
Martin Kernan
Holger Kirchmann
James Kirchner
Pirkko Kortelainen
Pavel Krám

Veronika Kronnäs
Meredith Kurpius
James Labaugh
Rattan Lal
Leon Lamers
Hjalmar Laudon
Ahti Lepistö
Greg Lewis
Gunnar Lischeid
Graeme Lockaby
Stephen Lofts
Gary Lovett*
Volker Lüderitz
Lars Lundin
Chris Maier
Daniel Markewitz
Egbert Matzner
Bernhard Mayer
Megan McGroddy
Peter B. McMahon
Patrick Megonigal
John Melack
Henning Meesenberg*
Robert Michel
Myron Mitchell
Filip Moldan
Florencia Montagnini
Eduardo Morales
Scott Neubauer
Tiina Nieminen
Steve Norton
Martin Novák*
Roch Oiumet
Aisling O'Sullivan
Tomáš Pačes
Steve Perakis
William Peterjohn
Hans Rudolph Pfeifer
Maximilian Posch
Jennifer Powers

Jerry Qualls
Olivier Radakovitch
David Raftos
Hannu Raitio
Jane Radford
Katri Rankinen
Heinz Rennenberg
Brian Reynolds
Eva Ritter
Cornelia Rumpel
Fred Scatena
Sherry Schiff
Peter Schleppi
William Schlesinger
Peter Schuster
James Shanley
Walter Shortle
Keith A. Smith
Ondřej Šráček
Michael Starr*
Eiliv Steinnes*
John Stoddard
Eric Strauss
Vince St. Louis
Merritt Turetsky
Liisa Ukonmaanaho
Noel Urban
Caroline van der Salm
Sjoerd van der Zee
Alain Véron
Josef Veselý
Melanie A. Vile*
Peter Vitousek
Jan Vymazal
Leonard Wassenaar
Paul Whitehead
Gary Whiting
R. Kelman Wieder*
Robert Wright

* Reviewed more than one manuscript.